Periodic Table of the Elements

period	group Ia	IIa	IIIb	IVb	Vb	VIb	VIIb	VIII	VIII	VIII	Ib	IIb	IIIa	IVa	Va	VIa	VIIa	0
1	1 1.008 H hydrogen																	2 4.003 He helium
2	3 6.94 Li lithium	4 9.01 Be beryllium											5 10.81 B boron	6 12.01 C carbon	7 14.01 N nitrogen	8 16.00 O oxygen	9 19.00 F fluorine	10 20.18 Ne neon
3	11 22.99 Na sodium	12 24.31 Mg magnesium											13 26.98 Al aluminum	14 28.09 Si silicon	15 30.97 P phosphorus	16 32.06 S sulfur	17 35.45 Cl chlorine	18 39.95 Ar argon
4	19 39.10 K potassium	20 40.08 Ca calcium	21 44.96 Sc scandium	22 47.90 Ti titanium	23 50.94 V vanadium	24 52.00 Cr chromium	25 54.94 Mn manganese	26 55.85 Fe iron	27 58.93 Co cobalt	28 58.71 Ni nickel	29 63.54 Cu copper	30 65.37 Zn zinc	31 69.72 Ga gallium	32 72.59 Ge germanium	33 74.92 As arsenic	34 78.96 Se selenium	35 79.91 Br bromine	36 83.80 Kr krypton
5	37 85.47 Rb rubidium	38 87.62 Sr strontium	39 88.91 Y yttrium	40 91.22 Zr zirconium	41 92.91 Nb niobium	42 95.94 Mo molybdenum	43 (99) Tc technetium	44 101.1 Ru ruthenium	45 102.9 Rh rhodium	46 106.4 Pd palladium	47 107.9 Ag silver	48 112.4 Cd cadmium	49 114.8 In indium	50 118.7 Sn tin	51 121.8 Sb antimony	52 127.6 Te tellurium	53 126.9 I iodine	54 131.3 Xe xenon
6	55 132.9 Cs cesium	56 137.3 Ba barium	57 138.9 La lanthanum	72 178.5 Hf hafnium	73 181.0 Ta tantalum	74 183.9 W tungsten	75 186.2 Re rhenium	76 190.2 Os osmium	77 192.2 Ir iridium	78 195.1 Pt platinum	79 197.0 Au gold	80 200.6 Hg mercury	81 204.4 Tl thallium	82 207.2 Pb lead	83 209.0 Bi bismuth	84 (210) Po polonium	85 (218) At astatine	86 (222) Rn radon
7	87 (223) Fr francium	88 (226) Ra radium	89 (227) Ac actinium	104 (261) Rf rutherfordium	105 (262) Ha hafnium													

atomic mass — 26
55.85
Fe
iron — atomic number / symbol

58 140.1 Ce cerium	59 140.9 Pr praseodymium	60 144.2 Nd neodymium	61 (147) Pm promethium	62 150.4 Sm samarium	63 152.0 Eu europium	64 157.3 Gd gadolinium	65 158.9 Tb terbium	66 162.5 Dy dysprosium	67 164.9 Ho holmium	68 167.3 Er erbium	69 168.9 Tm thulium	70 173.0 Yb ytterbium
90 (232) Th thorium	91 (231) Pa protactinium	92 (238) U uranium	93 (239) Np neptunium	94 (239) Pu plutonium	95 (243) Am americium	96 (245) Cm curium	97 (247) Bk berkelium	98 (249) Cf californium	99 (254) Es einsteinium	100 (253) Fm fermium	101 (253) Md mendelevium	102 (255) No nobelium

For an unstable element, the mass number of the most stable isotope is in parentheses.

TEACHING INTRODUCTORY PHYSICS

A SOURCEBOOK

TEACHING INTRODUCTORY PHYSICS

A SOURCEBOOK

CLIFFORD E. SWARTZ
State University of New York, Stony Brook
Stony Brook, New York

THOMAS MINER
Associate Editor
The Physics Teacher
1972–1988

American Institute of Physics   **Woodbury, New York**

In recognition of the importance of preserving what has been written, it is a policy of the American Institute of Physics to have books published in the United States printed on acid-free paper.

©1997 by American Institute of Physics
All rights reserved.
Printed in the United States of America.

Reproduction or translation of any part of this work beyond that permitted by Section 107 or 108 of the 1976 United States Copyright act without the permission of the copyright owner is unlawful. Requests for permission or further information should be addressed to the Office of Rights and Permissions, 500 Sunnyside Boulevard, Woodbury, NY 11797-2999; phone: 516-576-2268; fax: 516-576-2499; e-mail: rights@aip.org.

AIP Press
American Institute of Physics
500 Sunnyside Boulevard
Woodbury, NY 11797-2999

Library of Congress Cataloging-in-Publication Data
Swartz, Clifford E.
   Teaching introductory physics : a sourcebook / Clifford E. Swartz and Thomas Miner.
     p. cm.
  Includes bibliographical references and index.
  ISBN 1-56396-320-5
  1. Physics--Study and teaching (Higher) I. Miner, Thomas D. II. Title.
QC30.S87 1996   96-17141
530' .071--dc20   CIP

Contents

Preface vii

CHAPTER 1
Fatherly Advice 1

    A. Textbooks 1

    B. Class Discussions 5

    C. Demonstrations 8

    D. The Laboratory 17

    E. Evaluation: Tests and Grades 32

CHAPTER 2
Error (Uncertainty) Analysis 53

CHAPTER 3
Units, Dimensions, Vectors, and Scaling 71

CHAPTER 4
Friction 87

CHAPTER 5
Gravitation 101

CHAPTER 6
Reference Frames and Relativity 117

CHAPTER 7
Newton's Laws of Dynamics 156

CHAPTER 8
Angular Momentum 178

CHAPTER 9
Work and Energy 196

CHAPTER 10
Internal Energy 221

CHAPTER 11
Second Law of Thermodynamics 264

CHAPTER 12
Fluids 278

CHAPTER 13
Vibrations 305

CHAPTER 14
Wave Transmission 323

CHAPTER 15
Complex Waves and Wave Interactions 349

CHAPTER 16
Electrostatics 381

CHAPTER 17
Electric Current 404

CHAPTER 18
Magnetism 426

CHAPTER 19
Currents and Fields that Change with Time 451

CHAPTER 20
Electromagnetic Radiation 476

CHAPTER 21
Microstructure 503

Index 549

Preface

This is a reference book, a review book, and a book that contains some highly biased advice. It started out with a decision by the two authors to gather some of the background material concerning introductory physics that we had been accumulating during our years of editing *The Physics Teacher*. In writing the book we faced the same dilemma that we faced in choosing articles for the magazine. About one-third of the subscribers to *The Physics Teacher* teach high-school physics, two-thirds teach in colleges or universities. About 1000 teach outside the United States. In each issue we try to have topics and a level of treatment that will be useful to such a broad spectrum of interest and technical training. In writing this book we assumed that some readers would be new to the profession, and some would have taught for years. We have tried to include material that each group will find useful. Some chapters are concerned with classroom techniques. Most chapters are tutorials on specific topics from the standard introductory physics course. Much of this will be familiar to experienced teachers, but we hope that it will be useful to have explanations and background data gathered together for easy reference.

Some of the material in the tutorial chapters consists of excerpts and revisions of the college text *Phenomenal Physics* written by Swartz (John Wiley & Sons, Inc., New York, 1981). The English version of that text has been out of print since 1987 and the text is now available only in a Russian translation. Much of this tutorial material can be found in standard university introductory texts, but our presentation has some novel features, particularly in its emphasis on order-of-magnitude approximations and the analysis of everyday phenomena.

Tom Miner and I gave up editing *The Physics Teacher* in 1985. When I became editor again in 1989, Tom's health did not permit him to travel regularly to Stony Brook to join me in the editorial office, but we decided to write this book together. We—particularly I—didn't write fast enough. Tom died in 1991. He had written most of the chapters on classroom techniques and had started an annotated bibliography for all of the chapters. After his death I abandoned the project for several years, but have concluded it now as a testimonial of respect and admiration for my co-author. Besides being good friends, we were good critics of each other. The book would have been much stronger and more useful if Tom could have analyzed and reshaped the remaining chapters. In part, of course, he had already done that by teaching and influencing me in the years we worked together.

The diagrams were skillfully and patiently drawn by Arthlyn Ferguson.

In finishing this book I have sought advice from many people about physics and explanations of physics, but have not always taken their advice. In the early stages, Ernie Kuehl and Jim Gerhart critiqued some chapters. Later, Mario Iona reviewed part of the book, and Bob Bauman read it all. If there are faults remaining, dear Reader, be assured that it is not due to these stars, but to myself.

Cliff Swartz

1
Fatherly Advice

PART A. TEXTBOOKS

There is little agreement among teachers about the role of a text in the physics course. Some plan the entire course around the text, adopting its organization and subject-matter development, and using its exercises for homework. To have the greatest possible sales appeal most texts include more topics than any class can deal with in an academic year, so even teachers who follow the book must omit some subjects if they wish to get to the usual end chapters on quantum and nuclear physics. Poorly prepared and inexperienced teachers may be well advised to follow a text quite closely.

At the other end of the text-usage scale we find teachers who are so secure in their knowledge of physics and self-confident as teachers that they ignore the adopted text, writing worksheets and assignments to serve their own ideas of a proper course, personally presenting and explaining all the physics the students are expected to learn.

It is likely that most teachers and students do not use the text as much as they should. Today's texts, at least at the college level, are generally correct in physics and clear in style. The authors know physics and are themselves successful in the classroom. The explanations, worked over as they have been to put them in writing, are better than a teacher can come up with in front of a class, and the figures are more informative than most teachers can draw on the chalkboard. Good texts, at least at the college level, have been reviewed and critiqued by expert referees. Students can thus take a superior tutor to their study desk when they use a textbook.

Yet it is a common experience for teachers to find that few students make real tutorial use of their texts. For most of them the book is where you go to find the assigned homework ("Do all the odd-numbered problems.") and the formulas to solve the problems, with perhaps a quick scanning search for a solved sample problem resembling an assigned one.

Teachers should insist that their students read the text. In making an assignment, refer to text pages as well as problem sets. Ask students to compare the text's development of a tricky topic with the approach used in class. Point out a particularly successful figure or ask a question about it. Require each student to provide a real example of a phenomenon different from those used in the text. ("I saw beats the other day. It was raining and I was stopped at a traffic light. The wipers of the car in front of me worked at a different frequency than mine, and the two got in and out of step with each other!") One teacher makes his point when he announces that every test will include two items that he has not taught but which are well-treated in the textbook. Another requires that each student write an original test item on a text topic, the best to be incorporated in the next test.

One good reason for designing high-school instruction so that students will learn to use their texts properly and efficiently is that college instructors assume that students have already learned such skills. Professors make assignments to read text chapters, knowing full well that many students don't know how to read a physics text. Nevertheless, it's a rare college class where such instruction is given. High-school physics teachers would be doing a great service for their students by giving explicit instruction in how to read a physics text, and then following it up with assignments or tests that require such reading.

■ HELPING STUDENTS TO USE THEIR TEXTS

Here are some suggestions that can be given to students to help them get the most out of their texts:

1. Before starting to use the text try to familiarize yourself with its general features. Who is the author and where does he teach? In the preface what kind of course does he say the book was written for? Try to get an idea of the organization of the book as shown in the table of contents. Are sample problems solved in most chapters? Are there chapter summaries? Is there a glossary? This quick overview need not be detailed or time-consuming but will be very helpful when you start to use the book.

2. If possible, before a written assignment has been made, start reading the chapter that will be involved. "Speed reading" is not very helpful in physics. Read slowly and reflectively, asking questions that your reading suggests. Relate figures and their captions to the text. If the book is your own, write marginal comments and questions. Underlining can be a mindless process, but if you have found it useful, then underline very sparsely; most students underline too much—trivial ideas as well as essentials. Take notes on your reading, putting the notes in your own words.

3. When a problem assignment is given, get to it as soon as possible. Try to associate each problem with an idea or several ideas covered in the text and in the classroom. This will probably require more reading, again questioning thoughtfully. Actively try to relate each problem to the relevant text.

4. This procedure of careful reading, question asking, problem solving, and making sure that your own questions are answered will help you avoid one of the most serious traps for the beginning physics student. Too many students think that the most important point of a homework exercise is to get the answer. They can do this by plugging numbers into formulas that contain the right symbols to match the information in the problem. Physics is not learned this way! The primary purpose of a problem is to illustrate a principle and to give practice in applying the principle.

■ CHOOSING A PHYSICS TEXT

Many teachers are in a position to influence the adoption of a text either by making the choice or serving on an adoption committee. Most physics texts differ very little in subject matter and sequence, although the levels of presentation may vary widely. Some authors have experimented by starting the course with a topic other than mechanics. Such innovations have usually either expired or returned to a more conventional organization. An example is the PSSC (Physical Science Study Committee) text. The second half of the typical physics text is nearly always dominated by electricity and magnetism, with so-called "modern physics" coming in at the end of the book.

On the principle that physics books are so nearly the same that the problem of choice is trivial, some very odd criteria have been used in text selection. One teacher used the same book year after year because his name was mentioned in the preface as a reviewer of the manuscript. Another picked the "Naval Academy Edition" of a standard book, not because of any consideration of content but because he found that his (at that time) all-male students were impressed. Some books have been shunned because of their size of page; others because more males than females were depicted in the figures.

Here are some practical questions that can be considered in choosing a text [a number of authors have developed more formal checklists on this topic—see, for example, Owen (1962) and Vogel (1951)]:

1. Subject content
    a. Does the book deal with the subject matter you plan to teach?
    b. Is the material presented at the right level?
    c. Are the basic concepts of physics illustrated by practical examples without being smothered by such applications?
    d. Is the subject matter correct? [For a study on this point see Lehrman (1982).]
    e. Is the development of content based on student experiences, either from everyday life or from laboratory experiences?

2. Style
    a. Is the book readable for the type of student you will be teaching? [See Dukes and Kelley (1979) for readability testing.]
    b. Do you think it will appeal to student interests?
    c. Does the author's style invite thought rather than memorization?
    d. Are topics described and explained clearly? (Note the use of topic sentences, freedom from run-on sentences, continuity of thought, summarizing sentences, and an appropriate vocabulary.)
    e. Are words defined and explained at the point where they are first used?

3. Organization
    a. Is there an attempt to show relationships between parts of the course, or is the subject matter compartmentalized or fragmented?
    b. Are basic matters introduced before the more complex relationships depending on them are considered?

4. Illustrative materials
    a. Are photographs chosen to enlighten and to engage the reader's interest rather than merely to decorate the text?
    b. Has color (if used) been used effectively?
    c. Do illustrations appear on the same page as the text to which they refer?

d. Are captions complete and helpful?

   e. Are drawings clear, uncluttered, accurate, and well-labeled? Are the drawings artistically acceptable?

5. Teaching aids

   a. If you want chapter-end reviews, does this book have them, and are they informative summaries, not mere lists of definitions and formulas?

   b. Do you want a glossary, and if so, does the glossary in this book define technical terms understandably and correctly?

   c. Are chapter-end exercises useful, graded in difficulty, and numerous enough to give the teacher a good selection? Are there some questions requiring recall only, and others calling for thought, imagination, and ability to synthesize? Do the problems encourage comprehension, with only a few of the "plug-in" type? Are answers supplied for some problems? [Not all teachers consider this desirable. See Bork (1983).] Are some of the exercises of a project nature, requiring the students to experiment on their own, with a few of the tasks challenging enough to intrigue the superior student?

   d. Is the index adequate? (This is not a trivial point.)

   e. Is a Teacher's Edition, or Teacher's Guide, available? Is it really helpful or does it just "go through the motions"?

■ POSSIBLE CHECK POINTS

As a final test, study the presentation of a few difficult topics. Here are some suggestions of topics that many authors find troublesome to write about clearly and correctly:

1. The distinction between weight and mass
2. Centripetal and centrifugal forces
3. Diffraction
4. Electric potential
5. Induced electromotive force (emf) and Lenz's law
6. Binding energy and nuclear mass

■ FOR READING AND REFERENCE

Black, J. T., and Harding, I. W., "Developing reading skills in the high school physics class," Phys. Teach. **19**, 106 (1981).
   A reading program integrated with high-school physics. A "minilibrary" of physics texts, graded according to reading difficulty, is available to students.
Bork, A., "Letter to the editor," Am. J. Phys. **51**, 491 (1983).
   Argues strongly that supplying answers to problems in a text is a deterrent to learning. See also correspondence in later issues of *The American Journal of Physics*, for example, Whitaker, M. A. B., "Physics students, authority and the real world." Am. J. Phys. **51**, 1065 (1983), which maintains that supplying answers to problems is very useful.
Dukes, R. J., Jr., and Kelley, S., "The readability of college astronomy and physics texts," Phys. Teach. **17**, 168 (1979).
   A report on the subject of the title, covering a large number of books. Tells also how to apply the Flesch readability scale.

Dukes, R. J., Jr., "The readability of college astronomy and physics texts II," Phys. Teach. **18**, 290 (1980).
   A follow-up article on Dukes and Kelley (1979) giving further information about the Flesch readability formula and ratings of a number of new and overlooked texts.
Field, G., "Textbook reading difficulties in the calculus-based physics course," Phys. Teach. **22**, 174 (1984).
   Suggestions for helping college physics students get the most benefit from their textbooks.
Kennedy, K., "The reading levels of high school physics texts," Phys. Teach. **17**, 165 (1979).
   An analysis of the reading levels of ten texts. Shows the graph for estimating readability and tells how to use it.
Lehrman, R., (Chairman of committee), "Physics texts, an evaluative review," Phys. Teach. **20**, 508 (1982).
   The chairman of The Physics Teacher's text-evaluating committee makes specific critical remarks about 14 high-school physics texts.
Owen, A. M., "Selecting science textbooks," Sci. Teach. **1962**, Nov., 20. Quoted by Anderson, H. O., in *Readings in Science Education for the Secondary School* (Macmillan, New York, 1969).
   Practical details involved in making a selection, including a table telling how to go about it as well as a suggested rating sheet to be used in evaluating books.
Romey, W. D., *Inquiry Techniques for Teaching Science* (Prentice-Hall, Englewood Cliffs, NJ., 1968), pp. 44–48, 50–51.
   Describes an elaborate numerical method for rating textbooks and supplies a form for the analysis.
Symposium, "On Selecting a New Text," Phys. Teach. **20**, 524 (1982).
   Six high-school teachers tell about their experiences in changing from one text to another.
Vogel, L. F., "A spot-check evaluation scale for high school science textbooks," Sci. Teach. March (1951), quoted in Thurber, W. A. and Collette, A. T., *Teaching Science* (Allyn and Bacon, Boston, 1959), pp. 232–233.
   Provides a very detailed check list to be used when evaluating a text. Fifty features are covered and the result is expressed numerically.

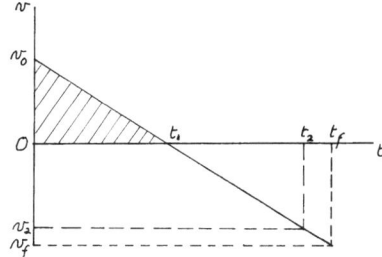

Figure 1.1. Velocity versus time graph for an object tossed vertically. Air resistance is ignored. At time t_1 the object is at its highest point, the velocity changing from upward (positive) to downward (negative). The distance the object has risen is represented by the shaded triangle: $h=(v_0/2)(t_1)$. To return to the thrower's hand it drops the same distance, and is caught at t_2, at which point $v_2 = -v_0$. If the thrower fails to catch the object, it falls to the ground at t_f with velocity $-v_f$. It would be interesting, with a more sophisticated class to introduce the effect of air resistance.

PART B. CLASS DISCUSSIONS

■ LET'S THINK ABOUT A CASE TOGETHER

A class is usually more engaged and teaching is more effective if points are made by "think-about" questions rather than by simple declarative statements. The think-about question considers a simple situation whose analysis can lead to more subtle ideas than are at first apparent. The thinking is ideally done by the students, prodded usually by questions and provocative comments from the teacher.

Say, for example, that your class has studied uniform acceleration using an air track and has been introduced to the acceleration due to gravity, perhaps measuring g by timing an object as it falls a measured distance. At this point you ask the class, "Can you think of a case in which an object is accelerating, but has no velocity?" Where you go from here is up to you and your students, but, given an inquiring class attitude, it is safe to predict an interested and perhaps passionate debate. The students who come up with a correct answer may be hooted down by others ("How could an object's velocity be changing if it has no velocity?").

The best answer will probably cite some version of a ball thrown vertically upward at the top of its flight, or a glider projected up an inclined air track at its highest point. You then propose to think further about this situation by drawing a graph of velocity against time, and you develop on the board the graph shown in Fig. 1.1. Start with the axes and v_0 and let the class tell you how to draw the graph. What is the significance of positive and negative on the graph? Why is the graph linear? Why is the slope negative? What's happening at the instant when the line crosses the time axis? (Of course, this answers the original question, but keep going!) How long a period of time does the object spend with no velocity? What's

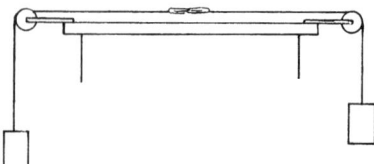

Figure 1.2. The two hanging objects have the same weight. What is the tension in the horizontal cord? Two paper clips are hooked together at the center of this section and can be separated to allow insertion of a spring balance.

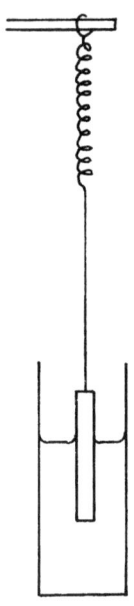

Figure 1.3. A metal rod hanging from a spring dips into water. Describe the motion when it bobs up and down. Consider changing the water level so that the rod is fully immersed or pulls out of the water for part of each cycle.

the significance of the area that's shaded? How fast is the object moving when it returns to the starting point? Then you can add a wrinkle: The ball or glider is not stopped at its starting position—the ball thrower, for instance, misses catching the ball where it was released and the ball continues to the ground. Add this complication to the graph at the student's direction.

Questions of this sort will occasionally arise spontaneously in your class, but more usually you will propose them yourself, to fit the topic being considered. Here are some examples of other questions that are worth thinking about in detail:

1. Galileo reported that the distance moved by an object rolling down an incline is, in successive equal time periods, in the ratio of the odd numbers 1, 3, 5, 7, etc. Is this generally true, or just a coincidence?

2. You are running in a rainstorm. Could you hold a cardboard tube while you're running so that the raindrops could enter the upper end and leave at the lower end without wetting the inside of the tube? How would you do this? Next imagine that there's a north wind and you're running east to west. How would this change your answers?

3. You are in a boat on a small pond. If you throw the anchor into the water will it affect the water level of the pond? Suppose that you jump out of the boat and stand on the bottom of the pond in water to your waist, and then you swim; how will each action affect the water level?

4. A solid wooden mast tapers uniformly from a thick base to a thin top. If you balance this pole horizontally, and then saw through it at the balance point, which portion of the pole weighs more, or are the weights the same?

5. Why is the base of a ladder more likely to slip as the painter climbs higher on the ladder?

6. Suppose that an experiment revealed a situation in which the distance an object moved from rest was proportional to the cube of the time ($s = kt^3$). How does this differ from the usual case? What can you say about the force acting on the object? If the force is actually not unusual, what other factor may be causing the phenomenon?

7. Hang equal masses over pulleys as shown in Fig. 1.2. What is the tension in the horizontal cord? (Have a couple of paper clips hooked together in this section so that a balance or two can be inserted.)

8. How can you best explain the direction and speed of precession of a gyroscope? (Try to keep the discussion conceptual, and be sure to give students, eventually, the mnemonic "The axis of spin chases the axis of torque.")

9. An early astronaut reported that, while in orbit and trying to get some sleep, his arms kept floating out in front of him. Explain why. (He finally hooked his thumbs into his helmet straps to hold his arms in place.)

10. How would an astronaut in a space ship catch up and dock with a satellite ahead of him in the same circular orbit? How does this case differ from that of a racecar driver catching up with a car ahead of him on the same track?

11. A rod hangs from a spring and dips into water (Fig. 1.3). The spring is made to vibrate with a small amplitude. Is the motion of the rod simple harmonic? Describe its motion if the water level is lower so that the rod is pulled out of the water for part of each cycle. What about the motion if more water is used

so that the rod is fully immersed for a portion of each cycle? If a liquid of different density were used how would it affect the motion?

12. A playground swing is mounted so that it swings at a right angle to the wall of a nearby building. When a child blows a whistle while he is swinging he says it sounds "funny." What might be affecting the sound of the whistle? How might the effect be changed by increasing the height he swings, or by increasing the distance between the swing and the building?

13. The mean free path of a gas molecule is the average distance it travels before colliding with another molecule. How would you expect the mean free path to be affected by (a) raising the temperature of the gas while keeping its volume constant, (b) reducing the pressure by cooling the gas, (c) reducing the pressure by allowing the gas to expand, (d) using a gas with larger molecules, all other conditions being the same?

14. Referring to a table of coefficients of linear expansion and a table of melting points, determine whether there is any connection between these two properties of a metal. Pick a group of metals listed in both tables and plot a graph of their melting points against their coefficients of linear expansion. Is the result to be expected? Explain. Try to predict and then confirm the relationship, if any, between melting point and tensile strength, heat of fusion, heat conductivity, and specific heat.

15. When a simple circuit, containing, for example, a lamp and a switch, is connected to a battery and the switch is closed, the electrons in the circuit drift from negative to positive at speeds typically about 0.01 cm/s—about the speed of the tip of the minute hand of a small clock. In view of this very low speed, how do you explain the fact that when the switch is closed the light is turned on so quickly?

16. Two lamp sockets are mounted on a panel and clearly wired in series. If two 40-W, 120-V lamps are used in the sockets, and the circuit is connected to 120-V ac, will the lamps light? Try it. If you substitute two 20-W lamps for the 40-W lamps, how would you expect them to work? If you use a 20- and a 40-W lamp in the sockets, what result do you predict? Try switching the lamps to see whether the position of the socket determines the result. Explain each step.

17. Compare the deflection of a charged particle as it crosses a magnetic field with the deflection of an identical particle crossing an electric field. Assume in each case that the particle enters at right angles to the field lines. How does the deflection depend on the particle's velocity? Describe the particle's path in each case. Describe the path of a charged particle entering a magnetic field at some angle other than 90°.

18. Self-inductance has been called "electrical inertia." Why?

19. How could you measure the focal length of a diverging lens?

20. How does a spectrum formed by a prism differ from a spectrum from the same source formed by a transmission grating? If you were given a photograph of a spectrum with wavelengths and angles of deviation indicated, how could you tell whether it was formed by a prism or a grating?

21. Sometimes a public address system makes a howling or ringing sound. What can cause the self-oscillation? What is a common source of feedback? What determines the frequency of the sound? How can this behavior be prevented?

■ A REMINDER ABOUT SOME OBVIOUS FACTORS THAT ARE SOMETIMES FORGOTTEN

When you ask a student a question, wait for the answer. That may seem obvious, but most of us find it awkward to wait long enough. In 1974, Mary Budd Rowe published a paper showing that the average delay time for most teachers was only about one second. If you increase that time even to three seconds, the student reaction is remarkable. They start to talk, and sometimes have thoughtful things to say. If you aren't used to waiting for a student answer, you may find the delay torturous. Try counting the time to yourself with "one thousand and one, one thousand and two, etc." up to 10 seconds.

If you toss a question out to everyone in general, no one in particular will answer. Call on students by name. No matter what the answer, try hard to use the response to advance the discussion and to make the student feel that he or she has made a contribution. *Never* let an honest attempt to answer one of your questions be turned into an embarrassing situation for the student.

Make a deliberate attempt to engage everyone in discussions at some time throughout the week. Call on the girls as well as the boys who are speaking out without being asked. Call on the shy students, boys or girls. If you need a third hand with demonstrations, be impartial. Girls can lift, time, count, fasten, and hold as well as boys.

■ REFERENCE

Rowe, Mary Budd, J. Res. Sci. Teach. **11**, 263 (1974).

PART C. DEMONSTRATIONS

The demonstration experiment is one of the most powerful teaching tools that a physics teacher can call on. In a demonstration an individual, usually the teacher, handles objects in a planned way to illustrate or clarify the physics to be taught. The objects can be as simple as a child's ball or as complex as a laser. As for the plan, it is essential. The principle to be featured and the facts of its physics are known, so the desired behavior of the apparatus is determined in advance. Although the whole event is rehearsed and staged it has vital elements that bring to the students a level of reality much higher than a lecture, a slide show, or a film. Teachers often find that returning alumni may say, "I remember when you showed us ...," but seldom or never reminisce about a lecture or film.

■ SEEING IS BELIEVING

We shall not linger here on the reasons for using demonstrations because they are widely described in the literature. (See, for example, a symposium on demonstrations in the September 1981 issue of *The Physics Teacher* and the introductory essays in Meiners' "Physics Demonstration Experiments.") It all boils down to the ancient adage "seeing is believing." Do students readily believe that a ball thrown horizontally and another dropped at the same instant will hit the floor simultaneously? They will accept it as a fact if you show them by any one of a number of techniques that it really works that way. (See, for example, Dem. M-9, p. 5 in Freier and Anderson.) You can then build the abstract ideas of the independence of velocities on the evidence supplied by the students' own senses.

Students (with the exception of some very grade-ambitious individuals who want only to know the "right" answer) generally find demonstrations interesting and rewarding. The discussion may go on long after the class. Students want to try it themselves. (Let them!) The reality of a demonstration may make students conscious of absurdly misfit values that they may come up with in problem solutions: a force of 2 newtons accelerating a car, or a charge of 50 coulombs on a capacitor.

Demonstrations combat rote learning just as formulas and definitions invite it; let's have the demonstrations first. They are also an antidote to the all-too-common feeling that physics is too difficult. When students see how things really "go" the analysis that follows makes more sense.

Demonstrations have their detractors. Many administrators find the cost in both money and time to be excessive. It is true that apparatus and materials for demonstrating can be expensive, but they need not be so. Some of the most graphic demonstrations use only everyday objects; indeed, in an extreme case a whole course could be devised using demonstrations that do not call on any expensive commercial apparatus. (See, for example, the third demonstration source on p. 15, and the column of the same name in *The Physics Teacher*.) It is also true that a teacher who uses demonstrations must have time to set them up, try them out, and put them away. Another objection related to time is that it takes more class time to show a principle in action than merely to tell about it. This is actually an advantage. Demonstrations do slow the pace of the teaching, giving students needed time to organize and digest the ideas, in contrast to the "drinking from a firehose" aspect of too many lectures. Experienced and successful teachers of introductory physics are nearly unanimous in using demonstrations. They are willing to invest the necessary effort and time because they know that the method works.

■ MAKING THE BEST OF A FAILURE

This brings us to a troublesome point: what if a demonstration *does not* work as planned? It should not happen very often. It is absolutely essential to practice a demonstration before showing it to students, and the outcome and possible pitfalls should be well known in advance. But it happens to all of us that occasionally some unanticipated factor throws the whole performance awry. Indeed, this possibility is one of the reasons for the effectiveness of demonstration-experiments. The risk of failure is a part of all real life, and this reality keeps students alert.

It is a great temptation for a teacher whose demonstration has come to an unexpected result to tell the class how it "should have worked" and move along to the next idea. But things *always* work as they should for the actual conditions; it is just that we are not always in control of all the conditions. There is little place in a physics class for talk about the way things "should" happen. Instead, the teacher has a splendid opportunity to show students how a physicist handles a problem and to invoke their help. This on-the-spot research can be a fine experience for students and must not be neglected.

■ HOW TO DO IT

Here are some additional guidelines for demonstrators:

1. Choose demonstrations that illuminate specific ideas that you want to teach. A great teacher (Harald Jensen of Lake Forest College, Illinois, now retired) once said, "If you can't show it, don't teach it!," and we find that turning this epigram around is just as significant, "If you can't teach it, don't show it!" In both forms the statement exaggerates, but there ought to be a clear link between the physics course and the demonstrations. Otherwise the motive becomes mere entertainment. More on this later.

2. Try to obtain some quantitative results in most demonstrations. There is seldom a need to make detailed measurements (unless that's the point of the demonstration). Instead, the demonstrator can call attention to and estimate comparative sizes and quantities. Based on these estimates the class can be encouraged to forecast results when a variable is changed, to appraise the plausibility of a result, or to draw a freehand graph showing how one quantity depends on another. A coupled-pendulum demonstration, for instance, can be used to define or clarify several fundamental ideas (the transfer of energy, resonance, phase, and phase relationships in particular); this is its qualitative use. But it is also valuable to estimate the amplitude of the response of the driven pendulum as the length of the driver is changed and draw a quick resonance curve on the chalkboard, thus revealing a quantitative component of the phenomenon.

    Here are a few examples of demonstrations in which rough measurements or estimates can help to teach a basic idea:

    a. In a first approach to the concept of velocity, pace across the classroom, having students estimate both the distance and the time in order to calculate the velocity. Teach them the "one thousand and one, one thousand and two, etc." method of timing. (Each syllable takes $\frac{1}{5}$ second.)

    b. Drop a ball from as high as you can reach, or toss it to near-ceiling height, and ask the class to "measure" the time it takes to hit the floor, using either their own watches or the "one thousand and one" method of timing. From the estimated distance and time, calculate rough values for the average.

    c. Calculate roughly the expected cost (as it would appear on a monthly utility bill) of operating a desk light or a heater on a plausible schedule. Have the device on display and working.

    d. Determine the approximate focal length of eyeglass lenses or a projector lens by forming images of a bright distant source. Try it with strong lenses used for the correction of farsightedness. The farsighted astigmatism correction is especially instructive.

    e. Demonstrate the concept of beats formed by the periodic interference of waves by using two pendulums of slightly different lengths. Measure roughly the frequency of each pendulum, and then observe the frequency with which they get out of step (interfere).

3. Prefer apparatus that is simple. Pedagogically a stopclock is more effective as a timing device than a photogate and scaler combination. For showing the characteristics of a simple circuit, a dry cell is a less distracting source than a rectifier power supply.

4. Make sure that your apparatus and what it is doing is visible from the last row in the room. This means, of course, that the equipment should be large. A golf ball is a better falling object than a ball bearing. Visibility can also be improved by using contrasting colors. It is also a fact that a demonstration is easiest to follow when the bench is uncluttered. Move previously used apparatus out of the way so that what you are showing occupies center stage. Neatness contributes to visibility.

5. Shun trickery. It may be fun to be fooled, but a teacher who deliberately sets out to fool students will teach them to look for tricks rather than physics. The art of the magician depends on misdirection, and directing attention away from relevant features is exactly the opposite of one of our goals as physics teachers—encouraging accurate observation. There are so many mind-challenging demonstrations available in physics that we do not need the magician's trickery to make physics fun. One suspects that a teacher who hides a wire or component in a circuit to bring about an amazing result is merely indulging his own ego.

6. There are times when a physics teacher must be a showman. In fact, experiments should generally be presented with a flair that speaks of the teacher's own enthusiasm. The purpose of adding an element of drama is to engage students' interest and attention, not to entertain. Students these days are very sophisticated about their entertainment, and teachers (with certain well-known exceptions) cannot really compete with the professional entertainers presented every night on television. A student who recently returned from his freshman year at an ivy-league college said, when questioned about his physics course (his final grade was B+), "They tried too hard to make it interesting!" His teacher's efforts to add frills to the demonstrations to make them more entertaining had been transparent. Students are not fooled by this patronizing approach. Good physics, in tune with the course, and genuine enthusiasm for the subject are enough. The teacher should let his or her enjoyment show.

   No matter how good the physics, however, it is usually not wise to show a demonstration that cannot be handled as a part of the course. For example, a bicycle-wheel gyroscope, fascinating as its behavior might be, is not a fit demonstration device for a class that will not study angular impulse and momentum.

7. Do not use so many demonstrations that there is insufficient time to let each make its own impression on the students. There is no place in physics teaching for a rapid-fire series of poorly comprehended demonstrations, no matter how relevant. Such a program may be fine for meetings of physics teachers, as several of our colleagues have shown us, but not for instruction. A worthwhile experience must be deliberate enough to be well-understood, and in many cases should be repeated.

8. Demonstrations must be rehearsed. This is time-consuming, but absolutely necessary. Part of the rehearsal is to make sure that the experiment really shows what is intended. The apparatus used in the trials ought to be exactly the same as that to be used in class. A second purpose of the rehearsal has to do with visibility. Where should the demonstrator stand so that he or she does not hide the apparatus or the chalkboard? Where should the chalkboard drawing be placed, and how oriented so that it is best related to the actual apparatus? Most experienced demonstrators have developed their technique to the point where their handling of the problem is nearly automatic. But the habits do have to be developed.

9. When possible it is a good idea to set up the demonstration with the class watching. Even though you have already assembled the apparatus for your own practice, take it apart! Consider a demonstration using a cart to be dragged up an incline. Don't have it set up in advance, but when you need it move the incline to a clear area of the bench, set the angle, hold the cart in place on the incline, pull the string (precut and knotted, of course) over the pulley, and hang the balancing object on the string. During the short time this is going on the teacher talks about what he or she is doing, so that when the arrangement is complete the class needs no more description of the apparatus. If you are going to project a spectrum, place the light source, lenses, and prism in position, calling attention to their locations and the orientation of the prism while the class watches.

   One of the most glaring mistakes made by many inexperienced demonstrators, as they make adjustments with the class watching, is to work on the apparatus in silence. It has been said that most teachers talk too much, and this is certainly quite often so, but the class deserves to know what the teacher is doing when he or she adjusts apparatus for a demonstration. Tell them, even if it means that you are just thinking out loud.

10. Keep records of your demonstrations on 5×8 in. cards or in a special notebook. You will save much time next year if you record the details of a successful demonstration. For each one write down the principle demonstrated, list the apparatus, and tell where it is stored. Attach a Polaroid photo of the setup ready to go, and make a simple sketch of the arrangement. Be sure to include a record of the values (voltage, current, force, mass, etc.) that work well.

■ SPECIAL KINDS OF DEMONSTRATIONS

1. Much of what we have presented above applies in particular to short, simple, single-concept demonstrations. These are sometimes descriptive, but often can be made to reveal important functional relationships, as described above. Another type of demonstration resembles a laboratory experiment in which data are taken and recorded, precision is noted, graphs are drawn, and calculations worked out to arrive at a specific conclusion. This type of demonstration might take an entire class period and make the basis of a homework assignment. Almost any laboratory experiment can be handled this way, if necessary, although it must be noted that a demonstration, no matter how skillfully executed, cannot really yield the benefits that we expect from a student experiment. The procedure is, however, particularly useful for courses

that do not include a laboratory program, and for such courses may be the only real contact that students have with the experimental basis of physics. When instruments cannot be projected or are not otherwise legible to students it is best to have readings taken by student assistants.

2. Some experiments should be shown as demonstrations because they are too difficult, dangerous or costly for student use. For example:

   a. A beautiful method of measuring the wavelength of sound requires too much apparatus and know-how for students to do it. Knowing the wavelength and the frequency permits calculation of the speed of sound. The experiment requires a meter stick, a calibrated audio signal generator, a speaker, a microphone, and an oscilloscope. The tone generator is connected to the speaker and the horizontal input of the scope; the microphone to the vertical input (see Fig. 1.4). The phase delay of the sound traveling in air from the speaker to the microphone produces, with the original signal, a Lissajous figure whose shape depends on the distance the wave has traveled in air. Shifting the relative positions of the speaker and microphone causes a continuous phase change and can bring about a succession of 360° changes, permitting a very accurate measurement of the wavelength. This is a delightful experiment, and every aspect of it is worth extended discussion.

   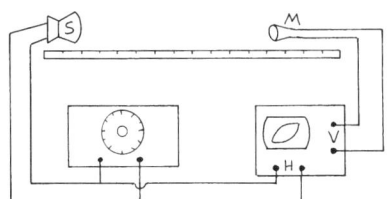

   Figure 1.4. Measuring the wavelength of sound.

   b. The power developed by an electric motor is measured by the usual bandbrake method. Apparatus problems make this more successful as a demonstration than as an experiment.

   c. Any experiment involving an element of danger should be done as a demonstration. This includes experiments involving the handling of metallic mercury, high pressures that may be encountered in, for example, the change of boiling point with increased pressure, lethal voltages, including the power line and the output of neon transformers, and nonencapsulated radioactive sources.

3. There is one form of demonstration that is really just a three-dimensional illustration. Most phenomena that we deal with in physics can be described with algebra and with graphs and diagrams. When either representation is shown, the experienced physics teacher can easily visualize the real-world phenomena represented. A slanted chalk line on the board, for instance, *is*, to the teacher, an inclined plane. A cross product takes on a direction perpendicular to the chalkboard, represented by a line at an angle, and a right-hand thread direction is intuitively felt in the hand muscles. Not so with many students! They are remarkably naive and inexperienced with simple devices and phenomena and frequently have trouble perceiving three-dimensional perspective drawings.

   Whenever possible, accompany algebra and graphs or diagrams with real-world three-dimensional objects. The inclined plane should also be shown by an inclined board, even if no measurements are to be made. Hold up "tinkertoy" assemblies or models made by sticking stiff wires into a styrofoam ball to show vector relationships, and manipulate a real screwdriver as you talk about the right-hand rule. Displacements with chalk arrows on the board can also be shown as real movements of teacher or students at the front of the room. Step off three strides to the north and four to the west. Freely falling objects should become dropped golf balls, and two-dimensional solenoids or lenses on the board should become the real objects held up for the class to see. Even when the real object is too small for students in the back row to see in detail, at least they can get a sense of the size and shape of the thing.

Sometimes you can easily make large-scale models of small devices out of wire, cardboard, and sticky tape. In other cases hold the models aloft in lecture, and then display them in a temporary "museum" at your desk or in the laboratory. If possible, make all the demonstration apparatuses or models available in some monitored place such as the student laboratory. Encourage (or require) your students to observe, handle, and manipulate the devices.

Here is a list of helpful things to have on hand:
- Sticky tape (single- and double-sided)
- Styrofoam in various shapes
- Stiff wire
- String and thread
- Cardboard
- Hand drill (manual)
- 1/4-in. dowel (various lengths)
- Quick-drying cement
- Boards (several lengths)
- Tinker-toy set
- Screwdriver
- Wrench
- Golf ball
- Tennis ball
- Foam ball (Nerf)
- Modeling clay
- Lead shot
- Aluminum foil
- Paper clips
- Drinking straws
- Rubber bands
- Thumb tacks
- Balloons (fresh)

■ SOURCES OF DEMONSTRATION EXPERIMENTS

1. *The Physics Teacher* is a membership journal of the American Association of Physics Teachers (AAPT) and is published nine times a year. Regular membership in AAPT (every serious physics teacher should be a member), with *The Physics Teacher* as journal of choice, costs (at this writing) $75 a year. For more information write to Membership Department, AAPT, One Physics Ellipse, College Park, MD 20740-3845.

The Physics Teacher focuses on the teaching of introductory physics at all levels. Every issue contains descriptions of useful teaching demonstrations, as well as other material of lasting value.

2. *The American Journal of Physics* is another membership journal of AAPT and is published monthly throughout the year. Membership in AAPT with *The American Journal of Physics* as journal of choice costs (at this writing) $96 a year. For more information write to the address given in item 1.

   The American Journal of Physics focuses on the instructional and cultural aspects of physical science. It is written at a higher technical and mathematical level than *The Physics Teacher*. Many issues include demonstration experiments suitable for use in elementary courses, and other experiments for intermediate and advanced studies, as well as articles and notes from the entire field of physics.

   Regular membership including both journals costs (at this writing) $140. Nonmember subscriptions are available for both journals; write to AAPT for details. Also write to AAPT for rates outside the United States, as well as arrangements for optional air freight or airmail special rates.

3. Edge, R. D., *String and Sticky Tape Experiments* (AAPT, College Park, MD), $24. Edge is the author of the "String and Sticky Tape" column in *The Physics Teacher*. His experiments use only the simplest sorts of materials: soda straws, aluminum foil, string, paper clips, and the like. Many are appropriate for demonstrations.

4. Freier, G. D., and Anderson, F. J., *Demonstration Handbook for Physics* (AAPT, College Park, MD, 1981), Publication Department (address given in 1 above), $24. This volume contains descriptions of over 800 demonstrations, both new and old. Each is illustrated by a line drawing. Many of the demonstrations use inexpensive, everyday materials.

5. Meiners, H. F., *Physics Demonstration Experiments*, original edition 1970, reprint (Krieger, Melbourne, FL, 19XX), $96.50. This is a two-volume work given mainly to detailed descriptions of demonstrations, ranging from simple to very complex, in all areas of physics. The first section of Vol. I contains papers on the purposes and uses of lecture demonstrations, written by teachers who are famous for their skills. Volume II starts with essays discussing various lecture techniques, including the use of closed-circuit TV, films, corridor demonstrations, the overhead projector, and stroboscopic effects. Throughout the book the apparatus is specified clearly. Construction details and materials lists for some of the more specialized apparatus appear in appendixes at the end of each volume.

6. Sutton, R. M., *Demonstration Experiments in Physics* (McGraw-Hill, New York, 1938). A classic, describing nearly 1200 demonstrations, that is now out of print. Many copies were sold over a long period of years, however, and it is probably in your library. A teacher lucky enough to own a copy will guard it zealously, so, unless reprinted, it will be hard to acquire. Reprinting it is unlikely because of the primitive nature of some of the apparatus (much emphasis on vacuum tubes, for instance) but most of the book is of timeless value.

7. Harris, J., and Ahlgren, A., "Apparatus, lecture-demonstration and laboratory: Some simple experiments and demonstrations." *Phys. Teach.* **4**, 314 (1966). Here are about a dozen ideas developed by the people working on the

Harvard Project Physics Program. Some of them later were incorporated in that course.

8. Francon, M., *et al*, *Experiments in Physical Optics* (Gordon and Breach, New York, 1970). The "experiments" in this manual are really intended as demonstrations and include a wide range of topics. Written for the teacher, it specifies readily available apparatus and gives many tips to aid the instructor.

9. Levinstein, H., "The physics of toys" *Phys. Teach.* **20**, 358 (1982). This article describes the use of toys in a demonstration course. The author has collected about 600 toys and uses 30 or 40 in each lecture. Specific examples are described and analyzed.

10. Carpenter, D. Rae, Jr. and Minnix, Richard B., *The Dick and Rae Physics Demo Notebook*, (Dick and Rae, Lexington, VA, 1993), $40. This is the definitive collection of demonstrations created or recreated in the many summer workshops that Dick and Rae have hosted. The book is well illustrated and indexed—indispensable!

■ FOR READING AND REFERENCE

A symposium on the use of demonstrations in teaching introductory physics appeared in the September 1981 issue of *The Physics Teacher*. It included the following papers:

Carpenter, D. R. and Minnix, R. B., "The lecture-demonstration: Try it, they'll like it," Phys. Teach. **19**, 391 (1981).
Freier, G., "The use of demonstrations in physics teaching," Phys. Teach. **19**, 384 (1981).
Hilton, W. A., "Demonstrations as an aid in the teaching of physics." Phys. Teach. **19**, 389 (1981).
Johnston, J. B., "The lecture demonstration, a developing crisis," Phys. Teach. **19**, 393 (1981).
   All the authors are famous for their skill and commitment as demonstrators. They all take strong and well-documented stands in favor of the technique.
Pinkston, E. R., "The use of demonstrations in teaching introductory physics," Phys. Teach. **19**, 387 (1981).
Davis, J. A., "Physics lecture demonstrations, an annotated bibliography, Part I," Phys. Teach. **12**, 523 (1974).
   This paper gives dozens of references to pre-1974 sources of demonstrations. Many of the suggestions are now out of print, and others are out of date. Part II was never published.
Davis, J. A., "Lecture-demonstrations are consistent with Piagetian theory," Phys. Teach. **12**, 299 (1974).
   A short note in which the author supports the contention of his title.
Lloyd, J. T., "Lord Kelvin demonstrated," Phys. Teach. **18**, 16 (1980).
   William Thomson, Lord Kelvin, one of the great physicists of his day, was also an innovative teacher. Many of his demonstrations have become standard in physics classrooms. This article tells about a number of his lecture-demonstrations, some simple and others on the grand scale.
Nicklin, R. C., "Notes of an itinerant demonstrator," Phys. Teach. **12**, 72 (1974).
   The author describes a series of demonstration lectures that he presented in the high schools of western North Carolina. In addition to organizational details he identifies and gives directions for the demonstrations that he found to be most useful.
Prigo, R., "A new addition to the homework assignment: demonstration-problems." Am. J. Phys. **45**, 433 (1977).
   In addition to textbook questions and problems students are assigned problems using demonstration apparatuses in the author's university-level Physics Learning Center. Eight examples are given. Education theory justifies the approach, and the results are positive.
Semper, R., *et al*, "Use of interactive exhibits in college physics teaching," Am. J. Phys. **50**, 425 (1982).
   The use of exhibits that can be operated by the student and can be displayed in the laboratory, hallways, special rooms, or public places. This also describes an internship program at the Exploratorium, the San Francisco Science Museum.

Willis, J., and Kirwan, D. F., "Physics demonstrations for the public," Phys. Teach. **14**, 210 (1976). An article describing the public lecture-demonstration series presented by the authors. The most useful demonstrations are listed, organized as to topic, and accompanied by references and comments.

PART D. THE LABORATORY

There is a difference between the experiments discussed in this chapter and demonstrations covered in Chap. 1.3. A demonstration is intended to make memorable to the students some particular fact, principle, or device, but the basic purpose of an experiment is to enable students to find out for themselves, by manipulating appropriate apparatus, how scientists solve problems and to impress on them the experimental basis of our knowledge of nature. The outcome of a successful demonstration is known to the demonstrator in advance, and the procedure is planned to bring about that result. For the participant in an experiment, however, a question is asked, and a procedure is designed to answer the question. Spend some time raising the question and involving the students in defining it. Make sure the question becomes *their* question. Discuss with them possible ways to answer the question, including the degree of error (uncertainty) that they will require. (Is it a 1% experiment?, 10%?, order of magnitude?) What sort of apparatus is needed, and how does it work?

In the laboratory the student should learn how the ideas of physics describe the reality of phenomena. The setting should provide practice in handling problems based on real situations. In addition to being a link with the real world there are many other hoped-for results. We think that students gain from the laboratory a deeper and better-remembered understanding of the principles investigated. We want them to appreciate the experimental method, and to realize that, in part because of the uncertainty of all measurements, our knowledge of physics must always be tentative. One desirable feature of laboratory experiments is the opportunity for the students to meet the unexpected, giving them a lifelike challenge in fitting this unanticipated result into their growing pattern of knowledge. The chance to make mistakes and then to correct them resembles the processes of learning encountered outside the school. Another reason for our conviction that introductory physics courses should all have a laboratory component is the belief that many, if not most, people learn with their hands as well as by seeing and hearing. In addition, for some courses the students should learn laboratory skills.

■ THE STRUCTURE OF EXPERIMENTS

The degree to which an experimental procedure is controlled can vary from the cookbook-recipe experiment at one extreme to the free-wheeling project at the other. In the first type the students are told in advance what the result must be. They study the theory before coming into the laboratory, where they find a specified apparatus ready for use. Tightly defined procedures are spelled out in detail. Results are good or bad according to how closely they agree with the previously stated law or with a numerical value given in a handbook. Most thoughtful teachers reject the cookbook laboratory as nonscience, unlikely to bring about the hoped-for results of the laboratory. The rigid structure, however, makes it easy on instructors, and it has some useful aspects that we shall discuss shortly.

The project laboratory is, in nearly all aspects, the opposite of the cookbook

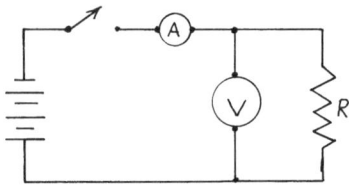

(a) Connect the meters this way if R is much smaller than the resistance of the voltmeter.

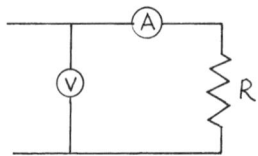

(b) Connect the meters this way if R is not a great deal smaller than the voltmeter resistance.

Figure 1.5. *What determines the current in a resistor?*

laboratory. Students investigate topics that may be of their own choosing. They design the experiment, assess its feasibility, requisition or construct the apparatus, and stay with the same problem for indeterminate periods of time. The instructor is used as a consultant and necessarily is a very busy person. Procedures have not been tested, so unexpected things happen. Problems and plans differ from group to group. To keep things moving in such a laboratory is indeed a challenge. It is easy to understand that this idealized situation has not been found to be practical as a complete laboratory program except in small and favorable settings. In addition to the demands made on the instructor, and the requirement of nearly limitless apparatuses organized for quick access, it is not realistic to expect beginning students to have the skills and background needed to work completely on their own. To use this kind of laboratory we must usually sacrifice the great advantage of coordinating the laboratory program with the class work, and instead accept the fact that most students participating in a project laboratory will emerge with glaring gaps in their backgrounds. A way of achieving some of the great benefits of the project laboratory without most of its difficulties is described below.

■ THE OPEN-END EXPERIMENT

What we want is order without rigidity and freedom without chaos. This ideal cannot be achieved by either of the extremes, but a program using the best features of both works well and the result of such a blend is the "open-end" experiment. The first part of the experiment, in which the scene is set and necessary techniques are taught, may be as firmly programmed as a cookbook recipe. The students, once set on the path, are then left to choose between alternative goals. They invent procedures, decide for themselves how far to go and with what results to be content. The outcome, it is hoped, is both order and freedom.

Let's see how something of the sort could be done with the students' introduction to the electrical circuit. The instructor has decided that the class should do laboratory work for which the organizing themes are the nature of a circuit and the relationship between potential difference and current. (In a cookbook laboratory this might be titled "Ohm's Law Experiment.") With students using meters for the first time some instruction will be necessary. The laboratory session begins, then, with careful and specific directions for wiring an ammeter, voltmeter, resistor, variable source, and switch in a circuit (Fig. 1.5). This part of the experiment is as closely controlled as any recipe.

Next the question is posed, "What factors determine the current in a given resistor?" The students are thus pointed in a specific direction and are free to collect evidence on which to base an answer. The next question is, "What is the relationship between these quantities?"; then, "Does it work the same way for a different resistor of the same nature?", and, "What happens if we replace the resistor by an incandescent lamp? ...a crystal diode? ...an electrolytic cell?" The student who comes to the laboratory with no knowledge of circuits may get no further than the linear current–voltage relationship for a wire-wound resistor and the definition of resistance; better-prepared students can be encouraged to move quickly into the investigation of diode and lamp. Each one is having an encounter with a physical phenomenon whose features are not known in advance, in a situation requiring choices and decisions, and a final evaluation of evidence.

An example of a question that could be answered in the laboratory, but will not be unless the question is carefully phrased, is how the current varies in successive resistors in series. We know that most students adhere to the "tired electricity" model, whereas most instructors reject the model out of hand. Unless the instructor accepts the question as important, students will not explore it in the laboratory.

■ USING PROJECTS

It is possible to use a modified project approach along with a conventional laboratory program. The students are told that each must carry out one experimental project (or more if desired) during the semester. These projects are independent of the regular laboratory course. The subject of a project is to be chosen from a list provided by the instructor, or it can be proposed by the student, in which case it must have the instructor's approval. The projects listed are nonroutine, call for apparatus that is known to be available, and are of such a nature that they can be approached by several routes at several levels of difficulty.

The student, before starting the experimental work, must hand in for approval a feasibility study, giving a brief description of the planned procedure, telling what apparatus will be needed and where it is to be obtained. The recommended source is home, but some specialized pieces can be loaned from the laboratory stores. An analysis of the expected precision is to be part of the feasibility study, important because many of the projects can be either trivial or extremely difficult depending on the expected precision.

The final report is to be neat, detailed, and formal, without a requirement that it be in any specific form.

Here are several examples of project topics used by one of the authors:

1. Without using a timing device, find the maximum speed with which you can throw a stone or ball.

2. Measure the specific heats of several metals to check the law of Dulong and Petit.

3. Measure $T(t)$ for water (ice) from -10 to $100\ °C$, starting with a sample of ice being supplied with heat at a constant rate. Determine from your result the heat of fusion and the heat of vaporization of water.

4. Measure and plot the diameter of a human eye pupil as a function of light intensity.

A list of other such projects will be found in Appendix 1, p. 27.

■ SCHEDULING LABORATORY EXPERIMENTS

With regard to the progress in the classroom there is a just-right time for each laboratory experiment. When laboratory work, classroom lectures, and exercises get out of step we may lose most of the sought-after benefits of the experiments.

In most instances, the best use of the laboratory is to *introduce* the physics concepts. The most productive timing is to do the laboratory work before the class discussion of the same ideas. The class work that follows is thus based on the reality of the laboratory, and much of the ensuing learning comes from the natural need to understand what was observed. On the other hand, an experiment that *follows* an introduction given by assigned reading, lecture, and class discussion will generally have the dry and sterile purpose of confirming what is already known, and can have very little justification. Even less likely to be useful is the experiment that is so unrelated to the students' studies that they have no background to support the work or motivation to learn from it.

In addition to introducing new ideas, an experiment can be used as a unifying device, as a review, as practice in interpreting data, to teach certain laboratory skills, or for some other purpose that may not relate directly to a single topic in the subject content of physics [the electric calorimeter (Fig. 1.6), for instance, brings together the fields of heat and electricity]. It is apparent that the scheduling of individual laboratory investigations must be based on the nature of each one; it is generally best to use an experiment to introduce course content.

■ EVALUATING EXPERIMENTS

Here are some of the desirable characteristics of experiments. Not all can be realized in every worthwhile experiment.

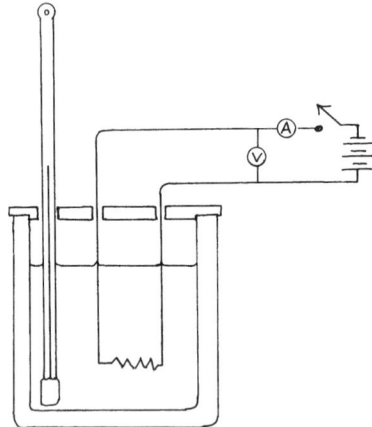

Figure 1.6. The electric calorimeter.

1. The subject must be worthy, at the level of the students for whom it is intended. We should not spend valuable laboratory time on dead-end, peripheral, or irrelevant topics. Compare, for example, the value of an experiment to measure Hooke's law for springs with that of an investigation of conservation and momentum. The first is a byway, the second mainstream. While the level and nature of the course may make a study of Hooke's law appropriate, if there is a need to choose between the two it is the truly basic concepts of physics that deserve the attention.

2. The experiment should teach skills appropriate to the course. Whereas one should stress uncertainty of measurement and the proper handling of error in a course for physics majors and engineers, the approach in a physics course for nurses should be much less technical, with an emphasis on such skills as reading instrument scales, measuring liquid levels and volumes, and estimating elapsed time.

3. An experiment should introduce an element of the unknown. "Hard thinking must take place in the laboratory" proclaims an early statement about the PSSC course, and one cannot expect much hard thinking if the result is known in advance.

4. An experiment must be capable of some expansion or extension. This is the open-end characteristic discussed above.

5. Where a result is numerical it should be "checkable." That is, there ought to be some way to evaluate its validity. Perhaps the check is internal (one part of an experiment agrees with another) or perhaps the check is made by testing a prediction based on the result. Sometimes a consensus test is useful: "How does your value of Planck's constant with its range of uncertainty compare with the values obtained by other members of the class?" In many cases, including this last one, we can turn to a handbook or the text to find the "accepted" value and its range of uncertainty.

6. An experiment should yield some insight into the uncertainties of measurement. The precision sought should be appropriate to the goals of the experiment and the course, and students must always be discouraged from thinking of a result as "right" or "wrong." Whether or not it is best in your course to make a detailed study of error, uncertainties of measurement should always be recognized and accounted for as much as possible.

7. An experiment should leave some decisions to the experimenter.
8. It is desirable for an experiment to be amenable to analysis. Most experiments, even those that are largely descriptive, should show the interplay between theory and experiment that is characteristic of all sciences.

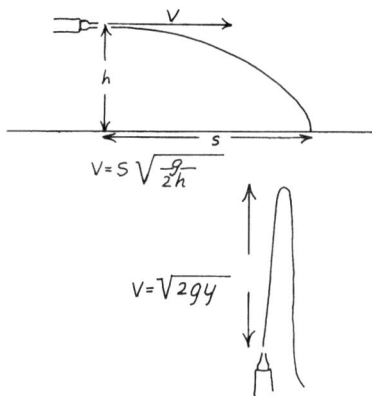

Figure 1.7. Measuring the velocity of a water jet.

■ TYPES OF EXPERIMENTS

When laboratory work falls into a set pattern, following a routine that has become familiar with use, we lose the freedom that is so necessary. Variety is not just the "spice" of laboratory life, it is an essential ingredient. There are enough different kinds of experiences available in a physics laboratory to make change of pace the rule. Here is a breakdown of kinds of experiments, with a few examples, that have different motivations and approaches.

1. *The empirical experiment.* Every physics student should have some experience with the process of collecting data and then manipulating it to reveal a relationship. The ideal topic is one that will be studied analytically later, but with which the student has had little experience before starting the experiment. The motion of a mass bobbing up and down on a coiled spring is an example. In determining the relationship between the period of oscillation and the mass, the student uses methods that are characteristic of actual scientific work, including the "inspired guess," and winds up with a result that can easily be tested. An examination of the graph of the raw data of period (T) and mass (m) reveals a nonlinear relationship, and suggests that T must be proportional to some power of m other than one. The choice of which power to test may be done by a thoughtful and informed "guess." The intuitive result is then checked by plotting another graph such as T versus $\sqrt{m}$. Whether this is the time to use logarithms and logarithmic graph paper to straighten the curve must depend on the mathematical preparation of the students.

   This experiment can be greatly expanded if desired by investigating the part played by the elasticity of the spring, and finally by evaluating the constant in the equation, which turns out, of course, to be 2π. Determining the effect of the mass of the spring is probably too difficult. When simple harmonic motion is studied later in the course the picture is complete, and the interplay between theory and experiment has been realistically illustrated.

2. *The analytical experiment.* In this sort of experiment (perhaps it should be termed an exercise), one starts from principles and from them makes a statement about a special case. It amounts, in the laboratory, to testing a prediction. For example, after a study of free fall and the vector nature of velocity and acceleration, the trajectory of a water jet is used as an illustration. One asks, "Is it possible to make measurements on the water jet and use what we have learned about falling bodies to tell the initial velocity of the water?" Two situations are identified for which we can easily write equations: one in which the initial velocity of the water is vertically upward, and the other in which the water is directed horizontally (Fig. 1.7). If we have applied our theory correctly the two results for the same flow of water will agree. When a constant-level supply tank is used the agreement is good, in part because critical measurements occur in the calculations under a radical!

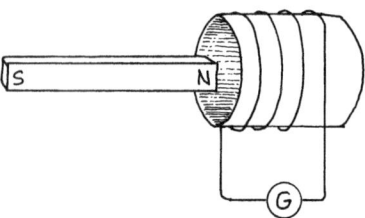

Figure 1.8. *A Lenz's law experiment.*

3. *The basic-concept experiment.* Many experiments are designed to reveal or reinforce basic concepts. They will generally involve an empirical-style approach, but will be aimed at the result rather than at the process. Newton's second law is well-handled this way. The hydrogen spectrum is studied in the laboratory to reinforce ideas about atomic-energy levels. The electric calorimeter is useful as a link between two areas of physics, but can serve more specifically to illuminate the basic concept of potential difference and the definition of the volt.

4. *The descriptive experiment.* Most experiments will be of a quantitative nature, with numerical measurements leading to graphs and numerical results. There is, however, especially with lower-level courses, a real place for an occasional descriptive experiment. In many cases this valuable qualitative experience can be enlarged into a quantitative study. No amount of listening or reading can supply as convincing a background for Lenz's law as manipulating a magnet near a coil connected to a galvanometer (Fig. 1.8). Students should be encouraged to play with a prism, white light, and filters, and certainly every physics student should personally observe Brownian motion. When handling apparatus this way the emphasis is on the personal experience, on becoming acquainted with the phenomenon. After the first encounter with, for example, total internal reflection, in which it is shown that the phenomenon actually does take place, the student can measure the critical angle and relate it to the indexes of refraction.

5. *The analog experiment.* As a general rule, we should not study a substitute for a phenomenon if the real thing is available. There are many situations, however, for which we can take advantage of the simplification possible using a model. At some point students should certainly use a counter and scaler to determine the half-life of a radio isotope, but we can introduce the topic with an analog experiment to reveal the statistical nature of what's going on. Whether we choose to pick marked cubes out of a box or let water drain from a column through a capillary tube, the resulting data will give a nearly exponential curve that can be analyzed for "half-life." The way the relationship applies to reality is more easily understood than the phenomenon we shall eventually study.

   Ripple tanks are really analog devices. We are not that interested in water ripples, but we can see and count and measure them, and they follow many of the same rules of behavior as the radiation that we actually want to learn about.

6. *Simulations.* The availability of microcomputers has made the simulated physical phenomenon very popular in the student laboratory. It is true that events that develop too slowly, or are dangerous, or require apparatus too expensive or delicate for student handling can be represented for the students by a computer simulation. But we must be careful not to use a computer in such a way that we violate the cardinal rule about never using a substitute when reality is available. The computer's ability to handle a great quantity of data and to permit studying a large number of special cases does not justify its use for simulating simple trajectories, for example.

   In the references at the end of this chapter we have listed a number of authoritative papers on the use of the microcomputer in the student laboratory (see, for example, Tinker, 1981). In each case, the reader, thinking of his or her own situation, should ponder the question whether the undeniable advantages of using the computer outweigh the loss of the direct experience with real things

■ THE LABORATORY REPORT

In deciding on the nature of the laboratory report we must consider several factors. The full-dress, extended write-up of every experiment, so extensively used in the past, has some merit, but generates many problems. To a certain extent, it is true that "every teacher is an English teacher," and requiring a good deal of use of the language is part of the job. But we are primarily interested in teaching physics and must make sure that time spent in writing does not overbalance the time devoted to physics. This time allotment refers not only to the students, but also to teachers; detailed writing by students means thorough criticism by teachers, with a follow-up of corrections. Are the benefits compatible with the time demands? There is no evidence that a detailed analysis and description of every experiment serves a worthwhile purpose for the student, but there is much reason to believe that the requirement can stifle the delight and creative thought that should accompany laboratory work, and a strong belief that the practice diverts a physics teacher from teaching physics.

On the other hand, we do want students to learn some things that are associated with the traditional formal report. What is needed is a way of keeping records in the lab so that the emphasis is on the physics, not the records; something that students do not see as "busy work" and that does not distract physics teachers from their proper concern.

1. One answer lies in the laboratory diary, augmented by a small number of detailed reports (say, two a semester). The diary is kept in a specified notebook; one useful form is spiral bound, with alternate pages of graph paper and lined paper. Each experiment record starts with essential facts: the date, names of other students in the laboratory group, a title, and a one-sentence statement of the problem. An apparatus diagram is drawn only if it will save words. Procedures can often be covered by descriptive headings on data columns, and data are written directly in the notebook. Graphs are drawn in the laboratory period, often while the data are being taken. Graphing data while they come in is a valuable skill to develop. Students should learn that it usually saves time. If there is a mistake in a particular reading, its position on the graph will stand out, indicating that something is wrong. The matter can be investigated immediately, while the apparatus is still assembled and working. If the student discovers such a problem after the laboratory session is over, there is little chance of recovery. Besides, maybe the out-of-line datum is not a mistake. Maybe it's a clue to the Nobel prize! What a shame to lose the prize because you failed to graph as you go.

   For the record, the students should calculate and state the results of the experiment, including a comment about the significance and reliability of the results. In many cases the entire record is completed during the laboratory class; in fact, some teachers require that this notebook remain in the laboratory for inspection at the instructor's convenience. This style can be modified to suit the situation.

2. Students are told in advance about the limited number of detailed reports to be required. The instructor selects the experiments to be reported because they seem to provide a good opportunity to realize the benefits of such an effort. The write-up can follow the form of a research report and is written in detail to be handed in for review by the instructor. Whatever the requirements the teacher should convince the class that the formal report has a useful purpose and anticipated outcomes and does not involve the "make-work" that busy students quite rightly resent.

   In at least one case, the teacher requires a "bare-bones" report to be completed in the laboratory and makes the traditional formal laboratory report optional for students who want an A or a B for the final course grade.

3. Some teachers have lightened the burden but retained the merits of the full report by having each working laboratory group present a single report on each experiment. The groupings are permanent, and the students take turns in writing up the experiments. All members of the group receive the same grade on the write-up, and students are strongly motivated to work together and help each other. In some programs one member of each group is formally designated as "principal investigator" with the others as assistants (see Murray, 1975). The principal investigator directs the work of the others and writes the final report. Each student acts as the leader in turn.

In summary, although it seems certain that some sort of written record ought to be made for each experiment, the record can take various forms. There are probably as many ways of tackling the problem as there are teachers who have given the problem some thought.

■ THE FOLLOW-UP OF LABORATORY WORK

Even more important than the write-up is the follow-up, in itself a sort of report. A major chunk of class time after the experiment should be devoted to a discussion of the results. In this session, various students state their results, offer interpretations of graphs and calculations, and try to understand and explain the inevitable disagreements. The teacher must beware of assigning the labels "right" or "wrong" to the students' results. Of course, there are mistakes in the laboratory as well as the usual errors. Mistakes can be corrected and can be said to be wrong. Errors, however, are to be expected. We try to reduce such uncertainties, but final results, based on uncertain data, are not "wrong" (except as actual mistakes may have intervened) even though far removed from the consensus or a book value. The class review of an experiment, analyzing difficulties and accounting for error, can be one of a physics teacher's most fruitful teaching opportunities.

■ "HOMELY LABS"

It is possible to provide experiences in experimenting and measuring away from the formal school surroundings, and there are good reasons for doing so occasionally. Some students maintain a psychological wall between the school and what they view as the real world. There is one kind of explanation that satisfies the teacher; there's another that is really useful at work or play. Particularly in science laboratories there is a danger that the special, small-scale delicate school apparatus may seem irrelevant. Why not, therefore, take science laboratories out of the classroom and measure real-world effects in the real world? Students can take their activities home with them or to the dormitories. Science becomes more visible, less foreign, and usually more fun. There is a healthy public relations benefit for the physics course. Furthermore, most home activities require more initiative and creativity than those used in school. Frequently they also involve more physical activity and larger muscles. One ancient maxim of education is that the larger the muscle, the more secure the learning.

How are students going to do experiments without special apparatus and precision instruments? There are many useful and interesting phenomena to explore that require only homely apparatus. For many cases a laboratory experiment can be coupled with an allied take-home project. Usually the precision that can be obtained at home is not as good as that in the laboratory, but so much the better! Large uncertainties in measurement emphasize the fact that there *are* uncertainties, and that they must be taken into account. Even more important, when errors are large it becomes obvious that judgment must be used to determine how much precision is needed, and how much time or effort or money it will cost.

Here are some examples of homely experiments that we have assigned in various courses. A more extensive list is given in Appendix 2 on page 30. Whether a particular experiment is feasible for a particular class depends on the school level, whether home is home or a dormitory, and in some cases on how easily you can loan simple apparatus overnight.

1. Measure the floor area of a house, dormitory, bedroom, school, or physics building. Limit the "expense" in terms of measurement time: for example, stipulate that the actual measurement must take no more than five minutes (thus ruling out, in most cases, the use of precision sticks or tapes, and encouraging the use of pacing or secondary standards).

2. Measure the diameters and volumes of several balls and graph the dependence of volume on diameter. Balls that might be available are Ping-Pong, golf, handball, tennis ball, and baseball. Volume can be measured by water displacement, using a kitchen measuring cup and a pot. (Perhaps the athletic department can furnish old balls that are too scuffed or worn for use but still have the same diameter and volume.)

3. The standard experiment of finding the functional dependence of pendulum period on other variables can be done at home as well as in the laboratory. It is necessary, however, to provide some instruction about the need to have the pendulum swing from a fixed point (a knot and a firm support), and there might be demonstrations and hints about various timing procedures. If the standard experiment is done in the laboratory, then as a home exercise students can find rough values for the variation of period with starting amplitude. A stopwatch and protractor are needed.

■ FOR READING AND REFERENCE

Chonacky, N., "Microcomputer data management in an introductory physics laboratory," Am. J. Phys. **50**, 170 (1982)
  The case for using a microcomputer in the laboratory, with examples, and a discussion of the probable problems to be faced by the neophyte developer.
Cook, G. P. "Grading labs: the self-checking questionnaire," Phys. Teach. **23**, 95 (1985).
  To avoid undue stress on the formal laboratory report and to reduce the time for grading these reports, the author uses a questionnaire, a week after the laboratory experiment. Each student gets

answers from his or her laboratory notebook, which must be in good shape to supply the information.

Corbridge, J. N., and Clay, A. R., "Laboratory teachers and legal liability," Phys. Teach. **17**, 449 (1979).

A detailed and scholarly analysis of a laboratory instructor's liability for negligence, including descriptions of several typical cases and their results. A teacher should determine to what extent insurance coverage either already exists or is available.

DeJong, M. L., and Layman, J. W., "Using the Apple II as a laboratory instrument," Phys. Teach. **22**, 291 (1984).

Game paddles and joysticks provide a microcomputer with the ability to make measurements and displays of physical quantities. This article tells how to use an Apple computer to measure resistance, temperature, light intensity, elapsed time, and more.

Denson, J., and Jones, G. F., "An equipment loan program for high schools," Phys. Teach. **23**, 31 (1985).

A university physics department in a rural setting loans kits of apparatus to local high-school teachers.

Kammer, D. W., and Shaltis, L. W., "Some ideas for the 'poets' physics' laboratory," Am. J. Phys. **41**, 178 (1973).

A description of a laboratory program suitable for a course in physics for nonscience majors.

Kuehl, E. (committee chairman), "An evaluation of high school physics laboratory manuals," Phys. Teach. **22**, 222 (1984).

A committee of teachers here describes and evaluates twelve of the leading laboratory manuals. An exhaustive and valuable study.

Lindsay, J. G., Jr., "Modified, self-paced, limited-option laboratory," Phys. Teach. **12**, 230 (1974).

Students choose experiments from a limited list and come to the laboratory whenever they wish. The laboratory experiment is supervised at all times by a faculty member whose task is tutorial. Written reports are required. The note discusses advantages and disadvantages.

Lochhead, J., and Collura, J., "A cure for cookbook laboratories," Phys. Teach. **19**, 46 (1981).

The design of laboratory activities that teach students to translate between English and mathematics, with a detailed consideration of the way each problem was solved, without using predetermined formulas.

Long, D. D., *et al*, "The influence of physics laboratories on student performance in a lecture course," Am. J. Phys. **54**, 122 (1986).

The authors show that taking a laboratory course has a positive influence on students with intermediate grade averages, less effect on students with high grades, and some indication that laboratory work may actually hinder students with low overall grades.

Long, R., II, "Laboratory learning modules," Am. J. Phys. **43**, 340 (1975).

Laboratory learning modules are self-teaching units that enable a student to learn measurement techniques and instrumentation independently of a laboratory course.

Murray, F., "Designating one laboratory partner as 'principal investigator'," Am. J. Phys. **43**, 734 (1975).

To reduce the report writing and reading load on students and instructors, a single member of each laboratory group is appointed principal investigator. This student directs the work of the others and writes the report.

Oppenheimer, F., and Correll, M., "A library of experiments," Am. J. Phys. **32**, 220 (1964).

A classic paper describing a laboratory organization with experiments assembled permanently, one to a table, and available to students in the same fashion as books in a library. The library-laboratory is open twelve hours a day, five days a week. Regular laboratory classes meet there, and students can also work or browse on their own time.

Patterson, J. R., and Prescott, J. R., "Self-paced freshman physics laboratory and student assessment," Am. J. Phys. **48**, 163 (1980).

A laboratory program that includes a strong element of mastery learning along with the self-pacing. Much attention is paid to assessment (grades). Source: Australia.

Peterson, R. W., *Teaching Physics Safely* (AAPT, College Park, MD, 1985). $10.

A 30-page booklet, written by an expert for other teachers of physics. Covers seven different areas of concern, and includes directions for heart–lung resuscitation, first aid, and selected references. A must!

Price, R. M., and Brandt, D., "Walk-in laboratory: a laboratory for introductory physics," Am. J. Phys. **42**, 126 (1974).

A self-instructional laboratory, making use of audio tapes. Students work through the experiments without assistance from an instructor.

Quisenberry, D. R., "Evaluating the laboratory experience," Phys. Teach. **19**, 549 (1981).

Describes one way to run a laboratory performance test with a small section. The note gives several sample performance-test tasks.

Roberts, D., "Errors, discrepancies, and the nature of physics," Phys. Teach. **21**, 155 (1983).

The principal concepts of error analysis for the introductory laboratory.

Roeder, J. L., "Physics and the amusement park," Phys. Teach. **13**, 327 (1975).

An analysis of many amusement park rides that might serve as the basis of a class "laboratory" experience. (See Taylor *et al.*, 1984).

Reif, F., and St. John, M., "Teaching physicists' thinking skills in the laboratory," Am. J. Phys. **47**, 950 (1979).

The authors have developed a one-quarter mechanics laboratory for the purpose expressed in their title. It uses minilaboratories (single concept) and group laboratories (more extended and complex). Some minilaboratories can be done at home.

St. John, M., "Thinking like a physicist in the laboratory," Phys. Teach. **18**, 436 (1980).

An extension and follow-up of the paper listed above (Reif and St. John, 1979).

Taylor, G. *et al.*, "A physics laboratory at Six Flags Over Georgia," Phys. Teach. **22**, 361 (1984).

Taylor and his colleagues run a laboratory session at an amusement park. In addition to some pedagogical background the authors describe six "problems," each based on one of the rides. (See Roeder, 1975).

Tinker, R., "Microcomputers in the teaching lab," Phys. Teach. **19**, 94 (1981).

Written by a leader in the field, this general article describes a number of applications of computers in physics laboratories and discusses hardware requirements and educational implications.

Toothacker, W. S., "A critical look at introductory laboratory instruction," Am. J. Phys. **51**, 516 (1983).

The author finds that introductory physics laboratory classes do not do very well what is claimed for them. He proposes to eliminate all freshman and sophomore physics laboratory courses.

■ APPENDIX 1: EXPERIMENTAL PROJECTS

1. Measure the period of a simple pendulum as a function of the angular amplitude. This problem starts out being trivial but can and should be carried to the point where sophisticated techniques are required, and the results are complicated. Results should be obtained and plotted for angles at least up to 60°. Precision should be good enough so that the form of the deviation from constancy is evident.

2. Measure and graph the position, velocity, and acceleration as a function of time for a simple pendulum. The three graphs should be plotted against each other and the precision should be good enough so that the relationships among the graphs can be derived.

3. Without using a timing device, find the maximum speed with which you can throw a stone or a baseball. In your feasibility study estimate the precision that you will be able to obtain.

4. Demonstrate experimentally the equivalence of inertial and gravitational mass to high precision.

5. Determine the inertial masses of four different objects by performing recoil experiments on the air track. Compare these masses with their gravitational masses.

6. Set up a gyroscope and measure its precessional angular velocity. Compare this velocity with that expected from your measurements of the other variables.

7. Measure the local g field in a rotating system by growing plants at various points on a turntable. A variation of this project might be to use plumb bobs at various radii, making precision measurements of the local g fields.

8. Produce and record sine-wave patterns starting out with the motion of a spring.

9. Set up an Atwood's machine of your own design and use it to measure the local gravitational field.

10. Compare the period of a simple pendulum swinging in a vertical plane with

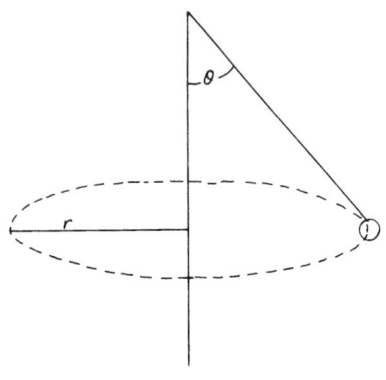

Figure 1.9. A conical pendulum.

its period when the bob is moving in a horizontal circle. Make measurements to find how the period depends on the radius of the circle (r), or on the angle (θ) at the apex of the cone described by the string (Fig. 1.9).

11. Measure the acceleration and maximum speed of an elevator for a and v in both directions.

12. Measure the efficiency of a commercial jack screw or chain hoist.

13. Measure the coefficient of restitution of various balls, including glass and steel.

14. Using two springs from discarded window-shade rollers, or any others available, measure the spring constant for each, then when they are connected end to end, and again when they are connected side by side. Explain the results.

15. Make a hydrometer and calibrate it without actually floating it. Then use the hydrometer to measure the specific gravity of a solution and check with the value of the specific gravity measured by a different method.

16. Measure the moment of inertia of a bicycle wheel. Then make whatever assumptions are necessary to calculate the value from the dimensions and masses of the parts. Compare the two results and analyze the uncertainty contributed by each assumption.

17. Calibrate a thermometer using a blank thermometer and the fixed points of ice water and boiling water. Analyze the accuracy and compare with a commercial thermometer.

18. Devise and perform a precision experiment to determine the mechanical equivalent of heat.

19. Measure the specific heats of several metals with sufficient precision to check the law of Dulong and Petit.

20. Make a careful measurement of P versus T for a gas at constant V and plot the data to determine the functional relationship.

21. Measure $T(t)$ for hot water left in an insulated jar (such as a styrofoam cup or thermos). Graph the data to find the functional dependence.

22. Measure $T(t)$ for water (ice) from -10 to 100 °C, starting with a sample of ice being supplied with heat at a constant rate. Graph your data and determine the heat of fusion per gram and the heat of vaporization per gram.

23. Measure $T(t)$ for hot water in two metal cans. Using identical cans paint one dull black and the other glossy white. Devise methods to reduce convection and conduction. Graph the data and determine the functional dependence.

24. Measure the heat of vaporization of liquid nitrogen, using the techniques described in *The Physics Teacher* in the May 1969 issue on page 289.

25. Measure the density of ice.

26. Measure the temperature–resistance characteristics of a thermistor.

27. Measure the temperature–length coefficient of a rubber band while it is under stress.

28. Make a precision measurement of the speed of a transverse wave in a length of clothesline. Compare your result with the calculated value.

29. Measure the speed of sound in aluminum by producing compressional vibra-

tions in an aluminum rod and measuring the wavelength and frequency when the rod is vibrating.

30. Examine and analyze a slide projector, opaque projector, or overhead projector. Measure the optical properties of all the mirrors or lenses and also the geometry of the instrument. Draw a ray diagram showing how it works and compute and test the magnification.

31. Measure and plot the index of refraction of two pieces of glass or plastic as a function of wavelength. For one material use a glass prism, and for the other use a plastic prism. Make the measurements with appropriate precision.

32. Design a pinhole camera (Fig. 1.10) and take a good picture with it.

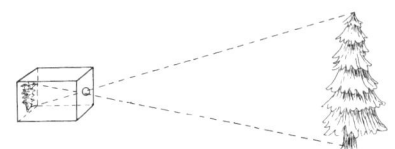

Figure 1.10. A pinhole camera.

33. Produce a series of pictures demonstrating the properties of a camera that has a range of shutter speeds and lens apertures. Demonstrate, among other things, the extent to which shutter speed and lens opening (as indicated by f/number) are interchangeable parameters.

34. Measure and plot the diameter and area of a human eye pupil as a function of light intensity.

35. Measure the optical properties of an eyeglass lens that has a strong effect, including at least one other correction besides that for myopia or hyperopia. In your report give a ray diagram illustrating the use of the lens.

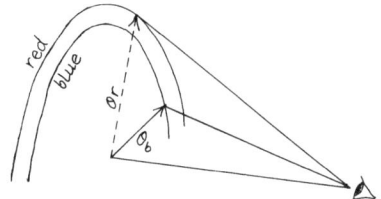

Figure 1.11. A rainbow.

36. Produce a rainbow and measure the angular displacement of the red and blue portions from the center of the bow (Fig. 1.11). In your report explain all, with diagrams.

37. Measure, plot, and find the functional dependence of light intensity on distance from a point source, a cylindrical source, and a plane source.

38. Produce a real image so that it can be viewed as if it actually existed in space. Diagram the light rays for the effect.

39. Devise an experiment to measure the way the index of refraction of sugar solutions depends on the amount of dissolved sugar. Present your results in a graph with an indication of its precision.

40. Make a hollow triangular prism by cementing three microscope slides together edge to edge (see Fig. 1.12). Cement a fourth slide to the bottom of the prism to complete the container. Use the hollow prism in measuring the dispersion of two different liquids. (Caution: Some liquids dissolve some cements!)

41. Measure the resistance–temperature coefficient of iron. (Construct a dc bridge.)

42. Use a cathode-ray tube to demonstrate quantitatively the Lorentz force on electrons.

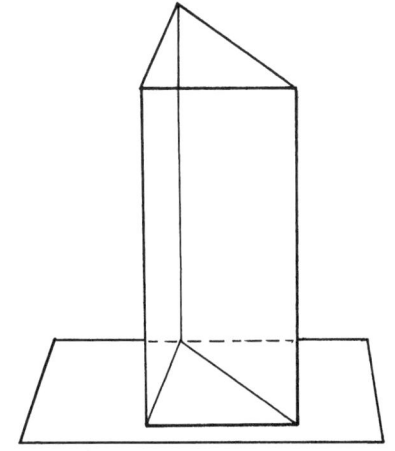

Figure 1.12. A hollow prism.

43. Build a thermocouple, explain why it works, and check it against a mercury thermometer.

44. Build a battery and measure its emf and internal resistance.

45. Construct a capacitor of several tenths of a microfarad, using aluminum foil and plastic wrap. Calculate the appropriate dimensions in advance and measure the capacitance when you have finished.

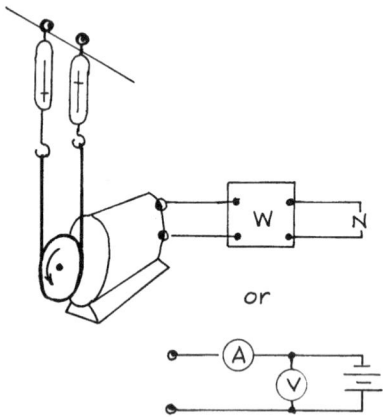

Figure 1.13. Measuring the efficiency of a motor.

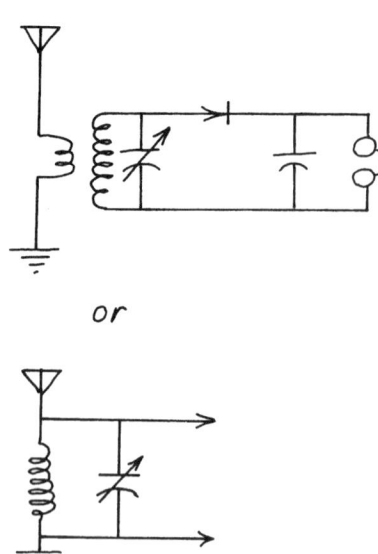

Figure 1.14. A crystal radio.

46. Measure quantitatively the magnetic field in the vicinity of a long straight wire with a current in it and plot the field as a function of the radial distance to the wire.

47. Construct an audio-frequency oscillator with its frequency controlled by a variable capacitor.

48. Build a transistor audio-frequency amplifier with a gain of 10. Calculate the values of components in advance.

49. Measure the output power of a small electric motor (Fig. 1.13). Use this result and your measurement of the input power to calculate the motor's efficiency.

50. Build a crystal radio using individual components obtained from the department's stock room or ordered separately from an electronics supplier (Fig. 1.14). Demonstrate it and explain how it works.

■ APPENDIX 2: "HOMELY" EXPERIMENTS

1. Measure in metric units all sorts of body sizes: height, circumference at waist, foot length, index finger width, index fingernail thickness, distance between elbow and wrist when hand is at right angles to arm, distance between eye pupils (using a mirror), distance between nose and fingertip when arm is held out to the side, and length of the pace (double step). These lengths can be measured with string and a plastic metric ruler. Measuring them provides exercise in the use of the metric scale and familiarity with the units when they appear in calculations.

2. Find human density by measuring mass (converting a bathroom scale reading) and volume by estimating the height, width, and depth of a box you could fit into (or height and diameter of cylinder). The volume calculated this way will probably be about twice the actual volume—a fact that will show up when the density is calculated and you compare your density with that of water. For a more accurate measurement of your volume, fill a bathtub half full. Mark the level, then get in and submerge yourself. Have a friend mark the new level. Calculate the volume from the dimensions of the tub.

3. Measure the mass, in kilograms, of common objects using a bathroom scale, a baby scale, a kitchen scale, or a spring scale borrowed from school. The objects might be a book, pen, nickel, shoe, cup of water, quart of water, liter of water, baseball, Ping-Pong ball, or yourself. With most scales conversion of units will be required. Some cases will call for ingenuity, such as weighing many nickels to find the mass of one or subtracting the mass of the container.

4. With a protractor, ruler, and string, various home-surveying exercises are possible: distance between buildings, height of buildings, width of a street, etc.

5. Measure reaction time by having a partner drop a ruler while you attempt to seize it between thumb and forefinger (see Fig. 1.15). The whole family or dormitory group will enjoy this one.

6. Determine the average speed of a runner, a bike rider, or a car between measured fixed marks. The time can be measured with a stopwatch or by using a counting method such as "one thousand and one, one thousand and two, etc." (With a little practice this method can be surprisingly precise down to one-fifth of a second—the time for one syllable.)

7. Measure g to within ±35% by dropping (or throwing) a ball. To get the time use only a pocket stopwatch, or better yet, counting fifths of a second (see item 6). If the ball is tossed to near-ceiling height, the time of fall will be greater, as will also be the uncertainty of the height measurement.

8. If it is not to be done as a class experiment or a demonstration, students can study the motion of a marble rolling down an incline. Prop up one end of a flat table to make a slope. Mark the position of the marble (or other ball) at the end of one second, two seconds, three seconds, etc. The average velocity, change of velocity, and acceleration can be computed directly. A graph will show the relationship between distance and time.

9. Adjust a water faucet so that it drips steadily. Take periodic measurements using a watch and a measuring cup to find the way the total water loss depends on time. Get the rate of the leak in liters per second from a graph of the data.

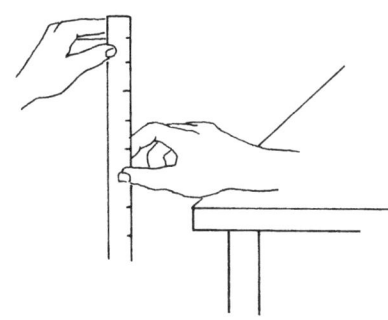

Figure 1.15. Measure reaction time.

10. Measure the angular width in radians of your index finger, palm, and hand-span when seen at arm's length. Use them as measuring tools to find the angular height of the north star above the horizon. Compare with the local latitude.

11. When you are a passenger in a car, hold a plumb bob and watch it as the car speeds up, slows down, moves at constant speed, and turns a corner. If you can obtain a helium-filled balloon, try the same experiment using the balloon as an inverted plumb bob.

12. Study resonance. Hang two pendulums from near the center of a string that is stretched horizontally between two fixed points, such as two chairs. (See Fig. 1.16.) Call one pendulum the driver and the other the oscillator. Keep the length of the oscillator constant and vary the length of the driver. When the driver is a bit shorter than the oscillator and is set swinging, how does the oscillator behave? Change the length of the driver in regular steps. When is the oscillator response strongest? Compare the phase of the two pendulums under various conditions. Observe the energy transfer.

Figure 1.16. Mechanical resonance.

13. Describe the length and apparent linear density of piano strings for high, middle, and low pitches. How does the length and linear density of a stretched rubber band affect its pitch when plucked? Note that as you increase the length of a rubber band by stretching it, the linear density decreases.

14. Find the speed of sound using an echo. You will need to be able to get about 50 m away from the flat side of a large building or a cliff. Clap your hands and listen for the echo. If you clap rhythmically you should be able to find a distance and clapping frequency for which the echoes arrive between the claps. Have a colleague use a watch to determine the frequency of clapping, pace off the distance to the wall, and solve for the speed of sound.

15. Cook some ice. Obtain a thermometer, a pan, ice from a freezer, and a steady source of heat (hot plate or stove). Working quickly, crush the ice, put the pan of ice on the hotplate, stir with the thermometer, and take the temperature of the contents. Continue taking the temperature once a minute until the pan and its contents are at about 20 °C. Plot a graph and estimate the heat needed to melt the ice.

Figure 1.17. "Truth mirror."

16. Use a large cookie tin or a bathtub as a ripple tank. Make circular waves by dipping a finger in the water and plane waves with a piece of wood or cardboard. Reflection is easily seen, but refraction, which depends on the depth of the water, is more difficult to observe. Use two fingers, dipping rhythmically in the water to see an interference pattern. The Doppler effect can be observed by moving the dipping finger steadily in one direction. To see diffraction, let a plane wave pass through a gap between two pieces of wood or cardboard.

17. Make a "truth mirror." Mount two rectangular mirrors at a right angle with their edges together (Fig. 1.17). If you look at yourself in this corner reflector, you will see yourself as others see you. Try combing your hair while looking into this double mirror. What happens to your image if you rotate the combination so that the connected edge is horizontal? Draw a ray diagram to explain the effect.

18. Study the diffraction of light. Punch a pinhole in a piece of aluminum foil and use the foil to cover a bare light bulb so that when you move away you can see a tiny, bright point of light. Stand a few feet away from this point source of light and look at it through pinholes of various sizes in another piece of foil held close to the eye. Try looking at the source through a pair of pinholes placed as close together as possible. Make and use other pairs of pinholes with different separations.

19. Find the focal lengths of eyeglass lenses used by several farsighted people by casting images of distance objects on a screen. If one of the lenses has a strong astigmatism correction, you may find two focal lengths.

20. A very sensitive electroscope can be made by hanging a 1-cm scrap of aluminum foil by a piece of light thread. Most of the usual electrostatic experiments can be done with this simple electroscope: identifying chargeable objects, attraction and repulsion, distinguishing between charges, conductors and insulators, induction, etc.

21. The current produced by a "lemon cell" (a lemon containing a copper penny and a galvanized nail in separate knife cuts) can be detected by a homemade tangent galvanometer. To make the galvanometer, wind a dozen or so turns of insulated copper wire (enameled No. 26 is good) around a magnetic compass. If the projecting penny and nail are simultaneously touched to the tongue, the current can also be detected by taste.

22. A portable AM radio can serve as a direction finder. The signal received is weakest when the built-in antenna points at the transmitter. Use such a radio to find the directions of several moderately weak stations (not local ones) and compare with the directions shown on a road map.

PART E. EVALUATION: TESTS AND GRADES

Most teachers have mixed feelings about tests. On the one hand, the organization of our system of education demands that student achievement be evaluated; thus we must invent some tasks for our students to help us rate them in their learning of the subject. On the other hand, making tests that reliably reveal which students have done well and which ones poorly presents a problem that bristles with technical difficulties, requires a lot of a teacher's time and effort, and nearly always brings more blame than credit. Most teachers and students have dreamed wistfully of a physics course without tests and some teachers claim to have devised just such a program (see Lucido, 1982). The trouble with this effort is that tests serve purposes more important than merely ranking students and too few teachers plan their test programs to take these goals into account.

The literature on tests is enormous; it has been a favorite topic of education research for years. Much of this research has developed ideas that are far more technical than could be useful to a classroom teacher. In what follows we shall stay close to matters that can be applied by teachers, ignoring many of the scholarly advances that (no matter how true) may be too complex to be usable or only marginally useful for the actual teaching of physics.

■ WHAT ARE THE REASONS FOR TESTING?

1. The most obvious reason for using tests is to help arrive at a grade for each student. But just as this is not the sole reason for testing, test results should not be the sole factor in determining grades. This topic will be addressed more fully later.

    The grade-setting function of tests, with the resulting feeling that test marks will help determine students' futures, is the main reason for the test anxiety that every teacher has observed. It is our duty as teachers to lower the pressure by making clear to students that tests have other purposes, by writing tests that students will think fair, by interpreting test results with humanity, and by making it known early in the course that factors other than test results will contribute to final grades.

2. A test is an important instructional opportunity. Test questions have a unique impact on the student mind. The selection of topics to be tested and the organization of questions provide significant clues to the nature of the course. We must write items that will not only test but also help students to understand both physics and our expectations of them.

3. The prospect of a test is a powerful motivating agent for students. Putting one's best foot forward ("looking good") and a sense of competing successfully with others are elements of human nature. A favorable test result satisfies both these basic urges. This purpose, however, will not serve all students; it goes wrong when the student feels threatened by the test. Motivation by fear is real enough, but unproductive in the long run.

4. Not only do tests spur students to greater efforts, but taking a test and studying its results are helpful to students. Misconceptions and gaps in knowledge are often revealed and can be corrected. Besides, student morale requires an answer to the question "How'm I doing?" Tests for this purpose are called "mirror exams" by Rogers (1969).

5. Someone has said, "Tests test teachers." There are two aspects of this idea: one invaluable, the other unjustifiable. A teacher needs to appraise his own success. Sometimes the results of a test are discouraging or even humiliating. But we all need to face this reality, and it is the poor results, the omissions, and the discovered misconceptions that are the most instructive.

But the idea of rating teachers professionally by using student test results is unacceptable. Good results do not necessarily indicate good teaching nor bad results bad teaching. The results of tests that were devised to rate and teach students should be only a minor piece of evidence of the competence of the teacher.

6. One characteristic of a good teacher is flexibility, a willingness to face new situations as they arise, never becoming locked into a fixed program, but using clues provided by students and classes to suggest the need for occasional "midcourse corrections." Test results enable a teacher to spot specific difficulties. Perhaps immediate reteaching taking a different approach is indicated, or a change in order, emphasis, or development for the next time the subject is taught.

■ TEACHERS SHOULD WRITE THEIR OWN TESTS

1. Note that of the six purposes of a testing program described above, only one, the first (determining grades) is implicit in the whole idea of testing. The others may or may not be achieved depending on the teacher's convictions and efforts. A teacher has the best chance to make the tests truly helpful by using homemade tests. The course may well culminate in a professionally written published test; such tests are designed to achieve purpose (1) noted above and may succeed very well. But students deserve tests that take into account their preparation, not only the content but their interaction with the teacher and even their local social backgrounds. As for the teacher, it is an important part of professional experience to write valid tests. The act of producing the tests will help with teaching.

2. There may be large-course situations in which the work of several teachers must be coordinated so that they give the same test at the same time. But even in this case every teacher who is directly responsible for students should have a hand in test preparation.

■ TYPES OF TEST QUESTIONS

1. The free-response item ranges in freedom from "tell what you know about..." to the relatively restricted "fill in the blank...," neither of which extreme forms has much to recommend it. Among typical free-response questions are the familiar essay, problem, and completion items, together with a variety of ingenious variations to be described later.

Among the advantages of using free-response test items is the important fact that they reveal to the instructor more about the student than do objective questions. Essay and problem questions can be written so that the answer requires fresh thinking, imagination, and a demonstration of a "feeling" for physics. Such items also test a student's grasp of interrelationships in course content. They may be easier to compose and thus take less time in the writing than objective items covering the same subject area, but when prepared with meticulous attention to possible student misinterpretation and similar technical details their development can be demanding indeed.

There are three major drawbacks to free-response items. Marking them takes time, effectively canceling any advantage they may have in being quickly written. Free-response items have a serious scoring unreliability. This is true even if scoring is to be done by just one person because the scorer's judgment is likely to change as more and more papers with varying answers are read. The problem is most severe if a question is marked by different readers. The story is told of one examiner giving a failing grade to a paper that served another examiner as a key of ideal answers. It could happen!

The third trouble with essay or problem questions is the limited range of topics that can be covered in a test of reasonable length. A side result of the narrow coverage is that this sort of question can unduly reward or punish a student who has a deep but specialized knowledge of a topic.

2. The common characteristic of objective test items is that the testee selects a response that has been stated in the question. The most common form of the objective question is the multiple-choice item, with its variants. True–false items and matching sets are also objective but the first is now regarded as weak, and the second, although acceptable, has also seen its day. The multiple-choice question can be applied effectively to most testing goals, and we shall, in what follows, emphasize this form of objective question.

Among the advantages of using objective test items is that the element of personal bias is removed from the scoring process. Ideally each item has only one acceptable answer. This feature means that the task of scoring is easily and quickly done, and does not require professional attention. A broad sampling of subject matter is possible, and as a further result of the use of many items per test each item has a relatively low value. Thus an inadvertent ambiguity or a misreading by the student has less influence. In physicists' terms, if there are enough data a few points that fall off the curve need not invalidate the conclusion.

On the other hand, objective test items have been justly criticized as unrealistic. In life we do not often solve problems by picking an answer from a set of stated alternatives. It is also true that exclusive use of objective items may make students immune to certain desirable educational outcomes, for instance, expository skill and value judgment. Nor can objective questions test some abilities, such as creativity, expression, organization, and manual skills. Another drawback of the multiple-choice item is the penalty it places on students of low reading ability, such as may be the case with students of disadvantaged backgrounds, or foreign students.

Some students develop a "quiz skill" for objective tests, seeming, without having real knowledge, to be able to select a correct choice from four or five alternatives by using subtle clues found in the wording of the stem and the interrelationships between the choices. Of course, the test writer tries to eliminate such "tips." This is just one reason why good objective test items are hard to write. Many multiple-choice questions written by experienced professionals can be shown to be flawed [see Lehrman (1977) and Iona (1978)].

After weighing the points stated above it must be clear that free-response items and objective items have their characteristic merits and drawbacks, and that both types of questions should be used, each for the purpose it serves best.

3. A very different sort of test, or assessment, is being proposed by many educators, particularly those in high administrative positions. Since there are so many faults to be found with the standard tests—ethnic and gender bias, arbitrariness of grading, narrow sampling of accomplishment, psychological pressures on students—there should be *authentic* tests. The very word implies that our traditional tests are artificial and wrong. In the best of all worlds, an authentic test would be a demonstration that the student had accomplished certain prescribed tasks and could now perform in new ways. The fad word for the nature of this proof is *portfolio*. In physics the portfolio might consist of a number of essays showing understanding of particular concepts, open-book solutions of interesting real-world problems, laboratory reports of investigations, or perhaps a Nobel prize.

The system has its adherents, but of course there are authentic problems. The very word *portfolio* encourages students to "gussy up" their offerings with cutesy covers, poems, and art work. As with any take-home assignment, there is the question of who did the work. There has been no large-scale demonstration that the system is practical under general conditions or that it would yield better instructional results.

■ WRITING A TEST

The first step in writing any form of test is to identify the subject matter to be covered. This is best done by listing the essential ideas. The written list is used as a checklist while composing the test, making sure that no basic concept is either neglected or overemphasized. This means that the selection and weight of items on the test will reflect what the teacher considers to be most important.

As suggested above, writing essay questions and problems can be done very quickly, but a perfunctory approach to so important a teaching task will cause trouble later in uncertainty of scoring, resulting student resentment, and if done as a general practice is likely to bring about disrespect by the students for the teacher, the tests, and the course.

Here are some suggestions for writing essay and problem questions:

1. Be specific in making clear what is expected as an answer. Indicate the proper length of the answer to an essay question, or suggest the time to be spent. Break the question into parts, stating the credit to be assigned for each part.

2. Try to avoid the common flaw of narrow coverage by having each question cover a broad but related field.

3. While writing the question, decide on the ideal answer and the distribution of credit. Making yourself answer the question while writing it will reveal the questions with multiple correct answers (and hence scoring problems) and the seemingly broad question that really belabors just one point.

Two examples of free-response items follow:

1. Taken from an exam written for university physics majors; 20 minutes, 20 points. A ball of *mass 0.5 kg* hangs on a spring of negligible weight, having a spring constant k. The level at which the ball hangs at rest is called $y = 0$. (See Fig. 1.18.) The ball is pulled down to $y = -4$ cm and *released at rest at* $t = 0$, as shown in Fig. 1.18. It then oscillates with a *period of 0.2 seconds*.

   a. Find the value of y for the highest point reached by the ball, and *find the time* at which the ball just reaches this high point. (2 points)

   b. Carefully sketch a graph of y versus time t for the ball from $t = 0$ to $t = 0.4$ seconds, using the axes provided. (4 points)

   c. Give the equation for y as a function of time t, putting in the values of all the constants and specifying units. (4 points)

   d. Find the equation for the acceleration a of the ball as a function of t. (4 points)

   e. Show how you can use Newton's second law and the results in (c) and (d) to obtain the value of the spring constant k, and give the value (little credit for just the value). (6 points)

2. From a high-school physics exam: A block of wood weighing 12 N is given a kinetic energy of 8 J, moving horizontally. The block slides across a surface and stops in a distance of 1 m.

   a. Calculate the force of friction that acts on the block. (2 points)

   b. How much thermal energy was developed? (1 point)

   c. Is this energy conversion a reversible process? Explain. (3 points)

(Note that this item provides more information than is needed to arrive at the answers. Thus selection of the relevant data is part of the problem.)

Here are some examples of essay questions in which college students were told that they would have to explain one or more phenomena or concepts from a list of five or six given in advance. All of these topics had been emphasized in class. Each question was broken down into a series of four or more sequential subquestions, both to help the student write a logical explanation and for uniformity in grading.

1. Explain to the horse why he can successfully pull the cart, in spite of Newton's third law. (Draw diagram, show forces.)

2. Explain with the aid of a diagram how astronomers measure the distance to a local star. Be quantitative.

3. Explain with diagrams what happens to the scale readings and why, as you stand on a bathroom scale in an elevator.

4. Explain, including the use of a diagram, why astronauts are apparently weightless in space.

5. Describe Newton's first law, and explain why it is not simply a special case of the second law.

6. Explain, using a diagram and equations, why a 2-kg ball does not fall faster in a vacuum than a 1-kg ball made of the same material. Then explain why, in air, it does.

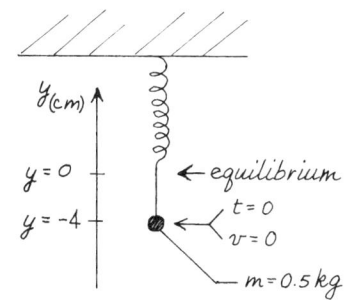

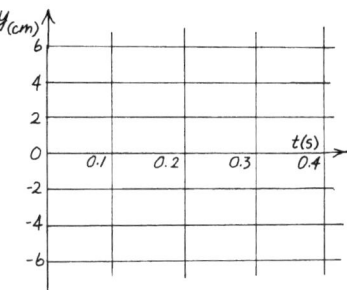

Figure 1.18. Exam problem for simple harmonic motion.

7. Use diagrams to help you explain exactly what a skater does to go into a twirl, and explain the result in terms of angular momentum.

8. Describe in detail how to calibrate a blank mercury thermometer. Why is there such a large change in the length of the mercury column when the volume expansion coefficient of mercury is only 1.7×10^{-4}?

9. The first law of thermodynamics is $Q = \Delta U + W$. Explain the meaning of this equation, including the meaning of each symbol. In the case of an ideal gas, describe how ΔU and W can be defined by P, V, and T.

10. In terms of the microstructure of matter, describe what happens to Q as it enters (a) an ideal monatomic gas, (b) an ideal diatomic gas, (c) water vapor, or (d) a metal.

Students hate these explanation questions at first, but most end up liking them and working hard to prepare answers in advance. They are encouraged to do so, and to get help and advice in advance. Of course, that's when a lot of teaching gets done!

Here are some suggestions for writing multiple-choice items:

1. Try for a good sampling of the ideas to be tested, but make sure that you do not overstress testing for the technical vocabulary of physics.

2. Keep the vocabulary and sentence structure as simple and direct as possible. Try to write many short items. Avoid negatives ("no" or "not"), and exclusives ("always" or "never"). Do not try to be clever or cute; as the testwriter's wittiness goes up the readability goes down.

   Avoid particularly writing the kind of question pinpointed by Hoffman (1961) in which one of the choices, although not obviously correct, may have more fundamental truth than the "right" answer, thus penalizing the thoughtful student.

3. Make each item internally consistent: a single concept. Make items independent of each other.

4. Try to write items that deemphasize memorized facts, but instead are designed to cover specific objectives. Examples are given below.

5. Use at least four choices for each item. Make one choice unambiguously the best choice, and make the distracters (the wrong answers) plausible.

6. As you write a series of multiple-choice items be sure that the correct answer does not too frequently appear as the same-numbered choice. It is easy, for instance, to favor the last choice.

Examples of multiple-choice test items are as follows:

1. From a university physics test: A potential difference V is imposed across the ends of a cylindrical copper wire that has a resistivity ρ, radius r, length l, and temperature (in kelvin) T (about room temperature). Which change would result in the largest percentage increase of the current in the wire?

    a. A 1% increase in V.

    b. A 1% decrease in ρ.

c. A 1% increase in r.

d. A 1% decrease in l.

e. A 1% decrease in T.

(Comment: Even though each student was given a reference sheet of formulas and useful numbers and one of the given relationships was $R = \rho l / A$, the solution of this item requires that the student apply several concepts, and it is a relatively difficult question. It is an example of an item in which no distracter can be dismissed without thought.)

2. Taken from the same university physics test: "Two systems in thermal equilibrium with a third system are in thermal equilibrium with each other." The statement above is

   a. The zeroth law of thermodynamics.

   b. The first law of thermodynamics.

   c. The second law of thermodynamics.

   d. The third law of thermodynamics.

   e. The definition of thermal equilibrium.

(Comment: This question is more straightforward, requiring only recall of a definition. The words quoted were lifted right out of the textbook. Items like this that demand recognition of vocabulary invite last-minute cramming, an effect enhanced in this case by using a statement repeated verbatim from the text.)

3. Taken from an article on achievement testing, where it served as an example of an item stressing a basic concept: A ball is thrown vertically upward with an initial speed of 9.8 m/s. At its highest position the ball has

   a. No acceleration.

   b. A downward acceleration of 9.8 m/s^2.

   c. An upward acceleration of 9.8 m/s^2.

   d. An instantaneous acceleration of 0, which quickly becomes 9.8 m/s^2.

   e. An acceleration change from 9.8 upward to 9.8 m/s^2 downward.

(Comment: The topic is excellent, the lively class discussions promised by the author will surely ensue, but not all will be for constructive reasons. Students should properly object to being told in the stem that the initial velocity of the ball was 9.8 m/s. Giving the number appears to be an effort to lead the student's thinking astray; it has no bearing on the correct answer to the question.)

4. From a high-school test: Mud flies off a spinning bicycle wheel. From the point of view of the cyclist the best explanation is the following:

   a. Centrifugal force threw it off.

   b. The centrifugal force equaled the centripetal force so the mud was in equilibrium and traveled in a straight line.

c. The centripetal force was less than the centrifugal force.

d. The centripetal force was not great enough to make the mud continue in its circular motion.

[Comment: This is a tricky question, even though the stem does identify the frame of reference (the cyclist). In the frame of reference of the *wheel* the preferred answer would be a.]

■ WRITING TEST ITEMS FOR SPECIFIC KNOWLEDGE AND SKILLS

Nedelsky (1965) has done a great deal of pioneering in the area of testing. Some of the material that follows was inspired by this valuable, but excessively detailed, work. The purpose of a physics course is not merely (a) to provide students with straight knowledge; that is with definitions, facts, and principles, but also (b) to give them an understanding of these basics and of interrelationships between them, and (c) to give them an appreciation of the part that experiment plays in the development of physics, and (d) to provide some skill with, and comprehension of, mathematical relationships and graphs. We shall use these representative but incomplete goals in the following analysis.

We must not be satisfied with writing a test that deals with just one of these ideas. Too many test questions concentrate on the facts and laws of physics without attempting to assess our success with other less direct goals.

Here are some examples of multiple-choice items, mostly at the high-school level, that have been written to test specific objectives.

I. Testing facts, definitions, and vocabulary

1. Waves spread out after passing through a gap. This is called

    a. Refraction.

    b. Diffraction.

    c. Dispersion.

    d. Diffusion.

    (Comment: This is a routine item calling only for recognition of a word. This type of question makes up too great a portion of many physics tests.)

2. Of the following, the frequency that is audible to the human ear is

    a. 0.5 Hz.

    b. 5.0 Hz.

    c. 5.0×10^3 Hz.

    d. 5.0×10^5 Hz.

    (Comment: Another single-fact question, requiring, in addition, some understanding of exponential notation.)

II. Laws and principles

1. When a photon collides with an electron the photon undergoes an increase in its wavelength. This shows that

    a. The photon has lost energy.

    b. The electron has wavelike properties.

c. The electron has partially canceled the photon wave.

d. The photon's speed has been reduced.

(Comment: The student need not have studied the Compton effect, which is not always taught in introductory physics, to recognize the truth of the first choice, based on the idea of the particle nature of light. Perhaps a penalty of one point should be exacted for any students selecting d.)

2. A lump of putty and a steel ball each weighing 0.5 N are dropped from a height of 2 m onto a massive steel plate. The putty sticks to the plate and the steel ball bounces. The resulting thermal energy is

   a. Greater for the putty than for the steel ball.

   b. The same for both but not zero.

   c. Greater for the steel ball than for the putty.

   d. Zero.

   (Comment: It might have been better to omit the numbers from the stem, specifying merely that the weights and heights were the same in both cases.)

III. Understanding of experimental techniques

1. If you wanted to measure the speed of sound you might use only an air column of variable length, a source of sound of known frequency, such as a tuning fork, and

   a. An oscilloscope.

   b. A stopwatch.

   c. A meter stick.

   d. An amplifier.

   [Comment: The experiment might work better if the sound were amplified (choice d) but could not be done at all without measuring the length of the resonant column. Thus c is the only correct answer.]

2. The heat of vaporization (L_v) of water is measured by bubbling steam into cool water in a calorimeter. L_v is then calculated after measuring the mass of the water, its temperature change, and the mass of the condensed steam. The result might be much higher than the accepted value if

   a. Hot water from the boiler entered the calorimeter along with the steam.

   b. A lump of ice used to cool the original water was not completely melted when you started the experiment.

   c. The water in the calorimeter changed temperature from 15 to 90 °C.

   d. The steam delivery tube removed some water from the calorimeter at the end of the experiment.

   (Comment: This item would be very useful for students who have done the experiment themselves. It would be inappropriate for students unacquainted with the methods of calorimetry, but illustrates a type of question that demands understanding of an experimental procedure. Note how much more difficult the question would have been if the last sentence of the stem were changed to "all of the following procedural mistakes would have made the result lower than the accepted value except". Nothing further would have been tested, but the reading difficulty would have been much greater.)

IV. Skill with measurements and units

1. A student uses a ruler with millimeter markings on it to measure the width of a 4-in. card. If he or she is careful and has an accurate ruler the percentage uncertainty in the measurement will be about

    a. 0.02%.

    b. 0.5%.

    c. 2%.

    d. 20%.

    (Comment: This question could have been made easier by giving the width of the card as 10 cm. The guessers will choose c, not appreciating the precision with which a simple ruler can be used.)

2. In solving a physics problem a student finds the units for the answer to be kg m^2/s^3. This represents

    a. Force.

    b. Energy.

    c. Power.

    d. Momentum.

    (Comment: This brief and seemingly direct-knowledge item requires manipulation of the units and has the merit that none of the choices can be dismissed without thought.)

V. Understanding general principles

1. Joule found in various experiments that 1 foot-pound of work always produced about 780 British thermal units (BTUs) of thermal energy. However, no cyclic engine has ever succeeded in converting 780 BTUs of thermal energy into 1 foot-pound of work because

    a. It would violate the law of conservation of energy.

    b. Thermal energy cannot be converted into work.

    c. Some thermal energy must be delivered by the engine to a low-temperature sink.

    d. Of the equivalence of mass and energy.

    (Comment: The question suffers the fault of not having reasonable distracters.)

■ ASSEMBLING THE TEST

1. Consider first the time available and make sure that the test is not too long. There is little justification in requiring speed in answering test questions and much reason to believe that many students are frightened by the prospect of a test that they cannot complete. Reflective thinking of the sort required by many good test questions, is seldom helped if done under pressure.

A multiple-choice item that does not depend mainly on recognition of a name or fact should be allowed about a minute. A multiple-choice item that is really a problem may take much longer, perhaps four minutes. For a test with a reasonable range of difficulty among the items, allow about a half-hour for 20 multiple-choice questions. If all multiple-choice items are problems, then eight items per half hour is about right.

It is much harder to plan a time allotment for essays and problems. A rule of thumb that might be helpful is to allow students about four times as much time as you yourself or a colleague need to write a complete answer to the question.

2. Both free-response and objective items should be used, but not necessarily on the same test.

3. Do not bother to plan the test so that credit totals just 100 points. Using the method of grading to be recommended below, a test totaling 23 points can be evaluated just as easily as one with 20, 50, or 100 credits.

4. When a test is being given to "cover" a subject and the students have studied the same material under the same teacher, there is no need to give any choice of questions.

5. Put the most straightforward items first and those requiring extended or imaginative reasoning last.

6. Many students, including some whose records in other courses have been brilliant, come to the physics class with the belief that success will be assured by memorizing formulas, definitions, and laws. Try to combat this by supplying at the head of each test whatever numerical information will be needed, and even a collection of commonly useful formulas. We shall deal later with open-book exams and the "cheat sheet."

■ HOW OFTEN SHOULD YOU TEST?

The extremes are, of course, the daily quiz along with unit and quarterly tests, and the single test that covers the work of a whole semester. As in most matters moderation is the best policy.

The daily quiz wastes valuable class time and puts too much emphasis on marks. One suspects that teachers who give a quiz every day are more concerned about discipline than teaching and are using the test to call the class to order or to keep students busy until the period-ending bell. There are better ways to motivate students to do their assignments.

The single do-or-die course test puts too much pressure on students whose entire grades will be determined by that unique experience, no matter what their personal situation may be on that fateful day.

The most reasonable schedule is to give a major test at the end of each well-defined unit of work, for example, "Electric Currents" or "Dynamics." A test of this sort must be announced in advance and the subject content to be covered should be carefully defined. Some teachers adhere to this rule, but also have an understanding with the class that a minor portion of the test will include material covered earlier in the course. This helps students appreciate the coherence of the subject and encourages a repeated (if perhaps superficial) brush-up review of what came before the current topic.

Unannounced or "snap" quizzes have a perfectly legitimate place in this testing program but have more a diagnostic than evaluative function and should not be allotted much time or credit.

The Open-Book Test and Its Variants

Although memory is recognized as an essential component of scholarship, it is less important in the study of physics than in most other disciplines. Physics teachers place less emphasis on the memorization of laws, in mathematical form or otherwise, than on their relevance to real-world situations and to each other. "$F=ma$" is not a useful answer to a physics question, but it certainly points to a procedure for obtaining answers to a variety of problems. Many students have found in other studies that memorization is the key to success. They arrive in the physics class to find that rote learning is relatively powerless. There are several ways of using tests to discourage memorization without understanding.

Many test writers deemphasize memorization by supplying a reference sheet with each test. The student need not memorize a complex formula or a difficult definition because he knows that it will be given to him when he takes the test, and the test questions will ask him to apply it or demonstrate his understanding of its consequences.

Another way of making the same point is the so-called "cheat sheet" (see Williams, 1980). Students are allowed, or in some cases required, to bring to the test a single page in their own handwriting containing information of any kind that they think may be useful. This sheet must then be handed in with the test paper. Having each student write a reference sheet encourages (or requires) a disciplined review of the physics to be tested.

The open-book test also has its advocates. One of the great advantages of this practice is that it rewards the student who has used the text enough to become thoroughly familiar with it. Another justification is its close resemblance to reality; outside the classroom one goes to references for help in solving problems. The open-book test is another way to take the premium off memorization.

Students sometimes complain that insufficient time was allowed for referring to the text, and the answer is a matter for the individual teacher's judgment. We certainly don't want students to have enough extra time to do the studying that should have been done last week, nor do we want the student to look up the answer to every test question. A judicious way to handle the timing problem is to caution students not to use the text until after they have answered all the test items they can without it. In writing an open-book test, the teacher should probably not write more items or more difficult ones than might be used on a closed-book test.

■ GUESSING

Students have nicknamed the multiple-choice item the "multiple-guess" item with some reason. Guessing should be encouraged because in most cases it is based on knowledge. The student who succeeds in eliminating two of the four choices in a question by recognizing that they are inappropriate has raised his chances in guessing from one in four to one in two, and if several of his guesses have been correct, he or she deserves some credit.

A good preliminary instruction for students is to answer all items they can, skipping those of which they are uncertain, and making sure that they do not devote excessive time to any one question. Then they can return to the more difficult items and attempt to eliminate choices that seem obviously wrong, making a thoughtful choice between those that remain. Even in the case of problems used as multiple-choice items, a common-sense consideration of answers based on estimates of reasonable magnitudes can help, especially if the time remaining in the test period is short.

■ HOW DIFFICULT SHOULD A TEST BE?

Any test for which the class average is 15% right, or 90% right, has been poorly matched to the class ability and preparation. In addition to separating the sheep from the goats (which, in a sense, almost any test will achieve), we want to separate the bright sheep from the brilliant sheep and the slow goats from the hopeless goats. We do not obtain a broad-scale discrimination if a test is either too easy or too difficult. Nor do we achieve the best discrimination by using a test made up of items that are all equally difficult. The item difficulty should range from easy (80+% of students answer correctly) to very difficult (80+% of students choose the wrong answer) with the average difficulty falling somewhere just below the 50% level. That is, if the test paper with the median score earns about 50–60 % of the possible credits, the test has been about right in difficulty for the group tested.

This does not mean that the median *mark* on the test should fall in the 50–60 % range. Teachers who give grades on tests so that two-thirds of their students fail often pride themselves on their "rigor" and bewail the stupidity and indolence of the students. Students, on the other hand, are outraged, drop the course as soon as possible, and the public image of physics has received another black eye. A wise old-timer once remarked rather cruelly to a young physics teacher who was complaining about his students, "Where there's no learning there has been no teaching."

A test composed with the proper difficulty to discriminate over the whole range of achievement should never be marked on a percentage basis. Turning raw scores into grades will be discussed later.

■ CHEATING

Everyone wishes that the problem of cheating would go away, but it has not and it will not. Even schools with elaborate honor codes administered by students themselves sometimes have serious troubles. The old saw, "Every man has his price," seems to be the controlling principle. If the penalties of failure are great enough most people will cheat to get by. Aside from the unusual individual who cheats as a way of life, there appear to be two principal causes of cheating on a troublesome scale.

1. *The system.* When a student is trapped in a situation in which the price of failure seems disastrous (for example, rejection by college or professional school, or loss of a coveted and necessary scholarship), the pressure may be intolerable. Some of our most publicized cheating scandals have taken place

in just such circumstances. There is little the individual teacher can do about this but stay alert to the possibility.

2. *The teacher.* Most of us know of situations in which a group of students who are earnest, industrious, and honest in one classroom will routinely cheat on tests in a different course. A teacher who has regular troubles with cheating should take the problem seriously as a student comment on the course and the teaching methods. The students are stating in their own way that they have little respect for the teacher or the course. They do not find the course worthwhile or the tests fair, and do not believe that the teacher has much regard for them. This sort of cheating is a serious indictment and calls for reassessment of the various factors involved.

   a. Is the teacher interested in his own course? Is it evident to students that the teacher knows and likes the subject? Has the course been planned with students' interests in mind? Do students believe that mastery of the course content will actually serve them well in their future lives and careers? Every teacher should try to get a student's-eye view of the course and what is going on in the classroom. By listening closely to questions and comments, and by watching facial expressions while the class is in progress, one can tell much about student attitudes, and, not surprisingly, they often reflect the teacher's attitude.

   b. Is the teacher interested in the students? Is the class atmosphere aloof or friendly? Are there chats between teacher and students in an informal way when there is an opportunity, as, for example, during a laboratory period? Is the teacher's office door open to students when they want help, and is the laboratory open to students after scheduled hours when they need more time for an experiment or project? No teacher should actively seek popularity among students, but a teacher can make clear that he or she cares for them. Sometimes a harsh drillmaster, apparently unpopular, gains student respect just because they realize that this teacher is making the extra effort for their sake.

   c. Is the teacher scrupulously fair in evaluating students? Do tests represent the subject as it was taught and as homework was assigned and discussed? Has the teacher taken the time to explain in detail and to justify his or her method of giving grades? Does the teacher listen to student questions about test items and admit a fault when a student has a just complaint? Are careful, honest evaluations and recommendations written when they are called for?

   d. Does the teacher make it difficult to cheat? Is it clear to students that working together for the purpose of learning is not cheating, but working together to get a higher mark without learning physics is against the rules? Does the instructor, or a deputy, proctor tests alertly so that students who are honest will not feel threatened by cheaters?

   If, as a physics teacher, you ever say of a class, "They're such awful cheats!" stop and ask yourself the questions above. Students who are not caught in a system that places a high premium on dishonesty will not be dishonest in a course they like and appreciate, taught by a teacher who is fair and who has their interests at heart.

■ MARKING THE TEST

There are a few apparently trivial techniques that can take some of the horror out of a test-marking session. If you have machine-scoring facilities you will, of course, use a separate answer sheet. Otherwise you should lay out the objective test questions so that a copy of the test can be used as the key. The best arrangement is to line up the answer spaces in the left margin of the paper, each space immediately adjacent to the question it serves. The key bearing the correct answers is then placed on the student's paper so that the answers and the key are lined up next to each other. Use a red pencil or pen for making the key and marking the test, and forbid use of red by students.

If there are several pages to the exam and a sufficient number of papers to present a significant chore, the separate keys can be cemented to the backs and fronts of cardboard strips, each labeled with its page number.

Essays and problems require a key, too. Identify the best and hoped-for answers, but also try to predict the probable variations. Decide in advance on the relative credit, making no effort to have the total credits add up to 100. Ideally the teacher should mark the first free-response question on all papers before going on to the second, but the result is too much paper handling for an individual. If there are several graders available, each one can be responsible for a given free-response answer on all papers.

Identify and mark first a few good papers. Answers written by able students may show up defects in the questions or in the key. In the case of essay questions, read a few good papers first without grading them.

■ TURNING RAW SCORES INTO GRADES

One of the handicaps with which physics teaching labors is the reputation of physics as being "tough," and the too-often-earned reputation of physics teachers as being self-assured and inhuman taskmasters. As a general guiding principle the grades of students in physics should not be too different from the marks the students and their peers receive in other academic subjects. If notably different, physics grades should be *higher* than grades in required courses. Because of the elective nature of physics in most schools the students are, at least in part, self-selected, and will be more competent than their fellows in quantitative thinking.

Here's a way of dealing with this problem:

1. Count the total credits on each paper to determine a raw score. Some teachers prefer to use only the raw scores throughout a whole marking period. Students, however, deserve some kind of running report on their relative achievement, and find it easiest to interpret a letter grade.

2. For a letter grade, list the raw scores on the test from high to low and the number of papers receiving each score. (See Table 1.) Note that if this test had been graded on a percentage basis half the class would have failed.

TABLE 1. *Raw scores and grades on a 24-item objective test taken by 52 students.*

Number right	Number of papers	Grades, as adjusted
24	1	A+
23	0	
22	2 ⎫	A
21	1 ⎭	
20	5 ⎫	B+
19	3 ⎭	
18	4 ⎫	
17	6 ⎬	B
16	4 ⎭	
	Median	
15	7 ⎫	C+
14	5 ⎭	
13	2 ⎫	C
12	3 ⎭	
11	4 ⎫	D
10	2 ⎭	
9	0 ⎫	
8	2 ⎬	F
7	1 ⎭	

3. At this point the objective test receives a nonobjective set of marks. Consider the conditions peculiar to this test, the topics covered, your own evaluation of its difficulty, and the level and apparent aptitude of the class as a whole, and decide on reasonable marks for the median, the high, and the low. Identify these grades on the list. The class whose results are tabulated had been interested, generally serious in purpose, and conscientious in work habits. The test covered the subject (as taught) adequately and after marking it was judged not unfair, but difficult. There were no "throwaways"; each item had a useful point. Two weak, nearly faulty items were revealed by the results.

  On the basis of such considerations (others might also be applied), the teacher decided that the median grade should be at the dividing line between B and C+. The single paper with all items right is obviously an A+ and the teacher decided that the three lowest were failing papers.

4. Once the high, median, and low grades have been identified, then the rest of the list can be marked in letter grades, which are written in the teacher's record book (along with the median) and on the students' papers.

5. It is essential that the students understand and accept this procedure. It enables us to give tests that are difficult enough to yield useful discrimination between students without getting too much out of step with the school's academic standards and the students' own expectations. To help students find it acceptable, show them that it always results in raising grades.

As we pointed out above, the recommended procedure for determining grades on tests contains a highly subjective element. It is difficult and probably wrong to exclude a subjective judgment in grading a class or student. Many students seem to think that there is a difference between an absolute grading scale and "grading on a curve." Of course, any absolute scale depends on the questions that have been written, and the difficulty of these questions is based on a curve determined by our expectations or experience with other classes. Any of us could prepare exams similar to the New York State Regents Examinations that have an inflexible passing level of 65%. We could then assign A to those getting better than 90%, etc. Then, depending on our questions, we could flunk everyone, or have most everyone get A. Somebody's subjective judgment always enters into the decisions about grades.

As for subjective judgment in grading a particular student, there are obvious problems. What do you do with the borderline cases if you have established a numerical formula for the final grade, composed of various percentages for various exams, laboratory reports, and projects? What do you do with the student who, you think, may need the encouragement of the benefit of the doubt, or the student who started well, but has slumped, and, you think, needs goading? And if you bend the grading formula for special cases, how clear is your conscience that perhaps these are favored cases? Perhaps you were conned.

■ GRADES IN THE LARGE-ENROLLMENT COURSE

These problems are particularly complicated in a large course given by a staff. If a university uses faculty, instead of graduate students for recitation teaching, then the faculty will truly be teachers only if they have some judgment about the grades of their own students. Without that power the faculty will be only guest performers and the students will sense that they are just part of the large lecture group. In the ideal arrangement, the recitation teacher gets to know the students because they come to him or her for tutoring and advice. Then how does the recitation teacher exercise judgment in grading when the students are taking course-wide exams? Clearly the main grade determinant must be a course-wide formula of weighted grades.

One solution to this staff problem is to use the numerical total only as a first approximation to the final grade. In one such situation our formula calls for 10% for each of three one-hour course-wide exams, 20% for the final exam, 25% for the recitation quiz average, and 25% for laboratory performance. Since the weekly recitation quizzes are designed by each instructor and laboratory grades are subject to individual instructor practices, these grades are normalized across the sections in terms of the average grade for that section in the course-wide exams. The process sounds complicated, but works smoothly. On grade-giving day, all faculty and teaching assistants gather to examine a distribution curve showing the numerical grade totals for the whole class. The staff then makes decisions about the numerical dividing lines for various letter grades. However, for a particular student the numerical total is only the first approximation to the letter grade. Each faculty staff member then goes over each of his or her students' records with the assistance of the graduate laboratory teaching assistant. They use specific criteria to decide whether borderline cases should be raised or lowered, usually by no more than a plus or minus letter grade. The first criterion is the gradient of each student's grade. A strong recovery on the final exam can, in effect, increase the weighting factor of the final exam. Clever, thorough, or poor performance on the semester laboratory project is the second criterion. Invariably, intangibles of student interest and prospects are also taken into account as criteria.

We have used this general system for many years at State University of New York at Stony Brook. Of course, there are inequities as there are with any system. However, both staff and students have experienced satisfaction, and as far as we know the process has not been abused by unreasonable exercise of judgment. The great advantage of the system is that it helps to break up large lecture courses into small classes with real student–faculty interchange and responsibility.

■ ANALYZE THE TEST RESULTS

The amount of time that a teacher can spend on studying the results of a test must depend on the teacher's personal goals and schedule. But if we accept the role of a test as a teaching tool the analysis is most important. If you have marked the errors on the key as well as on the test papers, as suggested above, you already have the most important information to guide you in a discussion of the test.

To help with future test-making and teaching, spend a little more time with the items that drew many errors by counting the popularity of each distracter. Consider also the items that almost everyone got right. Try to decide which items were really flawed. Write a note to yourself on the key so that when you write a new version of this test you will be guided by your experience with this one.

■ USE THE TEST FOR INSTRUCTION

Test papers should be returned at the earliest opportunity, if possible the next class meeting. When the students receive the marked test paper the test experience should be fresh in their minds. The scored paper shows not only the incorrect responses but also the adjusted letter grade. This is the time to announce the median grade.

Discuss the test fully, paying particular attention to items that were answered incorrectly by the greatest number of students. Reteach material that seems to need it. Since you know in advance (from your test analysis) what topics are going to raise the most questions, you can have a quick demonstration or two ready to keep the discussions realistic. Encourage students to take notes on their papers and permit them to keep the papers. Each student's collection of marked test papers will form a useful study guide in preparing for a semester or final exam.

Note that disregarding the security of test questions implicitly assumes that a new test will be written the next time around, and this is as it should be. Keep a few copies of old tests to use as "make-ups" after a student has been disappointed by a test result. It is wise to allow students to make up a poor test result on their own time. Keep the original grade and simply add the new one to the record. The result will almost always be higher than the original grade. Part of this effect is undoubtedly caused by the presence of earlier tests in notebooks of older students, but there seems also to be a good deal of physics learning in the result.

When discussing a test with a class, listen carefully to student complaints: they may be justified, and if not, your own interpretations of the questions can be instructive. Of course, tests should be written with such care that flawed items are rare, but allow credit for very poorly conceived test items that misled students.

Repeat: this should be sufficiently rare so that an argumentative student is not encouraged to raise problems needlessly.

After a few years of this kind of testing program a teacher will have file copies of a number of tests on standard topics and combinations of topics. Each file folder contains the original key, the distribution of marks, and whatever item analysis the teacher found time for. These make splendid review instruments at the end of the semester or the end of the course.

■ FOR READING AND REFERENCE

Adams, R. C., "Is physics a laughing matter?," Phys. Teach. **10**, 265 (1972).
   A note describing tests that incorporate humorous examples. The author claims that fear pressures are reduced and test results improved.

Applebaum, D. C., "A two-variable grading scheme," Phys. Teach. **17**, 110 (1979).
   This note describes a type of exam and a grading formula that allows the student some control of the analytical and descriptive portions of the exam.

Aubrecht, G. J., II and Aubrecht, J. D., "Constructing objective tests," Am. J. Phys. **51**, 613 (1983).
   One need not agree with all that is stated in this major article, or admire all the sample test items, to profit from the many sound principles outlined here. See also a follow-up letter to the editor, Iona, M., Am. J. Phys. **52**, 201 (1984), and the authors' reply, Am. J. Phys. **52**, 201 (1984).

Boedeker, R. R., "Procedure for written test follow-ups in large sections of university physics," Am. J. Phys. **51**, 859 (1983).
   Students are permitted to rework two test problems each quarter, turning them in one day after the return of the graded test.

Ferris, F. L., Jr., "Testing for physics achievement," Am. J. Phys. **28**, 269 (1960).
   The PSSC (Physical Science Study Committee) tests caused turmoil among teachers, students, and parents because they required a student to demonstrate an understanding of process and method as well as knowledge. This paper tells about the development of the PSSC tests.

Glanz, P. K., and Brown, R. S., "The evaluation of students in introductory physics needs a new outlook," Phys. Teach. **14**, 107 (1976).
   How should test grades be used in arriving at a final mark? This paper describes three approaches.

Golden, D. E. *et al.*, "Repeatable testing—a tool for learning physics," Am. J. Phys. **42**, 941 (1974).
   The authors write two or three versions of a test and encourage students to repeat the test if they wish, after a suitable interval spent restudying the material.

Greenslade, T. B., Jr., "An alternate form of examination question," Phys. Teach. **23**, 160 (1985).
   The teacher provides a poorly written or erroneous statement about some aspect of physics. This is given to students for correction and rewriting.

Guenter, J. M., "Testing techniques," Phys. Teach. **13**, 495 (1975).
   The author describes several unusual kinds of test questions that he has used with success.

Halloun, I. A., and Hestenes, D., "The initial knowledge state of college physics students," Am. J. Phys. **53**, 1043 (1985).
   The authors have written and evaluated a test designed to measure students' qualitative understanding of mechanics. It is a 35-item diagnostic test, and is given in full in an appendix.

Hoffmann, B., "Testing," Phys. Today, **14** (10), 38 (1961).
   Hoffmann wrote several articles and a book sharply criticizing multiple-choice questions. His basic point is that some questions have distracters that have more basic truth than does the preferred choice. See also the strong reply, Fornoff, F., "Objective tests—all bad?," Phys. Today, **15** (4), 36 (1962).

Iona, M., "The physics Advanced Placement exams," Phys. Teach. **16**, 150 (1978).
   The author, an expert critic of the failings of physics teachers and textbook writers, here gives advice on taking the Advanced Placement exam, along with some sharp comments on various questions.

Kruglak, H., "Resource letter AT-1 on achievement testing," Am. J. Phys. **33**, 255 (1965).
   This gives an annotated list of articles on physics achievement testing up to about 1965.

Kruglak, H., "Testing for achievement in physics," Phys. Teach. **4**, 199 (1966).
   A splendid comprehensive article on the subject.

Lehrman, R., "Sloppy thinking in the physics regents," Phys. Teach. **15**, 526 (1977).
   A number of New York State physics exam questions are analyzed and criticized in this paper.

Lewis, R. R., Jr., "Random self-paced examinations," Phys. Teach. **12**, 21 (1974).
   At the start of a unit students receive 30 test questions, of which five will be used on the exam. The student can take as much of the three-hour exam period as needed. Test pressures are lowered and make-ups are available.

Lucido, H., "Does this count? Teaching physics without grades," Phys. Teach. **20**, 455 (1982).
   The author says he has shifted his students' concern from grades to learning physics. Evaluation is done in an interview each quarter.

Lunde, B. K., and Jones, J., "Let the students write the exam," Phys. Teach. **10**, 270 (1972).

    The authors use student-written questions for more than 50% of a test. This note describes the results and states the advantages.

Nedelsky, L., *Science Teaching and Testing* (Harcourt, Brace and World, New York, 1965).

    Nedelsky is a physicist, and his book deals almost entirely with physics teaching and testing. It is a very personal statement, in parts too detailed and theoretical to help a busy teacher, but it is a useful reference and gives hundreds of examples of test questions, with comments on most.

Nelson, J., "AAPT/NSTA high school physics examination," Phys. Teach. **21**, 100(1983).

    The story of the development of a standardized physics exam for use in high schools.

Pfeiffenberger, W., "The 1974 Advanced Placement examination in physics—Part I," Phys. Teach. **14**, 344 (1976); "Part II," Phys. Teach. **14**, 425 (1976).

    Part I presents an entire Advanced Placement (AP) Physics B exam: 70 multiple-choice (MC) items (key supplied) and seven free-response questions. Part II gives the MC and free-response questions for the 1974 AP Physics C exam, with answers supplied for the MC items.

Quisenberry, D. R., "Evaluating the laboratory experience," Phys. Teach. **19**, 549 (1981).

    This paper describes one way to organize and administer a laboratory-performance test.

Rogers, E. M., "Examinations: powerful agents for good or ill in teaching," Am. J. Phys. **37**, 954 (1969).

    In his Oersted Medal response a famous teacher discusses the effect of exams on students. A final course exam he calls an "exit" exam, for which he sees little purpose. The "entrance" exam, given before a new course is tried seems to him more justifiable.

Williams, W., "A 3″ by 5″ crib," Phys. Teach. **18**, 461 (1980).

    Students are told to bring to a test a 3×5-in. file card bearing any formulas, definitions or other notes that they feel might be needed in answering test questions.

2
Error (Uncertainty) Analysis

The introductory physics course is the first, and in many cases the only, contact most people have with quantitative methods. Certainly the math courses seldom supply this experience. Mathematicians and math teachers are usually concerned with symbolic relationships. The idea that numbers have units and represent real values and measurements is not the prime concern of mathematicians. Furthermore, the introductory physics course is the first place where students must take seriously the fact that all measurements have uncertainties, and that these uncertainties have consequences concerning the significance of the measurements. We claim that learning how to determine the uncertainty of data and how to deal with it is one of the most important lessons that we physics teachers have to offer. For the purposes of a liberal education, learning the exact form of Newton's laws is not so important as being able to understand the difference between one part in a thousand and one part in a million or the difference between data with a 20% uncertainty and data good to 1%.

When a value is not known precisely, the amount of uncertainty is usually called *error*. This has given the whole business of uncertainty analysis a bad name because in common usage *error* implies sloppiness, very likely caused by sinfulness (from Victorian days: "She saw the error of her ways."). At the very least, most people think of error as a mistake. Students frequently describe any discrepancy between their results and the textbook answer as an "error," and when asked to explain, assign the cause to "human error."

This chapter deals with the analysis of errors—how to judge their magnitude, how to describe them in conventional ways, and how to take them into account in calculating numerical values based on a number of individual measurements, each containing error. As we shall emphasize, *error* represents *uncertainty* and has nothing to do with mistakes or sloppiness. Indeed, reducing the amount of error in a measurement beyond the immediate need is usually a mistake and a sign of poor judgment.

We categorize the reporting and handling of uncertainty in four stages, each involving a particular degree of precision. The first is concerned only with order of magnitude of a number. Consider this to be a zeroth approximation. Next are the conventions regulating the use of significant figures, a first approximation of limited usefulness, but obligatory politeness. The second approximation deals with the maximum and minimum range of measured quantities. The rules of manipulation of such error limits are simple, and this system should be the one most often used by students of introductory physics. Indeed, the system is so simple that if its use is causing the student a lot of extra work, the student does not yet understand the system. Finally, there is a third approximation to error citation and analysis, involving rules derived from probability and statistics. This system is frequently misunderstood and misapplied. Its use is justified only when the primitive data fulfill certain requirements of quantity, distribution, and probability. *There is essentially no need for students in any introductory physics course to use this third level of statistical error analysis.*

■ THE EXPONENTIAL NOTATION

The exponential notation is a means of indicating and keeping track of the decimal point position in a number. However, use of the method provides a powerful aid in doing arithmetic problems, particularly as a check of the approximate correctness of a calculator solution. Furthermore, the very nature of the formalism displays an attitude, a frame of mind, about numbers and data. The method emphasizes the importance of the *order of magnitude*, and then explicitly specifies the number of significant figures. The physics class routine should be to solve arithmetic first with exponential notation to order of magnitude, or to one significant figure, and then to turn to the hand calculator. Because students often meet—or at least, use—exponential notation for the first time in science classes, the system is sometimes called "scientific notation." Fortunately, this terminology is unknown in the real world outside the classroom.

Here is the system and a review of its use:

0.001	0.01	0.1	1	10	100	1000	10 000
10^{-3}	10^{-2}	10^{-1}	10^0	10^1	10^2	10^3	10^4

Note that the 3 in 10^3 can be thought of as either the number of zeros in 1000 or the number of times that 10 is multiplied by itself—10 cubed. Negative exponentials indicate the reciprocal power of the number: $10^{-2} = 1/10^2 = 1/100 = 0.01$.

To multiply the powers of 10, add the exponents: $100 \times 1000 = 10^2 \times 10^3 = 100\,000 = 10^5$.
$0.1 \times 0.001 \times 100 = 10^{-1} \times 10^{-3} \times 10^2 = 0.01 = 10^{-2}$.

This property justifies the assignment of $10^0 = 1$. We must have this equality since if $10 \times 0.1 = 1$, then $10^1 \times 10^{-1} = 10^{1-1} = 10^0 = 1$. Or, in general, $10^a \times 10^{-a} = 10^a \times 1/10^a = 10^{a-a} = 10^0 = 1$.

Here are some examples of numbers written in exponential notation:

$$15 = 1.5 \times 10^1, \quad 5\,380\,000 = 5.38 \times 10^6, \quad 0.0032 = 3.2 \times 10^{-3},$$

$$9 = 9 \times 10^0, \quad 90 = 9 \times 10^1, \quad 90.0 = 9.00 \times 10^1.$$

Notice the difference between writing 90.0 and 90. By convention the first number indicates that the value is known to be between 89.95 and 90.05. The extra zero is significant and must be retained. The form of the second number may be ambiguous. The value may be known to be between 89.5 and 90.5 or perhaps only between 85 and 95. The exponential notation provides a way of avoiding the ambiguity. To indicate the former value, we would write 9.0×10^1; to indicate the latter value, we would write 9×10^1.

Here is an example of how to set up a multiplication and division problem. Students should use this procedure *before* turning to their calculators:

$$\frac{5832 \times 0.051}{68\,000 \times 32} = \frac{5.832 \times 10^3 \times 5.1 \times 10^{-2}}{6.8 \times 10^4 \times 3.2 \times 10^1} = \frac{5.8 \times 5.1}{6.8 \times 3.2}\frac{10^1}{10^5} = 1.4 \times 10^{-4}.$$

Using this technique, a problem like this can quickly be solved to one or two significant figures by inspection, and the placement of the decimal point determined. For instance, the third step above sets up the problem so that it can be solved approximately by looking at it. The numerator product must be about 30 (6×5); the denominator product must be about 20 (7×3). Therefore, the answer must be about 1.5×10^{-4}. If our hand calculator does not read close to 1.5×10^{-4}, we should suspect that we have made a mistake in one method or the other.

Notice that in this example, no more significant figures are retained in any one of the numbers to be multiplied than belong to the number having the least number of significant figures. Since 32 and 0.051 each have only two significant figures, 5832 is rounded off to 5.8×10^3.

Raising a number to some power, or taking a root, is easy to do with exponential notation. For example, $(10^2)^3 = 10^2 \times 10^2 \times 10^2 = 10^6 = 10^{2 \times 3}$. When raising to a power, multiply the exponents. When taking a root, divide the exponents: $\sqrt{10^6} = 10^{6/2} = 10^3$. Here are some other examples:

$$(5800)^2 = (5.8 \times 10^3)^2 = (5.8)^2 \times 10^6 \approx 36 \times 10^6 \quad (= 34 \times 10^6), \tag{1}$$

$$\sqrt{0.47} = \sqrt{47 \times 10^{-2}} = \sqrt{47} \times 10^{-1} \approx 7 \times 10^{-1} \quad (= 0.69) \tag{2}$$

(Arrange the exponent of 10 so that it is easily divisible by the root. For instance, it would not be sensible to write $\sqrt{0.47} = \sqrt{4.7 \times 10^{-1}} = \sqrt{4.7} \times 10^{-1/2}$.)

$$\sqrt[3]{6 \times 10^{23}} = \sqrt[3]{600 \times 10^{21}} = \sqrt[3]{600} \times 10^7 \approx 8 \times 10^7 \quad (= 8.4 \times 10^7), \tag{3}$$

$$\frac{14000}{383 \times 1.0\%} = \frac{14 \times 10^3}{3.83 \times 10^2 \times 1.0 \times 10^{-2}} = \frac{14}{3.83}\frac{10^3}{10^0} \approx 3\tfrac{1}{2} \times 10^3 \quad (= 3.7 \times 10^3). \tag{4}$$

By leaving 14 000 as 14×10^3 instead of 1.4×10^4, we made the digits numerator larger than the denominator, and so ended with a quotient greater than 1—an unnecessary but convenient trick.

■ FERMI QUESTIONS—ORDER OF MAGNITUDE

Fermi questions were named after Enrico Fermi, the great physicist who contributed to both experiments and theory concerned with atomic nuclei and fundamental particles. He was a master at computing approximations to answers when it seemed that no information was available. The point of such questions is that reasonable assumptions linked with simple calculations can often narrow down the range of values within which an answer must lie. The *order of magnitude* refers to the power of 10 of the number that fits the value. To increase an order of magnitude means to increase by a factor of 10. Very often an order of magnitude calculation is all that the interest in a problem justifies. Even if more precision is needed, an order of magnitude calculation done first may indicate whether or not it is worthwhile to pursue the problem, and sometimes may indicate how the next approximation to the required value can best be obtained.

Here are some Fermi questions. The first one is the classic example. Note the method of problem solving. You zero in. You do not know the answer immediately, but you define quantitatively what you do know.

1. How many piano tuners are there in New York City? Assume 10^7 people in New York and 2×10^6 families. Assume 1 piano for every 5 families; therefore, 4×10^5 pianos. Assume each piano is tuned once every 2 years; therefore, 2×10^5 piano tunings each year. Assume each tuner tunes 2 a day for 250 days a year. (At \$50 per tuning, the tuner makes a living—a factor of 2 could make a big difference to the tuner.) Therefore,

   $(20\times10^4$ tunings per year$)/(500$ tunings per year per tuner$)=400$ tuners.

   It is unlikely that New York City has fewer than 40 piano tuners or more than 4000. If you do not like the assumptions made, choose your own reasonable guesses and see if your answer is not of the same order of magnitude. (Better yet, do this for a class and let them make the assumptions—the procedure never fails.) Note that sometimes in this calculation we carried along one significant figure. The extent to which you do this depends on the problem and your style; rules would be cumbersome and probably useless. For instance, whether 400 is of order of magnitude 10^2 or 10^3 is a silly question, because a reasonable answer depends on the meaning of the number and the way it is going to be used. When in doubt, carry an extra figure along. Note, incidentally, that a factor of 2 in one of the assumptions makes a big difference to the piano tuners but not to the final result of our order of magnitude calculation.

2. How many golf balls will fit in a suitcase? Assume that the suitcase is 30 in.$\times$8 in.$\times$24 in. The volume is 30 in.$\times$8 in.$\times(100/4)$ in.$\approx6\times10^3$ in.^{3}. Assume that each golf ball is a sphere 1 in. in diameter. The volume of the ball is a little less than 1 in.^{3}. The order of magnitude of the number of golf balls that will fit in a suitcase is 10^4.

   This question is not worth more extensive calculations unless it were asked by a traveling golf ball salesman. Since the size of the suitcase is not specified, there is no point going to the closet to measure a real one. Imagine a reasonable size. Nor is there any point in worrying about whether the balls are close packed as nested spheres; the packing factor could not be greater than 1.5. Surely the diameter of a golf ball is closer to 1 in. than 2 in. Doubling the diameter would increase the ball volume by a factor of 8. That would reduce the number of balls in the suitcase by an order of magnitude. Consider the reasonableness of our answer. Surely the number of golf balls that would fit in a standard suitcase must be greater than 1000 and less than 100 000. Our answer is good within a factor of 10.

3. How many cells are there in a human body? Assume that the average cell diameter is 10 microns or μm (micrometers)$=10^{-5}$ m. Then the volume is 10^{-15} m^3. The order of magnitude of a human volume is 10^{-1} m^3. Therefore, there are 10^{14} cells in a human body.

   This question illustrates again how some information can be obtained out of very little definite knowledge. Living cells come in a great range of sizes. However, they can all be seen with an ordinary light microscope and therefore must have a diameter larger than the wavelength of light. They can scarcely be seen with the unaided eye and so must have a diameter smaller than 0.1 mm. We assumed that the diameter was the geometric mean between these values. (The geometric mean of A and B is $\sqrt{A\times B}$.) In this case, it is

$\sqrt{(10^{-6})(10^{-4})} = 10^{-5}$. The arithmetic mean would be practically the same as 10^{-4}.) Notice that for these calculations it makes no sense to differentiate between the volume of a sphere and that of a cube. The volume of a human body could be estimated by assuming a reasonable height, width, and thickness of a column that is human size. An alternative method requires knowing that 1 liter of water has a mass of 1 kg and a volume of 1×10^{-3} m^3. A cubic meter of water (or flesh—we float) would therefore have a mass of 1000 kg and would weigh about a ton. The assumed volume for the body was 0.1 m^3, yielding a mass of 100 kg.

Here are some other Fermi questions to try with your students:

1. How many hairs are on a human head? (Assume that the spacing between hairs is 1 mm. Then there are 10/cm along a line or 100/cm^2. We figure 2×10^4 hairs per human head.)

2. How many individual frames of film are needed for a feature length film? (We get 1.5×10^5.)

3. What is the ratio of spacing between gas molecules to molecular diameter in a gas at standard temperature and pressure (STP)? [A mole (6×10^{23}) of gas molecules at STP occupies 22.4 ℓ. A molecular diameter might be 2×10^{-10} m. Compare available volume per molecule with the volume of a molecule. We get (spacing between molecules)/(molecular diameter)≈ 10.]

4. How many seconds are in a year? [We get 3×10^7 s/yr, or $10 \times 10^6 \pi$ s/yr ($\pi \times 10^7$ s/yr). Needless to say, it is not fair or profitable to use a calculator to do this problem.]

5. If your life earnings were to be doled out to you at the rate of so much per hour for every hour of your life, how much is your time worth?

6. What is the weight of the solid garbage thrown away by American families each year?

7. How many molecules are in the air in a standard classroom?

8. How much does it cost to toast a slice of bread?

9. How many photons of visible light are emitted each hour by a 60-W bulb? (The efficiency of emission of visible light by such a standard bulb is only about 4%!)

10. What is the population density (people/m^2) of the United States?

■ SIGNIFICANT FIGURES—A FIRST APPROXIMATION TO ERROR ANALYSIS

The number of significant figures in a numerical value is a first approximation to showing the limits within which the value is known. There are more precise ways of indicating the error limits, as we shall see in the next section.

Here are the generally accepted conventions for writing significant figures.

1. When we say that an experimental quantity has the value 3, we mean—by convention—that the value could actually be anywhere between 2.5 and 3.5:

$$2.5 < 3 < 3.5.$$

However, if we say that the value is 3.0, then we mean that it lies between 2.95 and 3.05:

$$2.95 < 3.0 < 3.05.$$

2. Note the ambiguity of a number such as 300. Does it imply $250 < 300 < 350$ or $299.5 < 300 < 300.5$? A superior method of writing such a number is to use power of 10 notation. In that form the number is 3×10^2 or 3.00×10^2, depending on which precision is intended.

3. 0.000 01 has one significant figure. (It could be written 1×10^{-5}.) 1.000 has four significant figures. 1.000 01 has six significant figures.

Here are examples of the rules for the proper use of significant figures in addition or subtraction:

$$
\begin{array}{cccccc}
5.2 & 6.843 & 6.843 & 6.843 & 500 & 5.00 \times 10^2 \\
+3.1 & +1.2 & +1 & +0.001 & -4 & -4 \\
\hline
8.3 & 8.0 & 8 & 6.844 & 500 & 4.96 \times 10^2
\end{array}
$$

In addition or subtraction, the sum or difference has significant figures only in the decimal places where *both* of the original numbers had significant figures. This does not mean that the sum cannot have more significant figures than one of the original numbers. In the examples, note that 0.001 has only one significant figure, but the sum properly has four. It is the decimal *place* of the significant figure that is important in addition or subtraction. In the last two cases there is another example of the ambiguity of final zeros. If you estimate that there are 500 students in a lecture, implying a number between 450 and 550, your estimate is not changed if 4 people leave. On the other hand, if you draw out $500 from the bank and spend $4, you have $496 left.

Here are examples of the rules for the proper use of significant figures in multiplication or division:

$$
\begin{array}{c}
5.2 \\
\times 3.1 \\
\hline
52 \\
156 \\
\hline
\end{array}
$$
$\dfrac{156}{16.12} = 16,$

$5.243 \times 3.1 = 16,$

$5.243 \times 0.0031 = 0.016,$

$\dfrac{37}{9} = 4 \qquad \dfrac{37}{9.1} = 4.1.$

In multiplication and division, the product or quotient cannot have more significant figures than there are in the least accurately known of the original numbers. Here is the justification for this rule. Do the calculation with the extremes of values allowed by the definition of "significance." Consider the first example: the product might be as large as $(5.25)(3.15) = 16.5375$, or as small as $(5.15)(3.05) = 15.7075$. The rule for significant figures in multiplication is evidently justified in this case. Usually during multiplication or division, an extra significant figure is carried along, and the final answer is then rounded off appropriately.

Here are some examples of the use of significant figures that illustrate particular points. You can check out the reasonableness of the answers by doing the calculation with extreme values as we did in the preceding paragraph.

1. $2.000 + 0.01 = 2.01$.
2. $2.000 \times 0.01 = 0.02$.
3. The volume of a piece of chalk that is 10.5 cm long and has a diameter of 1 cm is 8 cm^3.
4. The volume of a rectangular box with length 3.025 cm, width 2.5 cm, and height 2 cm is 15 cm^3. (This variation of the rule applies when the first digit is 1. To see the validity of this extra rule, do the calculation with extreme values. Such a special rule and the indeterminancy point up the need for a more precise way of specifying and dealing with error limits.)

■ ABSOLUTE AND PERCENTAGE ERRORS—A SECOND APPROXIMATION TO ERROR ANALYSIS

This second approximation to error statement and analysis is based on maximum pessimism. The *absolute error*, the ±2 cm in the first example in item 1, defines the maximum excursion of the main value. The implication is that all the lengths measured and the "true" length fall between 95 and 99 cm.

1.
   Examples of absolute error: 97 ± 2 cm, 12 ± 2 cm,
                                 1.03 ± 0.01 s, 104.89 ± 0.01 s.
   The corresponding errors in percent: 97 cm $\pm 2\%$, 12 cm $\pm 20\%$,
                                          1.03 s $\pm 1\%$, 104.89 s $\pm 0.01\%$.

   The rules for compounding measurements through addition or multiplication are based on the assumption that the worst possible coincidence of errors will occur. For instance, in adding 97 ± 2 cm to 12 ± 2 cm the sum could be as large as 113 or as small as 105 if the original values were off by the maximum amount *in the same direction*. Often there will be cancellation of errors in arriving at a compound quantity, because some of the original measurements will have values that are too high and others will have values that are too low. In the next section we will see how and when this effect can be taken into account, but these second approximation rules are simpler and often satisfactory. Indeed, for most introductory laboratory work, they are the only valid rules.

2. To find the compound error in addition or subtraction, *add* the absolute errors:

   $$(97 \pm 2 \text{ cm}) + (12 \pm 2 \text{ cm}) = 109 \pm 4 \text{ cm},$$

   $$(104.89 \pm 0.01 \text{ s}) - (1.03 \pm 0.01 \text{ s}) = 103.86 \pm 0.02 \text{ s},$$

   $$(88 \pm 2 \text{ kg}) + (3.26 \pm 0.02 \text{ kg}) = 91 \pm 2 \text{ kg}.$$

3. To find the compound error in multiplication or division, *add* the *percentage* errors:

   Area of square with sides 97 cm $\pm 2\% = 9400$ cm $\pm 4\% = 9400 \pm 400$ cm^2,

   $$(97 \text{ cm} \pm 2\%) \times (12 \text{ cm} \pm 20\%) = 1200 \text{ cm}^2 \pm 22\% = 1200 \pm 200 \text{ cm}^2.$$

(Note that it would be inappropriate to use 22%, implying that we know the error to two significant figures.)

$$(100.89 \text{ s} \pm 0.01\%)/(1.03 \text{ s} \pm 1\%) = 100 \pm 1\% = 100 \pm 1.$$

The determination of the absolute error is a matter of judgment. How would one estimate that a length is 97 ± 2 cm? The error being assigned to this value has nothing to do with any mistake. There is no mathematical treatment that can automatically compensate for mistakes. The error limits of ± 2 cm may be assigned as the result of the observation of numerical data. Several people may have measured the length or one person may have done it several times, and all the values were found to be between 95 and 99 cm, with 97 cm as the average. Perhaps only one measurement was made with a meter stick marked every 10 cm, and the experimenter estimated that the length was about 97 cm and could not be more than 2 cm different. In this case, the instrument itself was the limiting factor. Perhaps the measurement was done with a tape that had markings down to millimeter size, but the experimenter did not trust the shrinking or stretching of the tape closer than ± 2 cm. Perhaps the measuring instrument was precise and trustworthy, but the object being measured was moving and difficult to measure, or irregular so that a closer measurement was not justified. In all these cases, the assignment of reasonable error limits depends on the *judgment of the experimenter* based on the *instrument*, the *object*, and the *need for precision*. The absolute error represents the *personal guarantee* of the experimenter that the "true" value of the measurement lies within the error bounds. Note that it is not generally true that the absolute error is equal to some arbitrary fraction of the smallest scale division of the measuring instrument. In particular, it is wrong and obviously silly to promulgate a custom of measuring lengths to one-half or one-fifth of the smallest scale division of a meter stick. For instance, you could use a vernier caliper to measure the diameter of a piece of chalk, but because of the unevenness of the chalk the error would be closer to 2 mm than to 0.1 mm.

Notice in the examples given in item 1 that the same absolute error in two measurements may yield radically different percentage errors. Note also that if only one significant figure is given in the absolute error, only one significant figure is justified in the percentage error. The third example in item 2 illustrates the same sort of principle in the addition of errors. If there is only one significant figure in one of the errors, it seldom makes sense to end up with a compound error that contains two significant figures. As pointed out on page 00, an exception to this rule is justified if the first digit of the final number is 1. Thus $(62 \pm 10 \text{ cm}) + (21 \pm 5 \text{ cm}) = 83 \pm 15$ cm.

Although the numbers used as examples here display symmetric errors, there is no necessity for the given value to lie halfway between the extremes. For example, suppose you were sure that a mass that was determined to be 102 g could not be less than 100 g but might be as large as 108 g. Its mass should then be given as $102 \binom{+6}{-2}$ g.

Justification for the rule for compounding errors in multiplication can be obtained algebraically, graphically, and by the trial of extreme values.

Error (Uncertainty) Analysis

Suppose that an area with true length L and true width W is measured to have values of $L+l$ and $W+w$, where l and w are the absolute errors.

Algebraic

$$\text{true area} = LW,$$

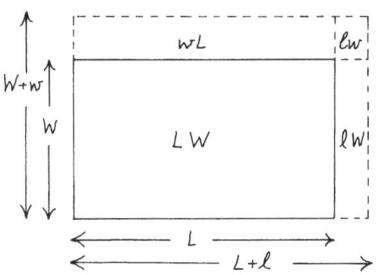

Figure 2.1. *Graphical explanation of error propagation for products.*

$$\text{calculated area} = (L+l)(W+w) = LW + lW + wL + lw,$$

$$\% \text{ error in area} = \frac{(\text{calculated area} - \text{true area})}{\text{true area}} \times 100 = \frac{lW + wL + lw}{LW} \times 100$$

$$= \frac{l}{L} \times 100 + \frac{w}{W} \times 100 + \frac{lw}{LW} \times 100 = (\% \text{ error in } L) + (\% \text{ error in } W)$$

$$+ (\text{comparatively small term}).$$

If l/L and w/W are $\ll 1$, lw/LW is small compared with l/L or w/W.

Graphical

The same derivation is shown in Fig. 2.1.

Numerical Trial

If $L = 100 \pm 10$ cm and $W = 50 \pm 10$ cm, then $L = 100 \pm 10\%$ and $W = 50 \pm 20\%$. The rule gives area $= 5000$ cm$^2 \pm 30\%$ or ± 1500 cm^2. If $L = 90$ cm and $W = 40$ cm, $A = 3600$ cm^2. If $L = 110$ cm and $W = 60$ cm, $A = 6600$ cm^2. Therefore, the rule is justified.

Here are some examples of the application of these rules:

1. The volume of a sphere that has a diameter of 6.2 ± 0.2 cm is 1.2×10^2 cm$^3 \pm 10\%$. The percentage error in the diameter (and hence the radius) is 3%. Since the volume is proportional to the diameter (or radius) cubed, the percentage error of the volume is three times that of the diameter.

2. If an event started at 3.4 ± 0.2 s and ended at 5.0 ± 0.2 s, the event lasted for 1.6 ± 0.4 s or $\pm 30\%$. Notice the large error introduced by the subtraction of large numbers resulting in a small number. Each datum has a percentage error of only about 5%. The subtraction yields a smaller number and the absolute error is the sum of the original errors. Thus the final percentage error is large.

3. The volume of a cylinder with length 6.24 ± 0.01 cm and diameter 2.1 ± 0.1 cm is 22 ± 2 cm^3 or 10%. In this case the percentage error in the length is only 0.2%. The percentage error in the diameter is almost 5%, which produces a 10% error in the square of the diameter. It appears that the precision of the

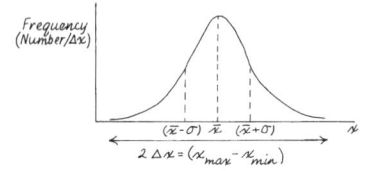

Figure 2.2. Gaussian or normal curve and error limits.

length measurement is overkill. If greater precision in the volume is required, it is the diameter measurement that must be improved.

■ DATA DISTRIBUTION CURVES—A THIRD APPROXIMATION TO ERROR ANALYSIS

The use of this third approximation to error analysis is justified only when certain experimental conditions and demands are met. If the formalism is applied blindly, as it often is, sophisticated precision may be claimed when it does not exist at all. The situation is aggravated by the easy availability of statistical programs on many hand calculators. Just enter a few numbers, press the keys, and standard deviations and correlations will come tumbling out to 10 insignificant figures. The mathematical techniques are derived from and are concerned with statistical laws of probability. The underlying assumption of their use is that the actual measurements are clustered around an average value and vary from that average in a way specified by the bell-shaped Gaussian distribution.

The Gaussian distribution function occurs under very general conditions of chance determination. The major requirement is that if the probability of getting a particular value i is p_i and the probability is getting j is p_j, then the probability of getting the values i and j in sequence is the product $p_i p_j$. It is also necessary that the probability be maximum for one particular value. These conditions are frequently met by data subject to ordinary *random* errors of measurement.

There are many reasons why the measurements may not fit the Gaussian function; there may be some cutoff or prejudice against low readings as opposed to high readings, producing a skewed distribution; there may be mistakes in readings or in instruments; the quantity being measured or the instruments may have periodic variations; the data sample may be so small that statistical fluctuations cause a warped distribution. Even when the *precision* (the scatter of data) determined by the statistical analysis of random errors is correctly stated, the *accuracy* may be poor because of systematic errors or other types of mistakes. If anyone, especially an inexperienced student, measures some static quantity over and over again with the same instrument, he will probably keep on getting the same prejudiced, nonrandom, results.

If elaborate analysis of experimental error is necessary, it may be helpful to consult Barford (1967), Beers (1958), Swartz (1993), Taylor (1982), and Youden (1962).

The only sure way to determine whether data can sustain statistical analysis is to plot it. If the distribution graph is approximately bellshaped, like the one in Fig. 2.2, then the third approximation is valid—though it still may not be worth the trouble. However, if the data *are* Gaussian, the maximum range is too pessimistic for the error limits. The precision is really better than that. A reasonable criterion for the width of the uncertainty is σ, the standard deviation of the Gaussian function. Two-thirds of the data points fall within the range between $\bar{x}+\sigma$ and $\bar{x}-\sigma$. The mean $\bar{x}$ is the arithmetic average of the readings.

The standard deviation is sometimes called the *root-mean-square deviation*. It can be calculated from the data in a straightforward, though sometimes laborious, way, using the following formula:

Error (Uncertainty) Analysis

$$\sigma = \sqrt{\frac{\sum_{i=1}^{n}(x_i - \bar{x})^2}{n}}.$$

(For practical purposes, σ is usually calculated by accumulating the values in a calculator using the Σ key and then pressing the σ key. The calculator gives you a readout whether you have entered 2 numbers or 200, and regardless of whether the distribution is Gaussian!)

The Gaussian function is given by

$$n(x) = \frac{N}{\sigma\sqrt{2\pi}} e^{(\bar{x}-x)^2/2\sigma^2}.$$

The constant $N/\sigma\sqrt{2\pi}$ is a normalizing factor that makes the integral over all values of x yield the total number of trials, N. Note that for $x = \bar{x}$, the exponent is 0, making the exponential 1, yielding the maximum of the distribution. Because of the squared terms in the exponent, the curve is symmetric about the mean $\bar{x}$ and falls off rapidly on either side, scaled, however, by the denominator. The total number of events between $\bar{x} - \sigma$ and $\bar{x} + \sigma$ is $0.68N$, or about $\frac{2}{3}N$.

■ PROPAGATION OF ERRORS IN THE THIRD APPROXIMATION

In the second approximation to error analysis we made the most pessimistic assumption about the way that errors in separate measurements might combine to yield a total error. If $x \pm \Delta x$ is to be added to $y \pm \Delta y$, it is possible that the true value could be as large as $x + y + \Delta x + \Delta y$, or as small as $x + y - \Delta x - \Delta y$. In some cases, however, it is possible that the errors will cancel to some extent so that the range of final error will not be $\pm(\Delta x + \Delta y)$. Certainly this would be the case if the errors were randomly distributed. *If* there are sufficient data to justify the use of the standard deviation to express the error probabilities in the measurement of the separate variables, then the following rules can be used in calculating compound errors. One additional requirement is that the variables and their measurement be independent of each other. (For instance, the percentage error in the volume of a sphere is still 3 times the percentage error in the measured diameter.)

1. The usefulness of repeated measurements of a quantity is specified in the example given below. Note that this relationship is valid only if each of the measurements contains sufficient data to justify the use of the standard deviation.

$$\bar{\sigma} = \frac{\sigma}{\sqrt{n}}.$$

For example, here are results of four measurements of p:

$$p_1 = 83 \pm 2.5,$$
$$p_2 = 84 \pm 2.4,$$
$$p_3 = 81 \pm 2.7,$$
$$p_4 = 85 \pm 2.0.$$

Then the average value is $\bar{p} = 83 \pm 1.2$.

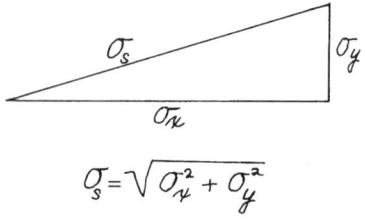

Figure 2.3. Propagation of errors when use of standard deviation is justified.

2. The plausibility that there will be cancellation in combining errors can be seen by considering what happens if two data distribution curves are added to each other. The Gaussian peaks will overlap. A revealing way to think of the error combination is in terms of right-angle geometry, which the formula below suggests. This representation is shown in Fig. 2.3 and emphasizes the way in which a larger error dominates a smaller one.

If $s = x \pm y$, $\sigma_s = \sqrt{\sigma_x^2 + \sigma_y^2}$, e.g.,

$$s = x+y, \quad x = 18.4 \pm 0.2, \quad y = 16.2 \pm 0.2,$$
$$\sigma_s = \sqrt{(4 \times 10^{-2}) + (4 \times 10^{-2})} = 0.3,$$
$$s = 34.6 \pm 0.3, \tag{5}$$
$$s = x-y, \quad x = 42 \pm 2 \text{ cm}, \quad y = 20.12 \pm 0.02 \text{ cm},$$
$$\sigma_s = \sqrt{(4) + (4 \times 10^{-4})} = 2,$$
$$s = 22 \pm 2 \text{ cm}. \tag{6}$$

3. The rule for finding errors of products is just an extension of the rule used in the second approximation, but with the probability of cancellation of errors taken into account. The derivations of these particular rules plus descriptions of how to deal with more complicated situations (such as when the variables are not completely independent) are given in the references cited on p. 00.

If $p = xy$ or x/y, then $\sigma_p/p = \sqrt{(\sigma_x/x)^2 + (\sigma_y/y)^2}$, e.g.,

$$p = xy, \quad x = 21 \pm 0.5 \text{ cm}, \quad y = 10 \pm 0.5 \text{ cm},$$
$$\frac{\sigma_x}{x} = 2.5 \times 10^{-2}, \quad \frac{\sigma_y}{y} = 5 \times 10^{-2},$$
$$\frac{\sigma_p}{p} = \sqrt{(6.3 \times 10^{-4}) + (25 \times 10^{-4})} = 5.6 \times 10^{-2},$$
$$p = 210 \pm 12 \text{ cm}^2, \tag{7}$$
$$p = \frac{x}{y}, \quad x = 2.3 \pm 0.1 \text{ radian (rad)}, \quad y = 1.002 \pm 0.001 \text{ s},$$
$$\frac{\sigma_x}{x} = 4 \times 10^{-2}, \quad \frac{\sigma_y}{y} = 1 \times 10^{-3},$$
$$\frac{\sigma_p}{p} = \sqrt{(16 \times 10^{-4}) + (1 \times 10^{-6})} = 4 \times 10^{-2},$$
$$p = 2.3 \pm 0.1 \text{ rad/s}. \tag{8}$$

Here are some examples of applications of these rules:

1. If the diameter of a sphere is measured to be 1.000 ± 0.002 cm, the volume is 0.524 ± 0.003 cm^3. In this case, $\sigma_d/d = 2 \times 10^{-3}$. Although the volume is proportional to the product $d \times d \times d$, these are obviously not values for three independent variables. Consequently, there is no chance for cancellation in compounding them, and $\sigma_v/V = 3(\sigma_d/d) = 6 \times 10^{-3}$.

2. The number of photons per second coming through a filter is $1.06 \times 10^4 \pm 1 \times 10^2$. If a detector is gated with an on time of 10.0 ± 0.1 ms, it will detect 106 ± 1.4 photons/gate. In this case $\sigma_N/N = 1 \times 10^{-2}$, and $\sigma_t/t = 1 \times 10^{-2}$. Consequently, $\sigma_p/p = \sqrt{1 \times 10^{-4} + 1 \times 10^{-4}} = 1.4 \times 10^{-2}$.

3. The measured masses of five blocks are 390±10 mg, 460±10 mg, 270±10 mg, 540±10 mg, and 420±10 mg. The total mass is 2080±22 mg. In this case, $\sigma = \sqrt{10^2 + 10^2 + 10^2 + 10^2 + 10^2} = 22$ mg. However, consider how unlikely it would be to have an actual experimental situation where this method would be appropriate! The statistical method is valid only if the data for each block fit a Gaussian curve. Why would anyone want to do that? What instrument could produce such a distribution?

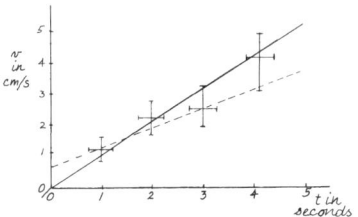

Figure 2.4. Graphical representation of errors using error bars.

■ THE GRAPHICAL REPRESENTATION OF ERROR

On a graph, a datum "point" should not be represented as a point. If displacement x is being plotted as a function of time t, then there is an error in the measurement of x at each uncertain time t. Instead of $x(t)$, the experimental situation is $x \pm \Delta x (t \pm \Delta t)$. The error range should be represented by an error bar. Each datum will then be shown as a cross of a vertical bar and a horizontal bar, faithfully presenting the assigned errors in x and t. The cross outlines a region of possible values within which the actual value may exist. This graphical representation of the data and errors provides an immediate check on the consistency of the data and indicates if any points should be measured again.

A curve through the data regions need not go through the center "points"; in fact if it does, something is wrong with the original assignment of error ranges. The usual criterion for linking the data is to draw a smooth curve that passes somewhere through all the rectangular data regions. If the functional dependence is linear, then you can probably draw various straight lines that pass through all the regions. By doing so you can find the maximum and minimum slope that will fit and thus determine the error in the slope. It is also easy to see how precision of the slope value might be improved by measuring only certain points with increased precision.

In the graph shown in Fig. 2.4, the velocity was measured each second as well as could be gauged, which was ±0.2 s. The velocity measurements were accurate to ±20%, making the absolute errors greater for the larger velocities. Notice that the data fit a model for constant acceleration; i.e., the data regions can be linked with straight lines. On the other hand, the graph makes clear in a way that a data chart never could that the acceleration may not have been constant.

In Fig. 2.5 we show another feature of drawing error bars on graphs. In Fig. 2.5(a) there is a graph of $y(t)$ for the displacement of a stone falling from rest at $t=0$. The time was measured to ±0.1 s and y was measured with a precision so good that the error in y cannot be shown on a graph with this scale. Since the absolute error in t was independent of the value of t, all the error bars have the same length. Looking at this graph you could not tell whether it is a plot of a parabola or of a higher power curve. Furthermore, it is hard to notice if any particular datum region is inconsistent with the assumed function of $y \propto t^2$.

To resolve these problems, and to test the assumed power dependence, plot y versus t^2. This graph is shown in Fig. 2.5(b). Since the data appear to fall on a straight line, the assumed functional dependence is demonstrated. A linear or straight-line function can be identified very accurately on a graph. The straightness of a line can be judged by eye with great precision, particularly if you sight along the line with the eye as close as possible to the paper and to the projection of the line. Notice that in this graph the error limits for t^2 are larger than they are for t. If there is an absolute error of 0.1 s in each measurement, then there is

(a)

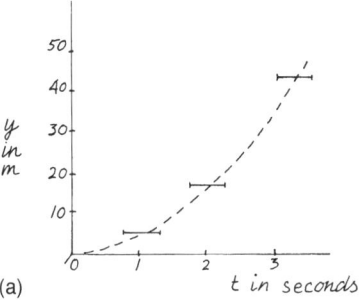

(b)

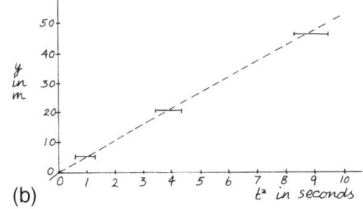

Figure 2.5. Graphical method of determining functional dependence of two variables.

±10% error in t at $t=1.0$ s, and therefore ±20% error in t^2. At $t=3.0$ s, there will be a ±3% error in t and ±7% error in t^2. When $t=1$ s, the absolute error (and the width of the error bar) for t^2 is ±0.2. At $t=3.0$ s, $t^2=9.0$ s^2, and the absolute error is ±0.6. The error bar extends from $t^2=8.4$ to $t^2=9.6$ s^2.

Should we bother introductory students with error analysis, particularly on graphs? Absolutely! The lessons to be learned are more important than memorizing all three of Newton's laws of motion. In the real world there are uncertainties, there are errors. Only in the physics class will students learn how to describe these and analyze them.

■ ACCURACY, PRECISION, AND COMMON SENSE

1. Does an average of 10 readings, each with n significant figures, yield a value with $n+1$ significant figures?

$$x = \frac{\Sigma x_i}{n} = \frac{18.4+19.2+18.5+18.9+18.8+18.5+18.9+18.6+18.7}{10} \stackrel{?}{=} 18.75.$$

It is commonly assumed that this is the case. The standard argument is that since the first figure after the decimal is significant for each reading, the first figure after the decimal should be significant for the sum (in this case, 187.5). Dividing by 10 does not change this significance, and therefore the average has an additional significant figure.

This situation illustrates the difficulty of laying down hard and fast rules about the analysis of data. In the example given, it may be that the experimenter could judge each reading within ±0.05, but for one reason or another the average values do not always lie within that range. The spread of readings indicates that the figure after the decimal point is not significant in the original technical sense. Circumstances such as this are common in actual experiments. In general, there is no justification for citing an additional significant figure for an average value. The main value in repeating a measurement of a quantity that is not subject to random fluctuations is to check against accidental mistakes. For that purpose, it is best to do the measurement by a different method, by using a different instrument, or by having another person do it. If an individual reads too large a length on the first measurement, that person will probably read too large a length the second time.

2. The error in a student experiment is *not* the difference between the student's experimental value and the textbook value! Error limits of individual measurements should be determined and cited by the experimenter. The compound error due to errors in the individual variables should then be calculated. If the error limits of the experimental value do not overlap the textbook value, there is evident reason to reassess the original judgments of possible error limits or to look for a mistake. An experimental value of 84±20 is not in disagreement with a handbook value of 100. (Of course, if the assignment was to determine the value within 5%, the experimenter has other problems.) Ideally, every experimenter (including a student) should make a preliminary estimate of the expected error limits before starting an experiment. Otherwise, how do you know what instruments to use and how much time to spend?

3. For a compound value made up of several individual values, do not seek more precision for any one value than is justified by the precision with which you know the others. If one value has a 10% error, there is usually no point in going to any trouble to obtain another value to within 1%. Notice especially how unimportant small errors become if they express standard deviations, linked with others through the square root of the sum of their squares. (In this case, the product of two values, one with a 10% error and the other with a 3% error, has a compound error of $\sqrt{10^2+3^2} = 10.4\%$, and not 13%.)

   If the diameter of a cylinder is measured to within 2% and the height to within 1%, then the volume is known to within 5%, assuming that the data were not sufficient to justify statistical procedures. The use of three significant figures for π (3.14) yields $\frac{1}{6}$% error margins, which do not add to the overall error margins for the volume. The use of four significant figures for π would not be justified. There would also be no point in improving the precision of the measurement of the height, since the major contribution to the error is provided by the diameter measurement.

4. You have no doubt heard that venerable saw, "If a thing is worth doing, it is worth doing well." That, of course, is nonsense. If a thing is worth doing, it is worth doing *well enough for the purpose at hand*. To do it any better than that is surely silly and probably wrong. Do not think, however, that this realistic view makes life easier. Far from it. Judgment is required. The purpose at hand may require years of devoted and meticulous work. Furthermore, the individual is faced with the awful responsibility of using his head to determine the requirements of the problem. No rules exist.

   Precision is expensive. In 1954 a particular cross-section value for a high-energy particle reaction was measured during the course of an afternoon to a precision of 15% or 20%. During the following two years, three scientists and numerous technicians and engineers spent a considerable fraction of their time and over $100,000 of government money to obtain that cross section to ±4.4% error. There was good reason for obtaining that precision, and the probable cost and difficulty were carefully considered in advance. The very first response that anyone should make when faced with a task is: "For what purpose is this information required?" If you do not know the answer, you do not know how to begin. This lesson is one of the most important ones that we can teach our students, preferably in the laboratory.

   What is the difference between precision and accuracy? One way to demonstrate the difference to students is to have one of them measure the height of a table with an expensive, precision meter stick, while another student measures the same height with a cheap, beat-up yard stick. (The good one we use has millimeter markings and a fancy plastic surface; the cheap one is left over from an elementary-school project and only measures to the nearest centimeter.) With only a little care, the first student can get four significant figures for the height of the table by using the expensive stick. The other student can quickly obtain two or three significant figures with the cheap one. Invariably, however, the two readings differ by a whole centimeter. It turns out that the first centimeter of the expensive meter stick has been sawed off. It is precise, but not accurate. It should be emphasized that there is a moral to this demonstration, making it more than just a trick. *All* instruments have errors in their zero settings. Also, even the most expensive meter may have been dropped the day before you use it.

What is the area of your front yard? Your choice of measuring instrument depends on the purpose for which the information is required. If you want to know the area so that you can buy lime in 80-lb bags at the local hardware store, then you can measure the yard by looking at it. Lime is cheap, you cannot buy a fractional bag, and the exact dosage is unimportant anyway. If you want to know the area in order to buy grass seed at $3.00/lb, then you should pace out the yard and perhaps use a pencil and the back of an envelope to sketch the geometry and check your arithmetic. Whether the yard is exactly rectangular, or whether your pace is 5 ft or 5 ft 3 in. is unimportant. You certainly would not use a meter stick. (Where is the seed salesman who can advise you in terms of lb/m^2?) If you want to measure the area of your yard to pay the taxes, the assessor will probably insist on knowing the area to within 0.01 acre. A surveyor's transit and chain are the appropriate measuring instruments. Note that each of the three methods was a measurement, and each was appropriate for the purpose. The surveyor's technique would have been as wrong for buying lime, as pacing would be for paying taxes.

■ PROPER CITATION OF STANDARD DEVIATION AND/OR ERROR LIMITS

How do we know what is meant when we read that a particular datum equals 10.0 ± 0.1 cm? Is this a personal guarantee of the experimenter that the results of all skillful measurements of this value will lie between 9.9 and 10.1 cm? Or does the writer mean that repeated measurements of the value yield a Gaussian distribution with a mean of 10.0 cm and a standard deviation (root mean square) of 0.1 cm? In that case, about $\frac{2}{3}$ of skillful measurements would produce values between 9.9 and 10.1 cm.

There is another possibility, frequently used in research reporting. The experimenter may have taken 100 measurements, plotted the distribution, and found that it had a mean of 10.0 cm and a standard deviation of 1 cm. The experimenter would then be justified in reasoning that another run of 100 measurements would yield a similar distribution with a mean and standard distribution very close to the first one. According to the Central Limit Theorem, if the means of repeated runs are plotted, the *means* will have a Gaussian distribution with a standard deviation equal to $\sigma/\sqrt{n}$, where σ is the standard deviation of the original data distribution (in this case equal to 1 cm) and n is the number of measurements in one of the runs. Therefore the *standard deviation of the means* is equal to (1 cm)/$\sqrt{100}$ or 0.1 cm.

But which of these possibilities is meant when the ± value is given? Unless the author specifies the meaning, or it is clearly implied by the experimental procedure, no one knows! How should *you* cite the error in reports? Define your error in words. If you are giving the maximum and minimum limits, say so, and explain how you chose those limits. Presumably the systematic errors (which do not produce a Gaussian distribution) were larger than the random errors. Perhaps you took only one or two measurements because there was no need for precision analysis. In that case you probably judged the error limits based on your experience and knowledge of the instruments and the procedure. If you did take enough measurements to yield a distribution curve, if that curve is roughly bell-shaped, and if you judge that your systematic errors are smaller than the random error, then you are justified in using the standard deviation. If you cite the standard deviation of the distribution and define it as such, then you are claiming that if one more measurement is made it stands a $\frac{2}{3}$ chance of having a value between $\bar{x}-\sigma$ and $\bar{x}+\sigma$. However, if you define σ to be the standard deviation of the mean ($\sigma_{\text{mean}}=\sigma_{\text{distribution}}/\sqrt{n}$), then you are claiming that if another set of n measurements is made, the mean of that set stands a $\frac{2}{3}$ chance of falling within the limits $\bar{x}\pm\sigma_{\text{mean}}$.

The standard deviation of a distribution is not necessarily related to the value of the mean or to the number of measurements in the sample. Distributions can be very sharply peaked or very broad. The standard IQ curve, for instance, has a mean of 100 and a standard deviation of 16 regardless of how many people are measured, providing that the sample is large enough to minimize statistical fluctuations. In a standard population, $\frac{2}{3}$ of the people tested will have IQ scores between 84 and 116. If you tested a class of 100 students and found that one of them had an IQ of 116, you would have no reason to think that there was anything unusual about the class. However, if the *mean* IQ of the class were 116, you would surely be dealing with a very special group since the standard deviation of the mean for this number of students should be $16/\sqrt{100} = 1.6$.

With the binomial distribution, the mean and standard deviation are linked. The standard deviation is $\sqrt{np(1-p)}$ where n is the number of "throws" or readings, p is the probability of "success" in any one throw, and np is the most likely number of successes, usually assumed to be the number that is measured. (For instance, if you throw a die 36 times, and the probability of getting a 1 on any throw is $\frac{1}{6}$, the most likely number of 1's in 36 throws is 6.) Since p is usually small, $(1-p)\approx1$, and $\sigma \approx \sqrt{np}$. For instance, if a Geiger counter measures 100 counts in one second, then the datum should be entered on the graph as $100 \pm \sqrt{100}$, or 100 ± 10. How does this work in the case of an IQ measurement? Suppose you test just one student and measure an IQ of 100. That tells you nothing about the standard distribution, but—assuming that the student's mistakes are random (an unlikely assumption)—suggests that $\frac{2}{3}$ of repeated measurements *of the same student* would yield test scores between 90 and 110. (Of course, it would be a major research project to construct a large number of truly duplicate exams in order to test the validity of such a conjecture.)

For binomial distributions, increasing the number of counts increases the absolute error, which is the standard deviation, since $\sigma \approx \sqrt{np} = \sqrt{\text{number of counts}}$. For instance, if your counter registers 100 counts, your datum is 100 ± 10. If you count 10 times as long and get 1000 counts, your absolute error is $\sqrt{1000}$ or approximately ±30. The fractional error has decreased, however, from $100\pm10\%$ to $1000\pm3\%$. Would you do better to take 10 runs, each taking $\frac{1}{10}$ the time, getting about 100 counts each time? Then your error for each is ±10. Combining 10 such runs, the combination error (as given on p. 000) is $\sigma/\sqrt{n} \approx 10/\sqrt{10} = 3$. We gain nothing by dividing up the data runs, except perhaps a check on the consistency of the apparatus.

■ REFERENCES

Barford, N. C., *Experimental Measurements: Precision, Errors, and Truth* (Addison-Wesley, Reading, MA., 1958).
Beers, Yardley, *Introduction to the Theory of Error* (Addison-Wesley, Reading, MA, 1958).

Swartz, C. E., *Used Math* (AAPT, College Park, MD, 1993).
Taylor, J. R., *An Introduction to Error Analysis* (University Science Books, Oxford University Press, New York, NY, 1982).
Youden, W. J., *Experimentation and Measurement* (Scholastic Book Service, New York, NY, 1962).

3
Units, Dimensions, Vectors, and Scaling

The subject of units and dimensions is usually thought of as one of the most necessary but least glamorous topics in our syllabus. Of course we have to choose a uniform system of units, and yet our students must be able to switch from one system to another. The general acceptance of *Système International* (SI) in college science studies has not been matched by its general use in the rest of life in the United States, or even in engineering classes. In universities, research physicists who take their turn teaching the introductory course have to relearn the SI values for many familiar constants that they have been happily using in more convenient units. Both the technical world and our everyday world are bedeviled with a multitude of units used to measure the same quantity. Length is measured in meters, feet, centimeters, inches, kilometers, yards, miles, furlongs, spans, fathoms, microns, Ångstroms, fermis, millimeters, rods, chains, light-years, parsecs, barleycorns, and probably many others. Some of these units survive as accidents of history, but many continue to be used because they are convenient for some particular trade. It is useful to have a standard unit about the same size as the object being measured.

■ CONVERTING UNITS

It is often necessary to change units for consistency in calculations. For simple cases this can be done without formal bookkeeping, but there is a foolproof way of keeping track of the changes. Students should be explicitly taught this method, and the instructor should be ostentatious in its use. The quantity to be converted should be written in the original numbers with all units specified in words or symbols, e.g., 60 miles/hour. The units are then treated like algebraic quantities that can be multiplied, divided, and canceled. The guiding rule for conversion of the units is to multiply the original value by fractions that are equal to 1. For example, let us change 60 miles/hour to feet/second. We use the following identities: 5280 ft = 1 mi; 60 min = 1 h; 60 s = 1 min.

$$\frac{60 \text{ mi}}{1 \text{ h}} \times \frac{5280 \text{ ft}}{1 \text{ mi}} \times \frac{1 \text{ h}}{60 \text{ min}} \times \frac{1 \text{ min}}{60 \text{ s}} = \frac{60 \times 5280 \text{ ft}}{60 \times 60 \text{ s}} = 88 \text{ ft/s}.$$

This conversion factor is frequently useful both in the technical and the everyday world. Notice that each of the multiplying fractions has the value 1. The fraction 5280 ft/1 mi was used, rather than 1 mi/5280 ft because we wanted to cancel the miles in the original numerator. The only units left are the ones desired: feet in the numerator and seconds in the denominator. Spelling out the units in this fashion keeps an automatic check on whether a conversion factor (such as 5280) should be multiplied or divided.

For another example, let us convert furlongs per fortnight to speed in cm/s:

$$\frac{1 \text{ furlong}}{1 \text{ fortnight}} \times \frac{220 \text{ yd}}{1 \text{ furlong}} \times \frac{3 \text{ ft}}{1 \text{ yd}} \times \frac{12 \text{ in}}{1 \text{ ft}} \times \frac{2.54 \text{ cm}}{1 \text{ in}} \times \frac{1 \text{ fortnight}}{14 \text{ days}} \times \frac{1 \text{ day}}{24 \text{ h}} \times \frac{1 \text{ h}}{3600 \text{ s}}$$

$$= 1.66 \times 10^{-2} \text{ cm/s}.$$

Evidently, 1 furlong/fortnight is a snail's pace. This conversion factor is little used.

We have gathered together a large number of conversion factors in Appendix 00. Learning to transform units is one of the basic skills that we must teach our students. To be sure, if we stick only with SI it may seem that we avoid headaches for ourselves and confusion for our students. But since these students are going to use technology in many trades in the real world where SI is not known or not honored (including advanced physics courses!), they must learn how to use and obtain other units. It is an enormous advantage to have the world settle on SI and we should use it as much as possible. However, we should also make sure that our students get a feel for the quantitative size of things. In order to do so, we must frequently interpret final SI units in terms of familiar quantities such as miles/hour. The baseball diamond is still calibrated in feet. When it comes to units, fanaticism as well as consistency should be eschewed.

■ DIMENSIONS

Even if we are referring to bride-price, we cannot justify an equation that says

$$4 \text{ horses} = 5 \text{ cows} + 5 \text{ sheep}.$$

It is conceivable, however, that the *cost* of these animals might be equated:

$$4 \text{ horses} \times (\$500/\text{horse}) = 5 \text{ cows} \times (\$300/\text{cow}) + 5 \text{ sheep} \times (\$100/\text{sheep}).$$

Each term in this second equation is a certain number of dollars. In every equation each term must describe the same quantity as every other term.

Frequently, the terms may not look alike. For instance, the final velocity of an object starting out with zero velocity and subject to constant acceleration a for a distance s is

$$v_f^2 = 2as.$$

The left-hand side is a velocity squared; the right-hand side is the product of an acceleration and a distance. Each of these, however, is composed of more primitive *dimensions*. The basic three dimensions are length (L), mass (M), and time (T). Velocity consists of a length divided by a time: L/T. Acceleration has the dimension L/T^2. The dimensions of the terms in the equation above are

$$\left(\frac{L}{T}\right)^2 = \frac{L}{T^2} L.$$

Dimensionally, the terms are the same.

This simple requirement provides a quick check on the consistency of results in a complicated derivation and also is the basis of powerful methods of analysis, particularly in complex situations requiring gross parameters, such as in hydrodynamics. As a trivial example of the method of dimensional analysis, what is the final velocity under constant acceleration if there in an initial velocity, v_0? The extra term must also have the dimensions of $(L/T)^2$. The simplest possibility is to introduce v_0^2:

$$v_f^2 = v_0^2 + 2as.$$

This makes sense because of an additional argument. If $a=0$, then v_f^2 must equal v_0^2.

Here are two examples of dimensional analysis used to derive the functional dependence of one variable on others.

1. Under constant acceleration, the distance traveled from rest is a function of powers of a and t. What is the function?

$$s = f(t,a), \quad s = Kt^n a^r.$$

Dimensionally, $L^1 = T^n(L/T^2)^r \rightarrow L^1 T^0 = T^{n-2r} L^r$. This can be satisfied if $r=1$, and $n-2r=0$. Therefore, $s = Kat^2$.

The method does not provide the value of the (dimensionless) constant.

2. Is the period of a simple pendulum

$$T = 2\pi\sqrt{\frac{l}{g}} \quad \text{or} \quad 2\pi\sqrt{\frac{g}{l}} \ ?$$

$$T = f(l,g) = K l^n g^r \rightarrow T^1 = L^n (L/T^2)^r = L^{n+r} T^{-2r}.$$

This can be satisfied if $2r = -1$, and $n + r = 0$. Therefore, $T = K l^{1/2} g^{-1/2} = K\sqrt{l/g}$. Once again, the method is incapable of yielding the constant.

Dimensional analysis can be used to discover much more complicated functional relationships. On p. 324 we use the technique to find the functional dependence of wave velocity on properties of the medium. For an elementary but very complete treatise on the subject, see Huntley (1967).

Routine use of dimensional analysis simply as a check on derivations can pay off in time saved. For instance, note that the arguments of all sines, exponents, etc., must be dimensionless. (Angles are dimensionless, since θ = arc/radius = L/L.) For a traveling wave,

$$y = A \sin 2\pi \left(\frac{t}{T} - \frac{x}{\lambda}\right) = A \sin \omega \left(t - \frac{x}{v}\right).$$

In the first expression each term is clearly dimensionless: T/T, L/L. In the second expression each term in the parentheses must have the dimension of T, since ω is a frequency with the dimension T^{-1}. The dimension of x/v is L/LT^{-1}.

The dimensions of most quantities are just multiples of L, M, and T. For work in thermodynamics it is convenient to consider temperature a separate dimension, denoted by Θ. Electromagnetic quantities have more complicated dimensions that include fractional values unless charge Q is taken as a basic dimension. The system has considerable arbitrariness connected with it, since other dimensions could be chosen as basic. It might be argued, for instance, that velocity is a basic dimension, since the velocity of light is a fundamental constant and has been *assigned* an exact value (without error) from which the standard of length is derived.

The following tables list the dimensions of some common quantities:

Quantity	Dimension	Quantity	Dimension
Angle	0	Frequency	T^{-1}
Area	L^2	Viscosity	$ML^{-1}T^{-1}$
Density	ML^{-3}	Acceleration	LT^{-2}
Moment of inertia	ML^2	Force	MLT^{-2}
Momentum	MLT^{-1}	Pressure	$ML^{-1}T^{-2}$
Angular momentum	ML^2T^{-1}	Surface tension	MT^{-2}
Torque	L^2MT^{-2}	Heat	ML^2T^{-2}
Energy	ML^2T^{-2}	Heat conductivity	$MLT^{-3}\Theta^{-1}$
Power	ML^2T^{-3}	Heat capacity (mass)	$L^2T^2\Theta^{-1}$
Temperature	Θ	Entropy	$ML^2T^{-2}\Theta^{-1}$
Specific heat	0		

With Q as a basic dimension	Quantity	Dimension in terms of M, L, T only
Q	Electric charge (q)	$L^{1/2}M^{1/2}$
$T^{-1}Q$	Electric current (i)	$L^{1/2}M^{1/2}T^{-1}$
$ML^2T^{-2}Q^{-1}$	Potential difference (V)	$M^{1/2}L^{3/2}T^{-2}$
$ML^2T^{-1}Q^{-2}$	Resistance (R)	LT^{-1}
$M^{-1}L^{-2}T^2Q^2$	Capacitance (C)	$L^{-1}T^2$
ML^2Q^{-2}	Inductance (L)	L
MLQ^{-2}	Permeability (μ)	0
$M^{-1}L^{-3}T^2Q^2$	Permittivity (ϵ)	$L^{-2}T^2$
$L^2T^{-1}Q$	Magnetic dipole moment (m)	$L^{5/2}L^{1/2}T^{-1}$
$MT^{-1}Q^{-1}$	Magnetic field (B)	$M^{1/2}L^{-1/2}T^{-1}$
$MLT^{-2}Q^{-1}$	Electric field (E)	$M^{1/2}L^{1/2}T^{-2}$
$M^{-1}L^{-3}TQ^2$	Conductivity (σ)	$L^{-2}T$
$L^{-2}T^{-1}Q$	Current density (j)	$M^{1/2}L^{-3/2}T^{-1}$

■ DIMENSIONLESS NUMBER—"SPECIFIC" QUANTITIES

Some quantities are specified in terms of their ratio to some common standard. In naming these quantities the word *specific* is often used. Thus we have *specific gravity* for the ratio of density of a material to the density of water. Since this is a ratio of densities, the value is dimensionless. In the cgs system of units, the density of water is 1.00 g/cm^3, by definition of the gram, and the density of iron is 7.86 g/cm^3. Consequently the specific gravity or specific density of iron is 7.86, numerically the same as its density.

Other such quantities are specific heat (the ratio of heat capacity of a substance to the heat capacity of water), but see also p. 231 in the chapter on internal energy, and atomic *weight* [the ratio of the nuclear mass of an element to 1/12 the mass of the carbon (twelve) nucleus].

■ DIMENSIONLESS NUMBERS—REYNOLDS NUMBER

In a number of cases, particularly in engineering, phenomena can be characterized by the values of dimensionless constants. These serve as scaling factors; as the value of the constant changes, the behavior of the phenomenon changes from one regime to another. The best known of these is the Reynolds number. It characterizes the behavior of an object moving through a fluid in terms of the relative importance of the inertial reaction of the fluid and the viscous drag. The ratio of these two terms is called the Reynolds number, named for Osborn Reynolds (1842–1912), an Anglo-Irish scientist. (For further information about the role of the Reynolds number in fluid behavior, in modeling, and in friction, see p. 83 in this chapter, p. 301 in the chapter on fluids, and p. 96 in the chapter on friction.)

To set up the dimensionless constant, analyze the forces acting on a tiny chunk of fluid that is being swept into a current from an initial speed of zero, as shown in Fig. 3.1. The driving force is equal to the product of the cross-sectional area of the chunk and the pressure difference from one end to the other. A viscous friction force reduces the effect of the driving force. The viscous stress on the small chunk of fluid is equal to

$$\eta \frac{dv}{dy} = \eta \frac{v}{L} \text{ (see p. 298)},$$

where η is the viscosity, v the velocity, and L the height of the chunk. The viscous force is equal to the product of this stress and the cross-sectional area of the fluid chunk. Let us assume for the moment that the chunk is cubic with a side length of L. The total viscous force acting on the cube of fluid is thus equal to $\eta(v/L)L^2 = \eta v L$.

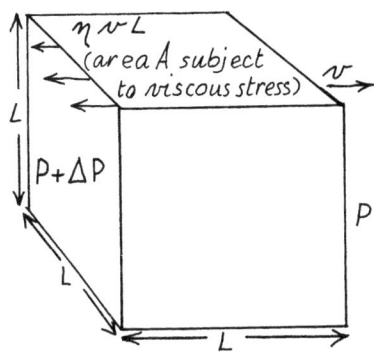

Figure 3.1. Block of fluid subject to viscous stress.

The net force acting on the fluid cube will produce an acceleration

$$L^2 \Delta P - \eta v L = ma.$$

The term on the right can be considered (as engineers like to do) the "inertial response" to the net force. We can express this inertial response, at least approximately, in terms of the density, size, and velocity of the fluid cube. To begin, note that $m = \rho L^3$. Next, $a = \Delta v/\Delta t$. Since the cube starts from rest, the change of velocity Δv is equal to v. The acceleration cannot occur in less time than it takes the cube to move through its own length: $\Delta t = L/v$. Thus the "inertial reaction" is

$$ma = (\rho L^3) \frac{v}{L/v} = \rho L^2 v^2.$$

To form a dimensionless constant, take the ratio of the two force terms, the "inertial reaction" and the viscous drag:

$$\text{Re} = \frac{\text{(inertial reaction)}}{\text{(viscous drag)}} = \frac{\rho L^2 v^2}{\eta v L} = \frac{\rho L v}{\eta}.$$

The density of the fluid is ρ, its viscosity is η, the velocity of the object is v, and L is its "characteristic" length. This length is approximately the maximum size of the object that is disturbing the fluid. For instance, for a dust particle or a raindrop falling through air, the diameter of the particle or drop is about the size of the disturbed chunk of air under consideration. An airplane will disturb air all around it in a region with a characteristic dimension of the length of the wing (actually, the disturbed air behind the plane extends over much greater distances but that is due to more complicated effects). As you can see, the assignment of the proper size for L is inexact by a factor of 2 or more. Nevertheless, the order of magnitude of the Reynolds number is useful in characterizing the type of motion that the fluid undergoes. For instance, if Re is between 0.1 and 10, both the inertial reaction of the fluid and its viscous drag play an important role in the motion. However, if Re<0.001, we can ignore the inertial reaction. Viscous drag dominates the motion, and there is essentially no acceleration of the fluid. If Re>1000, the viscous drag is negligible although the viscosity may still produce changes in the boundary region close to the object, thus influencing the flow pattern which may dominate the phenomenon. Usually, with Re>3000, the flow is turbulent.

We called the Reynolds number dimensionless. Here is the analysis:

$$\text{Re} = \frac{\rho L v}{\eta} \rightarrow (ML^{-3})(L)(LT^{-1})(ML^{-1}T^{-1})^{-1} = M^0 L^0 T^0.$$

The numerical value of the Reynolds number is necessarily also independent of units. For instance, compare your Reynolds number swimming in water in SI and in English units:

$$\text{Re} = (1 \times 10^3 \text{ kg/m}^3)(0.5 \text{ m})(1 \text{ m/s})/(1 \times 10^{-3} \text{ N s m}^{-2}) = 5 \times 10^5,$$

$$\text{Re} = (1.9 \text{ slugs/ft}^3)(1.6 \text{ ft})(3.3 \text{ ft/s})(2.1 \times 10^{-5} \text{ lb s ft}^{-2}) = 5 \times 10^5.$$

Another 19th-century Anglo-Irish scientist, George Stokes (1819–1903), analyzed the laws of motion for small Re. In this region objects move with constant velocity with a drag force proportional to $\eta v L$. The friction force is proportional only to the first power of the characteristic size. The drag on a spherical droplet, for instance, is proportional to the diameter and not to the cross-sectional area. The friction force is proportional to the first power of the velocity and is also proportional to the viscosity of the fluid. The drag does not depend, however, on the density of the fluid, since the fluid is not accelerated.

Note that viscous behavior depends on the *combination* of factors represented in Reynolds number and not just on the viscosity of the fluid. The viscosity of air, for instance, is very small and yet tiny dust particles or droplets travel at terminal velocity in air with zero acceleration *of the air* and hence are in the viscous regime. It is also interesting and surprising that the drag force in this regime does not depend on the density of the fluid. The terminal velocity of dust particles in air is the same at a pressure of 1 or 0.01 atmosphere. Although the fluid density is different by a factor of 100 in these two cases, the viscosity of a gas is independent of pressure.

Let us compute the Reynolds number for a dust particle and a raindrop. Assume that the dust particle has a characteristic size L of 0.1 mm and that for the rain droplet, $L=5$ mm. The dust particle is usually not solid and therefore has a density less than that of water. Its velocity might be 10^{-3} m/s. The velocity of a large raindrop would be about 10 m/s. The density of air is 1.3 kg/m^3. The viscosity of air at room temperature is 180 micropoise (μP)=1.8×10^{-4}, P=1.8×10^{-5} N s m^{-2}. For the dust particle

$$\text{Re} = \frac{\rho L v}{\eta} = \frac{(1.3 \text{ kg/m}^3)(10^{-4} \text{ m})(10^{-3} \text{ m/s})}{1.8 \times 10^{-5} \text{ N s m}^{-2}} = 10^{-2}.$$

According to the assumptions that we made, this particular dust particle is in the viscous regime. The Reynolds number for the rain droplet is larger by a factor of 5×10^5 (L is larger by a factor of 50 and v is larger by a factor of 10^4). Since, according to our assumptions, Re for the raindrop is equal to 5×10^3, this means that only the inertial reaction of the air that is shoved aside is important for the falling rain drop. Viscous drag is negligible.

Units, Dimensions, Vectors, and Scaling

For Reynolds number >100 or so, the drag force is proportional to $v^2 \rho L^2$. The viscosity does not enter, although it can produce very important effects that act like drag forces, because it influences the skin effect and the region where turbulence starts in. The drag is proportional to the density of the fluid since an object moving through a fluid must shoulder the fluid aside and so accelerate it. The drag is proportional to the cross-sectional area of the disturbed fluid L^2 and is also proportional to v^2.

■ DIMENSIONLESS NUMBERS—THE FINE-STRUCTURE CONSTANT

The fine-structure constant received its name because it characterizes the strength of a correction term in atomic spectra that is responsible for multiple lines, i.e., the fine structure. Its form is

$$\alpha = \frac{e^2}{4\pi\epsilon_0 \hbar c} \rightarrow \frac{(L^{1/2} M^{1/2})^2}{(L^{-2}T^2)(ML^2T^{-1})(LT^{-1})} = L^0 M^0 T^0.$$

The numerical value of α is approximately 1/137. In Dirac's relativistic treatment of the hydrogen-atom energy levels, the energies are

$$E = -\frac{\mu e^4}{(4\pi\epsilon_0)^2 2\hbar^2 n^2}\left[1 + \frac{\alpha^2}{n}\left(\frac{1}{j+\frac{1}{2}} - \frac{3}{4n}\right)\right].$$

The correction term that takes account of the angular momentum j term depends on the square of the small (1%) fine-structure constant. The j term is the total angular momentum of the atom. It is the vector sum of the orbital angular momentum l and the electron spin s. For instance, note that for $n=1$, where l must equal 0, and $j=s=1/2$, the correction term is equal to $\frac{1}{4}\alpha^2$. In quantum electrodynamics, the strength of successive approximations depends on α; hence these corrections are small.

One way to construct a dimensionless constant is to take the ratio of two relevant quantities, each of which has the same dimensions. In the case of the Reynolds number, the two quantities were forces. In the case of the fine-structure constant, we can form the ratio of the electrostatic self-energy of an electron that has a radius equal to its (reduced) Compton wavelength, to its relativistic total energy:

$$E_{\text{electrostatic}} = \frac{e^2}{4\pi\epsilon_0 r_c} \quad \text{where} \quad r_c = \lambda_c = \frac{\hbar}{mc},$$

$$E_{\text{total}} = mc^2,$$

$$\frac{E_{\text{electrostatic}}}{E_{\text{total}}} = \frac{e^2}{4\pi\epsilon_0 \hbar c}.$$

The (reduced) Compton wavelength is formed by dividing the unit of angular momentum $\hbar$ by the product of electron mass and speed of light. The ratio mvr/mv has the dimensions of length. (Planck's constant is "reduced" by expressing it in terms of radians: $\hbar = h/2\pi$.)

$$\lambda_c = \frac{\hbar}{mc} = \frac{1.06 \times 10^{-34}}{(9.11 \times 10^{-31})(3.00 \times 10^8)} = 3.88 \times 10^{-13} \text{ m}.$$

In Compton scattering of an x ray off a (free) electron, the *change* in wavelength of the x ray depends only on the deflection angle of the x ray, regardless of the original energy of the x ray: $\Delta\lambda = \lambda_0 - \lambda_f = \lambda_C(1 - \cos\theta)$. λ_C is the Compton wavelength:

$$\lambda_C = \frac{h}{m_0 c} = 2.4 \times 10^{-12} \text{ m}.$$

Note the units; an angular momentum h is divided by a linear momentum $m_0 c$, yielding a length.

■ DIMENSIONS OF COMBINATION VARIABLES

It is often useful to manufacture variables or constants that have specific dimensions. One example that we have just seen is the Compton wavelength.

1. The Rydberg constant for atomic spectra has the dimensions of reciprocal length:

   $$\text{Re} = \left(\frac{1}{4\pi\epsilon_0}\right)^2 \frac{me^4}{4\pi\hbar^3 c}.$$

   To check on this, we could appeal to Coulomb's law,

   $$F = \frac{1}{4\pi\epsilon_0} \frac{e^2}{r^2}.$$

   The Rydberg constant thus transforms *dimensionally* to

   $$\frac{F^2 m r^4}{(mvr)^3 v} \rightarrow \frac{(F^2/m^2) r^4}{r^3 v^4} \rightarrow \frac{a^2 r}{v^4} \rightarrow \frac{1}{L}.$$

2. Although temperature could be defined in terms of energy units, the common practice is to give it a separate dimension, Θ. The Boltzmann constant must therefore have the units of energy/degree. The combination, kT, has the units and dimensions of energy.

■ DIMENSIONAL CONSTRAINTS OF GRAPHS

The area under a graph curve must have the units of the product of the units on the two axes. Thus in a graph of power emitted by a filament as a function of time, the area under the curve between two times (the integral) represents the energy emitted during that time interval. It would be meaningless to plot energy as a function of time—what would that product be? However, you could graph the total energy expended up until a certain time, as a function of that time. The area under the curve would have no significance, but the slope (derivative) at any time would represent the power delivered at that time [(energy)/(time)].

Similarly, you cannot plot radiated power as a function of wavelength. Instead, the standard radiation curves show power *per wavelength interval* as a function of wavelength. The product is then the total power radiated between two wavelength limits, $(P/\Delta\lambda)(\lambda_2 - \lambda_1)$.

If you plot current versus time, the area under the curve between two time limits represents the charge that has flowed during that time interval. The slope (di/dt) is also physically meaningful. However, if you graph the charge on a capacitor as a function of time, $q(t)$, the area under the curve is meaningless. The slope, of course, represents the current in the circuit ($i = dq/dt$).

■ VECTORS

We usually define vectors as quantities that have direction and magnitude. There are several things wrong with this definition. First, we ought to make some attempt to distinguish between mathematical vectors and the physical quantities that they can represent. For instance, it is not quite valid to say that force *is* a vector; the question is, can force be represented by a vector? Do forces behave like vectors? That leads to the second point. How do vectors behave?

Mathematicians have rigorous ways of specifying vector properties. The simplest three dimensional vector is described as a number triplet (x_1, x_2, x_3), or (x,y,z), or (r,θ,ϕ), with specific rules of combination. To be sure, these numbers can represent line segments of certain lengths with particular directions in space. Whether they are vectors, however, depends on how they combine with each other. To combine ("add," "subtract," "multiply"), vectors follow special defined rules.

For physics classes, the best definition of the combining rule for "vector addition" is in terms of the process for displacements. If quantities combine or "add" like displacements, then they are described by vectors.

Are we being pedantic? Do not all quantities that have magnitude and direction add like displacements? No. Mountains, for instance, have magnitude (height or volume) and direction (up). But there is no useful way to represent a mountain with a vector. You cannot add two mountains. As we will see, there are less fanciful examples of quantities that have magnitude and direction but cannot be represented by vectors.

In his delightful and very instructive book, *About Vectors*, Hoffman (1966), Hoffman tells of an Indian tribe that accepted the early missionaries' doctrine that vectors were arrows, and therefore arrows were vectors. So well did they learn vector arithmetic that thenceforward they sent their braves out in pairs to shoot arrows in perpendicular directions and thus bring down the fallow deer with the deadly resultant. Hoffman argues that logically the story might be true, since had there been such a tribe they would surely have died of starvation, and it is a matter of record that there is now no such tribe.

How do displacements add? You can join them head to toe like arrows, and then draw the resultant from the first toe to the last head, as shown in Fig 3.2. However, except for sketching approximate results, this pictorial process has limited usefulness. (For the first approximation analysis of a problem, students should indeed learn how to draw such diagrams to scale.) For detailed calculations, it is better to resolve all vectors into perpendicular components and add in-line components separately. By doing this you avoid the conceptual problem of where the vector is "acting" if you move it around. For instance, compare the two diagrams in Fig 3.3. The head-to-toe combination of forces requires an unnatural displacement of the normal force away from the object and into a spatial limbo. On the other hand, this drawing does make it clear that the block will accelerate uphill only if $F - f > mg \sin \theta$.

Very few high-school texts show how to resolve vectors into components. It is not clear why they do not do this. The combination of trigonometry and geometry

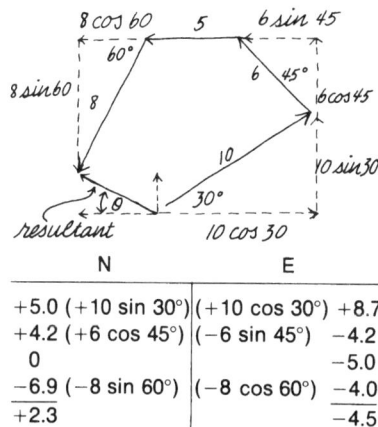

N	E
+5.0 (+10 sin 30°)	(+10 cos 30°) +8.7
+4.2 (+6 cos 45°)	(−6 sin 45°) −4.2
0	−5.0
−6.9 (−8 sin 60°)	(−8 cos 60°) −4.0
+2.3	−4.5

Length of displacement

$$= \sqrt{(2.3)^2 + (4.5)^2} = 5.1$$

$$\tan \theta = \frac{2.3}{4.5} = 0.51$$

$$\theta = 27° \text{ N of W}$$

Figure 3.2. Vector addition by resolution of vectors followed by addition of components.

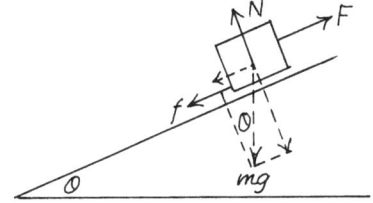

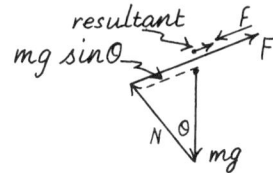

Figure 3.3. Forces perpendicular to plane: $N - mg \cos \theta = 0$. Force along the plane: $F - F_{friction} - mg \sin \theta = ma$.

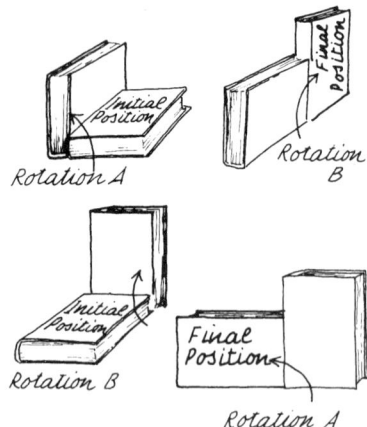

Rotation A + Rotation B ≠
Rotation B + Rotation A

Figure 3.4. Angles cannot be represented by vectors.

is simple and all of it has been taught in earlier math classes. Numerous high-school teachers have told us that their students can learn the technique easily.

Displacements, and therefore vectors, have simple associative and commutative properties of addition:

$$\vec{A}+(\vec{B}+\vec{C})=(\vec{A}+\vec{B})+\vec{C},$$

$$\vec{A}+\vec{B}=\vec{B}+\vec{A}.$$

These properties are not intuitively obvious to everyone in terms of the geometric representation and operations. In terms of components, however, the properties are manifest:

$$\vec{A}+\vec{B}=A_x+A_y+B_x+B_y=(A_x+B_x)+(A_y+B_y)=(B_x+A_x)+(B_y+A_y)=\vec{B}+\vec{A}.$$

Components in the same direction, or along the same axis, simply add like numbers.

Some college texts introduce the full vector notation and combination rules in an early chapter. That makes a grim introduction to our subject and profession. It is more natural and far more efficient to introduce each vector operation when it is needed for the description of some physical phenomenon. For first-year students it is never necessary to use the unit-vector notation:

$$\vec{A}=A_x\hat{i}+A_y\hat{j}+A_z\hat{k}.$$

Certainly problems of vector addition do not require this extra notation. At the introductory level the rules and procedures for multiplying the unit vectors are not really needed. Students are having enough troubles with the physical concepts without having to learn the rules for mathematical tricks that are needed only for problems more complicated than those appropriate at the introductory level.

What properties can be represented by vectors? Classification of the physical quantities into scalars or vectors is a time-honored classroom exercise. Be sure to emphasize the language. It is not force that *is* a vector; rather, force can be *represented* by a vector. In the case of force, incidentally, there is no necessary *logic* that requires force to be represented by a vector. It is a matter of experimental fact, to be tested with some arrangement like a force table. Of course, the vector nature of force is a consequence of Newton's second law if acceleration is a vector and mass is a scalar. But that law requires experimental justification in the usual logical scheme of definition. Velocity and acceleration are vector quantities because of their definitions in terms of the derivatives of displacement, which is the prototype of a vector quantity. However, just to make the situation nonobvious, remember that an angle cannot be represented by a vector. An angle has magnitude and we can assign a direction using the right-hand rule. However, angles not in the same plane do not follow combination rules like vectors, namely, $\vec{A}+\vec{B}=\vec{B}+\vec{A}$. This problem is illustrated in Fig 3.4. Nevertheless, the time derivative of an angle, which is the angular velocity, *can* be represented by a vector, as can the angular acceleration.

In listing vector and scalar quantities there are other surprises. Weight, being a force, goes in the vector column, but is mass a scalar? Certainly in everyday life inertial and gravitational mass are independent of any spatial orientation. However, in the general theory of relativity, mass is a component of a tensor, and therefore is not a scalar. Mass both produces and is affected by the curvature of space. In recent years, Dicke and co-workers have done highly precise experiments to see if a vibrator's mass is affected by the direction of oscillation. Perhaps when moving in line with the direction to the center of the galaxy, an object's mass is different than when it is moving perpendicular to that line. So far no such effect has been detected.

We usually treat angular speed, torque, and angular momentum as vector quantities, but strictly speaking they are not. Once again we appeal to the prototype behavior of displacements. If all the components of a displacement are reversed in direction, the displacement is reversed. If $A'_x \to -A_x$, $A'_y \to -A_y$, and $A'_z \to -A_z$, then $\vec{A}' \to -\vec{A}$. Such a transformation describes a reflection of the vector through the origin. In the case of torque and angular momentum, the quantities are themselves vector products of two other vectors:

$$\vec{\tau} = \vec{r} \times \vec{F}, \quad \vec{L} = \vec{r} \times m\vec{v}.$$

Upon reflection through the origin—reversing the sign of all components—

$$\vec{r}\,' \to -\vec{r}, \quad \vec{F}' \to -\vec{F}, \text{ and } \vec{\tau}\,' \to +\vec{\tau}.$$

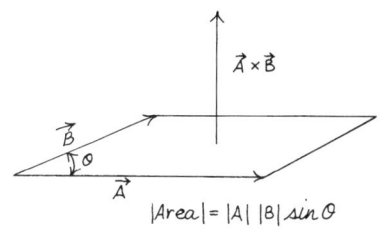

Figure 3.5. Representation of area by a pseudovector.

Because of this unvectorlike behavior, torque and angular momentum are referred to as pseudovectors. It is not something to worry students about, but some students may be beguiled by finding that simple things can be complex.

Most students are willing to put volume in the scalar column, but will be surprised to learn that area can be represented by a (pseudo-) vector. The area of a parallelogram is conveniently calculated as the vector product of two adjoining sides (Fig. 3.5):

$$\vec{A} = \vec{L} \times \vec{W}.$$

The assigned direction is given by the right-hand rule definition of the vector product and yields the direction of the perpendicular, or normal to the surface. In the natural sequence of introducing vectors, students will not have seen the vector product at the time that they will be classifying quantities, which is all right. Forget the vector-product nature of area (but remind them when they meet it later in the course) and merely point out that a surface area clearly has magnitude and can be oriented in different directions. As to whether areas combine like vectors (remember the problem with mountains), let the students consider the meaning of combining one acre of the vertical wall of a cliff with one acre of horizontal ground. Is there any useful sense in which the combined area can be considered as 1.4 acres with a direction of 45° to the vertical? Yes, if you are interested in projections of the area, such as in the problem of effectiveness of intercepting sunlight. Later in the standard course, at least at college level, this use of vectorial area is necessary in dealing with Gauss's law.

When introducing the way displacements combine, there is a temptation to start drawing arrows on the board right away. It may be apparent to you that the arrow is a scaled displacement, but some of your students may see only a chalk line with an arrow head. It might help to dramatize the problem by pacing out displacements, or better yet, having one of your students move around in front of the class. Two steps forward plus (combined with) two steps in a perpendicular direction obviously requires a journey of four steps, but also obviously leaves you only about three steps from your starting point. Do a number of these human displacements. Accompany each with the corresponding chalk arrow model on the board.

We have claimed that the unit-vector notation is an unnecessary conceit in an introductory-physics course. But without unit vectors, how do you deal with dot

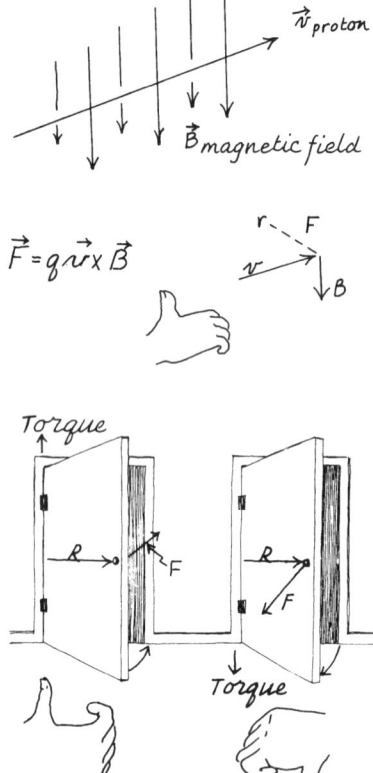

Figure 3.6. Right hand rule for vector multiplication.

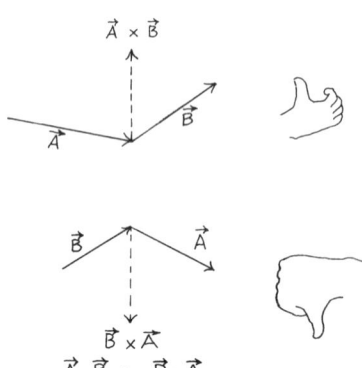

Figure 3.7. Demonstration, using right hand rule that $\vec{A} \times \vec{B} = -\vec{B} \times \vec{A}$.

and cross products? In terms of phenomena that require and illustrate them! The traditional first use of the scalar product is in the definition of work. Here the effective force in moving an object is clearly the component of the force in the direction of the displacement. The reason for the cosine of the subtended angle is apparent in the geometry. In the case of the vector product, the right-hand rule is expressive and almost tangible. You can use it to describe the opening of doors or the motion of protons in magnetic fields, as shown in Fig. 3.6. Furthermore, this simple right-hand rule of "turning the first vector into the second" makes it obvious that $\vec{A} \times \vec{B} = -\vec{B} \times \vec{A}$, as shown in Fig. 3.7.

■ SCALING

If you peel and chop two apples, one with a diameter of 4 in. and the other with a diameter of 2 in., how much more fruit do you get from the larger one? Fewer than 20% of college freshmen preparing to take introductory calculus-based physics realize that the bigger one yields 8 times the fruit. The principles of scaling and its consequences are not part of any curriculum. In the first PSSC book, back in the early 1960s, scaling was included in the first chapter, using examples from Gulliver's Travels. In later editions the unit was dropped because teachers reported that students found it too difficult. No doubt we are running into a Piagetian threshold problem. If 2/3 of a standard human population are incapable of doing simple algebra, regardless of their training (and surveys seem to indicate this), then problems involving proportionality and scaling are clearly beyond many of our students. This is a pity, because such notions are basic to our understanding of many physical phenomena.

Let us consider some of the consequences of the dependence of volume on length cubed and surface area on length squared. Note first of all that in scaling we are usually concerned with the comparison of objects with the same shape. If you double all the linear dimensions, it makes no difference whether the object is a sphere or a turtle; the volume is increased by 2^3. The surface area, or a cross section, is increased by 2^2.

1. The weight of an animal is proportional to its volume. The strength of its legs or arms is approximately proportional to their cross-sectional area. This is true whether the strength is that of legs used for support or arm muscles used for pulling. If we compare similar types of animals—mammals, for instance—then we conclude that an elephant's legs must be thick and those of a gazelle can be slender. Since the weight to be supported is proportional to L^3, and the strength is proportional to L^2, then leg diameter must be proportional to $L^{3/2}$. If you double the scale, the leg diameter must increase by $2^{3/2} = 2.8$, so that the cross-sectional area of the leg will increase by $(2^{3/2})^2 = 2^3 = 8$, in order to support the weight which has increased by 2^3.

2. Warm-blooded creatures must maintain their temperature above that of the environment. The heat energy lost is roughly proportional to the surface area, and hence to L^2. Consider the length ratio of human to mouse: perhaps 13:1. The volume and/or mass ratio would thus be 2000:1 (human mass of 60 kg; mouse mass of 30 g). In order to maintain body temperature, the human must eat 170 (13^2) times as much as a mouse. The mass of food for a day eaten by an adult human is about 2 kg, about 3% of body mass. Therefore for a mouse it would be 12 g, almost half its body mass. Clearly, the mouse must spend a

lot of time eating. It would be impossible for mammals to be much smaller than a mouse. (This game can be played with other information commonly available, such as 2 pounds of milk each day for a baby that weighs 10 pounds.)

It is generally the case that the smaller the creature, the more it must eat in comparison to its size. The food/body mass ratio is proportional to $L^2/L^3 = 1/L$. Hummingbirds must eat steadily to keep alive. The basal metabolism (energy consumption per body mass) of a week-old human is about twice that of an adult. Of course, each type of mammal or bird has its own proportionality constant for this size dependence, determined by the shape and nature of the surface. Since insects and fish are not warm blooded, they are not subject to the same law and so can exist in the form of very tiny species. On the other hand, insects cannot get very large because they do not have a circulatory blood system to transport oxygen to the inner cells. Consequently, muscles must stay close to surfaces where air is provided through tiny tubes called trachae.

3. Physical structures are also subject to volume–area scaling constraints. Consider a cube of granite, 1 cm on a side. Its mass is 2.5 g, the pressure on its base is $(2.5 \times 10^{-3}$ kg$)(9.8$ N/kg$)/(10^{-2}$ m$)^2 = 2.5 \times 10^2$ N/m^2. ($P_{atm} = 1 \times 10^5$ N/m^2.) Scale the cube to 1 m on a side. Its volume increases by a factor of 10^6, the area of its base increases by 10^4. The pressure is now 2.5×10^4 N/m^2. The pressure is proportional to $L^3/L^2 = L$. Granite crushes at a pressure of 1.6×10^8 N/m^2. Consequently, the cube cannot become greater than 6.4 km on a side—explaining well enough the limit to mountain height on earth. ("Granite" is a variable mix of minerals, and the values of density and compressibility are not good to better than 20%.)

4. For objects smaller than humans by a factor of 1000 or greater, the world behaves very differently than it does for us. For instance, insects and small pieces of metal (such as needles) can be supported by the surface tension of water. The support is proportional to the perimeter of the object. The surface tension of water is 7×10^{-2} N/m. For a needle 5 cm long, the perimeter is 10 cm. The surface tension would support 7×10^{-3} N, the weight of about 1 g. If all linear dimensions doubled (a 4 in. nail, for instance), the support would double but the weight would increase by a factor of 8 and the needle (nail) would sink.

The (mass)/(surface area) of objects is proportional to the scale factor. If the linear size is reduced by a factor of 1000, we are dealing with dust particles and tiny gnats. For them the Reynolds number is much less than 1. Their world is dominated by the *viscosity* of air, not its density. Stoke's law applies: drag is proportional to $h v L$. Note once again that in this regime the drag is proportional to the first power of the length of the object and the first power of the velocity.

5. Scale model planes do not fly at all like their real counterparts. There is a different ratio of surface to weight, and lift and drag are proportional to v^2. However, the real plane and its model would have approximately the same flight characteristics if their Reynolds numbers were equal: $Re = \rho L v / \eta$. If the model plane is flown in air, then ρ/η is the same for real plane and model. The problem is that compared with the real plane, L for the model is usually

small and so is v. This is especially true for homemade student wind tunnels. There are ways to get around this, usually expensive and impractical for amateurs. Notice that it is not very helpful to substitute water for air since

$$\frac{\rho}{\eta} = \frac{1 \times 10^3 \text{ kg/m}^3}{1 \times 10^{-3} \text{ N s m}^{-2}} = 10^6$$

for water, and

$$\frac{\rho}{\eta} = \frac{1.3 \text{ kg/m}^3}{1.8 \times 10^{-5} \text{ N s m}^{-2}} \approx 10^5$$

for air.

For boat models there is another problem. A major drag on surface boats (as opposed to submarines) is caused by the interference of their bow and stern waves. At the end of the 19th century, William Froude (rhymes with rude) investigated the behavior of small boats towed in tanks. He discovered another dimensionless scaling number, now called the Froude number: v^2/gL. For boats with geometrically similar hulls and equal Froude numbers, the wave patterns are the same. Consequently, a model with scale factor of 1/10 would have to be towed at speeds smaller by $\sqrt{10} \approx 3$. Note the dilemma. If a model matches the Froude number by traveling more slowly, it completely mismatches Reynolds' number which contains the *product* of v and L. Usually in (surface) boat model testing, the Froude number is matched and theoretical corrections are made for viscosity effects.

As all yachtsmen know, there is a practical limit to the speed of their sail boats. At *hull speed* the bow wave piles up ahead, leaving the boat in a small valley of its own making. To go faster requires an impractically large amount of energy. From Froude's number we can surmise that hull speed is proportional to $\sqrt{gL}$ for all boats of the same basic hull design. That's why longer boats can sail faster than shorter boats. The $\sqrt{L}$ is a standard handicapping factor for boat races. You can see this effect on the nearest mill pond by watching water fowl swim. Ducks and swans often paddle at hull speed, piling a ridge of water ahead of them. Swans, of course, can swim faster.

6. We can determine the scaling effects on falling objects by using dimensional analysis. Of course, if there is no air, there is no scaling effect; all objects fall with the same acceleration. However, if we require velocity to be a function of air density and the weight of the object and its cross section, then $v = f(mg, \rho, A)$:

$$L^1 T^{-1} = (MLT^{-2})^a (ML^{-3})^b (L^2)^c,$$
$$L^1 T^{-1} = M^{a+b} L^{a-3b+2c} T^{-2a},$$
$$a+b=0, \quad a-3b+2c=1, \quad -1=-2a.$$

Therefore,

$$a = \tfrac{1}{2}, \quad b = -\tfrac{1}{2}, \quad c = -\tfrac{1}{2},$$
$$v = K \sqrt{\frac{mg}{\rho A}}.$$

This agrees with the conditions for terminal velocity. The weight of the object equals the drag due to air friction:

$$mg = C\tfrac{1}{2}A\rho v^2.$$

C is the streamline coefficient; for a smooth sphere it equals 0.5. (The factor of $\tfrac{1}{2}$ stems from the derivation which takes into account the kinetic energy of the air swept out of the path of the ball.) The density of the fluid is ρ.

Let us calculate the terminal velocity of a standard baseball, which has a mass of 142 g and a diameter of 7.3 cm:

$$v_T = \sqrt{\frac{2(0.142 \text{ kg})(9.8 \text{ N/kg})}{0.5(42 \times 10^{-4} \text{ m}^2)(1.3 \text{ kg/m}^3)}} = 32 \text{ m/s} = 71 \text{ mph}.$$

The actual terminal velocity of a baseball is about 95 mph. Baseballs are not smooth. The roughness enhances turbulent flow, which twirls into a smaller wake (and creates less resistance) than part laminar, part turbulent flow.

We can calculate v_T for other smooth spheres by making some dimensional and unit transformations: $v_T = \sqrt{(8g/3C\rho)rD} = 6.3\sqrt{rD}$, where D is the mass density of the sphere and r is its radius. For instance, a sphere of iron with the same radius as a baseball would have a terminal velocity of $6.3\sqrt{(0.0365 \text{ m})(7.9 \times 10^3 \text{ kg/m}^3)} = 107$ m/s. The larger the ball, and the more dense, the greater its terminal velocity. The formula works well for smooth spheres, but as we have seen, the texture of the surface plays a large role. For instance, a golf ball has a mass of 0.046 kg and a radius of 2.1 cm. Its density is 1.2×10^3 kg/m^3 (golf balls sink!). Our formula gives $v_T = 32$ m/s, the same as a baseball. (The radius is smaller, but the density is higher.) The actual terminal velocity is 40 m/s = 90 mi/h, a little less than that of a baseball. Let us apply the formula to a Ping-Pong ball, which *has* a smooth surface, but a very small density. The mass is 2.7 g and the radius is 1.9 cm. Therefore the density is 94 kg/m^3. Our formula gives $v_T = 8.4$ m/s = 19 mi/h, in remarkable agreement with experiment.

Investigation of terminal velocity makes a good student research project. Unless you have a skydiver in your class, the experiments are most feasible with spheres dropping through liquids or very low density objects dropping in air. Check out the Reynolds number to determine the friction regime. For most human-size objects and speeds, the Reynolds number will be greater than 10, and thus the friction will be proportional to v^2. Remember that buoyancy reduces the effective force. In the case of the Ping-Pong ball with density of 94 kg/m^3, the buoyant effect of air with density of 1.3 kg/m^3, makes only a 1% reduction in net weight. A great reference for such school experiments is Shapiro (1961).

If the surface area producing the drag is unrelated to the weight of the object, then a different version of the formula ensues. For instance, a person falling with a parachute will ideally have terminal velocity of 15 mi/h. (If it is faster you break your ankle upon landing; if it is slower you drift too much.)

$$v_T = K\sqrt{\frac{\text{(weight)}}{\text{(area)}}}.$$

Parachutes must be matched to the weight of the parachuter. If you double the weight, you must double the area. If you do not have a parachute, your

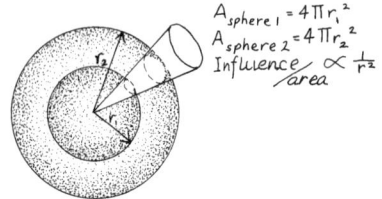

Figure 3.8. Functional dependence of fields (influence/area) on distance from source to observer, depends on source geometry.

terminal velocity (aptly named) will be between 150 and 200 mi/h, depending on how you tuck or flap your arms. Note that this speed roughly matches that obtained from the formula for a smooth sphere, assuming the density of water and some reasonable value for equivalent radius.

■ GEOMETRICAL FUNCTIONAL DEPENDENCE

In our introductory-physics course, a student meets the inverse square law in connection with gravity and later with electrostatics. Sometimes students end up thinking that this geometrical dependence is a characteristic of gravity and electromagnetism. The functional dependence on geometry is stressed in the study of Gauss's law, but students often find this analysis mysterious.

The inverse square law arises because of the geometry of a point (or homogeneous sphere) source, and a force influence that spreads out spherically, isotropically, and without decay or absorption. The law is a consequence of Euclidean three-dimensional space. The areas of concentric spheres surrounding a point source increase as r^2. If any "influence," such as electric field, gravitational field, or water spray, radiates away from the source equally in all directions, then the "influence" passing through each concentric sphere will be the same. Consequently, the "influence *per area*" must be proportional to $1/r^2$. (However, if the "influence" decays or has limited range, such as happens with virtual mesons, then an exponential fall-off must be added to the inverse square term. This would also be the case with a point source of light in a fog.)

If the source is a line source, such as an electrostatically charged rod, or a long fluorescent lamp, then the "influence" (electric field, or light intensity) falls off as $1/r$. That's because the influence is now traveling radially away from a line (or cylinder). The areas of concentric cylinders are proportional to r. Since the same amount of "influence" passes through each concentric cylinder, the "influence *per area*" must be proportional to $1/r$. Since we cannot have infinite line sources, the $1/r$ dependence is only an approximation valid in the region where $r \ll \frac{1}{2}L$, where L is the length of the source.

On an infinite flat earth, or in a parallel-plate capacitor, the lines of force are perpendicular to the surface, uniformly spaced, and parallel to each other. The "influence *per area*" is thus constant.

The three cases are shown in Fig. 3.8 and Fig. 16.12 on p. 398.

All three of these principal geometries can be explored by students with a light-intensity meter using light sources that are approximately points or spheres, lines or cylinders, or flat and uniformly illuminated. For practical purposes, such experiments must be conducted in surroundings that will not reflect much light. It is hard to get good results in classrooms, but the problem makes a good home project, to be done outdoors at night.

■ REFERENCES

Hoffman, Banesh, *About Vectors* (Prentice-Hall, Englewood Cliffs, NJ, 1966).
Huntley, H. E., *Dimensional Analysis* (Dover, New York, 1967).
Shapiro, Ascher, *Shape and Flow* (Doubleday Anchor, Garden City, NY, 1961).

4
Friction

Friction can be a bit of a drag. It robs us of useful energy and makes simple physics formulas only poor first approximations. On the other hand, it allows us to walk on most flat surfaces and it creates variations of Newton's second law with which we can test students. The subject occupies only a page or so of most introductory texts, indicating both its apparent simplicity and its relative unimportance, at least for the study of physics. As for simplicity, friction appears to involve only a table of coefficients and a few empirical laws, easily grasped. As for unimportance, note that friction is an embarrassment when we try to demonstrate the conservation of momentum and energy, and it can hardly compete with the excitement of lasers and subatomic physics.

As is the case with other folklore, the actual situation is quite different. Friction is very complicated. The simple laws, when they apply, are clues to the microstructure of matter. When they do not apply, and this happens in many everyday situations, the deviations supply even better clues. One of the happy features of this apparently dull subject is that the standard classroom study is only one step away from current research. The nature of surfaces has become a hot topic in the last few decades, both in pure and applied physics.

■ HISTORY

An Egyptian mural from 1900 B.C. shows a heavy sledge being drawn by many men while one man pours lubricant (perhaps milk) on the runners. Leonardo da Vinci, in the middle of the fifteenth century (two centuries before Newton), noted that the effort needed to pull an object was proportional to its weight but independent of the area of the object in contact with the surface. These simple rules were rediscovered by Guillaume Amontons in France at the end of the seventeenth century. In 1781 Charles Coulomb published a treatise on simple machines, and attempted an explanation of friction in terms of surface roughness. As shown in Fig. 4.1, in order to move the upper block to the right, we must exert a force equal to $W \tan \theta$. The ratio of F to W is thus $\tan \theta$, which might plausibly be less than 1 for many surfaces. This would explain the curious fact that the coefficient of dry sliding friction is between 0.1 and 1.0 for many pairs of materials. However, the theory gives us no guide as to how to calculate the average angle of the sawtooth planes, or how that angle would depend on the gross roughness of the surfaces or on the type of material. Furthermore, the model would not explain many of the exceptions to the general rules. For instance, the friction between very smooth surfaces can be extremely large. Finally, the simple inclined-plane model does not explain why friction produces energy loss. To be sure, it would take work to pull

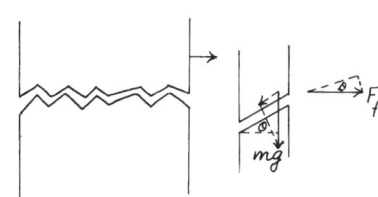

Figure 4.1. Coulomb's sawtooth version of surfaces and friction.

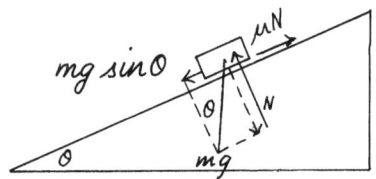

When block is sliding with constant speed,

$$mg \sin \theta = \mu N$$
$$= \mu\, mg \cos \theta$$

Therefore: $\mu = \tan \theta$

Figure 4.2. Inclined plane method of measuring coefficients of friction.

the upper block up the inclined planes, but that energy should be recovered when the block slides down into the next position of the sawtooth geometry. To claim that the energy would be lost due to the friction of sliding down or bumping along just begs the original question and calls for a more detailed molecular model.

Although many important practical facts were learned about friction during the nineteenth and early twentieth centuries, it was not until after World War II that solid-state theory and electron-beam probes were brought to bear on the subject. By the 1980s the general nature of friction was agreed upon by research people (though not always by introductory-physics texts), but detailed applications could still be mysterious. We will see why this is the case when we take a look at the nature of surfaces and the several causes of friction force and energy loss.

■ THE SIMPLE, GENERAL RULES OF DRY SLIDING FRICTION

1. The friction force is proportional to the normal force at the surface. However, this is true only if the objects in contact do not deform appreciably as the load increases. With rubber tires, for instance, where the tire spreads out as the load increases, the sliding (not rolling) friction force is proportional to W^n, where $n<1$.

2. The force necessary to start an object moving from rest is usually more than the force needed to keep it moving at constant speed. In terms of the coefficients of static and kinetic friction, $\mu_s > \mu_k$. However, μ_s is even more dependent than μ_k on small effects of surface conditions, such as cleanliness. Indeed, with freshly milled surfaces of hard metals under vacuum conditions, there is no difference between μ_s and μ_k, and both can be much larger than 1.

3. For a particular object, the friction force is independent of the apparent area in contact with the supporting surface. With a rectangular block, for example, the force needed to drag the block along its small end face is the same as that needed to drag it on its broad base. However, this is true only for hard materials where there is no appreciable compression or grooving.

4. For laboratory speeds of cm/s to m/s, μ_k is approximately independent of speed. Obviously there must be a transition region between μ_s and μ_k as speed increases from 0. At very high speeds (which can occur with certain types of machinery), frictional heating may change the surfaces sufficiently to change the coefficient of friction, usually lowering it because of local melting which provides lubrication.

The simple general rules can be demonstrated easily by students. Blocks made of various materials can be pulled along surfaces of various materials with spring force scales. There is pedagogical value in doing such a straightforward measurement but the traditional inclined-plane method shown in Fig. 4.2 may also provide a useful exercise in vector resolution. Whether one method is more accurate or yields more precision than the other is immaterial; there is almost never justification in a school laboratory for finding a coefficient of friction to more than one significant figure. More precision than that would be vitiated by slight changes in geometry or surface conditions in some other application with the same materials. Avoid using blocks of wood to demonstrate that the friction is independent of apparent surface area. The difference in grain structure on the several sides will usually produce different coefficients of friction.

It is particularly instructive to show that friction does not always depend in obvious ways on smoothness or roughness of the surfaces. For instance, a smooth board will glide easily on bristles or a pebbly plastic, but two pieces of smooth, flat glass may well stick together. One startling demonstration of this latter effect is to fit a string harness around a drinking glass filled with water and drag it along the surface of a sheet of window glass. Prepare the flat glass by rubbing your hands over the surface. You should be able to drag the drinking glass with only a small force. Then pour a little of the water on the flat glass. A normal expectation is that friction will now be even lower because of the water film. Instead the glass will stick, and often scratch the window glass. The oil left by your hands has floated to the surface of the water, thus allowing glass-to-glass molecular bonding.

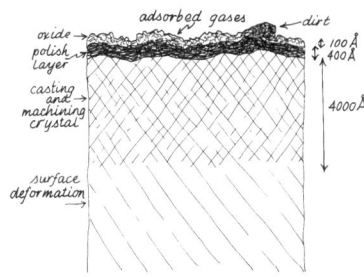

Figure 4.3. Surface structure of a solid.

■ THE SURFACES OF SOLIDS

Most solid chunks of things have surfaces that look rough or warped even to the eye. It is apparent that the actual contact area of a wooden box on a wood floor is less than the nominal area of the bottom of the box. It is not so obvious that this is also true of almost any object lying on another one. Flatness over regions longer than a centimeter or so is rarely better than a few wavelengths of light. A wavelength of light is equal to several thousand atomic diameters. If the surface is polished it will have scratch marks equal in size to the diameter of the polishing powder, perhaps micron size. Heat or fire polishing may fill in the scratches, but except for certain crystals like mica there will still be hills and valleys on the surface that are many atoms high. In Fig. 4.3 we show a profile of the surface of a typical piece of polished steel. The length unit of scale is the angstrom, 1×10^{-10} m, which is about an atomic radius. The peaks in the surface roughness are less than the wavelength of light; the surface would appear mirrorlike. The peaks and valleys are coated with gases. An oxide layer has worked its way in to a depth of about 100 atoms. Other gases, primarily nitrogen but also water and carbon dioxide, are not bound chemically, but are adsorbed and piled up several molecules thick. If the surface has been handled and has been in air for a few minutes then there are also big globules of dirt from the air and patches of oil from hands. Underneath this mess there are layers of iron alloy that may differ markedly from the interior steel. The polishing and the original machining or casting will produce crystal structure and flaw patterns at the surface that are different from those in the interior. The freezing process, for instance, usually leaves the surface harder and more brittle than the region several microns deep.

Consider what happens when you put one such surface on top of another one. As one of the great research workers in friction, Frank Bowden, said, "Putting two solids together is rather like turning Switzerland upside down and standing it on Austria—the area of intimate contact will be small." The actual area of contact, much smaller than the apparent area, will be in regions where mountain tops or rolling hills are crunching into each other. There is a limit to the pressure that any particular solid material can stand before it starts flowing. This is called the yield pressure. When two objects are forced together, the high peaks on their surfaces will crumble and the hill tops will flow, bringing more and more area into intimate contact. When the product of yield pressure and area equals the applied force, the two objects will resist further motion into each other.

Let us calculate an approximate value for the actual contact area of a sample

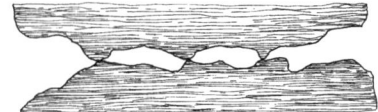

Figure 4.4. Weight is supported by many micro regions, some of them crushed and exerting their yield pressure. Where the metal has flowed under this pressure, welds have formed. The supported weight (or normal force) is equal to the product of actual contact area, A, and the yield pressure, P, N=AP. In order to slide the two surfaces over each other, the bonded micro regions must be sheared. The required force is F=AS, where S is the shear modulus.

case. When stress is proportional to strain, $F/A = -Y\Delta L/L$, where Y is Young's modulus. Young's modulus for steel is 1.9×10^{11} N/m². Yield pressure is considerably smaller than this—about 5×10^9 N/m². [Note the significance: $\Delta L/L = -(5 \times 10^9)/(1.9 \times 10^{11}) = 1/40$. The material cannot stand elastic stretching or compression greater than about 1/40 of its length.] Since the density of steel is 7900 kg/m³, the weight of a 10-cm cube of steel is about 80 N. That weight could be sustained by an actual contact area:

$$A = (\text{weight})/(\text{yield pressure}) = (80 \text{ N})/(5 \times 10^9 \text{ N/m}^2) = 1.6 \times 10^{-4} \text{ cm}^2.$$

Since the apparent surface area is 100 cm², the fraction of real to apparent surface area in contact is about 10^{-6}. (For a cube the weight increases as the cube of the length, but the surface area increases as the square. Therefore the real contact area would approximately equal the apparent area for a cube of steel 10^6 times larger, i.e., 100 km on a side.)

One of the simple rules of friction of nondeformable objects is now explained. Friction is independent of the apparent surface area because under most circumstances the actual area bearing the load is much smaller than the apparent area. The *actual* contact area is always proportional to the applied (normal) force.

Notice that the force required for dry sliding friction is in many cases about the same or a little less than the normal force. That is to say, $\mu \lesssim 1$. Let us see why this is approximately true. We assume that the compressed contact regions supporting the load are also providing the bonds that must be overcome if the object moves. The motion is perpendicular to the compression and therefore involves shear forces. The friction force must be $F = AS$, where S is the force per unit area to shear a junction, and A is the *actual* contact area. Since $A = (\text{normal force})/(\text{yield pressure}) = N/P$, and since for most materials the shear modulus is about 1/2 the compression modulus, it is plausible that the friction force should be about 1/2 the normal force: $F = (N/P)S \rightarrow N(S/P) \approx \frac{1}{2}N$.

According to this model, dry sliding friction is caused by the breaking of a large number of tiny spot welds that are continually made as two objects are pressed together and slide over each other (Fig. 4.4). There are some obvious questions that can be raised about such a model. First, is there any evidence that such welding takes place? Indeed, microscopic inspection of the surfaces of dissimilar materials that have been dragged across each other reveals that fragments of the materials have been interchanged. Similar studies have been done with radioactive tracers built into one of the objects; they transfer to the other object. Second, if you place one object on another and the surface hill tops flow and weld together, why can't you pick up the top object and lift the bottom one with it? Under the right circumstances you can. Take two cylinders of lead with freshly machined end faces. Press them strongly together, by hand. The upper one will then be able to support the weight of the lower one. If you break them apart, you can see the scars where bonding had taken place. Of course, special conditions are necessary in order to produce this effect. The surfaces should be machine flat and recently cleaned. Lead is soft enough so that the surfaces will mold into each other over small areas. Why can't you do this just by setting a book on a table and then lifting the book? Consider the nature of the surfaces. There are surely films of gas molecules, and assorted oils preventing molecular bonding. Furthermore, much of the load is being supported by springlike compressions without bonding. If the load is raised, these springs help to snap any molecular bonds that have formed. When you drag one object across the other, however, you scrape away much of the surface dirt and film, allowing more intimate contact between the surfaces and providing the energy for localized heating.

A third objection to the simple model of spot welds is that the same process ought to take place during rolling, and then rolling friction would be as large as sliding friction—which it is not. True enough, there must be as much actual area supporting a ball or cylinder as a flat plate of the same weight. Once again, however, because of the surface films most of these contact regions are not molecularly bonded. In rolling, the surfaces do not scrape over each other. Instead, hilly regions are lowered onto other hilly regions, compressing them, while the back surface is lifted or peeled off. Notice that you cannot polish a surface by rolling things on it.

The model of many spot welds cannot be the complete story, especially considering the messiness of real surfaces. During the sliding process, some of the rough spots will actually collide and be broken off. Whether through microscopic bond breaking or the fracturing of individual tiny peaks, the work done in pulling an object against friction ends up as thermal energy dispersed throughout the surfaces. (Some texts worry about the work done by the friction force. Since that force is opposite to the displacement, one might think that friction does negative work. That's nonsense. Friction uses no fuel and does no work, positive or negative.)

The coefficient of friction can be much larger than 1. Some of the cases are obvious, e.g., very rough surfaces, or sticky surfaces. Under some circumstances molecular bonding may take place over large regions of the surface. We have already mentioned the cases of a glass tumbler sliding over window glass, fresh surfaces of lead pressed together, and metal with milled surfaces in a vacuum. Optical glass can be made flat enough to bond, although that is not what usually holds microscope slides together. With these, as with machinists' gauges, there are viscous and surface tension effects of thin surface films. (The precision gauge blocks used by machinists are oiled to prevent corrosion and to prevent the scratching and pitting that would occur if they were wrung together and molecular bonds formed.)

A source of tricky freshman physics problems is the difference between static and kinetic friction. The difference is real, but more accidental than profound. If one messy surface is placed on another messy surface, various oils and dirt will flow and mix and bind the two surfaces together. The value of the coefficient of static friction depends not only on the chemical nature of the two materials but also on the recent history of the surfaces. With surfaces freshly prepared in a vacuum, there is little or no difference between static and kinetic coefficients, at least with hard metals. With softer materials, which can flow, the actual contact area may increase with time if two objects are left in contact with each other. With some materials there can also be diffusion of atoms through regions that are already bonded, thus strengthening the bonds. With lubricated surfaces a different sort of time-dependent effect may occur. Under weak tangential stress, the film may break down, allowing molecular bonding. Once the objects are in motion, however, they may slide over each other without allowing enough time for the film to separate, thus reducing the friction.

Since for most surfaces $\mu_s > \mu_k$, there arises a common form of motion called stick-slip. As a force is applied to move an object, the object is motionless until the applied force is greater than that needed to move the object with constant speed against kinetic friction. When the object breaks free it will accelerate, catching up with whatever was pulling it and then bouncing back, momentarily coming to rest. But then the whole process will repeat itself. The frequency of the stick-slip usually depends on the dynamics of the system. For instance, when chalk screeches against a blackboard, the frequency is determined by the time taken for the bending pulse to travel up the length of the chalk and back again. One of the most profound findings in educational theory is that this problem can be avoided by breaking the standard piece of chalk and using about 2/3 of the standard length. Another example of stick-slip is provided by the bowing of a violin string. In this case the release of the string that has been pulled by the bow, held by static friction, is triggered by the round-trip return of the kink in the string produced by the previous slip.

■ EFFECT OF HARDNESS ON METAL–METAL FRICTION

An interesting test of our model for friction is to compare the coefficients of friction of hard and soft metals. The shear strength of hard steel is 50–100 times that of tin, but $\mu_{\text{tin-tin}} = 3\mu_{\text{steel-steel}}$. Indeed, within a factor of 2, $\mu_k = 0.4$ for almost all dry metal–metal contacts under everyday conditions. Remember that the friction force is $F = AS$, where A is the actual contact area and S is the shear strength in N/m^2. The actual contact area is $A = N/P$, where N is the normal force and P is the yield pressure. Consequently, the friction force $= N(S/P)$. Hard metals have both large yield pressure and large shear strength, while soft metals have relatively low values for both. For hard–hard metal contacts, friction is about the same as for soft–soft contacts. However, if a hard metal can be covered with a thin layer (no more than a few hundred atoms thick) of a soft metal, then the substrate can support the normal force with only small actual area A in contact, while the soft surface film permits the weld spots to be sheared with less force. For instance, $\mu_{\text{steel-steel}} = 0.8$. For steel sliding on indium, a very soft metal, $\mu_{\text{indium-steel}} = 1.6$. In this case, the hard steel "mountains" on the surface settle into the soft indium, increasing A. When the two surfaces move across each other, these buried mountains plow through the indium. However, if hard steel is covered with a thin film of indium, then $\mu_{\text{steel-(indium)steel}} = 0.1$. This phenomenon is used in industry to reduce friction when drawing steel rods through dies. The rods are first coated with a microscopic layer of indium.

■ THE EFFECT OF NONRIGID SURFACES

The simple law that friction is proportional to the normal force and independent of the surface area certainly does not apply when one of the surfaces can be deformed. A skier experiences less friction while standing on the skis than while standing in the snow and holding the skis. When deformation occurs, the main determinant of the amount of friction is the "plowing" effect. Since the normal force determines the extent of deformation, as well as the required contact area for support, the friction is not proportional to the normal force.

For *elastomers*, soft substances such as rubber or plastics that can yield and then recover, friction depends on the shape of the objects. For these soft materials, the actual contact area is approximately equal to the apparent area. As a hard sphere or wheel sinks into a yielding surface, the apparent contact area increases with load, but because of the geometry, doubling the load does not double the area. The true area of contact, and hence the sliding friction, is proportional to N^n, where $n<1$. (For the case of a metal sphere on rubber, $n=2/3$.)

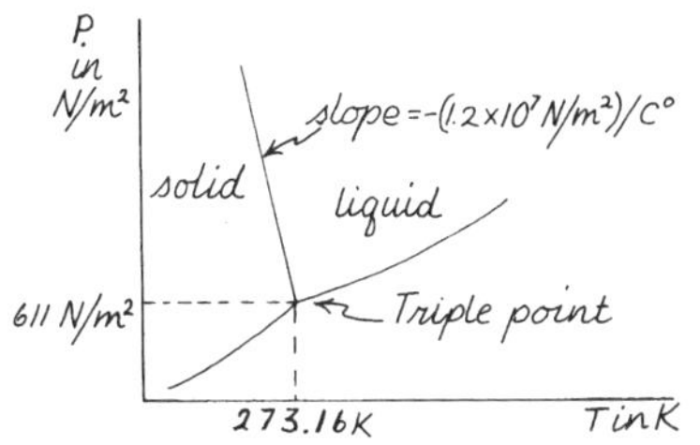

■ THE LOW FRICTION OF TEFLON

Another test of the molecular bonding model of friction is provided by the behavior of teflon–polytetrafluoroethylene, or PTFE. Teflon does not bond well with most things under most conditions. Therefore a surface coated with teflon should have very low friction, and this is indeed the case: $\mu<0.1$. Teflon and similar materials have been used as coatings on some skis, instead of wax. For most ski surfaces the coefficient of friction depends strongly on the temperature of the snow, since the low friction is provided by a thin water film. For teflon-coated skis, however, μ is almost independent of temperature.

Figure 4.5. Pressure-Temperature curves for water. Scales are non-linear to exaggerate the negative slope of the boundary line between liquid and solid.

■ SLIDING FRICTION ON ICE

Ice skates experience low friction because a thin water film is produced between blade and ice. A popular legend has it that this effect is caused by the lowering of the melting point due to the pressure exerted by the blade. It is true that one of the unusual features of water is that it expands during freezing. Consequently, if we prevent the expansion by increasing the pressure, we prevent the freezing and thus lower the melting point. The usual pressure–temperature diagram for water, Fig. 4.5, shows the boundary line between solid and liquid arching back to the left from the triple point. However, the actual effect is very small. The slope of the boundary line is $-(1.2\times10^7 \text{ N/m}^2)/°\text{C}$. It would take an increase of 120 atmospheres to lower the melting temperature 1 °C.

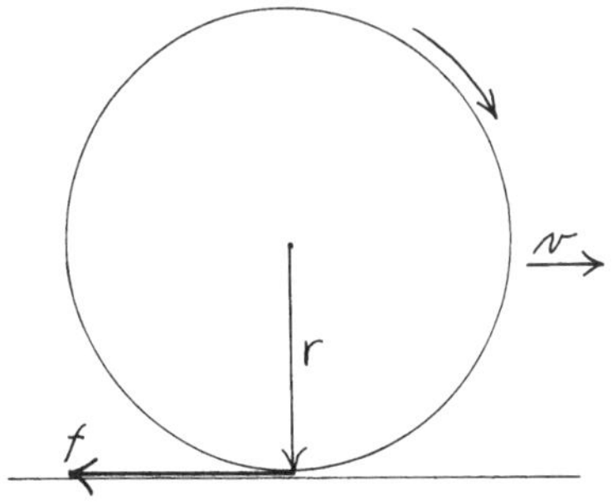

For the typical skate blade, the area is $27 \text{ cm}\times4 \text{ mm}=11 \text{ cm}^2$. If the full weight of a skater with mass 65 kg is exerted on one blade, the increased pressure would be $(650 \text{ N})/(11\times10^{-4} \text{ m}^2)=6\times10^5 \text{ N/m}^2\approx6$ atm. Sharpening the blades does not decrease the contact area appreciably, since the blade sinks down into the relatively soft ice.

Figure 4.6. Retarding friction force evidently provides torque to accelerate wheel.

What does produce the water film between blade and ice? There are two plausible explanations. When the leading edge of a blade strikes the ice, the resulting friction energy can melt a trail for the rest of the blade. A more important effect stems from a phenomenon first noticed by Faraday and then largely ignored because it was not understood. We now know that at the interface between ice and air there is a thin film of water. The thickness increases from monomolecular to several hundred molecules as the temperature rises from -10 to 0 °C. Since the reduction of friction depends on the water film, you might conclude that the fastest skating could be done on ice close to the melting point. However, warm ice is soft ice, allowing the blades to sink in more. On the other hand, cold ice, which is hard, has only a thin film of surface water. These two competing effects yield a minimum of friction for speed skating at about -7 °C. (For further details see the reference by James White at the end of the chapter.)

■ ROLLING FRICTION

In spite of the fact that we have a civilization that moves on wheels, rolling friction is seldom discussed in introductory-physics texts. Perhaps it is because of the embarrassment of an apparent paradox concerned with rolling. In Fig. 4.6 we

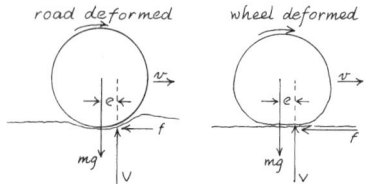

Figure 4.7. Exaggerated sketches of deformation occurring during rolling.

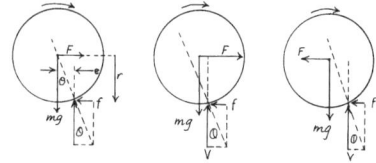

Figure 4.8. Geometry of friction and vertical forces on wheel rolling with constant velocity, with positive acceleration, and while being braked. Note that resultant of vertical support and friction maintains roughly constant angular direction. At constant velocity, resultant points to axle, producing no torque. The relative magnitudes of f and V are not drawn to scale. The geometry depends more on the offset, e, than on the magnitude of f.

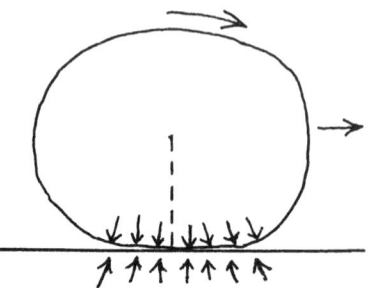

Figure 4.9. Because of the asymmetry in the bulge of the tire, the tire strikes the ground in the forward direction, causing the ground to recoil with a backward thrust.

represent an undriven wheel rolling in the direction shown and slowing down because of the retarding friction force. Note, however, that that force is exerting a torque about the axis tending to make the wheel speed up!

Evidently the geometry shown cannot be valid. Indeed, unless there is some compression in the wheel or the road, there can be spinning but not real rolling motion. The actual geometry, greatly exaggerated, is shown in Fig. 4.7.

When a wheel is rolling, the deformation of road or wheel is always such that the effective vertical force of the road is not in line with the center of mass of the wheel. The magnitude of the offset, e, depends on whether the wheel is accelerating, traveling at constant speed, or decelerating. Experiment shows that the magnitude of the retarding friction, f, is approximately independent of speed or acceleration. Therefore, the resultant of V and f always has the same angular direction. However, *where* the resultant points depends on the offset e. For constant speed, the resultant must point toward the axle, thus producing no torque. For acceleration, e must decrease, allowing the resultant to point behind the axle, thus producing an accelerating torque. For deceleration, e must increase, allowing the resultant to point ahead of the axle producing a braking torque. These cases are shown in Fig. 4.8 for a wheel that is being pulled by an external force.

We can calculate the magnitude of the offset e for a typical case. If $f/V = 0.01$ (compared with 0.5 for many materials sliding over each other), then $\phi = \tan^{-1} 0.01$ and $\phi = 0.6°$. At constant v, $\theta = \phi$, and $\sin \theta = e/r$. Therefore, $e = 0.01r$.

Why should the road exert a backward friction force? If the wheel–road contact region were symmetric about the vertical through the center of mass, then the reaction forces of the road would be symmetric, front to back, resulting only in an upward normal force to support the weight. However, because of the forward motion of the wheel, there is a backward horizontal component of the road reaction on the leading surface of the tire as it strikes the road. This is a reaction to the forward motion of the tire bulge created by the slightly higher pressure at the leading surface. This effect is shown in exaggerated form in Fig. 4.9.

In analyzing rolling friction, it is important to separate the friction force f from the traction. Energy is consumed by the friction, equal to the product of f and the distance traveled. Some of this energy is lost in the wheel–road interface, but in a pneumatic tire most of it is expended in the flexing of the tire structure. The traction force does not move anything through a distance, and so does no work. When you launch yourself from the edge of a swimming pool, for instance, the reaction force of the pool wall does not consume energy. When radial tires were first introduced, many people thought that they must create a lot of friction because they look partially flat. They actually create less friction because the flexing of radial fibers consumes less energy. They also provide greater traction because more rubber is in contact with the road.

If a wheel is being pulled at constant velocity there is no traction. The pulling force at the axle is equal to the friction force at the rim. There is no torque and no angular acceleration because the pulling force is acting through the center of the wheel, and the resultant of the normal force from the road and the friction is directed toward the center.

If a wheel is driven by an internal torque, that torque may be represented by a force couple as shown in Fig. 4.10. The driving torque τ transforms to the force couple F and F acting through the axle and the rim. These behave as if they were *external* forces acting on the wheel. The rim force shoves back against the road through the traction of the tire and the road. The road exerts an equal and opposite forward force on the tire. Note that no work is done by this force of traction; there is no displacement. Nor is there any energy loss unless there is slippage between tire and road, which does occur slightly during acceleration. Since the road force cancels one of the torque component forces, the case of the applied torque reduces to the previous case of a single force applied to the axle.

Some explanations of rolling friction point out that at the instantaneous contact point between wheel and road, the speed of the wheel is 0, and thus the coefficient of static friction should apply. Such an argument must be confusing to a student who might then wonder why rolling friction is nevertheless much smaller than sliding friction. This explanation has nothing to do with the energy loss of rolling friction. It does not even give a good account of the traction forces, because the contact region between tire and road is not a point and the area of the region is a function of the dynamics of the accelerating or braking wheel. Furthermore, the notion of static friction and kinetic friction and the difference between them arises from analysis of dry sliding friction of nonflexible solids. There is no reason to expect that the analysis applies to the momentary contact between a hard surface and a flexible rubber tire. Note, incidentally, that the reaction force of the road during braking or acceleration produces a torque that is in the wrong direction for the expected angular acceleration of the wheel. If the wheel accelerates forward, the traction reaction from the road is forward—but that force *by itself* would produce a torque that would slow the wheel. Remember, however, that the traction force arises as a reaction to one part of the force pair representing the driving torque. In this representation, the two forces at the rim cancel each other (both act as external forces on the tire), leaving the other half of the torque component at the axis to exert the forward force. The accelerating torque arises from the shift of the offset e between the vertical weight through the center of gravity and the normal force from the road.

Rolling friction is usually smaller than dry sliding friction by a factor of 10 or more. The surface contact between road and wheel changes by a peeling motion, not a shearing or plowing action. This is true at both molecular level and on the visible scale of tire treads. Any energy loss takes place because of hysteresis in the compression of wheel or road. For instance, tires get hot due to continual flexing.

With soft rubber, the actual contact area between tire and road is approximately the apparent area. This is clearly not the case with most car tires that have tread. The tread is there to cut down into snow or mud if necessary, and also to squeegee water away from the surface on wet roads. For clean, dry roads the greatest traction for acceleration or braking is obtained with wide, low pressure, soft rubber tires. That is what drag (or acceleration) racers use. In fact, just before starting a run they spin their drive wheels while standing still in order to get the rubber hot and sticky. Once started, however, they avoid wheel slippage if they can. (Some classes of vehicles lack the transmission that makes such control possible.) The dry sliding friction provided by slipping the wheels usually yields less traction (whether for acceleration or braking) than rolling friction just below the onset of slippage. "Patching out" reduces acceleration.

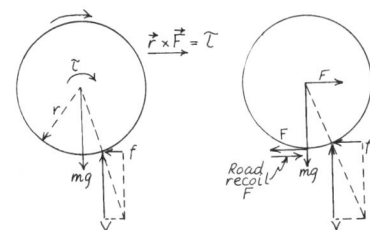

Figure 4.10. Internal torque, τ, is equivalent to force pair, $F-F$. F at the rim gives rise to road reaction, which provides net external force to counteract rolling friction. At constant velocity, $F=f$. For purposes of analysis, this decomposition of the torque reduces the problem to the previous case of a wheel being pulled with force, F. When a wheel is being pulled, there is rolling friction but no traction. When a wheel is being driven by an internal torque, there must be traction.

■ ROLLING FRICTION PRODUCED BY WHEEL AND AXLE

In an article in the November 1982 *The Physics Teacher*, Richard Stepp pointed out that reduction of friction in a wheel and axle arrangement is really due to a

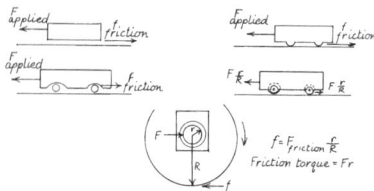

Figure 4.11. Analysis of role of bearing friction on rolling friction of wheel.

lever effect. Consider the four diagrams in Fig. 4.11. In the first there is dry sliding friction that depends only on the surface materials and the normal force. In the second diagram, in which the block has on its lower surface two round, fixed legs, the apparent contact area has changed. As we have seen, however, for rigid bodies this change of apparent area does not affect dry sliding friction. In the third diagram we assume that the materials are still the same, but the rounded legs have now been freed to revolve as wheels. Although these wheels roll along the floor, they do not turn on axles but slide against their upper bearings. The friction force remains the same as before because the wheels are still sliding in their bearings and are supporting the same normal force. In the fourth diagram, in which the wheels have been mounted on axles, the friction force *at the axles* remains the same (assuming the same materials) but now there is indeed an advantage in using the wheels. Remember that with a wheel turning on an axle the bearing is undergoing *sliding* friction, not rolling friction! (Ball bearings would change the situation.)

The force with which you pull the wagon *at constant speed* is translated as a torque to turn the wheel by the equal and opposite force exerted by the ground *at the rim of the wheel*. This is a traction force. The net force on the wagon must be 0: 10 N exerted by you in the forward direction; 10 N exerted by the ground in the backward direction. That small force at the large radius provides the necessary large friction force at the axle which is operating through a small radius. A wheel is a circular lever. If the ratio of wheel diameter to axle diameter is 10, then the force necessary to pull the wagon is smaller than the sliding friction force in the axle by a factor of 10.

Of course, with the sliding friction taking place in a small region, the materials in the axle can be made especially hard and lubricated, or can be made of a material with low sliding friction such as teflon, and that reduces axle friction and thus pulling force even further.

The advantage of such an arrangement can also be seen by considering the work done. Suppose it takes 10 N to pull the wagon; then it requires 10 J to pull the wagon 1 m. With a ratio of diameters of 10, the rubbing friction at the axle must be 100 N, but for a wheel travel of 1 m this axle force is exerted through a distance of only 0.1 m. The energy dissipated is 10 J. Note that this friction energy is dissipated at the axle where there is rubbing and not at the rim where the traction force of 10 N makes the wheel turn but does not move anything through a distance to do work.

■ FLUID FRICTION

When an object moves through a fluid, whether a gas or a liquid, it loses energy by imparting energy to the fluid. There are two distinct processes. The first is the viscous effect of parting the fluid in order for the object to slip through. The force required for this is proportional to the fluid viscosity η, the "breadth" of the object, L, and the *first* power of the speed v. $F_{\text{viscous}} = \eta L v$. The "breadth" has the dimensions of a length, rather than an area, and its magnitude depends on the shape and alignment of the object.

The second process of energy loss is due to kinetic energy being given to the fluid as the object shoulders its way forward. This process can be very complicated, depending on the amount of turbulence produced. However, for speeds below the speed of sound in the fluid, the inertial reaction force of the fluid is approximately proportional to the density of the fluid, ρ, the square of the "breadth" (thus, the "area"), and the square of the speed: $F_{\text{inertial}} = \rho L^2 v^2$.

The ratio of the forces involved in these two effects is a dimensionless constant known as the Reynolds number:

$$\text{Re} = \frac{(\text{inertial reaction})}{(\text{viscous drag})} = \frac{\rho L^2 v^2}{\eta L v} = \frac{\rho L v}{\eta}.$$

(For a derivation of the Reynolds number and its role in scaling, see p. 74. For its applications in fluid behavior, see p. 301.)

Two objects of the same shape but of different size will have similar lift and drag characteristics if their Reynolds numbers are about the same. For instance, a small-scale model of an airplane will have similar flight characteristics if its speed is greater than the speed of the large plane, satisfying the requirement that $(Lv)_{\text{model}} = (Lv)_{\text{large}}$, since the density and viscosity of air will be the same for both. Unfortunately for model airplane builders, this requirement is contrary to practice since the small models usually also have low speeds.

The density, viscosity, and speed in the Reynolds number are specific attributes of the fluid and the object moving through it. But the significance of L is not so obvious. It is equal to the breadth of the disturbed fluid, which is related to the size of the object and its shape. *Usually* the assignment of the proper size for L is inexact by a factor of 2 or more. Nevertheless, the order of magnitude of the Reynolds number is useful in characterizing the type of motion that the fluid undergoes. For instance, if Re is between 1/10 and 10, both the inertial reaction of the fluid and its viscous drag play an important role in the motion. However, if Re<1/1000, we can neglect the inertial reaction. Viscous drag dominates the motion and there is no acceleration of the fluid. If Re>1000, the viscous drag is negligible. Usually with Re>3000, the flow is turbulent. In general, for Re<1/10 the viscous drag dominates and the friction force is proportional to v. For Re>1000, the inertial force is more important, and the friction is proportional to v^2.

In analyzing the friction effects of everyday objects moving through air or water, it is pertinent to note that ρ/η for water is only about 10 times larger than for air:

$$\rho_{\text{air}} = 1.3 \text{ kg/m}^3, \quad \rho_{\text{water}} = 1.0 \times 10^3 \text{ kg/m}^3,$$

$$\eta_{\text{air}} = 1.8 \times 10^{-5} \text{ N s m}^{-2} = 1.8 \times 10^{-2} \text{ cP},$$

$$\eta_{\text{water}} = 1 \times 10^{-3} \text{ N s m}^{-2} = 1.0 \text{ cP (at 20 °C)}.$$

Since your characteristic "breadth" when you are swimming or are in a small boat is about 1 m, and any reasonable motion would have a speed of the order of 1 m/s, your Reynolds number would be equal to or greater than 10^6. The same argument holds for the motion of most human-size objects in air. Consequently, the fluid friction force in air is proportional to v^2 for baseball, cars, skiers, and even raindrops. For millimeter-size objects (such as mist) moving at speeds of a few millimeters per second, the fluid friction is proportional to v. These conditions

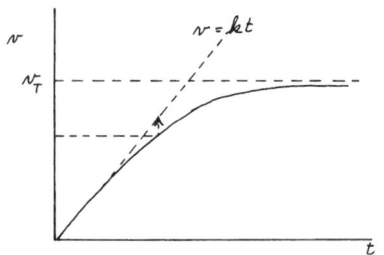

Figure 4.12. Terminal velocity relationships.

characterize the Stoke's law regime and apply, for example, to the motion of oil droplets in the Millikan experiment.

■ VELOCITY-DEPENDENT FRICTION

If friction does not depend on speed, then a constant force applied to an object will produce constant acceleration, regardless of the friction. However, if the friction increases with speed, an object subjected to a constant applied force will experience decreased acceleration as the speed increases and eventually will reach a terminal velocity. The most common example of this phenomenon is the motion of an object falling through air. The formula for the motion is: $mg - kv^2 = ma$. The constant k depends on the geometry of the falling object and will be small for a streamlined object. When fluid friction is equal to the weight, $mg = kv^2$, and $a = 0$. Fig. 4.12 shows the typical pattern of this kind of motion. Notice that $v \propto t$ within 10% up to a speed of 80% of the terminal velocity. For rule-of-thumb estimates, the terminal velocity for a baseball is 95 mph; for a tennis ball, 70 mph; for a basketball, 45 mph; and for a Ping-Pong ball, 20 mph. The terminal velocity for an adult human is between 100 and 200 mph, depending on whether the body is in a deliberate dive or is spread-eagled. For a *successful* parachutist, the standard terminal speed is 15 mph.

The retarding friction experienced by an automobile is composed of two major parts. Rolling friction is almost independent of speed, but air friction is proportional to v^2. For most cars, the two friction forces are equal at about 30 mph. As the speed increases beyond that, the air friction rapidly dominates. Because friction increases with speed, there must be a terminal velocity for each car. Floor the gas pedal and you will soon reach the maximum speed for that particular car.

■ FOR READING AND REFERENCE

Armenti, A., Jr., "How can a downhill skier move faster than a skydiver?," Phys. Teach. **22**, 109 (1984).
  An interesting study of terminal velocities, showing in particular the effect of the area that the moving object presents to the air.
Barker, R. E., Jr., "Power to overcome air resistance at highway speeds," Am. J. Phys. **44**, 108 (1976).
  A follow-up of the article by Shanebrook (q.v.) using data for four real cars.
Barret, R. P., "Friction on the air track," Phys. Teach. **12**, 297 (1974).
  An experiment in which students study the dependence of air-track friction on velocity, using a card on the glider to change the air resistance. Loss of energy through magnetic induction braking is included. See also the follow-up letters: Phys. Teach. **13**, 68 (1975).
Bartels, R. A., "Braking distance versus mass for automobiles," Am. J. Phys. **45**, 398 (1977).
  Using data from measurements made by *Consumer Reports* this note shows that mass has no effect, but that the stopping distance is less in the "wheels locked" position than when wheels are not locked. See also the papers by Logue (1979) and Smith (1978).
Bettis, C., "Capstan experiment," Am. J. Phys. **49**, 1080 (1981).
  Describes a simple student experiment using a capstan. See also Hazleton (1976) for the theory.
Bowden, F. P., and Tabor, D., *Friction, an Introduction to Tribology* (Robert E. Kreiger, Malibar, Florida, 1982).
  This is one of the PSSC Science Study Series, originally published in 1973 by Doubleday. It covers authoritatively most aspects of friction at a level suitable for bright young students.
Chaplin, R. L., and Miller, M. G., "Coefficient of friction for a sphere," Am. J. Phys. **52**, 1108 (1984).
  A study of the problem, culminating in measurement of the coefficient of kinetic friction for various materials. See also Shaw and Wunderlich (1984).
Drouin, B., and Gagnon, R., "A graphical solution to some problems involving dry friction," Phys. Teach. **15**, 422 (1977).
  A note presenting a graphical method that avoids differential equations in solving some problems.
Farr, J. R., "Determining the air drag on a car," Phys. Teach. **21**, 320 (1983).

A "take-home" experiment utilizing students' cars and wrist watches as the apparatus. Tables show data for actual makes and models of cars.

Flowers, K., "Friction and mechanical energy loss," Phys. Teach. **16**, 173 (1978).
Describes an experiment, using standard apparatus, to evaluate energy dissipation from friction.

Frohlich, C., "Aerodynamic drag crisis and its possible effect on the flight of baseballs," Am. J. Phys. **52**, 325 (1984).
This long paper may tell more than you want to know about the topic, but it reviews the literature, presents some new results, gives fascinating figures for use as examples, and gives a sensible answer to the question, "Does a pitched curve ball really 'break'?"

Goehl, J. F., Jr., "The effect of frictional force on the time of motion," Am. J. Phys. **52**, 271 (1984).
A short note dealing with the motion of a block projected up an inclined plane, then sliding back down.

Goldberg, F. M., "Friction in a moving car," Phys. Teach. **13**, 234 (1975).
A parking-lot experiment in which students measure the "coefficient of friction" for a rolling automobile, and the dependence of this friction on the tire pressure.

Grimm, R. D., "A classroom demonstration of automobile stopping distances," Phys. Teach. **16**, 559 (1978).
A simple demonstration to show that the deceleration of a sliding object is independent of its mass, other factors being the same.

Haber-Schaim, Uri, and Dodge, John, "There's more to it than friction," Phys. Teach. **29**, 56 (1991).
An analysis of the table cloth stunt, including experimental confirmation, showing that coefficients of sliding friction are major determinants of the effect. Disagrees with analysis of Hudson (1985).

Hafemeister, D., "Science and society test for scientists: Transportation," Am. J. Phys. **44**, 86 (1976).
Suggests a "coast-down" method for students to measure the drag force on a car. (See p. 89, answer to question 5a.)

Hazelton, G. L., "A force amplifier: the capstan," Phys. Teach. **14**, 432 (1976).
Describes and gives the theory of a device in which friction is essential. This note also contains a suggestion for a students experiment, said to give consistent results. [See Bettis (1981).]

Hart, J. B., "Frictional force rotator," Am. J. Phys. **50**, 631 (1982).
A friction vector can be rotated to make its component in a given direction as small as desired. The author gives practical applications and describes a piece of demonstration apparatus.

Hudson, H. T., "There's more to it than inertia," Phys. Teach. **23**, 163 (1985).
A note considering the role of friction in the "tablecloth snatch," and suggesting a variation in procedure. See Perez (1977), Jones (1977), and Haber-Schaim and Dodge (1991).

Jones, E., "The tablecloth pull," Phys. Teach. **15**, 389 (1977).
A letter, giving an analysis of the experiment, showing that friction is the principal factor. [See Hudson (1985), Perez (1977), and Haber-Schaim and Dodge (1991).]

Kwasnoski, J., and Murphy, R., "Determining the drag coefficient of an automobile," Am. J. Phys. **53**, 776 (1985).
An experiment using a real car, some simplifying assumptions, and easily obtained apparatus gives a value in good agreement with published values.

Lock, J. A., "The physics of air resistance," Phys. Teach. **20**, 158 (1982).
The drag force on a spherical object is shown to have two components: a linear part at low velocities and a quadratic part at higher velocities. See also a comment on this article in Mironer's letter Phys. Teach. **20**, 400 (1982) and the original author's response on the same page.

Logue, L. J., "Automobile stopping distances," Phys. Teach. **17**, 318 (1979).
The coefficient of friction between tire and road depends on the temperature, percent slip, normal force, and contact area.

Macomber, H. K., "Normal and frictional forces: an experiment," Phys. Teach. **17**, 593 (1979).
An experiment or demonstration to measure static friction.

Nelson, J., "About terminal velocity," Phys. Teach. **22**, 256 (1984).
This note gives an analog demonstration and some qualitative theory about terminal velocity.

Palmer, F., "What about friction?," Am. J. Phys. Part I **17**, 181 (1949); Part II **17**, 327 (1949); Part III **17**, 336 (1949).
This three-part article gives a good review of the topic as of 1949.

Perez, J., "The tablecloth pull," Phys. Teach. **15**, 242 (1977).
The author jerks a cloth out from under a table setting as a demonstration of inertia. He includes practical suggestions. [See also Jones (1977), Hudson (1985), and Haber-Schaim and Dodge (1991).]

Rabinowitz, F., "Resource letter F-1 on friction," Am. J. Phys. **31**, 897 (1963).
A list of the important publications and other resources available at the time of writing. The author is a distinguished contributor to the field.

Rochlin, G. I., "Drag force on highway vehicles," Am. J. Phys. **44**, 1010 (1976).
Raises serious objections to paper by Barker (1976) and recommends practical precautions in using the procedure suggested by Hafemeister (1976). See also Farr (1983).

Shanebrook, J. R., "Power to overcome air resistance at highway speeds," Am. J. Phys. **42**, 1028 (1974).

The requisite power is shown to be proportional to the cube of the speed. See also Barker (1976), Hafemeister (1976), Rochlin (1976), and Farr (1983).

Shaw, D. E., and Wunderlich, F. J., "Study of the slipping of a rolling sphere," Am. J. Phys. **52**, 997 (1984).

An experimental study of the speed of a ball rolling down an incline as the inclination is changed. This is an involved experiment, suitable for the better students. See also Chaplin and Miller (1984).

Smith, R. C., "General physics and the automobile tire," Am. J. Phys. **46**, 858 (1978).

A note pointing out the many ways that tire–road friction differs from that predicted by the usually accepted "laws" of friction. Considers the effect of temperature, speed, and tread. See also Logue (1979) and Bartels (1977).

Stepp, R. D., "Why wheels work," Phys. Teach. **20**, 550 (1982).

An unconventional explanation of the effectiveness of wheels in reducing the force needed to move an object. See also a letter commenting on this article by Macomber, Phys. Teach. **21**, 350 (1983).

White, J. D., "The role of surface melting in ice skating," Phys. Teach. **30**, 495 (1992).

Whitmire, D. F., and Alleman, T. J., "Effect of weight transfer on a vehicle's stopping distance," Am. J. Phys. **47**, 89 (1979).

Friction between tires and road produces a torque that changes the normal force at the wheels. This paper analyzes the effect and considers the situation for nonskidding stops as well as for stops with either or both the front and rear wheels locked.

Witters, J., and Duymelinck, D., "Rolling and sliding resistive forces on balls moving on a flat surface," Am. J. Phys. **54**, 80 (1986).

A theory on rolling friction is compared with experimental results on rolling billiard balls. Sliding friction for a ball is also measured.

5
Gravitation

Gravitation is a marvelously rich subject for introductory physics. It involves the concept of mass and the necessity of distinguishing between weight and mass. It provides a constant force for our demonstrations, experiments, and homework problems. As Newton first realized, the same law governs the fall of an apple and the motions of moon and planets. The simple fact that in a vacuum all objects fall with the same acceleration turns out to have profound implications and becomes the cornerstone of the general theory of relativity, which is our current explanation of gravity.

Newton dominates the subject of gravitation. He not only proposed the formula, he applied it to a vast range of phenomena. There may be a grain of truth in the legend that Newton discovered the law of gravitation upon seeing an apple fall from a tree. Many years after Newton's stay in the country to escape the plague, a friend, William Stukeley, wrote memoirs of Sir Isaac's life. Stukeley says that while he and Newton were having tea in the shade of some apple trees, Newton told him that it was in just the same situation long ago that he noticed the fall of an apple and became convinced that the same force that caused the apple to fall must also cause the moon to fall toward the earth.

There may also be more to this anecdote than meets the casual historian's eye. At the time when Newton recounted this story he was still concerned with establishing his priority over Hooke in discovering universal gravitation. In an article in *Scientific American*, I. Bernard Cohen points out that while Newton may have known about the effect of an inverse square law in his early twenties, he did not seem to understand the importance of the interaction of both masses in a system until he wrote the *Principia* 20 years later. Cohen suggests that Newton remembered and rewrote history in order to suit his own claims of priority. At any rate, there is surely more to the anecdote than meets our modern eyes. Newton was saying that celestial phenomena are subject to mundane laws. A generation earlier, Galileo was censured for a similar heresy.

■ NEWTON'S LAW OF GRAVITATION

In many introductory courses, the gravitation law is taught at about the same time as Newton's three laws of motion. Pedagogical steps should be taken to separate these very different topics. Students become confused because Newton's name is attached to all these equations, all of which are concerned with mass and force. Indeed, the appearance of mass in both the gravitation and inertia law is a harbinger of a more profound connection, as well as student confusion. To begin

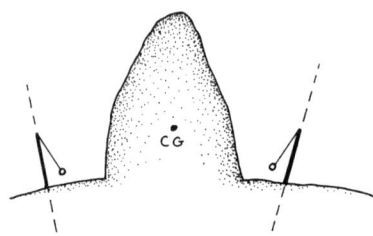

Figure 5.1. The principle of Maskelyne's mountain experiment.

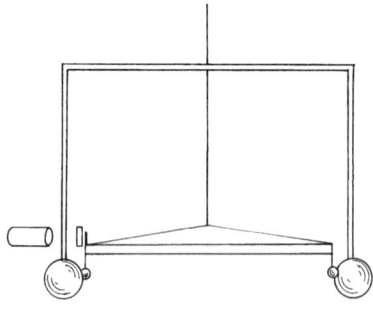

Figure 5.2. Cavendish's torsion balance.

with, however, the dynamics laws apply for any force. Gravitation is just one of many phenomena that may provide the "F."

Let us consider a number of aspects of Newton's famous formula:

$$F = G \frac{m_1 m_2}{r^2}.$$

1. The universal gravitational constant G is the same on earth as it is in the heavens. The realization of this fact was one of Newton's triumphs; G is not simply a constant that makes the units come out all right!

   In order to measure G you must be able to measure the force between two known masses separated by a known distance. Since the value of G is 6.67×10^{-11} N m^2/kg^2, the force of attraction between two 1-kg masses separated by 1 m is only 6.67×10^{-11} N. Clearly, experimental work with this law demands large masses or very sensitive force detectors. One very large mass that might be used is the earth. Unfortunately, there is no independent way to measure the mass of the earth except through this equation. Newton knew the approximate radius of the earth and knew the density of surface rocks, which is about 2500 kg/m^3. That would lead to a mass of 2.7×10^{24} kg. As we now know—through the measurement of G—the average density of the whole earth is 5500 kg/m^3. There is a dense core containing elements such as iron. The resulting mass of the earth is 5.98×10^{24} kg.

   In 1774, Nevil Maskelyne, the English Astronomer Royal, determined G to within about 20% of our present value. He compared the attraction exerted by a mountain on a plumb bob with that of the earth itself on the plumb bob. In Scotland the Schiehallion peak rises abruptly from a plain and has contours such that its volume can be measured with some precision. Its mass and the position of its center of gravity were calculated after an elaborate survey and numerous drillings to determine the density of the interior rocks. Two observation stations were selected to be at equal distances from the mountain's center of gravity and on a north–south line. A telescope lined up with the plumb bob at these two points showed a star to be shifted slightly, the total relative deflection being about 12 seconds of arc. See Fig. 5.1.

   Using half this angle (because the 12 seconds was the sum of two deflections), Maskelyne calculated the ratio of the mountain's and the earth's attraction on the plumb bob; this ratio was the tangent of the angle. From these measurements he concluded that the value of G was 7.4×10^{-11} N m^2/kg^2.

   The classic experiment to measure G was first done by Henry Cavendish in 1798, using apparatus that had originally been designed by the Rev. John Michell. Cavendish, one of England's wealthiest men, was something of a recluse, although active in the affairs of the Royal Society and British Museum. He had a morbid aversion to women, or for that matter, the female of any species. He was a skilled scientist and did major research in the nature of heat, the constitution of gases, electricity, astronomy, and gravitation. However, he published very little. It was not until many years after his death that it was discovered from his laboratory notes that he had anticipated many of the discoveries of others such as Faraday, Davies, Coulomb, and Priestley.

   To measure G, Cavendish used a torsion balance, as shown in Fig. 5.2. The suspension wire was silver-coated copper, about 0.01 in. in diameter. This wire held a horizontal wooden truss about 6 ft long from which lead balls were hung, each about 2 in. in diameter with a mass of 3/4 kg. This apparatus was enclosed in a thin, shallow wood box in order to reduce the effects of air currents. The oscillation period of the rig was about 7 min. Outside the box two very large lead spheres could be positioned close to the suspended spheres, first to produce a torque in one direction and then moved to produce a torque in the opposite direction. Each outside sphere had a diameter of 8 in. and a mass of 48 kg. Cavendish could alter the positions of the outside spheres with a pulley system so that he did not have to be in the room. He had found that his proximity to the box would produce thermal movements of the air which disturbed the suspension. Instead of measuring the angular deflection with an optical lever, such as modern versions employ, Cavendish used a telescope to view the movement of an ivory tab attached to the end of the suspended rod. This tab moved past a stationary tab, with vernier scales (invented just a few years earlier) inscribed on each. With this apparatus, Cavendish measured G to be 6.75×10^{-11} N m^2/kg^2 (in our present units), a value that is high by only 1%. His experiment and results caught the popular imagination at the time and became widely known. Cavendish had weighed the earth!

In the years since, numerous attempts have been made to reduce the uncertainty in our knowledge of the value of G. The torsion balance method was improved by Sir Charles Boys (who wrote the delightful book on soap bubbles). He perfected methods of producing and using quartz fibers for the suspension. Other types of balances have been used. With the outside spheres in line with the suspended ones, the suspended rod will oscillate (in the field of the large spheres) with a period that depends on G. If a large sphere is placed directly under a small one suspended from a spring, the spring will stretch a small additional amount. In recent years there have been attempts to sense the rotation of a torsion balance optically and feed back the signal to control a servo motor that maintains constant separation distance between the spheres. In spite of all these efforts, G remains the least well known of the important natural constants. Only four significant figures are justified: $G = (6.670 \pm 0.005) \times 10^{-11}$ N m^2/kg^2.

2. The exponent in the denominator of Newton's law of gravitation is 2.000 000 000. In other words, the number is known to be exactly 2 to within one part in a billion. Astronomical orbit calculations provide this certainty. However, the value is 2 only if each mass is concentrated at a point or is distributed in a radially symmetric sphere. Newton applied his mathematics of fluxions—the calculus—to prove that the gravitational attraction of a sphere is the same as if all the mass were concentrated at the center. (For a more sophisticated proof, see p. 108.) For the formula to be valid, one object cannot be inside the other. In a later section we will analyze gravity inside a sphere.

If the mass is distributed in an irregularly shaped object, such as a human body, then Newton's law is a good approximation as long as the separation distance r is large compared with the size of the object. For instance, we can use the inverse square formula to find our weight, since the earth is a sphere and since the distance to the center of the earth is a lot larger than our human dimension.

The exact value of 2 for the exponent is a reflection of the Euclidean geometry of space. As the influence from a point (or spherical) source spreads out isotropically in space, the amount of influence *per unit area* decreases with the square of the distance. That is because the surface area of a sphere is proportional to the square of the radius. The same amount of influence spreads out through larger and larger spherical shells; the area density is therefore proportional to $1/r^2$. To the extent that space is not Euclidean (and it is not), then the simple law does not hold and we must go to Einstein's general theory. We will mention such circumstances later, but note that the guidance of our Voyager and Explorer rockets requires only Newtonian laws.

3. To a first approximation, the force exerted on an object by the earth is called the weight of the object:

$$\text{Weight} = G\frac{mM}{R^2} = mg.$$

The mass of the earth is M, and its radius is R;

$$g = (6.67 \times 10^{-11} \text{ N m}^2/\text{kg}^2)\frac{(5.98 \times 10^{24} \text{ kg})}{(6.37 \times 10^6 \text{ m})^2} = 9.83 \text{ N/kg}.$$

We said that to a first approximation that this formula gives the weight, because there is a complication. Centrifugal force reduces the gravitational attraction. (In the reference frame of our spinning earth, the force is indeed *centrifugal*—see Chap. 6.) On the equator the magnitude of this effect is

$$\omega^2 R = \frac{4\pi^2}{T^2}R = \frac{4\pi^2}{(24 \times 3600 \text{ s})^2}(6.37 \times 10^6 \text{ m}) = 0.034 \text{ m/s}^2.$$

This number is $\frac{1}{3}\%$ of the value of g. Note also that we expressed g in units of N/kg, but the centrifugal effect in terms of m/s^2. Indeed, one set of units can be converted into the other:

$$\frac{\text{N}}{\text{kg}} = \frac{\text{kg m/s}^2}{\text{kg}} = \frac{\text{m}}{\text{s}^2}.$$

One way to specify which effect we are talking about is to call the Newtonian formula effect *gravitation* and the actual value that is measured at any point *gravity*. The centrifugal effect is a reference-frame, or inertial, force. As Einstein insisted, such forces are equivalent to gravitational forces and experimentally are indistinguishable, at least at any point. We usually express the inertial actions in terms of acceleration units. The constant g is usually called the *acceleration due to gravity*. This may cause pedagogical problems because a student may wonder how you can get the weight of an object by multiplying its mass with an acceleration, even though the object is probably standing still. It is possible to introduce g from the beginning as a field strength, with units of N/kg. This terminology will pave the way for the analogous concept with the electric field, where $E = F/q$. Students should think of mass as being the "charge" or source of the gravitational field.

The value of g, the experimental local field strength, varies with location on the earth. There is the variable effect of rotation, which is maximum at the equator and zero at the poles. There is also a slight effect because the earth's surface is about 14 miles closer to the center of the earth at the North Pole than at the equator. One might think, therefore, that the gravitational pull at the North Pole would be larger than that at the equator by a factor of $(4014/4000)^2$. This would yield an increase of 0.7%. The actual value is 0.18%. The shape of the earth is not that of a sphere but rather an oblate spheroid. During the formation of the earth centrifugal force created a slight girdle in the equatorial region. Consequently, although the North Pole is 14 miles closer to the center, more mass is distributed further from the Pole than would be the case with a sphere. The value of g at the pole is 9.8322 N/kg; at the equator g is 9.7805 N/kg. A good approximation for most regions at sea level is $g = 9.7805(1 + 0.005\ 29\ \sin^2\phi)$, where ϕ is the latitude.

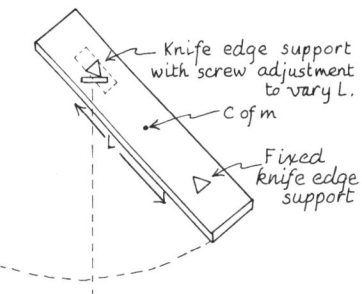

Figure 5.3. *Kater's reversible pendulum.*

The value of g also varies with height (or depth in a mine shaft) and with local geological features. This information is useful for research about the earth and for prospecting for minerals and oil. Modern instruments can measure relative gravity (from one location to another) with a sensitivity of 10^{-8} the earth's field. The unit for such measurements is the galileo: 1 Gal = 1 cm^2/s. Note that 1 mGal = 1×10^{-5} N/kg. Portable gravimeters in the field routinely measure to 0.01 mGal. Anomalies of interest in petroleum exploration are usually between 0.5 and 5 mGal. An increase of height of 3 m produces a decrease of 1 mGal. Some permanently maintained research instruments can measure to 0.001 mGal, and so can detect the change in weight if an object is moved up or down by half a centimeter.

For relative measurements of g, gravimeters consist of a weight hung from delicate spring assemblies. Until the middle of the twentieth century, geologists used devices called Kater's pendulums as field gravimeters. In the early 1800s, Henry Kater in England devised a compound pendulum with knife edges at the conjugate points as shown in Fig. 5.3. The period is the same for the upside-down suspension as for the right side up, and the length L of the pendulum is the distance between the knife edges. This distance can be measured very precisely, as can the period T, and so the value of g can be determined precisely. The formula for the period has the same form as for the simple pendulum: $T = 2\pi\sqrt{L/g}$. In recent decades, commercial gravimeters have been available that measure g by timing the fall of a plummet, with timing sensitivity of the order of a microsecond. In current tests of the equivalence of inertial and gravitational mass, falling plummets are being timed with an error of 0.1 ns (1×10^{-10} s).

In discussing the weight of an object, there is always the question as to whether astronauts in earth orbit are weightless. Or, equivalently, is a person in a freely falling elevator weightless? Clearly, in both cases, the value of g is about the same as the standard value on the surface of the earth. [In an orbit 100 miles above the earth, g is smaller by a factor of $(4000/4100)^2$, or in other terms, smaller by about 5%.] Since all objects in the capsule or elevator are falling with the same acceleration, regardless of mass, everything and everyone float around without any sensation of falling or weight. The normal sensation of weight comes from the tension of our muscles that keeps us from collapsing to the floor, and the feeling of falling is a combination of lack of muscle tension and the sight of things flying past us. As for the weightless-

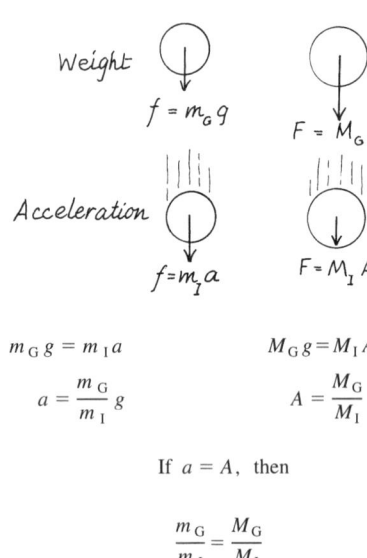

Figure 5.4. Proportionality of gravitational and inertial mass.

ness of astronauts, the question is clearly one of semantics and not physics. To be sure, it is not semantics that make some astronauts throw up under these conditions; this is due to a physiological and psychological problem of lack of muscle tension on the stomach and disorientation.

4. Every part of Newton's law of gravitation contains a revelation. Note that in the numerator there is the product of two masses. Gravity is not simply the force that the earth exerts on the apple; it is an interaction between earth and apple. If the apple weighs 2 N in the earth's gravitational field, then the earth weighs 2 N in the apple's gravitational field. When the apple falls toward the earth, the earth falls toward the apple. In the case of moon and earth, the moon does not revolve around the center of the earth. Both earth and moon revolve around their common center of mass which is about $\frac{2}{3} R$ from the center of the earth.

There is another profound problem connected with the dependence of gravitational force on the masses of the interacting objects. What is mass? In this equation it appears to act as a "charge" or source of the influence. But in Newton's law of motion, mass plays an entirely different role. In these it is a measure of the inertia of an object, of the reluctance of the object to change its velocity. The defining experiment for gravitation is a balance—a static measurement. The defining experiment for Newton's three laws involves acceleration or a recoil. Since these two phenomena—universal attraction and sluggishness—are so unrelated to each other, there is no prima facie reason to expect that inertial mass has anything to do with gravitational mass. Unfortunately, that is not obvious because we brainwash ourselves and our students by using the word *mass* and the symbol *m* for both cases.

If we want students to see the importance of this problem, we have to be dramatic about it. Most students are not at all bothered by either the identity or possible difference between the two aspects of mass. Mass is mass. Try the Galileo demonstration. Drop a golf ball and baseball together from as great a height as you can manage, and have the students witness that they fall at the same rate. Galileo, who died the year Newton was born, was not trying to prove the equivalence of inertial and gravitational mass. He was just arguing that Aristotle was wrong in saying that an object with twice the weight would fall with twice the speed of a smaller object. But the demonstration does prove the equivalence of inertial and gravitational mass. If an object has twice the gravitational mass of a smaller object, then it has twice the weight. If that weight provides the force in $F = ma$, then the acceleration of the heavy object will be the same as the acceleration of the light body if, and only if, inertial mass is proportional to gravitational mass. The argument is shown in Fig. 5.4.

Newton understood the implications of this experiment, and worried over the dual nature of mass. He compared the periods of pendulums with bobs made of different materials. Gravity and inertia interact in the same way with pendulum bobs as they do with a falling object. If inertial mass is proportional to gravitational mass, then the acceleration of an object in free fall is independent of the mass, and the period of a pendulum is independent of the mass of the bob. Newton could have detected a very small dependence of the period on mass because it takes only a small discrepancy between frequencies to produce a detectable phase shift.

Many great scientists have mulled over these strange facts. Mach realized that the implications had cosmological significance. He claimed that the mass of an object was not some intrinsic attribute. Without other matter in the universe the mass of an isolated object would be a meaningless concept. There would be no reference frame, such as the fixed stars, in which to measure its acceleration. Consequently, the mass of any object must be determined by all the rest of the mass in the universe, and, in particular, by the material at the far reaches of the universe, where most of the mass resides. At the end of the nineteenth century, Roland, Baron Eötvös of Vásárosnamény, performed the definitive test of the equivalence of inertial and gravitational mass. He used a Cavendish–Coulomb type torsion balance that had objects at the ends made of material from different regions of the periodic table. The end objects were subject to the earth's downward gravitational pull and a centrifugal outward force due to the earth's spin. The magnitude of the centrifugal force depends on inertial mass. By looking for slight twists of the suspension, Eötvös claimed that he had shown the equivalence of inertial and gravitational mass to five parts in 10^9.

Very bright people marveled over the fact that experimentally $m_G = m_I$. It took the genius of Einstein to set the matter in the right perspective. He said that it was not remarkable that the two values were equal, because gravitational and inertial mass were *identical*! The identity and equivalence of inertial and gravitational mass is the cornerstone of the general theory of relativity, and hence of our modern theory of gravity.

The testing of equivalence still goes on, challenging the general theory. Within the last two decades R. H. Dicke of Princeton did an Eötvös type experiment and increased the precision by a factor of 100. At the beginning of 1986 an article in *Physical Review Letters* challenged the conclusions of the original Eötvös work, and proposed that at short distances (up to a few hundred meters) there might be a new type of force that would have the effect of slightly reducing gravitational force. The source of this force would be the nucleons in atomic nuclei. Their number (the atomic weight number A) is not quite proportional to the atomic mass, because when nucleons combine in a nucleus some of the mass-energy shoots away. Most subsequent experiments have not found this proposed effect.

■ THE UNIT OF MASS

When the metric system was established, during the French Revolution, the unit of mass was chosen to be the mass of one cubic centimeter of water at its lowest density, which occurs at about 4 °C. This much mass was named the gram. For practical purposes, the official standard of comparison was chosen to be a kilogram. Our international standard today is the mass of a platinum–iridium cylinder that was constructed in 1889 to be equal to the mass of a cubic decimeter of water. The volume of water could only be measured to one part in 10^6. Weights can be compared with balances to two parts in 10^8.

Nations and their industries take standards of measurement very seriously. This point is emphasized in the eleventh chapter of the Book of Proverbs: "A false balance is abomination to the Lord, but a just weight is His delight." In 1889 many copies of the kilogram were made and calibrated against the official one and each other. These were then distributed by lot to the participating countries. When

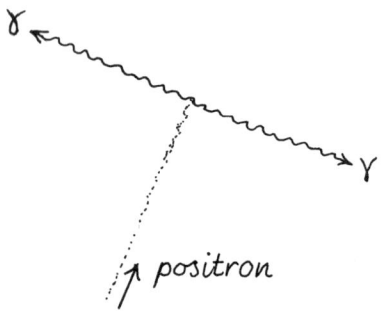

Figure 5.5. Positron annihilation. The positron loses energy as it ionizes atoms in its path. It leaves a more and more dense trail and is more buffeted by these collisions as it slows down. When it comes to rest it combines with an electron in an atom. They annihilate each other. Their rest mass becomes the energy of the two photons.

the secondary standard was delivered to the United States, President Benjamin Harrison officially received it at the White House and certified it as the United States standard mass.

■ MASS IS ENERGY IS MASS

Starting on p. 148, in the chapter on relativity, there is a discussion of the meaning of the famous equation: $E = mc^2$. The most important moral of that story is that mass cannot be turned into energy. *Mass is energy*! (Note Fig. 5.5.)

■ GRAVITATIONAL FORCE INSIDE A HOMOGENEOUS SPHERE

If you could drill a hole through the center of the earth starting in the United States, you would come out near Australia—not China. (After all, China is also north of the equator.) The antipode for New York City is just to the west of Perth. If the earth were not rotating, had uniform density, and were not hot, and if you could drop a bowling ball down the hole without any air friction, what would happen? When the ball is at r, where $r < R$, the gravitational pull of the interior sphere of radius r is equal to $-G(mm_r/r^2)$, where m_r is the mass of that interior sphere: $m_r = \rho \frac{4}{3}\pi r^3$. But what is the influence of the spherical shell that is at a larger radius than the ball? As shown qualitatively in Fig. 5.6, the gravitational pulls from the various parts of the outer shell exactly cancel. This cancellation can be proved with a rather complicated calculus integration, but a much nicer and more sophisticated proof is given by Gauss's theorem. In the more familiar statement of Gauss's law, applied to electric fields, there are $4\pi k$ electric field lines (lines of E) coming out of each unit charge. Similarly, there are $4\pi G$ gravitational field lines (lines of g) coming out of each unit mass. Gauss's law states that the number of field lines coming out of a closed surface is proportional to the amount of charge enclosed by the surface:

$$\oint \mathbf{E} \cdot d\mathbf{A} = 4\pi k q, \quad \oint \mathbf{g} \cdot d\mathbf{A} = 4\pi G m.$$

Because of the spherical symmetry of the case we are considering, all field lines must be radial, and the field has the same value at all points at a given radius. Therefore, for a spherical surface at radius r:

$$\oint \mathbf{g} \cdot d\mathbf{A} = g_r 4\pi r^2 = 4\pi G m_r \rightarrow g_r = G\frac{m_r}{r^2}.$$

The mass of the enclosed sphere of radius r is $\frac{4}{3}\pi r^3 \rho$. Therefore, $g_r = \frac{4}{3}\pi G \rho r$. Since the mass of the whole sphere is $M = \frac{4}{3}\pi R^3 \rho$,

$$g_r = G\frac{M}{R^3}r = g_R \frac{r}{R}.$$

Gauss's law tells us first of all that in a radially symmetric sphere, the field at r depends only on the mass inside the surface of radius r. Second, in a homogeneous sphere, the field is proportional to the radius. Note that when $r = R$, the familiar Newton's law for spheres reappears. Also note that at $r = 0$, $g_r = 0$.

The graph of $g(r)$ is given in Fig. 5.7. The field is proportional to r from 0 to R, and then is proportional to the inverse square of r. It is plotted as negative

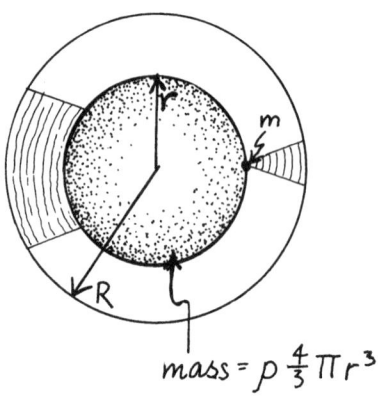

Figure 5.6. The small cone of mass close to the object pulls to the right. This pull is exactly balanced by the larger mass which is further away on the opposite side. The pull of all the mass at a radius larger than r cancels out. The net force on the object comes only from the inner sphere of radius r.

because its direction is toward negative r.

As for the bowling ball in the hole of our hypothetical earth, it is subject to a restoring force proportional to its displacement from zero:

$$F_{\text{restoring}} = -m\frac{g_R}{R}r = -m(\tfrac{4}{3}\pi\rho G)r,$$

where m is the mass of the bowling ball. Such a force produces simple harmonic motion, with a period equal to $T = 2\pi\sqrt{m/k}$, where k is the spring, or proportionality, constant. In this case

$$T = 2\pi\sqrt{R/g_R} = 2\pi\sqrt{(6.37\times 10^6 \text{ m})/(9.8 \text{ m/s}^2)} = 5.1\times 10^3 \text{ s} = 1.4 \text{ h}.$$

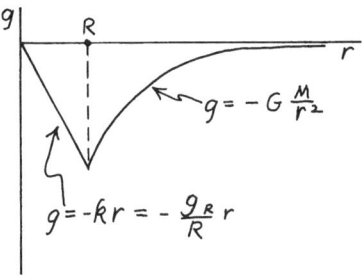

Figure 5.7. Gravitational field strength g as a function of the distance from the center of sphere with radius R.

The ball will shuttle back and forth between New York and Perth with a round-trip time of 1.4 h. Surprisingly, it appears that the period of such an oscillation is independent of the size of the sphere, since the spring constant can also be written as $\tfrac{4}{3}\pi\rho G$. We do not need a hypothetical earth. The experiment could, in principle, be done with a marble dropped through a baseball. Since the spring constant is proportional to the density of the sphere, the period for a marble dropped through a gold sphere of any radius would be $[\sqrt{(5.5)/(19.3)}](1.4 \text{ h}) = 0.75 \text{ h}$.

■ ORBITAL MOTION IN A SPHERICALLY GRAVITATIONAL FIELD

On a flat (infinite) earth, the vertical acceleration of a projectile in free fall is constant, and its horizontal velocity is constant. the resulting trajectory is parabolic. On a spherical earth, however, the trajectory should be elliptical with the center of the earth as one of the foci. For most objects moving through the air, the actual trajectories are not symmetric and neither the parabolic nor elliptical orbits are good approximations. This generalization applies to Ping-Pong balls, baseballs, and bullets.

For the special case of circular orbital motion, gravitation provides the centripetal force. Take the case of a satellite with mass m at radius r in the field of a much larger object with mass M:

$$G\frac{mM}{r^2} = \frac{mv^2}{r} = m\omega^2 r = m4\pi^2 f^2 r = m\frac{4\pi^2}{T^2}r.$$

From this we get Kepler's relationship between period and radius:

$$T^2 = \left(\frac{4\pi^2}{GM}\right)r^3.$$

Note that the mass of the satellite cancels out. For an orbit close to the surface of the earth, where $r = R$, we have:

$$T^2 = \frac{4\pi^2}{(GM)/R^2}R = \frac{4\pi^2}{g_R}R \quad \text{and} \quad T = 2\pi\sqrt{R/g_R}.$$

This is the same expression that we obtained for the period of a ball dropping down a hole through the center of the earth. The period is 1.4 h.

For a "hovering" satellite, the period must be 24 h:

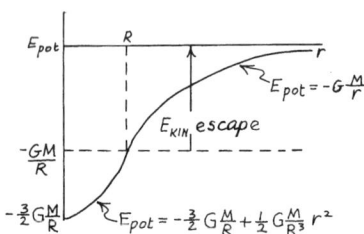

Figure 5.8. Gravitational potential as a function of r for a sphere.

$$(24\times 3600)^2 = \frac{4\pi^2}{(6.67\times 10^{-11})(5.98\times 10^{24})} r^3 \rightarrow r = 4.23\times 10^7 \text{ m} = 26\,300 \text{ miles}.$$

The geosynchronous satellites are a little over 22 000 miles above the earth's *surface*, necessarily above the equator.

The mean radius of the moon's orbit is 3.84×10^8 m. This calls for a period of 2.37×10^6 s = 27.4 days. Note that this is the sidereal period. New moon to new moon takes $29\frac{1}{2}$ days.

■ GRAVITATIONAL POTENTIAL ENERGY

We live in a potential well here on earth. This is evident enough if we try to get out; we have to provide positive kinetic energy from rockets to get free. The zero of gravitational potential energy is usually assigned to $r=\infty$, at the limit of the influence of the gravitational source. Since it takes work to get there, we must be in a negative potential region to begin with. The work that we must do to lift unit mass from r to ∞ is

$$(\text{work})/m = \frac{1}{m}\int_r^\infty \mathbf{F}\cdot d\mathbf{r} = GM\int_r^\infty r^{-2}dr = \frac{GM}{r}.$$

Since it takes that much work to bring the potential energy of the system to zero, the original potential energy of the system with the unit mass at r must have been $-GM/r$.

The gravitational potential is inversely proportional to the distance to the center of a spherical source, down to the surface of the sphere. As we have seen, at that radius the force changes from inverse square to proportional to the radius. If we had a hole through the diameter of the earth, or a bowling ball, the force on unit mass would be equal to $F/m = \frac{4}{3}\pi\rho Gr = g_R(r/R)$. If we had to extract unit mass from the center of the sphere to radius r inside the sphere, we would have to do additional work equal to:

$$\int_0^r \left(\frac{g_R}{R}r\right)dr = \frac{1}{2}\frac{g_R}{R}r^2.$$

The work done in taking unit mass from the center to the surface is

$$\tfrac{1}{2}g_R R = \frac{1}{2}\frac{GM}{R}.$$

For 1 kg, this work equals 3.12×10^7 J. The work done in carrying the kilogram from the earth's surface to infinity is $GM/R = 6.26\times 10^7$ J. Lest this seem like a lot of energy, consider that it is about what you would get by burning half a gallon of gasoline. Or, in other terms, it is about 17 kW h.

The graph of gravitational potential as a function of r is shown in Fig. 5.8. Note that from 0 to R, the curve is a parabola; from R to ∞ the potential is proportional to $1/r$.

Rather than pulling objects off the surface of the earth, we usually just shoot them off, giving them a large amount of initial kinetic energy. In order to escape, the object's initial positive kinetic energy must equal the magnitude of its negative binding energy, so that its total energy is zero:

Gravitation

$$\tfrac{1}{2}mv^2 - m\frac{GM}{R} = 0, \quad v_{\text{escape}} = \sqrt{\frac{2GM}{R}} = \sqrt{2g_R R} = \sqrt{2(9.8 \text{ m/s}^2)(6.4 \times 10^6 \text{ m})}$$

$$= 11 \times 10^3 \text{ m/s} \approx 25\,000 \text{ miles/h}.$$

Note that the potential energy of an object is proportional to its mass; it is the product of its mass and the gravitational potential at that point. However, the escape velocity is independent of mass. It is the same for a baseball, a rocket ship, or a molecule.

Let us compare *escape* velocity with *orbital* velocity when $r = R$. (To be sure, the orbits are usually at least 100 miles above the surface.) For circular orbits:

$$G\frac{mM}{R^2} = \frac{mv^2}{R} \rightarrow v = \sqrt{\frac{GM}{R}} = \sqrt{g_R R} = 7.9 \times 10^3 \text{ m/s} \approx 17\,700 \text{ miles/h}.$$

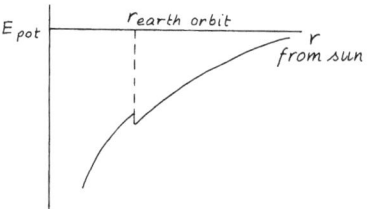

Figure 5.9. Gravitational potential as a function of distance from the sun. Potential well at earth's surface, due to earth, is only 1/15 of depth of well due to sun at earth's orbit.

Even if you escape from the earth, you are still bound in the sun's gravitational field. Let us compute the binding energy of a 1kg object that is motionless, not on the earth but at the orbit of the earth:

$$E_{\text{grav. pot.}} = -G\frac{mM_{\text{sun}}}{r_{\text{earth-sun}}} =$$

$$-(6.7 \times 10^{-11} \text{ N m}^2/\text{kg}^2)\frac{(1 \text{ kg})(2 \times 10^{30} \text{ kg})}{(1.4 \times 10^{11} \text{ m})} = -9 \times 10^8 \text{ J}.$$

Here we have used the mass of the sun and the orbital radius of the earth. Evidently, we are in a hole due to the sun's pull that is 15 times deeper than the local well caused by the earth, which has a depth of 6.26×10^7 J. The comparison is shown in Fig. 5.9.

One of the confusing features for students in learning about gravitational potential energy is that we teach them two different formulas. If you lift an object from the surface of the earth, you increase its potential energy by mgh. This equation is appropriate for an infinite flat earth and a constant weight, mg. Let us demonstrate how we get this approximation on a spherical earth, with weight equal to $G(mM/R^2)$ and gravitational potential energy equal to $-G(mM/R)$. Remember that only *changes* in potential energy are measurable:

$$\Delta E_{\text{gray. pot.}} = -GmM\left(\frac{1}{R+h} - \frac{1}{R}\right) = -GmM\left(\frac{R-(R+h)}{(R+h)R}\right) = GmM\left(\frac{h}{R^2+Rh}\right).$$

For small vertical displacements, hR is negligible compared with R^2. Since $g = GM/R^2$, $\Delta E_{\text{gray. pot.}} \approx mgh$. If the earth were an infinite plane, and the weight of an object independent of its height, then it would take infinite energy to escape and we would never get those rockets up and away.

The potential energy of a satellite in circular orbit has twice the magnitude of the kinetic energy but is negative. The total energy is

$$E_{\text{total}} = E_{\text{kin}} + E_{\text{pot}} = \frac{1}{2}\frac{GmM}{r} - \frac{GmM}{r} = -\frac{1}{2}\frac{GmM}{r}.$$

The total energy is negative, meaning that the satellite is bound, which, of course, it is. The smaller the radius of the orbit, the larger the magnitude of the negative total energy; that is, the satellite is deeper in the potential well and is more tightly bound. However, the smaller the radius, the larger the kinetic energy. As a satellite loses energy and descends, its speed gets greater and greater. The velocity of a

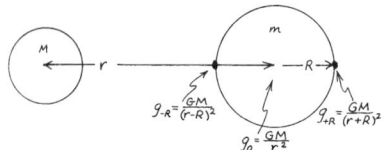

Figure 5.10. Differential pull of Moon on Earth.

satellite close to the earth, for instance, is greater than the velocity of the moon.

If one satellite is chasing another and they are both in the same orbit, the hunter might fire its tail rockets to catch up. In so doing it would be gaining total motion energy. As the magnitude of its negative total energy is reduced, the satellite moves to larger r. But at larger r its kinetic energy is reduced and it slows down, falling further behind its quarry. To catch up, the rockets must be fired in the backward direction.

The binding energy for a hydrogen molecule on the surface of the earth is

$$-m\frac{GM}{R} = -\frac{(3.3\times 10^{-27}\ \text{kg})(6.67\times 10^{-11})(5.98\times 10^{24}\ \text{kg})}{(6.37\times 10^6\ \text{m})}$$

$$= -2.1\times 10^{-19}\ J \rightarrow -1.3\ \text{eV}.$$

The molecular thermal energy is equal to $\frac{3}{2}kT \approx \frac{1}{26}$ eV at room temperature. It appears that hydrogen is tightly bound to the earth since the gravitational potential well is about 34 times deeper than the average kinetic energy resulting from thermal motion. Nevertheless, hydrogen does not stay in the earth's atmosphere. In spite of the mixing of gases within the atmosphere, the lighter hydrogen gradually works its way upwards. Above the ionosphere there are regions where the equivalent temperature is very high because of the bombardment of particles coming in from the sun. Hydrogen molecules in the high-energy tail of the thermal energy distribution occasionally reach escape speed in the right direction and are lost from earth.

The average thermal kinetic energy of nitrogen or oxygen is the same as it is for hydrogen. The depth of the gravitational potential well, however, is greater by a factor 28/2 or 32/2, for nitrogen or oxygen, respectively. A nitrogen molecule, for example, is in a gravitational potential well with a depth of about 18 eV. Almost no nitrogen escapes from this much deeper well, and so the earth has an atmosphere. (An explanation of the escape probability is given on p. 244.)

■ TIDES

In everyday life we are not aware of the inverse-square attraction of sun and moon on the earth. We tend to think of the sun pulling the earth, and the earth pulling the moon, and the earth pulling on us. Since all of these effects are constant we do not notice them. Nevertheless, the solid earth rises and falls under our feet twice a day by as much as 20 cm, and on the shores the ocean tides wait for no one. These effects depend on the inverse third power of the distance between gravitational bodies. Even more peculiar, at any given time there are high tides on opposite sides of the earth.

Forget the rotational motion of moon and earth around each other and consider them both in free fall toward each other. This is the actual case, since the mutual force and accelerations are along the line connecting them. Concentrate on the sphere with mass m and radius R, which is falling toward the sphere with mass M. The distance between the centers of mass of the two spheres is r. Then the gravitational field at the center of m, due to M, is $g_0 = GM/r^2$. However, at the near side of the sphere with mass m, the gravitational field due to M is $g_{-R} = GM/(r-R)^2$, and at the far side the field is $g_{+R} = GM/(r+R)^2$. The geometry is shown in Fig. 5.10. An object on the near side has a larger acceleration than the (solid) sphere itself, which in turn has a larger acceleration than an object on the far side. During the free fall, an object on the near side would speed ahead of the sphere while an object on the far side would lag behind. If these objects were tied to the sphere with springs, each would pull outward radially. If we call weight the measured pull toward the center of the sphere, then the weight of both objects is reduced.

The difference in g (the field produced by the distant sphere), and hence the reduction in weight, is

$$\Delta g = GM\left(\frac{1}{r^2}-\frac{1}{(r+R)^2}\right) = GM\left(\frac{(r+R)^2-r^2}{r^2(r+R)^2}\right) = GM\frac{2rR+R^2}{r^4+2r^3R+r^2R^2}.$$

For $R \ll r$, $\Delta g \approx 2GMR/r^3$. The difference in the field depends on the inverse third power of the distance between centers of the two spheres. (The same expression can be obtained by taking the radial differential of g: $dg = -2GMr^{-3}dr$.)

The effect of the moon on the earth is to pull the surface water toward the moon on the near side and to pull the earth away from the surface water on the far side. In our formula, M would be the mass of the moon, r the distance between centers of earth and moon, and R the radius of the earth. As the earth spins on its axis, the two high tides surge around the globe. The time between high tides is 12 h and 25 min. The time is not 12 h because the moon is moving around the earth with an apparent motion of about 1 h (15°) per day.

The *sun* also produces a differential gravitational effect—a tide—on the earth. Obviously the gravitational pull of the sun on the earth is stronger than that of the moon. Why then does the moon dominate the tides? This is the effect of the inverse third power of the separation distance:

$$\frac{\text{(Tidal effect of moon)}}{\text{(Tidal effect of sun)}} = \frac{2GM_{\text{moon}}R_{\text{earth}}/(r_{\text{to moon}})^3}{2GM_{\text{sun}}R_{\text{earth}}/(r_{\text{to sun}})^3} = \frac{M_{\text{moon}}(r_{\text{to sun}})^3}{M_{\text{sun}}(r_{\text{to moon}})^3}$$

$$= \frac{(7.34\times 10^{22} \text{ kg})(1.5\times 10^{11} \text{ m})^3}{(1.99\times 10^{30} \text{ kg})(3.8\times 10^8 \text{ m})^3} = 2.3.$$

When the sun and moon are lined up with the earth, at new moon or at full, their tidal effects add and produce the so-called spring tides. The high tides are at their highest and the lows at their lowest. The maximum vertical difference is called the range or the reach. When the sun and moon are in quadrature, 90° apart, at half-moon, the tidal effects of sun and moon partially cancel each other, leading to neap tides. The height of the tides also changes with the changing radial distance of the moon in its elliptical orbit around the earth. The mean radial distance is 384 400 km; at apogee it is 407 000 km, about 6% larger than mean; at perigee it is 357 000 km, about 7% smaller than mean. Because of the inverse-third-power dependence, the tidal height fluctuations produced by these changes is about ±20%. When the moon is at perigee and new or full, the high tide is particularly high, and the low is particularly low.

Another general factor influencing the height and timing of the tides is the orientation of the plane of the moon's orbit. That plane is tilted at 5° to the ecliptic (the plane of the earth's orbit), which in turn is tilted at 23.5° to the equator. Superimposed on the annual motions of the moon through the sky (higher at night during the winter, lower during the summer) there are monthly changes. During half of each lunar month the moon is slightly higher in the northern hemisphere and lower in the southern hemisphere—and vice versa during the other half of the

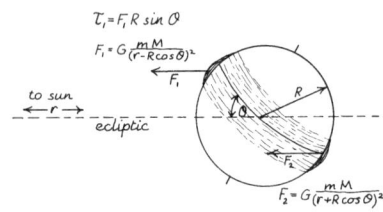

$$F_1 - F_2 \approx GmM_s \frac{2R\cos\vartheta}{r^3}$$

Precession frequency

$$= \frac{\tau}{J} = \frac{GmM_s 2R^2 \cos\vartheta \sin\vartheta / r^3}{(0.34)m_E R_E^2 \omega}$$

$$= 2 \times 10^{-12} \text{ rad/s}$$

$M_s = 1.99 \times 10^{30}$ kg

$m_E = 5.98 \times 10^{24}$ kg

$\omega = 2\pi/(24 \times 3600)$

$m_{\text{girdle}} = 1 \times 10^{22}$ kg

$\vartheta = 23.5°$

$r_{E-S} = 1.5 \times 10^{11}$ m

For Earth: $I = (0.34)m_E R_E^2$

$R_E = 6.37 \times 10^6$ m

Actual precession frequency

$$= 1.3 \times 10^{-12} \text{ rad/s}$$

Figure 5.11. *Geometrical cause of torque exerted on Earth by Sun, and consequent precession of Earth's axis.*

month. This geometry produces a diurnal (daily) fluctuation so that one of the two daily tides is higher than the other.

So far we have been describing tides that might be produced on a planet completely covered by water. The actual tides depend strongly on geometry of the land. Consider that in the Great Lakes, tidal effects are practically undetectable. To produce tides you must have a body of water large enough so that there is a difference of gravitational field from one end to the other. Simply reducing the weight of the water on the whole surface slightly twice a day will not produce any sloshing motion. At the beginning of this section on tides we spoke of the high-water bulges surging around the world as the earth turns. The speed of long-wavelength waves in deep water is $v = \sqrt{gh}$, where h is the water depth. The average depth of the oceans is 3800 m. The speed of the tidal surge is thus 190 m/s = 430 miles/h. In 12 h the wave would travel 5000 miles; hence the natural oscillation period of an ocean is comparable to the tidal forcing period.

The tidal range in the open ocean is only a couple of feet, although there are many exceptions and peculiarities. Because there are reflections and combinations of currents, there are places that have essentially no tide, and there are other places that have only one tide a day. More dramatic phenomena occur where the ocean tides act on the mouth of a long channel, such as a sound or a river estuary. Where the incoming tide enters a narrowing channel, such as at the mouth of the Colorado River, a wall of water called a bore piles up and surges upstream. In the Colorado River this wall may be as high as 10 ft. Where the channel does not narrow, the ocean tides at the throat may act as a forcing oscillation in a resonant system. For a rectangular channel, driven at one end and closed at the other, the fundamental period is $T = 4L/v = 4L/\sqrt{gh}$.

The most famous example of such a resonant tide occurs in the Bay of Fundy between New Brunswick and Nova Scotia. The bay is forked at the end, but can be approximated by a rectangular basin 170 miles long with average depth of 240 ft. This yields a resonant period of $11\frac{1}{2}$ h, close to the driving period of $12\frac{1}{2}$ h. The range of the tide sometimes reaches 70 ft. Another example is Long Island Sound between Long Island, N.Y. and Connecticut. The natural period is a little under 10 h, leading to a tidal range of 8 ft. The Gulf of California has a period of 25 h and experiences a single very high tide each day, resonant with the diurnal asymmetry of the free ocean tide.

The atmosphere experiences tides, less affected by the distribution of land masses. They have an average amplitude of about half a meter. The solid earth also flexes and bulges with a tidal amplitude of about 20 cm. We do not feel it because everything around us is going up and down at the same time.

■ LOCKED PHASES AND PRECESSION

The combination of centrifugal force and tidal force produces bulges or girdles on otherwise spherical planets and moons. If a planet is spherical, then of course the sun exerts no torque on it. If a planet has a girdle around its equator, but is spinning with the equator in the orbital plane, then the sun exerts no torque. However, the earth has an equatorial bulge from centrifugal force during its formation, and is spinning with its equatorial plane at an angle of $23\frac{1}{2}°$ from the ecliptic or orbital plane. As shown in Fig. 5.11, the sun exerts a torque on the earth trying to straighten it up, because there is a stronger gravitational pull on one side than the other. Instead of straightening up the earth, the torque makes the spinning earth precess. The period is about 26 000 years. Spring occurs each year when we are on the same side of the sun with respect to the fixed stars. In 13 000 years, spring will occur when we are on the other side of the sun. At present our pole star, or North Star, is α Ursae Minoris; in 13 000 years, or 13 000 years ago, Vega will be close to the pole position.

The order of magnitude of the precession frequency can be found by assuming that the equatorial bulge is a ring with a width of about 6000 km and a thickness of 21 km—which is the difference between equatorial and polar radius. The geometry and a crude calculation are shown in Fig. 5.11.

The moon's orbit is inclined at an angle of 5° from the ecliptic, and hence the moon may be anywhere from $18\frac{1}{2}°$ to $28\frac{1}{2}°$ from the plane of the earth's equator. The lunar orbit precesses with a period of 18.6 years. In this case also the sun forces the precession. The geometry is shown in Fig. 5.12.

The tidal motions of the oceans turn enormous amounts of rotational energy into heat. The rate of dissipation is 2.6×10^6 MW, which is about equal to the world's power consumption. The dynamical effect is to slow the earth's spin and increase the angular momentum of the moon, which means increasing its radial distance. The length of the day increases by 20×10^{-6} s per year, and the moon's orbital radius increases by about 3 cm per year. Such a small increase in the length of each day may seem insignificant. Consider, however, that at the end of a century, the day has lengthened by 20×10^{-4} s. The average day in the century is therefore longer by 10×10^{-4} s. Since the effect is cumulative, the total lag is $100 \times 365 \times 10 \times 10^{-4}$ s = 36.5 s. Such a delay is easily seen in comparing astronomical records that depend on the earth's timing.

While it is plausible that sloshing tides must dissipate energy, and that this energy must come out of the earth–moon–sun rotational system, it is not obvious how the earth can exert a torque on the moon in order to transfer its spin angular momentum to the moon's orbital angular momentum. The tidal action produces its own bulges on which it can act to produce a torque. As shown in Fig. 5.13 the tidal bulges are not instantaneously in line with the line from earth to moon. Due to friction, high tide lags the moon by about 3°, on the average. (Coast and channel geometry affects this lag at any particular place. The time between the moon's crossing of the meridian and the next high tide at a place is called the lunitidal interval. The average such interval at full and new moon is known as the *establishment* of the port.) The moon can act on this induced bulge to slow the spinning earth. Since the tidal effect to produce the bulge depends on the inverse third power of the satellite distance, and the torque exerted depends on the inverse third power and the size of the bulge, the overall effect depends on the inverse sixth power of the radial distance.

In Fig. 13 we did not show the induced tides in the moon. Our moon is now locked in phase with its rotation around the earth, so that it always presents the same side to the earth. It was surely spinning to begin with and so would have had the same kind of lagging tidal bulges in its solid crust. In a few million years the tidal torque from the earth would have reduced the spin to zero. There is a slight asymmetry of the centrifugal girdle of the moon, with a larger bulge toward the earth, thus making the locked phase stable.

■ ELLIPTICAL ORBITS

In analyzing orbital dynamics, we have assumed circular orbits. Actually, gravitational orbits are elliptical, but for most planetary and satellite problems, circles

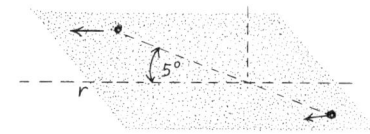

Figure 5.12. Precession of the moon's orbit. The period is 18.6 yr.

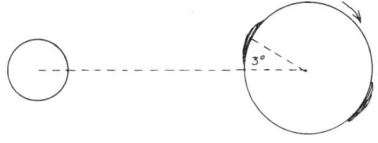

Figure 5.13. Tidal bulges are not in phase with moon. The resulting drag slows the earth.

are a good approximation. More information is given about elliptical orbits on p. 181. There is an expanded treatment of ellipses in *Used Math* by Swartz, pp. 110–133, pp. 139, and pp. 146–148.

■ FOR READING AND REFERENCE

Chaffee, F. H. Jr., "Discovery of a gravitational lens," Sci. Am. **243**, 70 (Nov. 1980).
 Two images of a quasar are formed by a galaxy between us and the quasar as it acts as a gravitational lens.
Cohen, I. B., "Newton's discovery of gravity," Sci. Am. **244**, 166 (March 1981).
 The development of Newton's thinking in working out the concept of gravitation.
Crandall, R. F., "Electronic Cavendish device," Am. J. Phys. **51**, 367 (1983).
 A description of a demonstration device for measuring G. The torsion beam is prevented from rotating by the magnetic field of a solenoid. G is calculated from the current needed in the coil to prevent rotation when a large external mass is moved into position.
Diamond, J. B., "The deflection of light by gravity," Phys. Teach. **20**, 543 (1982).
 The author takes a physical approach to the problem, and in an appendix outlines the calculation of the effect.
Dicke, R. H., "The Eötvös Experiment," Sci. Am. **205**, 84 (Dec. 1961).
 Describes the Eötvös experiment and Dicke's modern version to test the equivalence of inertial and gravitational mass.
French, A. P., "Is g really the acceleration due to gravity?," Phys. Teach. **21**, 528 (1983).
 An analysis of the latitude effect on the weight (apparent weight) of an object, and its acceleration in free fall.
MacKeown, P. K., "Gravity is geometry," Phys. Teach. **22**, 557 (1984).
 An article describing Einstein's answer to the question "What is gravity?"
Mentzer, R. .G., "Measuring the acceleration due to gravity," Phys. Teach. **22**, 580 (1984).
 Useful expansion cf a *Project Physics* experiment.
Raman, V. V., "A background to Newtonian gravitation," Phys. Teach. **10**, 439 (1972).
 A historian of science concludes that, in addition to Newton, Hooke, and Halley deserve mention in any historical discussion of gravitation.
Sayer, G. A., "A game of gravity," Phys. Teach. **14**, 359 (1976).
 A note by a high-school teacher describing a number of curiosity-arousing questions about gravitation. Used as a class activity.
Stauffer, F. R., "Seeds of general relativity," Phys. Teach. **22**, 27 (1984).
 This note shows the equivalence of inertial and gravitational mass, the equivalence of an accelerating reference frame to a gravitational field, and the gravitational bending of light. The mathematical level is moderate.
Swartz, C. E., *Used Math* (American Association of Physics Teachers, College Park, MD, 1993).
Van Flandern, T. C., "Is gravity getting weaker?," Sci. Am. **44** (Feb. 1976).
 There is some evidence that the force of gravity diminishes as the universe expands.
Walters, J., "Time dilation and the Lorentz contraction," Phys. Teach. **20**, 42 (1982).
 This note describes and analyzes a string model said to be useful for clarifying the ideas of the title.
Worden, F. W., Jr. and Everitt, C. W. F., editors, *Gravity and Inertia* (a reprint book) (American Association of Physics Teachers, College Park, MD, 1983).
 The book starts with a "Resource Letter" (an annotated bibliography on the topic) and contains eight reprints of papers on current research topics in gravitation and the general theory of relativity.

6
Reference Frames and Relativity

Some elements of the special theory of relativity are usually taught in introductory-physics classes, particularly in these days of emphasis on "modern" physics. Often the subject is reduced to the memorization and use of $E = mc^2$. The subject is modern all right—modern, classical, and fundamental. The nature of our entire system of dynamics is based on the structure of our reference frames. A straight-line path in one system becomes a parabolic path in another. A force in the forward direction in one system is a force in the backward direction in another. Indeed, the very definition of force depends on the reference frame. (The content of Newton's first law is basically a definition of reference frame.) At the introductory level we usually avoid these subtle but vital problems. As teachers we must be aware of them, however, so that at least we do not tell untruths. We should also be prepared to introduce our better students to the problems. Good students will not be frightened by the transformation algebra and may be fascinated by the complexities and insights.

Relativity is simply a matter of describing things from someone else's viewpoint. This is particularly difficult if the other person is moving past or circling around you. The other person has a different reference frame, or a different coordinate system. If you are stopped at a red light beside another car, you might notice that the other car is rolling backward; the other person, in the other car, might claim that your car is rolling forward. With a smooth motion, it is often hard to tell which is happening. A third person beside the road could tell which car was moving, but *with respect to her*.

Suppose that while in the moving car you drop a coin; it will fall straight down into your lap, not into the seat behind you. According to you, the path of the falling coin is a straight line down. An observer beside the road sees both you and the coin traveling forward. Then she sees the coin drop along a parabolic path. Is the true path of the falling coin a straight line or a parabola? Note that the question is meaningless; there is no *true* path. (Remember this conclusion when we get to centrifugal–centripetal force!)

If the car in which you are sitting accelerates suddenly in the forward direction, you will be thrown back against the seat. From your reference frame you were evidently accelerated backward, presumably by some backward force. (After all, $F = ma$.) The observer on the side of the road will claim, however, that your car accelerated forward and so did you, presumably because there was a forward force acting on you. Is the true direction of the force on you forward or backward? If you go around a sharp curve while riding in a car, you will be thrown to the outside, presumably by an *outward* radial force. An observer watching from a roadway above (which is fastened to a rotating earth) would conclude that you started to travel along a line tangential to the car's curved path, but were then pulled into the circular path by an *inward* radial force. Is the force on you inward or outward?

Ask your students if the earth goes around the sun, or does the sun go around the earth? They have all known the correct answer since third grade. In spite of all the evidence of their senses (the sun rises and sets), they will have learned the *truth* in school. Nevertheless, for 99% of the purposes of humans, the ancient Ptolemaic model is more convenient. Navigation, after all, is based on a stationary earth. Try to use the Copernican model in a ship far from shore and you will be lost at sea. If you are aiming a rocket for Mars, you will want to use the model where the planets go around the sun. The choice between these two models is not a matter of one being true and the other therefore false. The choice is determined by the convenience of the reference frame.

There are a number of demonstration apparatuses that give a feel for reference-frame phenomena. For instance, there is a cart with a spring-loaded vertical cannon. You can roll this cart horizontally, triggering the cannon with an attached string. A ball shoots out and executes a path that lands it back in the barrel of the cannon. For constant velocity of cart along the horizontal the effect is plausible but not obvious. Galileo describes experiments to determine the result of releasing a ball from the top of the mast of a sailing vessel. (As we point out on p. 132, the ball will *not* fall straight down to the foot of the mast.) What happens if the cart is accelerating, either along the horizontal or down an incline? Students should find out. If you and your students have access to a car, driver, and quiet road or parking lot, you can try these experiments by having a passenger lob balls up or drop them down from the window of a slowly moving car.

You can determine the direction of "down" with a plumb line. (It is educational to have a builder's plumb line in class; many students will never have seen one.) Of course you can use any weight tied to a string. Have your students use homemade plumb lines to observe the direction of "down" when they are passengers in accelerating cars—either forward or braking in a straight line or going around curves. Whenever students have helium-filled balloons, they should use them in the same car-acceleration experiments. In this case, they have "anti–plumb lines", whose direction defines "up." The anti–plumb line behavior of helium balloons surprises and delights even the jaded. Incidentally, another completely valid way to describe and understand the behavior of a submerged floating object is to assign it *negative* mass. Since gravitational and inertial fields are, of course, locally equivalent, an object with negative mass experiences negative acceleration in either kind of field.

■ THE GALILEAN TRANSFORMATION

In spite of the formidable name, the problem that we face is very simple. Two people, with measuring instruments, observe the same phenomenon. It might be a ball shooting through the air, the collision of two objects, or a weight on a spring moving up and down with simple harmonic motion. If the two observers are moving with constant velocity with respect to each other, what differences do they observe in the phenomenon? To make it easier to talk about, we give each observer his own (x,y,z) coordinate system. One of them will move in the x

direction at constant velocity v with respect to the other one. In Fig. 6.1 the (x',y',z') frame is moving to the right with respect to the (x,y,z) frame. We must resist the temptation of thinking that the (x,y,z) frame is stationary while the (x',y',z') frame is the one actually moving. All that is important is that they are moving with respect to each other. The observer in the (x',y',z') frame assumes that her frame is stationary while the (x,y,z) frame is moving to the left with a velocity $-v$.

In order to transform a description of motion from one frame to the other, we must write down the coordinate transformation. Since the motion is only in the x direction, the y and z coordinates in the two frames are the same. Furthermore, time is the same in both reference frames. After all, why should your clock run fast or slow just because you are traveling along a highway? (Of course, your clock *does* change, but for now we will use the Galilean assumption.) The Galilean transformations are

$$t=t', \quad y=y',$$

$$z=z', \quad x=x'+v\Delta t.$$

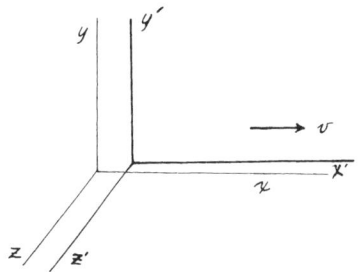

Figure 6.1. Reference frames moving with respect to each other. The primed frame is moving to the right with velocity v, as seen by an observer in the x, y, z frame. But the x, y, z frame is moving with velocity v to the left, as seen by an observer in the primed frame.

For instance, suppose that at 10 o'clock in the primed frame an object is at the point $x'=4$ miles, $y'=1$ mile, and $z'=2$ miles. If the origins of the coordinate frames were coincident at 9 o'clock, and the primed frame has been moving in the $+x$ direction with a speed of 10 miles/h, then the coordinates of the (x,y,z) frame are $t=10$ o'clock, $y=1$ mile, $z=2$ miles, and $x=4+(10 \text{ miles/h})(1 \text{ h})=14$ miles, as shown in Fig. 6.2.

Example 1

Let us use the Galilean transformation to confirm the observations we have already assumed about the path of a falling stone. Let the observer by the side of the road take the unprimed reference frame and the person in the car use the primed system. The easiest reference frame to use is the one in the car where the stone was stationary to begin with:

$$y'=\tfrac{1}{2}g(\Delta t')^2.$$

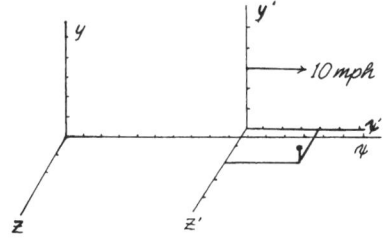

Figure 6.2. Location of object in two different reference frames.

If we simply substitute the expressions for y and t, we find what we already know. The y component of a moving object is unaffected by its motion in the perpendicular x direction:

$$y=\tfrac{1}{2}g(\Delta t)^2.$$

However, since the time is also related to x and v, let us make another substitution for t':

$$\Delta t'=\Delta t=(x-x')/v.$$

Substituting this expression for Δt into the formula describing y, we get

$$y=\frac{1}{2}g\left(\frac{x-x'}{v}\right)^2.$$

If we choose, we can claim that the stone is dropping from the origin of our moving reference frame, which means that $x'=0$. In that case,

$$y = \frac{1}{2}\frac{g}{v^2}x^2.$$

The equation of motion in the unprimed frame describes a parabola ($y = Ax^2$). This is the path as seen by an observer at the side of the road. The formula describes an object thrown with horizontal velocity v subject to vertical acceleration g. We have learned no new physics by using the Galilean transformation, but note the different mental reference frame in considering the problem in this new way. Before, we threw a ball and knew very well that it was traveling independently in both the x and y directions. However, in terms of a reference frame problem, the ball's motion in one reference frame is straight down and in another reference frame it follows a parabolic path.

Example 2

Now let us transform a motion in the (x,y,z) frame into the (x',y',z') frame. Suppose that the motion is simple harmonic, so that at $x=0$ an object is oscillating up and down with a motion:

$$y = A \sin(2\pi ft).$$

What does this motion look like in a reference frame that is moving along the x axis with velocity v? You can demonstrate the answer approximately by moving a chalk up and down a blackboard, using a motion that is as close as possible to simple harmonic motion (SHM), while walking slowly and steadily beside the board. Or, you can do the same thing with a pencil and paper while someone else slowly moves the paper. Note the relative difference, if any, between these two demonstrations. In the first case, you are in the moving system; in the second case the paper is moving.

To transform the equation for simple harmonic motion to the primed system (the "moving" paper or the "stationary" blackboard), we use the substitutions

$$y = y', \quad \Delta t = \Delta t' = (x - x')/v.$$

If the SHM always takes place at the origin of coordinates where $x=0$, the transformation is

$$y' = A \sin[2\pi f(-x'/v)] = -A \sin[2\pi(f/v)x'].$$

The transformation has changed a sinusoidal motion in *time* into a sinusoidal motion in *distance*. The wavelength of this sine wave is λ, which is a function of both the frequency of the SHM and also of the velocity between the two reference frames:

$$y' = -A \sin[(2\pi/\lambda)x'], \quad \text{where } \lambda = v/f.$$

The reference-frame change that we have just described changes an oscillation in time into an oscillation in space. The best-known example of this transformation is the phonograph record. Both the recording needle and pick-up needle oscillate along a horizontal line with a motion that is vibratory in time, if not SHM. The groove in the record, however, snakes along with a sinusoidal-type wave in space. Note that the wavelength along the groove in the record is $\lambda = v/f$. In this case, v is the speed of the record groove past the recording needle, and f is the frequency of the recording needle. In the playback arrangement, the coordinate systems must be transformed again. The sinusoidal motion in space along the groove turns into a sinusoidal vibration in time of the playback needle. If the velocity of the playback is v_1 instead of v, the oscillations of the playback needle are

$$y = -A \sin[(2\pi/\lambda)x'] = -A \sin[(2\pi/\lambda)(-v_1 t)].$$

We obtained this substitution by choosing $x=0$ in the expression $x' = x - vt$. Since $\lambda = v/f$, our formula for y becomes

$$y = A \sin[2\pi(v_1 f/v)t] = A \sin[2\pi f_1 t].$$

The frequency that we hear is not necessarily the frequency that was recorded. The playback frequency is $f_1 = (v_1/v)f$. The lesson is dramatic if a $33\frac{1}{3}$ rpm record is played on an old 78 rpm turntable.

Example 3

We have seen that a change in reference frame can change the description of the path of a moving object and can also transform displacement as a function of time into displacement as a function of position. How do velocities transform? If you are walking at 2 m/s in a train that is moving at 30 m/s, how fast are you moving with respect to the ground? ("Nonsense," said the inspector as the train hurtled past the station, "I have been sitting motionless in my seat the entire time.") We know intuitively (that deceptive prejudice) that if you are walking toward the front of the train, your speed with respect to the ground must be the sum of the train's speed and your speed with respect to the train. If you are walking toward the back of the train, your resultant speed with respect to the station is the difference between the train's speed and your speed. Let us prove this analytically and consider the consequences.

At time t:

$$x = x' + vt \quad (\text{where } t_0 = 0).$$

At time $(t + \Delta t)$:

$$(x + \Delta x) = (x' + \Delta x') + v(t + \Delta t).$$

Subtracting the first equation from the second:

$$\Delta x = \Delta x' + v \Delta t.$$

Therefore

$$\frac{\Delta x}{\Delta t} = \frac{\Delta x'}{\Delta t} + v.$$

If we call the resultant speed in the unprimed frame w, and the speed in the primed frame u, then

$$w = u + v.$$

We have been dealing only with the component of velocities in the x direction. The components of velocities in the other directions do not depend on the motion along the x axis. In the direction of motion, however, the velocity in one reference frame is the algebraic sum of the velocity in the other reference frame plus the relative velocity between the two frames. This conclusion may seem obvious but remember that it is only a low-velocity approximation to the complete relativistic law of addition of velocities. As for the extent to which the results are intuitive or obvious, ask your students what happens if you fire a BB gun with a muzzle velocity of 30 m/s on a train traveling 30 m/s. If you shoot in the forward direction, what is the speed of the BB with respect to somebody on the train that it hits and with respect to somebody watching from the station platform? If you stand at the back of the train and fire backward, what is the motion of the BB as seen by an observer at the station? In each case, what happens if the BB hits someone on the platform instead of someone on the moving train?

Example 4

As usual, when we start probing into nature, simple and obvious rules sometimes lead to paradoxical conclusions. It seems reasonable that velocities should add algebraically in a change of coordinate systems. What happens to the kinetic energy of the BB as seen in the two reference frames? *Energy must also be relative*. In the moving train, the BB has the same kinetic energy regardless of which direction it is fired: $E_{kin} = \frac{1}{2}m(30 \text{ m/s})^2$. The observer at the station, however, sees quite a different situation. If the BB is fired in the forward direction, then the observer measures the kinetic energy to be $E_{kin} = \frac{1}{2}m(60 \text{ m/s})^2$. That is four times the kinetic energy that is measured inside the train. To make matters worse, if the BB is fired in the backward direction, the observer at the station is correct in claiming that both the velocity and the kinetic energy of the BB are zero. Does the BB have kinetic energy or does it not? We might decide by measuring the size of the hole it makes in something. If the gun is fired backward at someone standing at the station, the BB will simply drop to the ground (according to the observer at the station) and leave no mark. If the gun is fired backward at someone on the train, it will leave a mark corresponding to a collision at 30 m/s. In this case the observer at the station would claim that the BB stood still and the person on the train ran into it. If the gun is fired forward and the BB hits someone at the station, the damage is four times that produced by a BB traveling at 30 m/s.

Collision problems must be handled very carefully and thoroughly as you transform from one coordinate system to another. In each reference frame you must set up the equations for conservation of energy and momentum within that reference frame, and you must include the motions and recoils of all the participating objects. As an example of the problems encountered when you do not take these precautions, here is a simplified version of a problem that was described years ago in the magazine *Astounding Science Fiction*. The author claimed that he had discovered an insurmountable obstacle to the establishment of space stations. Suppose that a supply rocket must link with a space station that has a relative velocity of 1 (1 m/s or any other unit—it makes no difference). The kinetic energy of the supply rocket relative to the station is therefore $\frac{1}{2}m(1)^2$. That much energy must be absorbed by the shocks of the space station. These shock springs are tested on earth so that they can slow down and stop the supply rocket without damage. Now let us put the space station in orbit where it is traveling at a velocity of 10 with respect to the earth (10 of the same units, whichever you have chosen). The supply rocket must have a velocity of 11 units with respect to the earth if its relative velocity with respect to the station is to remain 1 unit. Therefore, the kinetic energy of the supply rocket, with respect to earth, is $\frac{1}{2}m(11)^2$. If it slows down and stops in the space station, its final velocity will be 10, and its final kinetic energy will be $\frac{1}{2}m(10)^2$. Look at the disastrous consequences! The kinetic energy that must be lost to the shocks in the space station is $\frac{1}{2}m(11)^2 - \frac{1}{2}m(10)^2$

$= \frac{1}{2}m(21)$. That is 21 times the energy than we calculated would be lost on earth. The actual situation with the space station would be much worse, because the actual velocity of a satellite in orbit around the earth is so large that taking the difference between two large squares yields an enormous difference. As The Astounding Science Fiction author said long ago, "The supply rocket will gently ease its way into the shock springs and slowly but inexorably push them aside and continue plowing its way through the space station and out the other side."

Where did we go wrong? We did not take into account the necessity for applying the law of conservation of momentum. The supply rocket cannot just enter the space station and come to rest. There must be a recoil of the merged system. The recoil velocity is small because the mass of the station is much larger than that of the supply ship. Nevertheless, the kinetic energy associated with the recoil accounts for the extra energy that the supply ship appears to lose. The arithmetic is easy to work out. Assume $m=1$ and $M=10$.

Example 5

So far we have been comparing descriptions of events as seen from two different reference frames. The one containing ourselves we usually consider to be stationary; the other we assign to some system moving past us. Sometimes it is useful and revealing to analyze events from a reference frame attached to the center of mass of interacting objects. The laws of conservation of energy and momentum are particularly simple in the reference frame of the center of mass. In Fig. 6.3 we compare the collision of two protons as seen from two reference frames. In the first, we assume that one of the protons is at rest with respect to us. In the second part of the diagram, we see the same event from the point of view of the center of mass. Since the protons have identical mass, the center of mass must always be halfway between the two protons. From our point of view, if the bombarding proton is traveling to the right with velocity v_0, the center of mass is traveling to the right with velocity $\frac{1}{2}v_0$. From the reference frame of the center of mass, however, the proton from the left is approaching it at $\frac{1}{2}v_0$ and also the proton to the right is approaching it with a velocity of $-\frac{1}{2}v_0$. Necessarily, in this privileged reference frame, the net momentum of the participants is 0. At all times, including after the collision, the net momentum must remain 0. That can happen only if the protons in their glancing collision shoot out in directions opposite each other. The speed of each proton with respect to the center of mass will remain the same: $\frac{1}{2}v_0$. Otherwise kinetic energy would not be conserved in the center of mass frame, which must be the case since the collision is elastic.

In this reference frame, it is easier to calculate the probability of a proton leaving at any particular angle. The protons can shoot off at any angle as long as they leave opposite each other. It might seem therefore that there would be an equal number of protons at each angle. Let us assume that there is equal probability for a proton to pass through any small region on the surface of a concentric sphere, as shown in Fig. 6.4. This would mean that there is equal probability for a proton to be emitted into any solid angle. In terms of the angle θ, however, there is a much larger solid angle available in the region of $\theta=90°$ than there is in the region $\theta=0°$. Therefore, more protons would be emitted at 90° in the center-of-mass system, which becomes 45° in the laboratory system.

The transformation back to the laboratory system is easy. Simply add vectorially the velocity of the center of mass to the velocity of each of the protons. That

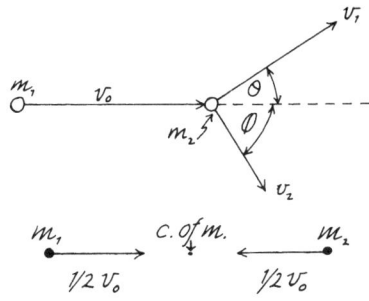

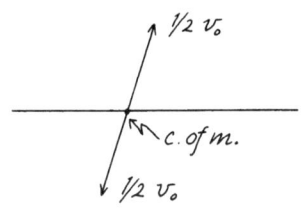

Figure 6.3. Two-particle collision in a center of mass reference frame.

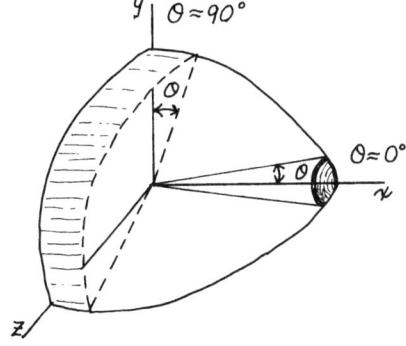

Figure 6.4. Angular distribution of particles in a two-body collision.

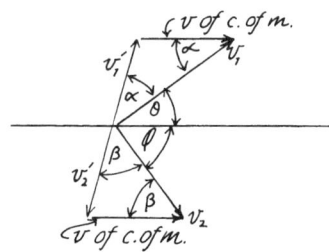

Because: $v_1' = \frac{1}{2} v_0 = v_{c.of m.}$
and: $v_2' = \frac{1}{2} v_0 = v_{c.of m.}$
∴ each triangle is isoceles
and $\alpha = \alpha$
$\beta = \beta$
Because of equality of interior angles:
$\theta = \alpha$ and $\phi = \beta$
Since: $\alpha + \theta + \phi + \beta = 180°$
$2\theta + 2\phi = 180°$
$(\theta + \phi) = 90°$

Figure 6.5. Opening angle between paths of identical particles after collision.

geometrical construction is shown in Fig. 6.5. The transformation is particularly easy in the case of identical particles and is not very difficult even when the particles have different masses. In the case of identical particles, note that the vector addition of velocities for each particle forms an isoceles triangle. The simple geometrical relationships shown in the diagram demonstrate that in a free elastic collision between two particles with identical mass, one of which was motionless with respect to us, the opening angle between their final directions must be 90°. This fact is frequently observed and exploited in identifying subatomic particles.

Example 6

Every introductory text derives the Doppler shift, either with diagrams or algebra. Since the Doppler shift arises because of a difference between reference frames of the source and the observer, the easiest way to describe the effect is to use the Galilean transformation. This treatment was described by Eli Maor. For the transformation we use $x' = x - vt$ for two systems whose axes coincide at $t = 0$. The primed system moves along the x axis in the positive direction with constant velocity v. We will use the symbol V for the velocity of sound in air, so as not to confuse that velocity with the velocity of the moving reference frame, v.

Let us start by assuming that the waves are generated in our (unprimed) reference system, perhaps by a factory whistle. As they spread along the x axis, we describe the *excess* air pressure P as

$$P = P_0 \sin\left[2\pi\left(\frac{x}{\lambda} - \frac{t}{T}\right)\right] = P_0 \sin\left[2\pi f\left(\frac{x}{f\lambda} - \frac{t}{fT}\right)\right] = P_0 \sin\left[2\pi f\left(\frac{x}{V} - t\right)\right].$$

The air is stationary with respect to us in the unprimed system. Therefore, the velocity of sound that we measure is the "true" velocity but not the velocity measured by the observer in the primed system, who is in a car. The observer in the car will find the speed of sound to be equal to $V' = V \pm v$. (The minus sign is used if the car is moving in $+x$ direction, away from the factory; the plus sign if the car is moving in $-x$ direction, toward the factory.)

We can now transform the formula for the traveling wave into the primed reference system, moving away from the factory, by simply substituting the transformation for x:

$$P = P_0 \sin\left[2\pi f\left(\frac{x' + vt}{V} - t\right)\right].$$

The primed observer in the car hears a sinusoidal wave that she would describe with the same formula we have used in the unprimed system. Of course, the primed observer would use variables from that system including a different frequency and a different velocity for the wave:

$$P' = P_0 \sin\left[2\pi f'\left(\frac{x'}{V'} - t\right)\right].$$

The two expressions must be identical. Therefore,

$$P_0 \sin\left[2\pi f\left(\frac{x' + vt}{V} - t\right)\right] = P_0 \sin\left[2\pi f'\left(\frac{x'}{V'} - t\right)\right].$$

Since V' equals $V - v$, the equality becomes

$$f\left(\frac{x'+vt}{V}-t\right)=f'\left(\frac{x'}{V-v}-t\right).$$

We now have an expression for the frequency observed in the primed system. Let us reduce the algebra:

$$f'_{obs}=f\,\frac{(Vt-x'-vt)/V}{(Vt-x'-vt)/(V-v)}$$

$$=f\,\frac{V-v}{V}\quad\text{(observer moving away from source)}.$$

Now that we have the algebra taken care of, consider the physical situation that is described. The source of sound (such as a factory whistle) is stationary with respect to us in the unprimed system and with respect to the air. The traveling observer is moving away from the source of sound. Consequently, the wave crests are not moving past her as rapidly or as frequently as they would if she were standing still in the air. Therefore, the frequency that she observes is lower. That is just what the formula tells us.

If the observer is traveling toward the source of sound instead of away from it, her velocity would be $-v$. Everything in our derivation would remain the same except that $+v$ should be replaced by $-v$. The final formula becomes:

$$f''_{obs}=f\,\frac{V+v}{V}\quad\text{(observer moving toward source)}.$$

According to this formula, if the observer is moving toward the source that is stationary in the air, the observer hears a higher frequency. The waves in the air are moving out toward the observer, and the observer in turn is moving in toward them. Consequently, the wave crests move past the observer faster and more frequently than they would if she were standing still. She hears a higher frequency. It is instructive to check out both cases to see what the formulas predict if $v=0$, $v=\frac{1}{2}V$, and $v=V$.

Now let us take up the case of the stationary observer (stationary with respect to the medium) and a traveling-wave source. This is the train problem. In the train's reference frame, it is producing a signal

$$P'=P_0\sin\left[2\pi f'\left(\frac{x'}{V'}+t\right)\right].$$

If the train is rushing away from us, $x'=x-vt$. This situation corresponds to our previous calculation where source and observer were traveling away from each other. Note that the train observes that the speed of sound is V' since the train is moving through the medium. If the waves in the medium are moving with a velocity V, then the train, looking back toward the observer, sees the crests to be moving away from itself with a velocity $V'=V+v$.

Now let us transform the variables of the wave equation from the train's reference frame to that of the stationary observer:

$$P=P_0\sin\left[2\pi f'\left(\frac{x-vt}{V+v}+t\right)\right].$$

The stationary observer hears sine waves passing over him but with a frequency f. Therefore he observes

$$P = P_0 \sin\left[2\pi f\left(\frac{x}{V} - t\right)\right].$$

Once again we can find the relationship between frequencies in the moving and the stationary reference frame by equating the arguments of the sine functions:

$$f\left(\frac{x}{V} + t\right) = f'\left(\frac{x - vt}{V + v} + t\right) \rightarrow \frac{f}{f'} = \frac{Vt + vt + x - vt}{V + v} \frac{V}{Vt + x} = \frac{V}{V + v},$$

$$f_{\text{obs}} = f' \frac{V}{V + v} \quad \text{(source moving away from observer)}.$$

This expression agrees with our experience that a train whistle moving away from us will have a lower pitch. If the train is moving toward us, everything in the derivation would remain the same except that we should replace $+v$ by $-v$. Our equation then becomes

$$f_{\text{obs}} = f' \frac{V}{V - v} \quad \text{(source moving toward observer)}.$$

Once again, it is instructive to note the ratio of f/f' for the cases of $\pm v = 0, \frac{1}{2}V$, and V. The two cases, moving observer or moving source, are not symmetrical. The waves move in a medium and when we say "moving" or "stationary" we mean with respect to the medium. Let us compare the two sets of equations for the situation where the source and observer are moving away from each other. Whether the source or the observer is stationary with respect to the air, the observer will hear a lower frequency.

If the source is stationary with frequency f, the observer hears f':

$$f'_{\text{obs}} = f \frac{V - v}{V} = f(1 - v/V) \quad \text{(factory whistle, car moving)}.$$

If the source is moving with frequency f', the observer hears f:

$$f_{\text{obs}} = f' \frac{V}{V + v} = f' \frac{1}{1 + v/V} \quad \text{(train whistle, listener stationary)}.$$

This last expression can be changed with a binomial expansion. If $v/V \ll 1$, then only the first two or three terms of the expansion are needed:

$$f_{\text{obs}} = f'[1 - v/V + (v/V)^2 - \cdots].$$

Now we can see the extent to which there is symmetry between a moving source and moving observer. If v/V is much less than 1, there is little difference in the frequency changes between the moving source and moving observer.

We claimed that a common example of the Doppler shift is the change in pitch when a train rushes past you. Let us use our equations to calculate the frequency shift that might be observed. If the train is traveling at 60 miles/h, its speed is 27 m/s. Let us take the velocity of sound in air to be 340 m/s and find out what happens if the train's whistle has a basic frequency of 1000 Hz. Since $v/V = 27/340 \ll 1$, we could use either formula. However, using the one for a moving source:

$$f' = 1000, \quad f = (1000 \text{ Hz}) \frac{1}{1 \pm (27)/(340)}.$$

We used the expression $\pm$ in the denominator to account for the train moving toward us (minus) or away from us (plus). Since $27/340 = 8\%$, the frequency that we hear as the train approaches is $(1000 \text{ Hz})(1.08) = 1080$ Hz. If we are standing right by the crossing, the shriek of the whistle gets louder as it comes toward us but does not change pitch appreciably. As soon as it passes us, however, it is moving away from us and the frequency drops to 920 Hz. In musical terminology, the basic frequency of 1000 Hz is a little higher than high B (an octave above middle B on the piano). The approaching train sounds like high C sharp and the departing train sounds like B flat. This abrupt shift of pitch is dramatic and easily heard.

In our derivation of the Doppler shift, we made crucial use of the fact that the waves exist in a medium that we can use as an absolute reference frame. It was furthermore crucial to the derivation that the wave velocity measured by anyone moving through the medium would be more or less, depending on whether they were traveling with the waves or against them: $V' = V \pm v$. Consequently, this Galilean transformation cannot be used for light signals. There is no medium in which electromagnetic waves propagate and which serves as an absolute reference frame for all other kinds of motion. The measured velocity of light for any observer in any reference frame, regardless of the relative velocity of the source reference frame, is always $c = 3 \times 10^8$ m/s. If we apply the correct Lorentz transformation to our previous analysis, we get just one formula for the Doppler shift. All that matters is the *relative* velocity between source and observer: there are not two special cases depending on who is moving. The Doppler shift for electromagnetic radiation is

$$f_{obs} = f_{source}[1 \pm v/c + \tfrac{1}{2}(v/c)^2 \pm \cdots].$$

Note that the second-order correction term places the frequency shift just half way between the shift for moving observer and moving source in our previous derivation.

The Doppler shift is a prime tool in determining velocities. Traffic radar depends on it. When a beam of microwaves strikes a moving car and bounces back to the police radar observer, the frequency of the waves has changed because they have been reflected from a moving mirror (the car). The instrument is calibrated to measure the frequency shift in terms of miles per hour of the moving car. (Because of the moving mirror effect, the shift is twice that which would be obtained from a source traveling at the car's speed.) When moving atoms or subatomic particles give off light or x rays, the radiation frequency is affected by the motion of the source.

Perhaps the most dramatic use of the Doppler shift is in astronomy. When light comes to us from a distant star or galaxy, it may contain very little direct geometrical information. With a few exceptions, all stars are truly point sources for telescopes on earth. We cannot see their diameter or structure or internal motions. However, we can spread the light out into a spectrum so that different frequencies of the light appear at different places on our photographic plates.

Much of the light is concentrated in particular discrete frequencies corresponding to certain atomic processes. Each atom produces its own "fingerprint" of certain frequencies. If the light from a star or galaxy contains this same pattern, then we know that contains that particular atom. However, if the star or galaxy is moving away from us, every frequency will be shifted lower and we will see the same pattern but in a different frequency range. Sometimes the spectral pattern from a star will shift from lower frequencies to higher frequencies and back again. We interpret this to mean that the star is revolving around another one so that sometimes it is moving toward us and its frequencies are increasing and later it is moving away from us and its frequencies are decreasing.

The most sensational finding using the Doppler shift has been that the distant galaxies are all moving away from us. Since the spectral patterns move to lower frequencies for a source moving away from an observer, and since in the visible range this means a shift toward redder light, this effect has become known as the redshift. Not only are the distant galaxies moving away from us, but their speed of recession is proportional to their distance from us, as determined by other methods. We have detected systems at the far reaches of the universe whose Doppler shift indicates that they are moving away from us with v/c as large as 0.8. Many attempts have been made to interpret the redshift in terms of some other phenomenon but so far the Doppler shift seems the only reasonable explanation. The Doppler shift has very particular characteristics. Note that $\Delta f \propto f$ as well as $\propto v/c$. In the "redshift," for instance, the photons do not all lose the same energy in traveling toward us from fleeing galaxies. The blue photons have a greater shift, and thus lose more energy, than do the red photons. (The *fractional* loss of energy is the same for all photons.)

■ ACCELERATED REFERENCE FRAMES AND LOCAL FIELDS

Newton was very concerned about the problems of describing motions from different reference frames. His three laws apply strictly only within the reference frame that he proposed—a reference frame at rest or moving with constant velocity with respect to the fixed stars. Such a frame is called *inertial*. Newton's first law can be used as a definition of a reference frame, but not necessarily an inertial one! If *with respect to the observer* an object suffers no acceleration, then the object is subject to zero force *with respect to the observer*. On the surface of the earth, we observe that objects fall down with increasing speed. Nothing is visibly touching the object and so we blame the effect on the force of gravity. We also observe that in the northern hemisphere objects moving along the surface of the earth accelerate to the right with respect to straight lines on the surface of the earth. We blame the Coriolis force for this effect, although it is not a force between "charges" like gravitation or electromagnetism. Alternatively, we can describe the Coriolis force as an "inertial" force arising because the earth is rotating with respect to the fixed stars and does not constitute an inertial reference frame.

Actually, it is not at all obvious what constitutes an "inertial frame." If we claim that an inertial frame is one that is at rest with respect to the fixed stars, we are closing our mental eyes to the fact that no stars are fixed. The term is meaningless. Aside from the rotation of the stars within our own galaxy, and the motion of galaxies with respect to each other, we know that the universe—space itself—is expanding. If we claim that an inertial frame is one in which Newton's first law holds, then what do we say to astronauts in orbit? In their reference frame an apple does not fall. They either claim that there are no forces acting on the apple (and so Newton's first law is satisfied), or if they choose to assign a gravitational influence coming from the distant earth then they can also assign reality to the *centrifugal* field in which they find themselves. If you protest and say, "Ah, no! The *centrifugal* force is fictitious," the astronauts respond, "You think that's fictitious! What about this mysterious inverse square gravitational force?" You see, the astronauts really do believe in the general theory of relativity, which has been our "accepted" theory of gravitation for 70 years. (It is called *modern physics*.) They really believe that "inertial" and "gravitational" fields are really equivalent. Instead of thinking that point objects are attracted by an inverse square force, they think that each object is falling naturally in the curved space-time produced by the mass of the other object.

Now, you may not want to confuse either the astronauts or your students with these subtle problems. Is it not interesting, however, to know that in this most elementary aspect of beginning physics, we are only a step away from profound and ancient and modern questions? And is it not fun to realize that after all these years Newton's laws are not so obvious after all? They never were.

Newton claimed that you could tell whether you were in an inertial reference frame by looking at the surface of liquids. For instance, when you are at rest or traveling with constant velocity, the surface of a liquid is horizontal and flat. Of course, there must be a uniform gravitational field perpendicular to the surface to keep it flat. (Actually, on our spherical earth the field lines are radial; the water surface is *not* flat.) Although you cannot measure absolute velocity, you can determine *acceleration* in the inertial reference frame simply by observing a liquid surface. If you are accelerating forward, the water piles up toward the back, creating a flat surface at an angle to the horizontal. If you spin a pail of water, the surface becomes parabolic. If you were to station an observer on the surface of the rotating water and let her claim that she is at rest, then she would have to explain why all the fixed stars were rotating rapidly around her. As a matter of fact, Ernst Mach (1838–1916) pointed out that it was not so far-fetched for an observer to assume that the universe rotating around her produces centrifugal force. That would explain why the water is pulled outward producing a parabolic surface. After all, the source of inertia is not understood. Perhaps the inertial mass of any object is produced by the gravitational influence of all of the rest of the mass in the universe. An object without any universe around it would have zero mass. Perhaps in the reference frame of an object, a rotating universe produces centrifugal force.

Aside from these real and fascinating philosophical problems, it is frequently convenient to describe phenomena from reference frames that are not inertial. We are frequently in reference frames that are accelerating with respect to the fixed stars. We ride on elevators, go around corners in cars, and map the movement of the winds on our rotating earth. In all these cases, we may find it convenient to deal with inertial forces that are caused by our acceleration with respect to inertial reference frames. These inertial forces cannot be described in terms of "source charges" such as gravitational mass, electromagnetic charge, or nuclear hypercharge. Instead, the inertial forces act on the mass of an object as if they were

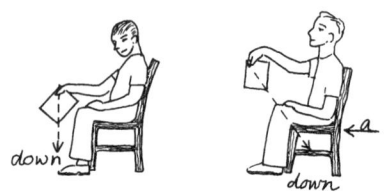

Figure 6.6. The direction of "down" in an accelerated reference frame.

local gravitational fields. These inertial fields—which are very real—are functions of the position or motion of the object within the noninertial reference frame. Here are some examples.

■ NONINERTIAL REFERENCE FRAMES

Example 1

The direction of "down" is determined experimentally by a plumb line. A plumb line does not necessarily point toward the center of the earth. There are local anomalies created by nearby mountains. Besides, the earth is not quite spherical. For most purposes, the plumb line points approximately toward the center of the earth which defines the vertical direction. However, if you hold a plumb line in a car or airplane that is accelerating forward, the line is no longer vertical; it points slightly backward.

This effect is particularly easy to see as you take off in a jet liner. As Betty Wood pointed out in her delightful book, *Science for the Airplane Passenger*, you do not need a builder's plumb line to see the effect of the acceleration field produced by the jet. Instead, you can use the airline magazine from the seat pocket in front of you. Simply hold the magazine at one corner between your thumb and index finger. You will see the corner that is diagonally opposite swing back toward you as the plane accelerates.

One way to describe this phenomenon is with the force diagram shown in Fig. 6.6. The string attached to the bob exerts a force sufficient to balance the downward gravitational force on the bob and also to provide a horizontal force that produces a horizontal acceleration. As you can see from the diagram, the tangent of the angle is equal to the ratio of the horizontal acceleration a and the gravitational field strength g. (Note that it is best to refer to g as a field strength—and not to use the confusing and useless expression "acceleration due to gravity." The units of gravitational field strength are N/kg.) However, suppose that your plane is completely sealed so that you do not know that the airport is rushing backward away from you. The direction of the plumb line *is* the direction of "down." That direction is apparently produced by the vector sum of two gravitational-type fields. The gravitational field of the earth produces a force of mg; a local inertial acceleration field produces a force of ma. Their resultant is the net field. Not only does the plumb line point in a new direction, but the weight of the bob is also slightly greater, as would be indicated by a spring balance.

Example 2

In an elevator, the acceleration is in line with the gravitational field, although its direction may be plus or minus. The direction of the gravitational field is conventionally taken as minus. You have experienced the result of these local fields in elevators that start or stop abruptly. When the elevator accelerates upward you feel heavier; just before it stops at a floor you feel lighter. The effect is easy to measure quantitatively by standing on a bathroom scale while riding an elevator. If other passengers get on, you can devise many interesting explanations of what you are doing. We have had other passengers ride extra floors with us just to observe the effect.

Once again, we have two ways of describing the readings on the bathroom scales in the elevator. First, we can simply use Newton's second law on the person being weighed. As you can see in Fig. 6.7, there are two forces in the vertical direction that act on the person standing on the scale. The gravitational attraction of the earth is pulling down with a force equal to mg. The springs on the scale are pushing up with a force that can be read from the dial. Since it is well known that bathroom scales never lie, that force is equal to the person's weight W (but in the opposite direction). According to Newton's second law:

$$F_{net} = ma,$$

$$F_{spring} + mg = ma.$$

If you know your mass m and the value of the local gravitational field g you can determine your upward acceleration by reading your weight on the scale: $W = m|\mathbf{a} - \mathbf{g}|$. For instance, some elevators can produce an upward acceleration of 1 m/s^2. If your weight while standing still is 500 N, your weight when the elevator starts up is $W = (500/g)[+1.0 - (-9.8)] = 550$ N.

The other way of explaining the experimental results is to conclude that the local gravitational field changes from time to time. At the moment when the first floor of the building leaves you and hurtles downward (as seen in your reference frame), the gravitational field increases. As the second floor arrives opposite you, slowing down to stop, your local gravitational field decreases. The net field acting on your mass is the sum of the earth's gravitational field and the acceleration field. If the outside world has an acceleration, a, the resultant field is $\mathbf{g} + \mathbf{a}$. If the outside world is accelerating downward (while you go up relatively), your weight will be $W = m[(-1) + (-9.8)]$. As the outside world slows down opposite your elevator, its acceleration must be positive. According to the bathroom scales, and your own feeling, your weight is now only $W = m[(+1) + (-9.8)]$. Notice that the quantitative conclusions are the same regardless of which mental model we use to describe the elevator trip.

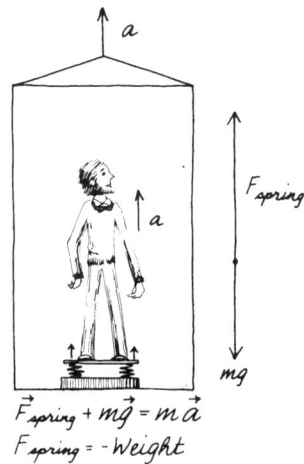

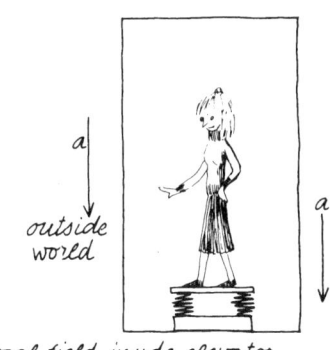

Figure 6.7. Weight in an elevator.

Example 3

Physics courses and physics textbooks frequently warn against the fallacy of thinking that there are centrifugal forces. Centrifugal forces are alleged to be merely fictitious forces. Indeed, you can describe the events in a rotating system by removing yourself from the action and standing above it all in an inertial reference frame. Take the case of the passenger in a car that is careening in a left-hand turn. The car door flies open, the passenger flies out. From your sublime and inertial vantage point, you can explain that the car door no longer provided sufficient centripetal force to keep the passenger going in a circle. From the passenger's point of view, of course, the door flew open and he shot out radially. You would have a hard time convincing the coroner's jury that the death was caused by a fictitious force.

Which reference frames we use for our descriptions of phenomena is a matter of convenience. There is never any conflict between descriptions from different reference frames *as long as we do not mix them up*. For instance, as we on earth observe a satellite in orbit around us, we say that it is gravitational force that provides the centripetal force needed to maintain the orbit. There are no other forces and the astronauts feel weightless simply because they and their surround-

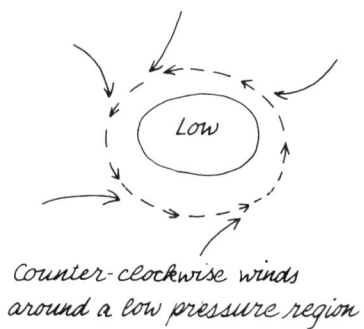

Counter-clockwise winds around a low pressure region

Figure 6.8. The effect of Coriolis force on wind circulation in the northern hemisphere.

ings are all in continual free fall with common acceleration toward the earth. On board the closed satellite, the astronauts observe that they and their test objects are not accelerating. Therefore they may conclude that they are not subject to any external forces. However, if they know that the earth is down there exerting a gravitational pull, they can propose that there must be an equal and opposite centrifugal force out the other way. That is why they are weightless. It is a perfectly reasonable and consistent explanation. Right here on earth we live on a giant merry-go-round, and it is common practice to name and use the inertial forces just as if they were real in the Newtonian sense. For instance, we lose weight because of the centrifugal force due to the earth's rotation. It is not much, of course. On the equator it is only $\frac{1}{3}$%.

Example 4

On the rotating earth there is an inertial force that plays a far more important role than the centrifugal force. The Coriolis force acts on any object that moves in the rotating system. In the northern hemisphere the Coriolis force acts to the right on an object moving along the surface of the earth, and always to the left on an object moving in the southern hemisphere. It is the Coriolis force that creates the great cyclonic motions of the atmosphere as the winds move from regions of high pressure to low pressure. In Fig. 6.8 we show how winds moving from all directions toward a low-pressure region in the northern hemisphere produce a counterclockwise circulation; around a high-pressure region, a clockwise circulation is set up. Hurricanes follow the same directional rule, although the sense of rotation of the smaller and more intense tornadoes is largely determined by chance.

The origin of the Coriolis force can be understood in terms of the requirements of conservation of angular momentum. Suppose for instance that a cannon shoots a shell vertically upward from the equator. If the shell is lobbed up only a few feet, we would expect it to fall right back down into the mouth of the cannon. To be sure, the earth is rotating, but the shell shares in that tangential velocity as it leaves the mouth of the cannon. If the shell rises to a great height, however, another effect begins to be important. The shell must maintain its angular momentum with respect to the earth. That angular momentum is given by $mvr = mr^2\omega$. As the distance r from the axis *increases*, the angular velocity ω must *decrease*. Consequently, the angular velocity of the shell is less than that of the cannon throughout the entire trajectory. It will fall to the west of the cannon. On the other hand, a ball dropped from the top of the mast of a sailing vessel will fall slightly to the east of the mast.

Now consider what happens when the wind blows due north in the northern hemisphere. Since the wind follows the curvature of the earth, it is getting closer to the axis of the earth. Therefore, the angular velocity must increase, since r is decreasing. This increase of angular velocity will divert the wind to the east, which is to the right of its northward motion. Similarly, if the wind is attempting to blow due south, it will follow the curvature of the earth and move to regions of greater r from the axis. Therefore, its angular velocity must decrease, making it slip to the west. Once again this direction is to the right of its main motion. For motion in the east direction, the surface velocity adds to that of the rotational velocity due to the turning of the earth. In the reference frame of the earth that creates additional *centrifugal* force. That forces the winds out radially, which on the surface is translated to a force up to the right (south), where the radial distance to the axis is larger. For motion in the westerly direction, the surface velocity reduces the centrifugal force, letting the winds slip down to the north, thus closer to the axis.

A complete analysis of the Coriolis force on the rotating earth shows that it is equal to

$$\mathbf{F}_{\text{Coriolis}} = 2m\mathbf{v} \times \boldsymbol{\omega},$$

where $\mathbf{v}$ is velocity (with respect to earth) and $\boldsymbol{\omega}$ is the angular velocity *of the earth*. Notice in this formula that the Coriolis force is 0 for *tangential* velocities north or south on the earth's equator. For such motion the velocity is parallel to the direction of rotation, $\boldsymbol{\omega}$. If the two vectors are parallel, their cross product is **0**. At the poles the Coriolis force is maximum, since any tangential velocity is automatically perpendicular to the direction of the rotation. Heading northward in the northern hemisphere, there is a component of tangential velocity in the direction of the axis and a component perpendicular to the axis that produces the force to the right. Similarly, in moving southward, the perpendicular component of velocity is away from the axis. The resultant force is once again toward the right.

There is a legend that wash bowls in the northern hemisphere drain counterclockwise while wash bowls in the southern hemisphere drain clockwise. Let us use the formula for the Coriolis force and some reasonable estimates to see whether the Coriolis force in a wash bowl is sufficient to produce such rotational motion. Let us compare the Coriolis *field strength* with the gravitational field strength. The ratio is $2\omega v/g$. The angular velocity is that due to the earth. Its magnitude is 2π radians per 24 hours, which is equal to 7.3×10^{-5} rad/s. The tangential velocity of water in a wash bowl in the United States is not perpendicular to the axis of rotation of the earth as it would be at the North Pole. Let us estimate that the perpendicular component is about one-half the velocity. That takes care of the factor of 2 in the Coriolis force equation. Now we must estimate the velocity of water in an average wash bowl as it moves in toward the center. Let us say 1 m/s? In that case, the ratio of the Coriolis force to the gravitational force on any chunk of water is about equal to

$$\frac{F_{\text{Coriolis}}}{F_{\text{grav}}} \approx \frac{(7 \times 10^{-5} \text{ rad/s})(1 \text{ m/s})}{9.8 \text{ m/s}^2} = 7 \times 10^{-6}.$$

The relative importance of the Coriolis force in wash bowls is negligible. The actual drainage circulation, clockwise or counterclockwise, is determined by asymmetries in the shape of the bowl or by the long-term memory of the water concerning the direction in which it swirled as it filled. Very careful experiments were done in the 1960s by a group from the Massachusetts Institute of Technology to observe the wash bowl effect. They used special hemispherical bowls that were machined to be very symmetrical. They also let the water rest for long intervals to eliminate any motions remaining from the filling process. Under these special laboratory conditions, they observed that the drainage circulation was generally counterclockwise in Boston. They shipped the device to Australia and, sure enough, the drainage there was clockwise.

Example 5

Serious proposals are being made to establish space colonies that can house many thousands of people. These colonies would be located in orbits between the

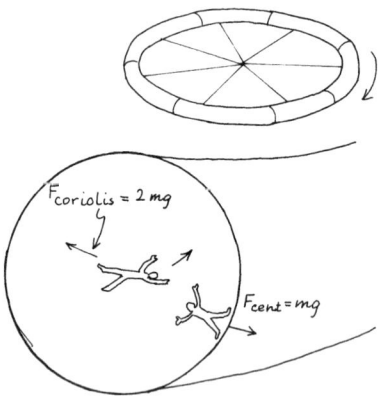

Figure 6.9. Effect of Coriolis force as seen by two different observers in rotating torus.

earth and the moon. Since the inhabitants would have to live in the space colonies for many years, if not permanently, living conditions would have to be pleasant and in many ways earthlike. In particular, humans and other living things adapted to the earth expect to live in a gravitational field equal to 9.8 N/kg. To produce this effect in the colonies, the habitations would be giant cylinders, toruses, or spheres rotating at constant velocity. The centrifugal force experienced by the inhabitants—who would live on the inside of the shells—would seem to them like normal earth gravity. A cylinder or torus with a diameter of 1 km would have to spin at a frequency of 0.14 rad/s to yield a centrifugal field of 9.8 N/kg. The period is 45 s.

One of the advantages of living in such a colony would be that you would have easy access to regions near the axis of the rotating system where "gravity" would be less. This feature opens up whole new realms of sports possibilities. On the other hand, because of the comparatively large value of ω ($\omega_{earth}=7.3\times10^{-5}$ rad/s) the Coriolis forces would introduce other, larger complications that we are not used to on earth. For instance, tangential motion along the perimeter of the habitat at only 10 m/s (22 miles/h) would produce a radial Coriolis field of $2(0.14$ rad/s$)(10$ m/s$)=2.8$ m/s$^2=2.8$ N/kg.

There is an instructive reference-frame problem, illustrated in Fig. 6.9, associated with the strong Coriolis fields in a rotating space colony. Suppose the 1-km torus has been made and has been set spinning but has not yet been filled with air. An astronaut is inside the tube and is at rest *with respect to an inertial reference frame*. To an observer standing still on the inner wall (in a comfortable field of 9.8 N/kg), the astronaut appears to be balanced in space inside the tube and racing along the circumference at a speed of $(0.14$ rad/s$)(500$ m$)=70$ m/s. But why, the observer wonders, does not the astronaut fall? He is in a centrifugal field of 9.8 N/kg. However, because of his tangential motion he is in a self-created Coriolis field of $2(0.14$ rad/s$)(70$ m/s$)=19.6$ N/kg, directed radially inward. That explains it, the observer reasons. The net field on the astronaut is 9.8 N/kg directed inward, and that field provides the *centripetal* force needed to keep the astronaut going in a circle along the circumference of the torus. Note that in this case the easiest reference frame to use to explain the astronaut's motion is inertial. However, the rotating reference frame provides consistent explanations. In the colony, centrifugal and Coriolis forces would be as real as gravitation. Here on our rotating earth, hurricanes are very real, and are most easily explained on weather maps by appealing to the Coriolis "counterclockwise around a low."

■ THE GALILEAN TRANSFORMATION IS ONLY RELATIVELY TRUE

The Galilean transformation depends crucially on the assumption that time intervals are the same in all reference frames: $\Delta t=\Delta t'$. This does not mean that if it is 10 o'clock in New York, it must be 10 o'clock in Milwaukee; but it does mean that if an event takes 10 seconds in one reference frame, it takes 10 seconds in the other reference frame. This assumption is so obviously true that it would not be worth mentioning if it were not indeed false. Note that this assumption led directly to the conclusion that relative velocities add algebraically. But they do not! Maxwell's equations call for a very specific speed for electromagnetic waves—for light. The equations say nothing, however, about the reference frame in which this velocity is to be measured. Apparently the equations hold in *any* reference frame and in *all* reference frames. Consequently, in a vacuum, light always travels at 3.0×10^8 m/s, whether the light has been emitted from a source down the hall or from a quasar moving away from us at half the speed of light. This unique speed is the same regardless of the speed of the light source or the speed of the observer who actually makes the measurement.

So much for the Galilean transformation. It is not really true. It is only an approximation to the more general case of the Lorentz transformation:

$$x' = \gamma(x - vt), \quad t' = \gamma\left(t - \frac{v}{c^2}x\right),$$

where

$$\gamma = \frac{1}{\sqrt{1 - v^2/c^2}}.$$

Note that for $v^2/c^2 \ll 1$, the Lorentz transformations collapse to the Galilean, low-speed approximation. The Lorentz transformations take us into all the paradoxes and fun of the special theory of relativity. On the other hand, the Galilean approximation has its charm and its relative usefulness. In our slow everyday world, our velocities add algebraically, the approaching train whistle has a higher, lonesome pitch, and in the northern hemisphere, the winds blow counterclockwise around the low.

■ EINSTEIN'S SPECIAL THEORY OF RELATIVITY

Many experiments have demonstrated that the speed of light is independent of the velocity between source and observer. For instance, we can see many double stars in the heavens. If twin stars rotate around each other at high speed, we may see them first as a double star, then as a single star when one passes behind the other, and then as a doublet again. In such cases, we have light sources swinging toward us and away from us at high speeds (as they go around each other), and these light sources are also turning off and on regularly (as they go behind each other). If the light from one of the stars swinging away from us travels slower, we would see a delay in its becoming obscured by the front star. If its light traveled faster as it swings toward us, the effect would be very strange indeed. The effect would depend on the distance from us of the twin stars. We might end up with "ghost" stars—sometimes the doublet would become a triplet. Note also that experimental observation of twin stars demonstrates that in a vacuum all colors of light travel at the same speed. Otherwise the eclipsing twins would be seen in rapidly varying colors—a remarkably pretty effect if it happened. It does not.

The explanation of the proper behavior of twin stars might be that light always travels at the same speed, independent of the speed of the source, because light is some sort of wave motion in stationary space. After all, that is the way sound travels in air. Regardless of the speed of a train toward an observer or away from him, an observer measures a constant speed of sound from the whistle (but the pitch may change). If this explanation applied to light, an observer on earth could detect a change in the velocity of light caused by the earth's motion through space. If the light waves exist in the medium of space, like sound waves exist in the medium of air, then the speed of light in the direction of the earth's motion should be different from the speed of light perpendicular to that direction.

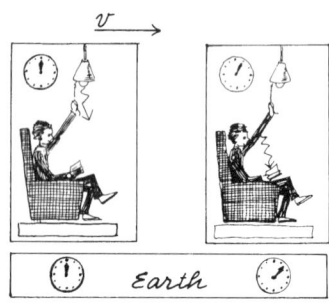

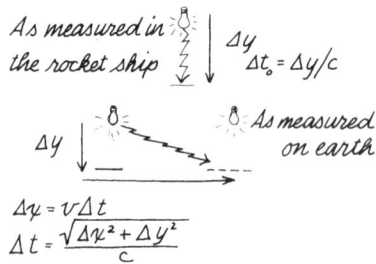

Figure 6.10. Time interval between two events as seen by observers in two different reference frames.

Suppose that we do an experiment to measure the effect of the earth's motion on light traveling in the same direction and also perpendicular to that direction. To guard against the possibility that during the first measurement we were at the moment relatively motionless in space, we should do the experiment again six months later. At any given time, we on earth are traveling 120 000 miles/h (6×10^4 m/s) opposite to the direction we were traveling six months before. If by chance we were motionless the first time, then surely we will be going 6×10^4 m/s with respect to space the second time.

The experiment that answered this question was first done almost 100 years ago in Cleveland, Ohio. This was the famous Michelson–Morley experiment (see p. 530). An optical interferometer was used to compare the flight time of light parallel and perpendicular to the earth's motion. So sensitive was this device that Michelson could have detected a difference of flight time over the several-meter path length of only 10^{-16} s (in 1881, when there were no electronics, no computers, and no National Science Foundation!).

Observations of double stars show that the speed of light is independent of the speed of the source; the Michelson–Morley experiment shows that the speed of light does not depend on the speed of the observer through space. Indeed, this latter experiment implies that it is meaningless to talk about velocity through space. The velocity of one object with respect to another can be measured, but there apparently is no reference frame attached to space itself. (Unless, perhaps, the background 2.7-K radiation left over from the Big Bang at creation constitutes such a reference frame. We can, indeed, detect our movement through it.)

There are numerous other experiments that demonstrate that light has a unique speed. We can also start out with this assumption as a theoretical hypothesis. That is the basis of Einstein's *special theory of relativity*.

A Thought Experiment

If you are sitting down reading a book and turn on an overhead lamp, how long does it take the light to reach your book? Naturally we are not concerned with the experimental situation exactly shown in Fig. 6.10. It usually takes much longer for the lamp to get bright than it takes for the light to travel from bulb to book. Still, light sources can be switched on rapidly enough to make this experiment feasible. Furthermore, modern electronic timing gear is quite capable of measuring such a short time:

$$\Delta t = \Delta y/c = (1 \text{ m})/(3 \times 10^8 \text{ m/s}) = 3 \times 10^{-9} \text{ s} = 3 \text{ ns}.$$

In this calculation we assumed that light was traveling the distance Δy straight down. Suppose that you observe the same experimental situation going on in a rocket ship that is flying past you at the speed v. Just as the ship passes overhead, the light is pulsed and you observe the time on your clock. Some distance along the flight path the light strikes the book and this particular time is noted on a clock directly underneath the ship at that time. The chain of clocks on the ground has been previously synchronized so that the time interval as measured on the ground is equal to the difference in the two clock readings. Once again, the experiment is straightforward and perfectly feasible. You now have two measurements of the time that it takes the light to reach the book—one was made in the rocket ship, just as we calculated above, and the other was made on the ground. The geometry of each measurement is shown in Fig. 6.10. In the rocket the light traveled the distance Δy. As observed from the ground, the light traveled the distance $\sqrt{\Delta x^2 + \Delta y^2}$. But if light always travels with speed c, then the time interval in one frame was different from that in the other!

The questions are: How far did the light *really* travel, and how long did it *really* take? In neither case, of course, does anyone see the light travel along the zig-zag path shown in the diagram. The only observations are that the switch is thrown to pulse the light and a short time later the book (perhaps a photocell) is illuminated. Both observers see the same event, and one set of measuring instruments is just as good as the other. In case you feel prejudiced in favor of the observer on the ground, remember that she and all her chain of clocks might be in another rocket ship with both ships out in space so that there is no way to tell which one is moving. Notice particularly the crucial role of our basic hypothesis. *The speed of light is the same in all reference frames.*

Since we observe the events in a ship going past us (regardless of whether *we* are moving with respect to the earth or the *ship* is moving with respect to the earth, we observe the ship moving past *us*), we must conclude that the light traveled further than simply Δy in reaching the book. Incidentally, we assume that the perpendicular distance Δy appears the same to either experimenter. Therefore, the time interval that we observe, Δt, must be greater than the time interval Δt_0 measured in the rocket ship.

$$\Delta t_0 = \Delta y/c, \quad \Delta t = \frac{\sqrt{\Delta x^2 + \Delta y^2}}{c} = \frac{\sqrt{v^2 \Delta t^2 + \Delta y^2}}{c}.$$

Since $\Delta y = c \Delta t_0$, we can substitute for Δy in the second equation and solve for Δt in terms of Δt_0:

$$\Delta t = \frac{\sqrt{v^2 \Delta t^2 + c^2 \Delta t_0^2}}{c},$$

$$c^2 \Delta t^2 = v^2 \Delta t^2 + c^2 \Delta t_0^2,$$

$$\Delta t = \frac{\Delta t_0}{\sqrt{1 - v^2/c^2}}.$$

The measured time interval *between the same two events* is longer in our system than it is in the rocket ship. The fact that time measured in one system is not the same as time measured in another system is contrary to all our everyday experience. After all, if you travel at high speeds on the highway, you do not expect to have your watch tick at a different rate from what it would if you were standing still. For that matter, the observer in the rocket ship does not notice his watch slowing down. From *our* reference frame, *everything* in the ship is going at a slower rate, including the biological processes such as the heartbeat. It is not just a case of clocks behaving strangely; time itself is different.

Only by comparison of time in two systems traveling past each other do we notice the effect. However, on a highway the effect will hardly be noticeable. The relationship between the two times depends on v^2/c^2. At 60 miles/h (27 m/s), the value of v/c is only 10^{-7}. The factor of v^2/c^2 is therefore only 10^{-14}, giving rise to a negligible difference between time intervals measured in a car and those measured on the road. Because of the dependence on the square of the velocity ratio, the factor remains negligible unless the velocity between the two observers

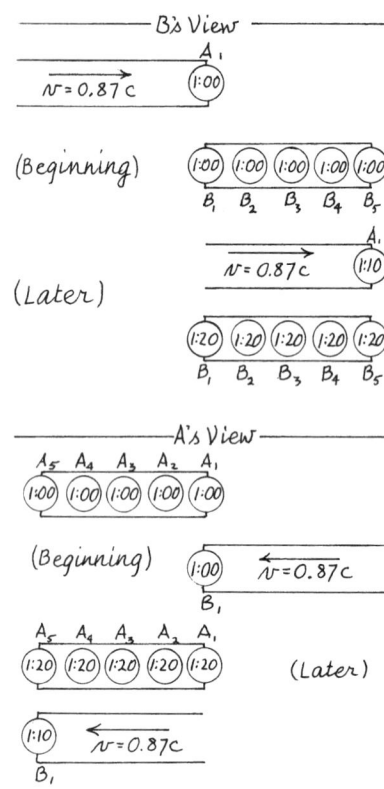

Figure 6.11. Time intervals as seen by two different observers and the problem of clock synchronization.

is very close to that of light. For a 10% effect, $\Delta t/\Delta t_0 = 1.1$ and $v/c = 0.4$. If $\Delta t/\Delta t_0 = 2.00$, $v/c = 0.867$.

An Apparent Paradox

There is an apparent paradox hidden in the situation that we have just described. If rocket ship A goes by rocket ship B, observer B sees that the clocks in ship A are going at a slow rate. For instance, if the relative velocity is 0.867, observer B would claim that, while his clocks ticked off a full hour, the clock in ship A advanced only half an hour. On the other hand, as far as observer A is concerned, she is standing still and ship B is hurtling by her. Observer A also has a chain of clocks strung out in the direction in which ship B is going, and she has carefully synchronized all these clocks. Then the derivation applies to her observations as well, and she observes that ship B's clocks are going slow. Can they both be right? They certainly can!

Let us suppose that just as ships A and B pass each other, both of their clocks (A_1 and B_1) read 1:00 o'clock, as shown in Fig. 6.11. Now as far as observer B is concerned, he will note the times read by his chain of clocks as clock A_1 rushes past. Suppose that when clock A_1 passes the last clock on observer B's chain, clock B_5 reads 1:20. If the velocity with which ship A is passing is about $0.87c$, it will be found by the clock B_5 observer that clock A_1 reads only 1:10. In other words, B observes that ship A's clock is going slower by a factor of 2. Now switch your thoughts over to observer A's system. She agrees that when clock B_1 passed clock A_1, both clocks read 1:00. In her system, ship B is rushing past to the left and when clock B_1 passes clock A_5, clock A_5 reads 1:20 but clock B_1 reads 1:10. As far as observer A is concerned, ship B's clocks are running slow by a factor of 2. What is observer B's response to this? After all, he knows that he has synchronized his five clocks, which is easy to do since they are at rest with respect to him. (This synchronization could be done by bringing all the clocks together and synchronizing them at the same place and time. Then carry them slowly to their stations. As a check when they are in position, station yourself at the midpoint between two of them, and arrange for them to emit light signals at their same clock times. If the signals arrive simultaneously, the clocks must be synchronized—in your reference frame.)

Observer B can also compare his clock B_1 when it passes clock A_5. Sure enough, observer B would agree that clock A_5 reads 1:20 while he reads only 1:10. Does he conclude that now A's clocks must be going *fast* compared with his? Not a bit. After all, as far as observer B is concerned, A_5 is not the clock which he is timing in his B system. (He is timing A_1.) His explanation is very simple; observer A did not properly synchronize her clocks. Up in the A system, however, observer A is making the same complaint about observer B. Neither one is actually to blame. The system of clocks synchronized in one system is not synchronized in another system traveling past it. No mistake is involved; synchronization in both systems at the same time simply is not possible.

Why not use the same synchronization scheme we proposed before? Place both observers at the midpoints of their respective systems: clocks A_3 and B_3. Clocks B_1 and B_5 emit light signals just at the moment when clocks A_3 and B_3 are opposite each other. Clock B_3 receives two signals at the same time a short time later. Observer B therefore says that clocks B_1 and B_5 are synchronized. Meanwhile, however, clock A_3 has been traveling toward the right and so receives the signal from clock B_5 before the signal is received from clock B_1. Therefore, observer A says that clocks B_1 and B_5 are *not* synchronized. It does no good to say that observer A should know she is moving. As far as she is concerned, she is at rest and observer B is moving. Because light travels at the same speed with respect to both ships A and B, both are equally good observers.

Experimental Evidence for Different Clock Rates

Can it really be true that time itself is different in different reference frames? The evidence is so commonplace that we are continually affected by it. Within the last minute several dozen muons have passed through your body leaving trails of ionized atoms and destroyed molecules. These muons are short-lived heavy electrons produced near the top of the atmosphere by incoming cosmic rays. When they are at rest, they decay with a half-life of about 2×10^{-6} s. In that time, traveling close to the speed of light, they could travel 600 m. But since most of them are generated 30 km or so (50 "half-distances") above the earth, essentially none would make it down here before decaying. Nevertheless, the muon intensity at sea level is high. Because they are traveling very fast with respect to the earth and us, their time (as we see it) goes more slowly. The experimental evidence for this effect is solid and quantitatively agrees with the time-relation formula that we have deduced. In every high-energy physics laboratory this time-relation formula must be used in dealing with the experimental effects of many other kinds of particles. It always works.

There is even more direct evidence for the slowing of clocks in systems moving with respect to us. Hafele and Keating describe a time-dilation experiment done with clocks on airplanes. They sent cesium-beam atomic clocks around the world in commercial airplanes. One clock traveled eastward, one stayed home, and one traveled westward. Many corrections were necessary to account for varying speeds and altitudes (an effect involving the general theory of relativity depends on the local value of g). Nevertheless, the effect was clearly measurable. For both traveling clocks time went more slowly than for the earth-bound clock. The eastward flying clock, which had a greater speed relative to the earth-centered inertial system than the westward flying clock, lost more time.

Another Apparent Paradox—Length Measurement

Like many other relativistic applications, there seems to be a paradox in this case of the muon. By realizing that the muon clock is ticking slowly, we can explain how it lives long enough to travel 20 miles. But in the muon's system it still lives only 2×10^{-6} s, and in both systems the velocity, close to the speed of light, is the same. How does the muon explain its ability to travel so far when it lives such a short time?

The apparent paradox with the muon could have been foreseen in our original example of the two clock chains passing each other. Instead of asking observers A and B to compare their clocks, let us see what they conclude about the *length* of each other's clock chains. If observer A wants to measure a length in ship B, she cannot hold a meter stick against ship B since it is racing past. However, both parties will agree that their relative velocity is v (in this case, equal to $0.87c$). If you want to measure the length of an object moving past you at constant velocity, simply time how long it takes to pass, and multiply by the velocity. Then B_1 will

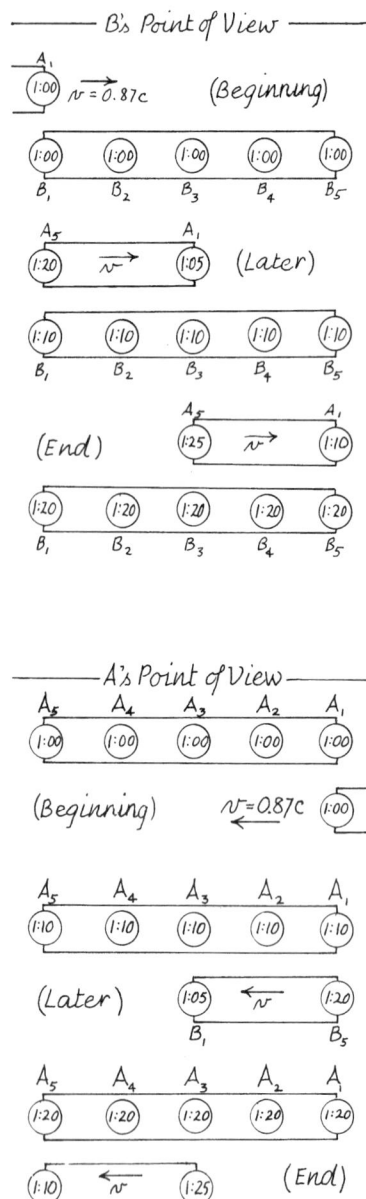

Figure 6.12. Effect of relative motion on length measurements.

observe that the leading edge of ship A arrives at 1:00 and that the tail end, clock A_5, passed at 1:10 (to be sure, the clock A_5 did not read 1:10, but that, according to observer B, is A's synchronization problem). Furthermore, observer B knows that the leading edge of ship A does not arrive at clock B_5 until 1:20. If it takes 20 minutes for the leading edge of ship A to travel the whole length of ship B, but only 10 minutes for the whole object to pass a particular point, then ship A must be only half as long as ship B. The action is detailed in Fig. 6.12.

On the other hand, observer A could use the same arguments and would come up with the conclusion that ship B is only half as long as ship A. Can they both be right? Certainly! The logic is straightforward and inescapable. Once again, the resolution of the paradox is that there is no way to synchronize clocks between moving systems. To compare a length with a standard you must line up both edges of the standard at the same time. When observer B performs his measurement, he would say that he lined up ship A against marks at clocks B_1 and B_3 at the very instant when the leading edge of ship A was at clock B_3 and the tail end of ship A was at clock B_1. The observer in ship A would say that observer B used those marks all right, but had lined up the mark at clock B_3 long before clocks A_5 and B_1 were together. According to observer A, when her clock A_1 read 1:10, clock B_5 was opposite clock A_1 (observer B agrees), and *at the same time* (observer B disagrees) clock B_1 was opposite clock A_3, which read 1:10. Observer B agrees that when clock B_1 passed clock A_3, A_3 read 1:10, but observer B claims that it was really 1:05, which is what the clock B_1 read.

Here is a summary of the times involved:

A_1 passes B_1	A_1 reads 1:00	B_1 reads 1:00
A_1 passes B_3	A_1 reads 1:05	B_3 reads 1:10
A_1 passes B_5	A_1 reads 1:10	B_5 reads 1:20

Both observers A and B agree about all these readings. However, observer B claims that his clocks are synchronized so that the time read at clock B_1 is the same as the time in B_2 and so on. Observer A claims that observers B's clocks are not synchronized.

B_1 passes A_1	A_1 reads 1:00	B_1 reads 1:00
B_1 passes A_3	A_3 reads 1:10	B_1 reads 1:05
B_1 passes A_5	A_5 reads 1:20	B_1 reads 1:10

Both observers A and B agree about all these readings. However, observer A claims that her clocks are synchronized. Observer B claims that A's clocks are not synchronized.

Observer B claims that at 1:10 (according to his clocks) clock A_1 was lined up with clock B_3 and clock A_5 was lined up with B_1. Therefore observer B claims that ship A is only half the length of ship B. On the other hand, observer A claims that at 1:10 (according to her clocks) clock B_1 was lined up with clock A_3, and B_5 was lined up with A_1. Therefore, observer A claims that ship B is only half the length of ship A. Each is right in his and her own system. There is no such quantity as absolute length. If you measure a length to be Δx in a system passing by you, then an observer in that system will measure a longer length, Δx_0. The relationship is

$$\Delta x = \Delta x_0 \sqrt{1 - v^2/c^2}.$$

Remember now the problem of the muon lasting long enough to penetrate the atmosphere. Here on earth we explain that the muon lived long enough to go 20 miles, because time is slowed in a system moving very fast with respect to us. The muon, decaying in only 2×10^{-6} s in its own system, would claim that it can live long enough to get to earth because the distance is really very short. The muon sees the earth traveling upward at close to the speed of light. Consequently the distance to the earth is shrunk to a length in the range of 600 m.

■ THE LORENTZ TRANSFORMATION

Clearly the Galilean transformations cannot describe events satisfactorily in a system moving past us with a velocity close to that of light. With the Galilean transformations, time intervals are the same in all reference frames, and the length of a meter stick remains 1 m. The true transformations between moving coordinate systems are named for the Dutch physicist Hendrik Lorentz (1853–1928). The Galilean transformations are merely approximations that are good if $v/c \ll 1$.

To determine the Lorentz transformations, we will assume once again that the primed coordinate system is moving to the right along the x axis with velocity v. The origins of the coordinate system touch when $t=t'=0$. We can no longer assume, however, that $t=t'$ at any other time or place. However, since the relative motion is along the x axis, we can argue because of symmetry that $y=y'$ and $z=z'$. Since the Lorentz transformations must reduce to the Galilean form at low velocity, it is a reasonable guess that

$$x'=\gamma(x-vt).$$

The constant γ must be some function of v and c such that when v/c goes to 0, γ goes to 1. In such a case, the Lorentz transformation would reduce to the Galilean transformation. The inverse relationship for x must be

$$x=\gamma(x'+vt').$$

To find the value of the constant γ, we now appeal to the basic hypothesis of the special theory of relativity: *the speed of light is a constant and equal to c in all reference frames*. If we start a burst of light at the origin when the origins of the two systems coincide, then the law for the spread of light must be the same in both reference frames:

$$x=ct, \quad x'=ct'.$$

Since we have transformations for x and x', we can substitute these, eliminating x and x' and also getting rid of t and t':

$$ct=x=\gamma(x'+vt')=\gamma(ct'+vt')=\gamma(c+v)t',$$

$$ct'=x'=\gamma(x-vt)=\gamma(ct-vt)=\gamma(c-v)t,$$

$$ct=\gamma(c+v)[\gamma(c-v)t/c],$$

$$\gamma^2=\frac{c^2}{c^2-v^2}=\frac{1}{1-v^2/c^2},$$

$$\gamma=\frac{1}{\sqrt{1-v^2/c^2}}.$$

Notice that the unknown constant of our transformation turns out to be the distortion factor that affects time and length intervals. For $v/c \ll 1$, $\gamma \approx 1$.

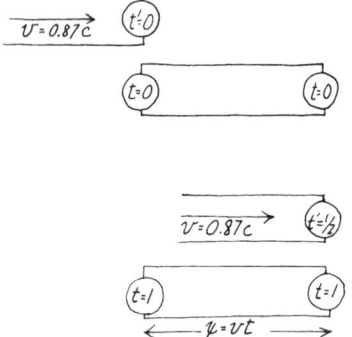

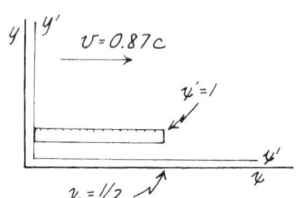

Figure 6.13. Explanation of space and time warping, using the Lorentz transformations.

Now we must find the transformation between t and t'. To find this, we use the original transformation for x and x' and solve for t or t':

$$x = \gamma(x' + vt') = \gamma[\gamma(x-vt) + vt'] = \gamma^2 x - \gamma^2 vt + \gamma vt',$$

$$\gamma vt' = x + \gamma^2 vt - \gamma^2 x = x(1-\gamma^2) + \gamma^2 vt,$$

$$t' = \gamma t + \frac{1-\gamma^2}{\gamma v} x = \gamma \left(t + \frac{1-\gamma^2}{\gamma^2 v} x \right),$$

$$t' = \gamma \left[t + \left(\frac{1}{\gamma^2} - 1 \right) \frac{x}{v} \right] = \gamma \left[t + \left(-\frac{v^2}{c^2} \right) \frac{x}{v} \right] = \gamma \left(t - \frac{v}{c^2} x \right).$$

A similar analysis leads to the corresponding formula for t:

$$t = \gamma \left(t' + \frac{v}{c^2} x' \right).$$

The Lorentz transformations for time show that clocks synchronized in one system are not synchronized in a different system, moving with respect to the first system. For instance, when the origins of our two systems coincide, $x = x' = 0$, and $t = t' = 0$. Suppose that we have a whole string of clocks in our unprimed reference frame that we have carefully synchronized with the method that we described earlier. When the origins of the two systems coincide, all the clocks in the unprimed reference frame will read 0. Not so in the primed frame. Out along the positive x axis, the recorders in *our* frame report that the clocks in the primed reference frame are reading earlier and earlier:

$$t' = \gamma \left(0 - \frac{v}{c^2} x \right).$$

Our recorders along the negative x axis find that the clocks in the primed reference frame register positive times. Yet at all the points along the x axis where we make these observations, our clocks at that instant read 0.

Do the Lorentz transformations account for the space and time warping that we calculated earlier? To see that they do, we must apply them very carefully. Rather than just dealing with the equations algebraically, it is a good idea to draw simple diagrams of the action. In Fig. 6.13 we show the two reference frames after the primed system has been moving to the right for a time, which according to *our* clocks is $t = 1$. All the clocks along the x axis read 1 at that instant since we can operationally define simultaneity within a system. The clocks in the primed system have various readings, but we are interested in the particular clock at the origin, since we know that it read 0 when $t = 0$. The time that it now reads is

$$t' = \gamma \left(t - \frac{v}{c^2} x \right).$$

The distance that the origin of the moving system has gone is $x = vt$. Since $t = 1$,

$$t' = \gamma \left(t - \frac{v}{c^2} vt \right) = \gamma \left(1 - \frac{v^2}{c^2} \right) = \gamma \left(\frac{1}{\gamma^2} \right) = \frac{1}{\gamma}$$

As we saw earlier, if $v/c = 0.87$, then $\gamma = 2$. In this case the clocks at the origin of the primed frame would read $t = 1/2$ when $t = 1$. The clock in the moving frame appears to us to be slow.

Reference Frames and Relativity

Now let us place a meter stick in the primed reference frame, with one end at $x'=0$ and the other at $x'=1$. We will measure the length of this meter stick in the unprimed frame at the instant when the origins coincide. Since $x'=x=0$, all we have to do is to find the position x of the other end at the same instant *in our reference frame*, when $t=0$. The Lorentz transformation is $x'=\gamma(x-vt)$. When $t=0$, $x'=1=\gamma x$. Therefore, $x=1/\gamma$.

For our familiar case when $v/c=0.87$, $\gamma=2$ and $x=1/2$ m. The meter stick in the moving frame appears to us to be shortened to only $1/2$ m. Evidently the Lorentz transformations yield the same shortening of length and slowing of clocks that we found with the earlier arguments.

■ PROPER TIME AND THE INVARIANT INTERVAL

If the time and the distance between two events depend on the reference frame in which we measure them, it may seem impossible to describe the separation between two events. One person claims that two bolts of lightning struck at the same time, only 1 m apart. A person traveling past him would deny that the lightning struck at the same time and would measure a different distance between the two marks where the lightning hit.

There is a somewhat similar problem in describing the length of a stick in ordinary two-dimensional geometry. In Fig. 6.14 we show the stick with one end at the origin of the coordinate system and the other end defined by $x=1$, $y=2$. Then $\Delta x = 1$ and $\Delta y = 2$. If we rotate the coordinate system but leave the stick lying still, we change both the x and y coordinates of the end of the stick. For instance, with the proper rotation, we can make the x component have the length 0. On the other hand, we know very well that the length of the stick itself is invariant and is given by the formula $L = \sqrt{\Delta x^2 + \Delta y^2}$. We can describe this invariance of length by writing $L^2 = \Delta x^2 + \Delta y^2 = (\Delta x')^2 + (\Delta y')^2$. There is an analogous invariant interval in space-time. It is

$$s^2 = \Delta x^2 + \Delta y^2 + \Delta z^2 - c^2 \Delta t^2 = (\Delta x')^2 + (\Delta y')^2 + (\Delta z')^2 - c^2(\Delta t')^2.$$

We should verify that the Lorentz transformations satisfy the invariance condition. Let us take the velocity in the x direction, and start our coordinates so that $x_0 = x'_0 = 0$ and $t_0 = t'_0 = 0$. The Lorentz equations are

$$x = \gamma(x' + vt'),$$

$$t = \gamma\left(t' + \frac{v}{c^2}x'\right),$$

$$x^2 - c^2 t^2 = \gamma^2(x' + vt')^2 - c^2\gamma^2\left(t' + \frac{v}{c^2}x'\right)^2$$

$$= \gamma^2(x')^2 + \gamma^2 v^2(t')^2 + 2\gamma^2 x'vt' - c^2\gamma^2(t')^2 - \frac{\gamma^2 v^2(x')^2}{c^2} - 2\gamma^2 t'vx'$$

$$= \gamma^2[(v^2-c^2)(t')^2 + (x')^2(1-v^2/c^2)] = \gamma^2[-c^2(t')^2(1-v^2/c^2) + (x')^2(1-v^2/c^2)]$$

$$= \gamma^2[-c^2(t')^2(1/\gamma^2) + (x')^2(1/\gamma^2)] = (x')^2 - c^2(t')^2.$$

All is well!

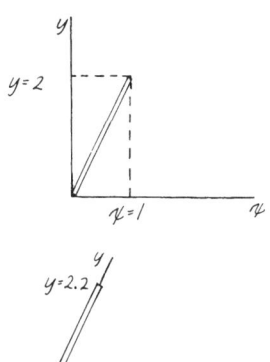

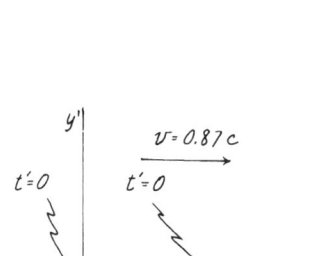

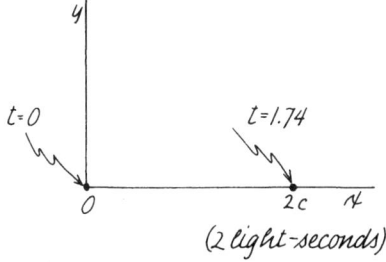

Figure 6.14. Satisfaction of the invariance condition by the Lorentz transformations.

Let us take a specific example of how this invariant interval works. Suppose that the observer in the primed frame claims that lightning struck in two places at the same time. She observes that at $t'=0$, the lightning struck at $x'=0$ and $x'=1$ (one light-second!). The space-time interval between the two events in her frame is given by

$$(\Delta x')^2 - c^2(\Delta t')^2 = (1)^2 - 0 = 1.$$

The observer in the unprimed frame does not agree about the sequence of events. He does agree that a bolt of lightning struck at the common origin $x=0$ at $t=0$, but he claims that the second bolt of lightning struck later in time and farther away. His coordinates for the second bolt of lightning are

$$x = \gamma(x' + vt') = \gamma(1+0) = \gamma,$$

$$t = \gamma\left(t' + \frac{v}{c^2}x'\right) = \gamma\left(0 + \frac{v}{c^2}\right) = \gamma\frac{v}{c^2}.$$

If we use our standard example where $v/c = 0.87$, and $\gamma = 2$, then

$$x = \Delta x = 2, \quad t = \Delta t = 1.74/c,$$

since $x_0 = 0$, $t_0 = 0$.

The invariant interval in the unprimed frame is

$$\Delta x^2 - c^2 \Delta t^2 = 4 - (1.74)^2 = 1.$$

The space-time interval is indeed the same in both reference systems.

Notice how the clock times and distance times compare in the two systems. In the primed frame the observer claims that the events were simultaneous. In the unprimed frame, they are separated by a considerable time. In the primed frame the two marks left by the lightning are separated by a distance of 1. In the unprimed frame the two marks are separated by a distance of 2. Do we have a paradox here? Should we not in the unprimed frame observe that the distance separation in the primed frame is only 1/2? Actually, we do. It is in *our* frame that we measure the separation distance of 2. As we observed the primed reference frame, we agreed that there was a mark made by a lightning bolt at $x'=0$ when $t=t'=0$. However, according to us, the next lightning bolt struck at $t=1.74/c$ and at $x=2$. We know that by that time the origin of the primed frame has traveled a distance of $vt = (0.87c)(1.74/c) = 1.5$. Since, according to us, the lightning struck at $x=2$ and the origin of the primed system was then at $x=1.5$, we claim that the distance between the two marks in the primed frame must be 0.5. As we have seen before, distances in a system moving past us appear to shrink.

The spatial distance between any two events *in* a system is larger than it appears to be when measured from any system moving past it. The time between any two events measured *in* a system (the "proper time") is smaller than it is measured in any other system moving past. The space-time interval between any two events is the same in any system.

■ VELOCITY ADDITION IN MOVING REFERENCE FRAMES

The simple Galilean formula for velocity addition cannot apply when the speeds are close to that of light. For instance, if a distant galaxy is moving toward us with a speed of $\frac{1}{2}c$ and sends light to us with a velocity of c with respect to itself, we still measure the velocity of light to be c and not $\frac{3}{2}c$. The Galilean formula for the addition of velocities must be just an approximation to the true one given by the Lorentz transformations. Let us derive the true law.

The velocity of a BB in the primed frame is found by measuring the distance it travels, $\Delta x'$, during a time interval $\Delta t'$. Its velocity then is $u = \Delta x'/\Delta t'$. We will use the Lorentz equations to find Δx and Δt:

$$\Delta x = \gamma (\Delta x' + v \Delta t'), \quad \Delta t = \gamma \left(\Delta t' + \frac{v}{c^2} \Delta x' \right).$$

The velocity that we measure in the unprimed frame is

$$w = \frac{\Delta x}{\Delta t} = \frac{\gamma (\Delta x' + v \Delta t')}{\gamma \left(\Delta t' + \frac{v \Delta x'}{c^2} \right)} = \frac{\Delta x'/\Delta t' + v}{1 + (v/c^2)(\Delta x'/\Delta t')} = \frac{u + v}{1 + \frac{uv}{c^2}}.$$

Let us see how this true formula for the addition of velocities applies to specific cases. First, note that for the case of BBs the true formula reduces to the approximate Galilean formula. If $u \ll c$ and $v \ll c$, then $w = u + v$. Consider next the galaxy that is flying toward us with a speed of $v = \frac{1}{2}c$. It emits light traveling with a speed in its own reference frame of $u = c$. The velocity that we observe is

$$w = \frac{c + \frac{1}{2}c}{1 + (\frac{1}{2}c)(c)/c^2} = \frac{\frac{3}{2}c}{1 + \frac{1}{2}} = c.$$

Let us take one other example of velocity addition that we can use later. We have already seen that in a reference frame moving past us with a velocity of $0.87c$ ($v^2/c^2 = 3/4$), the Lorentz transformation constant is $\gamma = 2$. Suppose that we have a third system, C, moving with a velocity of $0.87c$ with respect to system B. If system B is moving with a speed of $0.87c$ with respect to us in system A, what is the speed of system C with respect to system A? For ease in doing the calculation, we will work with fractions rather than decimals:

$$v/c = \sqrt{3/4} w = \frac{u+v}{1+uv/c^2} = \frac{2\sqrt{(3/4)}c}{1 + 3/4} = \frac{\sqrt{3}c}{1.75} = 0.9897c.$$

Note that both u and v are very close to the speed of light. Their sum is much closer, but is still less than c. While we are at it, let's compute γ for system C as seen by observer A:

$$\gamma = \frac{1}{\sqrt{1 - w^2/c^2}} = \frac{1}{\sqrt{1 - 3/(7/4)^2}} = \frac{1}{\sqrt{1 - (48/49)}} = \frac{1}{\sqrt{1/49}} = 7.$$

The factor of space warping and time dilation between systems B and C or systems A and B is simply 2; but the γ factor between systems C and A is 7.

■ THE FAMOUS TWIN PARADOX

First, there is no paradox. We will be able to calculate every detail of the event. Here is the story and the problem. There are two twins, a girl (G) and a boy (B). B stays at home, presumably on earth, while G hurtles off at a constant velocity to a distant star. As soon as she gets there, she turns around and returns at the same speed to earth. Ignoring all details of acceleration time and turnaround time, it is clear that when she gets back G is younger than B. Or is it? B argues that G has

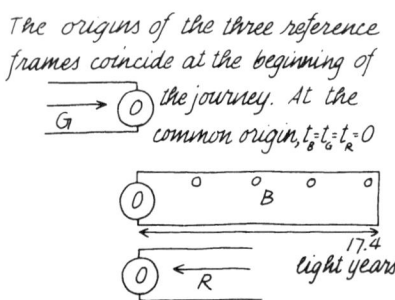

Figure 6.15. Journey of the twin in the twin paradox, the view at the beginning as seen by B.

been traveling continuously at high velocity, except for the one short break at turnaround, and that therefore her clocks have been slow. Therefore, G herself has not aged as much as B. The seeming paradox concerns the symmetry of the situation. Why does not G observe that B went shooting in the opposite direction and then eventually turned around and came hurrying back again? G would argue that B has been traveling at high velocity and therefore his clocks should be slower. Is there a physical difference between the two sequences? Yes, there is. B stays in the same reference frame at all times. G, on the other hand, must switch reference frames. First, she is in a reference frame that is traveling away from B, and then she must transfer to a reference frame that is moving toward B. The transfer between reference frames is a physical event (requiring acceleration, for instance) that can be measured. There is no symmetry between the experiences of G and B. B just sits there for the entire time.

Let us choose some simple numbers and actually calculate the various times and distances as seen by both G and B. Let us arrange for G to travel at a speed of about $0.87c$ so that the space-time warping factor is $\gamma=2$. Therefore, as B observes events in G's reference frame, he will see that her meter sticks are shorter by a factor of 2 and that her clocks run slow by a factor of 2. Of course, G will also observe the same effects as she measures events taking place around B. We will require G to travel for a time of 20 (years) to a distant star. The distance to that star as measured by B is therefore $x=vt=(0.87c)(20 \text{ yr})=17.4$ light-years. During this time G will only age by 10 years, since her clocks run slow by a factor of 2. If G turns around immediately, she will take another 20 years to return home as measured by B. In G's reference frame, however, she will age only another 10 years. At the end of the trip, G will be only 20 years older, while her stay-at-home twin will be 40 years older.

To start the action, we will arrange for the origins of the coordinate systems to be at the same place at $t=t'=0$. The situation *at the beginning* is shown in Fig. 6.15. We record the distances (in terms of an arbitrary unit: $1=17.4$ light-years) along the bottom of each reference frame. Along the top we place the reading of the clocks at each position. Notice that a third reference frame, R (return), also has its origins in common with the first two. This is the reference frame to which G must transfer when she starts her return journey home. The times in this reference frame are crucial because they will be compared with the earth clocks at the end of the journey. Both G and R are traveling with a velocity $v=0.87c$ with respect to B, but they are traveling in opposite directions. As we saw earlier, the relative velocity of G and R is $0.9897c$, and the γ factor between them is 7.

In Fig. 6.16 we show positions and clock times for the three reference frames at the time when G reaches the distant star. The event is shown as viewed from each of the three different reference frames. Note that they all agree about what the actual clock readings and position coordinates are *at the position of the star*. G says that her clock reads 10. She explains that it took her only 10 years to reach the star because from her reference frame the star's distance from earth has shrunk by a factor of 2. The clocks attached to the reference frame at rest with respect to the earth read 20 when the rocket ship arrives at the distant star. The clocks in R read 70. Note that each observer claims that all the clocks in his reference frame are synchronized, but the clocks in the other reference frames are not. Hence, as seen by B, his chain of clocks *all* read 20 at the time when G arrives at the distant star.

How do we know that a clock in R at the position of the star reads 70 when the

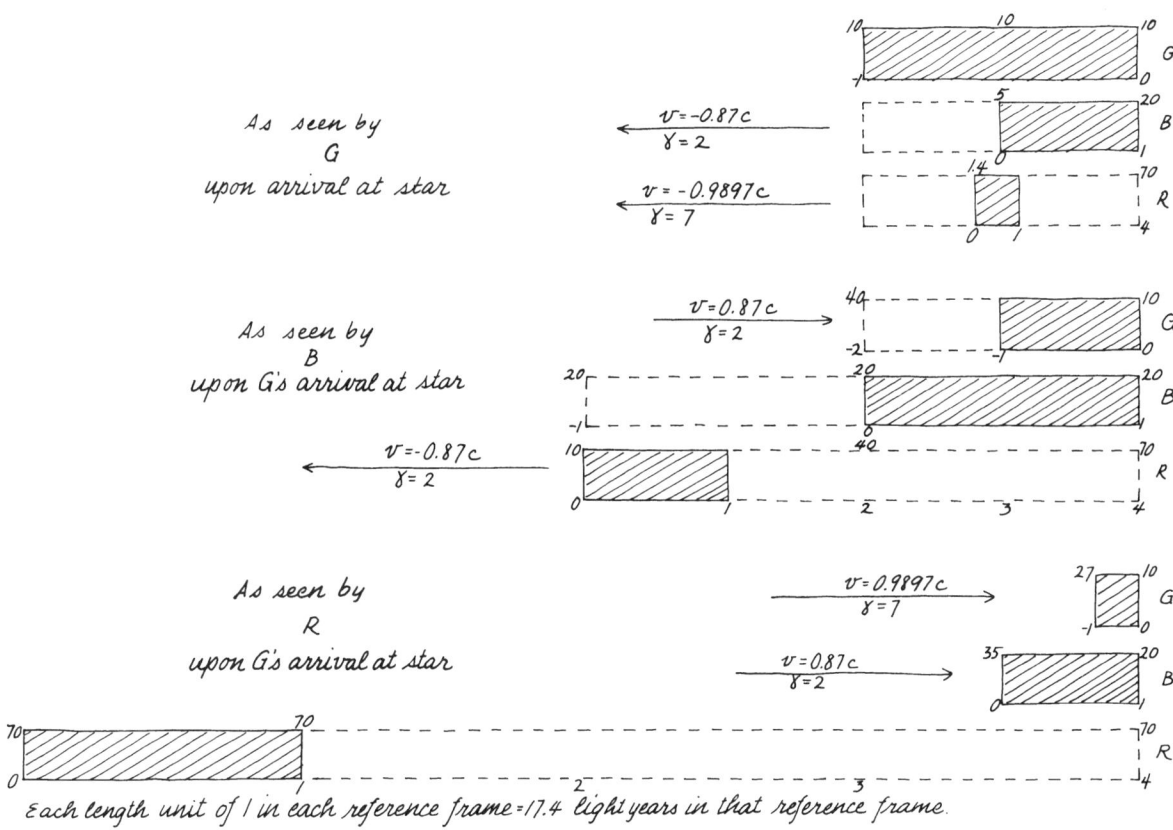

Figure 6.16. Journey of the twin in the twin paradox. Values of variables when G arrives at star. Each length unit of 1 in each reference frame = 17.4 light-years in that reference frame.

rocket ship arrives? We could argue either through symmetry or by using the Lorentz transformation. As seen by B, the clock *at the origin* of R must read 10 when the rocket ship reaches the star. On the other hand, at the origin of B where $x=0$, the time in either of the two moving frames must be $t' = \gamma[t + (xv/c^2)] = \gamma t$. Therefore, the clocks in both G and R passing the origin of B when $t=20$ must read $t'=2(20)=40$. Since in R's reference frame as seen by B, the clock at the left end reads 10 and in the middle reads 40, then at the right end it must read 70.

The Lorentz transformation also predicts that R's clock at the star should read 70 when the rocket gets there. Remember that the appropriate distance unit for x is not 1, but rather 17.4 light-years:

$$t' = \gamma\left(t + \frac{v}{c^2}x\right) = 2\left(20 + \frac{0.87c}{c^2}\,17.4c\right) = 70.$$

The nature of the paradox is seen by the views of G and B when G arrives at the star. B knows that it took G 20 years to get there, but agrees that G's clock reads only 10 years. Now look at the scene as viewed by G. Ten years have passed and G is at the distant star. But according to the string of clocks in her own reference frame, the clock back on earth where $x=0$ in B reads only 5 years.

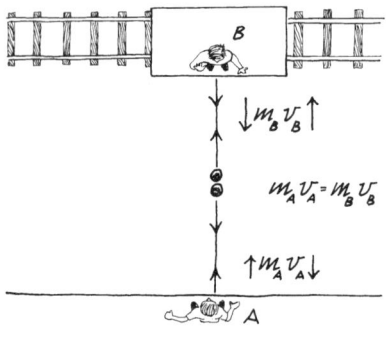

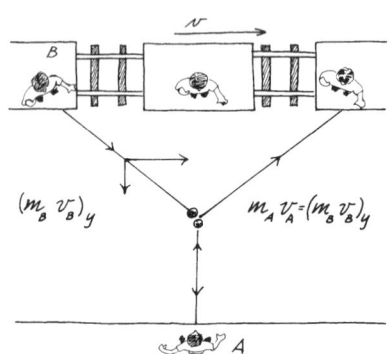

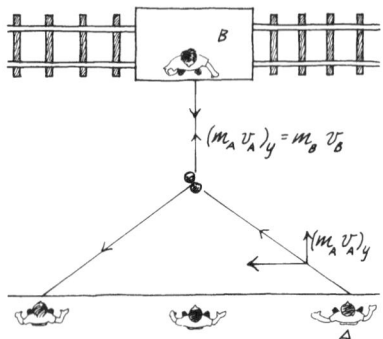

Figure 6.17. Reference frame effects on momentum, mass, and energy.

According to G, the fact that B's clock at the position of the star reads 20 is merely a matter of B's problem of synchronization. The resolution of the paradox comes when G starts home again. To do so she must transfer to the reference frame R. That reference frame had common origins with the others at $t=0$. It is the clock system that G must use on the return flight. Note that there is an abrupt shift of time as far as G is concerned. From a reading of 10 she suddenly jumps to a clock reading of 70. She can note with satisfaction, however, that at that same instant, as seen in her new reference frame R, the clock back on earth, where $x=0$, reads only 35. Since earth seems to be hurtling toward her with a γ of 2, the rate of subsequent clock time should once again be only half of that in her new reference frame.

The arithmetic of the rest of the journey is simply a repeat of the first part. As seen by B, G will return in another 20 years at which time B will be 40 years older than at the start. As far as G is concerned, in her new reference frame R, she need travel only another 10 years to get home. At that time her clock in R will read 80 as she observes the clocks on earth, which have advanced slowly from 35 to 40.

Although G changes reference frames and B does not, and although the necessary acceleration during that change creates a physical difference between the situations of G and B, nevertheless it is not the acceleration itself that produces the sudden change of clock readings between G and R. Acceleration does indeed produce changes in time intervals. This phenomenon is part of Einstein's general theory of relativity. However, in this case, we need appeal only to the rules of the special theory of relativity, which concern the relationships between inertial frames traveling at constant velocity with respect to each other.

■ MOMENTUM, ENERGY, AND MASS RELATIONSHIPS

We have already seen with Galilean relativity that the energy and momentum of an object depend on the reference frame in which we measure them. If we are traveling beside an object, its velocity with respect to us is 0 and so are its momentum and kinetic energy.

We might expect that the Lorentz transformations for momentum and energy would produce some surprises, since time also depends on the reference frame. In Fig. 6.17 we show an experiment that would be difficult to perform even at low speeds but that can be analyzed with both Galilean and Lorentz transformations. Two identical basketballs are chosen and thrown toward each other by two observers. One of the observers is on a train and the other is on the familiar station platform. If the train is standing still, the two skillful experimenters can shoot the balls toward each other with equal speeds and therefore with equal momentum since the basketballs have the same mass by definition. (At rest they are identical.) The balls will meet halfway and bounce back with equal speeds since these are ideal basketballs. The same experiment can be done if the train is moving past the station at constant velocity. Of course, the timing is more crucial. Each experimenter still shoots the basketball with the same speed that he produced in the first case, but this time the observer on the platform sees a triangular trajectory for the ball coming toward him from the train, as shown in the upper diagram. We know that the observer on the train sees the inverse of the situation as shown in the lower diagram. Nevertheless, in Galilean physics the y component of the ball's velocities both going and coming will be the same in each reference frame.

However, if the relative velocity between station and train is very great, time will be different in the two frames. Experimenter A on the platform observes that the clock of the moving experimenter runs slow. The y component of the ball's velocity coming from B is therefore smaller since everything in the B system is moving more slowly. Nevertheless, the balls hit as before, and A's ball returns with its usual speed. Since momentum must be conserved, it must be that the relativistic mass of B's ball as measured by A has increased by the same factor by which its transverse velocity has decreased. The true momentum of an object is not mv but rather

$$(\text{Momentum}) = \gamma(mv) = \frac{mv}{\sqrt{1-v^2/c^2}}.$$

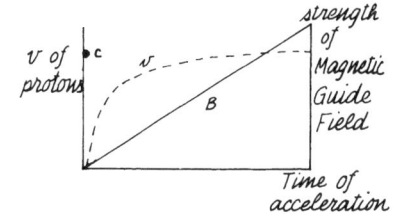

Figure 6.18. Velocity of protons subject to constant acceleration in forward direction (at each revolution) in synchrotron.

The relativistic increase of momentum is sometimes thought of as a relativistic increase in the mass of the object. However, the only way to measure the mass of a moving object is through a momentum exchange experiment. Experimentally, all that we can measure is the relativistic increase of momentum as a whole. And experimentally we do measure it—not, of course, by bouncing basketballs. Instead, we observe the effect in the equivalent experiments with subatomic particles. For instance, as electrons or protons are accelerated in circular guide fields, their velocity rapidly approaches that of light. Energy and momentum continue to be fed into them, but their velocity can increase very little. Nevertheless, it is necessary to keep right on increasing the strength of the magnetic guide field because the momentum of the particles continues to increase as γ increases. The situation is illustrated in Fig. 6.18.

The particles in an accelerator not only increase in momentum but also in energy. Evidently the formula for kinetic energy cannot be simply $E_{kin} = \frac{1}{2}mv^2$. We cannot account for the increased energy simply by multiplying the kinetic energy by γ. Instead, we must make use of a new formula, which we will not derive, relating the energy of an object to its momentum and its rest mass. The rest mass m is just the ordinary mass that is measured in a static experiment with the object at rest. The energy–mass–momentum relationship is

$$E^2 = p^2c^2 + (mc^2)^2,$$

where p is the momentum $= \gamma(mv)$.

The *form* of this equation is something like that of the Pythagorean theorem: $c^2 = a^2 + b^2$. It appears that momentum and the rest mass of an object are associated mathematically like two components of a vector. The magnitude of the resultant is the total energy of the object, including both its rest mass and its motion energy. Consider the consequences. According to our formula, an object has energy even if it is standing still with respect to us: $E = mc^2$.

Here we have the statement of the *equivalence* of energy and mass. Note that it is not a case of being able to turn energy into mass or mass into energy. *Energy and mass are the same thing.* We could and do measure energy in mass units of kilograms (or the equivalent), and we could and we do measure mass in energy units. The transformation equation is basically the same as the one transforming feet into yards: 1 yard = 3 feet. In the case of mass–energy, the conversion factor has the dimensions of velocity squared. However, there is nothing fundamental in this situation; it results from the units originally chosen for mass and energy.

Let us calculate the energy equivalent of 1 kg:

$$E = mc^2 = (1 \text{ kg})(3 \times 10^8 \text{ m/s})^2 = 9 \times 10^{16} \text{ J}.$$

The power complex at Niagara Falls produces about 10 000 MW. The system could produce 9×10^{16} J in about 10^7 s, which is about 4 months.

In an atomic bomb explosion less than 0.1% of the rest mass-energy turns into other forms of energy. But that is usually sufficient. When antimatter annihilates with matter, the conversion of rest mass-energy to other forms can be complete. For instance, when a positron, which is an antielectron, comes to rest in the vicinity of an ordinary electron, they momentarily form a hydrogenlike atom called positronium. When they mutually (because of their equal mass) descend to the $n=1$ configuration, their probability functions overlap and they annihilate each other. Their total mass–energy is converted to the energy of two γ rays. These shoot out in opposite directions, thus conserving momentum. Each γ ray shares equally in the available energy. The mass of an electron or positron is 9×10^{-31} kg. The energy equivalent is

$$E = (9 \times 10^{-31} \text{ kg})(3 \times 10^8 \text{ m/s})^2 = 8 \times 10^{-14} \text{ J} \left(\frac{1 \text{ eV}}{1.6 \times 10^{-19} \text{ J}} \right) = \tfrac{1}{2} \text{ MeV}.$$

We have calculated the mass, in energy terms, of one electron. This is also the energy of one of the γ rays produced in an electron–positron annihilation. We must not conclude, however, that "mass has been converted into energy." Mass and energy are names for the same thing. The original mass–energy of the electron and positron was concentrated in two small regions of space. After the annihilation the same amount of mass-energy existed in the two γ rays streaking away at the speed of light. To be sure, γ rays or photons have no rest mass since they cannot exist at rest. Nevertheless, they carry momentum and energy and, hence, the original mass has not been lost. The same thing is true of an atomic bomb explosion. The small fraction of mass that turns into other forms of energy has not been lost but, instead, is scattered throughout the countryside, at first in the form of kinetic energy of the exploding particles. The inverse process takes place when a high-energy x ray materializes into an electron–positron pair. In this case, the energy of the photon provides the rest mass-energy of the electron and positron and also their kinetic energy.

We have asserted that the total mass-energy of an object is composed of two parts: a term that depends on the momentum and a term that depends on the mass of the object measured when it is at rest. The total mass-energy can also be expressed in another way:

$$E = \frac{mc^2}{\sqrt{1 - v^2/c^2}} = \gamma m c^2.$$

Let us demonstrate that the two formulas for total energy are equivalent:

$$\gamma^2 (mc^2)^2 = E^2 = p^2 c^2 + (mc^2)^2 = \gamma^2 (mv)^2 c^2 + (mc^2)^2.$$

In the equation above, we have substituted on the left our alternative expression for E and on the right have made the substitution that the momentum of an object moving at high speeds is $\gamma(mv)$. Now let us gather terms and reduce the equation:

$$(\gamma^2 - 1)(mc^2)^2 = \gamma^2 (mc^2) m v^2,$$

$$\gamma^2 (mc^2 - mv^2) = mc^2,$$

$$\gamma^2 (c^2 - v^2) = c^2,$$

$$\frac{c^2-v^2}{1-v^2/c^2} = c^2,$$

$$1=1.$$

Evidently the two expressions for mass-energy are the same. *However*, if a photon has no rest mass since it can never be at rest, how can it have any energy if $E = \gamma mc^2$? In this form, E is an indeterminate quantity because it has 0 in the numerator and 0 in the denominator ($\gamma = 1/\sqrt{1-c^2/c^2}$). The energy of a photon is best expressed by the first relativistic expression for the total energy: $E^2 = p^2c^2 + (mc^2)^2$. Since $m=0$ for the photon, the energy of the photon is simply $E=pc$. Note that this formula gives us the magnitude of the momentum of a photon: $p = E/c = h\nu/c = h/\lambda$.

We can get some useful insights into the relative sizes of rest mass-energy and other forms of energy by performing an algebraic expansion on our second formula for E:

$$E = \gamma mc^2 = mc^2\left(1-\frac{v^2}{c^2}\right)^{-1/2}.$$

The standard expansion of a binomial is

$$(1-x)^{-n} = 1 + nx + \frac{n(n+1)}{2}x^2 + \cdots, \quad x<1$$

$$E = mc^2 + \tfrac{1}{2}mv^2 + \tfrac{3}{8}mv^2\left(\frac{v^2}{c^2}\right) + \cdots.$$

The total mass-energy of an object consists of its rest mass, mc^2, plus its ordinary kinetic energy, $\tfrac{1}{2}mv^2$, plus other terms that depend on higher powers of v^2/c^2. For ordinary velocities, we can neglect all of the kinetic-energy terms except $\tfrac{1}{2}mv^2$. Let us compare the sizes of the first three terms for the case of a very fast jet plane with $v = 10^3$ m/s. For 1 kg of the plane, the first term is 9×10^{16} J. The value of the second term is 5×10^5 J. The value of the third term is 4×10^{-6} J. The third term is less than the ordinary kinetic energy by a factor of 10^{11} and so can be neglected. But notice that the ordinary kinetic energy is smaller than the rest mass-energy by a factor of 10^{11} also. Why do we not just neglect the kinetic energy in ordinary motions? The reason is that in ordinary interactions the rest mass-energy remains essentially the same for all of the participants before and after the interaction. Therefore, it always cancels out.

Actually, the rest mass-energy *does* change during ordinary interactions. There are both mechanical and chemical methods of feeding energy into objects. A stretched spring, for instance, has extra energy and therefore extra mass compared with its unstretched state. Let us calculate the extra mass. A good spring with a mass of 1 kg might require an *average* force of 2000 N to compress it by 0.1 m. The stored energy is

$$\Delta E = \bar{F}\Delta x = (2 \times 10^3 \text{ N})(1 \times 10^{-1} \text{ m}) = 200 \text{ J}.$$

The mass increase of the compressed spring is therefore

$$\Delta m = \frac{\Delta E}{c^2} = \frac{200 \text{ J}}{9 \times 10^{16} \text{ m}^2/\text{s}^2} \approx 2 \times 10^{-15} \text{ kg}.$$

Clearly, no laboratory balance can measure such a slight increase in mass.

If you feed energy into an object and make it hotter, you can also increase its mass. Of course, you are producing higher velocities; the molecules or atoms are vibrating faster in their locked positions. We might expect therefore that their mass would increase, but note that such an increase affects the *first* term of the relativistic expansion for energy. The rest mass of the object as a whole is actually greater. For example, it takes about 2×10^5 J of thermal energy to make a 1-kg metal ball red hot. The extra mass of the ball is

$$\Delta m = \frac{2 \times 10^5 \text{ J}}{(3 \times 10^8 \text{ m/s})^2} \approx 2 \times 10^{-12} \text{ kg}.$$

Once again, the increase in mass is too small to be measured.

Let us calculate the mass loss for one of the most energetic chemical interactions that we can produce. The energy provided in the production of 1 kg of water, by burning hydrogen and oxygen, is 1.6×10^7 J. This energy release corresponds to a loss (or dissipation) of 1.8×10^{-10} kg. Even this much of a mass change is too small to be detected by the best precision chemical balance, although it is getting in the range.

■ MASS-ENERGY AND THE LORENTZ TRANSFORMATION

In space-time, as we have seen, there is an invariant interval between two events, regardless of which inertial reference frame you are in. The quantity $\Delta x^2 + \Delta y^2 + \Delta z^2 - c^2 \Delta t^2$ has the same value in your reference frame as it does in any other inertial frame. There is a similar invariant quantity concerning the momentum and energy of an object. This quantity is

$$E^2 - c^2 p^2 = (mc^2)^2.$$

The invariant quantity is just the rest mass-energy of the object. That is the mass-energy that we measure when the object is motionless with respect to us. From a reference frame moving rapidly past us, the object would seem to be in motion. Its total energy E would be greater, but it would also have momentum p. If the other observer measured both E and p for the object and calculated the quantity $\sqrt{(E^2 - c^2 p^2)}$, he would find just the value that we measure to be the rest mass-energy.

Total energy and momentum for an object are related in the same way as the space-time coordinates for the object are related. If you measure E and p in one inertial frame, you can transfer their values for any other inertial frame through the Lorentz transformations. The momentum variable, pc, transforms in the same way as the position coordinate x. The energy variable, E/c, transforms similarly to the time t. For instance,

$$x' = \gamma (x - vt) \rightarrow p'c = \gamma \left(pc - v \frac{E}{c} \right).$$

Therefore

$$p' = \gamma \left(p - v \frac{E}{c^2} \right).$$

The Lorentz transformations for the two sets of variables look like this:

$$x' = \gamma(x - vt), \quad p'_x = \gamma\left(p_x - \frac{v}{c^2}E\right),$$

$$t' = \gamma\left(t - \frac{v}{c^2}x\right), \quad E' = \gamma(E - vp).$$

Actually, there are four equations for each of these sets. For real and more complicated geometry we would have to add the transformations for y and z, and p_y and p_z. In each case the four variables of position and time, or momentum and energy, are called four-vectors. The Lorentz transformation can be considered a rotation in four-space, similar to the rotation we described in two-dimensional space on p. 143. In that case the length of the two-dimensional vector was an invariant, the same regardless of rotation of the coordinate system. We have seen that in four-dimensional space-time there is also an invariant time length, and in the four-space of momentum and energy the rest mass of an object is an invariant.

On p. 447 in the chapter on magnetism, we demonstrate the use of the special theory of relativity in linking the electric and magnetic fields. We have already mentioned that Maxwell's equations yield the speed of em waves without reference to any coordinate system. It is essential that Maxwell's equations—and all other equations describing physical events—maintain their form after being subjected to a Lorentz transformation. For electric and magnetic fields *perpendicular to the velocity v* these equations are

$$\mathbf{E}' = \gamma(\mathbf{E} + \mathbf{v} \times \mathbf{B}), \quad \mathbf{B}' = \gamma\left(\mathbf{B} - \frac{1}{c^2}\mathbf{v} \times \mathbf{E}\right).$$

(All of these transformation formulas are a little more complicated in the general vector case of motion in $\mathbf{r}$ and for $\mathbf{E}$ and $\mathbf{B}$ fields other than those perpendicular to $\mathbf{v}$.) When Maxwell's equations are transformed using the equations above, they retain their same form in the moving reference frame. However, notice the way in which *each* transformed field is made up of *both* electric and magnetic fields. An electrostatic field in one system could look like a combination electric and magnetic field in a system moving by.

■ GRAVITATION AND THE GENERAL THEORY OF RELATIVITY

In all our examples with Galilean relativity or with Einstein's special theory of relativity, we have compared observations between observers moving with constant velocity with respect to each other. That is an unfortunate restriction since so many interesting events happen in accelerating reference frames. At the beginning of this chapter we saw that "pseudo" or inertial forces arise in accelerated reference frames. With rotating systems these are the centrifugal and Coriolis forces. In acceleration along a straight line, there appears to be a gravitational-type force in the direction opposite to that of the acceleration. These inertial forces can be described in terms of local gravitational-type fields that can be added vectorially to any gravitational field of the normal Newtonian type.

Einstein proposed that these fields—gravitational and inertial—are completely equivalent in their action at any *point* in space. He went on from there to work out a theory in which gravitational mass warps space in such a way that gravitation becomes simply a consequence of geometry. In this model, planets go around the sun not because they experience a radial centripetal force but because the sun has distorted space in its vicinity. The planet is constrained like a marble rolling around on the inside of a bowl.

Einstein's union of geometry and gravitation reduces to Newtonian gravitation except for very large masses. Nevertheless, the differences in predictions concerning experimental effects are sufficiently large in several cases to be tested. The three following original predictions that Einstein made have been confirmed with considerable precision:

1. The elliptical orbit of Mercury precesses at a rate slightly different from the rate that could be accounted for by the known effects of other planets. The difference in precession rate as predicted by Einstein is only 43 seconds of arc *per century*. That small discrepancy is measurable and had been approximately known, but not understood, before Einstein made his prediction.

2. If space is warped near massive objects, then the path of light through such space should not be a straight line. One way to observe such an effect is to observe a star whose light has just grazed the surface of our sun. If the star's position seems to shift slightly as the sun moves close to the path, then the light from the star must have been deflected by the warped space near the sun. Until recently such an observation could be made only during a total eclipse of the sun. Shortly after Einstein predicted the effect, it was observed during the total eclipse of 1919 (to worldwide acclaim!). The measurement has been made many times since then with varying degrees of success and precision. Recently Einstein's prediction has been confirmed with great accuracy with radio wavelengths that can be used even when the sun is not eclipsed. It should be noted that, since the *special* theory of relativity requires that photons have mass-energy (but not rest mass), we would expect a gravitational deflection of the passing photons even with Newtonian gravitation. The *general* theory of relativity, however, predicts a deflection twice as great, and this is what is observed. During the 1980s, astronomers have observed numerous events that are most easily explained in terms of gravitational "lenses." The model is that light coming to us from certain distant galaxies passes through lens-shaped warped space created by large masses. Here on earth we see focused or distorted images.

3. A third prediction of the general theory is that time depends on the gravitational field, or in the Einstein model, the curvature of space. Clocks slow down in gravitational fields. If a particular element emits a characteristic pattern of different frequencies of light when it is incandescent in our laboratory, we would expect to see the same pattern from that element in a distant star. However, if time is slower in that star because of the strong gravitational field, each of the frequencies should be smaller because time is slower. We might expect to see the same *pattern* of frequencies, but shifted toward longer wavelengths. This frequency shift is mixed with other effects, including the Doppler shifts, and has not served as a critical test of the general theory. However, there is a frequency-dependent phenomenon (the Mössbauer effect) so sensitive that the gravitational effect on clocks has been measured here on earth. A clock (the Mössbauer source) in the basement of a building runs slow

compared with a similar clock in the attic. The effect has been measured and agrees with Einstein's prediction.

■ REFERENCES

Hafele, J. C., and Keating, R., *Around-the-World Atomic Clocks Observed Relativistic Time Gains* Science **177**, 166 (1972).
Maor, Eli, *Some Applications of the Galilean Transformation*, Phys. Teach. **13**, 399 (1975).
Wood, Elizabeth A., *Science for the Airplane Passenger* (Ballantine Books, New York, 1968).

7

Newton's Laws of Dynamics

Surely Newton's exposition of his three laws of dynamics is one of the grandest achievements of the human intellect. Not only did Newton clarify and summarize the hazy notions that were prevalent at the time, but he then proceeded to apply the laws to a vast array of phenomena, most importantly to a detailed explanation of solar-system dynamics. Small wonder, then, that the laws have been the subject of study, analysis, misunderstanding, and controversy ever since.

From the viewpoint of the learner, Newton's laws are formulas for memorization and plug-in exercises. With a little practice most students can learn to solve the standard problems, but comprehension of the implications in the everyday world is another matter. We all know what a struggle it is to persuade students that the vertical acceleration of a ball throughout its flight is constant, even at the top of the trajectory where the vertical velocity is zero. Yet clearly the weight of the ball must remain constant. It is also hard to persuade students that a table somehow exerts a force on a book that it supports. Furthermore, in the everyday world forces seem to be proportional to *velocity*, not acceleration. Double the oxen, double the speed of the ox cart. Everyone "knows" that it takes a lot of force to "overcome" the inertia of a heavy object.

From the viewpoint of the philosopher of science, Newton's laws present a continuing subject for logical analysis and argument. Here are some of the problems: Are these laws or definitions? Are there three laws or just one with special cases? In what reference frame do the laws hold, and does any such reference frame exist? Should the primary operational definition involve mass or force, and does this require circular reasoning? What is the relationship between inertial and gravitational mass? How do the laws transform for high velocities (are they invariant under Lorentz transformation)? What is the validity or use of the laws in the microworld where the operational definition of force—or the whole concept of force—is meaningless? One of the most comprehensive reviews of some of these problems is contained in an article by Eisenbud.

■ VIEWPOINTS ON THE LOGICAL RELATIONSHIPS OF NEWTON'S LAWS

The First Law

We start with a definition of "no force." If *with regard to an observer*, there is no acceleration of an object, then *according to that observer* there is no net force on the object. This is a good operational definition because there are good operational ways of defining and measuring acceleration (deflection of plumb bobs, measurement of change of velocity). However, this definition would mean that in an orbiting space capsule the astronauts would declare that there is no net force on the cup floating in front of them. Here on earth we would disagree with that conclusion, pointing out that the astronauts and the cup are subject to the earth's gravitational force, and that, indeed, the cup is accelerating. The astronauts would look out their window, see the earth, and agree with us about the gravitational field. However, they would argue that they are also in a centrifugal field which in their location just balances out the gravitational field, leaving a zero net force. We would point out that they are not at rest with respect to the fixed stars, but they might point out that we are not either, although more so than they are.

Perhaps we could take this definition: *If no net force acts on an object, the object will remain at rest if it was at rest. If it is moving with constant velocity, it will continue to do so.* Obviously, this definition leaves the crucial reference frame problems unanswered. Perhaps we can define the reference frame and state the law in this way: *If the net force on an object is zero, then it is possible to find a reference frame in which the object has zero acceleration.* To test such a definition, you would have to know that you have an object on which the net force is zero. How would you know? You can isolate the object from contact forces by suspending it (somehow) in a vacuum, and then you can isolate the object from electromagnetic forces by surrounding it with an electrically conducting shield (a Faraday cage). The only long-range forces we know about are electromagnetism and gravitation. As a practical matter, how will you isolate it from gravitation? How will you know that the object—like the astronaut's cup—is not subject to "inertial" fields, which Einstein assures us are completely equivalent (at any point) to gravitational fields. (This assertion is the cornerstone of the general theory of relativity, which is our theory of gravitation.)

Perhaps the cleanest solution is to start out by defining a reference frame at rest with respect to the fixed stars. This may be inconvenient in many cases, and it may be more practical—and perfectly legitimate—at times to name and use local inertial fields, such as centrifugal and Coriolis forces. But at least with the reference frame chosen, "no force" can be clearly defined in terms of the operational definition and measurement of "no acceleration." If an object is not accelerated with respect to a reference frame at rest with respect to the fixed stars, then the net force on it is zero. A philosophical problem remains. *There are no "fixed" stars.* Our solar system is moving around the center of the galaxy, along with all the other stars in the galaxy, and the galaxies themselves are not only moving around local centers but space itself is expanding carrying the galaxies with it. Never mind! A good first approximation here on earth—or within the solar system—is that the surrounding stars are fixed for most human purposes. An even better answer has appeared in 1992 as a result of the measurements made by the COBE satellite. The 2.7-K electromagnetic background radiation is indeed fixed, or at least uniform in all directions, and we mortals (or anyone else) can measure our motion relative to it.

The Second Law

Having defined "no force," we can now define force as $F = ma$. The trouble is, how do we define m? If we had a definition for force, we could say that m, inertial

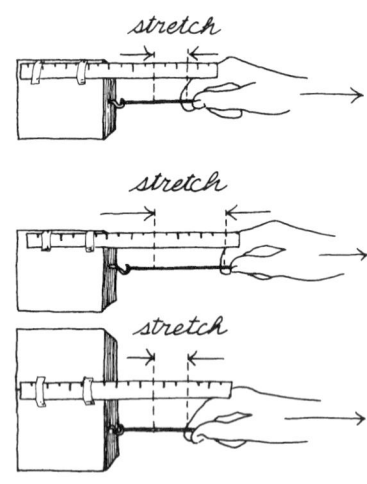

Figure 7.1. Simple method of applying constant force to sliding object.

mass, is the proportionality constant between F and a. But we cannot use m, defined that way, in order to define force. To do that would involve us in circular reasoning. There *is* a way to define inertial mass operationally through Newton's third law. We will spell out how to do that in the next section.

There is an alternative and logical scheme to define force. Force has two aspects, each enshrined in a classical equation, though they are two very different types of equations. In Newton's $F = ma$, force displays itself as a mover, a dynamic aspect. In Hooke's $F = kx$, force becomes a stretcher, a static behavior. Without regard to dynamics, we could set up units and standards of force in terms of the stretching of standard springs. The arguments are tricky and not simply a matter of logic. *In particular, we cannot assume the validity of Hooke's law in establishing our system.* Experiments must be performed to justify that for (some) springs two identical weights hung on the spring yield twice the stretch of one weight. We must assume that two identical weights acting in the same direction yield twice the force of one weight. Other experiments must be performed to justify the fact that forces combine like vectors. A self-consistent set of static spring demonstrations can generate a unit and scale of forces. This approach is a good one for introductory physics. Statics comes before dynamics, or even kinematics. There are many interesting phenomena to be studied that depend on the strength of materials and on structure.

With force independently defined, we can demonstrate experimentally that forces also accelerate objects, and that the acceleration is proportional to the force. In this sequence of arguments, $F \propto a$ is a law that requires experimental proof. The law requires this proof not only for logical consistency but also because it is not obviously true, or should not be! In everyday life it takes constant force (constant pressure on the "accelerator," constant pushing of the bike pedals) to make the car or bike go at constant *speed*. The acceleration experiments introduced long ago by PSSC are still valuable for beginning students. Create a constant (spring) force by attaching one end of a rubber band to the end of a meter stick and the other end to a wheeled cart, as shown in Fig. 7.1. Maintain a constant stretch of the rubber band, as indicated by the position of the meter stick, and feel how you and the stick and the cart accelerate. A student will learn the lesson best by performing the exercise, rather than just by watching a demonstration. The bigger the cart, the larger the rubber band or spring, the better the lesson.

Another way of generating an apparently constant force has been proposed by Robert Morse. He mounts a battery driven air propeller on low-friction carts or on air-track gliders, and then his students measure $x(t)$ in one of the usual ways, as a function of several different masses of the carts.

Once you are convinced that $F \propto a$, it takes only a series of calibrating experiments to obtain a unit and scale of inertial mass, the proportionality constant. There remains the problem of the relationship between inertial mass m_I and the gravitational mass m_G. This gravitational mass appears as the "charge" in another great law by Newton, one that is completely unrelated to the laws of dynamics: $F = Gm_G M_G/r^2$. In the teaching of introductory physics, the gravitation law should be separated in time and space and treatment from the dynamics laws. Otherwise, we confuse the students and they may miss the astonishing import of the identity of inertial and gravitational mass. This problem is discussed at more length on p. 106 in the chapter on gravitation (Chap. 5).

The second law, $F = ma$, is the centerpiece of the first semester of the introductory-physics course. How important it is, therefore, to let students know that it is only an approximation. This fact illustrates the nature of our subject; most of our explanations are successive approximations. The true statement of the second law is

$$F = \frac{d(mv)}{dt} = \frac{dp}{dt}.$$

Force is the time rate of change of momentum. The usual form is valid for velocities much less than the speed of light. The general form is useful not only for velocities close to that of light, but also for cases where the mass of the object is a function of time, such as rocket problems.

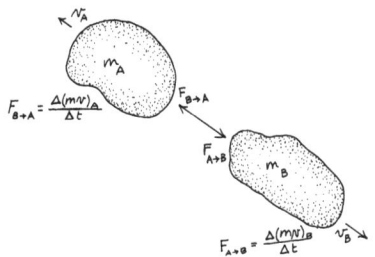

Figure 7.2. Equivalence of Newton's Third Law and Conservation of Momentum.

The Third Law

We think it best to avoid the action-reaction statement. It has led to endless confusion and misrepresentation. (Why did the stock market rebound?—because for every action there is an equal and opposite reaction.) The word "action" has many different technical meanings in different professions. In physics it is usually related to the product of momentum and displacement. In Chap. 9 we will discuss the Principal of Least Action. We do not mean any of these things in the well-known statement of the third law. We are talking about *forces*. (Besides, have you ever seen an "action" meter?)

If object A exerts a force on object B, then B exerts an equal and opposite force on object A:

$$F_{A \to B} = -F_{B \to A}.$$

This statement is equivalent to the law of momentum conservation, and this is a useful way to think of the third law. In Fig. 7.2, we show two objects influencing each other. If the third law holds at all times and for any time interval, and if we appeal to the general form of the second law,

$$\left(\frac{\Delta p_B}{\Delta t}\right)_{A \to B} \Delta t = -\left(\frac{\Delta p_A}{\Delta t}\right)_{B \to A} \Delta t.$$

Therefore, in this two-body system, the total change of momentum must be 0:

$$\Delta p_B + \Delta p_A = 0.$$

For any object, the internal forces between each and every pair of particles must always behave in this way. Consequently, the *internal* forces produce no change in momentum of the object.

The experimentally determined fact of momentum conservation can be used to define inertial mass. The sequence of arguments runs like this.

1. Use gliders on an air track to approximate isolated objects. Collide glider A into stationary *identical* glider B.

2. Repel A and B from each other (with a spring) from a stationary start. Observe that speed v is not conserved, but total **v** is.

3. Repel A from double-length glider C and observe that **v** is no longer conserved. Define a new quantity, momentum=$m\mathbf{v}$, such that $(mv)_A = -(mv)_C$. Then vector momentum is conserved in these one-dimensional explosions, as long as $m_A/m_C = v_C/v_A$.

4. Choose m_A as unit inertial mass. Perform repulsion measurements with other objects to measure their masses, and thus establish mass scale.

5. Carry out other collision (interaction) experiments, including in three dimensions, and demonstrate experimentally that vector momentum is always conserved.

6. From the momentum-conservation law, deduce Newton's third law with the inverse of the argument illustrated in Fig. 7.2. In this sequence of arguments, force is a *defined* (and secondary) quantity:

$$F_{\text{on B}} = \frac{dp_B}{dt}.$$

Let us stress several features of this approach to Newton's laws. First, it provides an *operational* definition of inertial mass. There is no need for vague claims about the "quantity" of matter. The operations described can be used for air-track gliders or for electrons and photons. Second, the approach emphasizes the primacy of a conservation law—in this case, linear momentum. Third, the logic of the arguments is simple and straightforward. Newton's laws are not all definitions, and are not all the same statement. The third law is an experimentally determined conservation law, yielding consistent results if mass is defined in terms of the law, and providing an operational way to define that mass. The second law is a definition of force derived from the third law. In conjunction with the first law, the relationship between force and acceleration defines a reference frame.

Regardless of the sequence of logic concerning Newton's three laws, it is important to emphasize to our students that "forces" are *interactions between two objects*. Therefore, of course $F_{A \to B} = -F_{B \to A}$. There are not *two* forces. There is only *one* interaction. In particle physics, force does not even show up. We measure only changes of momentum.

■ MASS AS THE "QUANTITY OF MATTER"

Even as energy is sometimes erroneously defined as "the ability to do work," mass is sometimes referred to as the "quantity of matter." The trouble is that "quantity," like "ability," is not defined, or depends on circumstances. For that matter, the meaning of "matter" is not always obvious. Are mesons matter? They have mass.

Clearly, in connection with mass the quantity of matter should not be determined by the volume. There is more mass in 1 cm^3 of lead than in 10 cm^3 of styrofoam. The quantity of matter is frequently measured in terms of its weight, but it is not obvious that weight is proportional to the inertial property described in $F = ma$. (See p. 106.) For instance, both a bicycle and a train have their weight fully supported by the earth. If they were to roll slowly toward you, however, you could more easily change the velocity of the bike than the train. It would not be the vertical force between earth and train (the weight) that would bowl you over. You could not stop the train in far outer space, where it would have no weight at all.

If the inertial property of matter is not proportional to the volume of objects and exists whether or not the objects have weight, perhaps it is proportional to the number of atoms in an object. Not so. Different atoms have different inertial responses. Perhaps, however, you could count the total number of protons, neutrons, and electrons in an object. These constituents of the atom are the same in every atom, but each type of atom has a different number of them. While it is not practical to count every particle in large objects, we can actually compare the inertial response of subatomic systems where we can count the individual protons, neutrons, and electrons. One easily performed experiment, with a mass spectrometer, allows us to measure the relative acceleration of a deuterium nucleus and a helium nucleus, when each is subjected to a known electric force. The nucleus of deuterium contains one proton and one neutron. The nucleus of a helium nucleus (an α particle) contains exactly twice as much "matter" (or so it would seem). It consists of two protons and two neutrons. We might expect, therefore, that if the two nuclei were subjected to the same force, the acceleration of the deuterium nucleus would be exactly twice that of the helium nucleus. But this is not quite true. The inertial property, the mass, of the helium nucleus is a little less than twice that of the deuterium nucleus. (For details of a similar problem, see p. 533.) It appears that as far as inertial effects are concerned, the only way to define the "quantity of matter" is in terms of an inertial experiment using Newton's laws.

■ UNITS OF FORCE AND MASS

The *Système International* (SI) units for mass and force are convenient and foolproof. Students should use their muscles to heft and maneuver objects with various masses so that the units mean something to them. They should know the masses of themselves, their textbooks, common coins, and automobiles. Throughout the world, except in the United States, people know the "weight" of things in terms of "kilos." Newtons, the unit of force, is not much used anywhere except in scientific work. Nevertheless, our physics students must have a feel for the size of the newton in terms of their muscles. They should know their own weight in newtons and should have engaged in strength contests with spring scales calibrated in newtons.

Some physics research is still done and reported in dynes, grams, centimeters, and seconds. 1 dyne=(1 gram)(1 cm/s^2). Consequently, 1 N=1×10^5 dyn. Introductory-physics texts tend to weigh about 10 N, or about a megadyne. The weight of a gram is about 1×10^{-2} N. [$(1\times10^{-2}$ N$)\approx(1\times10^{-3}$ kg$)(9.8$ N/kg$)$]. The weight of 1 mg is about 1 dyn. [1 dyne$\approx(1\times10^{-3}$ g$)(980$ cm/s$^2)$]. The gravitational field strength at the earth's surface in this system is 980 dyn/g=980 cm/s^2.

Unfortunately, pounds and feet are still used in everyday life in the United States and in a lot of engineering practice. We should prepare our students to translate. The only tricky part is that the pound is the unit of *force*, not mass. If you drop an object at the earth's surface with a weight of 1 lb, it will accelerate at 32.2 ft/s^2. We describe this behavior with the second law, 1 lb=$(m)(32.2$ ft/s$^2)$. Evidently, the mass of the 1 lb weight must have the numerical value of 1/32.2. The unit is called appropriately enough, the *slug*. A slug is a large mass; its weight is 32.2 lb. Nobody, anywhere, refers to the mass of something in terms of slugs. (What is *your* mass in slugs?) When working dynamics problems in the "English" system you must adopt a simple artifice. *Wherever you see "m," use the weight in pounds and divide by 32.2.* For instance, here's how to find the centripetal force necessary to keep a 2-lb physics text traveling in a circle with a 3-ft radius at a speed of 15 miles/h (60 miles/h=88 ft/s):

$$F_{\text{cent}} = \frac{mv^2}{r} = \frac{[(2\ \text{lb}/32.2\ \text{lb})/\text{slug}](22\ \text{ft/s})^2}{3\ \text{ft}} = 10\ \text{lb}.$$

■ MOMENTUM CONSERVATION

Collisions between two objects, or explosions of one object into many, can be very complicated events. The details will depend on relative velocities and masses, and also on the types of forces involved and impact angles. The final products of the event and their distribution may look very different from the original situation. Rather than focusing on what changes, it is frequently helpful in solving such problems to ask what remains the same. We appeal to conservation laws. No matter how complex the internal forces, or the number of particles involved, an isolated system maintains its original momentum. Here are several examples of the application of the conservation of momentum.

1. In the collision of an air-track glider with an identical stationary glider, the bombarding glider stops and the target glider proceeds with the original velocity. The momentum before equals the momentum afterward:

   $$(m_A v_A)_0 = (m_A v_A)_{\text{final}} + (m_B v_B)_{\text{final}}.$$

   Since $m_A = m_B$, and $(v_A)_{\text{final}} = 0$, then $(v_B)_{\text{final}} = (v_A)_0$. The observed event certainly satisfies momentum conservation. However, many other combinations of $(v_A)_{\text{final}}$ and $(v_B)_{\text{final}}$ would also satisfy the conservation law. For instance, $(v_A)_{\text{final}} = (v_B)_{\text{final}} = \frac{1}{2}(v_A)_0$ is a solution. In fact, these are the final velocities observed if the two gliders have velcro on them and stick together. Evidently since we have two unknowns, $(v_A)_{\text{final}}$ and $(v_B)_{\text{final}}$ and only one equation, an infinite number of solutions is possible. If nature chooses only one of these solutions for an elastic collision, there must be another constraint equation. That constraint, for an elastic collision, is provided by the conservation of kinetic energy. We will take that into account and pursue the problem further in Chap. 9.

2. A .22 slug with a mass of 2 g can be shot with a muzzle velocity of 340 m/s. (Compare this with the speed of sound in air at room temperature, which is 343 m/s; or with the speed of an ordinary jet liner at 270 m/s.) Just before firing, the gun–bullet system has zero momentum. Immediately afterward, the lead slug is carrying off momentum equal to $(2 \times 10^{-3}\ \text{kg}) \times (3.4 \times 10^2\ \text{m/s}) = 0.68$ kg m/s. If the gun were floating by itself and had a mass of 3 kg, it would recoil with a velocity of 0.23 m/s:

   (momentum before) = (momentum after),

   $$0 = (3 \times 10^{-3}\ \text{kg}) \times (3.4 \times 10^2\ \text{m/s}) + (3\ \text{kg}) \times (-0.23\ \text{m/s}).$$

   The gun, of course, is not floating; it is braced against your shoulder. In the next section we will see how to estimate the effect of the kick.

3. Is momentum conserved when a tennis ball bounces from the ground? A new tennis ball has a mass of 56 g. When dropped to a hard surface from a height of 1 m it rises to a height of 63 cm. You can find its speed just before and after striking the ground by using $v_f^2 = 2gy$. Just before striking, $v = 4.4$ m/s. Just after, $v = 3.5$ m/s. At first thought, it appears that momentum is not conserved for several reasons. The ball starts out from rest with zero momentum. The

momentum increases steadily, but during the collision with the ground the momentum abruptly goes to zero, then suddenly becomes large in the opposite direction, but not so large as it had been. Momentum appears to be increasing, decreasing, and changing direction. Remember, however, that the conservation law is concerned with the momentum of a *whole system*. In this case the system includes both ball and the earth. While the ball is falling toward the earth and picking up momentum in one direction, the earth is "falling" toward the ball, picking up equal momentum in the opposite direction. The total momentum is zero at all times. When the ball and earth bounce, the ball goes one way, the earth the other, each with equal magnitude of momentum but in opposite directions so that the vector quantities always add to 0.

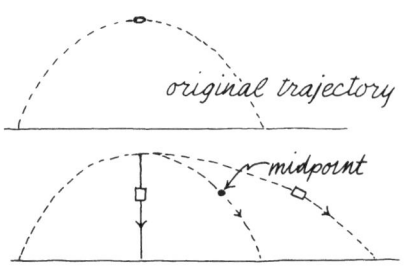

Figure 7.3. Trajectories of parts of object exploding at highest point.

Surely it is fantastic to think of the earth falling up toward a tennis ball, and then bouncing back. In order to conserve the momentum of the earth–ball system, it must be true at all times that $MV = -mv$. The capital letters refer to the earth; the lowercase letters refer to the ball.

Before bounce:

$$(6 \times 10^{24} \text{ kg})(V_b) = -(0.056 \text{ kg})(-4.4 \text{ m/s}),$$
$$V_b = 4.1 \times 10^{-26} \text{ m/s}.$$

After bounce:

$$(6 \times 10^{24} \text{ kg})(V_a) = -(0.056 \text{ kg})(+3.5 \text{ m/s}),$$
$$V_a = -3.3 \times 10^{-26} \text{ m/s}.$$

The *change* of velocity of the earth is -7.4×10^{-26} m/s, which is not enough to make anything spill!

4. When bombs are bursting in air, the fragments shoot out in such a way that the original momentum of the bomb remains unchanged. That condition imposes severe constraints on the relative speeds and directions of the fragments. Let us study a simple example of a shell exploding in midtrajectory into two equal pieces. Furthermore, let us take the special case where one piece is shot horizontally backward at a relative speed just equal to that of the shell when it exploded. That piece is suddenly left with zero horizontal velocity, with respect to the ground, and will start dropping straight down. The other half of the shell must be kicked forward with twice the horizontal velocity that it did have. In that way, the momentum of the shell before explosion, which was horizontal with the value mv, remains unchanged since the momentum of the fragment is $(\frac{1}{2}m)(2v) = mv$.

As the two fragments fall, their vertical heights are identical at all times; those heights are also the same as the height that the whole shell would have had at those same times. The first fragment falls straight down, and the horizontal distance covered by the second fragment is just twice the horizontal distance that the original bomb would have traveled. (Of course, all these statements ignore air resistance, which would actually play a major role.) Fig. 7.3 shows the trajectories and also shows the continuation of the original trajectory which contains the midpoint of the two fragments.

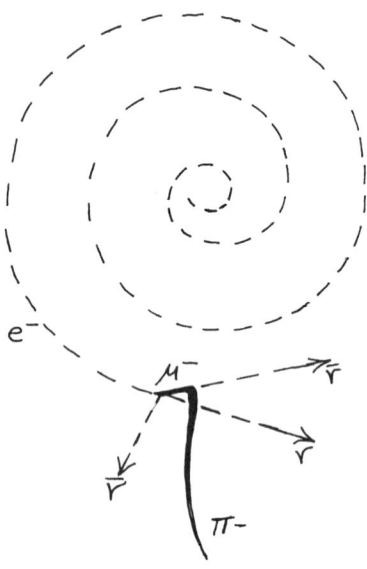

Figure 7.4. π^-, u^-, e^- decay.

5. Here is an example of how the law of momentum conservation can tell us that something must have happened even when we cannot see it. Figure 7.4 shows the bubble-chamber tracks of the decay of a π meson into a muon, and the subsequent decay of the muon into an electron. The actual photograph shows that the π meson is slowing down toward the end of its path. Its track is thicker and the curvature in the magnetic field is greater. It came to rest and then decayed. From other experiments we know that the decay time must have been about 10^{-8} s. The decay product is a muon, which had small kinetic energy and therefore left a short, thick trail. After losing its kinetic energy, the muon came to rest and then, after about 10^{-6} s, decayed into an electron that produced a very sparse, curved trail. The fact that the trail of the final decay product is very thin indicates that the electron was traveling very fast—close to the speed of light. The electron's spiral path shows that the electron continually lost momentum going through the liquid hydrogen and thus the sideways force exerted by the magnetic field had more and more effect on its path.

So far we have described most of the things that you can see in the picture. Some of the most interesting events cannot be seen, but the law of momentum conservation tells us that they must have taken place. Note that the π^- meson slowed down, stopped, and then decayed with the emission of a single particle going off to the side. What if you saw a hand grenade roll to a stop, explode, and shoot just one fragment to the side? It could not happen. Something must have shot out in the opposite direction in order to conserve momentum. In the case of the subatomic particles, we know that the invisible recoil particle was an antineutrino. Because an antineutrino has no electric charge, it leaves no trail in a bubble chamber.

The same argument can be used to describe what we do not see when the μ^- decays. Once again, there must have been recoil particles. The evidence of many similar pictures is that the μ^- track always has the same length, regardless of the angle between it and the π^- track. On the other hand, the electron track is sometimes long and sometimes short. In the case of the π–μ decay, the μ can get the same momentum each time only if it is recoiling against one—and only one—particle. In the μ–e decay, the momentum of the electron can vary from one time to the next only if it is recoiling from two or more particles. From other evidence, we know that when the μ muon decays it turns into an electron, a neutrino, and an antineutrino. We cannot see the neutrino and antineutrino, but we know that the vector sum of their momenta must be equal and opposite to the momentum of the electron.

■ IMPULSE

A special name is given to the product of the external force on an object and the time interval during which it acts: (impulse)$=F\Delta t=\Delta p=\Delta(mv)$. In this form the law makes it clear that if $F_{ext}=0$, $\Delta(mv)=0$, and the momentum of the object remains constant. This form is also particularly useful in analyzing the trade-off between force and time during collisions.

1. You get a "kick" out of firing a gun. For the .22 rifle firing a "short" slug, we calculated (on p. 162) the change of momentum to be 0.68 kg·m/s$^2=F\cdot\Delta t$. The size of the force (recoil) exerted on the gun evidently depends on the

length of the time interval Δt. That time interval depends on the length of the gun barrel, the muzzle velocity, and the nature of the explosion. Actually, the recoil force is greatest shortly after the explosion begins and falls off as the gases expand and the slug shoots down the barrel. However, because the slug starts out slowly, the slug is close to the explosion point during most of the time that it spends in the barrel. For our purposes, a good approximation is that the force is constant during the time that it takes the slug to leave the gun. If the muzzle velocity is 340 m/s, the average velocity with constant acceleration is 170 m/s. If the gun barrel is $\frac{2}{3}$ m long, the time interval is $\Delta t = (\frac{2}{3}\text{ m})/(170\text{ m/s}) = 0.004$ s, or 4 ms. Now we can find the force on the gun: $F = \Delta(mv)/\Delta t = (0.68\text{ kg m/s})/(0.004\text{ s}) = 170$ N. To find out whether that is much of a kick, compare this force with the weight of a typical .22 rifle which has a mass of 3 kg and a weight of 30 N. If the gun is against your shoulder, you momentarily feel a horizontal force of almost 6 times the gun's weight.

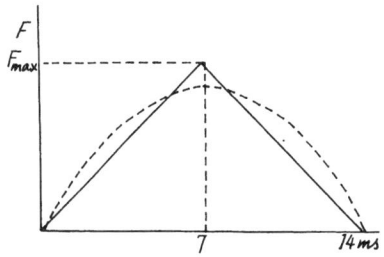

Figure 7.5. Model of $F(t)$ for bouncing tennis ball.

2. The force on a tennis ball as it bounces is certainly not constant. Whether the ball bounces against the ground or against the strings of a racket, the restoring force must be approximately proportional to the distortion. That requirement would produce a force-versus-time curve something like the dotted one shown in Fig. 7.5, *if* the bounce were symmetrical. (It is not, as we have seen on p. 162, since the upward speed is less than the downward speed.) For an easier calculation, however, let us use the triangular approximation shown with the solid line. The time it takes the ball to come to rest during the first half of the bounce can be estimated by assuming that the ball slows down with constant acceleration in a distance equal to some fraction of its diameter. Let us assume that when the ball is dropped from a height of 1 m, it has stopped by the time it is squashed one-fourth of its diameter—about 1.5 cm. Just before it hits the ground the ball is traveling with a speed of 4.4 m/s. If it were to slow down with constant acceleration, which is a very rough approximation, its average speed during the bounce would be 2.2 m/s. The time it would take to come to rest would be $\Delta t = (1.5 \times 10^{-2}\text{ m})/(2.2\text{ m/s}) = 6.8 \times 10^{-3}$ s, which is about 7 ms.

We can now calibrate the graph of F vs t in Fig. 5. The time scale must be such that it takes 14 ms for the complete bounce. The product $F\Delta t$ must be equal to the change of momentum, $\Delta(mv)$:

$$\Delta(mv) = \Delta(mv)_{\text{after}} - \Delta(mv)_{\text{before}} = (0.056\text{ kg})(+3.5\text{ m/s})$$
$$- (0.056\text{ kg})(-4.4\text{ m/s}) = +0.44\text{ kg m/s} = 0.44\text{ N s}.$$

However, we cannot simply set $F\Delta t = 0.44$ N s. The force is varying. The area under the force–time curve is equal to the impulse, which is the change of momentum. The area under this triangular curve is equal to $\frac{1}{2}F_{\max}\Delta t_{\text{total}} = \frac{1}{2}F_{\max}(14 \times 10^{-3}\text{ s}) = 0.44$ N s. Therefore, $F_{\max} = 60$ N. The ball itself, with a mass of 56 g, has a weight of only 0.55 N. Apparently, even for a drop of only 1 m, the restoring force on the tennis ball becomes more than 100 times the weight of the ball. In the language of test pilots of jets or rockets, the ball is subject to more than 100 "g's" (gravities).

3. In case you feel no empathy for an accelerated tennis ball, consider the forces involved when you yourself stop in a car. Assume that you drive at 40 miles/h into a concrete wall. The car will stop in a distance of less than 1 m, that

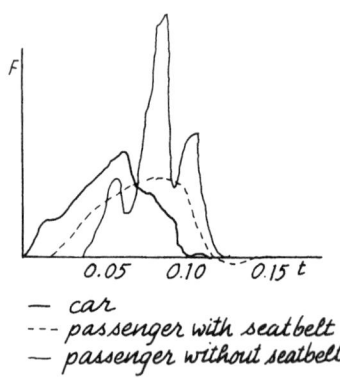

Figure 7.6. F(t) for car and driver during collision.

distance being taken up by the space where the motor and things used to be. Assuming that the car slows down with constant acceleration from an initial speed of 40 miles/h, which is about 18 m/s, the time it takes will be $\Delta t = (1 \text{ m})/(9 \text{ m/s}) = 0.11$ s. If your seat belts are well adjusted and hold, you will slow down in that same time. If you are not wearing seat belts, you will feel no strain or pain until your body hits the steering wheel. The wheel will not take long to collapse, giving you some more free time before you hit the dashboard and windshield. Then you will stop in a hurry unless you make it all the way through the windshield. On the graph of force versus time for this accident, shown in Fig. 7.6, we have drawn three curves: one for the car frame, one for a driver with seat belts, and one for the driver without seat belts. The time scale of events is realistic in terms of automobile manufacturers' tests with dummies. Actual events vary drastically, of course, depending on whether your stomach gets caught in the wheel and on which portion of your head hits the glass.

There is one important thing to keep in mind during the tenth of a second that all this action takes. The change of momentum of your body will be the same regardless of whether you are wearing a seat belt. You were traveling at 18 m/s, and about a tenth of a second later your speed is 0. Therefore, the total area under your $F(t)$ curve must be the same:

$$\Delta(mv) = (60 \text{ kg})(0 - 18 \text{ m/s}) = -1.1 \times 10^3 \text{ kg m/s} = -1.1 \times 10^3 \text{ N s}.$$

If you could stretch the force out evenly over the time of the collision, that average force would be $F = (-1.1 \times 10^3 \text{ N s})/(0.11 \text{ s}) = -1 \times 10^4$ N. The minus sign just indicates that the direction of the force is opposite your original direction. With your weight of 600 N, you would experience about 16 g's. That would not be comfortable, but if the belt is properly positioned you might live. Besides, the seat belt exerts its restraining forces on sections of the body that are more flexible than the skull, spreading the force over a larger area of the body.

Notice that it is not exactly the car collision that kills the driver. Instead, the fatal blow comes from the second collision when the driver strikes the inside of the car. The car has pretty much come to rest before the unbelted driver hits. The driver's stopping distance, and thus the stopping time, is shorter than that of the car by a factor of about 10. Consequently, the stopping force on the body is about 10 times what it would have been had the driver been fastened to the car. (In many cases, what kills is the third collision when the brain hits the front of the skull.)

4. In each of the examples so far, it made no difference whether we described the change in momentum, $\Delta(mv)$, or the product of mass and change of velocity, $m\Delta v$. The mass of the bullet, the tennis ball, or the human body did not change during the interaction. When a subatomic particle travels at speeds close to that of light, momentum becomes more meaningful than mass and is not simply mv. There is another case, perhaps less exotic, where the mass of an object changes during an interaction. A rocket accelerates by shooting away some of its own mass. The exhaust gases have a fairly constant velocity *with respect to the rocket*. If the rate of ejection of gas is constant during firing, the rate of change of momentum will be constant. However, the acceleration of the rocket will not be constant because the mass of the rocket will

be decreasing. Not only will the velocity of the rocket increase, its *acceleration* will increase. During the flights of the Apollo rockets, the astronauts experienced their largest g forces just before the end of the first stage firing. Let us take a look at the arithmetic of this situation.

In free flight, or neglecting gravitational influences, the momentum of the rocket–gas system must be conserved. If the rocket's mass at the beginning of a firing is M_0 and if it loses mass at the rate of $\Delta m/\Delta t$, then after a time t its mass will be $M_0 - (\Delta m/\Delta t)t$. During the next interval of time after t, the rocket will pick up an additional velocity, Δv, and will lose an additional mass, Δm. Meanwhile, an amount of gas with mass Δm has been shot out with an exhaust speed of u with respect to the ship. The forward momentum gained by the ship must be equal in magnitude to the backward momentum of the gas:

$$\left(M_0 - \frac{\Delta m}{\Delta t} t - \Delta m\right)\Delta v = \Delta m\, u.$$

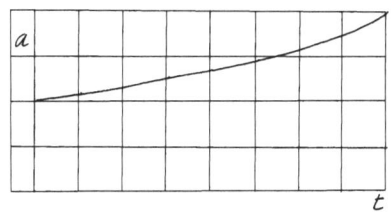

Acceleration of a rocket with constant thrust but with decreasing mass.

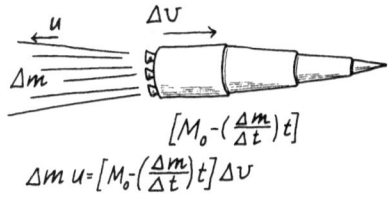

Figure 7.7. Acceleration as a function of time for rocket ship with constant thrust.

Notice that the gas is emitted with speed u with respect to the ship. With respect to the ground, both ship and gases are traveling in the same direction when $u < v$!

If the amount of mass shot out during this interval is small compared with the mass of the ship at that time, we can make the approximation that

$$M_0 - \frac{\Delta m}{\Delta t} - \Delta m \approx M_0 - \frac{\Delta m}{\Delta t} t.$$

Rearrange the variables and divide both sides by Δt:

$$\frac{\Delta v}{\Delta t} = \frac{\Delta m}{\Delta t} \frac{u}{M_0 - \frac{\Delta m}{\Delta t} t}.$$

The left-hand side of this formula is simply the acceleration of the rocket ship. From Newton's second law, we expect that an acceleration is equal to a force divided by a mass. In this case, the force is constant and equal to the product of constant speed u and a constant mass loss, $\Delta m/\Delta t$. The mass in the denominator of the right-hand side is a variable mass because the rocket ship is continually getting lighter. Even with a constant force exerted on a ship by the constant exhaust, the acceleration increases because the mass decreases, as shown in Fig. 7.7.

Let us try some actual numbers in the rocket equation. Consider the lift-off of a 100-ton rocket. At $t=0$, the mass is simply M_0. The thrust of the rocket must be at least equal to its weight or the rocket will never get off the pad. The thrust of the rocket is equal to the product of its mass and acceleration: $M_0(\Delta v/\Delta t)$. From the rocket equation, at $t=0$,

$$(\text{thrust}) = M_0 \frac{\Delta v}{\Delta t} = u \frac{\Delta m}{\Delta t}.$$

The exhaust speed of the gas is about equal to the average thermal speed of the molecules of gas at the combustion temperature. The average velocity of oxygen and nitrogen molecules at room temperature is about 500 m/s. On the

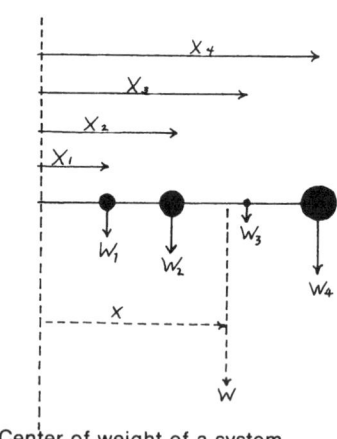

Center of weight of a system.

Figure 7.8. Center of weight for system of objects.

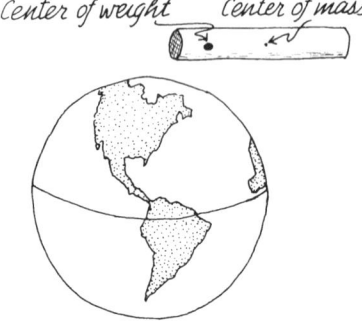

Figure 7.9. Example of difference between center of mass and center of weight.

absolute-temperature scale, the combustion temperature of rocket gas is about 10 times higher than room temperature. The molecular velocities are proportional to the square root of the temperature and inversely proportional to the square root of the mass of the molecule. Since hydrogen is part of the exhaust and since the temperature is very high, the exhaust speeds can be as high as 3000 m/s. If we require a thrust of exactly 100 tons, just enough to balance the initial weight of the rocket ship, then the rate of mass loss must be

$$\frac{\Delta m}{\Delta t} = \frac{(100 \text{ tons})(9 \times 10^3 \text{ N/ton})}{3 \times 10^3 \text{ m/s}} = 300 \text{ kg/s}.$$

To produce an upward acceleration of g, the rocket will have to shoot out twice this much, or 600 kg/s. (Note that an English ton is about equal to 900 kg.) Evidently the rocket cannot lose mass at this initial rate for many seconds.

■ CENTER OF MASS AND MOMENTUM CONSERVATION

The center of *weight* of an object is that point around which all the weight torques of the constituent particles add to 0. For a system of n point particles fastened together, as shown in Fig. 7.8, each with weight W_i, and each located a distance x_i, from some axis, the distance of the center of weight from that same axis is

$$x = \frac{\Sigma x_i W_i}{W}.$$

(It may not be obvious to a student, but the *location* of the center of weight is independent of the axis chosen. This point can be demonstrated by calculating the center of weight for two particles, using simple numbers for m_1 and m_2 and two different sets of values for x_i, corresponding to two different axes.) If the gravitational force on each particle is parallel to that of every other part (the flat-earth approximation), then the summation of torques, $x_i W_i$, requires just ordinary arithmetic instead of vector addition. We can substitute $m_i g$ for W_i and Mg for W and rewrite the equation for the center point:

$$x = \frac{\Sigma x_i m_i}{M},$$

where $M = \Sigma M_i$. With this change, x becomes a distance along the x axis from some origin to the *center of mass*. Similarly, the y component of the distance from the origin to the center of mass is

$$y = \frac{\Sigma y_i m_i}{M}.$$

The distances x and y are "weighted averages" of the distances from the origin to the individual particles. It may seem as if the center of mass is the same as the center of weight. For small systems of particles on earth that is generally true, but it is not true for a system of particles on the celestial scale. As an extreme example, consider the difference between the center of mass and center of weight for the uniform rod shown in Fig. 7.9.

The idea of the center of mass is vital to the law of conservation of momentum. If you follow the path of one point object moving all by itself without external

Newton's Laws of Dynamics

forces, it is simple to assert that its momentum remains constant. If the object bumps into another point object, it is not so obvious that the momentum of the system must be conserved, but we have seen that mass can be defined so that this is the case. If there is a whole system of objects, exerting all sorts of internal forces on each other, it is not even clear what momentum is being conserved. Certainly the momentum of any individual part changes. The claim of the law of conservation of momentum is that the vector sum of the x, y, and z components of the individual momenta must not change. Let us write out that sum for the x components:

$$\sum m_i(v_x)_i = \sum m_i\left(\frac{\Delta x}{\Delta t}\right)_i = \frac{1}{\Delta t}\left(\sum m_i(x_f - x_0)_i\right) = \frac{1}{\Delta t}\left(\sum m_i(x_f)_i - \sum m_i(x_0)_i\right).$$

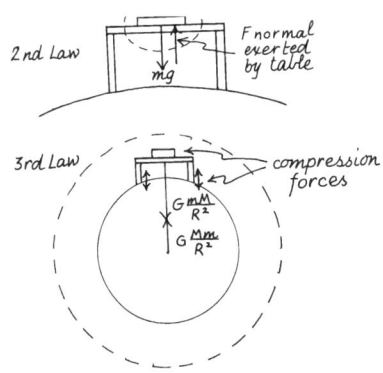

Figure 7.10. Analysis of the forces on a book resting on a table.

[We wrote the x component of the velocity of object i as $(\Delta x/\Delta t)_i$. Since the time interval Δt is common to all the objects, we took it outside the summation sign. The displacement Δx is equal to $x_f - x_0$.]

Now the definition of the center of mass enters the derivation. The weighted sum of the x components is $\sum m_i x_i = Mx$. Let us substitute this expression into our statement about the sum of momenta of the objects in the system:

(sum of the x components of momenta)

$$= \frac{1}{\Delta t}[(Mx)_f - (Mx)_0] = \frac{1}{\Delta t}M\Delta x = M\frac{\Delta x}{\Delta t} = Mv_x.$$

It appears that the sum of the x components of momenta of all the objects in a system is equal to the total mass times the x component of the velocity of the center of mass. The same argument holds for the y and z components. The law of momentum conservation for a system of objects states that the momentum *of the center of mass* remains constant. Furthermore, if the system is subject to external forces, the motion of the center of mass is the same as if the vector sum of all the external forces acted on the total mass situated at the center of mass.

■ APPLICATIONS OF THE SECOND LAW

To solve any dynamics problem, start out by drawing a diagram showing the objects, the forces, and the geometry closely to scale. The next step illustrates the difference between Newton's second and third laws. To apply the second law you must choose the particular object whose motion you want to determine and draw a dotted curve around it. *You must isolate the object.* To apply the third law you must decide which *two* objects are interacting and draw a dotted curve around *them*. The internal forces that each exerts on the other are equal and opposite.

1. For instance, you would think that one of the simplest problems would be to determine the forces that support a book on a table, but many students have trouble with this problem. In Fig. 7.10 we show both a second-law and a third-law approach to the analysis. In the second law, we first isolate the book. We are concerned only with forces acting *on* the book, not on any forces that it may exert. Gravity (the book's weight) is pulling the book downward; the table is shoving it upward. Up to the limit of the strength of the table, the upward force will be sufficient to make the net force, and thus the acceleration, zero. How can the table exert just the right force? Substitute springs or a bathroom scale for the table, and let the students see and handle the appa-

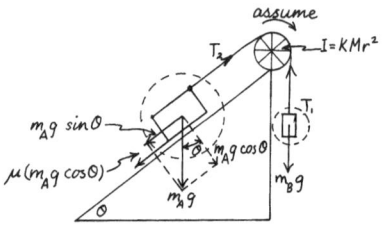

(B) $(m_B g - T_1) = m_B a$
(Pulley) $(T_1 - T_2)r = (kMr^2)\alpha = (kMr^2)\dfrac{a}{r} \to (T_1 - T_2) = kMa$
(A) $T_2 - \mu(m_A g \cos\theta) - m_A g \sin\theta = m_A a$

3 equations 3 unknowns: T_1, T_2, a

Figure 7.11. A complicated Atwood's machine problem and analysis.

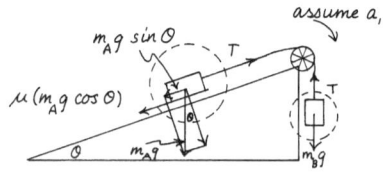

$m_B g - T = m_B a_1 \to T = m_B(g - a_1)$

$T - m_A g \sin\theta - \mu(m_A g \cos\theta) = m_A a_1$

2 equations 2unknown: T and a_1

Eliminate T: $m_B(g - a_1) = m_A g(\sin\theta + \mu\cos\theta) + m_A a_1$

$$a_1 = \frac{m_B g - m_A g(\sin\theta + \mu\cos\theta)}{m_A + m_B}$$

If $m_B g > m_A g(\sin\theta + \mu\cos\theta)$ assumed direction correct
However, if $m_B g < m_A g(\sin\theta + \mu\cos\theta)$ then a may be given by:
$$a_2 = \frac{m_A g(\sin\theta - \mu\cos\theta) - m_B g}{m_1 + m_2}$$

$a_2 \neq -a_1$

If a_2 is negative, then friction is too great and $a = 0$

Figure 7.12. Atwood's machine with sliding friction.

ratus. The heavier the book, the greater the compression of the spring, and thus the greater the upward force. To show that a solid table also compresses or deflects, try intermediate cases of supports, such as a flexible board.

For the application of the third law, the dotted curve surrounds the earth and the book. There is an interaction of attraction between them, yielding the weight of the book in the earth's gravitational field and the *equal* weight of the earth in the book's gravitational field. There are other internal forces within that system. The table between earth and book can be treated as if it were a large spring, holding the book and earth apart. There are repulsive interactions between book and table top, and between earth and table legs.

2. Variations of the Atwood's machine are favorites for examples of how to use Newton's second law. In Fig. 7.11 we show a complicated version and solution. It incorporates about all the complexities anyone would want in an introductory course. First, we draw a dotted curve around each of the three objects and analyze all forces acting on each object from outside its boundary. Second, we resolve all force vectors into perpendicular components. In the case of the forces acting on A, these components are perpendicular and parallel to the plane, and are not vertical and horizontal. Third, we make an arbitrary (but considered) assignment of direction of the resulting velocity. If our choice is wrong, the arithmetic will correct us by yielding a negative acceleration. Note that in this example, $T_1 \neq T_2$. In simpler cases we assume that the mass (or moment of inertia) of the pulley is 0. Then the pulley only changes the direction of the rope and tension and does not enter into the dynamics. In such a case we assume that the tension is constant in magnitude along the rope.

Shrewd students will rapidly figure out that some Atwood problems can be solved by considering the separate objects to be all one system. Because the objects are fastened together, they all have the same magnitude of acceleration. Frequently the in-line forces can be summed and used for the net force acting on the total mass. We think that students should be discouraged from using this method. For one thing, the method is tricky for complicated geometries, particularly ones involving moments of inertia such as the one shown in Fig. 7.11. More importantly, beginning students should learn to isolate simple objects and analyze the exterior forces acting on each object. Most dynamics problems, including these, can also be solved through energy considerations applied to the whole system. We show an example of this in Chap. 9.

The sliding friction force always opposes the relative motion of two surfaces. Once we choose a direction of motion of the objects on the Atwood's machine, we also choose the direction of friction. If we have chosen incorrectly, the arithmetic will not automatically adjust itself. We should consider the first solution a trial. If the direction is wrong, we must reverse the sign of the friction and solve the problem again. However, there is no guarantee that the second trial will work. Once again the arithmetic may yield a negative acceleration. That is because there are three possible outcomes: the system may accelerate to the right, the system may accelerate to the left, or the system may be stuck because of the friction. We illustrate this situation with the diagram and equations in Fig. 7.12.

When problems involving ropes are portrayed in free-body diagrams with arrows representing forces, the tension in a rope may pull in different directions at different ends. Worse yet, sometimes when the rope goes over a pulley the arrows point in the same direction. Most students find the concept of tension in a rope to be mysterious. Try the standard demonstration of pulling on a rope that is tied to a wall. Use two force meters, one inserted into the middle of the rope and the other at the end where you pull. Students will readily believe that if you pull with a force of 100 N, the tension in the rope, as read by the meter in the middle, is also 100 N. Now untie the rope from the wall, add a third force meter, and ask a student to pull on that end with the same force you exert at the other end. Before you start, have the class vote whether the meter in the middle will read, 0, 100, or 200 N. The 200 and 0 choices win most of the votes. There is no magic explanation to resolve this conceptual stumbling block. However, the problem should be recognized and talked over—not just by you, the teacher, but by the students. It seems to help the analysis if you substitute a long, tightly coiled spring for a rope in several geometries. The stretch and thus the tension all along the spring become more visible and thus, we hope, more plausible.

A beginning physics student might well be excused for thinking that the main reason for studying friction is to provide variations on Newton's second law. That is usually the only place the student meets the subject. To the extent that dry sliding friction is independent of velocity, the friction force simply reduces the applied force. $F_{net}=ma$ becomes $F_{applied}-F_{friction}=ma$. We continue to get constant acceleration.

If the friction is a function of velocity, we get very different results. As the velocity increases, the friction increases. The net force is not a constant, but continually decreases to zero as the velocity increases to the terminal velocity. If the friction is *proportional* to the velocity, Newton's second law is:

$$F_{app}-k\frac{dx}{dt}=m\frac{d^2x}{dt^2}.$$

This is a second-order, *linear* differential equation that can be solved analytically. Suppose, for instance, that the applied force is the net weight of a very small object, falling with a Reynolds number $\ll 1$. This would be the case for a mist droplet in air or the oil droplet in the Millikan experiment. We can reduce the equation one order by letting $dx/dt=v$:

$$m\frac{dv}{dt}+kv=W,$$

where

$$W=mg-(\text{buoyancy}).$$

Separate the variables and integrate:

$$mdv+kvdt=Wdt \rightarrow dv=(W-kv)\frac{1}{m}dt=\left(\frac{W}{k}-v\right)\frac{k}{m}dt,$$

$$\frac{dv}{v-(W/k)}=-\frac{k}{m}dt \rightarrow \int_{v_0}^{v}\frac{dv}{v-W/k}=-\frac{k}{m}\int_0^t dt$$

$$\ln\left(\frac{v-W/k}{v_0-W/k}\right)=\frac{k}{m}t \rightarrow \ln\left(\frac{W/k-v}{W/k}\right)=-\frac{k}{m}t \quad \text{if } v_0=0 \text{ when } t=0,$$

$$\frac{W}{k}-v=\frac{W}{k}e^{-(k/m)t} \rightarrow v=\frac{dx}{dt}=\frac{W}{k}(1-e^{-(k/m)t})$$

$$x=\frac{W}{k}t+\frac{Wm}{k^2}e^{-(k/m)t}+C.$$

If $x=0$ when $t=0$, $C=-Wm/k^2$ and

$$x=\frac{W}{k}t-\frac{Wm}{k^2}(1-e^{-(k/m)t}).$$

Note that at $t=0$, $x=0$, and as $t \rightarrow \infty$, $x \rightarrow (W/k)t-Wm/k^2$. The velocity for large t is essentially constant and equal to W/k.

For the motion of most human-size objects at human-size speed, the Reynolds number $\gg 1$, and the fluid friction is proportional to v^2. The resulting differential equation is nonlinear and is messy to solve analytically. The numerical solution of such problems is an ideal exercise for a student and computer, particularly as a spreadsheet exercise. For first approximations, the computation can be carried out with a hand calculator.

The equation for a large object falling through air is

$$\frac{d^2y}{dt^2}=g-\frac{k}{m}\left(\frac{dy}{dt}\right)^2.$$

First, note that the velocity will increase steadily until the acceleration is 0. At that time, $mg=k(dy/dt)^2$. The terminal velocity is $v_{\text{terminal}}=\sqrt{mg/k}$.

We show a sketch of $v(t)$ in Fig. 4.12, p. 98. At first the object falls with nearly constant acceleration g. As the velocity increases, the drag increases, and the acceleration decreases.

Let us calculate the velocity and displacement during the first 10 s, step by step, using 1-s intervals. Assume that the object is a human body with a mass of 60 kg and an approximate value for k of $\frac{1}{6}$ kg/m, and let $g=9.8 \approx 10$ m/s^2:

$$v_{\text{terminal}}=\sqrt{\frac{600}{1/6}}=60 \text{ m/s} \approx 135 \text{ miles/h}.$$

The calculation method is based on the approximation that the velocity at time $t=n$ is equal to the velocity at t_{n-1} plus the change of velocity during the time interval between $n-1$ and n: $v_n=v_{n-1}+\Delta v=v_{n-1}+a_{n-1}\Delta t$. Similarly, the distance fallen at $t=n$ is equal to the distance fallen at $t=n-1$ plus the product of the average velocity during the time interval and the length of the time interval: $y_n=y_{n-1}+v_{\text{av}}\Delta t$. The average velocity during this time interval is

$$v_{\text{av}}=\left(\frac{v_n+v_{n-1}}{2}\right)\Delta t.$$

Here is the second-by-second array of calculations, starting with $t=0$:

$$v_n=v_{n-1}+a_{n-1}\Delta t, \quad y_n=y_{n-1}+\left(\frac{v_n+v_{n-1}}{2}\right)\Delta t, \quad [y_{\text{vacuum}}=\tfrac{1}{2}gt^2].$$

t		$[v_n^2]$		$v_{\text{vaccum}}=gt$		$a_n=g-\dfrac{k}{m}v_n^2$
0	0	0	0	0	0	$g=10$
1	10	100	5	10	5	9.72
2	19.72	388	19.86	20	20	8.92
3	28.64	821	44.04	30	45	7.72
4	36.36	1320	76.54	40	80	6.33
5	46.29	1825	116.06	50	125	4.93
6	47.62	2265	161.21	60	180	3.71
7	51.33	2635	210.68	70	245	2.68
8	54.01	2920	263.35	80	320	1.89
9	55.90	3130	318.30	90	405	1.31
10	57.21	3275	374.85	100	500	0.90

It was assumed that during each interval the applicable acceleration was that calculated for the end of the previous interval. The extra significant figures were carried to prevent cumulative error. Substitution of 10.0 for g instead of 9.8 produces only a constant 2% error in both calculated and vacuum values. In Fig. 7.13 we show the graphs for v and y from the data calculated above. You can see the comparison between falling in air and falling in vacuum. In Chap. 4, p. 98, more information is given about the terminal velocities of various objects. There is a similar treatment of step-by-step calculations for a different problem, in Feynman's *Lectures on Physics*.

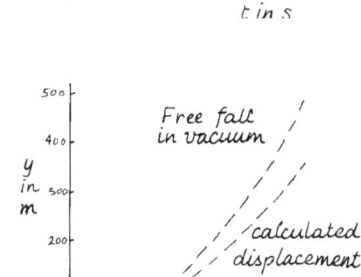

Figure 7.13. $v(t)$ and $y(t)$ for object falling a long distance in air.

■ CIRCULAR DYNAMICS AND NEWTON'S SECOND LAW

Each of the variables used to describe motion along a line has its counterpart in circular terms. The position x becomes the angle θ. The time rate of change of position, $dx/dt=v$, is linear speed. Similarly, the time rate of change of angular displacement, $d\theta/dt=\omega$, is the angular or rotational speed. The time rate of change of angular velocity is angular acceleration: $\alpha=d\omega/dt$.

The dimensions of these circular units are different from their linear counterparts. An angle has the dimensions of just a number, since the measure of an angle is the ratio of an arc (a length) to the radius (a length). The dimensions of ω are therefore T^{-1}, and the dimensions of α are T^{-2}. Three kinds of units for angular velocity are used, depending on the unit chosen for θ. The angle might be given in degrees, radians, or revolutions. The angular velocity would then be given in degrees per second, radians per second, or revolutions per second. Conventional usage is that ω refers to radians per second, and frequency f is the name for revolutions per second. The unit for revolutions per second, or vibrations per second, is Hz, hertz. If a wheel has a frequency f of 1 Hz, it turns 2π radians during one second. Consequently, $\omega=2\pi f$.

Velocity and acceleration, both in linear and circular motion, can be represented by vectors. It is not obvious how to represent the direction of angular velocity and acceleration, except in terms of clockwise and counterclockwise. Of course, when we say "clockwise", we refer to a rotation around an axis that itself has a particular direction. The custom is to give the direction of rotation with the right-hand rule. Let the fingers of your right hand curl in the direction of the circular motion. Your thumb points along the axis in the direction assigned to the motion. However, this system works only for small rotational displacements or instantaneous rotational velocities and accelerations. If you use the system for large displacements or long-term average velocities, you will find that vector addition does not work: $\mathbf{A}+\mathbf{B} \neq \mathbf{B}+\mathbf{A}$. This difficulty was illustrated in Fig. 4 of Chap. 3.

There are several other useful relationships to be noted. The period of rotation is equal to the reciprocal of the frequency: $T = 1/f$. If a wheel is rotating at 10 Hz, then the period is $\frac{1}{10}$ s. There is also a simple relationship between the tangential velocity of an object going in a circle and its angular velocity. With tangential velocity v the object would move through an arc, $v\,dt$, during a time interval dt. That arc length is equal to $r\,d\theta$. Since $d\theta/dt = \omega$, $v = r\omega$. (Since all these quantities are vectors, the equation really is $\mathbf{v} = \boldsymbol{\omega} \times \mathbf{r}$.)

This formula holds only if ω is expressed in units of radians per second. The same argument gives a similar relationship between the tangential acceleration and the rotational acceleration, $\mathbf{a} = \boldsymbol{\alpha} \times \mathbf{r}$. Finally, the formula for centripetal acceleration is usually derived in one form, but all of the following forms are useful:

$$a_{\text{centripetal}} = \frac{v^2}{r} = \omega^2 r = 4\pi^2 f^2 r = \frac{4\pi^2}{T^2} r.$$

The direction of the *rotational* or *tangential* acceleration is *axial*; the direction of the *centripetal* acceleration is negative *radial*. (For an analysis of that perennial problem about whether *centrifugal* force, which is positive radial, is real or fictitious, see p. 131.)

Let us apply Newton's second law to a point object with mass m rotating in a circle. We will exert a force that is constant in magnitude, but continually changing in direction so that it is always in line with the tangential velocity. If it were not for this changing direction, we could just use Newton's second law in its usual form: $F = ma$. To account for the continually changing direction, we can transform the equation by vector multiplication with the radius vector $\mathbf{r}$:

$$\mathbf{r} \times \mathbf{F} = \boldsymbol{\tau} = \mathbf{r} \times m\mathbf{a} = (mr^2)\boldsymbol{\alpha} \rightarrow I\boldsymbol{\alpha}.$$

This transformed equation is Newton's second law for rotational motion. Notice how parallel it is to the original. Force becomes torque τ. Tangential acceleration becomes angular acceleration. Mass becomes a combination of mass and its location. This particular combination (mr^2) is called the *moment of inertia* and is usually given the symbol I. The moment of inertia plays the same role in rotational motion that mass does in linear motion. The greater the moment of inertia of an object, the larger the torque needed to produce a change in its angular velocity. However, it is not just the mass that determines the moment of inertia; the radial location of the mass is even more important since I depends on the *square* of r.

■ MOMENTS OF INERTIA

Although we derived Newton's second law for rotational motion in terms of a point object at radius r, the same formula would apply to a thin hoop or rim with radius r. The moment of inertia of each little piece on the hoop would be mr^2. The moment of inertia of the whole hoop is Mr^2, where M is the sum of the masses of all the individual pieces. For any other kind of rotating object, each piece with mass m has its own particular radius and contributes its own particular value, mr^2, to the total moment of inertia.

If two rotating systems have the same mass, but the mass of the first is mostly around the rim while the mass of the second is concentrated at the hub, the first one will be much more sluggish in responding to torques speeding it up or slowing it down. Here is a way to demonstrate the phenomenon. Take four full soup cans (or any other convenient weights) and place them symmetrically on a record player turntable. With the turntable disengaged and free to rotate, try spinning it slowly by hand—first while the cans are near the axis, and then while the cans are near the rim. The mass is the same in both cases, but the torque required to start or stop the spinning is noticeably different.

Calculation of moments of inertia for various geometries is a standard calculus exercise. It involves the solution of the integral, $I = \int r^2 dm$, where the mass element, dm, must be expressed in terms of density and r. For instance, for a solid cylinder, the convenient mass element would be a ring with $dm = \rho L 2\pi r\, dr$. The integral would be

$$I = \rho L 2\pi \int_0^R r^3 dr = \tfrac{1}{4}\rho L 2\pi R^4 = \tfrac{1}{2}MR^2,$$

where $M = \rho \pi R^2 L$.

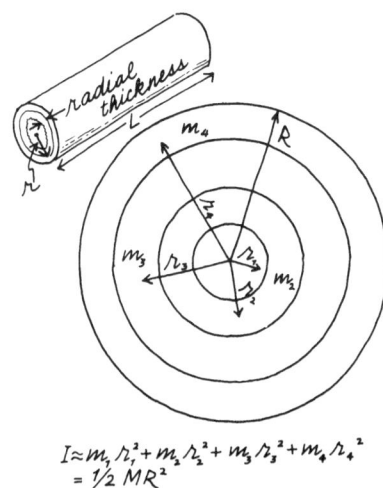

Figure 7.14. Approximation analysis of moment of inertia of cylinder.

Sometimes it is helpful for students to see approximate solutions to a simple problem. Here is such a numerical calculation for the case of the solid cylinder. Divide the cylinder into four thin rings as shown in Fig. 7.14. Each ring has a moment of inertia of $m_i r_i^2$. Add together the separate moments of inertia:

$$\sum m_i = M, \quad \sum I_i = I.$$

Choose rings with $\Delta r = 1$. Choose the average radius in the middle of each ring so that $r_1 = \tfrac{1}{2}$, $r_2 = \tfrac{3}{2}$, $r_3 = \tfrac{5}{2}$, $r_4 = \tfrac{7}{2}$, and $R = 4$. The mass of a ring is approximately equal to $\rho 2\pi r L \Delta r$. The mass of the ith ring is $m_i \approx \rho L 2\pi r_i$, and the moment of inertia of that ring is $\rho L 2\pi r_i^3$:

$$I_1 = \rho L \pi\left(\frac{1}{4}\right), \quad I_2 = \rho L \pi\left(\frac{27}{4}\right), \quad I_3 = \rho L \pi\left(\frac{125}{4}\right), \quad I_4 = \rho L \pi\left(\frac{343}{4}\right),$$

$$I = \rho L \pi (124).$$

Since $M = \rho L \pi R^2 = \rho L \pi (16)$, $I = M(7.8)$. The actual value for the moment of inertia of this cylinder is $I = \tfrac{1}{2} MR^2 = 8M$. Even with our crude approximation of only four rings, we get an answer correct to $2\tfrac{1}{2}\%$. Of course the approximation would get better as the rings chosen became thinner.

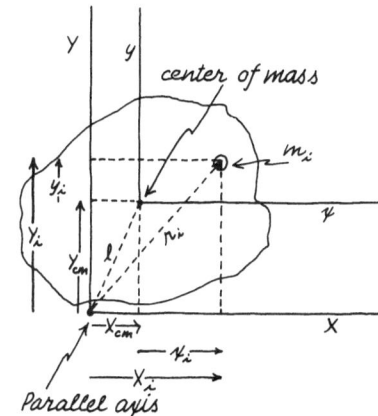

Figure 7.15. Analysis of parallel axis theorem for moment of inertia.

The general definition of the moment of inertia does not define the location of the axis or assume that the object has symmetry. In introductory physics the objects we deal with are usually symmetrical, and the calculations are easiest if we assume that the axis is through the center of mass. However, sometimes things roll on their rims rather than on their axes. It would be difficult to use the previous method to calculate the moment of inertia of a solid cylinder rotating about an instantaneous axis at its rim. Evidently $I_{\text{rim}} > I_{\text{c.m.}}$ because some of the mass is much farther away from the axis when it is rolling on the rim. Fortunately, there is a simple relationship called the *parallel axis theorem*.

The moment of inertia of the object shown in Fig. 7.15 is

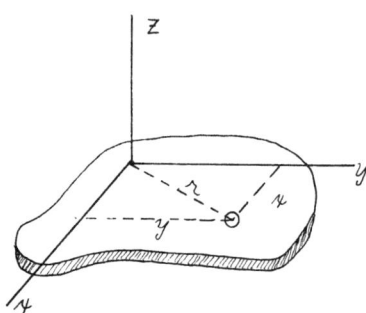

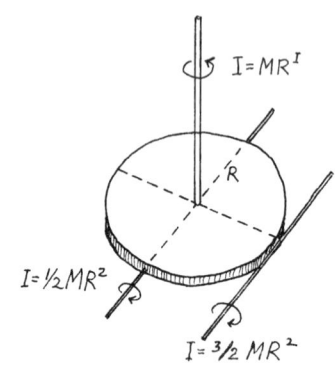

Figure 7.16. Application of perpendicular and parallel axis theorems to find moments of inertia of ring around several axes.

$$I = \sum m_i r_i^2 = \sum m_i (X_i^2 + Y_i^2)$$

The x and y locations of particle i can be written in terms of the x and y location of the center of mass:

$$X_i = X_{\text{c.m.}} + x_i, \quad Y_i = Y_{\text{c.m.}} + y_i$$

Substitute these values into the equation for I:

$$I = \sum m_i(x_i^2 + y_i^2) + \sum m_i(X_{\text{c.m.}}^2 + Y_{\text{c.m.}}^2) + 2X_{\text{c.m.}} \sum m_i x_i + 2Y_{\text{c.m.}} \sum m_i y_i$$

Each of the last two terms is 0, because the sum of the $m_i x_i$ to the right of the center of mass cancels the sum of the $m_i x_i$ to the left of the center of mass. The same argument applies to the y components. In the second term of the equation, $X_{\text{c.m.}}^2 + Y_{\text{c.m.}}^2$ is just the distance squared from the center of mass to the parallel axis around which the rotation is taking place. Call that distance l. The first term in the equation is the moment of inertia around an axis through the center of mass. Therefore, the moment of inertia of an object around an axis parallel to the center of mass and a distance l from it is $I = I_{\text{c.m.}} + Ml^2$.

There is another helpful general relationship about moments of inertia, called the *perpendicular axis theorem*. The sum of the moments of inertia of a plane (two-dimensional) object about any two perpendicular axes in the plane is equal to the moment of inertia about an axis through the point of intersection perpendicular to the plane. Here is the proof, following the geometry in Fig. 7.16. The moment of inertia of a particle about the x axis is $I_x = my^2$. Its moment of inertia about the y axis is $I_y = mx^2$. The sum of these is $I_x + I_y = m(x^2 + y^2) = mr^2 = I_z$. Since the moment of inertia of the whole object is just the sum of the individual moments, the theorem holds for the whole object.

For an example of the use of the perpendicular axis theorem, consider how you would calculate the moment of inertia of a ring around a *diameter* through its center. The moment of inertia of the ring around its axis is elementary—just MR^2. However, to calculate the moment of inertia around a diameter requires some careful analytical geometry and use of the calculus. Our theorem allows us to bypass these complications. According to the theorem the sum of the moment of inertia about any diameter plus the *equal* value for the moment about a perpendicular diameter is equal to the moment of inertia around the axis, which is equal to MR^2. Consequently, the desired moment of inertia about a diameter is equal to $\frac{1}{2}MR^2$. Taking the exercise one step further by using the parallel axis theorem, we deduce that the moment of inertia of a ring rotated about an axis at its edge, *parallel to a diameter*, is $\frac{1}{2}MR^2 + R^2M = \frac{3}{2}MR^2$.

Here are moments of inertia of some standard symmetrical objects:

	About the center axis	About a point on the rim
Ring	MR^2	$2MR^2$
Washer (inner R_1, outer R_2)	$\frac{1}{2}M(R_1^2+R_2^2)$	$\frac{1}{2}M(R_1^2+3R_2^2)$
Solid cylinder	$\frac{1}{2}MR^2$	$\frac{3}{2}MR^2$
Spherical shell	$\frac{2}{3}MR^2$	$\frac{5}{3}MR^2$
Solid sphere	$\frac{2}{5}MR^2$	$\frac{7}{5}MR^2$
The earth	$0.3444MR^2$	!
Uniform thin rod	$\frac{1}{12}ML^2$ (about the axis through center, perpendicular to the length)	$\frac{1}{3}ML^2$ (about the axis through the end, perpendicular to length)

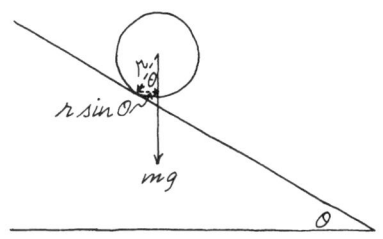

Figure 7.17. Dynamics of objects rolling down hill.

The moment of inertia of the earth is less than that of a solid sphere because the earth has a high density core.

The classical demonstration of moments of inertia involves a race downhill between various rolling objects. Students should consider the factors and place their bets before the race begins. If an object is rolling downhill, its weight provides a torque of $(mg)R \sin \theta$, as shown in Fig. 7.17. Newton's second law for rotation is $\tau = I\alpha$.

$$mgR \sin \theta = KmR^2 \alpha = KmR^2 \frac{a}{R} \rightarrow g \sin \theta = Ka.$$

The acceleration is independent of the mass and of the radius, but the acceleration is inversely proportional to the geometrical factor of the moment of inertia. The largest value for K is 2, for a hollow cylinder rolling on its rim. For a solid sphere, $K = \frac{7}{5}$. For a solid cylinder, $K = \frac{3}{2}$. The sphere will beat the solid cylinder, which will beat the hollow cylinder. Since the effect is independent of mass or size of the objects, you can race any kind and any size of solid ball against tin cans or solid pencils, and so on.

■ REFERENCES

Eisenbud, Leonard, *On the Classical Laws of Motion*, Am. J. Phys. **26**, 124 (1958).

Feynman, Richard, *Lectures on Physics*, Addison-Wesley Pub. Co., Vol. I, Sec. 9–6. Reading, MA (1963).

Morse, Robert, *Constant Acceleration: Experiments with a Fan-Driven Dynamics Cart*, The Physics Teacher **31**, 436 (Oct. 1993).

8

Angular Momentum

Angular momentum plays the same role for rotational motion that linear momentum plays for motion in a line. However, angular momentum is responsible for some special, startling effects. Spinning objects, such as tops, seem to move in directions perpendicular to what you would at first expect. The conservation of angular momentum is responsible for the fast spins of twirling skaters and dancers and also played a dominant role in forming our solar system. The conserved nature of angular momentum permits certain interactions among subatomic particles and forbids others. Strangest of all, the angular momentum of a bound system is quantized, much like the situation with electric charge. A rotating system cannot increase or decrease its angular momentum continuously in magnitude or direction, but only in unit amounts that are integral multiples of a basic unit. The topic of angular momentum is a marvelous one for easy, fascinating demonstrations and for tying together everyday phenomena with the microworld and the cosmos.

■ PRELIMINARY DEMONSTRATION

Every physics room should have a low-friction rotating stool and a bicycle wheel, loaded at the rim, with handles. Each student should experience (not just watch) the effect of spinning on the stool with a textbook in each outstretched hand, and then bringing in the books to the lap. You must make sure that there is lots of room around the stool, because sometimes the effect is so startling that a student may fall. Actually, you can experience the effect without a rotating stool and even if you are not a good skater or dancer. With ordinary shoes and lots of free room, anyone can start spinning on one toe. Hold a textbook or some other weighty object in each outstretched hand and then rapidly bring the books in toward your chest. You will have to bully students to try these stunts, but the effects are dramatic and worth the time.

Each student should feel the effect of trying to change the direction of a spinning bicycle wheel. They should hold the wheel in front of them with both hands while someone else gives it a spin. Then try to turn the wheel. Angular momentum is conserved in direction as well as magnitude. Have each student combine the rotating stool with the spinning wheel. While sitting on the stool with legs off the floor, hold the spinning wheel with axis vertical. Then turn the wheel over 180°. The student starts rotating so that the total system conserves its angular momentum.

■ THE FORMS OF ANGULAR MOMENTUM

An object barreling along in a straight line has momentum $m\mathbf{v}$. With respect to some axis that it is passing, it also has *angular momentum* equal to $\mathbf{r} \times m\mathbf{v}$. The cross product of the two vectors simply yields the product of mv and the distance of nearest approach, as shown in Fig. 8.1. So far this may seem like a useless definition. Just because an object is going past an axis does not mean that any rotational motion exists. However, if the object ran into a paddle of a wheel extending into its path, it would start the paddle wheel spinning. Therefore, it is consistent to say that angular momentum already existed in the complete system—whether or not it gets transferred from the projectile to the paddle wheel. Note that if the object is headed straight for the axis, it has no angular momentum about that axis. Note also that the angular momentum of this system is not quantized, since the system is not bound. The mass and radius and velocity can have arbitrary values.

It might seem that with this definition, the angular momentum would change continuously as the projectile goes past the axis. The distance $\mathbf{r}$ to the axis changes steadily. However, the magnitude of the vector angular momentum is $|r||mv|\sin\phi$, and $r\sin\phi$ is a constant equal to the distance of closest approach. Since $r\sin\phi$ is a constant, the angular momentum with respect to the axis is a constant.

In the case of a spinning bicycle wheel, the magnitude of the angular momentum about its own axis is approximately mvr. To get this value, we make the approximation that all the mass is at the radius r. The tangential velocity v of each part of the rim is automatically perpendicular to the radius. In terms of the rotation variables, the angular momentum is

$$mvr = m(r\omega)r = (mr^2)\omega = I\omega,$$

where I is the moment of inertia about the axis. This last form is valid, regardless of how the mass of the rotating system is distributed.

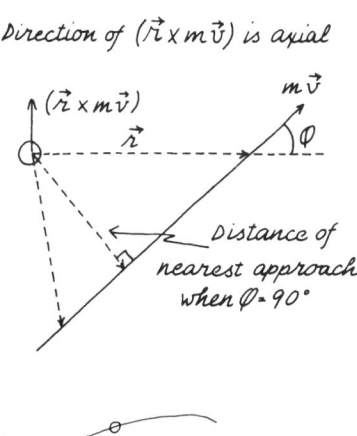

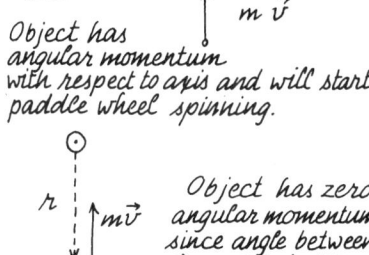

Figure 8.1. An object traveling in a straight line has angular momentum with respect to an axis through which it is not passing.

There is a fundamental difference between the angular momentum of an object rotating around its center of mass and the angular momentum of the object as it goes past some axis point. The former is frequently called *spin*, to distinguish it from the angular momentum caused by its rotation around or past an external axis. An object's spin is a feature of the object itself, without regard to any outside axis. In general, however, the magnitude of any angular momentum depends on the axis point chosen. For instance, the angular momentum of a bicycle wheel around its axis—its spin—depends only on the radius of the wheel, the mass and mass distribution of the wheel, and the speed. For some purposes, however, it is convenient to consider the angular momentum of the wheel around the momentary axis where it touches the road. That angular momentum is larger than the spin because it includes not only the spin but also the angular momentum of the center of mass going past the momentary axis.

In Chap. 7 we transformed Newton's second law for linear motion into the law for rotation: $F=ma \to \tau = I\alpha$. However, we also pointed out that the most general form of the second law is

$$F = \frac{dp}{dt} \approx \frac{d(mv)}{dt}.$$

Similarly, for rotation the general form is

$$\tau = \frac{dL}{dt} \approx \frac{d(I\omega)}{dt}.$$

The angular momentum $I\omega$ is given the symbol L.

■ MAGNITUDES OF SOME ANGULAR MOMENTA

Objects appear to be spinning or in orbit at every range of magnitude, from the subatomic realm to the domain of the galaxies. Let us calculate the angular momenta of objects in three different ranges of size. We should compare these angular momenta in terms of some common unit. Although, as we will see, nature has a basic unit of angular momentum, no special name is given to the *Système International* (SI), or humanmade unit. In terms of kilograms, meters, and seconds, the unit of angular momentum is 1 kg m²/s. These same units appear whether you define angular momentum as mvr or $I\omega$.

1. Let us start with an object that is human size. An ordinary bicycle wheel has a mass of about 2 kg and a radius of 30 cm. If we make the approximation that all of the mass is at the rim, then a wheel traveling at 10 miles/h ($4\frac{1}{2}$ m/s) has a spin (an intrinsic angular momentum about its axis) of $mvr = (2 \text{ kg})(4.5 \text{ m/s})(0.30 \text{ m}) = 2.7$ kg m²/s. If the wheel is rolling along the road, there is an equal but separate value of mvr with respect to an instantaneous axis where the rim touches the road.

2. In the Bohr model of the atom, proposed by Niels Bohr in 1913 and still very useful, the valence electron in an atom travels in a circular orbit around the central nucleus. The mass of an electron is about 1×10^{-30} kg; the radius of an atom (and therefore the radius of the orbit of the outer electron) is about 1×10^{-10} m, and the speed of the electron is about 1×10^6 m/s. (This last value can be obtained from the data on p. 517 in Chap. 21.) *On the basis of this model*, the angular momentum of such an electron is
$$mvr = (1 \times 10^{-30} \text{ kg})(1 \times 10^6 \text{ m/s})(1 \times 10^{-10} \text{ m}) = 1 \times 10^{-34} \text{ kg} \cdot \text{m}^2/\text{s}.$$

This seems like an unimaginably small quantity. Compare it with the angular momentum of a bicycle wheel. But look at the number—we have obtained the (reduced) Planck's constant, $h/2\pi$, which is the quantum unit of angular momentum!

3. Now let us calculate the angular momentum of something big. The earth spins on its axis with a frequency of 1 revolution per day. If the earth's density were uniform, which it is not, the moment of inertia would be $I = \frac{2}{5}MR^2 = 0.4MR^2$. Because the core of the earth is more dense than the mantle, the actual moment of inertia is $0.3444MR^2$. The angular momentum is $I\omega = (0.3444)(5.98 \times 10^{24} \text{ kg})(6.38 \times 10^6 \text{ m})^2 (2\pi \text{ rad})/(24 \text{ h})(1 \text{ h}/3600 \text{ s}) = 6.1 \times 10^{33}$ kg m²/s. This quantity of angular momentum is about as large in one direction as the size of the angular momentum of the valence electron is small in the other direction. Actually, as we shall see, this angular momentum of the spinning earth is small compared with other angular momenta in the solar system.

THE CONSERVATION OF ANGULAR MOMENTUM

For a closed system, with no external torques, the change of angular momentum, $\Delta(I\omega)$, must equal zero. That does not mean, however, that isolated rotating objects always have the same angular *speed*. Here are four examples that illustrate this peculiar nature of rotational motion.

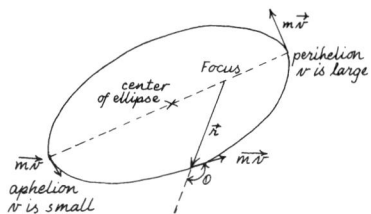

Figure 8.2. The constant angular momentum of an object in an elliptical orbit requires different speeds at different points in the orbit.

1. The orbits of the planets (and of most earth satellites) are not really circles, but ellipses. The eccentricity of the earth's orbit around the sun is so small that the elliptical nature can hardly be shown in a page-size drawing. (The eccentricity has a value, $e=0.0167$, where $e=0$ is a circle and $e=1$ is a parabola. Mercury has $e=0.2056$, the most eccentric of the planetary orbits. For the moon's orbit around the earth, $e=0.055$, producing easily observed differences in apparent size of the moon and noticeable differences in tides.) The ratio of the largest diameter to the shortest diameter of the earth's orbit differs from unity by only $1\frac{1}{2}$ parts per 10^4. Nevertheless, the difference between aphelion radius, when we are furthest from the sun, and perihelion radius, when we are closest, is about 3%. This is because the focal point of the ellipse, which is the sun's location, is off center by about $1\frac{1}{2}\%$ of the radial length. The focal distance is 1.6×10^6 miles from the center of the orbit. Since the sun is about 1×10^6 miles in diameter, the center point of the earth's orbit is only about the sun's radius away from the edge of the sun. What happens to the angular momentum in a noncircular orbit? If the radius is not always perpendicular to the velocity, perhaps the sun exerts a torque on the earth and changes its angular momentum.

   To simplify this problem, let us assume that the effects of the other planets are negligible, and that the sun–earth system can be considered to be an isolated two-body system. The sun does, indeed, exert a force on the planet, but it is a gravitational force directed along the radius, connecting the planet to the sun. The torque on the planet with respect to the sun as axis is $|\boldsymbol{\tau}|=|\mathbf{r}\times\mathbf{F}|=|\mathbf{r}||\mathbf{F}|\sin\theta$. Remember that θ is the angle between the radius arm and the force vector. Torque is maximum when θ is 90°; it is 0 when θ is 0° or 180°. For any *radial* force, F lies along r, either in the positive or negative direction, and therefore θ is 0° or 180° and the torque is zero. Consequently there can be no change in angular momentum. This is true for the orbits controlled by any central force. In the case of the gravitational force, which is proportional to $1/r^2$, the allowed orbits can be circles, ellipses, parabolas, or hyperbolas.

   In Fig. 8.2, which shows a very eccentric ellipse, we emphasize that the angular momentum of a planet is not just mvr, where v is the tangential speed and r is the radial distance between sun and planet. The amplitude of the angular momentum $=m|\mathbf{v}||\mathbf{r}|\sin\phi$, where ϕ is the angle between $\mathbf{v}$ and $\mathbf{r}$. At all points of the orbit, $\mathbf{v}$, $\mathbf{r}$, and ϕ must change in such a way that the angular momentum stays constant. At aphelion and perihelion, the velocity and radius are perpendicular to each other: $\phi=90°$. In those cases, the angular momentum is simply mvr. Since that product must be the same at both points, at aphelion the tangential velocity of the earth must be 3% smaller than it is at perihelion. For comets the effect is more dramatic. Their eccentricity is close to 1, and their speeds at aphelion are so small compared with

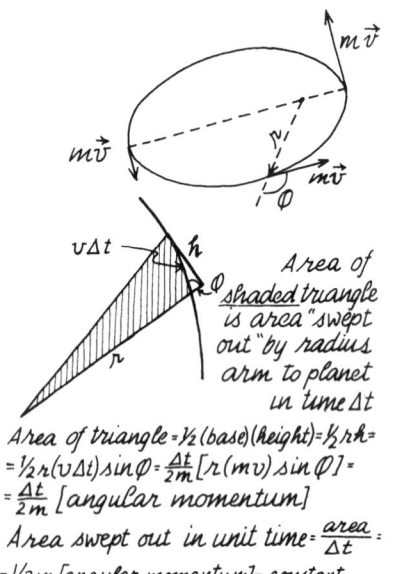

Figure 8.3. Proof of Kepler's Second law for object in orbit that conserves angular momentum.

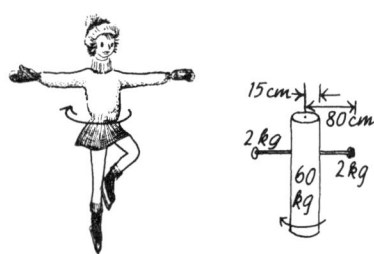

Figure 8.4. Analysis of a model of a twirling skater.

their speed at perihelion that they spend most of their time traveling slowly at the far reaches of the solar system.

A geometrical consequence of the conservation of angular momentum was discovered by Kepler back in 1609. He not only figured out that the orbits of the planets are ellipses, but he also found from his observations that the radius arm from sun to planet sweeps out equal areas in equal times. The derivation of this rule (which had to wait until Newton) is shown in Fig. 8.3.

2. Now we must explain how skaters can twirl so fast, and what happens when you spin on one foot while holding weights or textbooks. The person starts turning with arms outstretched. Once on tiptoe the twirler is not subject to much external torque; therefore, the angular momentum of the twirler must be nearly constant: Fig. 8.4 shows a crude model of the situation. The trunk of the body has become a solid cylinder with a mass of 60 kg and an outer radius of 15 cm. The arms and weights turn into lumps at a radius of 80 cm with a total mass of 4 kg. The total angular momentum of the model is

$$[\tfrac{1}{2}(60 \text{ kg})(0.15 \text{ m})^2]\omega + [(4 \text{ kg})(0.80 \text{ m})^2]\omega = [0.68]\omega + [2.6]\omega.$$

Note that the angular momentum of the much lighter arms and hands is about four times as large as the angular momentum of the core body. If the original frequency is 1 revolution every 2 seconds, $\omega = 3.1$ rad/s. The total angular momentum is $(3.3 \text{ kg m}^2)(3.1 \text{ rad/s}) = 10$ kg m^2/s.

Next the twirler brings her arms and the weights down beside her body. The angular momentum stays the same, but the radius of the arms and weights is now greatly reduced.

$$10 \text{ kg m}^2/\text{s} = 0.68\omega + [(4 \text{ kg})(0.15 \text{ m})^2]\omega = 0.68\omega + 0.09\omega = (0.77 \text{ kg m}^2)\omega,$$
$$\omega = 13 \text{ rad/s}.$$

The angular frequency has increased by over a factor of 4. The twirler is now going over 2 revolutions per second.

Divers create a torque that produces angular momentum as they spring off the board. At the moment their feet are thrown into the air by the board, their center of mass is already beyond the edge of the board. From that point on, their angular momentum is approximately constant, although air resistance can exert a torque in a high dive. How fast divers spin depends on how they distribute their mass on the way down. If they stay stretched out, their feet will rise relative to their center of mass and they will enter the water after half a revolution, more or less, depending on the height of the board. If divers tuck their arms and legs together, they will go into a fast spin, still conserving angular momentum, but with a smaller moment of inertia.

If a cat starts falling without angular momentum, and if air effects are negligible, the cat cannot generate net angular momentum. However, the cat can rotate one part of its body one way and its legs in the opposite direction and thus can change its angular orientation. Spinning the tail is not necessary, but arching its back is. Even a Manx cat, which has no tail, always lands on its feet. There is a discussion of this complicated problem, which cats and some divers can solve easily, by Galli in *The Physics Teacher* (Sept. 1995).

3. Rotational motion is one of the most common and striking features of the universe. Planets, moons, and stars spin on their axes; the planets have moons and the sun has planets revolving around it; a large fraction of stars appear to be binaries—rotating double stars with a complete range of size of partners; the stars and their satellites revolve around the centers of their galaxies; and many galaxies are parts of swirling clusters of other galaxies. On a more homely scale, you have probably seen dust devils of wind pick up dust or leaves and swirl them into the air as if there were a miniature tornado. In rowing a boat or in swimming, you may have noticed tiny whirlpools spinning off in the wake. Most of these rotations start with motion of material along straight lines. What makes everything swirl?

We gave an example earlier of why an object hurtling past an axis really has angular momentum with respect to that axis. If the object hits the paddle of a wheel centered on the axis, the wheel will start turning. If two currents of fluid, either air or water, are flowing side by side in opposite directions, they have angular momentum with respect to each other. If the conditions of viscosity and relative velocity are sufficient, whirlpools will be generated at the boundary between the two currents.

On a cosmic scale, if atoms were uniformly distributed in space and motionless to begin with, they would all fall toward the center of mass along straight, radial lines. Eventually they would collide and form a star with no rotational motion at all. However, the original condition of uniform distribution of the raw materials just never occurs. In a region of the galaxy where a star is going to form, the atoms are distributed with varying density and varying initial velocities. The motions of the atoms are not radial toward the center of mass and, therefore, angular momentum exists. As the atoms condense into a star, the radius of the system shrinks from a distance of a light-year or so to the star's final size of about 10^9 m, a reduction by a factor of 10^7. If a new star had no companion or satellite system to absorb the angular momentum, in most cases it would be left spinning so fast that it would not be stable.

Let us see where the angular momentum in our solar system resides. We have already calculated the spin angular momentum of the earth. The sun is also spinning on its axis, although since it is not a rigid body the spin period varies with latitude. The period for the equatorial region is $24\frac{2}{3}$ of our days, and at a latitude of 75° the rotation period is about 33 days. The assumption that the sphere has uniform density is even less accurate for the sun than for the earth. The density of the sun as a whole is 1.4×10^3 kg/m^3; the density at the center is about 100 times greater. If we assume that for such a body the moment of inertia is $I = (0.1)MR^2$, then the angular momentum of the sun is about

$I\omega = (0.1)(2 \times 10^{30}$ kg$)(7 \times 10^8$ m$)^2(2\pi$ rad/25 days$)(1$ day/8.6×10^4 s$)$

$= 3 \times 10^{41}$ kg m^2/s.

The absolute value of this enormous quantity is unimportant; just compare it with the angular momentum of the spinning earth, which is smaller by a factor of 5×10^7. But these quantities are still much smaller than the *orbital* angular momenta of the planets.

Let us calculate the orbital angular momenta of earth and Jupiter. Here are the data:

$$M_{earth} = 6 \times 10^{24} \text{ kg}, \quad M_{Jupiter} = 1.9 \times 10^{27} \text{ kg},$$
$$\text{orbital } r_{earth} = 1.5 \times 10^{11} \text{ m}, \quad \text{orbital } r_{Jupiter} = 7.8 \times 10^{11} \text{ m},$$
$$T_{Jupiter} = 11.9 \text{ earth years}, \quad 1 \text{ earth year} = 3 \times 10^7 \text{ s}.$$

For earth:
$$I\omega = (6 \times 10^{24} \text{ kg})(1.5 \times 10^{11} \text{ m})^2 (2\pi \text{ rad})/(3 \times 10^7 \text{ s}) = 3 \times 10^{40} \text{ kg m}^2/\text{s}.$$

For Jupiter:
$$I\omega = (1.9 \times 10^{27} \text{ kg})(7.8 \times 10^{11} \text{ m})^2 (2\pi \text{ rad})/(11.9 \text{ yr})(3 \times 10^7 \text{ s/yr})$$
$$= 2 \times 10^{42} \text{ kg m}^2/\text{s}.$$

As you can see, Jupiter in its orbit possesses much more angular momentum than the spinning sun itself. Is it possible that some other planet, perhaps Neptune with its great orbital radius, carries more angular momentum than even Jupiter? Here is a table of the needed data.

Planet	Distance from sun (10^6 km)	Mass (10^{24} kg)	Period in days
Mercury	57.9	0.3	88
Venus	108.1	4.9	225
Earth	149.5	6.0	365
Mars	227.8	0.6	687
Jupiter	777.8	1900	4 333
Saturn	1426	569	10 760
Uranus	2868	87	30 690
Neptune	4494	103	60 190
Pluto	5908	5.4	90 740

Rather than repeating the detailed calculations, we can compare the angular momenta of the planets by manipulating one of Kepler's laws: $T^2 = Kr^3$. The orbital angular momentum is then equal to

$$mr^2\omega = mr^2 \frac{2\pi}{T} = \frac{mr^2 2\pi}{\sqrt{Kr^3}} = 2\pi m \sqrt{\frac{r}{K}}.$$

It appears that the orbital angular momentum of a planet is proportional to the mass of the planet and the square root of its orbital radius. Comparing the angular momenta of Neptune and Jupiter, we see that the orbital radius for Neptune is 45×10^{11} m, and the orbital radius for Jupiter is 7.8×10^{11}. The ratio of these distances is 5.77 and the square root of the ratio is 2.4. On the other hand, the ratio of the mass of Jupiter to the mass of Neptune is 18.4. Therefore, the orbital angular momentum of Jupiter is greater than that of Neptune by a factor of $18.4/2.4 = 7.7$.

As viewed looking down at our North Pole, all of the planets revolve about the sun in a counterclockwise direction. Their orbits are all approximately in the same plane, and the orbits of most of their moons are also in that plane.

Pluto's orbit is the most out of the plane. It is inclined at an angle of 17° to the plane going through the sun's equator, and Mercury's orbit is inclined at an angle of 7°. The axes of most, but not all, of the planets are perpendicular to the rotation plane, and their spins are mostly in the counterclockwise direction. Most of the angular momenta therefore add up arithmetically. Presumably, the sum is the angular momentum of the original condensing cloud of atoms that formed the solar system.

The spin angular momentum and orbital angular momentum of a satellite can be linked through tidal forces with the parent body. The tidal motions (within the solid sphere as well as in any surface oceans) drain energy from the spinning body, reducing the spin. In order to conserve angular momentum, the distance between satellite and parent increases, thus increasing the *orbital* angular momentum. Apparently this happened long ago to our moon, because now the period of the moon's spin is the same as its period of revolution. The motions are locked so that we always see the same face of the moon. The earth's spin continues to be drained by the friction of the tides. The day is lengthening by 0.0016 second per century and the moon–earth distance continues to increase in order to conserve angular momentum.

4. The prime evidence that we offer in our classes for the conservation of linear momentum comes from the collisions of two objects. You can provide a similar homely demonstration of the conservation of angular momentum. All you need are two old music records that are already scratched or that can be sacrificed. Get a thread or string, at least a meter long, and tie one end to a large paper clip. Thread the other end of the string through the center holes of the two records, and then suspend the array from some overhead support. Tape a small piece of paper to the rim of the bottom record, so that you can time its rotation. Most likely the records will not hang in a horizontal plane, but will tip. Get someone else to hold the top record some distance up the thread, while you give the bottom record a spin in the horizontal plane. Immediately you and your students will see one effect of angular momentum. The spinning record will remain in the plane in which you spun it. You have a gyroscope. (With only the one record on the thread, you can let the whole system swing back and forth like a pendulum while the record is spinning. The record will remain flat, which looks remarkable.)

Time several revolutions of the bottom record with a watch or by counting steadily. Then have your assistant drop the upper record a short distance onto the bottom one. There will be some grinding of the grooves, but then the two will spin together. Immediately time the rotation period for the two records together. Regardless of the dissipated energy, angular momentum must be conserved. If the two records have equal moments of inertia, the combination period must be twice the single one.

If the records rotate too many times in one direction, they will wind up the thread so much that the restoring torque will interfere with your observations. To avoid this, use a long thread, and reverse the direction of spin each time. You can make an experimental exercise out of this apparatus by using nonidentical records, differing both in mass and in radius. Experimentally

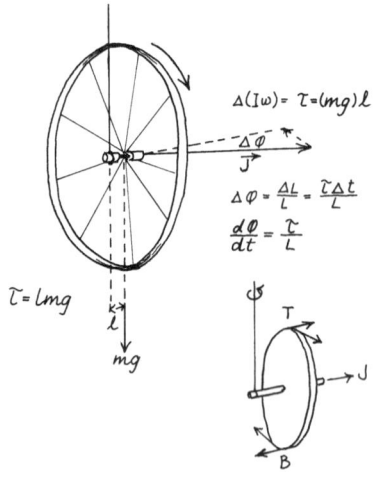

Figure 8.5. Bicycle wheel gyroscope.

determined relative moments of inertia can then be compared with calculated values.

■ THE DIRECTION NATURE OF ANGULAR MOMENTUM

For most purposes in introductory physics, assigning an axial direction to the angular-momentum vector is more a matter of convenience than for actual use. After all, usually you can describe a rotation perfectly well by calling it clockwise or counterclockwise as seen from a particular direction. Now we will see a practical use for saying that the appropriate direction for torque, angular velocity, and angular momentum is along the axis of rotation. Without such a description, it is hard to explain the motion of ordinary tops. In the next section we will see that this assignment is valid only in certain symmetrical situations, and that the complete description of the real world is, as usual, more complicated.

You can create a dramatic example of a kind of top, or gyroscope, with a bicycle wheel. Every classroom should have at least one such wheel, preferably with handles, and preferably with its rim loaded with wire instead of air inside the rubber tire. Support one end of the axle with a vertical string. Then let go of the handles, holding the wheel only by the string on one handle. Immediately there is an unbalanced torque acting on the wheel, as shown in Fig. 8.5. Gravity exerts a force equal to the weight of the wheel, which acts downward through the center of the wheel. The radius arm of the torque is the distance between the center of the wheel and the suspension point of the string. According to the right-hand rule for the cross product, $\boldsymbol{\tau} = \mathbf{r} \times \mathbf{F}$, the direction of the torque is horizontal and parallel to the face of the wheel. As you let go of the opposite handle, the wheel will naturally flop over.

Now start out the same way, but this time have someone else spin the wheel before you let go. There will still be the same unbalanced torque due to gravity, but the wheel will not flop over. Instead, it will remain essentially vertical, and the whole wheel will precess slowly in a horizontal circle around the string. You have an excellent, low-friction gyroscope, whose geometry is manifest and whose properties are easy to measure and calculate.

For a qualitative explanation of why the spinning wheel does not flop over, consider what happens to a point, T, at the top of the wheel. As seen in Fig. 8.5(a), the top of the wheel is tipping over to the right. Consequently, T receives an impulse to its left. However, T is also moving rapidly toward us. The tipping impulse provides a small momentum (to our right) that adds vectorially to the main momentum of T. The resultant momentum is in the direction of a turn to the left for T, and thus a counterclockwise turn for the whole wheel as seen from above. Instead of tipping over, the wheel, and its plane of rotation, turns counterclockwise. Try the same argument for point B at the bottom of this wheel. The tipping impulse provides a small momentum to our left, which must be added vectorially to the main momentum of B away from us. The resultant momentum of B is also in a direction of a turn of the whole wheel counterclockwise.

In Fig. 8.5(b) we show the angular momenta and torques with arrows representing axial vectors. The largest angular momentum in the system is that of the spinning wheel, and that is in the direction of the horizontal axle. The torque due to gravity is in the horizontal plane also, but is perpendicular to the axle. During each interval of time, the torque produces a change of angular momentum: $\boldsymbol{\tau}(\Delta t) = \boldsymbol{\Delta}(I\omega)$. This change of angular momentum is in the direction of the torque and is perpendicular to the main angular momentum of the wheel. Therefore, it does not change the magnitude of the main angular momentum. But the torque does change the *direction* of the angular momentum of the wheel. The axle moves in the horizontal plane and continues to do so as long as the gravitational torque continues.

Let us find the angular velocity of the bicycle wheel's precession. So as not to get confused with the angular velocity of the spinning wheel itself, call the wheel's angular momentum J. We will assume that the magnitude of J stays constant for the duration of our experiment—no friction in the axle bearing or from the air. Then, as shown in Fig. 8.5, a small change in the precession angle is given by

$$\Delta \phi = \frac{\Delta(I\omega)}{J} = \frac{\tau(\Delta t)}{J} = \frac{mgl(\Delta t)}{J}.$$

The precession frequency is

$$\frac{d\phi}{dt} = \frac{mgl}{J}.$$

Let us assume some reasonable values for the variables and get an approximate value for the precession frequency of a bicycle wheel under these circumstances. To first approximation we assume that the mass of a bicycle wheel (2 kg) is concentrated at its outer radius ($r=30$ cm). The angular momentum J is therefore $mr^2\omega$. You can make a chalk mark on the wheel and measure its rotation period when it is spun by hand. For $\omega=10$ rad/s, $J=(2$ kg$)(0.3$ m$)^2(10$ rad/s$)=1.8$ kg m^2/s. If you tie the support string on the handle so that $l=6$ cm, the precession frequency will be

$$\frac{d\phi}{dt} = \frac{(2\text{ kg})(9.8\text{ N/kg})(0.06\text{ m})}{1.8\text{ kg m}^2\text{/s}} = 0.65\text{ rad/s}.$$

The period of the precession will be $T = 2\pi/\omega = 9.7$ s.

The phenomenon of precession is actually more complicated than the simple explanation we have given. Note that the wheel also has angular momentum in the *vertical* direction, because it is precessing around a vertical axis. Where did *that* component of angular momentum come from? You can see the answer qualitatively by noticing that, even though the axle starts in the horizontal plane, it dips below as it goes around. Consequently, part of the angular momentum of the axle is vertical.

Spinning objects react to torques in directions perpendicular to what we would naively expect. On the other hand, an experienced bike rider makes use of the effects almost intuitively. If you are riding no-hands, and want to turn right, you *lean* to the right. Instead of tipping the front wheel over to the right, that direction of torque *turns* the wheel to the right. This gyroscopic effect must be taken into account in flying a single-engine plane. If you try to climb abruptly, the resultant *horizontal* torque applied to the rotating propeller will make the plane swing sideways.

In the musical *Hair* a song called prematurely for the dawning of the Age of Aquarius. Aquarius is one of the twelve astrological houses, or constellations, marking the ecliptic (sun's path) in the sky. Each 24 hours the sun, moon, and stars appear to rotate once around the earth, all moving from east to west. However, as the earth makes its annual trip around the sun, the sun appears to move *eastward*

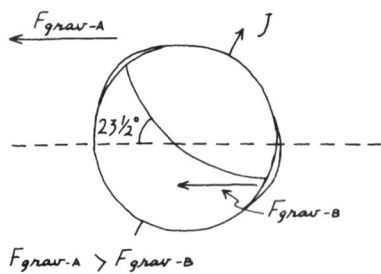

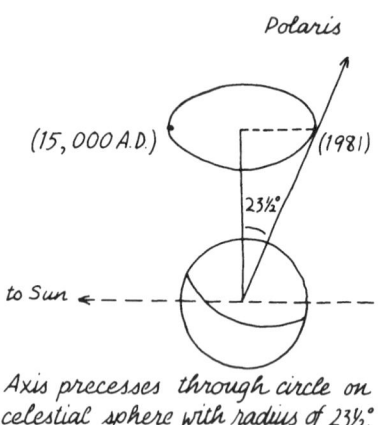

Figure 8.6. Explanation of Earth's precession.

with respect to the stars—about 1° per day. In the course of one year the sun moves through each of the twelve regions, or "houses," along the ecliptic. The astrological "age" depends on which house the sun is in at the beginning of the spring (the vernal equinox). That position changes very slowly from one year to the next. Starting some hundreds of years from now, each spring the sun will be entering the ill-defined region called Aquarius.

The reason the sun's position on a given date keeps changing is because our spinning earth precesses. The tilt of the axis remains the same, but the direction of the axis slowly changes with respect to the fixed stars. When spring comes now, we are always on a particular side of the sun (with respect to the stars). Eventually spring will come when we are on the opposite side of the sun. The axis now points toward Polaris, our North Star, but it did not point in that direction when the pyramids were built in Egypt. The period of precession is about 26 000 years—about 2200 years per house!

Precession requires a torque. In the case of a spinning toy top, the torque comes from the weight of the top acting through the lever arm that exists because the top is tilted. In the case of the earth, the torque must come from the pull of the sun trying to tilt the earth in some way. If the earth were truly spherical, there would be no lever arm and no torque—you cannot "straighten up" a sphere. However, the earth is not quite spherical. There is a slight bulge around the equator, no doubt produced by the centrifugal force as the spinning earth formed. The equatorial radius is 6378.2 km, and the polar radius is 6356.7 km. This slight difference of 21.5 km is enough to provide a girdle, which the sun can twist. As we show in Fig. 8.6 there is a greater force on the bulge toward the sun than on the side opposite. Instead of straightening up the earth, the resulting torque makes the spinning earth precess. There is a quantitative treatment of this on p. 114.

■ THE QUANTIZATION OF ANGULAR MOMENTUM

So far as we know, the length of an object can be as large or as small as we choose. No minimum length is known, not even for subatomic particles. The same condition holds for mass and time. There are no basic units, m and t, such that some other mass or time must be exactly $1083\, m$ or $2\, t$. As shown in Chap. 6, our measured values of length, mass, and time for objects depend on our speed with respect to those objects. The standard kilogram itself would have a larger effective mass (in terms of momentum, mv) if it went hurtling by us, and we could change that mass by an infinitesimal amount just by changing the speed slightly. Consequently, length, mass, and time cannot be integral multiples of some small basic unit; they are not quantized.

On the other hand, the angular momentum of a bound system is quantized. The basic, ultimate unit of angular momentum is called *Planck's constant*. When the frequency is given in terms of cycles or revolutions per second, Planck's constant is

$$h = 6.626 \times 10^{-34} \text{ kg m}^2/\text{s}.$$

We use Planck's constant in this form when we link it to the frequency (in Hz) and hence the energy of photons. For now, however, we have been describing angular velocity in terms of radians per second. In those units, Planck's constant is called h-bar:

$$\hbar = \frac{h}{2\pi} = 1.054 \times 10^{-34} \text{ kg m}^2/\text{s}.$$

This is a very small amount of angular momentum. We found a quantity about that size when we calculated the orbital angular momentum of a valence electron in an atom. The order-of-magnitude calculation on p. 180 yielded the value 1×10^{-34} kg m²/s. Our calculation was based on a very crude and now rather old-fashioned model of the atom. The remarkable thing is that such a model, picturing electrons in planetlike orbits, should work so well. In many cases, the subatomic particles act as if they were just tiny versions of the large-scale objects that we can handle. In many other cases, however, such mechanical models break down, and we have to describe the particles in terms of the probability functions of quantum mechanics.

As an example of how a mechanical model for subatomic particles sometimes works and sometimes does not, consider the analogy between electrons in atoms and planets in orbit. The planets not only revolve in orbit; many of them rotate on their own axes. It turns out that many of the subatomic particles also act as if they were spinning on their own axes. However, their spin angular momentum can only be zero, or $\frac{1}{2}\hbar$, $1\hbar$, $\frac{3}{2}\hbar$, or $2\hbar$, and so on. Immediately, there appears to be a contradiction! How can a particle have half of the basic unit? Perhaps the basic unit is $\frac{1}{2}\hbar$. Actually, the quantization rule is that angular momentum must be either integral or half-integral, but cannot *change* except in integral multiples of $\hbar$.

The angular momentum of a particle or a system of particles must have a particular direction, usually determined by the direction of the electric or magnetic fields. For instance, an electron may have its spin aligned with a magnetic field and then suddenly flip so that it is lined up in the opposite direction. At all times the electron keeps the magnitude of its angular momentum of $\frac{1}{2}\hbar$, but the *change* of angular momentum is one whole unit of $\hbar$—from $+\frac{1}{2}\hbar$ (up) to $-\frac{1}{2}\hbar$ (down).

We know that an electron or an atom cannot just change its angular momentum without something else happening. Some other particles must be given a recoil angular momentum. Angular momentum must be conserved. Frequently when an atomic system changes angular momentum, the recoil is carried off by a photon. Photons not only carry linear momentum, but also angular momentum. Both effects have been demonstrated in very direct mechanical experiments. If an intense light is reflected from one side of a tiny vane suspended in a high vacuum, the vane will recoil, thus demonstrating that light carries linear momentum.

In an ingenious experiment in 1935, Richard Beth produced rotation in a disk by shining polarized light along the axis of the disk. (Technically, the light was circularly polarized, which means that the photons were selected so that the direction of the angular momentum of all of them was the same. In ordinary, untreated light, there is a 50% chance that the angular momentum of a photon will be in the direction of travel, and a 50% chance that it will be in the opposite direction.) The spin, or angular momentum, of a photon is $1\hbar$.

If the angular momentum of any rotating system must equal $n\hbar$, the value of n for any human-size object is enormous. For instance, for our bicycle wheel with $m = 2$ kg, $r = 30$ cm, and $v = 10$ miles/h ($4\frac{1}{2}$ m/s), the angular momentum is 2.7 kg m²/s $= n(1 \times 10^{-34}$ kg m²/s). Therefore, n is equal to some integer with 35 digits, approximately equal to 2.7×10^{34}. Evidently, the quantization of angular momentum has no effect on the motion of things that we handle in everyday life. If Planck's constant were larger by a factor of 10^{34}, a bicycle wheel would behave

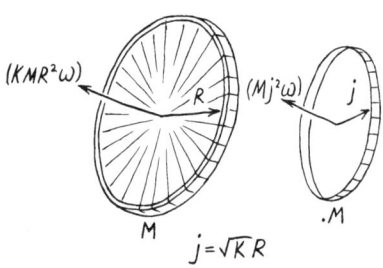

very strangely. It could rotate only at a certain speed, or twice that speed, or three times that speed, and so forth, and its axis could change direction only abruptly and only by certain amounts.

■ RADIUS OF GYRATION

The vector sum of all the forces acting on a system of objects will accelerate the center of mass of the system as if all the mass were located there and the resultant force were applied there. However, there may also be rotation of the system, and that is not described by assigning all the mass to the center-of-mass point. For instance, the center of mass of a wheel rotating on its axis is right at the center of the wheel. If you throw the wheel, no matter how it tumbles, the center of mass will move in a parabolic trajectory. However, the effect of mass of a rotating object depends on the *location* of the mass. In order to calculate rotation around the axis, the mass of the wheel must be considered to be divided up and distributed at various radii.

It is sometimes convenient to describe a radius for a rotating object where all the mass could be concentrated without affecting the angular momentum. This distance is called the *radius of gyration*, j.

For any rotating object, the angular momentum is given by $I\omega$, where I is the moment of inertia. For a system composed of parts at various radii, $I = \Sigma m_i r_i^2$. As we saw on p. 177, the moment of inertia of common symmetrical objects can be written as $I = KMR^2$. The total mass is M, the outside radius is R, and K is a constant that depends on the geometrical distribution of mass of an object with a particular shape. We can produce the same moment of inertia by rotating a point object or a hoop with mass M at a distance j from the axis. To find this distance j, the radius of gyration, we equate the moments of inertia of the two systems:

$$KMR^2 = Mj^2.$$

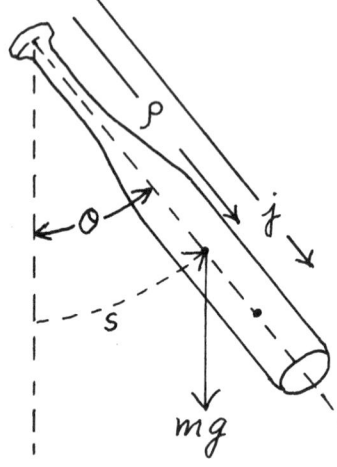

Figure 8.7. Radius of gyration and the physical pendulum.

Evidently, $j = \sqrt{K}R$. For instance, the radius of gyration of a solid disk of radius 10 cm, rotating on its own axis, is $j = \sqrt{\frac{1}{2}}$ (10 cm) = 7 cm.

■ THE PHYSICAL PENDULUM

The *simple* pendulum consists ideally of a bob with its mass concentrated at a point, swinging on a massless string. Any solid object supported from a point above its center of mass will also swing back and forth with a pendulum motion. However, the relationship between its period and its length is different from that of the simple pendulum.

The moment of inertia of a solid swinging object, called a "physical pendulum" would be the same if all its mass were concentrated at its radius of gyration. Perhaps its period is the same as the period of a simple pendulum with a string as long as the radius of gyration? It is not that simple. The trouble is that placing the mass at the radius of gyration produces the right moment of rotational inertia, but the restoring torque provided by gravity acts through the *center of mass*—which is at a different radius. Here is the proper analysis; refer to Fig. 8.7:

$$\tau = I\alpha,$$

$$(mg\sin\theta)\rho = I\frac{a}{\rho}.$$

The distance from the axis to the center of mass is ρ. The acceleration of the center of mass along the arc is $a = \alpha\rho$. The distance along the arc from the equilibrium line is called s. We use the small-angle approximation so that $\sin\theta \approx \theta = s/\rho$. The pendulum equation then becomes

$$mgs = I\frac{a}{\rho} \rightarrow a = \frac{mgs\rho}{I}.$$

The restoring force on the center of mass is $F = ma = -(m^2 g\rho/I)s$. The negative sign indicates that the restoring force is opposite in sign to s, the displacement along the arc.

We have arrived at the condition for simple harmonic motion. The restoring force is proportional to the negative of the displacement. For $F = ma = -kx$, the period of oscillation is $T = 2\pi\sqrt{m/k}$. Therefore, the period of the physical pendulum is

$$T = 2\pi\sqrt{\frac{m}{m^2 g\rho/I}} = 2\pi\sqrt{\frac{I}{mg\rho}}.$$

Remember that the period of a simple pendulum is $T = 2\pi\sqrt{L/g}$. For a simple pendulum to have the same period as a physical pendulum, the length of the simple pendulum would have to be $L = I/m\rho$. (Note that this formula reduces to that for the simple pendulum. If $I = mL^2$, then $L = L^2/\rho$, and $\rho = L$.)

In terms of the radius of gyration, j, the moment of inertia is $I = mj^2$. The length of the simple pendulum with the same frequency as a physical pendulum is therefore

$$L = \frac{mj^2}{m\rho} = \frac{j^2}{\rho}.$$

Evidently, the length of the equivalent simple pendulum is *not* the same as the radius of gyration.

Let us calculate the period of a meter stick swinging from one end. The moment of inertia of a uniform rod, swinging about one end, is $I = \frac{1}{3}ML^2$. Its period of oscillation is

$$T = 2\pi\sqrt{\frac{I}{Mg\rho}} = 2\pi\sqrt{\frac{\frac{1}{3}M(1\text{ m})^2}{M(9.8\text{ N/kg})(0.5\text{ m})}} = 1.6\text{ s}.$$

The length of a simple pendulum with that period is

$$L = \frac{I}{M\rho} = \frac{(\frac{1}{3}M)(1.0\text{ m})^2}{M(0.5\text{ m})} = \frac{2}{3}\text{ m}.$$

■ THE RADIUS OF PERCUSSION

There is a very familiar type of collision involving angular momentum of one object and linear momentum of the other: a bat hits a ball. The very least that we can do in teaching introductory physics is to explain the dynamics of the great American pastime. As every schoolboy knows, a bat has a diamond imprint about 4/5 the distance from the handle to the end. If the bat strikes the ball at that point

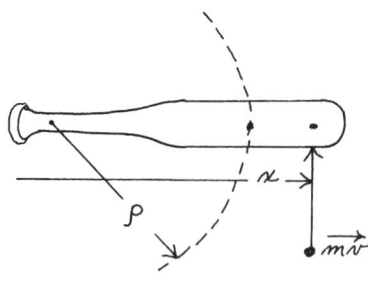

Figure 8.8. Radius of percussion and the baseball bat.

(but on the opposite side), there will be no kick or backlash to the hands. How does the maker know where to put that imprint?

Before presenting the analysis of the problem, have your students handle the phenomenon in the following simple way. Get a ruler, or better yet a meter stick, and place it flat on a table or other smooth surface. Put your finger or some marker next to one end, and snap the stick at the far end with the fingers of your other hand. Notice what happens to the end near the marker. Now line up the stick again and snap it close to the marker end. The end at the marker will move again, but in the opposite direction from the first time. Clearly, there must be some point along the stick where you could snap it and not make the marker end move. Find that point. Is it at the half mark? The two-thirds mark? The three-fourths mark? Notice that a blow at this special point rotates the stick around the end point as an axis, but does not make the end point move forward or backward. If the stick were a bat, and you were holding the bat at the end point, your hands would feel no pain. The distance from the axis to that point is called the *radius of percussion*.

We will assume that the only force acting on the stationary bat comes from the blow when the ball strikes it. In other words, we assume that the bat is only loosely held by the hands, and we require a condition such that the hands do not have to exert any force as a result of the blow. The ball is traveling perpendicular to the bat and strikes it at a distance x from the pivot point held by the hands. The geometry is shown in Fig. 8.8. The force delivered to the bat is

$$F = \frac{d(mv)}{dt}.$$

The change of momentum of the ball is $d(mv)$. That force will produce a torque on the bat around the pivot point: $\boldsymbol{\tau} = \mathbf{x} \times \mathbf{F} = I\boldsymbol{\alpha}$. The moment of inertia of the bat around the pivot point is I.

The force delivered by the ball will also make the center of mass of the bat move. If no force is exerted by the hands, then the center of mass must accelerate in the same direction as the force applied by the ball. That direction is perpendicular to the bat. The only way that that can happen (if the pivot point does not move) is for the center of mass to start moving on an arc around the pivot point, since momentarily that direction is perpendicular to the bat. The equation of motion for the center of mass must be $F = Ma = M\alpha\rho$, where M is the mass of the bat. The angular acceleration of the bat around the pivot point is α, and the distance from that pivot point to the center of mass is ρ. Now let us equate the force expressions from the torque equation and the force equation: $M\alpha\rho = I\alpha/x$. The special distance x which allows the bat to swing around the pivot point without any reaction force by the hands, is $x = I/M\rho$. We can change this expression one step further, by writing the expression for the moment of inertia in terms of the radius of gyration: $I = Mj^2$. Then,

$$x = \frac{Mj^2}{M\rho} = \frac{j^2}{\rho}.$$

This is a remarkable result! The radius of percussion of an object is the same as the length of a simple pendulum with the same period. To find the radius of percussion of a bat (or any other solid object), let it swing from the pivot point and measure the period of oscillation. (In the case of a bat or golf club, the pivot point is midway between your two hands.) Then, find (or calculate) the length of a simple pendulum that has the same period. That length is j^2/ρ, which is the same as the radius of percussion.

Let us find the radius of percussion of a meter stick. The radius of *gyration* of a uniform rod swinging around one end is given by $\frac{1}{3}ML^2 = Mj^2$. Therefore, $j^2 = \frac{1}{3}L^2$. The radius of percussion is

$$x = \frac{j^2}{\rho} = \frac{\frac{1}{3}L^2}{\frac{1}{2}L} = \frac{2}{3}L.$$

Note that the radius of percussion is indeed the same as the length of the simple pendulum that has the same period as the meter stick. In Fig. 8.9 we summarize the geometry of these various radii for a bat and for a meter stick

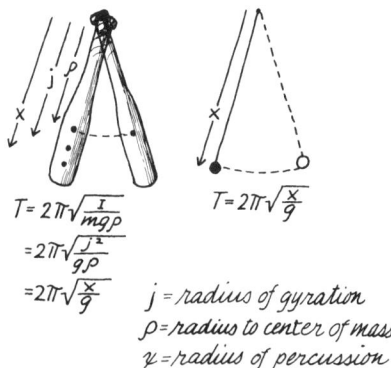

■ WOBBLY ROTATIONS

The rotation of solid objects is a much more complicated phenomenon than we have seen so far—and much more complicated than we can reasonably present in an introductory course. Nevertheless, students should be aware that there are complications, some of which are fun to demonstrate and play with. Some of the effects make themselves painfully evident in the everyday world. Physics at the introductory level can at least give qualitative explanations.

A wheel on an axle spins smoothly as long as the bearings have low friction, the wheel is symmetrically "balanced," and the plane of the wheel is perpendicular to the axle. As for the effect of static balancing, suppose that the wheel has a high-density spot or is not quite round. When allowed to spin freely it will come to rest with the heavier side down. While it is spinning, the asymmetrical masses will create centrifugal forces that will not cancel out. There will be a rotating pull on the axle, leading to a resonant vibration when the rotation frequency matches the frequency of the wheel suspension. Even when the wheel is balanced statically—so that there is no preferred stopping direction—it may not be balanced dynamically. A mass of 2 at a radius of 1 can statically balance a mass of 1 at a radius of 2. But the small mass would have a larger moment of inertia because of the r^2 dependence. This is why car wheels should be balanced with a machine that can spin them and indicate where the lead slugs should be placed for balance.

There is another potential problem with asymmetries of rotation. If the plane of the wheel is not perpendicular to the axis, there will be a wobbling that puts a severe strain on the bearings. The bearings experience a rotating twist to which they must respond by exerting a rotating torque. This is not an axial torque to keep the wheel rotating, but a torque at an angle to the axle.

These wobbling effects can be demonstrated nicely with a Tinkertoy set. (No physics classroom is complete without one.) In Fig. 8.10 we show several simple geometries of rods and connectors. As usual, students should individually feel these phenomena with their hands; demonstrations cannot convey the effect. In Fig. 8.10(a), there is a symmetrical rotator. The fingers spinning and holding the axle feel no wobbling torques. In Fig. 8.10(b) we show an unbalanced rotator. Now the fingers feel the unbalanced, rotating centrifugal force. The effect of the geometry shown in Fig. 8.10(c) is even more dramatic. The finger bearings must exert continuous torques as the tilted rotator wobbles about the axis.

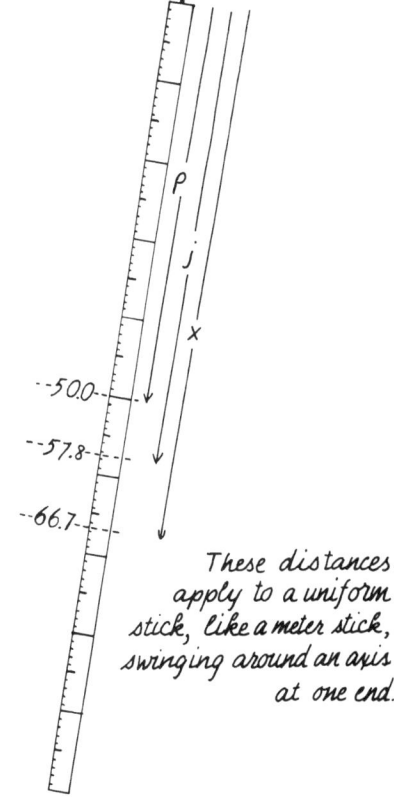

Figure 8.9. Relationship of the radius of percussion, the radius of gyration, and the radius to the center of mass

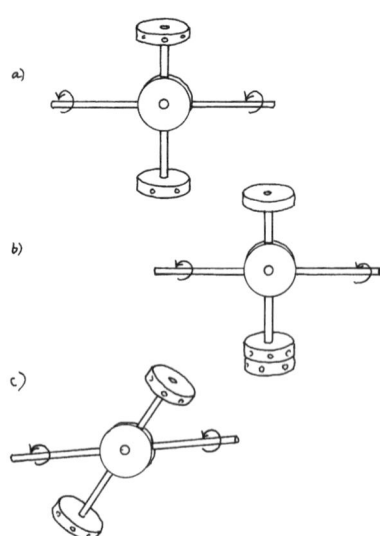

Figure 8.10. Wobbly motions produced by asymmetrical rotators.

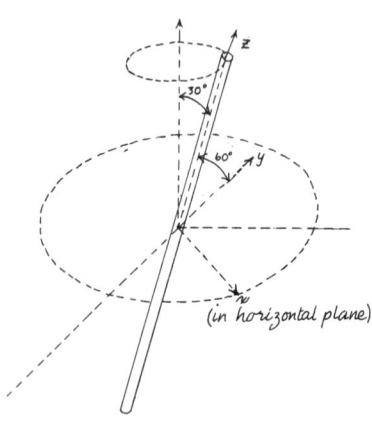

Figure 8.11. Angular momentum need not be in same direction as angular velocity.

There is a surprising implication of the phenomenon felt in Fig. 8.10(c). If the fingers (axle) must supply a nonaxial torque, then there must be angular momentum that is nonaxial. But ω, the angular velocity, *is* axial. We must conclude that angular momentum is not necessarily in the same direction as angular velocity. *Linear* momentum is mv, and is necessarily in line with the velocity since m (for these purposes) is a scalar. Angular momentum is $I\omega$. The moment of inertia, I, is defined as $\int r^2 dm$. Its value depends on the axis about which the object is rotating, and from which r is calculated. The moment of inertia is not a scalar; it is a tensor. Suppose we insert an axis through the center of mass of an object. The moment of inertia of each element of mass depends on the direction of the axis. For three dimensional space, defined by three perpendicular axes, the moment of inertia tensor is a 3×3 array:

$$\mathbf{I} = \begin{pmatrix} I_{xx} & I_{xy} & I_{xz} \\ I_{yx} & I_{yy} & I_{yz} \\ I_{zx} & I_{zy} & I_{zz} \end{pmatrix}.$$

The diagonal components of **I** are the familiar moments of inertia about each of the three axes.

$$I_{xx} = \int\int\int \rho(y^2+z^2)dV, \quad I_{yy} = \int\int\int \rho(z^2+x^2)dV, \quad I_{zz} = \int\int\int \rho(x^2+y^2)dV.$$

The off-diagonal elements are called *products of inertia*:

$$I_{xy}=I_{yx}=-\int\int\int \rho xy\, dV, \quad I_{yz}=I_{zy}=-\int\int\int \rho yz\, dV, \quad I_{zx}=I_{xz}=-\int\int\int \rho zx\, dV.$$

The angular *velocity* is a three-dimensional vector, represented by a column array. The angular *momentum* is the tensor product of **I** and $\boldsymbol{\omega}$:

$$\mathbf{L} = \hat{\mathbf{x}}(I_{xx}\omega_x + I_{xy}\omega_y + I_{xz}\omega_z) + \hat{\mathbf{y}}(I_{yx}\omega_x + I_{yy}\omega_y + I_{yz}\omega_z)$$
$$+ \hat{\mathbf{z}}(I_{zx}\omega_x + I_{zy}\omega_y + I_{zz}\omega_z).$$

To show how the angular momentum can be in a different direction from the angular velocity, consider the thin rotating rod shown in Fig. 8.11. In order to calculate the moments of inertia easily, we choose the axes fastened to the rod with the center at the center of mass. The z axis is along the rod, and we assume that the rod has small radius so that x and y never have large values to contribute to the integrals for the moments of inertia. The rod is tilted at 30° from the vertical, and is rotating around a vertical axis. As you can imagine from playing with the tinker toys, the rotating rod will wobble. From the diagram, you see that $\omega_x=0$, $\omega_y=\omega\sin 30°$ and $\omega_z=\omega\cos 30°$. Because we assume that x and y values for the rod are negligible, many of the components of the moment of inertia are also negligible. We make the approximation that $I_{zz}=I_{xz}=I_{zx}=I_{zy}=I_{yz}=I_{xy}=I_{yx}=0$. All of the off-diagonal elements have an x or y in their integrals, and I_{zz}, which is the moment of inertia of the rod about its own z axis, has both x and y in the integral. Since $\omega_x=0$, we are left with only one component:

$$\mathbf{L} = \hat{\mathbf{y}}(I_{yy}\omega_y).$$

The moment of inertia, I_{yy}, is the moment of inertia of a thin rod about an axis through its center of mass, perpendicular to its length: $I_{yy}=\frac{1}{12}ML^2$. Notice the crucial point. The angular velocity ω is vertical. The angular momentum is in the

y direction ($\hat{y}$ is the unit vector in the y direction), which rotates about the vertical axis at an angle of 60°.

■ PRINCIPAL AXES OF ROTATION

For any solid object, no matter how irregular, there is a set of mutually perpendicular axes through its center of mass that has special properties. When the object rotates around any one of these axes, the angular momentum is in the same direction as the angular velocity. Furthermore, the moment of inertia around a particular one of these axes has the maximum value for any rotation around the center of mass; the moment of inertia around another of the perpendicular axes is the minimum; and the moment of inertia around the third has an intermediate value. For the axes that yield the largest and smallest moment of inertia, rotation is stable. Rotation around the third axis is unstable. If the object is set spinning freely around the third axis, it will wobble into a mixture of rotations around the other two axes. These *principal* axes are also the symmetry axes of objects that have three dimensional symmetry.

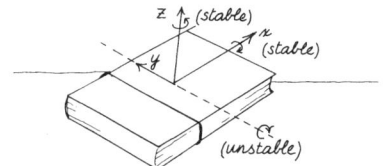

Figure 8.12. Demonstration of principal axes.

An easy and instructive demonstration of the role of principal axes can also be a party stunt. Put a rubber band around a book so that it will not fly open, or use a rectangular board. If the book or board is lying flat, imagine the three perpendicular axes with the origin at the center of mass. Define the z axis to be upward, the x axis to be along the length, and the y axis to be along the width, as shown in Fig. 8.12. It is best to have the length considerably larger than the width, and both much larger than the thickness. Spin the book in air around the z axis, catching it before it falls too far. The spin is stable; there is no tendency for the book to flop over. Notice that the moment of inertia about this axis is maximum. The mass at the corners of the book is further away from the axis than can be arranged with any other geometry. The smallest moment of inertia must be produced by rotation about the x axis. Try spinning the book about that axis and observe that this rotation is also stable. Now spin the book about the y axis. It will twist in midflight, apparently trying to find a new axis.

You can make the demonstration more dramatic but more dangerous by flipping a claw hammer. (A tennis racket will also work, but it is easier to see the directional effects with the asymmetry of the claw and head.) Hold the hammer by the handle with the claw and head *horizontal*, and flip it up so that it rotates in a circle whose plane is vertical. Throw it so that you can catch the handle after one full rotation. If there is nothing in the way immediately overhead, you will see the hammer go through an astonishing twist at the top of its trajectory. That particular axis is the unstable one. If the claw was on the left when you threw the hammer, it will be on the right when you catch it. You can demonstrate that there is stable rotation around either of the other two axes. It is not obvious, but the axis through the center of mass with maximum moment of inertia is perpendicular to the head-claw line. Try spinning the hammer by holding the handle with the head and claw in the *vertical* plane. For the axis with minimum moment of inertia, spin the hammer around the axis running down the handle.

■ REFERENCE

Galli, John Ronald, *Angular Momentum Conservation and the Cat Twist*, Phys. Teach. **33**, 404 (1995).

9
Work and Energy

■ THE PEDAGOGICAL PROBLEM

Everyone knows what work is—until they try to define it. The problem exists at several levels. For beginning students, the standard definition does not seem to agree with common usage of the word. Homework, for instance, is a lot of work. Clearly, holding a heavy object motionless is work. After we work out those problems, there is the question of who is doing the work. If I do work on a system I say that I have done positive work, but does the system think that it is negative? For instance, if I work to pull a boat trailer through the sand, does the friction in the sand perform negative work? Serious physics teachers can argue for hours on this subject. At stake are all the systematics of potential-energy theory and the formalities of the definition of a conservative force. On yet another level, energy, not work, is the fundamental quantity to be defined. Work is a way to transfer energy from one system to another. For interactions between subatomic particles, force is a derived and rather useless quantity, and the concept of work is meaningless.

Every fifth grader learns that energy is conserved. But of course they do not know what energy is and neither did anyone else until about 150 years ago. In the first edition of *Encyclopædia Brittanica*, 1771, there are fourteen pages on electricity, with numerous references to the researches of Dr. Benjamin Franklin. However, the entire entry under *Energy* was: "Energy, a term of Greek origin, signifying the power, virtue, or efficacy of a thing. It is also used, figuratively, to denote emphasis of speech." Our modern concept of energy was not formulated until the middle of the nineteenth century. For all their sophistication, Galileo, Newton, and Franklin did not know that a quantity called energy could be defined in such a way that it is always conserved. The idea is not at all obvious. Energy appears in many different forms. A rolling car has energy. A motionless flashlight battery has energy. A stone at the top of a cliff has energy. A pat of butter has energy. A kettle of boiling water has energy. Sunlight has energy. Energy in all these different forms can be defined in such a way that the total energy is conserved as one system transforms into another. However, so long as a system never changes in any way, its energy content is meaningless (although measurable as mass). It is only during transformations from one shape to another or from one place to another that the concept of energy becomes useful as a bookkeeping device.

Energy is something like money. It is meaningful only in terms of exchange processes. Transformations occur when money is spent, not when it stays in the bank. It is the same with energy. A car battery does not start the car until a switch is thrown. A change in the internal structure of the battery takes place, and an electric motor starts turning. We can account for the transformations in terms of the battery losing energy and the motor gaining it.

Pedagogically, we can define work done on a point object as $\int \mathbf{F} \cdot d\mathbf{x}$, or $|F||\Delta x|\cos\theta$, or "the product of the force on the object and the distance the object moves in the direction of the force." After playing with examples of these definitions, we can ask how much work it takes to accelerate an object to a particular speed and then define kinetic energy to match that quantity of work. *Or*, we can start out with kinetic energy as a conserved quantity under certain conditions, and then ask how to produce that kinetic energy with a force in the first place. Either approach is valid. We will present the latter method here because it is less well known and also because it emphasizes the nature of the energy conservation law and specifies that work is a particular type of transfer of energy.

■ THE SECOND CONSTRAINT ON CERTAIN COLLISIONS

In Chap. 7, we found that in certain types of air-track glider collisions, momentum can be defined in such a way that it is a conserved quantity. There is a problem, however. Momentum conservation is a necessary condition for explaining the results, but is not sufficient. For instance, if a moving glider with a spring bumper collides with a stationary identical one, the target glider goes forward with the original speed while the bombarding glider stops completely. Momentum is conserved, but momentum would also be conserved if the two gliders stuck together and went off with half the original speed. The momentum equation is

$$m_1 v_0 = m_1 v_1 + m_2 v_2.$$

There is an infinite number of ways in which the two unknowns, v_1 and v_2, can satisfy the one equation. Nature chooses one. There must be another equation of constraint.

It turns out that one and only one other combination of m and v is also conserved. That is the product, mv^2. This particular quantity of motion is written as $\frac{1}{2}mv^2$ and is called *kinetic energy* or E_{kin}. There are several points to be noted about the nature of this defined quantity. First, kinetic energy is a scalar and not a vector like momentum. The velocity, which *is* a vector, is squared in the expression, making the whole term just a directionless scalar. Second, the existence of the constant, $\frac{1}{2}$, implies that there must be some more basic definition of the unit of energy. This is right and we will soon get to it. In the meantime, note that if an object with a mass of 1 kg moves with a speed of 1 m/s, then its kinetic energy is equal to one-half the basic unit of energy, or one-half *joule*. The unit is named after James Joule (1818–1889), an English brewer turned scientist who first demonstrated the energy equivalence of heat. Finally, note that we now have two different quantities to define the quantity of motion: the momentum $m\mathbf{v}$ and the kinetic energy $\frac{1}{2}mv^2$. One is a vector; the other is a scalar. Why do we need them both? Three hundred years ago, people argued about this question; the Germans championed the preeminence of mv^2, which was called *vis viva*, or living force, and the English were sure that the quality of motion was best described in terms of the momentum $m\mathbf{v}$. Note that in air-track collisions, momentum is conserved regardless of the nature of the bumpers on the gliders—spring or Velco. However, E_{kin} is conserved only when the gliders are equipped with good spring bumpers. Such collisions are called, appropriately enough, *elastic*.

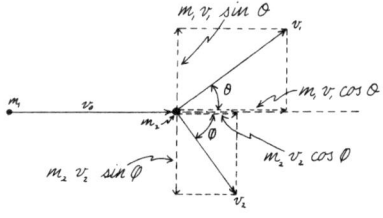

Figure 9.1. Elastic collision in two dimensions. $m_1 = m_2$, and initial velocity of $m_2 = 0$.

If two quantities must be conserved in the air-track collisions, two equations must be satisfied.

Conservation of momentum $\quad m_1 v_0 = m_1 v_1 + m_2 v_2$
Conservation of E_{kin} $\quad \frac{1}{2} m_1 v_0^2 = \frac{1}{2} m_1 v_1^2 + \frac{1}{2} m_2 v_2^2$

Now we have two equations for our two unknowns. There must be a unique solution. Take the case of identical gliders, so that $m_1 = m_2$. Our two equations become

$$v_0 = v_1 + v_2 \rightarrow v_0^2 = v_1^2 + v_2^2 + 2v_1 v_2,$$

$$v_0^2 = v_1^2 + v_2^2.$$

These two equations (the first one has been squared), are consistent with each other only if $2v_1 v_2 = 0$. Apparently there is not one unique solution; there are two! Either $v_1 = 0$ or $v_2 = 0$. What we actually see happen on the air track is that the first glider stops after the collision; $v_1 = 0$. The equations then require that $v_2 = v_0$, which is just what happens. The two conservation laws completely constrain the outcome of the collision so that only one combination of velocities is possible, and the prediction matches the experimental results. But what about the other possibility? The mathematics would be satisfied if $v_2 = 0$ and $v_1 = v_0$. To satisfy these conditions, the original glider would have to continue on its way with its original velocity. Do we call this "extraneous solution" an artifact of the mathematics and throw it away? No, our faithful mathematics has accurately described a situation where the bombarding glider misses the target and continues forward. This is not likely to happen on an air track, but we did not tell the mathematics that it was describing an air-track collision. For that constraint we would have needed a third equation.

Let us apply both conservation laws to the collision of identical spheres in *two* dimensions, as shown in Fig. 9.1. Once again, since the spheres are identical, the masses will cancel. Now we have three equations, two for momentum and one for kinetic energy:

Momentum in x $\quad v_0 = v_1 \cos \theta + v_2 \cos \varphi$
Momentum in y $\quad 0 = v_1 \sin \theta - v_2 \sin \varphi$
E_{kin} $\quad v_0^2 = v_1^2 + v_2^2$

To solve these three simultaneous equations, square the first and second equations:

$$v_0^2 = v_1^2 \cos^2 \theta + v_2^2 \cos^2 \varphi + 2v_1 v_2 \cos \theta \cos \varphi,$$

$$0 = v_1^2 \sin^2 \theta + v_2^2 \sin^2 \varphi - 2v_1 v_2 \sin \theta \sin \varphi.$$

Add these two equations, remembering that $\sin^2 \theta + \cos^2 \theta = 1$, for any θ:

$$v_0^2 = v_1^2 + v_2^2 + 2v_1 v_2 (\cos \theta \cos \varphi - \sin \theta \sin \varphi).$$

Compare this equation with the kinetic-energy equation. They are consistent only if $2v_1 v_2 (\cos \theta \cos \varphi - \sin \theta \sin \varphi) = 0$. If either $v_1 = 0$ or $v_2 = 0$, the term would always be 0. Those conditions would correspond with either a head-on collision or a complete miss. For any other kind of collision, both spheres have nonzero velocities afterward. Consequently, the expression in the parentheses must be identically zero. That expression can be transformed by using the following trigonometric identity:

$$\cos \theta \cos \varphi - \sin \theta \sin \varphi = \cos(\theta + \varphi).$$

But if $\cos(\theta+\varphi)=0$, then $\theta+\varphi=90°$. The opening angle between the paths of the two spheres after the collision must be exactly 90°. Several examples are shown in Fig. 9.2. Notice what a remarkable constraint is imposed by the conservation laws on the collision of identical particles when one starts at rest!

This same condition about particle collisions was deduced on p. 124 from an analysis using only momentum conservation and a transformation to the center-of-mass coordinate system. The assumption that in the center-of-mass system the particle speeds are the same before and after collision is equivalent to the constraint that there is no loss of kinetic energy. Without this constraint, the opening angle between identical particles is less than 90°.

Notice that in two dimensions we have three equations, but that there are *four* unknowns: v_1, v_2, θ, and φ. The results are undetermined unless one more condition is imposed. However, if you measure just one of the four final variables, the other three are then determined. For example, if you observe that the target sphere goes off at 30°, then there is just one possible set of values for the speed of the two spheres and the angle of the bombarding one. Moreover, if you know the original offset between the two spheres, perpendicular to the forward direction, you know the angular momentum of the system. Since angular momentum must be conserved, there is a fourth constraint, or equation, so that the values of all four final variables can be predicted.

Identical mass collisions with kinetic energy conserved can be seen in low-to-medium-energy collisions of protons in hydrogen cloud chambers. When the proton speeds are close to that of light, relativistic dynamics must be used and the 90° opening angle is not produced. The 90° effect can be produced fairly well in the introductory laboratory with marbles colliding in mid-air. There is a standard laboratory apparatus that holds one marble on a tee as a target for another marble that rolls down an incline to hit it. The apparatus sits on a table, and the two marbles make carbon-paper marks when they hit the floor. The 90° effect is *not* produced when two spheres, such as billiard balls, collide while rolling on a surface. The two balls do not form an isolated system. During the collision the bombarding ball must change its axis of rotation and to do so it must dig into the surface and thus change its momentum. Any billiard player is familiar with such spin–table-top interactions, sometimes called *English*.

The conservation laws can be used to derive a general expression for head-on collisions, where $m_1 \neq m_2$. As usual, we start out with m_2 at rest and m_1 coming in from the left with positive velocity, v_0.

Momentum conservation $\qquad m_1 v_0 = m_1 v_1 + m_2 v_2$
E_{kin} conservation $\qquad \frac{1}{2} m_1 v_0^2 = \frac{1}{2} m_1 v_1^2 + \frac{1}{2} m_2 v_2^2$

Divide the first equation by m_1 and the second equation by $\frac{1}{2} m_1$:

$$v_0 = v_1 + \frac{m_2}{m_1} v_2, \quad v_0^2 = v_1^2 + \frac{m_2}{m_1} v_2^2.$$

Square the first equation:

$$v_0^2 = v_1^2 + \left(\frac{m_2}{m_1}\right)^2 v_2^2 + 2\left(\frac{m_2}{m_1}\right) v_1 v_2.$$

Subtract the second equation from this one:

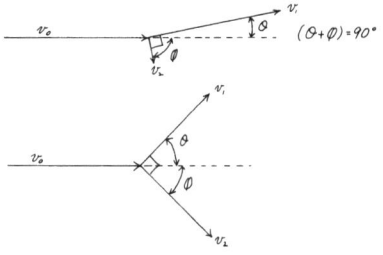

Figure 9.2. In a collision between identical particles, one starting at rest, the opening angle between the resulting trajectories is 90°.

$$0 = \left(\frac{m_2}{m_1}\right)v_2^2\left(\frac{m_2}{m_1}-1\right)+2\left(\frac{m_2}{m_1}\right)v_1 v_2.$$

Divide through by $(m_2/m_1)v_2$:

$$0 = \left(\frac{m_2}{m_1}-1\right)v_2 + 2v_1.$$

Notice that after all this algebra, we have ended up with one equation for two unknowns. At least, however, we can solve for the ratio v_1/v_2:

$$\frac{v_1}{v_2} = \frac{1}{2}\left(1-\frac{m_2}{m_1}\right) = \frac{m_1-m_2}{2m_1}.$$

If $m_1 = m_2$, $v_1 = 0$, agreeing with what we have already deduced. If $m_1 = \frac{1}{2}m_2$, then $v_1/v_2 = -\frac{1}{2}$. In terms of v_0 as given by the momentum equation, $v_0 = -\frac{1}{2}v_2 + 2v_2$. Therefore, in this case where a glider strikes one with twice as much mass, $v_2 = \frac{2}{3}v_0$. The more massive target glider goes off with a velocity of two-thirds that of the original. The bombarding glider bounces back with a velocity of $v_1 = -\frac{1}{2}v_2 = -\frac{1}{3}v_0$.

When $m_1 = 2m_2$, then $v_1/v_2 = 1/4$. In terms of the momentum equation, $v_0 = \frac{1}{4}v_2 + \frac{1}{2}v_2$. Therefore, when a glider strikes another one with one-half its mass, the target glider shoots forward with a velocity $v_2 = \frac{4}{3}v_0$. The more massive bombarding glider continues in the forward direction with a velocity $v_1 = \frac{1}{3}v_0$.

If $m_1 \gg m_2$, then $v_1/v_2 \to 1/2$. The target object goes shooting off with a velocity twice that of the bombarding object. As you can see from the momentum equation, the velocity of the bombarding object is approximately the same as its original velocity. For instance, a car hitting a tennis ball does not slow down appreciably. Note that the tennis ball would be knocked forward with a speed twice that of the car. If the car—perhaps a van with flat windshield—were traveling at 30 miles/h, the driver would see the ball coming at 30 miles/h and then bouncing off at 30 miles/h. (Actually, it would probably bounce off at about 20 miles/h because of the low coefficient of restitution.) An observer by the roadside would see the ball take off at 60 (or 50) miles/h.

■ THE WORK NECESSARY TO PRODUCE KINETIC ENERGY

How does a point object get to have kinetic energy? The energy can be transferred to the object in a collision, but it can also be produced if some outside system shoves the object until it has the required velocity. Starting from rest, an object will reach the velocity v if it is subject to a constant acceleration a over a distance x satisfying the condition $v^2 = 2ax$. The constant acceleration is produced by a constant force, given by $a = F/m$. If we substitute this formula for acceleration into the formula for final velocity, we get

$$v^2 = 2\frac{F}{m}x \to \tfrac{1}{2}mv^2 = Fx.$$

The product of force, and the distance through which the force is exerted, is called *work*. Work is a method of transferring energy from one system to another. The system that exerted the force F on the point object lost energy.

This simple definition of work raises all sorts of problems. Both force and displacement are vectors. Is their product a vector (cross) product, like torque?

Does the simple definition of work agree with the everyday meaning of the word? Is the definition any good if the applied force is not constant? Are work and energy part of the same conservation law—so that if you do some work on an object and produce kinetic energy, you can make the object exert a force and perform the same amount of work thus returning the energy? Finally, we must look into the relationships that exist among force, displacement, time, momentum, and kinetic energy.

First, what about that product of force and displacement? Certainly the maximum acceleration and hence the maximum velocity will be produced if the force is applied in the direction of the displacement. In fact, any force applied perpendicular to the displacement produces no useful effect at all—unless it reduces friction. In the case of the product of force and a lever arm to produce torque, the situation is exactly the opposite. Only the component of force *perpendicular* to the lever arm is effective. Apparently we need another kind of product of two vectors. This one is called the *dot product* or *scalar product*:

$$\text{Work} = W = \mathbf{F} \cdot \mathbf{x} = |\mathbf{F}||\mathbf{x}|\cos\theta.$$

The magnitude of the force component *in the direction of the displacement* is given by $|F|\cos\theta$, where θ is the angle between $\mathbf{F}$ and $\mathbf{x}$.

The *vector product* of two vectors yields another vector, perpendicular to the first two. The *dot product* of two vectors yields a scalar. Work and energy have no direction properties. The amount of work done or energy transferred is the same, regardless of whether the object is headed north, south, east, or west.

According to this definition, no work would be done on a heavy box if you carried it in your arms across the room. The force that you exert to hold it is directed upward, but the displacement is horizontal. Therefore, the angle between force and displacement is 90° and the work done on the box is 0. You would not have transferred any energy to the box. That is reasonable! In terms of our everyday meaning of the word, work is something that somebody gets paid for. Why should anyone pay you for carrying a box across a room when the job could be done just as well by sliding the box along a low-friction runway? Such a method would require no fuel, cost no money, and hence involve no work.

Even if no work is done *on* the heavy box as you carry it across the room, surely you *do* work and have dissipated energy. If you had to do that work all day, you would get tired out and your muscles would ache. Even if you just held the box all day, you would think that you had done a lot of work. Actually, there really is a displacement through which your muscles are exerting forces when you hold an object or press against something. Two muscles control each lever arm in your body. When you exert a force, both muscles are in tension and opposed to each other. There are continuous tightenings and relaxations of the opposing muscles, creating tiny movements. Your muscles are doing a great deal of work according to our standard definition, just trying to maintain a constant force, even though the net displacement is zero.

To get the relationship that we did between $\mathbf{F} \cdot \mathbf{x}$ and $\frac{1}{2}mv^2$, we made use of one of the formulas for constant acceleration (and also assumed $x_0 = 0$). What happens if the force, and hence the acceleration, is not constant? We can always make the approximation that the force is constant over a short distance. Our definition of work then becomes:

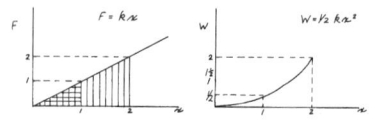

Figure 9.3. $F(x)$ and $W(x)$ for a spring obeying Hooke's law.

$$W_{a \to b} = \mathbf{F}_{av} \cdot (\mathbf{b} - \mathbf{a}) \to \int_a^b \mathbf{F} \cdot d\mathbf{x}.$$

We can also make a graph of the effective force exerted at every point: $F(x)$. If the force is constant, clearly the product of force and displacement is equal to the work done. It is also equal to the area under the curve of the $F(x)$ graph. For example, Fig. 9.3 shows the graph of $F(x)$ for a spring that obeys Hooke's law: $F = kx$. In this case there is no negative sign in Hooke's law, because F is the force that we must exert in the direction of x in order to stretch the spring—as opposed to the restoring force of the spring, which is in the opposite direction. The work done in stretching the spring to a distance x is the triangular area under the $F(x)$ curve, as shown in Fig. 9.3. That area, which is equal to the work, is

$$W = \tfrac{1}{2}(\text{base})(\text{height}) = \tfrac{1}{2}(x)(F_x) = \tfrac{1}{2}(x)(kx) = \tfrac{1}{2}kx^2$$

or

$$W_{0 \to x} = \int_0^x kx \, dx = \tfrac{1}{2}kx^2.$$

It takes four times the work to stretch a spring 2 cm than it takes to stretch it 1 cm.

It appears that we have pulled a fast one in logic here. Applying a force to a spring will stretch it, and that takes work according to our definition, but the spring does not pick up any kinetic energy. Of course, we can then let the spring expand against an air-track glider and produce kinetic energy of the glider. The spring appears to be a device for "storing" work and then transmitting it to an object that can take on kinetic energy. In a later section we will use this train of logic to define various forms of potential energy.

Now let us consider that historical question about whether the true quantity of motion is best described by mv or $\tfrac{1}{2}mv^2$, by momentum, or by *vis viva*. If you exert a force F on any object for a *time* interval Δt, then you have exerted an impulse that produces a change in momentum of the object:

$$\mathbf{F}\Delta t = \Delta(m\mathbf{v}).$$

If you exert a force F on an object through a *distance* interval Δx then you have done work on the object. If all the work goes into changing the kinetic energy,

$$\mathbf{F} \cdot \Delta \mathbf{x} = \Delta(\tfrac{1}{2}mv^2).$$

The area under the curve of a graph of $F(t)$ is equal to the change of momentum. The area under a curve of $F(x)$ is equal to the work done. If the object is free to accelerate, and F is the net force in the direction of displacement, then the area under the $F(x)$ curve is the change in kinetic energy of the object.

As an example of the different uses of momentum and kinetic energy in a problem, suppose that you want to accelerate an air-track glider with a mass of 200 g to a velocity of 1 m/s by exerting a force of 0.1 N. How long will it take?

$$\Delta(mv) = F\Delta t, \quad (0.2 \text{ kg})(1 \text{ m/s}) = (0.1 \text{ N})\Delta t,$$

$$\Delta t = 2 \text{ s}.$$

How far does the glider go before its speed has reached 1 m/s?

$$\Delta(\tfrac{1}{2}mv^2) = \mathbf{F} \cdot \Delta \mathbf{x}, \quad \tfrac{1}{2}(0.2 \text{ kg})(1 \text{ m/s})^2 = (0.1 \text{ N})\Delta x, \quad \Delta x = 1 \text{ m}.$$

Notice that the two methods agree with each other. Since the acceleration is constant, the average velocity is just half the final velocity. At an average velocity of $\frac{1}{2}$(1 m/s), the glider will travel 1 m in the first 2 seconds.

To figure out the damage a car can do in a collision, and the force that it exerts while it stops, should you consider the car's momentum or its kinetic energy? The question is not entirely academic! At 60 miles/h the car has twice the momentum that it does at 30 miles/h, but it has four times the kinetic energy. Either momentum or kinetic energy can be used as a criterion for the damage done *if* the other variable used is the right one. If momentum is used, then the *time* of stopping is the critical factor in finding the forces produced. If kinetic energy is used, then the *distance* of stopping should be used. With a constant force being exerted to stop the cars, a car starting at 60 miles/h will take twice the *time* to stop as a car at 30 miles/h, but the 60 miles/h car will require four times the stopping *distance*. For a given distance of stopping—perhaps 1 m provided by the displacement of the front end of the car as it runs into a concrete wall—the energy equation calls for four times the average force for the 60 miles/h car compared with the 30 miles/h car. This accounting also agrees with the momentum equation. The 60 miles/h car has twice the momentum and takes half the time to travel 1 m. Once again, the average force for the 60 miles/h car is larger by a factor of 4.

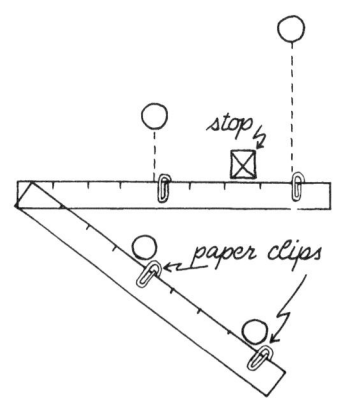

Figure 9.4. Tabletop demonstration that stopping distance (with dry sliding friction) is proportional to the square of the original speed.

Braking distances for cars are roughly proportional to the square of the original speed. There is an easy tabletop simulation of the road test. Get a ruler or flat stick, some paper clips, and several projectiles such as aspirin pills or thumbtacks. Place one projectile at twice the radius of the other along the ruler, as shown in Fig. 9.4. Then swing the ruler around one end, through an arc of about 30°, so that it hits a finger and stops abruptly. The pills or tacks will be swept along until the ruler stops, and then they will continue along the table in a straight line. Since the radial distance of the outer projectile is twice that of the inner, the outer one will be given twice the speed of the inner one. Students can try this game several times and with other ratios of radii, such as 3:1. There is an assumption that the retarding force on the projectile is constant, and that is approximately true for dry sliding friction.

■ WHO DOES THE WORK?

Work involves the scalar product. We have assumed that the effective force is in the direction of the displacement so that the work done is positive. Consider the following paradox that is sometimes proposed. You exert a force of 20 N on a box, pulling it along the floor at constant speed. There must be an equal but opposite force from friction, since $a=0$. After pulling the box 1 m, you have done 20 J of work, since $\theta=0°$ and $\cos 0°=+1$. $W=|F||x|\cos\theta$. The friction must have done -20 J of work, since for the friction force $\theta=180°$ and $\cos 180°=-1$. The total work done on the box is therefore 20 J -20 J$=0$ J. Perhaps that is all right. After all, there is no increase in kinetic energy. There is, however, an increase in temperature. Where did that extra thermal energy come from?

This is an unnecessary paradox, a result of sloppy bookkeeping. For students at the introductory level there is never any need to consider negative work. Instead, simply keep track of who is doing the positive work. The person (or agent) doing the positive work needs fuel since work is a transfer of energy. When in doubt, figure out who gets the fuel. In the case of pulling the box, obviously the friction mechanism gets no food, gas, or electricity. The person pulling the box uses up food in order to perform positive work. In this case with constant speed, all the transferred energy goes into the disruption and agitation of molecular bonds, which we describe in the chapter on friction. But will we not need the concept of negative work when we get to thermodynamics or electrostatics? No. There are other ways to keep track of energy flow. Why ensnare the innocent?

■ ROTATIONAL KINETIC ENERGY

An object traveling in circles or spinning on its axis has rotational kinetic energy. For a *point* object, at a distance r from the axis, the angular speed and tangential speed are related by $v = \omega r$. The angular speed ω is in radians per second, and the moment of inertia I is equal to mr^2. The rotational kinetic energy of the object is therefore

$$E_{\text{kin}} = \tfrac{1}{2}mv^2 = \tfrac{1}{2}m(\omega r)^2 = \tfrac{1}{2}I\omega^2.$$

This last expression for rotational kinetic energy holds good for any kind of rotational motion, whether or not it is a point object, and whether the object is circling an outside axis or spinning on its own (as long as the angular momentum is in the same direction as the angular velocity—which is usually the case in introductory-physics problems.)

On p. 185 in the chapter on angular momentum (Chap. 8) we described a method of dropping a music record onto one that is already spinning. Angular momentum is conserved: $(I\omega)_0 = (I\omega)_f \rightarrow \omega_f = \tfrac{1}{2}\omega_0$. However, the kinetic energy of the system is not conserved. As you can tell by the sound when the records mesh, the collision is not elastic. The final rotational energy is

$$E_f = \tfrac{1}{2}(2I)\omega_f^2 = \tfrac{1}{2}(2I)(\tfrac{1}{2}\omega_0)^2 = \tfrac{1}{2}(\tfrac{1}{2}I\omega_0^2) = \tfrac{1}{2}E_0.$$

The result is just like that for the totally inelastic linear collision of two objects with equal masses. Exactly half the kinetic energy disappears.

Let us compare the rotational kinetic energy of a rolling bicycle wheel with its translational kinetic energy. Assume that the 2-kg mass is concentrated on the rim at a radius of 33 cm. At 15 miles/h, the linear speed is 6.7 m/s; the angular speed is

$$\omega = \frac{v}{r} = \frac{6.7 \text{ m/s}}{0.33 \text{ m}} = 20 \text{ rad/s}.$$

The rotational kinetic energy of the wheel about its axle is therefore

$$E_{\text{rot kin}} = \tfrac{1}{2}I\omega^2 = \tfrac{1}{2}(2 \text{ kg})(0.33 \text{ m})^2(20 \text{ rad/s})^2 = 45 \text{ J}.$$

The translational kinetic energy of the wheel is

$$E_{\text{tran kin}} = \tfrac{1}{2}mv^2 = \tfrac{1}{2}(2 \text{ kg})(6.7 \text{ m/s})^2 = 45 \text{ J}.$$

It appears numerically in this case that the rotational kinetic energy is equal to the translational kinetic energy. We could have deduced the same result from the general formulas. For any rolling wheel *with the mass on the rim*, the rotational kinetic energy is equal to the translational energy:

$$\tfrac{1}{2}I\omega^2 = \tfrac{1}{2}(mr^2)\omega^2 = \tfrac{1}{2}m(r\omega)^2 = \tfrac{1}{2}mv^2.$$

The total kinetic energy of the rolling wheel is $\tfrac{1}{2}I\omega^2 + \tfrac{1}{2}mv^2$.

Serious research has been done on the design of super flywheels for the storage of very large amounts of energy. Since the energy is proportional to the square of the rotational velocity, obviously it pays to increase the rotational frequency of the flywheel. Unfortunately, the centripetal force needed to keep the outer rim going in a circle is also proportional to the square of the rotational velocity: $F_{\text{cent}} = m\omega^2 r$.

Let us calculate the energy storage in one proposed type of flywheel. Suppose that the wheel is made of radial wires, anchored at a hub. Then the centripetal force must be exerted along the length of each wire. So long as the wires do not pull out of the hub, the only effect of the rotation is to stretch the wires. A giant circular wire brush of this kind is not the most efficient design for energy storage. With the wires packed closely together at the hub, they spread out at larger radii, occupying only a fraction of the volume. Furthermore, the part of the system most subject to stress is the anchor point of each wire. A better design consists of cylindrical layers of materials bonded together. Still, the order of magnitude of energy storage in a radial wire wheel is easy to calculate.

Assume that the radius of the wheel is $\frac{1}{2}$ m, so that it could fit easily into a car. If the thickness of the wheel were $\frac{1}{2}$ m, the volume would be about $\frac{1}{3}$ m³. If the total volume were filled with a material of the density of steel, the mass would be 2000 kg. Since the wires occupy only a fraction of the volume, let us assume a mass of 500 kg, composed of individual radial wires, each with a moment of inertia equal to $\frac{1}{3}Mr^2$. (This is the moment of inertia of a thin rod rotating about an axis at one end.) Such devices can be spun in vacuum at frequencies of 4000 rpm. The energy storage would be

$$E_{\text{rot}} = \tfrac{1}{2} I \omega^2 = \tfrac{1}{2}(\tfrac{1}{3})(500 \text{ kg})(\tfrac{1}{2} \text{ m})^2 4\pi^2 [(\tfrac{2}{3})(10^2) \text{ Hz}]^2 \approx 3\tfrac{1}{2} \times 10^6 \text{ J}.$$

We can compare this energy with other types of storage by noting that 3.6×10^6 J = 1 kilowatt-hour (kW h), and that a standard car battery can store about 1 kW h. The flywheel would thus provide the equivalent of about one car battery.

Can ordinary wires exert the necessary centripetal force without flying apart? The mass of one of the wires is equal to its density times the cross-section area times the length: $M = \rho A r$. Exaggerate the effect of the force by assuming that all that mass is located at the outer radius. Assume a density of 6×10^3 kg/m³ for the wire. Young's modulus Y for steel is 1×10^{11} N/m².

$$F = m\omega^2 r = (\rho A r)(4\pi^2 f^2) r = [6 \times 10^3 A \tfrac{1}{2}][4\pi^2 (\tfrac{2}{3} \times 10^2)^2] \tfrac{1}{2},$$

$$\frac{F}{A} = \frac{6 \times 10^3 \times 4\pi^2 \times 2 \times 2 \times 10^4}{2 \times 3 \times 3 \times 2} = 24 \times 10^7 = Y \frac{\Delta L}{L},$$

$$\frac{\Delta L}{L} = \frac{24 \times 10^7}{1 \times 10^{11}} = 2.4 \times 10^{-3}.$$

If the strain of an ordinary steel wire is much greater than about 2×10^{-3}, it will break. It appears that with the assumptions that we have made, the wire in the flywheel would be at the breaking point. Actually, much stronger materials and better designs are available.

■ SIMPLE MACHINES

The lever, the inclined plane, and the pulley, all used to be part of introductory physics. In most syllabi we have banished the simple machines to earlier grades and have put in their place quarks and unified field theories. Unfortunately, the lower grades are usually too busy teaching quantum mechanics. A sampling of 15 senior physics majors at a well-known research university revealed that only two had ever played with a pulley and none could remember learning about the principles of simple machines in any class. All we can do here is to urge that large, working examples of the machines be available in physics laboratories, and that students be required to perform work with them. In each case, the machine should be large enough so that its advantage is clearly evident. That is to say, the pulley should have enough wheels and be strong enough to lift a bag of sand that is too heavy to lift by hand.

The simple machines save no work. With large-size devices, the students can demonstrate this by maneuvering heavy objects, measuring forces with newton scales and displacements with meter sticks. Except in the case of levers, the real mechanical advantage is always less than the ideal. With all machines, (work input)$\geq$(work output), $(F_{in})(x_{in}) \geq (F_{out})(x_{out})$, where x is the displacement produced by the force. Most machines are really devices allowing us to trade a smaller input force and a longer input displacement for a larger output force and a shorter output displacement, or vice versa. Such a trade is frequently useful. Let us examine a few special cases.

With levers, we usually assume that the lever bar itself has no weight, and that the forces are exerted at right angles to the lever. Actually, in the case of a heavy crowbar dislodging a rock, the weight of the crowbar may add important leverage, and neither the weight nor the applied force may be perpendicular to the bar. At any rate, if the applied force is perpendicular to the lever, then the distance through which that force is exerted is just the arc through which that end of the lever moves. The arc is proportional to the length of the lever from the fulcrum to the point where the force is applied. The same conditions are true for the output force and the output distance. Consequently,

$$(F_{in})(\mathrm{arc}_{in}) = (F_{out})(\mathrm{arc}_{out}) \rightarrow (F_{in})(L_{in})(\theta) = (F_{out})(L_{out})(\theta),$$

$$\frac{F_{out}}{F_{in}} = \frac{L_{in}}{L_{out}}.$$

With most levers—nutcrackers, can openers, or crowbars—we want to exert only a small force with our hands, even though we have to do so through a longer distance, in order to apply a large force on some object. Usually the lever provides other important advantages as well, providing an "impedance match" between large, soft hands and small, hard materials. The input force of your hand on a can opener is distributed over a much larger area than is the output force of the cutting edge. Not only is the input force smaller than the output force because of the leverage, but the input *pressure* is much smaller than the output pressure. It is well known (at least in Georgia) that it is hard to crack one pecan by itself in the hand, but it is easy to crack one pecan if two are pressed against each other in the hand.

There are other examples of the way simple tools provide impedance match between the low-pressure exertions of human hands or feet and the high pressure of cutting edges. For instance, have you ever tried cutting a steak by grabbing the sharp blade and cutting with the handle? A knife is not a lever. There is no mechanical advantage. You just exert a force downward with your hand, and that same force is exerted by the blade. But, the *pressure* on your hand is not the same as the *pressure* on the steak. For another example, compare the radius of the handle of a screwdriver with the radius of the blade. The ratio is the mechanical advantage, which is usually small. However, the handle is fashioned so that the whole hand can grasp it and exert pressure over a large area to provide the torque. Similarly, door knobs are designed to accommodate the size and softness of the human hand.

The prototype machine for amplifying force by transmitting pressure is the pneumatic lever, shown in Fig. 9.5. The pressure on both input and output pistons is the same: 10 N/cm². A small force on the input piston produces a large output force because of the large area of the output piston: $F = \text{(pressure)} \times \text{(area)}$. Of course, the input piston must be shoved a proportionately greater distance, so that $(\text{work})_{\text{out}} = (\text{work})_{\text{in}}$. This device is used in automobile break pedals and pneumatic lifts.

Some elementary-school texts still emphasize the classification of levers into three categories, depending on the placement of the fulcrum. (A class 1 lever has the fulcrum between the "effort" and the "load"!) Such classification has no use whatsoever and is unknown in the real world, but you may run into students who have memorized the scheme.

It may seem as if most lever arrangements that we use provide a greater output force in exchange for a greater input distance. Actually, the levers that we use most of the time are rigged just the other way. Most of the joints of our body consist of levers arranged to deliver large output displacement—and consequently high output speed—in exchange for large input force. Figure 9.6 shows the main bones and muscles of the human arm. For every body part that can move, there have to be two muscles. Muscles cannot push; they can only pull. When you raise your lower arm, the biceps contracts and the triceps expands. When you lower your arm, the opposite happens, as you can readily feel for yourself.

On a typical arm, the biceps muscle is fastened to the forearm at a point about 5 cm beyond the elbow. The hand is about 35 cm from the elbow. The biceps must act through a lever arm of 5 cm while everything being lifted in the hand has a lever arm seven times as long. If the point where the biceps is fastened moves upward 1 cm, the hand rises 7 cm. That is clever! It means that your muscle can make a short movement, which makes your hand move much further and thus very rapidly. On the other hand (on either hand, for that matter), a lever that trades distance for force demands a great deal of force. If you hold 10 kg in your hand, the force down is 100 N, and your biceps must pull up with a force of 700 N. That is why biceps have to bulge. A hand that can move very fast can throw a stone very fast, and a leg that can move fast allows its owner to run very fast. Maybe this effect provides an evolutionary advantage.

The class of machines called *wheel and axle* includes the screwdriver and doorknob. The principle is really the same as that for the lever. In most wheel-and-axle arrangements a small input force exerted on a long lever arm moves through a large arc. The output force is larger, but is exerted through a smaller arc. Note, however, that the output torque and angular displacement are the same as the input torque and angular displacement.

Gears are toothed wheels that are not on the same axle. One gear may mesh directly with another or they may be connected with an endless chain. The important point about gears is that they provide a way for a shaft turning at a particular angular speed to drive another shaft at a different angular speed, and therefore with a different torque. *Gears are torque converters.* A small gear with few teeth can rotate at high speed and low torque to drive a large gear with many teeth at low speed and high torque. The ideal mechanical advantage, the ratio of output

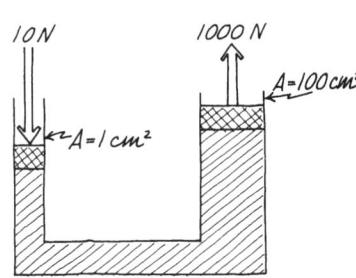

Figure 9.5. Hydraulic lever.

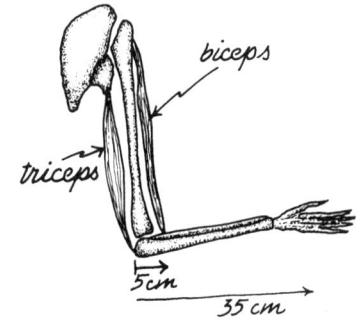

Figure 9.6. The lever in the arm.

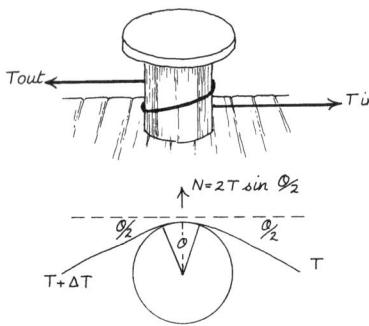

Figure 9.7. Geometry of the capstan.

torque to input torque, is the ratio of angular displacement of the drive wheel to the angular displacement produced in the final wheel.

A bike brought into the classroom makes a great laboratory apparatus. The drive system is a combination of gears and two wheel-and-axle arrangements (the pedal wheel and the back wheel). The way to measure the mechanical advantage is to measure the distance traveled by a point on the back-wheel circumference, while the pedals go around once. The ratio of pedal-circle circumference to wheel arc is the ideal mechanical advantage. The actual mechanical advantage is less because there is friction in the gears and chain drive. Friction does not affect the ratio of arcs, but it does require one to push harder on the pedals. The actual mechanical advantage can be measured by using two spring scales, one placed at the rim of the back wheel and the other pulling the pedal. Incidentally, even for 10-speed bikes, the mechanical advantage is less than 1 in all gears.

When a rope goes around a wheel that is free to turn without friction (and acquires negligible rotational energy), the tension in the rope on one side of the wheel is the same as the tension on the other side. Hence it is possible with a pulley wheel on the ceiling to lift something by pulling down. There may be a practical advantage in this, but there is no *mechanical* advantage. By using another wheel traveling with the load, you can suspend the load with two strands of the same rope, each pulling up with a force equal to the tension. The person pulling on the rope is exerting a force equal to the tension in the rope; thus the person's effective force is doubled. However, for every meter of rope that is pulled, the load moves through only half a meter. Once again, in an ideal case without friction, input work equals output work.

The ideal mechanical advantage of a simple pulley system can be calculated by counting the number of strands supporting the load. The real mechanical advantage is usually considerably less than the ideal, since pulleys under heavy load frequently have a lot of friction. To measure the real mechanical advantage, the ratio of output force (the load) to input force must be measured *while the load is actually being moved*.

The *capstan* is another much-used device that is simple, although it is not ordinarily included as one of the simple machines. That is because it is a friction amplifier that is used to prevent a load from moving. Consequently, there is no energy input or output. We show the capstan mechanism in Fig. 9.7. It is a particularly familiar device to anyone who has observed boats being warped to a pier or who has snugged down the sheets of a sailboat. By taking one or more turns of the rope around the capstan, a person can exert a tension T_{in} that yields a much larger tension T_{out} on the other end. Compare this arrangement with the one-wheel pulley where we assumed that the pulley was friction-free and therefore that the tension was the same on both sides of the wheel.

The difference in tension between load side and effort side of the capstan must be equal to the amount of the friction, $\Delta T = \mu N$, where N is the normal force exerted on the rope by the capstan. As you can see from the diagram,

$$N = 2T \sin \frac{\Delta \theta}{2} \approx T \Delta \theta.$$

Therefore,

$$\Delta T = \mu T \Delta \theta \rightarrow \frac{dT}{d\theta} = \mu T.$$

This differential equation is mathematically the same as the one for compound interest; the *increase* in the function is proportional to the function. The solution is the exponential:

$$T_{out} = T_{in} e^{\mu \theta}.$$

Consider how rapidly the holding tension increases. For $\mu = 0.30$ and for one turn of the line around the capstan, $T_{out}/T_{in} = e^{0.3 \times 2\pi} = 6.6$. For two turns of the line around the capstan, $T_{out}/T_{in} = e^{0.3 \times 4\pi} = 43$.

There is a whole other class of simple machines that does not appeal in any simple way to the conservation of energy or to the exchange of force and distance. These are *impulse machines*. With these devices, *force* and *time* are the variables of exchange. The prototype example is the hammer. The controlling principle is Newton's third law, framed to emphasize the product of force and time interval, which is known as the impulse:

$$F\Delta t = \Delta(mv).$$

When the hammer hits the nail, its change of momentum is equal to its momentum since it comes to rest. The magnitude of the force depends on the size of the time interval. The shorter the time, the greater the force.

One grim example of this increase of force with decrease of collision time is the sequence of events in car accidents, a subject dealt with quantitatively on p. 166. Consider more homely examples: slamming doors, dropping things on shelves, tramping upstairs, using a baseball glove.

■ POTENTIAL ENERGY

When you do work on a system, you can produce kinetic energy or, among other possibilities, you can distort the system. Objects move up hill, springs stretch, electric charges separate, molecules dissociate. To preserve a conservation law, we define the *potential energy* of the system in such a way that we can claim that the work done has transferred energy that is then stored. The energy of the system is *potential* in the sense that it can sometimes be transferred back again and do work as the distortion goes back to 0. Of course, in this sense, *kinetic energy* is also potential in so far as it can be turned into other kinds of energy, such as stretched springs. The point is that we define forms of energy so that as transformations occur we maintain a quantity that is conserved. Does that mean that energy conservation is just a human invention of clever definitions? Not really. The definition of a new form of energy may, in the first instance, be a human invention. The test of the usefulness of the invention, however, is the consistency with which the model describes a wide range of phenomena. For instance, if our definition of gravitational potential energy only conserved energy for air-track gliders, we should look further for a law of nature.

In general, what we mean by potential energy is energy that is a function only of position or displacement. With some systems and some types of changes, the system regains its original potential energy when it regains its original position or displacement. These are called *conservative potentials*. In other cases, such as the stretching of a spring beyond its elastic limit, some of the original work gets transformed into internal forms of energy, which sometimes show up as higher temperature. In this chapter we will analyze only a few forms of conservative potential energy.

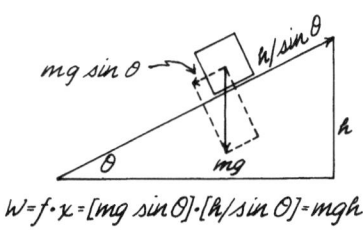

Figure 9.8. The work done to shove an object up an inclined plane is independent of angle (for no friction).

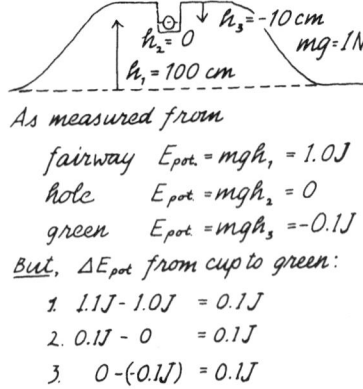

Figure 9.9. Potential energy of a golf ball depends on arbitrary location of zero potential.

There are many classic demonstrations of energy transformation from potential to kinetic and back again. It is instructive for students to handle devices where they must do work to store up energy and then see it turn into some other form. Perhaps the easiest to make and most fun to use are wind-up toys, particularly rubber-band boats or airplanes. Another simple demonstration consists of a pendulum that transforms kinetic and gravitational potential energy back and forth. You can revive interest in this familiar device by demonstrating coupled identical pendulums. An arrangement of this can be easily reproduced as a party stunt. Tie a string between the tops of two dining room chairs, about a meter apart. From the string suspend two other strings of equal length with some reasonable kind of weights attached as pendulum bobs. You can change the tautness of the suspension line, and thus the strength of the coupling, by moving the chairs slightly. Everyone is always fascinated to see energy flowing back and forth between the two pendulums, as one decreases in amplitude while the other increases and then back again.

In a constant gravitational field, the work done in raising an object through a height h is mgh, where mg is the weight of an object with mass m. The potential is conservative, so that the change in potential energy is independent of the route taken as the object rises through the height h. That is why inclined planes work as simple machines. To shove a box up an incline with angle θ requires a force equal to $mg \sin \theta$, as shown in Fig. 9.8. To reach the height h, you must shove the box a distance $L = h/\sin \theta$. The work you do is $(mg \sin \theta)(h/\sin \theta) = mgh$, an amount of work independent of the incline angle θ. That work has turned into gravitational potential energy of the earth–box system. As we point out in Chap. 5, the gravitational field is not really constant, but is inversely proportional to the square of the distance to the center of the earth. The actual gravitational potential energy is not mgh, but $-GmM/r$. On p. 111 we show the limitations in using the approximation that $h = \Delta r$.

When we use the approximation that $E_{\text{grav pot}} = mgh$, there is the problem of specifying h. For your own gravitational potential energy, do you assume that h is your height from the floor, the ground, or the center of the earth? The zero point of potential energy appears to depend on a rather arbitrary definition. Actually, only *differences* in potential energy are ever measured. When work is done to distort a system, the stored energy is the difference between the final and initial potential energy.

Consider the case of a golf ball in a hole as shown in Fig. 9.9. The potential energy of the ball appears to depend on whether the zero point is chosen at the level of the fairway, the green, or the hole. Depending on this choice, the ball has a potential energy that is positive, negative, or 0. The physical significance of the definition, however, involves only the *change* of potential energy as the ball moves from one height to another. As the diagram shows, the change of potential energy of the ball if it is raised from the cup to the green is the same positive quantity regardless of how the zero point is defined.

While the definition of the zero point of potential energy is arbitrary, there are some reasonable conventions. For instance, it is reasonable to assign zero potential energy to a spring that is undistorted. If two objects that influence each other through electric, magnetic, or gravitational forces are so far from each other that the influence is not measurable, then we might say that at that distance their potential energy is 0. For instance, $E_{\text{grav pot}} \to 0$ as $r \to \infty$.

Let us calculate the potential energy produced by two particular nonconstant forces. The work done in stretching a spring from zero distortion to a stretch length x is

$$W = \int_0^x \mathbf{F}(x) \cdot d\mathbf{x} = \int_0^x kx\, dx = \tfrac{1}{2}kx^2.$$

Since we can get that energy back when x returns to 0, we define the potential energy of the spring as $E_{\text{pot spring}} = +\tfrac{1}{2}kx^2$.

The work we do in liberating an object with mass m from the earth's inverse-square gravitational field is:

$$W = \int_r^\infty \mathbf{F}(r) \cdot d\mathbf{r} = GmM \int_r^\infty \frac{1}{r^2} dr = -G\left.\frac{mM}{r}\right|_r^\infty = +G\frac{mM}{r}.$$

(Note that an inverse-square force gives rise to an inverse-first-power potential.)

Since that is the work *you* must do on the object to raise it to zero potential, the object itself at position r must have been at the negative potential: $-GmM/r$. Note the similarity of the arguments in these two cases. In both cases you do positive work first. In the case of the spring you define the resulting potential energy of the spring as positive, because you can get that much energy out of it. In the case of gravity you define the resulting potential as 0, therefore assigning negative potential energy to the original situation when the object was at r. You can also get positive energy out of the object in the gravitational potential by letting it fall back down from ∞ to r. The difference in these two cases is caused by the arbitrary (but reasonable and convenient) assignment of zero potential.

What can it mean to have *negative* energy? It is just a matter of bookkeeping. If you must do work on a system to distort it, then in that distorted geometry the system has positive potential energy. If you do work on a system to restore it to a situation so that it is no longer distorted, then the system originally had negative potential energy. In either case, for an object free to move without friction in a potential field, the bookkeeping works out so that

$$\Delta E_{\text{kin}} + \Delta E_{\text{pot}} = 0.$$

If E_{pot} is positive and decreases, then ΔE_{pot} is negative and ΔE_{kin} is positive. When a ball rolls down a hill, the potential energy decreases and the kinetic energy increases. For the opposite case, consider what happens when a magnet trapped by another one is given a shove to separate them. E_{pot} was negative but becomes less so; therefore ΔE_{pot} is positive. As the magnet pulls away, its speed decreases so ΔE_{kin} is negative.

If an object has negative potential energy, it is trapped in a *potential well*. If you fall into a water well, ΔE_{pot} is negative and as you fall ΔE_{kin} is positive. As you land in the water, your kinetic energy disappears into other forms of energy (splashing of the water and thermal energy). Then you are trapped in a gravitational potential energy well. To furnish the positive quantity equal to $|E_{\text{pot}}|$ to raise you back to the surface, there would have to be an expenditure of energy from some other source. If you could jump out, your kinetic energy as you start the leap is greater than it is at the top: ΔE_{kin} is negative.

You do not have to fall into a water well to be trapped by the earth's gravitational field. In Chap. 5 we analyze $E_{\text{pot grav}}(r)$ for the earth and show how this is

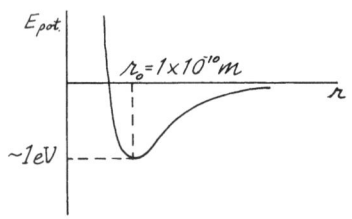

Figure 9.10. Plausible graph of potential energy of bound atom.

related to various phenomena such as escape velocity and the behavior of satellites.

■ MOLECULAR POTENTIAL WELLS

We can use the idea of potential-energy curves even in cases where we do not know the exact nature of the energy or the detailed shape of the curve. For example, we know that certain combinations of atoms stick together to form molecules, and so there must be a force of attraction between the atoms. On the other hand, the atoms do not fall completely into each other, and so at close range there must be a force of repulsion. When an atom is in its normal molecular position with respect to its neighbors, it must be in a potential-energy well. If the atom moves to a larger radius or to a smaller radius, there must be a restoring force to bring it back to its equilibrium position. In Fig. 9.10 we sketch a plausible shape for such a potential-energy well. The curve is a plot of $E_{pot}(r)$ for a bound atom as a function of its distance r from a neighbor.

When an atom moves too close to its neighbor, at small r, the potential energy becomes positive because of the repulsion. When the atom moves beyond the equilibrium point, the potential energy becomes less negative and finally goes to 0, indicating that the atom has been freed. Note that the potential-energy well is not symmetrical. The repulsion force that dominates when an atom gets too close to its neighbor usually rises more abruptly than the attraction force when the atom is pulled away. This asymmetry is vital for the explanation of thermal expansion of solids. We expect that for r larger than the equilibrium distance, the curve rises much more rapidly than would be given by a $1/r$ dependence. The molecular binding force is electromagnetic, but is much shorter range than that produced by isolated charges.

Even without knowing the source of the molecular forces, we can roughly calibrate the axes of the $E(r)$ curve. The radius of most atoms is about 1×10^{-10} m. In a molecule the atoms are essentially touching each other and so the equilibrium distance to the boundary between atoms is about 1×10^{-10} m. The binding energy of atoms (the depth of the potential well) has about the same value in most molecules. Within a factor of 3 or so this value is 1.6×10^{-19} J, an electron volt or eV. As an example of how the electron volt is a common magnitude of energy on the atomic scale, consider that most electrochemical cells (found in flashlight batteries, car batteries, or lemons) yield between one and two volts. These potentials are created by the rearrangement of molecules, each one of which contributes 1 to 2 eV. It is also the case that the energy of each photon of visible light is from $1\frac{1}{2}$ to 3 eV, depending on the color. Since light can cause many chemical reactions—bleaching dyes, photosynthesis, or tanning skin—it must be that the binding energy of chemical compounds is in the electron-volt range.

There is one more plausibility argument that we can make about the shape of the $E(r)$ curve for bound atoms. Although the whole well is asymmetrical, for small displacements from the equilibrium point any smooth-shaped well can be approximated by a parabola: $E_{pot} \approx \frac{1}{2}kx^2$, where x is the displacement from the equilibrium point. (See Chap. 13, p. 309.) That is the expression for the potential energy of a simple harmonic oscillator. If the bound atom is slightly disturbed, it should oscillate about the equilibrium point with simple harmonic motion. If enough energy is fed into the system so that the maximum kinetic energy is greater than E_0, the total energy is positive and the object is no longer bound.

■ RESTORING FORCES AND POTENTIAL ENERGY

The potential energy of a system, $E(x)$, is a function only of the final position or distortion of the system, regardless of how it got that way. A spring could be compressed rapidly or slowly, with a large force part of the way producing acceleration, and then a reverse force bringing the spring to rest. Extra force might also have been necessary to overcome friction. No matter what the history, once the spring is distorted by an amount x, its potential energy is $E(x)$.

The potential energy of a system is a scalar quantity expressed in joules and by itself does not provide information about the future action of the system. Look at the sketches in Fig. 9.11 showing $E(x)$ for three different springs, and notice the point for each of them where $E_{pot}=1$ J. The first one apparently represents the situation for a weak spring that has been stretched a long distance. The second one must represent a strong spring, which had to be stretched only a short distance in order to store 1 J. In the third case the spring has been compressed. Although the potential energy has the same value in each case, the actions of the springs if released will be quite different. The first one will pull back slowly to the left; the second one will snap back to the left; the third one will expand to the right. Although the single value of potential energy does not allow one to predict these different actions, the information is apparent in the shape of the whole $E(x)$ graph. It is the *slope* of the $E(x)$ curve at any point that is related to the restoring force *in the x direction* that the system exerts at that point. Let us see some examples.

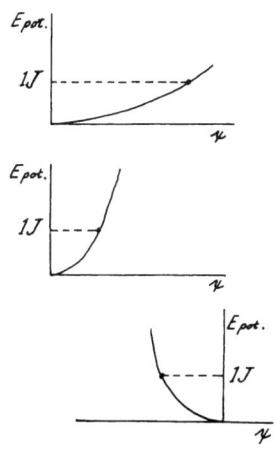

Figure 9.11. Potential energy graphs of three different springs.

The graph of $E(h)$ for an object lifted away from the earth's surface (for small distances) has a constant slope: $\Delta(mgh)/\Delta h = mg$. The slope is equal to the weight of the object. There is a catch, however. The restoring force of gravity is pointed downward, and so is negative. The slope of $E(h)$ is positive. If we want the *restoring* force of the system, we must take the *negative* slope:

$$F_{restoring} = -\frac{\Delta E(h)}{\Delta h} \rightarrow -\frac{dE(h)}{dh}.$$

The *external* force that you would have to exert to store energy in a gravitational system is in the opposite direction, upward and positive.

The same situation applies to energy stored in a spring. The *restoring* force is given by

$$F_{spring} = -\frac{dE(x)}{dx} = -\frac{d(\tfrac{1}{2}kx^2)}{dx} = -kx.$$

The restoring force obeys Hooke's law; it is proportional to the distortion and in a direction opposite the distortion. Notice that this definition agrees with what you would expect qualitatively in the cases of the three springs that we described before. In the first case the slope is small and positive; therefore the restoring force will be small and negative—toward smaller x. In the second case the slope is large and positive; the restoring force will be large and negative. In the third case the slope is negative; therefore the restoring force will be positive, making the spring expand.

We now have two relationships between force and other dynamical quantities.

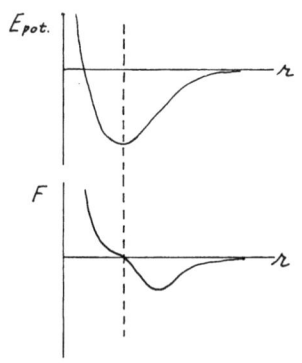

Figure 9.12. E(r) and F(r) for a bound atom.

1. The net force on an object is the *time* rate of change of its *momentum*:

$$\mathbf{F} = \frac{d\mathbf{p}}{dt} = \frac{d(m\mathbf{v})}{dt}.$$

Conversely,

$$\Delta(m\mathbf{v}) = \int_0^t \mathbf{F}\, dt.$$

The change of momentum is in the direction of the net force.

2. The restoring force provided by a (conservative) distorted system is the negative of the *spatial* rate of change of its *potential energy*:

$$F_r = -\frac{d(E_r)}{dr}.$$

Conversely,

$$\Delta E_{a \to b} = \int_a^b \mathbf{F}_{\text{ext}} \cdot d\mathbf{r}.$$

If E is a function of several variables, such as $E(x,y,z)$, then

$$F_x = -\frac{\partial E}{\partial x}, \quad F_y = -\frac{\partial E}{\partial y}, \quad F_z = -\frac{\partial E}{\partial z}.$$

■ MOLECULAR FORCES

Any time that we have a potential-energy system, we can find the restoring force associated with it. The force at any point is simply the negative of the slope of the $E(x)$ curve at that point. Let us take another look at our plausibility argument about the shape and size of the potential-energy curve of a bound molecule. In Fig. 9.12 we reproduce the curve of Fig. 9.10 and underneath have sketched the negative slope (the negative derivative) as a function of r. At distances smaller than the equilibrium point, the slope of $E(r)$ is large and negative. Therefore the restoring force on the atom is large and positive, toward larger r. At the equilibrium point, the slope of $E(r)$ is zero. Of course, the force should be zero when the atom is in equilibrium. For larger values of r, the slope of $E(r)$ is positive; the restoring force is negative, bringing the atom back toward equilibrium, but then, if the atom gets too far away, the force gets smaller and becomes zero.

Since we have a rough value for the binding energy and radial separation of atoms in a molecule, we can derive an approximate value for the forces involved. Let us assume that the half-width of the potential well is equal to an atomic radius. In other words, when the atoms are touching, shoulder to shoulder, they are in their bound equilibrium position, but if they separate by more than a distance of the atomic radius, they are free. For $\Delta r = 1 \times 10^{-10}$ m, $\Delta E = 1$ eV $= 1.6 \times 10^{-19}$ J. Then the average restoring force during this separation is

$$F = -\frac{\Delta E}{\Delta r} = -\frac{1.6 \times 10^{-19}\ \text{J}}{1 \times 10^{-10}\ \text{m}} \approx -2 \times 10^{-9}\ \text{N}.$$

That is a very small force, but then that is just the force on one atom. Let us find the force that it would take to pull apart one square meter of such atoms. There are

10^{10} atoms along each meter, and so there are 10^{20} atoms in 1 m². The force exerted per square meter is $F/A \approx (2 \times 10^{-9} \text{ N/atom})(1 \times 10^{20} \text{ atoms/m}^2) = 2 \times 10^{11}$ N/m².

The relationship between stress and strain for a metal rod is $F/A = -Y(\Delta L)/L$, where Y is Young's modulus. The values of Young's modulus for metals are in the range of 1×10^{11} N/m². The value of F/A that we obtained from the molecular forces was based on the assumption that the separation distance (in the direction of the force) between each atom was being doubled, from 1×10^{-10} to 2×10^{-10} m. In other words, $(\Delta L)/L = 1$. It appears that Young's modulus is approximately the molecular force per unit area required to stretch a wire to double its original length.

Of course, Young's modulus does not have the same value for all materials, nor do all atoms have the same radius and the same binding energy. Still, it is apparent that our plausibility arguments about the nature of the potential-energy curve for bound atoms must be qualitatively correct. Note how nicely introductory physics can explore first approximations in the microstructure. Actually, a wire cannot be stretched until the strain, $(\Delta L)/L$, is equal to 1. As a matter of fact, most metals will fracture if the strain gets as large as 0.002. From the shape of the molecular $E(r)$ curve, there is no reason to expect breakdown for separation distances of only 2/1000 of the atomic radius. Fractures occur because of impurities or imperfections in the crystal structure of the metal.

We can also use the values for the size of the molecular energy well to determine the frequency of oscillation of a bound atom. If the well were parabolic all the way to an extension from equilibrium of an atomic radius, the potential energy would be

$$E_{pot} = -\tfrac{1}{2}kx^2 = -1 \text{ eV} = -1.6 \times 10^{-19} \text{ J} = -\tfrac{1}{2}k(1 \times 10^{-10} \text{ m})^2,$$

$$k \approx 30 \text{ J/m}^2.$$

This value of k also agrees with Hooke's law:

$$k = \frac{3 \times 10^{-9} \text{ N}}{1 \times 10^{-10} \text{ m}} = 30 \text{ N/m}.$$

The frequency of a harmonic oscillator is given by

$$f = \frac{1}{2\pi}\sqrt{\frac{k}{m}}.$$

The mass of a carbon atom is about 2×10^{-26} kg. According to our model, the oscillation frequency of a carbon atom bound in graphite or diamond should be in the range

$$f = \frac{1}{2\pi}\sqrt{\frac{30 \text{ J/m}^2}{2 \times 10^{-26} \text{ kg}}} \approx 10^{13} \text{ Hz}.$$

This frequency is in the far-infrared range. We consider this problem again when we look at thermal energy in the microstructure in Chap. 10, and will find that

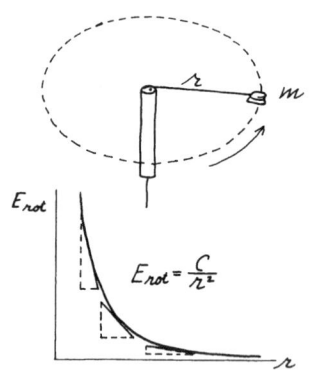

Figure 9.13. $E_{rot}(r)$ for object going in a circle with constant angular momentum.

with the addition of some quantum conditions we get another reasonable first approximation.

■ CENTRIFUGAL FORCE IN A ROTATING REFERENCE FRAME

A rotating object has rotational kinetic energy equal to

$$E_{\text{rot}} = \tfrac{1}{2} I \omega^2 = \frac{1}{2} \frac{(I\omega)^2}{I}.$$

If the rotating object is a point mass at the radius r, the moment of inertia is mr^2. Let us analyze the situation where no external torques act on the system. Then the angular momentum, $I\omega$, is a constant, and the rotational energy is

$$E_{\text{rot}} = \frac{1}{2m} \frac{(I\omega)^2}{r^2} = \frac{C}{r^2}.$$

The constant C ties together the constants of mass and angular momentum. It appears that under these conditions (constant angular momentum and circular motion) the rotational energy is a function of position only. Energy that is a function of position can be considered potential energy! Each time that we define potential energy, however, we have to choose an appropriate zero point and determine whether the energy is positive or negative. In this case, the convenient zero point is where $r \to \infty$. At first glance, this may seem like a poor choice, because an object rotating with a large radius usually has a large rotational energy ($E_{\text{rot}} = \tfrac{1}{2} m r^2 \omega^2$). Remember, however, that we are considering the case where angular momentum remains constant. For a system without external torques, as r increases, the kinetic energy decreases.

Whether we define the energy as being positive or negative depends on who does the work of changing position. If the system automatically tries to restore itself to zero energy, then the potential energy is positive. (This is like the situation with a stretched spring.) Take a look at the sketch and graph in Fig. 9.13. Once the bob has been set into rotation, the only torque on it is caused by the small amount of friction where the string rubs the edge of the tube. Since this torque can be made very small, the angular momentum remains constant for short periods of time. The only force that can be exerted within the system is radial. If you try spinning this device, you will find that the bob tries to move toward a larger radius. To make it move in toward a smaller radius, you must do work on the system by pulling on the string. Therefore, according to our convention, the rotational energy is positive. The graph of $E(r)$ is shown in the diagram.

Every system with potential energy provides a restoring force equal to the negative slope of the potential-energy curve. Since the energy of the rotating bob is a function of r, the force must be radial. The slope of $E(r)$ is negative, and so the restoring force must be positive—in the direction of increasing r. The magnitude of the force is

$$F = -\frac{dE}{dr} = -\frac{d(C/r^2)}{dr} = +\frac{2C}{r^3}.$$

Let us take a closer look at the value for the restoring force in this rotating system:

$$F=\frac{2C}{r^3}=2\left(\frac{1}{2m}\right)\frac{(I\omega)^2}{r^3}=\frac{1}{m}\frac{(m^2r^4)\omega^2}{r^3}=m\omega^2 r.$$

The restoring force has the same magnitude as the centripetal force, but is in the outward radial direction; it is a *centrifugal* force. Whether the radial force in circular motion is outward or inward depends on the reference frame in which the force is measured. If you are outside a rotating system, you observe that it is necessary to exert an inward radial force to keep the object rotating in a circle instead of flying off at a tangent. If you are inside the rotating system, such as in this case, then you observe that there is a force tending to throw objects out radially.

■ USE OF ENERGY CONSERVATION IN DYNAMICS PROBLEMS

Sometimes it is easier to solve dynamics problems by appealing to the conservation of energy rather than by setting up Newton's laws. At the next stage of the study of mechanics, these methods are transformed and formalized with Lagrange's and Hamilton's equation. Here is an example of how to solve a complicated Atwood's machine problem, using energy conservation. The situation was presented in Chap. 7 and is shown in Fig. 7.12 of that chapter on p. 170.

Assuming that B falls lower:

$$\Delta E_{\text{pot A}} - \Delta E_{\text{pot B}} = \Delta E_{\text{kin A+B}} + \Delta E_{\text{kin rot pulley}} + \Delta E_{\text{friction}},$$

$$m_A g h - m_B g(h\sin\theta) = \tfrac{1}{2}(m_A+m_B)v_f^2 + \tfrac{1}{2}(kMr^2)\left(\frac{v_f}{r}\right)^2 + \mu(m_A g\cos\theta)h.$$

For a given distance of fall, h, this one equation can be solved for the one missing variable, v_f, the final speed. In the solution in Chap. 7, the value of v would come out negative if the wrong direction of motion were chosen. In this formulation the wrong assumption will yield an imaginary value for v. This will also be true if the friction force is too large.

■ POWER

Power is the time rate of transferring energy: $(\text{power})=\Delta E/\Delta t \rightarrow dE/dt$. The unit of one joule per second is called the watt, symbolized by W. It is named after James Watt, a Scottish engineer and inventor who invented a new form of steam engine in the latter part of the eighteenth century. The watt is more familiar to most students than the joule, since electrical power is commonly measured in watts. A 60-W bulb, for instance, uses 60 J of electrical energy each second. We do not pay the electric company for watts—or power—however, we pay for energy. The common commercial unit of electrical energy is the kilowatt-hour (kW h). Notice that the unit is a product of power and time, yielding a unit of energy. 1 kW h$=3.6\times 10^6$ J. Another common unit of power is the horsepower. This one was defined by Watt himself, in 1783, as the average amount of work per second that could be done by a strong English dray horse, working steadily all day. (Actually, Watt deliberately chose a value about 50% larger than most horses could perform.) One horsepower equals 746 W.

If you had to work steadily all day, you probably couldnot generate even one-tenth horsepower. During brief bursts of effort, however, you may be as powerful as a horse. The classic way to find out is to time yourself while performing the work of lifting your own weight. You can lift yourself by chinning or by running upstairs. The upstairs run makes a fine exercise for a class. They must measure the vertical height and each person must measure or calculate his or her own weight and time their run. If you can chin yourself at a rate of once a second, you are producing a little less than one-half horsepower. You can probably produce more power by using your leg muscles in running upstairs.

In many cases, the time during which a certain amount of energy is produced or absorbed makes a tremendous difference in the consequences. The amount of energy contained in a lightning bolt is in the range of 10^9 J or 300 kW h. That is about as much electrical energy as a household would use in one or two weeks. When it arrives in about one-third second, however, the narrow path down which it passes does not have time to dissipate the energy harmlessly, and instead vaporizes or explodes. The average power during a stroke is 10^9 J/$\frac{1}{3}$ s$=3\times 10^9$ W, or 3000 megawatts (MW). Peak power during flashbacks that last a few microseconds reaches 10^{12} W. In a typical car collision, a car with a mass of 1000 kg going at 40 miles/h (19 m/s) might stop in 0.1 s. The power dissipated is

$$\frac{\frac{1}{2}(1000 \text{ kg})(18 \text{ m/s})^2}{0.1 \text{ s}} = 1.6\times 10^6 \text{ W}.$$

■ THE PRINCIPLE OF LEAST ACTION

In the middle of the sixteenth century, Pierre de Fermat (of Fermat's last theorem fame) devised his principle of "least time" for the passage of light rays. Whether reflecting from surfaces or refracting through different media, the path of light between two points will be such that the time taken will be a minimum—or a maximum—or at any rate, an *extremum*. Examples of this behavior are given on p. 335, and an explanation in terms of phase cancellations and reinforcement is given on pp. 355 and 356. In 1744 the French mathematician Pierre Maupertuis announced his discovery of a mechanical version of the least time principle. He called it the "principle of least action." His claim was that mechanical processes always occur in such a way that the product of mass, velocity, and distance (properly interpreted) is a minimum.

In modern terminology, if the product of momentum and displacement is summed for all elements of path as an object goes from one point in space to another, and if energy is conserved, then this sum is a minimum or a maximum for the natural path compared with any other path. To be mathematically precise:

$$\delta \int_{s_1}^{s_2} p \, ds = 0,$$

where δ means a small variation in path for the integral. The product of momentum and displacement is called the "action." A graph of p vs s is a plot in *phase space*. Note that the area under such a curve has the dimensions of (energy)$\times$(time).

A more complete and useful variational principle makes use of the Lagrangian function. In classical mechanics the Lagrangian is simply the difference between the kinetic and potential energy: $\mathscr{L}=E_{\text{kin}}-E_{\text{pot}}$. This equivalent formulation of the Maupertuis equation was given by William Hamilton in the nineteenth century:

$$\delta \int_{t_1}^{t_2} \mathscr{L}\, dt.$$

Conservation of (mechanical) energy need not be assumed in this formulation; it tumbles out of the equation along with Newton's laws (in the classical case). Note that the Maupertuis integral is over a spatial path while Hamilton's is over a temporal path. In either case, all the different paths for the integral must start at the same point in space and time and must end at another particular point, taking identical times. The principle is that the integral taken along the true path must be one where slight changes in path will not cause a first-order variation in the integral. This is a normal feature of a minimum or maximum—an extremum condition. A small departure of the variable from the equilibrium point makes only a second-order change in the function. For instance, at the bottom of the parabola $y = x^2$, if $x \to x + \Delta x$, then $y \to x^2 + 2x\Delta x + \Delta x^2$. Since at the bottom (minimum point) $x = 0$, then $\Delta y = 0 + \Delta x^2$, a second-order effect.

Feynman gives some simple examples of the application of the least action principle. For instance, for an object moving horizontally in the vertical gravitational field, the potential-energy term in the Lagrangian can be ignored since there will be no variation in it for any horizontal path. What remains is the condition

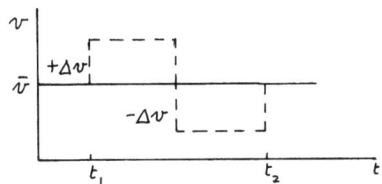

Figure 9.14. $v(t)$ curves for object moving horizontally.

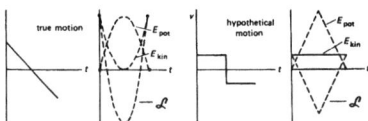

Figure 9.15. $v(t)$ curves (true and hypothetical) for object thrown straight up.

$$\delta \int_{t_1}^{t_2} E_{\text{kin}}\, dt = \delta \int_{t_1}^{t_2} \tfrac{1}{2} m v^2\, dt = 0.$$

The integration path (as opposed to the horizontal path of the object) could be one of constant speed or it could be one where the object first had large constant speed for a while and then small, as shown in Fig. 9.14. The only requirement is that the total time $(t_2 - t_1)$ be the same for all possible types of trips over the same distance. Therefore, the average speed must be the same. Suppose that the object travels at higher than average speed for the first half of the time, and then, necessarily slower for the second half. The integral becomes a sum of two parts:

$$\int_{t_1}^{t_2} \mathscr{L}\, dt = \tfrac{1}{2} m \left((\bar{v} + \Delta v)^2 \frac{\Delta t}{2} + (\bar{v} - \Delta v)^2 \frac{\Delta t}{2} \right) = \tfrac{1}{2} m \bar{v}^2 \Delta t + \tfrac{1}{2} m (\Delta v)^2 \Delta t.$$

The result is larger than the integral when the speed remains constant, and because of the squared term any variation of speeds greater or less than the average will produce an integral with a larger value. Therefore, the integral for the true path is a minimum. Note that energy is not conserved for paths other than the true one.

If an object is thrown straight up, the required motion involves changes in potential energy. Constant speed will no longer yield the smallest integral. To reduce the value of the integral, the object should rapidly move to a region where the potential energy is large, thus making $E_{\text{kin}} - E_{\text{pot}}$ small. The solution to the general problem yields Newton's second law, from which the parabolic nature $y(t)$ of this particular path can be calculated.

Rather than attacking the general solution, let us examine one particular alternative motion and compare it with Nature's solution. In Fig. 9.15 there are descriptions of two different paths for the flight of an object straight up and down. For the flight that actually takes place, the velocity linearly decreases from maximum (up) to maximum negative (down). The kinetic energy, potential energy, and Lagrangian are shown plotted as a function of time. The integral of least action corresponds to the area under the Lagrangian curve. The two positive regions cancel out only a small fraction of the negative area, leaving a large negative value. An alternative path might be one where the object climbs with constant speed to the same height as in the true path. The $v(t)$ graph is shown for this case. In the plot of E_{kin}, E_{pot}, and $\mathscr{L}$ for this path, it is clear that the (negative) area under the $\mathscr{L}$ curve is smaller than that for the true path. Therefore, the least-action integral is more negative (less) for the true path—in other words, a minimum.

Whether a variational principle exists for a particular phenomenon must be determined in the final analysis by whether or not the laws of motion (or equivalently for light, the law of refraction) can be deduced from the principle. Sometimes calculations are easier to do starting with the integral form; more often they are easier with the laws of motion in the more widely known differential form. There is, however, a major philosophical difference between the two forms. In the differential form (e.g., $F=ma$) if an object changes velocity at a point it is because there is a force acting on it at that point. In the integral form, it appears as if the object must somehow know which of all possible paths will produce the least (or maximum) action. In the case of light, we might imagine that the wavefronts probe various routes, calculate the times, and then choose the extremum. In anthropomorphic terms, it sounds as if Nature uses the end to choose the means—effecting some great economy in doing so!

It is not like that of course. The variational principles are clever for concise formulations, fruitful in suggesting analogies, and sometimes powerful in calculations. Always, however, their validity hinges on the deduced laws of motion. In the case of wave motion the choice of optimum paths (whether minimum or maximum) can be interpreted in terms of phase interferences. The similar relationships with classical mechanics take us into the complexities of the Hamilton–Jacobi formulation of mechanics, which in turn can form a base for generalization from classical to wave mechanics. All this, however, is far beyond the scope of introductory physics!

■ REFERENCES

Clifford E. Swartz, Editorial Phys. Teach. **10**, 516 (1972).
Feynman, R., *Lectures,* Addison-Wesley, Reading MA (1963), Vol. II, p. 19-1.
Goldstein, Herbert, *Classical Mechanics,* Addison-Wesley, Reading, MA, (1950), p. 308.

10
Internal Energy

A glider on an air track eventually loses its kinetic energy; after a few bounces a tennis ball rolls to a resting place; even a pendulum finally stops swinging. So much for the law of conservation of energy! We must either conclude that the law is very limited, or else we must discover other forms of energy and invent appropriate definitions so that the missing energy is accounted for. The realization that energy could be defined in this way did not occur until about 140 years ago. Through a methodical series of experiments done to high precision, James Joule in England showed that the mechanical energy used, and presumably lost, in stirring water was proportional to the temperature rise of the water. Apparently the lost mechanical energy had changed the internal properties of the material. The results were the same as they would have been had the water been heated on a stove.

In this chapter, we will explore the changes that take place in solids, liquids, and gases as energy is fed into them. We will see that many of these changes can be defined as a function of temperature, which, in turn, we will define in terms of one or more of the changes. These changes in materials occur in the same way whether the energy is fed in through friction effects or through exposing the material to a hotter object. The energy that flows between objects because one is hotter than the other is called *heat*. When heat enters a system, it can provide the energy to do work and can also raise the temperature or change other properties of the system.

In recent years there have been many calls to introduce "modern physics" into the introductory course. As we trace the "lost" energy into the microstructure, we must introduce a model of atoms and their binding to each other. A simple model dependent only on size, binding energy, and a plausible quantum condition can bring together and explain a vast range of phenomena. The thermometer becomes a prime tool for exploring the microstructure of matter. Before introducing the model, however, we must describe some gross properties of matter and make some comments about thermometers.

■ HUMAN THERMOMETERS

How good a thermometer is a human? The old, standard demonstration is still a good one, but only if the effect is experienced by each student. Arrange three pans of water, one hot, one cold, and one at room temperature. Soak the left hand in the hot water and the right hand in the cold water for 15 seconds or so. Then put both hands in the room-temperature water. The right hand will not believe what the left hand is "saying." On the other hand, we all know how exquisitely sensitive the body is to even a one-degree change from its normal temperature of 37 °C.

■ LIQUID THERMOMETERS

The most common thermometers exploit the fact that liquids usually expand as they get hotter. The liquid used in laboratory or clinical thermometers is mercury. A less expensive and potentially less dangerous liquid used in household thermometers is a red-colored alcohol, or toluene. A liquid thermometer can be used only in the range above the freezing point of the liquid and well below its boiling point. In the case of mercury, this practical range is from −38 to 260 °C.

The ordinary type of liquid thermometer is really a very ingenious instrument. If it consisted simply of liquid in a glass capillary tube, the temperature change would be very hard to detect. Liquids expand as they get hotter, but the percentage change of volume for a reasonable temperature change is very small. For mercury, it is only 0.018% per degree Celsius. From the melting point of ice at 0 °C to the boiling point of water at 100 °C, the volume of a column of mercury (and therefore its height in a tube) would change by only 1.8%. The actual construction of this type of thermometer makes use of a clever technique of amplification. Most of the volume of the thermometer liquid is in the bulb at the end. The column of liquid, which indicates the temperature, is in a capillary tube with a very small volume. A 2% change in the total volume of the liquid can change the height of the column by a factor of 10 or more. Note that as the temperature increases, the volume of the glass tube will also increase, and so does the volume of the cavity. Because the coefficient of expansion of glass is less than that of most liquids, the effect is small and the necessary corrections can be made during the calibration procedure.

■ THE THERMOCOUPLE

Temperature changes produce electrical changes in many materials. One kind of thermometer that makes use of such changes is called a thermocouple. Two fine wires of *different* metals are welded or soldered together at one end. The other ends are connected to a voltage-reading device. The voltage across the wires is a function of the temperature difference between the welded end and the voltmeter. In this case, the microstructure of the wires at the junctions acts like an electrical generator. The standard metals used for most thermocouples are copper and constantan. The latter is an alloy of copper and nickel. A table giving the voltage of this combination as a function of temperature is readily available in many handbooks. As an example of the size of the voltage generated in a copper–constantan thermocouple, if the voltmeter leads are at a room temperature of 20 °C and the junction of the wires is at 100 °C, the voltage produced is 3.48 mV. One of the virtues of thermocouples is their small size—no larger than the junction point of the fine wires. The wires can be embedded deep in the material and the internal temperature read remotely. Using several combinations of metals, thermocouples can be used to measure temperatures all the way from −269 °C to almost 2300 °C.

■ ELECTRICAL RESISTANCE THERMOMETERS

The resistance of metals generally increases with increased temperature. This is a very reproducible effect and is used in many precision thermometers. Platinum wire wound in a small coil is usually used because the characteristics of platinum remain so stable. The temperature measured can range from −258 to 900 °C.

■ THERMISTORS

Thermistor is the name for a class of semiconductors that have a temperature–resistance response opposite to that of metals. The electrical resistance decreases with increasing temperature. The thermistor materials, which are usually claylike metal oxides, can be made in the form of tiny beads. They can be much more sensitive than thermocouples or platinum-resistor thermometers, but are not so stable over long periods of time. Hospital thermometers with electrical readouts use thermistors.

■ BIMETALLIC STRIPS

The thermometer in most home thermostats consists of a coil of two pieces of different metals bonded together. The metals are chosen so that one expands much more than the other as the temperature increases. The only way that this can happen with the two pieces fastened together is for the bimetallic coil to change its radius of curvature. One of the metals usually used has the trade name Invar. It is an iron–nickel alloy whose volume changes very little over a large temperature range. You can see the coiled bimetallic strip in a home thermostat simply by taking off the cover cap. Usually the coil moves a needle to indicate the temperature and also trips a mercury switch to turn the heating system on or off. Thermometers of this type are usually accurate only to 0.5 °C and have a small, limited range.

■ GAS THERMOMETERS

When heat is fed into a gas, the pressure of the gas increases, or, if the pressure is kept constant, the volume increases. Either of these effects can be used to indicate temperature, but they are seldom used any more in actual thermometers. However, experiments done with constant-volume gas thermometers helped to define the meaning of temperature. All of the other thermometers depend critically on the nature and purity of the materials used. As we will see, under the right conditions the behavior of gas thermometers is almost independent of the type of gas used. Furthermore, under these conditions the theory of the temperature response of gas is simple and well understood.

■ OTHER THERMOMETERS

Many other physical properties have been used to measure temperature. Skaters and skiers can judge temperature crudely in terms of the slipperiness of ice or snow. Generations of fudge makers know that when the candy has reached the "soft ball" stage (when a little of the hot syrup dropped into a cup of cold water clings together) it is ready to take off the stove. A novelty thermometer item makes use of liquid crystals that can be mixed to change color at a particular temperature. For industrial purposes, with hot ovens and furnaces, there is a whole range of waxes, each of which melts at a particular temperature. The color of glowing metals is a function of temperature. There are special instruments—optical pyrometers—that allow the brightness of glowing metals to be judged against various standards.

■ CALIBRATING THERMOMETERS AND TEMPERATURE SCALES

We have not defined temperature in any formal way, but some of the difficulties in doing so will become apparent when we try to calibrate thermometers. Apparently no one measured temperature with instruments until Galileo designed an air thermometer in 1592. Many types of liquid-in-glass thermometers were made during the seventeenth century, each calibrated to the individual standard of its maker. In the early 1700s, Gabriel Fahrenheit standardized procedures for making mercury thermometers very much like our present ones. He also chose a scale that, with a few changes, we still use for household temperatures in the United States. The zero point of the scale was chosen to be the lowest temperature that Fahrenheit could conveniently produce in the laboratory. It is the temperature of salt and ordinary ice mixed together, the same mixture and process that is used in home ice-cream makers. On this scale, 32° was assigned to the melting point of ordinary ice and 96° to normal body temperature. Our present Fahrenheit scale maintains the 32° as the melting point of ice and 212° as the boiling point of water at normal atmospheric pressure at sea level but, with those points fixed, normal body temperature turns out to be 98.6°. During those same years, Anders Celsius, a Swedish astronomer, proposed that the range of temperature between the melting point of ice and the boiling point of water be divided into 100 degrees. (He originally specified 100° for the melting point and 0° for boiling.) The scale derived from his proposal was originally called "centigrade," but since 1960 has been known officially as the Celsius scale. Celsius degrees are standard in the Système International (SI) system of units. The old-fashioned Fahrenheit degree does have the advantage of being about the smallest change of temperature to which the human body is sensitive. (The actual history of how Fahrenheit, Celsius, and others fumbled their way to our present thermometric scales is complicated and murky.)

Calibrating a thermometer should be easy in principle. If you have an unmarked thermometer, place it in a dish of melting ice and mark the indicator position on the thermometer 0 °C. Then place the thermometer in a dish of boiling water and label the resulting point 100 °C. Divide the region in between on the thermometer scale into 100 equal parts. Unfortunately, there are all sorts of problems in doing such a simple thing, and even more problems when you are through in deciding whether or not you have a true temperature scale. Consider the following small, practical problems.

1. The thermometer must be small compared with the system whose temperature is being measured. If you have a small amount of water and a great big air

thermometer, you might disturb the temperature of the thing that you are trying to measure, or, during calibration, not have the sensitive part of the thermometer at a uniform temperature.

2. There is a subtle problem about when the thermometer should be read. When a thermometer is first moved to a new environment, it takes some time for the reading to settle down. (Every nurse or parent knows this.) Indeed, when we read the temperature *on* a thermometer, we are reading the temperature *of* the thermometer. We assume that the temperature and the environment have been together long enough so that the two have come to an equilibrium, which we define as being the same temperature.

3. There are problems in trying to measure the temperature of melting ice. An ice cube in water may have a temperature of 0 °C at its surface but −10 °C inside. Furthermore, the water only a short distance away from the ice cube may be at a temperature of +5 °C. To satisfy the definition of a calibration point, the ice–water mixture must consist of finely crushed ice that is kept in equilibrium with the water by continually stirring. Notice, incidentally, that we have repeatedly referred to the melting point of ice instead of the freezing point of water. Ice melts at 0 °C by definition, but water does not necessarily freeze at 0 °C. Pure water under clean conditions may go well below 0 °C before suddenly crystallizing. This happens frequently in the atmosphere. Ordinary tap water in an ordinary glass is usually sufficiently contaminated so that it will freeze at 0 °C or very slightly below.

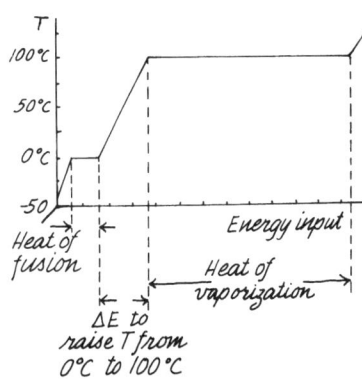

Figure 10.1. Graph showing theoretical data for temperature of water as function of constant heat input. An actual measurement to produce the sudden changes in slope would be hard to do.

In most elementary texts, there is a graph showing the temperature of water as function of time as it is heated from ice to water to steam. There is a typical graph of this kind in Fig. 10.1. Actually, it is extremely difficult to duplicate such a graph with real experimental data because of the problems that we have mentioned above. In most school laboratory situations, the straight lines would slither into each other along curved paths.

4. It is relatively easy to keep the thermometer within a degree of the boiling point of water, but there is no guarantee that the temperature is 100 °C. The boiling point and, to a much smaller extent, the freezing point depend on the atmospheric pressure. The boiling point is defined to be 100 °C at normal atmospheric pressure at sea level. At normal atmospheric pressure in Denver, about a mile above sea level, the boiling point is about 95 °C.

5. There is one other bothersome problem that we must confront when we calibrate a temperature scale. Suppose that you calibrate a mercury thermometer and an alcohol thermometer in the manner that we have described. Clearly, they will both agree at 0 and 100 °C, since they were in identical conditions when they were marked. In each case, then, the range between the two marks was divided into 100 equal parts. If the two thermometers are now placed in a glass of water and the mercury thermometer reads 50 °C, is there any guarantee that the alcohol thermometer will also read 50 °C? There is not. We have no reason to expect that the expansion rate of alcohol and mercury is the same with increasing temperature. In fact, the expansion rate is *not* quite the same. We cannot really expect that, as temperature changes, the changes of properties of various materials will all be proportional to each other. For one thing, as temperature increases, many different properties of each material are changing. Perhaps some of them depend on each other.

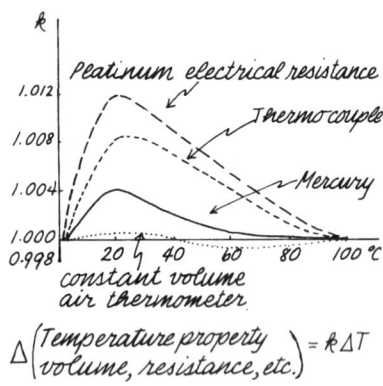

Figure 10.2. Responses of four different thermometric materials to temperatures between 0 °C and 100 °C. All are calibrated at 0 °C and 100 °C.

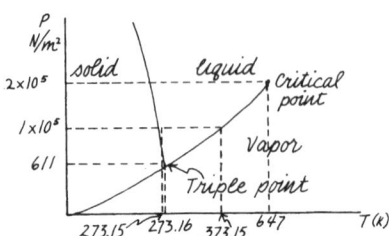

Figure 10.3. Pressure-temperature phase diagram for water (The slope of the line separating solid from liquid is greatly exaggerated.)

Electrical resistance, for instance, may depend in some way on the density of the metal, which also changes with temperature. Furthermore, we would not expect that the change of volume of a liquid would be the same as the change of volume of a gas of the same material. In fact, it is surprising, though very convenient, to find that certain properties of some materials are sufficiently linear with respect to temperature so that we can use them in thermometers. In Fig. 10.2 we have plotted the responses of several thermometric materials to various temperatures. It is assumed that all of the thermometers have been calibrated to agree with each other at 0 and 100 °C. They do not, however, agree at any intermediate point, although the differences are not great. Note that 1.01 on the vertical axis corresponds to a 1% difference. There is an underlying assumption in displaying such a graph; if we are plotting response versus temperature, we must have some way of measuring temperature! There are, indeed, several mutually consistent ways of defining temperature, independent of the properties of any particular material. We will soon analyze these. In the meantime, note that for most everyday purposes the thermometers shown in the diagram are almost linear.

■ THE KELVIN (ABSOLUTE) TEMPERATURE SCALE

For precise work, modern temperature scales are not defined in terms of the temperature interval between the melting point of ice and the boiling point of water. The Kelvin scale has a zero point at about -273 °C. That zero point, which has both theoretical and practical implications, is used as one of the fixed points of the Kelvin scale. The other point is the temperature at which ice, liquid water, and water vapor are present together. The temperature at this equilibrium of water, ice, and vapor is arbitrarily defined to be 273.16 kelvin, or 273.16 K. (Note that no degree symbol is used for kelvins.)

This particular state of water was chosen as the calibration point because it is relatively easy to reproduce it in the laboratory. In Fig. 10.3 we show the thermal behavior of water in terms of a pressure–temperature graph. Note the nonlinear scales of the axes, and the domain of the phases—vapor, liquid, and solid. At one atmosphere pressure (1×10^5 N/m^2), any of the three phases can exist within appropriate temperature ranges. The transition from solid to liquid, for instance, is shown at 273.15 K, and the boiling point is at 373.15 K. Below a pressure of 4.58 mm of mercury (611 N/m^2), however, the liquid phase cannot exist. Below that pressure ice would sublime, turning directly from solid to gas. At that pressure, and at a temperature of 273.16 K, all three phases can exist in equilibrium, and this is the triple point used for calibration. In laboratory practice, a vial of very pure water and water vapor (no air) is sealed. A standard arrangement is to enclose the water in the interior of a double-walled well. If a suitably cold object is put into the well, some of the water inside the vial will start to freeze. The cold object is then replaced by the thermometer to be calibrated. As long as the water, ice, and vapor are present together, the vapor pressure inside the vial is 611 N/m^2 and the temperature is, by definition, 273.16 K.

There is a *critical* point at a pressure of 218 atm (221×10^5 N/m^2) and a

TABLE 1. Boiling and melting temperatures of various substances at atmospheric pressure.

Substance	Melting point °C	Boiling point °C
Alcohol, ethyl	−114	78
Carbon, graphite		3827 (sublimes)
Gold	1063	2700
Helium		−269
Hydrogen	−259	−253
Lead	327	1737
Mercury	−39	357
Nitrogen	−210	−196
Oxygen	−219	−183
Sulfur	119	445
Sulfuric acid	9	326
Tungsten	3380	5555

temperature of 647 K, above which there is no difference between the liquid and the vapor. Therefore, there would be no heat of vaporization.

■ PHASE CHANGES IN MATERIALS AS A FUNCTION OF TEMPERATURE

We usually think of iron as a solid, alcohol as a liquid, and oxygen as a gas. Such a classification is based on our human prejudices about normal temperatures. Most materials can be solids or liquids or gases at particular temperatures and pressures. The most common material that can exist in all three phases within the range of human temperatures and pressures is water. Other examples are benzene and p-dichlorobenzene (mothballs). The boiling and melting points of a number of other materials at *atmospheric pressure* are shown in Table 1. Note that hydrogen turns into a liquid at 20.4 K and freezes solid at 14 K. Liquid hydrogen looks just like water and is routinely used as a target for high-energy particle beams in research on subatomic particles. Helium can also be liquified at 4.2 K, but cannot be solidified at atmospheric pressure, even at temperatures approaching absolute 0. Instead, helium starts undergoing a transition at 2.18 K, which turns it into a different kind of liquid with fantastic properties. It is a superfluid with zero viscosity. At the opposite extreme, at very high temperatures, note that carbon

TABLE 2. Pressures and boiling points of water at various heights above sea level.

Location	Pressure (atm)	Pressure (N/m^2)	Boiling point (°C)
See level	1	1.0×10^5	100
1 km	0.89	9.0×10^4	96.4
1.5 km (Denver)	0.83	8.5×10^4	94.6
2 km	0.78	8.0×10^4	92.8
4 km	0.61	6.2×10^4	86.4
8 km (Mt. Everest)	0.35	3.6×10^4	72.8
15 km (Concorde jet)	0.12	1.2×10^4	49.6

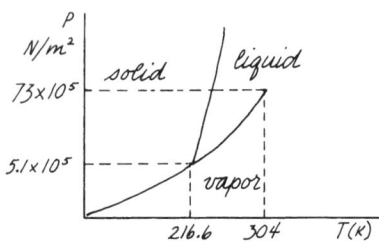

Figure 10.4. Pressure-temperature phase diagram for carbon dioxide.

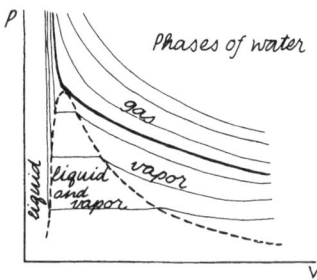

Figure 10.5. Pressure-volume phase diagram for water.

never melts at atmospheric pressure, but instead sublimes. Tungsten, from which light bulb filaments are made, melts at 3653 K and would boil at 5828 K.

At pressures below one atmosphere, the boiling point of water is less than 373.15 K, or 100 °C. In Table 2 we list some of the values of the boiling temperature of water for various heights above sea level.

Notice in Fig. 10.3 that the line between solid and liquid slopes upward steeply but is inclined slightly to the left. As the pressure increases, the melting point decreases slightly. This behavior is characteristic of water and bismuth, which both expand upon freezing. If you increase the pressure on ice, you decrease the volume, thus tending to create the condition of a liquid, which can then exist at a slightly lower temperature. In Fig. 10.3, the *slant* of the line between solid and liquid has been greatly exaggerated. Actually, on this scale, the line should appear vertical. The slope is 1.2×10^7 N/m^2 per K, or 120 atm per K. As we show on p. 93 in the chapter on friction, this thermal behavior of water does not explain the low friction of skate blades.

The pressure-versus-temperature graph for carbon dioxide, as shown in Fig. 10.4, is different from that of water in several crucial ways. The slope of the line between solid and liquid is positive; this is normal for most materials. The solid form is more dense than the liquid form. However, a major factor in the behavior of CO_2 is evident in the values for the triple point. That point for CO_2 is at -56.6 °C and 5.1 atm. The critical point, at which there is no longer a difference between gas and liquid, is at 304 K and 73 atm. At ordinary temperatures and pressures, the liquid phase cannot exist. The solid form of CO_2 is dry ice. If you have handled dry ice, you know that it has a very low temperature and that it does not melt into a liquid. Instead, it sublimes into the gas form.

The *pressure-versus-volume* graph for water is shown in Fig. 10.5. At temperatures well above the critical point of 647 K, the $P-V$ lines are hyperbolas, corresponding to the equation for an ideal gas: $P \propto 1/V$. Above the critical temperature, there is no difference between liquid and vapor. If gas below that temperature is compressed, its pressure will rise until it reaches a crucial pressure and volume. From that point on as the volume is decreased, the pressure stays constant. The gas is condensing into liquid form. Eventually, when all of the gas has turned into liquid, the $P-V$ curve rises abruptly, since the liquid is nearly incompressible and a small decrease in volume would require an enormous increase in pressure.

■ THE DEFINITION OF TEMPERATURE

We have been talking about many different temperature effects and have even described how to make thermometers to measure temperature. So far, however, we have avoided an actual definition of temperature. The use of a thermometer depends on the assumption that when a thermometer is placed in contact with an object and the thermometer reading settles down, the thermometer and the object are at the same temperature. This is an unusual property. If two objects with different masses or volumes or densities are connected, they will not end up with the same properties at some equilibrium point. But if two objects are arranged so that they exchange heat with each other but are isolated from everything else, then they will finally come to the same temperature. If you take a thermometer out of your mouth and it reads 100 °F, then you assume that the temperature in your mouth was 100 °F.

Another peculiar property of temperature is that if thermometer A and thermometer B are placed in contact and come to the same temperature, and then thermometer A is placed in contact with thermometer C and does not change its reading, not only are thermometers A and C at the same temperature, but so are thermometers B and C. That fact seems almost too obvious to mention. Remember, however, that thermometer A may be a thermocouple, thermometer B may be mercury in a tube, and thermometer C may be a curved bimetallic strip. When thermometers A and B are brought together, the electrical voltage produced by thermometer A and the length of the mercury column of thermometer B may change until they come to equilibrium. Then if the mercury column is put next to the bimetallic strip and neither one changes, the thermocouple can be put next to the bimetallic strip and not change either. This property of temperature is called the *zeroth law of thermodynamics*. If you think that the situation is obvious, consider what happens when you place a piece of iron, A, near a magnet, B. If B is attracted to another piece of iron, C, will A also be attracted to C?

All we have done with the zeroth law is to define what it means to be at the *same* temperature. The next question is: How do we know when we make up the temperature scale that there is the same difference of temperature between 10 and 11 °C as there is between 90 and 91 °C? We have already seen that the size of a degree on a mercury thermometer is not necessarily the same as the size of a degree on any other kind of thermometer. The expansion of a substance, or the electrical resistance, or the potential difference is not a linear function of temperature, or at least not a linear function of the corresponding changes in some other thermometer. For one class of substances, however, there is great uniformity. The graphs of pressure versus temperature or volume versus temperature for all simple gases are almost identical, as long as the gas pressures are low. For practical purposes we could make a constant-volume thermometer using low-pressure hydrogen or helium gas and *assume* that the pressure is proportional to the temperature: $P \propto T$. This assumption would define the intercept point in the gas-pressure straight-line graph to be zero temperature. If the triple point of water is then defined to be 273.16 K, the temperature scale is completely defined.

There are two theoretical reasons for thinking that this type of thermometer is a true indicator of temperature. First, as we will soon see, the simple theory of the microstructure of gases gives a satisfactory explanation of why the pressure of the gas is proportional to the temperature. As a matter of fact, that theory will also provide another interpretation of the meaning of temperature.

There is a second theoretical interpretation of temperature. This one comes from the second law of thermodynamics and is independent of any particular thermometer and thus of the properties of any particular material. This thermodynamic temperature scale agrees with the temperatures measured with a low-pressure volume gas thermometer.

There is one additional requirement we must consider in determining the significance of a temperature reading. *We can assign a temperature to a system only if the system is in equilibrium both internally and with its surroundings.* If you take an ice cube from the freezer and put it into a glass of water, the temperature of the system is not 0 °C. As a matter of actual experimental fact, as well as of theory, the idea of temperature for such a system is meaningless. However, if you put the ice cube with the water into an insulated glass along with a thermometer and stir the mixture until the thermometer reading remains constant for some time, then you can assign a temperature to the system. Similarly, if gas in a cylinder suddenly expands, neither the pressure nor temperature of the gas can be defined until the shock waves settle down and equilibrium has been established within the cylinder. Whether a system is in equilibrium is not always easy to decide. The question involves the length of time for thermal mixing within the system compared with the time required for the disruption. For instance, it *is* reasonable to assign temperature to the water in a glass that is slowly rising from 0 °C to room temperature. During the short time required for reading the thermometer and stirring the water, the temperature remains approximately constant and uniform throughout the glass.

■ THE FIRST LAW OF THERMODYNAMICS

We started this chapter by asking where the energy goes that seems to disappear as a result of friction. Our answer has been that the "lost" energy changes the properties of materials, just as the heat from flames can change materials. In many cases (but not all), the changed properties can be characterized in terms of the temperature of the material.

The phenomena that we have been describing and the declaration that heat is energy in transit can be summed up in the first law of thermodynamics:

$$Q = W + \Delta U,$$

$$\begin{pmatrix} \text{Heat entering} \\ \text{a system} \end{pmatrix} = \begin{pmatrix} \text{External work done} \\ \text{during the process} \end{pmatrix} + \begin{pmatrix} \text{Change of internal energy} \\ \text{of the system} \end{pmatrix}.$$

This law is really just a statement of the conservation of energy along with the claim that energy can be locked up in internal forms of materials. Heat energy Q is energy in transit from a hotter to a colder system. It is not a quantity that can be stored in the system into which it passes. Some of the heat energy may be converted immediately into external work. The rest turns into the internal forms measured by ΔU. Similarly, W is not energy that can be stored. It is a transfer of energy from one system to another, not involving a difference of temperature. Under some circumstances we might add a third transfer mechanism—radiation.

The symbol Q stands for the amount of heat entering a system. The measuring unit for heat could be the joule, although frequently in the United States it is still engineering practice to use the old-fashioned *calorie*. A calorie is approximately the amount of heat it takes to raise the temperature of 1 gram of water 1 °C. The present definition is 1 cal=4.1840 J. The kilocalorie, which is equal to 4184 J, is the amount of energy that it takes to raise the temperature of 1 kilogram of water 1 °C. The diet calorie that many people try to avoid is actually a kilocalorie.

In many engineering applications in the United States, heat is measured in yet another unit—the British thermal unit, or BTU. A BTU is the amount of heat required to raise the temperature of 1 pound of water 1 °F. 1 BTU=1055 J=252 cal=$\frac{1}{4}$ kcal.

The second term in the first law (W) accounts for the fact that heat entering a system may produce external work whether we want it to or not. For instance, when a gas is heated it may expand, pushing a piston in front of it and thus exerting a force through a distance. If the volume of any material expands by an amount ΔV, then the work done is $P\Delta V$, where P is the average pressure that the material exerts as it expands ($F\Delta x \rightarrow P\Delta V$). For an object being heated in the open air, P would be just the atmospheric pressure. When P is a rapidly varying function of V, $W = \int P \, dV$.

The third term in the first law of thermodynamics refers to the change of internal energy, ΔU, of the system. Note how cleverly we account for the disappearance of heat energy. We say that, except for the amount of energy used immediately to do work, the rest is stored in the form of changed characteristics of the material, including any change of temperature. The internal energy of a system may be a function of several variables, including its density, magnetic and electric strains, and also its temperature. *For an ideal gas, changes in the internal energy are a function of temperature changes only.*

Regardless of how heat is supplied or how external work is done, if a system starts out from some initial state (a particular pressure, temperature, etc.) and then has heat added and taken away, and work done on it and by it, the system will always end up in its original state as long as $(Q-W)_{net}$ is equal to 0. ΔU does not depend on the details or sequence of the changes, but only on the net value of $Q-W$. That makes U a potential. As is the case with any of the other potentials, we are free to choose the zero point in any convenient way. It is only the change of energy, ΔU, that has physical significance.

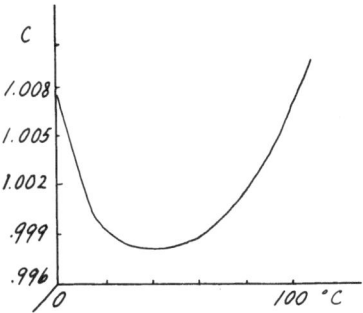

Figure 10.6. Specific heat of water.

When we defined potential energy that was the result of warped springs or gravity, we did it in such a way that kinetic energy could turn into potential energy and back again. Is there the same situation with respect to internal energy? If heat enters a system and no external work is done, is the increase in internal energy, ΔU, stored in such a way that it can be converted into useful work? Whether this is possible depends on how you want to use the energy. If you pour heat into a hot water bottle, then practically all of that heat energy is stored in the internal energy of the water. (Note that the *heat* is not in the water: Q turned into ΔU.) That internal energy can be completely recovered *as released heat energy* to keep your feet warm. However, if heat energy goes into the internal energy of a steam boiler, only a fraction of the stored energy can be used later to do external work in a cyclic machine. That problem is the subject of the second law of thermodynamics.

■ HEAT CAPACITIES

If we assume that no external work is done when a system is heated, then $Q = \Delta U$. Over small temperature ranges, providing that the phase of the system does not change, the change in internal energy is proportional to the change in temperature: $\Delta U \propto \Delta T$, and therefore $Q \propto \Delta T$. The proportionality constant must have something to do with the type of material and the amount of material in the system. One way to specify the amount of material is in terms of its mass. Then the proportionality constant becomes $Q = cm\Delta T$. This proportionality constant c is called the *specific heat* of the material. The values remain fairly constant over small temperature ranges, although we will see examples where the "constant" is not constant. For example, Fig. 10.6 shows the value of the specific heat of water as a function of temperature. From 0 to 100 °C, the value changes by less than 1%.

There is a nomenclature problem with the words *specific heat*. If the proportionality constant between Q and ΔT refers only to a particular entity, say a sample of diatomic gas, then we might refer to the *heat capacity* of that particular sample. If the proportionality constant concerns a *mole* of a substance (rather than a gram) then we specify a *molar heat capacity* or a *molar specific heat*. Since the

TABLE 3. *Some specific heats in the range from 0 to 100 °C.*

Material	cal/g °C=kcal/kg °C
Aluminum	0.215
Carbon	0.121
Copper	0.0923
Lead	0.0305
Silver	0.0564
Tungsten	0.0321
Water	1.0

specific heat of water is 1 cal/g °C, the specific heat in those units of any other substance is also the ratio of the heat capacity of that substance to the heat capacity of an equal mass of water.

Table 3 shows values of c for various common materials in the range from 0 to 100 °C. The units shown are derived from the units commonly used to measure the other quantities in the equation: $c=$cal/g °C.

There is no temperature change produced by heat during a change of phase. The amount of heat needed to melt 1 gram of material is called the *heat of fusion*. For water the heat of fusion is 80 cal/g. The *heat of vaporization* is the number of calories needed to turn 1 g of a liquid into a gas (see Table 4.) For water at atmospheric pressure, the heat of vaporization is 540 cal/g. Note how much larger the heat of vaporization is than the heat of fusion. It seems as if the internal energy of a gram of steam must be very large compared to the internal energy of a gram of water. That is why "live" steam can produce such serious burns. However, we will take another look at the nature of this internal energy on p. 256.

Notice that it always takes more energy to turn material from liquid to gas than it does from solid to liquid. A small part of the heat of vaporization goes into the work of expanding the material from liquid to gas form. For most materials, the volume of the gas is about 1000 times greater than the volume of the liquid from which it came. The amount of work done in the expansion at constant pressure is equal to $P\Delta V$. In the case of water, the volume occupied by 1 gram of steam at 100 °C and atmospheric pressure is equal to 1.7×10^{-3} m^3. (The volume of a mole of gas at standard temperature and pressure is 22.4 liters. One gram of water steam is $\frac{1}{18}$ mol.) The volume of the 1 gram of water is equal to only 1×10^{-6} m^3. Atmospheric pressure is equal to 1.01×10^5 N/m^2. Therefore, the work required for the expansion is equal to 171 J=41 cal. The other 499 cal/g of the heat of vaporization must be needed to free the water from the liquid state and convert it into a gas.

■ THE REGULARITIES AND IRREGULARITIES OF HEAT-CAPACITY VALUES

The specific heats of materials given in Table 3 appear to vary over a large range of values without any apparent uniformity. Water seems to have a very large value of specific heat compared with any other of the common materials, and most metals appear to have a very low specific heat. In the middle of the nineteenth century, DuLong and Petit noticed that a uniformity does exist for the specific heats of most solid elements. Note from the graph in Fig. 10.7 that the metals with

TABLE 4. *Heats of fusion and vaporization at atmospheric pressure.*

	Fusion			Vaporization		
Substance	cal/g	J/g	J/mol	cal/g	J/g	J/mol
Helium				5.0	21	5.0
Hydrogen	14	59	30	108	452	226
Alcohol, ethyl	25	104	2.3	204	854	18.6
Nitrogen	6.2	26	0.9	48	201	7.2
Oxygen	3.4	14	0.4	51	213	6.7
Mercury	2.9	12	0.06	65	272	1.4
Water	80	333	18.5	539	2256	125
Lead	6.0	25	0.12	208	871	4.2
Gold	15.5	65	0.33	378	1580	8.0
Sulfur	9.1	38	1.19	78	326	10.2

small atomic numbers have large specific heats, and the metals with large atomic numbers have small specific heats. If the specific heat, in cal/g °C, is multiplied by its atomic weight, the number turns out to be approximately 6. What we are really doing in multiplying by the atomic weight is finding the *molar* heat capacity—the number of calories required to raise the temperature of one mole of the material by 1 °C. Table 5 gives the molar heat capacities for many materials.

Whenever there is a regularity in nature, we suspect that there is some underlying cause. When we first suggested that $Q \propto T$, we said that the proportionality constant must be a function of the kind of material and the *amount* of material in the sample. Then we said that the amount of material might be defined by its mass.

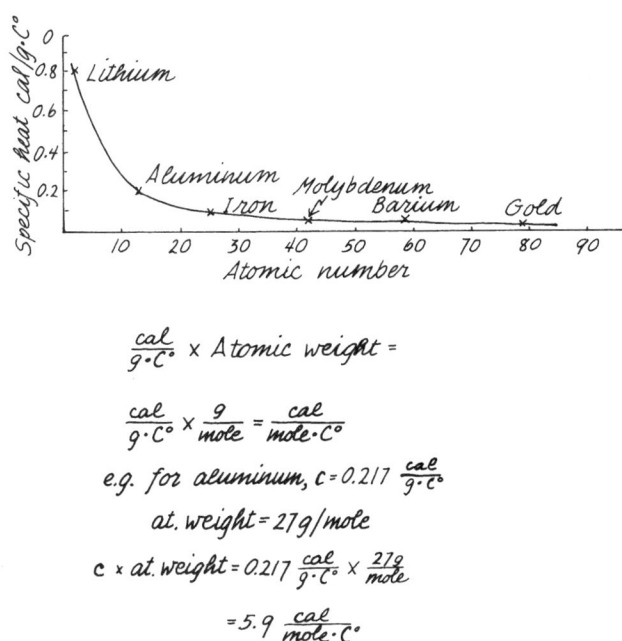

Figure 10.7. Law of DuLong & Petit.

TABLE 5. *Molar specific heats of solids and liquids at room temperature.*

Substance	Molar specific heat in cal/mol °C	in J/mol °C
Aluminum	5.82	24.4
Carbon (graphite)	1.46	6.1
Copper	5.85	24.5
Lead	6.32	26.4
Silver	6.09	25.5
Tungsten	5.92	24.8
Mercury	6.69	28.0
Uranium	6.7	28.0
Water	18.0	75.3
Bromine	18.1	75.7

Another way to describe the amount of material is to give the number of atoms or molecules in it. When we take a mole of material, we are using 6×10^{23} atoms (or molecules). The *molar specific heat* is thus the number of calories (or joules) that it takes to raise 1 mole of material 1 °C. For materials with the same basic structure—metals, for instance—there seems to be a simple relationship between the absorption of energy per mole (or per atom) and the change of temperature.

As a matter of fact, the regularity that we have just observed in specific heats of metals applies only at room temperature, and for the light metals not too well even there. The specific heat of a material is not constant but depends on temperature. Fig. 10.8 shows graphs of the specific heats of various elements as a function of temperature. Evidently what is going on in the microstructure of materials is simple in the first approximation, but more complicated when we take a closer look.

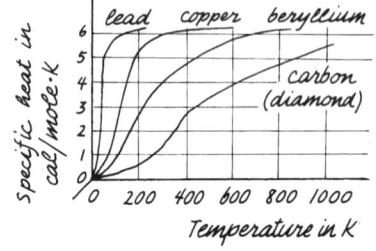

Figure 10.8. Specific heats of four substances as function of temperature.

The specific heats of *gases* also have both regularities and differences. Table 6 gives specific heats for several different types of gases; here they are grouped according to whether they are monatomic, diatomic, or polyatomic. Notice the uniformities within each group. Also notice that the molar specific heat for monatomic gases is almost exactly half that for the metals at room temperature.

These specific heats are given for room temperature. They change with temperature. One of the most revealing examples of this is shown in Fig. 10.9. The specific heat for hydrogen rises with temperature, leveling off at two plateaus. Our model of the microstructure must be able to explain the temperature dependence of specific heat for this particularly simple molecule.

■ **SPECIAL CONDITIONS FOR FEEDING ENERGY INTO GASES**

The specific heats that we have given so far are for cases where the volume of the material stays constant; thus no external work is done: $W = P\Delta V = 0$. The specific heat at constant volume should be labeled c_V, as opposed to the specific heat at constant pressure, c_P. It is very hard to obtain *experimental* values of c_V for solids or liquids. When heat is fed into a solid or liquid sample, it is hard to keep it from expanding. Usually, c_P is measured. Unfortunately, for *theoretical* considerations, the situation is just reversed. In terms of the microstructure, it is

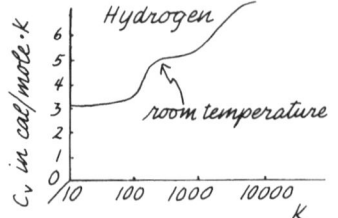

Figure 10.9. Specific heat of hydrogen.

TABLE 6. *Specific heats of gases at room temperature.*

Type of gas	Gas	c_P (cal/mol °K)	c_V (cal/mol °K)	$c_P - c_V$	$\gamma = \dfrac{c_P}{c_V}$
Monatomic	He	4.97	2.98	1.99	1.67
	A	4.97	2.98	1.99	1.67
Diatomic	H_2	6.87	4.88	1.99	1.41
	O_2	7.03	5.03	2.00	1.40
	N_2	6.95	4.96	1.99	1.40
	Cl_2	8.11	6.15	1.96	1.32
	HCl	6.45	5.00	1.45	1.29
Polyatomic	CO_2	8.83	6.80	2.03	1.30
	SO_2	9.65	7.50	2.15	1.29
	NH_3	8.80	6.65	2.15	1.31
	C_2H_6	12.35	10.30	2.05	1.20

more meaningful to consider c_V. For our purposes, the extra work done in the expansion of solids and liquids makes only a negligible difference between c_V and c_P. The work done in thermal expansion is $P\Delta V$ for a solid or liquid at constant pressure. But $P\Delta V = P(V\beta\Delta T)$, where the volume thermal expansion coefficient is given by $\Delta V = V\beta\Delta T$. Typical values for a solid are $V_{\text{mole}} = 10 \text{ cm}^3 = 10^{-5} \text{ m}^3$ and $\beta = 40 \times 10^{-6}$. For $\Delta T = 1 \text{ °C}$, $P\Delta V = (1 \times 10^5 \text{ N/m}^2)(10^{-5} \text{ m}^3)(40 \times 10^{-6})(1 \text{ °C}) = 4 \times 10^{-5}$ J. Compare this small extra heat capacity of 4×10^{-5} J/mol K with the molar heat capacity of copper which is 24 J/mol K.

In the case of gases, however, there is a large difference between c_P and c_V. Table 6 gives both sets of values and also shows the differences between the two as well as their ratios. Notice that the *ratio* c_P/c_V, which is given the special symbol γ, has special values for each class of molecule. The *difference* $(c_P - c_V)$ has approximately the same value regardless of the type of molecule. The difference between c_P and c_V corresponds to the extra energy that is required to expand 1 mole of the gas at constant pressure as its temperature rises 1 °C.

Of course, energy can be fed into a gas in such a way that the gas expands against *varying* pressure. For purposes of analysis, however, it is usually convenient to use only the limiting cases where either the pressure or the volume is kept constant. To find the difference between c_P and c_V we must appeal to the gas law. The "ideal" gas law is $PV = nRT = NkT$, where n is the number of moles of gas and N is the number of molecules. The gas constant is $R = 8.31$ J/mol K $= 1.99$ cal/mol K. Boltzmann's constant is $k = R/6 \times 10^{23} = 1.38 \times 10^{-23}$ J/molecule K. The remarkable feature of this law is that it is independent of the kind of gas or its mass. The only thing that counts is the quantity, measured in terms of moles or number of molecules.

We consider just two cases: constant volume (no external work done because ΔV is zero) and constant pressure (external work, $P\Delta V$, done, but easy to calculate because P is constant). We expect that $c_P > c_V$. The specific heat is

$$c = \frac{Q}{\Delta T} = \frac{\Delta U + P\Delta V}{\Delta T}.$$

For constant volume, $c_V = \Delta U / \Delta T$. For constant pressure, $c_P = c_V + P(\Delta V / \Delta T) = c_V + R$. [To find $P(\Delta V / \Delta T)$, at constant pressure, we appealed to the ideal-gas law for one mole: $PV = RT$.] Therefore, for any ideal gas (whether monatomic or

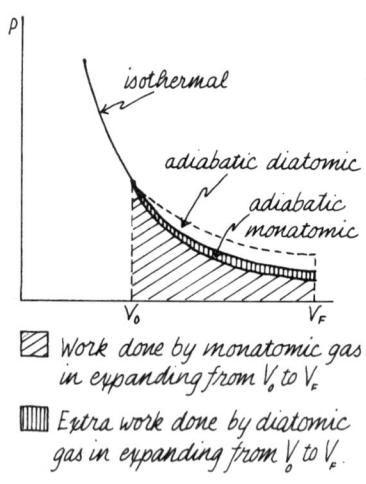

Figure 10.10. Pressure-volume graphs for expansion of two kinds of gases (monatomic and diatomic) under two different conditions (isothermal and adiabatic).

polyatomic), $c_P - c_V = R$. This result agrees with Table 6, which gives a value of about 2 cal/mol K for $c_P - c_V$.

Let us consider the $P-V$ relationship for gases under two limiting conditions: first, the temperature must be kept constant (isothermal) and, second, no heat can flow in or out of the gas (adiabatic). The experimental conditions for doing each measurement are illustrated in Fig. 10.10. In the first case, the gas is trapped in a cylinder with metal walls that can easily conduct heat in or out of the chamber. As the gas expands or contracts slowly to maintain temperature equilibrium with the walls, the piston is shoved out or in. For the second case, the cylinder and piston must be made of good insulating material, and the expansion must be fast so that no heat can flow in or out of the gas as it expands. In both cases, we will allow the gas to expand and we will plot the pressure as a function of volume. When the temperature remains constant, the relationship between P and V is simple; it is given by the ideal-gas formula, $PV = nRT$. With T constant, the equation reduces simply to $PV = \text{const}$, or $P = \text{const}/V$. This is the equation for the hyperbola plotted in the upper left part of Fig. 10.10.

Under most conditions of rapid change, the temperature of an expanding gas will decrease and the temperature of a gas being compressed will increase. This is because the expanding gas is doing work, and because work is being done on the gas to compress it. These phenomena are easy to demonstrate with a bike or car tire and a pump. As you pump up the tire, the hose gets hot; if you release the air in the tire, the air coming out and the nozzle will feel cold.

Under the particular and special conditions of adiabatic expansion, with insulating walls, what is the relationship between pressure and volume? As the volume increases, the pressure will drop, but so will the temperature. We cannot, therefore, set $PV = K$. Instead we must resort to the following derivation:

1. $Q = dU + W$ First law of thermodynamics
2. $Q = 0$ Adiabatic process—no heat in or out
3. $W = P\,dV$ Pressure is changing but has value P over small range dV
4. $dU = nc_V dT$ For ideal gas, U is function of T alone
5. Therefore $nc_V dT = -P\,dV$ Linking 1–4
6. $PV = nRT$ Ideal-gas law
7. $P\,dV + V\,dP = nR\,dT$ Differentiating 6
8. $P\,dV + V\,dP = nR[-(P/nc_V)dV]$ From 7 and 4
9. $P\,dV(1 + R/c_V) + V\,dP = 0$ Rearranging 8
10. $R = c_P - c_V$ For all ideal gases
11. $P\,dV[1 + (c_P - c_V)/c_V] + V\,dP = 0$ Combining 9 and 10
12. $P\,dV(c_P/c_V) + V\,dP = 0$ Rearranging 11
13. $(dV/V)\gamma + dP/P = 0$ Rearranging 12, $\gamma = c_P/c_V$
14. $\gamma \ln V + \ln P = 0$ Integrating 13 $[\int dx/x = \ln x]$
15. $PV^\gamma = K$ Taking antilogarithms

Surprisingly enough, the value of γ for a gas enters into several formulas concerned with gas behavior. One of these is the formula for the speed of sound in a gas. The speed depends on the relationship between the pressure and volume of a gas under conditions where the pressure changes so rapidly that practically no heat flows into or out of the gas. In the case of the $P-V$ diagram, notice the curves plotted in the lower part of Fig. 10.10 for both a monatomic and a diatomic gas.

For a given increase of volume, the pressure drops more in the adiabatic case than it does when the temperature remains constant.

Any change in pressure, volume, or any other variable that takes place at constant temperature is called *isothermal*. For an ideal gas in which the internal energy is a function of temperature only, an isothermal change implies no change in the internal energy of the system. If $\Delta T = 0$, then $\Delta U = 0$. In this case, the first law of thermodynamics becomes $Q = W$. At the other extreme where the system is insulated so that no heat can enter or leave, any changes that take place are called *adiabatic*. For an adiabatic change, $Q = 0$; therefore, the first law of thermodynamics becomes $0 = \Delta U + W$. In our example of the gas in a cylinder during isothermal expansion, the incoming heat must have been turned completely into the external work done by the piston. The area under that curve represents the work done during the expansion. In the adiabatic expansion, the work done by the piston must have come from the internal energy of the gas. The change in internal energy must have been negative, corresponding to a drop in temperature.

The graphs in Fig. 10.10 show an isothermal expansion of a gas followed by an adiabatic expansion. The curves for a monatomic gas and a diatomic gas are superimposed, along with a dotted line showing a continuation of the original isothermal expansion. In all three curves, $P \propto 1/V^x$. For the isothermal expansion, $x = 1$ and the curve is a simple hyperbola. The larger the value of the exponent x, the more steeply the curve will fall. Therefore, the curve for the adiabatic expansion of the monatomic gas ($x = \gamma = 1.67$) lies below the adiabatic curve for the diatomic gas ($x = \gamma = 1.4$). Since the work to expand the gas must come out of the internal energy, it appears that the diatomic gas can provide more work for a given expansion than can the same quantity of a monatomic gas. This result is plausible since the diatomic gas with its larger heat capacity can evidently store more internal energy than can a monatomic gas.

■ ASSUMPTIONS ABOUT OUR ATOMIC MODEL

All of our students believe deep in their hearts that there are atoms. They learned this in third grade. However, practically no high school senior or college freshman can offer any reasonable proof that atoms exist. In Chap. 21 we suggest several proofs that can be argued or demonstrated at an introductory level, while cautioning that the idea is not obvious. Atomicity was not fully accepted until the beginning of the twentieth century.

Here are the simple assumptions that we will make about atoms, leaving until Chap. 21 the explanations of how we determine the quantitative features. All the way from hydrogen to uranium, most atoms are about the same size. Within a factor of 2 their radius is 1×10^{-10} m. (There are a few exceptions, but even these are within a factor of 3. Actually, the "radius" of an atom depends on the method of measurement. For instance, the radius might be half the internuclear distance in a covalent molecule.) The mass of an individual atom can be calculated by dividing its molar mass by Avogadro's number, 6×10^{23}. For instance, the mass of 1 mole of H_2 is 2 g; a hydrogen atom has a mass of 1.6×10^{-24} g. At the other extreme, an atom of uranium has a mass of $(238 \text{ g/mol})/(6 \times 10^{23} \text{ atoms/mol}) = 4.0 \times 10^{-22}$ g.

In the solid phase, the atoms are shoulder to shoulder. If a mole of atoms is arranged in a cube, then along one edge of the cube there would be about 10^8

atoms. (The cube root of $6\times 10^{23} \approx 10^8$.) Since the diameter of each atom is about 2×10^{-10} m, the edge of this cube would be 2 cm. The molar volume is therefore about 8 cm^3. It may seem surprising that a mole of uranium atoms has about the same volume as a mole of carbon atoms. You can see how good the assumptions are for a number of elements by dividing the atomic weight number (the grams per mole) by the density. Keep in mind that a 10% change in diameter makes a 30% change in volume. Sample data are listed in Table 7.

There must be attractive forces between atoms that keep them bound together in the solid and liquid phases. In Chap. 9 we proposed that each atom must exist in a potential well created by the attractive forces of all its neighbors. The depth of the well is about 1 eV. At the bottom of the potential well, the atom is at an equilibrium distance from its nearest neighbors. At that point the restoring force on the atom is 0. We would expect, however, that the atom would slosh back and forth in the potential well. In the section on molecular forces in Chap. 9, we made an order-of-magnitude calculation concerning the spring constant k, responsible for small-amplitude oscillations of the atom in the well. That constant turns out to be 30 N/m = 30 J/m^2. Assuming that the atom oscillates in simple harmonic motion about its equilibrium point, the frequency is 6×10^{12} Hz for carbon and 1.4×10^{12} Hz for uranium. (As we shall see, the effect of the extra significant figure in comparing relative effects is important.)

■ SPECIAL ASSUMPTIONS FOR THE GAS MODEL

It is far simpler to create a kinetic theory of *gases* than it is to make a quantitative model of solids and especially liquids. The model that we will sketch here is commonly presented in introductory courses. In deducing the simple form of the gas model, we will assume some things about the atom that we know are not strictly true. First, let us assume that the volume of the atoms or molecules themselves is negligible compared with the volume in which the molecules are flying around. One way to demonstrate this is to boil water in a liter flask whose opening is covered with a piece of aluminum foil. Prick a hole in the foil to allow the water vapor to escape. The vapor will drive out the air, leaving a liter of water vapor when all the water has vaporized. Let the flask cool and observe the amount of condensed water. It will exist in droplets here and there on the walls, but enough will coalesce to make it possible to estimate its volume. You will usually get about 1 cm^3, about 1/1000 of the original volume of vapor. Or, given our primary assumptions about the size of atoms, calculate the volume of a mole of atoms, each of which has a diameter of 2×10^{-10} m. Each atom has a volume of 10^{-29} m^3. A mole of them would have a volume of 10^{-5} m^3 and would occupy 22.4 liters $\approx 10^{-2}$ m^3 at standard temperature and pressure (STP). Once again we get a ratio of 10^{-3} for the ratio of atomic volume to occupied volume of a gas.

TABLE 7. *Atomic masses and densities.*

Element	Atomic mass number	Density (g/cm^3)	Molar volume (cm^3)
Beryllium	9	1.8	5
Aluminum	27	2.7	10
Carbon (graphite)	12	2.3	5.2
Sodium	23	0.97	24
Iron	56	7.9	7.1
Lead	207	11.3	18.3
Gold	197	19.3	10.2
Uranium	238	19.1	12.5

Internal Energy

In constructing our model of gases, we will assume, at least at first, that the collision of the atoms or molecules with each other and with the walls of the container are elastic. No energy is lost due to friction and, more realistically on the atomic scale, no energy is lost due to the breaking up of molecules or setting them into internal vibration. You might wonder how atoms can collide elastically if there is an attractive force between them characterized by a potential well. As long as the total energy (kinetic plus potential) of two atoms is positive, they will collide by increasing their speeds as they fall into each other's attractive field, come very close together into the region of mutual repulsion, and then go shooting away from each other just as if they were two hard, impenetrable marbles. Indeed, our first model for gases will be completely classical, considering their momentum, energy, and flight times as if they were a swarm of marbles having no interaction with each other unless they collide, possessing no weight, and bouncing off each other according to the laws of classical mechanics.

■ THE KINETIC THEORY OF GASES

Let us start out with just one atom enclosed in a cubic box of length L. It is bouncing around randomly, striking the walls at various angles. Consider only the component of velocity in the x direction and what happens when the atom hits the y-z walls opposite that component. For the sake of visualizing the action, we could consider that the atom is simply moving back and forth in the x-direction. In that case, since it is moving to the right, it has momentum $+mv_x$. When it hits the wall and bounces back, its momentum has become $-mv_x$. The change of momentum, $\Delta(\text{mom}) = 2mv_x$. The wall suffered a recoil impulse to the right and will do so again every time that the atom bounces back. To find the time between bounces, divide the distance traveled across the box and back again by the velocity: $\Delta t = 2L/v_x$.

Remember that the general expression for Newton's second law is that $F = \Delta(\text{mom})/\Delta t$. The impulses on the wall can be equated to an average force,

$$\bar{F} = \frac{2mv_x}{2L/v_x} = \frac{mv_x^2}{L}.$$

Of course, one atom will not produce much of an average force but if there are N atoms in the box, all traveling in the same direction, then the force would be $\bar{F} = Nmv_x^2/L$.

The atoms will not be going in the same direction, however. Their motion is completely random and they are not only bouncing off the walls but off each other. The Pythagorean theorem gives a simple relationship between the square of a vector and the squares of its components in x, y, and z: $v^2 = v_x^2 + v_y^2 + v_z^2$. For any given atom at a particular time, one component may be much larger than either of the other two components. On the average, however, the components in the x direction will be the same as those in the y direction or the z direction. Therefore, for the average $\overline{v^2}$, $\overline{v_x^2} = \overline{v_y^2} = \overline{v_z^2}$. Consequently, $\overline{v_x^2} = \frac{1}{3}\overline{v^2}$. The force on the y-z wall of the box averaged over both time and for many different atoms is equal to $\bar{F} = \frac{1}{3}Nm\overline{v^2}/L$. Instead of describing the *force* on the y-z wall, we can make the calculation more general by describing the *pressure*:

$$P = \frac{\bar{F}}{A} = \frac{\frac{1}{3}Nm\overline{v^2}}{V},$$

where $V=LA$ is the volume of the box. The equation now transforms to

$$PV = \tfrac{1}{3}N\overline{v^2}.$$

There is yet another relationship that involves $m\overline{v^2}$ of the molecules. The average kinetic energy of the atoms or molecules in the box must be $\overline{E_{\text{kin}}} = \tfrac{1}{2}m\overline{v^2}$. Therefore, we can rewrite the equation for PV as

$$PV = \tfrac{2}{3}N\overline{E_{\text{kin}}}.$$

The experimental ideal-gas law is $PV=nRT=NkT$. If our model is correct, these two equations for PV should be identical. Therefore

$$\overline{E_{\text{kin}}} = \tfrac{3}{2}kT.$$

■ CONSEQUENCES OF OUR MODEL—AVERAGE ENERGY AND AVERAGE SPEED

Do we really prove with this sequence of arguments that temperature is proportional to the average kinetic energy of the molecules of the gas? We have proposed a theoretical model, deduced a consequence from the model, and *if* the model is correct, the temperature of a gas is a measure of the kinetic energy of its molecules. Such a prediction needs experimental testing. Let's put in some numbers and see what values we get for kinetic energy and speed of gas molecules at room temperature:

$$E_{\text{kin}} = \tfrac{3}{2}(1.38 \times 10^{-23} \text{ J/molecule K})(293 \text{ K}) \rightarrow \frac{6.1 \times 10^{-21} \text{J/molecule}}{1.6 \times 10^{-19} \text{J/eV}}$$

$$= 3.8 \times 10^{-2} \text{ eV/molecule} = \tfrac{1}{26} \text{ eV/molecule}.$$

The average energy of a gas molecule at room temperature is smaller than 1 eV by a factor of 26. At first glance that fact seems to explain why we and other chemically bound objects are stable at room temperature. On the other hand, we know that we are not stable at twice that temperature—at 586 K—where the average energy of a gas molecule is smaller than 1 eV by a factor of 13. Evidently, there is more to be learned about the phenomena.

The speed of the gas molecules with average kinetic energy can be found as follows: $\overline{E_{\text{kin}}} = \tfrac{1}{2}m\overline{v^2} = \tfrac{3}{2}kT$. The "root-mean-square" speed is

$$v_{\text{rms}} = \sqrt{\frac{3kT}{m}}.$$

For air at room temperature, we can use the mass of the nitrogen molecule and $T=293$ K:

$$v_{\text{rms}} = \sqrt{\frac{3(1.38 \times 10^{-23} \text{ J/molecule K})(293 \text{ K})}{4.7 \times 10^{-26} \text{ kg}}} = 5.1 \times 10^2 \text{ m/s}.$$

The average speed of air molecules at room temperature is about 500 m/s. This is over 1100 miles/h. The speed of sound in air at room temperature is about 340 m/s. Since sound in air must be carried by the collision of the molecules with each other, it is encouraging that the speed of sound is the same order of magnitude as the theoretical value for the average speed of air molecules.

■ THE DISTRIBUTION OF MOLECULAR ENERGY IN A GAS

In both gases and human societies, many phenomena are dominated by the nonaverage members of the population. Since the molecules in the gas are continually colliding with each other and the walls, some will be going very slowly and some very fast. The distribution of energies was worked out by Maxwell and Boltzmann a century ago. The distributions of speeds and energy are shown in Fig. 10.11. Although the formulas look complex at first, there are several simple points that can be seen.

First note that the distributions are not symmetrical about the average speed or the average energy. In fact, with these kinds of distributions, we have to be careful about what we mean by "average." In the speed diagram, we have indicated the most probable speed, the mean speed (the ordinary arithmetic average) and the root-mean-square speed, which is the average speed we calculated above. For very small values of speed or energy, the exponential term is close to unity because the exponent itself is close to 0. The distribution functions are therefore dominated by the v^2 term in one case or the $\sqrt{E}$ term in the other. Apparently it is very unlikely that a molecule being buffeted about in the gas will end up with zero speed or energy. In the distribution function the rising values of the v^2 or $\sqrt{E}$ terms are eventually overcome by the decreasing exponential terms. Indeed, once past the maximum, the distributions are quite completely governed by the exponential terms. Exponential terms decay very rapidly but they never decay away completely. Therefore it is possible to have some molecules with energies many times those of the average. Notice that the distribution functions are given in terms of the number of molecules *per speed or energy interval*. Consequently, the total area under each curve is equal to the total number of molecules present.

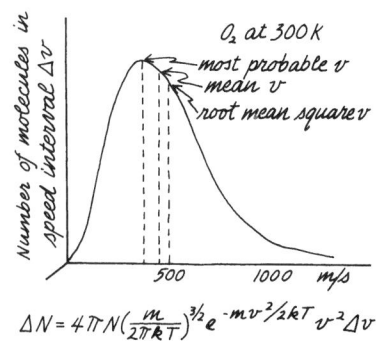

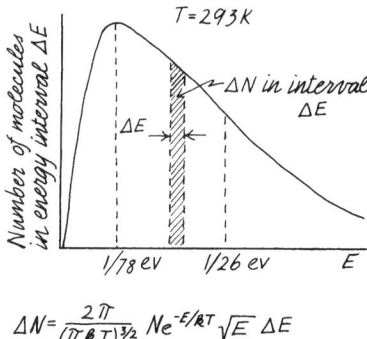

Figure 10.11. Distribution of molecular speeds and energies in a gas.

The exponential dependence produces surprising relationships. For instance, the average energy of a molecule is $\bar{E} = \frac{3}{2}kT$. Let us find the ratio of the number of molecules that have $2\bar{E}$ to the number that have $\bar{E}$:

$$\frac{\Delta N(2\bar{E})}{\Delta N(\bar{E})} = \frac{\sqrt{2\bar{E}}e^{-2\bar{E}/kT}}{\sqrt{\bar{E}}e^{-\bar{E}/kT}} = \sqrt{2}e^{-\bar{E}/kT} = \sqrt{2}e^{-3/2} = 0.32.$$

However,

$$\frac{\Delta N(10\bar{E})}{\Delta N(5\bar{E})} = \sqrt{2}e^{-15/2} = 7.8 \times 10^{-4}.$$

At first look, one might think that $2\bar{E}/\bar{E}$ is the same as $10\bar{E}/5\bar{E}$. However, in the first case we are comparing the number of molecules that have energy $3kT$ with the number that have energy $\frac{3}{2}kT$. The second comparison is for the number of molecules that have an energy of $15kT$ compared with those that have an energy of $7\frac{1}{2}kT$. Although the energy ratios are equal, the population ratios are not the same at all.

Molecular speeds can be measured in a very direct fashion. All you have to do is time the flight of molecules in the same basic way that you might time the throw of a baseball or the flight of a bullet. Starting back in the 1920s, Otto Stern and his students produced beams of molecules with apparatus shown schematically in Fig.

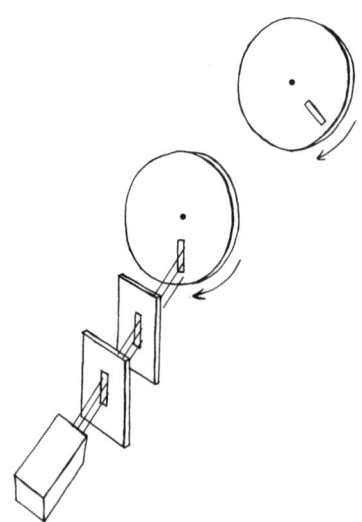

Figure 10.12. Schematic diagram of apparatus to measure molecular speed.

10.12. Many different kinds of atoms and molecules, even those of metals, can be produced in the gas phase. In some cases the source for the beam is an oven that is boiling off the atoms. The source box has a tiny hole through which the gas atoms can pass if they happen to be going in that direction. From that point on they are traveling in a vacuum. Those that can pass through the successive holes in the baffles form a narrow beam. The small group that gets through the opening in the first high-speed rotating shutter starts its flight at a particular time. Only those with a particular speed will arrive at the second shutter in time to get through it and on into the collector. By changing the angle by which the second shutter lags the first, we can accept groups of molecules with different speeds. The resulting distribution gives us ΔN versus v experimentally, and the results agree with the theory. (There is a subtle complication that at first confused the experimenters. The distribution of speeds *in the source* is proportional to v^2. However, the probability that any molecule will strike the hole in the source is proportional to v. Therefore, the distribution *in the beam* is proportional to v^3.)

Although we have been talking about the kinetic energy of molecules in a gas, the same exponential factor is involved in the distribution of molecules subject to a potential field. For example, the gravitational potential energy of molecules depends on their height, h, above the surface of the earth. Assuming that the temperature is independent of height (a poor assumption over any considerable distance), then the ratio of the number density of molecules at two heights is given by

$$\frac{n_1}{n_2} = \frac{e^{-mgh_1/kT}}{e^{-mgh_2/kT}} = e^{-mg(h_1-h_2)/kT}.$$

The number density n is the number of molecules per unit volume.

So far we have been talking about molecules in the gas form. A similar distribution occurs whenever particles can mix in a chaotic random fashion, subject to the conservation of overall energy and the conservation of the total number of particles. The generalized Maxwell–Boltsmann distribution for such a situation gives the ratio of number of particles in two different energy states, E_1 and E_2, as follows:

$$\frac{n_1}{n_2} = e^{-(E_1-E_2)/kT}.$$

Let us put some numbers into these distribution laws to see consequences concerning some familiar situations.

1. The atmosphere is less dense at high altitudes, and above 50 km is practically nonexistent in terms of what we know as air. Many factors influence the density of the air and also its composition. Since air is mostly transparent to solar energy, heating of the atmosphere is mainly a matter of convection from the earth's surface where the solar energy is absorbed. Consequently, temperature falls rapidly as altitude increases. (At about 50 km there is a layer of ozone created by ultraviolet radiation. This thin and fragile layer of chemically active oxygen has a temperature of about 10 °C. It acts as a heat source for the atmosphere slightly below and slightly above.) In the troposphere, below about 12 km, storms can affect air pressure (the "highs" and "lows" of weather patterns).

In spite of the factors of variable temperatures and storms, the Maxwell–Boltzmann distribution law yields a first approximation to the average density of the atmosphere as a function of height. The ratio of the number of molecules per unit volume at two heights n_1/n_2 is also the ratio of densities at those two heights, σ_1/σ_2, and therefore is also the ratio of pressures P_1/P_2. For P_2 use the pressure at sea level, where $h_2=0$. Then the pressure at any given height h is

$$P(h) = P_0 e^{-mgh/kT}.$$

Notice that this formula gives the right value for atmospheric pressure at sea level where $h=0$. To find the pressure at any other height, we have to evaluate the argument of the exponential:

$$\frac{mgh}{kT} = \frac{(4.7 \times 10^{-26} \text{ kg})(9.8 \text{ N/kg})(h)}{(1.38 \times 10^{-23} \text{ J/K})(290 \text{ K})} = 1.15 \times 10^{-4} h \approx 1 \times 10^{-4} h.$$

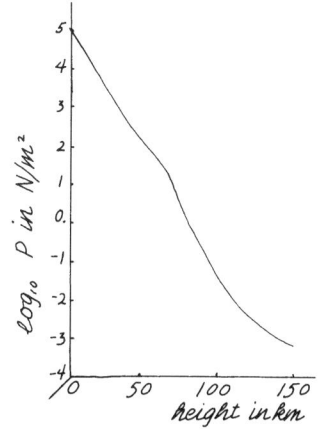

Figure 10.13. Pressure in atmosphere as measured in rocket flights. If the pressure followed the text formula, the curve would be a straight line.

At a height of 1.5 km, about the altitude of Denver, Colorado, the argument of the exponential would become $(1.15 \times 10^{-4})(1.5 \times 10^3 \text{ m}) = 0.17$. For small values of the argument of an exponential, $e^{-x} \approx 1 - x$. Consequently, the normal atmospheric pressure at Denver should be about $0.83 P_0$, where P_0 is the atmospheric pressure at sea level. Indeed, the normal atmospheric pressure in Denver is about 630 mm of mercury. The graph in Fig. 10.13 shows actual pressure readings in the atmosphere. If the pressure followed the exponential formula that we derived, the curve would be a straight line. Evidently the approximation is fairly good.

The pressure of the atmosphere is always just sufficient to hold the weight of a column of air of unit cross section going all the way to the top of the atmosphere. Atmospheric pressure at sea level is approximately 1×10^5 N/m². Since most of the atmosphere is in a region where g equals 9.8 N/kg, the mass of a column of air molecules with a cross section of 1 m² and a height going to the top of the atmosphere must be approximately 1×10^4 kg.

These molecules are bouncing around. It is not obvious how they exert a force just equal to what their weight would be if they were all lying quietly on top of each other. Nevertheless, let us use the distribution law to calculate how many air molecules there are in a column with a cross section of 1 m² reaching from $h=0$ to $h=\infty$. The situation is shown schematically and graphically in Fig. 10.14:

$$N/m^2 = n_0 \int_0^\infty e^{-mgh/kT} dh = \left. \frac{n_0 kT}{mg} e^{-mgh/kT} \right|_0^\infty = n_0 \frac{kT}{mg} = n_0 \frac{1}{1.15 \times 10^{-4}}$$

$$= (2.68 \times 10^{25} \text{ molecules/m}^3)(8.7 \times 10^3 \text{ m}) = 2.3 \times 10^{29} \text{ molecules/m}^2.$$

Since the mass of each air (nitrogen) molecule is 4.7×10^{-26} kg, the mass of the whole 1-m² column is $(2.3 \times 10^{29} \text{ molecules})(4.7 \times 10^{-26} \text{ kg/molecule}) = 1.1 \times 10^4$ kg, which is close to the mass deduced from the size of atmospheric pressure.

The same arguments apply to the energy distribution of gas in a box. The density of gas at the top of the box must be less than it is at the bottom. If the gas is at uniform temperature, then the pressure at the bottom must be greater than it is at the top by just enough to support the weight of the gas in the box.

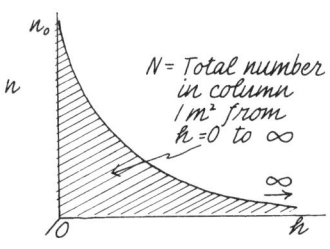

$n = n_0 e^{-mgh/kT}$

$n_0 = \dfrac{(6 \times 10^{23}\ \text{mol/mole})}{(22.4 \times 10^{-3}\ \text{m}^3/\text{mole})}$

$n_0 = 2.68 \times 10^{25}\ \text{mol/m}^3$ at STP

$N = 2.2 \times 10^{29}$

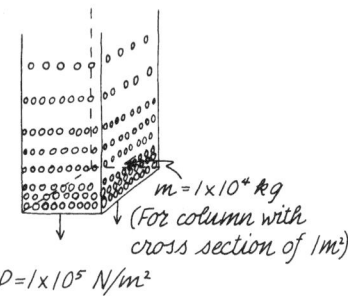

$m = 1 \times 10^4\ kg$
(For column with cross section of $1\ m^2$)

$P = 1 \times 10^5\ N/m^2$

Figure 10.14. Schematic and graphical representation of molecular density as function of height in atmosphere.

The number of molecules per cubic meter at STP is 2.68×10^{25}. At 290 K this number is reduced to 2.52×10^{25}. Therefore, for air molecules with a mass of 4.7×10^{-26} kg each, the mass of 1 m³ is 1.19 kg with a weight of 11.6 N. The pressure at the top of the box is

$$P = P_0 e^{-mgh/kT} \approx P_0 \left(1 - \frac{mg}{kT} h\right).$$

The extra pressure at the bottom compared with the top is

$$\Delta P = P_0 - P = P_0 \frac{mg}{kT} h.$$

For a box 1 m high, $\Delta P = (1.01 \times 10^5\ \text{N/m}^2)(1.15 \times 10^{-4}) = 11.6\ \text{N/m}^2$. The extra pressure at the bottom of the box is just enough to sustain the weight of the gas in the box.

2. Like everything else on the surface of the earth, an air molecule is in a gravitational potential well. The depth of that well is

$$E_{\text{grav. pot.}} = -G \frac{mM}{R} = -(6.7 \times 10^{-11}\ \text{N m}^2/\text{kg}^2) \frac{m(6 \times 10^{24}\ \text{kg})}{6.4 \times 10^6\ \text{m}}.$$

For a hydrogen molecule with a mass of 3.3×10^{-27} kg, the gravitational binding energy is 2.07×10^{-19} J $= 1.3$ eV. The molecular thermal energy is equal to $\tfrac{3}{2}kT$, which is about $\tfrac{1}{26}$ eV at room temperature. It appears that hydrogen is tightly bound to the earth since the gravitational potential well is about 34 times deeper than the average kinetic energy resulting from thermal motion. Nevertheless, hydrogen does not stay in the earth's atmosphere. In spite of the mixing of gases within the atmosphere the lighter hydrogen gradually works its way upwards. Above the ionosphere there are regions where the equivalent temperature is very high because of the bombardment of particles coming in from the sun. Hydrogen molecules in the high-energy tail of the thermal energy distribution occasionally reach escape speed in the right direction and are lost to earth.

The average *thermal* energy of nitrogen or oxygen is the same as it is for hydrogen. The depth of the gravitational well, however, is greater by a factor of 28/2 or 32/2 for nitrogen or oxygen, respectively. A nitrogen molecule, for example, is in a gravitational potential well with a depth of about 18 eV. This is why the earth has an atmosphere. The depth of the gravitational potential well on the surface of the moon is smaller than that on the earth by a factor of 22. (The radius of the moon is 1.74×10^6 m and the mass of the moon is 7.34×10^{22} kg.) Therefore a nitrogen molecule on the surface of the moon is in a gravitational potential well of less than 1 eV. That much binding is still greater than the average thermal energy but not great enough to prevent the escape of the molecules at the high-energy end of the thermal distribution.

3. The average thermal energy of different kinds of molecules is the same regardless of the molecular mass. Consequently at a particular temperature, lighter molecules must move more rapidly than heavy molecules:

$$(\tfrac{1}{2}mv^2)_{H_2} = (\tfrac{1}{2}mv^2)_{O_2},$$

$$\frac{v_H}{v_O} = \sqrt{\frac{m_O}{m_H}}.$$

Since the ratio of the mass of oxygen to hydrogen is 16, the ratio of speeds is 4.

This dependence of average speed on the mass of the molecule shows up in a variety of ways. Remember that we found that the average speed of sound in air is about two-thirds the average speed of air molecules. We should expect to find that the speed of sound in hydrogen or helium is much greater than it is in air. The effect can be demonstrated by flushing helium through an organ pipe (or bugle). The pitch of an organ pipe depends on its size, which determines the wavelength, and on the speed of sound in the pipe. Since the wavelength remains constant, the frequency of the sound rises as helium fills the pipe. The ratio of speeds, and hence frequency, for helium to air is

$$\frac{v_{He}}{v_{O_2}} = \sqrt{\frac{32}{4}} = 2\sqrt{2}.$$

The pitch rises by $1\tfrac{1}{2}$ octaves. There is a similar effect, but much more complicated, when you take a breath of helium and talk like Donald Duck. In this case the basic frequency is determined by the vocal cords but the emitted sound is influenced by several resonant cavities in the nose and throat. These resonances rise in frequency but the resultant combination sound suffers a nonlinear rise in pitch. Simply lowering the pitch electronically does not restore the original voice. This situation causes a serious technical problem of communicating with divers who breathe a mixture of helium and oxygen. If you try the helium breathing stunt as a demonstration, take only one breath. Helium is not poisonous but it is not oxygen. The lungs hold only about three breaths. If those three are helium, you pass out and may die.

Vacuum systems frequently leak. The leak is not caused by gas molecules being sucked into a tiny crack or hole in the vacuum wall. The molecules simply bombard all sections of the wall. If they happen to strike where the crack is, they may get inside. The chance of a molecule striking the wall right in the crack depends on how frequently the molecule hits the wall and therefore on its velocity. Whether the molecule gets through the crack may also depend on the relative size of crack and molecule. Since the lightest gas is molecular hydrogen, it might seem that hydrogen would leak into or out of a system more readily than any other gas. Indeed, a vacuum chamber or pressure chamber must be very tight to keep hydrogen in or out. It turns out, however, that helium is even harder to hold onto. Helium is monatomic, whereas hydrogen is diatomic. The molecular (atomic) mass of helium is twice that of hydrogen and so its average speed is slower by a factor of 1.4. Still, the helium gets through leaks more readily because the single atom of helium is considerably smaller than the double atom of the hydrogen molecule.

The diffusion of gases through air is somewhat similar to the leaking of gases into a chamber. Ammonia diffuses much more rapidly than perfume.

The ammonia has a molecular mass of only 17, while most perfume molecules consist of long chains of carbon, hydrogen, and other things and so have large molecular mass and size.

4. Our simple model of gases ought to apply to any particle that satisfies the assumptions of the model. If an object is suspended in a gas, even though it is large enough to be visible to the naked eye, it ought to take part in the random bombardment and motion. This effect had been seen but misunderstood in 1827 by an English botanist named Robert Brown. He was using a microscope to observe a water suspension of grains of pollen. The pollen grains seemed to jiggle and dash around by themselves as if they were alive. This effect, which can be seen with particles suspended in either gas or liquid, is called Brownian motion. When Maxwell and Boltzmann created the kinetic theory of gases, they assumed that the Brownian motion was due to molecular bombardment. The theory was not made quantitative, however, until Einstein did so in one of his three famous papers in the same issue of *Annalen der Physik* in 1905. J. Perrin applied the theory in 1908 to find an experimental value of Avogadro's number. Einsteins' theory and Perrin's experiment were historically the capstone of the arguments in favor of atomicity. Until that time, as late as 1908, such influential scientists as the physicist Mach and the chemist Ostwald felt that it was not necessary to hypothesize the existence of atoms.

A grain that is visible in a low-powered microscope must have a diameter of about 10 μm (micrometers, commonly called *microns*). The diameter of a gas molecule is about 5×10^{-10} m. The ratio of diameters of grain to molecule is therefore about 2×10^4. The ratio of volumes and therefore approximately the ratio of masses are about 10^{13}. The scale of the situation is comparable to that of an ocean liner surrounded and buffeted by a sea of corks. The astonishing feature of the phenomenon is not so much that the bombardment of the molecules (or corks) can move the grain (or ship) but that the *fluctuations* in the bombardment produce such motion. After all, at any given instant, the larger object is being bombarded on all sides.

The quantitative detection of Brownian motion consists of measuring the displacement of the particle as a function of time. The movement is a form of random walk. A pollen grain (or a smoke particle) with a diameter of 10×10^{-6} m has a volume of about 10^{-15} m^3. With an approximate density the same as that of water, the mass of the grain would be 10^{-12} kg. Let us find its root-mean-square speed:

$$\tfrac{1}{2}m\overline{v^2}=\tfrac{3}{2}kT,$$
$$(10^{-12}\text{ kg})\overline{v^2}=3(1.38\times10^{-23}\text{ J/K})(293\text{ K})$$
$$\langle v\rangle=1\times10^{-4}\text{ m/s.}=0.1\text{mm/s.}$$

The speed is small, but clearly large enough to explain the jiggling motion seen in the microscope.

Brownian motion is easily seen with a 50× power microscope. There are inexpensive commercial cells available to make the setup simple. These are cubes about 1 in. on a side with two glass ports opposite each other to let a light beam in and out, a glass port on top for viewing, and a rubber suction bulb on one side and a nozzle on the other for inhaling smoke. As long as you know what you are seeing, it's fascinating to see such direct evidence of the behavior of molecules.

5. At room temperature we are relatively stable, but if the temperature were doubled, from 293 to 586 K, we and our goose would be cooked. Even small differences in temperature can drastically affect cooking times. Eggs will not harden and potatoes will not cook (without pressure cookers) at elevations of 2 miles, no matter how long you boil them. If all the molecules participating in a chemical interaction had exactly the same thermal energy, then we might explain the need for particular temperatures in cooking in terms of a required threshold energy for an interaction. However, the Boltzmann distribution of molecular thermal energy applies approximately to solids and liquids as well as to gases. (In solids and liquids, the kinetic energy is vibrational about fixed positions.) There are always some atoms or molecules in the high-energy tail of the thermal distribution. It must be that a relatively small increase in temperature can dramatically increase the population in the high-energy tail of the distribution.

Let us put numbers in, starting out with the size of the energy scale factor, kT. In terms of electron volts, $k=8.63\times10^{-5}$ eV/K. At a room temperature of 293 K, $kT=2.53\times10^{-2}$ eV $\approx \frac{1}{40}$ eV. The *average* kinetic energy of this distribution turns out to be the same as the average translational kinetic energy of a gas molecule at this temperature: $E=\frac{3}{2}kT$, which is about $\frac{1}{25}$ eV. In the chemist's terminology, this is about 0.9 kcal/mol. These energies should be compared with the energies involved in chemical changes, which are in the range of 1 eV or 22 kcal/mol.

If $E_{\text{threshold}}=1$ eV, then the fraction of molecules in the tail of the distribution with energies greater than the threshold is

$$\frac{N(>E_t)}{N} = \frac{2\pi}{(\pi kT)^{3/2}} \int_{E_t}^{\infty} e^{-E/kT} \sqrt{E}\, dE.$$

This integral cannot be reduced in closed form, but for our present purposes it is a good approximation that $\sqrt{E}$, which is a slowly varying function compared with the exponential, remains constant at the value E_t. The fraction of molecules with enough energy to participate in the reaction thus becomes

$$\frac{N(>E_t)}{N} \approx \sqrt{\frac{E_t}{kT}}\, e^{-E_t/kT}$$

At room temperature, where $kT=\frac{1}{40}$ eV, the value of this crucial fraction is

$$\sqrt{40}e^{-40} \approx 3\times10^{-17}.$$

The small size of the crucial fraction is comforting unless we remember that there are 6×10^{23} molecules in every mole, and that means that in every mole of us there are 2×10^7 molecules with energies over the threshold. Still, we fall apart very slowly. Why is the effect so disastrous if the temperature is doubled? After all, the average kinetic energy in the distribution will then be only $\frac{1}{12}$ eV, still a safe factor of 12 below the threshold. However, the crucial fraction above the threshold is now

$$\sqrt{20}e^{-20}=1\times10^{-8}.$$

A factor of 2 in temperature (or of 2 in the threshold energy) changes the active population in the tail by a factor of 3×10^8. As those molecules above the threshold combine or decay, their places are taken by other molecules from the main body of the distribution. You're cooked!

■ THERMAL EFFECTS IN REAL GASES, SOLIDS, AND LIQUIDS

1. Few things in this world are ideal, including gases. The molecules in our ideal-gas model were supposed to have no volume of their own and no interaction until they hit each other. We know that both these assumptions are wrong. The volume of the molecules in a gas is about 1/1000 of the volume occupied by the gas at room temperature. We also know that the molecules attract each other at separation distances larger than that where sharp repulsion occurs. In the $P-V$ graph in Fig. 10.5, the hyperbolic curves called for by the ideal-gas equation applied only for temperatures well above the condensation point.

Many "equations of state" have been proposed to describe gas behavior over the complete range of pressure, volume, and temperature. The best known of these is van der Waals's equation:

$$\left(P+\frac{n^2a}{V^2}\right)(V-nb)=nRT.$$

The significance of b is straightforward; it is the *effective* volume occupied by 1 mol of the molecules of the gas. The actual volume available for the gas is less than V, the volume of the container, because part of the volume is taken up by the molecules themselves. At room temperature for most gases the correction is only 0.1%. (In practice, the correction term turns out to be about four times the actual volume of the molecules.)

The explanation of the correction to the pressure term is a little more complicated. According to van der Waals's equation, the pressure measured on a wall of the container, P, must be slightly less than the actual pressure in the gas, which is represented by the whole term $P+n^2a/V^2$. The standard explanation in many texts is that there are short-range forces between the gas molecules. For any particular molecule, these small attractive forces cancel out on the average as long as the molecule is surrounded by others. The argument then claims that as the molecule approaches the container wall, there will be an unbalanced force toward the interior of the gas. In effect, this will pull the punch of the molecule when it strikes the wall, reducing the pressure on the wall. There are some obvious problems with this theory. Why shouldn't the wall, which is much more dense than the gas, exert a strong attractive force on the advancing molecule and thus increase its speed and its impulse? Furthermore, if the speed is reduced then the kinetic energy will be reduced, and so the gas temperature would be less near the walls. Bob Bauman, in an article in *The Physics Teacher* (April, 1996) points out that in the elastic collisions of point particles, which we assume in the ideal gas derivation, the gas density does not interfere with the average time for a molecule to cross the container. If one molecule is slowed in a collision, the other is speeded up and the net momentum flow in any direction remains constant. Remember that the derivation of the ideal gas equation depends on the velocity in two ways. First, the pressure is proportional to the momentum transfer of the molecules with the wall, and that gives one power of v. Second, the momentum transfer (and thus the pressure) is proportional to the frequency with which the molecules make a round trip of the box and strike the wall. That frequency is also proportional to the velocity. However, with an ideal gas we assume that the average distance traveled between collisions with a particular wall is just the round trip distance. When there are short range attractive forces between molecules, the collisions can still be elastic but there will be extra time spent hovering around each other (perhaps even orbiting). That will increase the round trip time, decrease the frequency of hitting the wall and so decrease the pressure at the wall. This mechanism will not, however, decrease the average speed and therefore will not affect the temperature. The delay experienced by a molecule will be proportional to the density. Since the number of molecules striking the wall is also proportional to the density, the effect is proportional to the square of the density. Since we are dealing with a fixed quantity of gas, in this case n moles, the density is inversely proportional to the volume. Therefore the correction term is proportional to $1/V^2$. For air at normal temperature and pressure, the correction term for the pressure is about 0.3%.

There is an apparent paradox in the pressure term of van der Waals's equation. Consider that the lower the pressure of a gas, the more it is supposed to behave like an ideal gas. Note, however, that if the pressure is lowered, the correction term, $n^2 a/V^2$, would appear to become more and more important. The resolution of the paradox is that the equation and the correction terms are valid only for a fixed quantity of gas—namely, n mol. To lower the pressure of a fixed quantity of gas (at constant temperature) you must increase the volume. In order to lower the pressure by a factor of 2, you must increase the volume by a factor of 2. The correction term would then decrease by a factor of 4.

2. According to the kinetic theory, the temperature of a gas is related to the average kinetic energy of the molecules. Solids get hot too. According to our model, each atom is trapped in a potential well and oscillates about its equilibrium position. The energy relationships are shown in Fig. 10.15. As heat enters a solid, the energy spreads out among all the atoms, causing them to oscillate with larger amplitudes and a distribution of oscillation energy similar to that of the Maxwell–Boltzmann distribution.

What is the relative magnitude of the positive energy of oscillation to the negative potential energy of binding? As we have seen in many different circumstances, the chemical binding energy per atom is about 1 eV. The average thermal energy at room temperature is about $\frac{1}{26}$ eV. Consequently, the atoms of a solid are deep in potential wells and their oscillations, on the average, do not take them far up the wall of the well. The temperature of a solid is related to the kinetic energy of *oscillation* of the atoms in the same way that the temperature of a gas is related to the *translational* kinetic energy of the molecules. Heat passes through a metal bar from high temperature to low temperature because the atoms with large amplitude of oscillation at one end share their energy with their neighbors, thus spreading the thermal energy

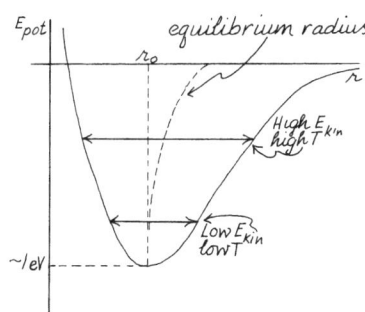

Figure 10.15. In a solid, each atom is trapped in an asymmetric potential well and oscillates about an equilibrium radius.

along the bar. There is similar contact between atoms in the surrounding gas and atoms in a solid. The bombarding gas molecules and the oscillating atoms of the solid come to energy—and hence temperature—equilibrium.

Good electrical conductors are also usually good heat conductors. To explain this situation, we must add one particular feature to our model. With electrically insulating materials the electrons are bound in position along with their parent atoms. With metals, however, one or more of the valence electrons from each atom are free to move around the whole expanse of the object. These electrical conduction electrons can also spread thermal energy and they do so much faster than when neighboring atoms share oscillation energy—with one exception. Diamond is not an electrical conductor, but it is the best heat conductor of any material at ordinary temperatures—better even than silver. The speed of a longitudinal pulse in a solid is proportional to $\sqrt{k/m}$, where k is the spring constant of the binding between atoms and m is the atomic mass. The carbon atoms in diamond have low m, and since diamond is the hardest material, the spring constant k is very large. Hence, oscillating atoms of diamond pass along their energy very rapidly, even without the aid of free electrons.

Our model must explain why most materials expand as they get hotter. At first thought, it might seem that the expansion could be explained in terms of the increasing amplitudes of oscillation of the bound atoms. After all, if each atom oscillates back and forth with a larger amplitude as its average kinetic energy increases, shouldn't the whole object become larger? Not if the potential wells are symmetric! The equilibrium positions would remain the same, and only the atoms on the outside walls might reach out slightly. The increase in volume would be negligible. As we show in Fig. 10.15, however, the potential well is asymmetric. As the amplitude of oscillation increases, so does the average separation of each atom from its neighbors. As the temperature increases, the average oscillation energy and amplitude of each atom increases and so does the average separation distance. Hence the solid expands.

Let us calculate on a classical basis the amplitude of oscillation of a bound atom. At 300 K the total kinetic energy of oscillation is $\frac{3}{2}kT$. The energy associated with oscillation along one dimension is $\frac{1}{2}kT \approx \frac{1}{80}$ eV$=2\times10^{-21}$ J$=\frac{1}{2}kA^2$. With a spring constant of 30 J/m^2, $A=0.1\times10^{-10}$ m. This is a reasonable amplitude, considering our assumption that the equilibrium distance between atoms is 1×10^{-10} m. If the temperature, and thus the oscillation energy, are doubled (from 300 to 600 K), the amplitude will increase by a factor of $\sqrt{2}$. Therefore, $\Delta A=0.04\times10^{-10}$ m$=0.04$ Å. To find the *actual* expansion, use the coefficient of volume expansion. A typical value for metals at room temperature is $\beta=(\Delta V)/V(1/\Delta T)=40\times10^{-6}$ K^{-1}. For $\Delta T=300$ K, $(\Delta V)/V=1.2\times10^{-2}$, and $(\Delta L)/L=\frac{1}{3}(\Delta V)/V=4\times10^{-3}$. Therefore, $\Delta L=4\times10^{-13}$ m$=0.004$ Å. Notice that the increase in equilibrium position due to the asymmetry of the potential well is only one tenth the increase in the amplitude of oscillation. But it is the increase in equilibrium position that causes the solid to expand. Notice also that the work involved in expanding the metal is small compared with the thermal energy fed in and stored in kinetic form. The stretching energy is $\frac{1}{2}kx^2 = \frac{1}{2}(30 \text{ J/m}^2)(4\times 10^{-13} \text{ m})^2 = 2.4 \times 10^{-24}$ J $= 1.5\times 10^{-5}$ eV.

3. The thermal behavior of a liquid is more like that of a solid than a gas. The atoms or molecules are just about as close to each other in a liquid as they are in a solid. After all, for most materials the density of the liquid is about the same as the density of its solid. Nevertheless, the binding of individual atoms or molecules in a liquid is sufficiently loose that they can move around each other. However, the energy associated with this meandering motion is much smaller than the energy of oscillation of each atom about its gradually moving equilibrium point. The period of oscillation of the atom in its temporary cell of neighbors is much shorter than the time it takes to drift past those neighbors.

A quantitative treatment of the liquid phase is still hard to do for most materials. Both the gas phase and the solid phase are much better understood. In liquids there are changing clusters of atoms or molecules, sometimes with sufficient order to their temporary arrangements so that they behave in some ways like crystals. Not only are these clusters continually forming and then breaking up, but there are fluctuations in the mass movement of microvolumes of the fluid, even when the average velocity of the liquid as a whole is 0.

4. As heat enters a solid, the material can turn into a liquid and then into a gas. (Under certain circumstances, the solid may change directly into a gas.) During the transition between phases, energy is absorbed without any change of temperature. Apparently the internal energy of the liquid phase is larger than that of the solid phase and the internal energy of the gas phase is greater than that of the liquid phase. However, this does not mean that the kinetic energy of the atoms in the liquid is greater than the kinetic energy of the atoms in the solid, or that the translational kinetic energy of the gases is greater than the vibrational kinetic energy of solids or liquids at their same temperature. Instead, the energy required for phase transition—the heat of fusion or vaporization—must go into freeing atoms from potential wells. Let us compare the depth of these potential wells with those produced by chemical binding.

For water, the heat of fusion is 80 cal/g and the heat of vaporization is 540 cal/g. Let us change each of these figures to joules per mole and then to electron volts per molecule.

Heat of fusion:
$$(80 \text{ cal/g})(4.18 \text{ J/cal})(18 \text{ g/mol}) = 6020 \text{ J/mol} = 1.4 \text{ kcal/mol},$$

$$\frac{(6020 \text{ J/mol})(6\times 10^{18} \text{ eV/J})}{6\times 10^{23} \text{ molecules/mol}} = 0.06 \text{ eV/molecule}.$$

Heat of vaporization:
$$(6020 \text{ J/mol})\frac{540}{80} = 40\,700 \text{ J/mol} = 9.7 \text{ kcal/mol},$$

$$(0.06 \text{ eV/molecule})\frac{540}{80} = 0.4 \text{ eV/molecule}.$$

Phase transitions are usually classified as physical changes as opposed to chemical changes. Such distinctions cannot be made solely on the basis of energy effects, although there is usually less energy per molecule involved with so-called physical changes than with chemical changes. In the case of water, the heat of fusion is small compared with the activation energy required by most chemical interactions. But the heat of vaporization is comparable to that of some chemical transitions. The energy required to break up a molecule of water into oxygen and hydrogen is 2.98 eV/molecule. But only 0.23 eV is needed for disintegration of H_2S into hydrogen and sulfur. Note that this "chemical" change takes less energy than the "physical" change of turning water into steam. Table 8 gives phase transition energies and chemical binding energies for a number of materials.

Liquids do not necessarily freeze at the freezing point nor do liquids necessarily boil at the boiling point. Similarly, in a dust-free atmosphere, it is possible to have humidity as high as 140% without condensation. Superheating, supercooling, and supersaturation are phenomena that are still poorly understood. The phase transitions proceed most rapidly when there are nuclei or "seeds" for the bubbles or crystals to form on. Each water droplet in saturated air, for instance, formed on a minute dust particle or on an electrical ion left by the passage of an ionizing particle. This allowed the droplet to grow beyond the threshold size where condensation onto the surface overcame the competition of evaporation from the surface.

■ HEAT CAPACITIES ACCORDING TO THE ATOMIC MODEL

Gases

As shown in Table 6, there are many regularities in the values for specific heats. At room temperature, monatomic gases have specific heats of about 3 cal/mol K. Diatomic gases have specific heats of about 5 cal/mol K. On the other hand, specific heats change with temperature. A good atomic theory should be able to explain not only the regularities but also the exceptions.

If we assume that the atoms or molecules in our model have no size and no internal structure, then the total internal energy of the gas is given by the translational kinetic energy of the particles.

TABLE 8. *Energies for the formation of molecules and for phase transitions in eV/molecule. The heat of vaporization is for standard atmospheric pressure. 1 eV/molecule = 23 kcal/g mol.*

	Formation of molecules from atoms	Heat of vaporization	Heat of fusion
HCl	0.96	0.17	0.021
H_2S	0.23	0.19	0.025
NaCl	1.89	---	0.29
CO_2	4.09	Sublimes	0.087
CO	1.15	0.063	0.0087
NH_3	0.48	0.243	0.059
H_2O	2.5	0.42	0.062

$$U = N\tfrac{1}{2}\overline{mv^2} = N\tfrac{3}{2}kT = \tfrac{3}{2}nRT.$$

Therefore, if there is a change in temperature, ΔT, the change of internal energy is equal to $\Delta U = \tfrac{3}{2}nR\Delta T$. If heat is fed into the gas at constant volume so that no external work is done, the specific heat is

$$c_v = \frac{Q}{\Delta T} = \frac{\Delta U}{\Delta T} + 0 = \tfrac{3}{2}R = 3 \text{ cal/mol K}.$$

As we showed on p. 236, for an ideal gas, $c_P - c_V = R = 2$ cal/mol K. The extra energy required at constant pressure goes into the work done in expansion, $P\Delta V$.

According to our simple model, all gases must have the same values for specific heat and the difference between c_P and c_V is a constant. Compare these theoretical conclusions with the experimental data from Table 6 on p. 235. To start out with success, note first that the difference $c_P - c_V$ does indeed have the value of 2 cal/mol K. Furthermore, argon and helium have exactly the specific heat values that we derived. They apparently act like our simple model. However, the gases that are diatomic or that have even more complicated molecules apparently require a more complicated model.

It must be that energy fed into molecules can be absorbed in some way other than translational kinetic energy. This is not surprising! Two atoms bound together ought to be able to rotate around their center of mass and also ought to vibrate back and forth as if they were held by a spring. Each mode of motion should be able to absorb energy. Perhaps the internal energy U should consist of $N\tfrac{1}{2}\overline{mv^2}$ (translation) + $N\tfrac{1}{2}\overline{I\omega^2}$ (rotation) + $N\tfrac{1}{2}\overline{kx^2}$ (vibration). The problem is, how does the energy divide up among the various modes? The translational kinetic energy of the whole molecule can be shared by motion in three perpendicular component directions. Surely in a gas the energy must be *equally* shared among these three directions. A diatomic molecule can rotate like a dumbbell around two axes perpendicular to its length. There should be as much energy used for one rotation as the other. The molecule can also spin around its longitudinal axis. Should that spin share in the energy distribution? As the diatomic molecule vibrates, it has both kinetic and potential energy; on the average there is an equal amount of energy in the kinetic and potential modes.

In the classical theory of mechanics (before quantum mechanics was known), it was assumed that the principle of *equipartition of energy* holds for all these energy modes, providing that the system can freely interact with all its components for a long enough time. Each such mode is called a "degree of freedom." For instance, translational motion has three degrees of freedom since motion can be broken down into three independent directions, such as the x, y, and z axes. Vibrational motion along a single axis has two degrees of freedom—for kinetic and potential energies. The principle held that in a system of many interacting objects, where all these motions could influence each other and share energy, each degree of freedom would, on a time average, share equally in the total energy. Not only would $\tfrac{1}{2}\overline{mv_x^2} = \tfrac{1}{2}\overline{mv_y^2}$ for the same molecule, but also $\tfrac{1}{2}\overline{mv_A^2} = \tfrac{1}{2}\overline{mv_B^2}$ for two different molecules in the same gas. Furthermore, $\tfrac{1}{2}\overline{I_A\omega_A^2} = \tfrac{1}{2}\overline{m_B v_B^2} = \tfrac{1}{2}\overline{k_c x_c^2}$ for all combinations of motions of the different molecules interacting with each other. The principle of equipartition of energy was a cornerstone of classical mechanics in the nineteenth century, even though the truth of the principle is not immediately obvious, nor easy to prove.

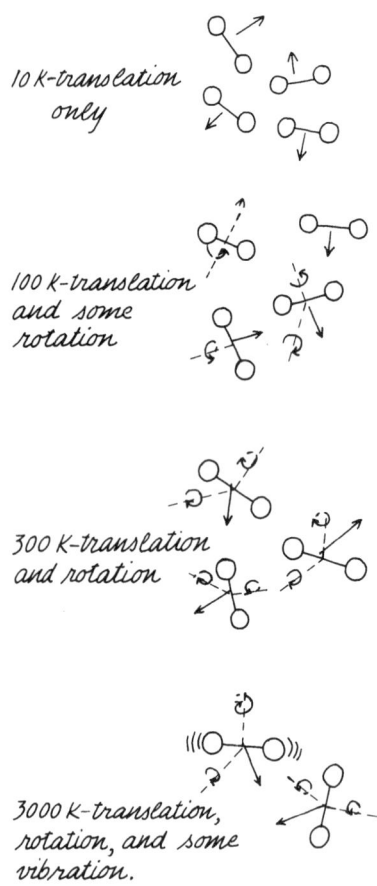

Figure 10.16. Types of motion energy possible at increasing temperatures for a diatomic molecule.

If there is really equipartition of energy in a gas, each of the possible modes of motion of molecules should share equally in the heat energy fed into the gas. It is at this point that our simple classical model begins to break down. The next step of explanation suddenly requires quantum mechanics. When we look at the experimental data for the specific heats at constant volume of diatomic molecules, we see that at room temperature H_2, O_2, N_2, and HCl apparently have five degrees of freedom. (Note that each degree of freedom is responsible for a contribution of 1 cal/mol K to the specific-heat value. (The happenstance that the value in calories is approximately 1 is the main reason for still using calories.) From other experiments and from quantum mechanics calculations, we know that the two extra degrees of freedom of these diatomic molecules are related to the possible rotation around two axes. It is reasonable enough that as the molecules run into each other, they set each other spinning.

The heat capacity of even the simplest molecule, H_2, cannot be explained on the basis of classical mechanics. As you can see from Fig. 10.9, it appears that at low temperatures some of the degrees of freedom of the hydrogen molecule are frozen out. Since the molecules are surely free to move about in space, it must be that for some reason their rotational and vibrational modes cannot share in the energy distribution. There appears to be a threshold effect with new degrees of freedom gradually becoming possible as the temperature (and energy) rises. That is just what happens. Rotations can only occur with certain discrete amounts of energy, because changes of angular momentum can occur only as integral multiples of Planck's constant. The minimum amount of energy needed for *vibrations* is usually even larger. The equipartition of energy theorem assumes that any mode of action can have *any* amount of energy, including very small amounts, and that the energy in that mode can change continuously. Since rotational and vibrational energies are quantized, they cannot share in the division of energy until the available energy in a collision is greater than the quantum threshold. In Fig. 10.16 we illustrate the types of motion energy possible at increasing temperatures for a diatomic molecule.

To understand why rotational and vibrational modes are frozen out at low temperatures, let us calculate the energies involved, subject to the quantum conditions. For rotations, $E_{rot} = \frac{1}{2}I\omega^2 = \frac{1}{2}L^2/I$, where L is the angular momentum $= n\hbar$, an integral multiple of Planck's constant. For the hydrogen molecule, with each atom having a mass of 1.67×10^{-27} kg, and assuming a molecular radius of 1×10^{-10} m:

$$E_{rot} = \frac{1}{2} \frac{(1 \times 10^{-34})^2}{2(1.7 \times 10^{-27})(1 \times 10^{-10})^2} = 1.5 \times 10^{-22} \text{ J} = 9 \times 10^{-4} \text{ eV}.$$

This energy corresponds to a temperature of

$$T = \frac{E}{k} = \frac{9 \times 10^{-4} \text{ eV}}{8.6 \times 10^{-5} \text{ eV/mol K}} = 10 \text{ K}.$$

Evidently the temperature and average energy must be even higher for a complete equipartition of energy with the rotational mode. Note the inverse dependence of the rotational energy on the moment of inertia. For more massive or larger diatomic molecules, the threshold energy for rotation is less and $c_V = 5$ cal/mol K starting at much lower temperatures. We can also see one reason why the diatomic molecule does not spin on its axis—or for that matter why the helium *atom* cannot share rotation energy by spinning on its axis. The moment of inertia of such system would involve a nuclear spin radius, smaller than molecular by a factor of

10^5. This would increase the threshold energy and temperature by a factor of 10^{10}. (Atomic nuclei do spin, involving energies in the MeV range.)

Changes in the vibration energy of a molecule must occur in multiples of $h\nu$, where ν is the vibration frequency. For hydrogen, with a reduced (center of mass) mass of $\frac{1}{2}(1.67\times10^{-27}$ kg), and a spring (binding) constant of 30 N/m:

$$E_{vib}=h\nu=(6.6\times10^{-34} \text{ J/Hz})\left(\frac{1}{2\pi}\sqrt{\frac{30}{0.83\times10^{-27}}} \text{ Hz}\right)=2\times10^{-20} \text{ J}=0.1 \text{ eV}.$$

This energy corresponds to a temperature of 1400 K. Indeed, above 1000 K, c_V begins to rise.

Solids

The molar specific heats of most metals have about the same value at room temperature: 6 cal/mol K. According to our simple model there must be six degrees of freedom for each atom. How can this be? Each atom is locked in place in the metallic crystal. It can oscillate, of course, and in three independent directions. These modes could provide three degrees of freedom. However, each mode of oscillation can absorb twice the energy allotted to an unbound translation motion. One unit of energy goes to the kinetic energy of oscillation, and the other unit goes to the potential energy. Unfortunately, the classical theory has trouble explaining why the specific heats of metals should depend on temperature. As you can see in Fig. 10.8, the regularity that we observe occurs at room temperature and even then only if we ignore the elements that are low in the periodic table.

We can make a first approximation to understanding the temperature dependence of the specific heats of solids by appealing to the formula for quantized vibration energies. In this semiclassical approach, we assume that each atom is vibrating in its potential well *and that these vibrations do not affect its neighbors*. Einstein first worked out a theory of specific heats of solids making this assumption. The energy levels of an atom bound in a parabolic well are evenly spaced by an energy $h\nu$. Let us calculate this energy for lead, which has an atomic mass of 3.5×10^{-25} kg and a "spring" constant of only 3 N/m:

$$E=h\nu=h\frac{1}{2\pi}\sqrt{\frac{k}{m}}=\frac{6.6\times10^{-34}}{2\pi}\sqrt{\frac{3}{3.5\times10^{-25}}}=3\times10^{-22} \text{ J}=2\times10^{-3} \text{ eV}.$$

This energy corresponds to a temperature of 20 K. The vibration modes of lead can be excited and share in thermal energy at low temperatures; therefore the specific heat rises rapidly to the full value of 6 cal/mol K. At the other extreme, consider diamond. Not only does carbon have an atomic mass smaller than lead by a factor of 17, but the spring constant of diamond is larger than for any other material. Consequently the quantized energy steps are very large and the specific heat of diamond approaches the full value only at temperatures above 1000 K.

There is an extra problem in explaining the specific heats of metals. According to our semiclassical model, we might expect that the conduction electrons could share thermal energy. Each such electron, acting like a gas particle, ought to be able to absorb $\frac{3}{2}kT$, adding an extra 3 cal/mol K to the specific heat of the metal. The actual contribution to the specific heat by electrons is very small. They can spread thermal energy but they absorb almost none for themselves. The electrons occupy the energy levels in pairs, filling up the valence band to a height of about 10 eV (the Fermi level). An electron down in the band cannot accept small amounts of thermal energy because the slightly higher energy level to which it would rise is already occupied. Only the few electrons within kT ($\frac{1}{40}$ eV) of the Fermi level can accept thermal energy.

It is revealing to calculate the specific heat of a metal per *atom*: 6 cal/mol K=25 J/mol K=1.6×10^{20} eV/mol K=2.6×10^{-4} eV/atom K. If the specific heat were constant all the way from 0 K to room temperature at 300 K (a good approximation for lead), the average thermal energy of each atom would be 7.8×10^{-2} eV= $\frac{1}{13}$ eV, and the average kinetic energy of oscillation would be half that, or $\frac{1}{26}$ eV. This value is the same as the average kinetic energy of gas molecules at 300 K. If we were to do the same calculation for diamond, we would have to use smaller values of specific heat, especially at low temperatures. From Fig. 10.8 it appears that the average value of specific heat for diamond from 0 to 300 K is less than $\frac{1}{2}$ cal/mol K. Therefore the average thermal energy per atom is less than $\frac{1}{150}$ eV and the average kinetic energy would be less than $\frac{1}{300}$ eV. But if the diamond is in temperature equilibrium with a gas at 300 K, in some way the average kinetic energy of the diamond atoms must be $\frac{1}{26}$ eV, the same as that for the gas molecules. The resolution of the paradox is that most of the diamond atoms are not sharing in the thermal energy distribution. The quantum thresholds require a different distribution with many of the atoms in low-lying levels. The resulting distribution does have an rms average of $\frac{1}{26}$ eV, but this is not the average value obtained by dividing the total kinetic energy of oscillation by the total number of atoms.

There are special problems involved with explaining the specific heat of water. We usually think that water has a very high specific heat compared with other substances, and it does—per gram. Molten lithium, because of its low mass, has a higher gram specific heat than water, but it is awkward to use molten lithium in hot-water bottles. However there are many other substances with larger *molar* specific heats. The molecules of these substances are usually large and can absorb energy in many forms of internal vibration and rotation. With water, the molecules are lumped together with polar and hydrogen bonding into large clusters. These clusters contain thousands of molecules at room temperature, with molecules continually joining and breaking off. It takes energy to free a molecule from the cluster binding. As energy is fed into the water it breaks up the clusters so that at 100 °C the average number of molecules per cluster is reduced almost to 1. The remarkable feature of the specific heat of water is not that it is large, but that it is almost constant from 0 to 100 °C.

Why does "live" steam at 100 °C produce more serious burns than water at 100 °C? At one level, the answer is obvious. As one gram of steam turns into one gram of water, it gives up 540 cal. At the molecular level, the explanation is a little trickier. The average kinetic energy of translation of the vapor molecules at 100 °C is the same as the average kinetic energy of oscillation of the water molecules at 100 °C. Loosely speaking, the vapor molecules do not "slow down" as they condense. However, a vapor molecule does fall into a deep potential well when it joins the condensed state. That means that it drops to negative energy, and to do so must give up positive energy. It gets that energy by falling into the well. That energy must be taken up by the rest of the liquid into which the molecule condenses. Note that if the temperature of the liquid is 100 °C, it cannot take up any energy from a molecule that is also at 100 °C, and therefore there cannot be any net condensation. (Because of the energy distribution of the molecules, some molecules are continually joining the liquid while others leave. We are talking about the average effect.) So as vapor molecules condense they do not give up their original kinetic energy. The energy that they release is obtained from their fall into the potential well. Conversely, when they escape to the vapor, the evaporation energy lifts them out of the well but does not provide extra kinetic energy of translation. One of the corollaries of this situation is that two vapor molecules at energies corresponding to 100 °C cannot bind together. If they fell into each other's potential well, there would be no agency to take up the released energy. This is different from the nuclear merger of a proton and neutron. In that case the released energy is carried off by a photon.

■ HEAT TRANSMISSION

In general science courses, everyone learns that the three methods of heat transfer are conduction, radiation, and convection. In the first method, two systems are in physical contact and thermal energy passes from one to the other because of molecular vibrations or electron excitations (for instance, a pot on the hot plate of a stove). In the second, electromagnetic radiation is emitted by all objects, and heat loss of an object due to radiation depends on the difference between radiation emitted and radiation absorbed from other sources (for instance, the earth cooling off on a cloudless night). In the third, an intermediary system absorbs heat from one system and then physically moves to another system to which it delivers the thermal energy (for instance, a home hot water heating system).

Newton experimented with heat flow. His name is attached to the primary equation for conduction through a system. Heat flow is proportional to the temperature difference across the system, proportional to the area of the conducting path, and inversely proportional to the thickness of the conductor:

$$\frac{Q}{\Delta t} = k \frac{A \Delta T}{x}.$$

Newton's equation can be written in an Ohm's law style, with the time rate of thermal energy passage corresponding to electric current; ΔT corresponding to voltage V, and with a thermal "resistance" equal to x/kA. There must be a word of caution about the measurement and application of ΔT. Unless special precautions are taken, the temperature difference across an insulation barrier (or heat conductor) is not necessarily the temperature difference between the source of heat on one side and the colder "sink" on the other. For instance, the temperature difference across a single-pane window in a room at 20 °C and the air outside at −10 °C is probably not 30 °C. If there is Jack Frost on the inside of the window, the temperature of the glass on the inside must be below freezing. As with all measurements in thermodynamics, it is difficult to establish well-defined systems. For another example, the popular measurement of the cooling curve for a cup of coffee (with cream added sooner or later) is very susceptible to the drafts in the room. If there is no air movement around the cup, a major share of the insulation will consist of the very thin air film clinging to the outer surface. Of course, it is hard to maintain a hot cup without convection air currents around it. For yet another example, an egg will cook faster in vigorously boiling water than it will in a pan with water at 100 °C—same temperature of the hot bath, but not the same thermal contact and therefore not the same ΔT across the shell.

The most common use of Newton's equation in everyday life is in calculating the insulation needed in house construction. Building suppliers provide information about insulation properties of materials in terms of R values. An R value is given for a particular thickness of material, such as one pane of glass, or 2 in. of styrofoam. In terms of our previous definition of thermal resistance, $R = \Delta x/k$, where Δx is the specified thickness of the material. To calculate the effect of combining insulators, we can by analogy make the same arguments as we do for combining electrical resistors, since the heat conduction equation has the same form as Ohm's law. Therefore, if one panel of insulation has an R value of 6, a sandwich of two panels of the same material (thus with twice the thickness) will have an R value of 12.

To calculate the heat loss through a wall, we must take the surface area into account. In terms of R, Newton's formula becomes

$$\frac{Q}{\Delta T} = \frac{A}{R} \Delta T.$$

If there are two panels of the same material side by side, with R values of R_1 and R_2, the heat loss through a wall with total area A will be

$$\frac{A}{R_{\text{total}}} \Delta T = \left(\frac{A_1}{R_1} + \frac{A_2}{R_2}\right) \Delta T.$$

It appears that combining thermal insulation values for adjacent panels follows the rules for combining electrical resistances in parallel, except that we must take into account the surface areas. If the two panels are the same size and made of the same material, then $A/R_{\text{total}} = (R_2 A_1 + R_1 A_2)/R_1 R_2 = R(A_1 + A_2)/R^2$. since $R_1 = R_2 = R_{\text{total}}$, and $A_1 + A_2 = A$. This result seems reasonable: if you use panels, each of which has $R = 6$, then the whole area has an R value of 6. However, note what happens if one section of the wall has a much lower R value than the rest, perhaps because there is a single-pane window, or a crack under the door. A single pane window has an R value of less than 1. Suppose you thoroughly insulate an outside wall with fiberglass having an R of 20. Suppose the windows have an area of 1/5 that of the wall. The effective R of the combination will be given by

$$\frac{A}{R_{\text{total}}} = \frac{\frac{4}{5}A}{20} + \frac{\frac{1}{5}A}{1} \approx \frac{1}{4} \rightarrow R_{\text{total}} = 4.$$

The window dominates the magnitude of the heat flow. Note that it would do little good to increase the thickness of the wall insulation.

In Table 9 we give the thermal conductivity k for various materials in both W/m K and BTU in./h ft^2 °F. The American system of units may seem very complicated, but they correspond to the units actually used when you buy building materials. Note the standard units for R values given in Table 10: h ft^2 °F/BTU.

As an example of the complexity of real world calculations of thermal phenomena, consider the listed R value of a single pane of window glass, which is about equal to 1. It seems reasonable that a double-pane window would have twice that R value. Actually, it does not under most conditions. Consider that the value of k for glass is about 5 BTU in./h ft^2 °F. For a $\frac{1}{4}$-in. pane of glass, the R value should be $\Delta x/k = 1/20$. We might as well have used a pane of aluminum! It isn't the glass that provides the insulation of a window: It is the barrier against air movement and the provision of two flat surfaces on which there are thin films of air. The air films do the insulating. With a single-pane window, the outside surface is often subject to wind which destroys the air film, leaving only one film on the inside. With a double-pane window, there are four air films. Even if the outside is

TABLE 9. *Thermal conductivities for various materials.*

	k in W/m·K	$k = \dfrac{Q}{\Delta t}\dfrac{x}{A}\dfrac{1}{\Delta T}$ k in (Btu·in)/(h·ft². °F)
Air (room temp)	0.026	0.18
Water (27 °C)	0.609	4.22
Ice	0.592	4.11
Aluminum	237	1644
Copper	401	2780
Iron	80	558
Silver	429	2980
Wood (maple)	0.16	1.1
Wood (pine)	0.11	0.78
Brick	0.7	5
Concrete	1.1	7
Glass	0.8	5
Glass wool	0.042	0.29
Plaster	0.5	3

eliminated, there are still three left, making the double pane three times as effective as a single pane.

We can calculate an approximate value of the thickness of the insulating air film on a surface by assuming that the total insulation is due to the air. Since the R value for a single pane is about equal to 1, the value for the film on each side must be about 0.5. From Table 9 we see that the value of k for air is 0.18 BTU in./h ft² °F. Then

$$R = 0.5 = \frac{\Delta x}{0.18} \rightarrow \Delta x = 0.09 \text{ in.} \approx 2 \text{ mm.}$$

Increasing the separation distance between window panes to more than about 1 cm is counterproductive. With a larger spacing, convection currents will be set up, destroying the thin air films.

As you can see from Table 9, air has the lowest heat conductivity of any of the common materials. That is why insulators made of materials that trap small pockets of air are so good—fiberglass, wool, styrofoam, goose down, etc. Materials with the largest values of heat conductivity are the metals, where the heat conduction is provided by electron motion. There is one exception, and that is diamond, which is an electrical insulator, but at room temperature has the highest thermal conductivity known—2000 W/m K. (See p. 250.) In the case of diamond, heat conduction occurs because of the high speed of molecular vibration waves (sound). The speed is high because the binding force between carbon atoms in the diamond crystal is very large and the mass of each atom is small (compared with most metallic atoms).

All objects radiate electromagnetic radiation. The emitted power (in watts) depends strongly on the absolute temperature: $I = e\sigma A T^4$. The area of the emitting surface is A, and σ is a proportionality constant, named for Josef Stefan who determined its value: $\sigma = 5.67 \times 10^{-8}$ W/m² K⁴. The factor e is a fraction between 0 and 1, depending on the nature of the surface. A good emitter of radiation is necessarily also a good absorber. For a perfect absorber (absolutely black), $e = 1$.

TABLE 10. R values for various building materials.

Material	Thickness	$R=\frac{\Delta x}{k}$ in (h·ft²·°F)/Btu	$R=\frac{1}{k}$ per inch in (h·ft²·°F)/(Btu·in)
Sheetrock	0.375	0.32	
Plywood	0.5	0.62	
Particle Board			1.06
Carpet			2.08
Tile		0.5	
Hardwood	0.75	0.68	
Insulating blanket	3	11	
Mineral Fiber	6	22	
Asphalt shingles		0.44	
Windows			
Single-pane		0.9	
Double-pane		1.8	
Triple-pane		2.7	

For a perfect reflector, $e=0$. If the fraction absorbed by a surface and the fraction emitted were not identical, you could arrange a geometry where two objects were exchanging radiation without loss to other objects (for instance, each could be at the focus of a reflecting ellipsoid). The object that could emit better than it absorbs would get colder while the other got hotter, thus violating the second law of thermodynamics.

Newton's law of cooling for conduction specifies that heat flow is proportional to the first power of temperature difference across the heat conductor: $Q/\Delta t \propto \Delta T$. It appears that this may not be true for heat transfer due to radiation, which depends on T^4. If you place a cup of coffee on a table, it radiates heat at its temperature, T_h, and absorbs energy from its surroundings at their temperature, T_c. The net heat transfer must be $Q/\Delta t \propto T_h^4 - T_c^4$. For many laboratory purposes, including monitoring the temperature of coffee, $T_h - T_c$ is small compared with either T_h or T_c. Then we can factor:

$$T_h^4 - T_c^4 = (T_h^2 + T_c^2)(T_h^2 - T_c^2) = (T_h^2 + T_c^2)(T_h + T_c)(T_h - T_c).$$

For $T_h \approx T_c$, we can use the mean temperature in the sums. Then, $Q/\Delta t \approx T^3 \Delta T$, and Newton's law of cooling holds approximately for $\Delta T \ll T$.

There is an interesting peculiarity in the radiation formula: the emitted power is proportional to the *exact* fourth power of the temperature. When functional dependence of one variable on another involves integral powers there is usually geometry involved. For instance, the inverse-square laws of gravity and electrostatics involve the three-dimensional nature of space and the fact that the surface area of a sphere is proportional to r^2. But what does temperature have to do with geometry? The relationships were worked out a century ago and are hallowed by such names as blackbody radiation, the ultraviolet catastrophe, the Stefan–Boltzmann law, the Rayleigh–Jeans formula, and finally Planck's quantum hypothesis.

The problem is this. Inside a cavity there must be electromagnetic radiation in temperature equilibrium with the walls. However, these radiation waves can have only certain frequencies, even as sound waves in an organ pipe can exist only as standing waves with wavelengths related by integer fractions of the length of the pipe. For a pipe closed at both ends, the first such mode is the fundamental with $\lambda = 2L$, the next with $\lambda = L$, etc. There is an infinite number of such modes, but the number *within a given frequency interval* remains constant in a one dimensional pipe. For a three-dimensional cavity, the number of frequency modes *within a frequency interval of $d\nu$* is proportional to ν^2. (The frequency is ν-nu.) Within a frequency interval of 1 kHz, there are 10^6 more modes at 10^9 Hz than there are at 10^6 Hz.

In classical thermodynamics, there was a fundamental theorem of *equipartition of energy*. If systems interact with each other, then on the average each possible mode (degree of freedom) must have the same energy. Consider the consequences, however, if this were true with the electromagnetic radiation in a cavity. Since there is an infinite number of modes, and their density increases as ν^2, if each mode has the same average energy there would be an infinite amount of energy in the short wavelengths—the ultraviolet catastrophe.

Planck resolved the issue by proposing that the energy of electromagnetic radiation must depend on the frequency: $E = h\nu$. This requirement violates the equipartition of energy, because a higher-energy mode (a higher frequency) could not be excited by interaction with the low-frequency modes that have low energy. The density of modes and the dependence of density on frequency remains unaltered; it is just that the higher modes cannot be excited. The number of modes in each frequency interval is

$$N(\nu)d\nu = \frac{8\pi\nu^2}{c^3} d\nu.$$

Planck's specification for the average energy of a given mode is

$$\bar{\epsilon} = \frac{h\nu}{e^{h\nu/kT} - 1}.$$

The energy density of the spectrum therefore is

$$\rho_T(\nu)d\nu = \frac{8\pi\nu^2}{c^3} \frac{h\nu}{e^{h\nu/kT} - 1} d\nu.$$

For the total energy emitted, the density must be integrated over all frequencies but the exponential in the denominator rapidly cuts off the contributions for $h\nu \gg kT$. By making substitutions ($x = h\nu/kT$), we can fit this integral to a standard form:

$$\int_0^\infty \frac{x^3 dx}{e^x - 1} = \frac{\pi^4}{15}.$$

The total radiation is

$$R_T = \int_0^\infty \frac{2\pi h}{c^2} \frac{\nu^3 d\nu}{e^{h\nu/kT} - 1} = \sigma T^4,$$

where σ is Stefan's constant,

$$\sigma = \frac{2\pi^5 k^4}{15 c^2 h^3}.$$

The mysterious integral fourth-power dependence of electromagnetic radiation on temperature is indeed due to geometry. In this case it is the geometry of fitting standing waves into a three dimensional cavity. The density of modes times the frequency-dependent modal energy yields a total emitted energy proportional to T^4.

To demonstrate the three kinds of heat transfer, it is hard to beat the use of an ordinary thermos bottle that can be disassembled. The inner container touches the outer container over only a small area, and these regions are made of poor conductors such as plastic or glass. Therefore there is very little contact heat conduction. The inner chamber is a double-walled glass container with a vacuum between the walls. (You can usually see the glass nipple that sealed off the vacuum.) Consequently there can be no convection from outer to inner wall, quite different from the situation with double-pane windows. Finally, the glass bottle is silvered both inside and out, thus making the surface a very poor absorber or emitter of radiation.

■ RECAPITULATION—WHERE DID THE ENERGY GO?

When you rub your hands together, or strike a match, where does the mechanical energy go? It spreads out into the microstructure of the material, but into the same forms that we study in the first part of the introductory course of physics. There is kinetic energy of translation, vibration, and rotation. There is potential energy due to electromagnetic binding. The vibration energies are quantized because the wave functions of objects in potential wells must satisfy boundary conditions. For parabolic wells, the energy levels are evenly spaced, with $E_0 = \frac{1}{2}h\nu$ and $\Delta E = h\nu$. The vibration frequencies are given by $(1/2\pi)\sqrt{k/m}$, and values for ΔE lie in the range 0.01–0.1 eV. The rotation energies are quantized because changes of angular momentum must be integral multiples of Planck's constant. Typical values for ΔE_{rot} are in the range 10^{-3}–10^{-4} eV.

Now that we have explored the ways internal energy can be stored in materials, we can take another look at the definition of temperature. We know that temperature is an *intensive* quantity. Like *density* it does not depend on the amount of material. The total internal energy of a piece of material does depend on the amount of material. It is an *extensive* property. Temperature is somehow related to the average internal energy possessed by each unit in the material. In the case of an ideal gas, the internal energy is equal to the total kinetic energy of translation of the atoms. The temperature is just proportional to the *average* kinetic energy: $\overline{E_{\text{kin}}} = \frac{3}{2}kT$. Why not, therefore, set the proportionality constant equal to 1, and define temperature as the average kinetic energy of a unit of the material? As a matter of fact, in advanced texts in thermodynamics, Boltzmann's constant is frequently assigned the value unity, and temperature then has the units of energy. In this system, entropy is dimensionless, since it is the ratio of energy to temperature, or is proportional to Boltzmann's constant and a logarithm.

With this system, however, we must be careful to define the kind of average energy per unit that counts. As we have seen, if the units of the gas are molecules, they may have internal motions of rotation or vibration. Worse yet, these motions are quantized and are not excited unless the interaction energy is above a threshold. Look again at the graph of specific heat of hydrogen on p. 234. The average kinetic energy corresponding to temperature is any kind of motion energy—translation, vibration, rotation—that can be communicated to the other units of the system. Each type of motion shares equally in the thermal energy, providing that the interaction energy is well over any quantum threshold. Therefore, the temperature corresponds to the average kinetic energy of any one of the modes—not the sum of the kinetic energies of the several modes. For instance, at room temperature, the two rotation modes of the hydrogen molecule are fully activated. The same average energy is possessed by each of the five modes of motion—three translational and two rotational. If hydrogen is in contact with a piece of metal, the average kinetic energy of any one of these modes of the gas atoms must be equal to the average kinetic energy of one of the three modes (x,y,z) of vibration of the metal atoms bound in their crystal potential wells.

To point out the problem of defining temperature as the average kinetic energy of the units of the system, consider that within any metal there is another and completely different system of particles in temperature equilibrium with the ions of the lattice—the electrons. Their energies do not follow the Maxwell–Boltzmann distribution. Instead, the conduction electrons (one or two from each ion) fill up quantum levels pair by pair with their spins opposed. The energy of each level increases by a minute but discrete amount. In a conductor there are empty levels available for the electrons at the high energy top of the distribution. This is called the Fermi level and has an energy of several electron volts. Except for the electrons in the conduction band, the rest are not free to accept the small amounts of thermal energy involved in room temperatures of $\frac{1}{25}$ eV, and therefore do not contribute to the heat capacity. (The whole distribution of electrons does shift with an applied electron field, however, and so all the conduction electrons take part in electric current.) So the two populations are in thermal equilibrium with each other but have very different energy distributions—Boltzmann for the ions, Fermi–Dirac for the electrons. When the conduction-electron distribution is in thermal equilibrium with an ion lattice of 300 K, the average energy of the electrons is half that of the Fermi level, or around a couple of electron volts. If that average energy were to be equated to temperature by using $\overline{E_{\text{kin}}} = \frac{3}{2}kT$, the electron temperature would be 15 000 K!

There is one other problem in accounting for thermal energy that is distributed among many components. If you lift a pendulum bob in a gravitational field, you can get the energy back in the form of kinetic energy of the whole bob. If you feed the energy into a vast number of interacting systems, you will usually need a mechanism for retrieving the energy into a coherent arrangement. The problem of bringing order out of chaos is the subject of the second law of thermodynamics.

11

Second Law of Thermodynamics

There is no trouble at all in turning "mechanical" energy into heat or into internal energy. Simply let a block slide down an incline at roughly constant velocity. Friction drains the gravitational potential energy, raising the temperature of the surfaces and causing heat to flow to the cooler surroundings.

Now let us try to reverse the process. Can we let heat flow into the block and make it slide back uphill? This is not so easily done, of course. The original organized motion of the block as a whole has been turned into the random chaotic motion of the molecules in the block and in the inclined plane.

There are ways to produce organized effort out of chaotic motions. In a balloon the gas molecules are moving equally in all directions and each one changes velocity as it runs into another one. Nevertheless, if the nozzle of the balloon is opened, the chaotic motions are turned into an organized surge of the gas out of the nozzle while the balloon goes in the other direction. This conversion of chaos into organization is caused by a selection process. Those molecules that happen to have a velocity in the direction of the nozzle escape. Is there some similar way to change the chaotic internal motions produced by heat into the organized motion that does work?

The first law of thermodynamics defines heat as a form of energy and asserts that heat shares in the energy conservation law: $Q = W + \Delta U$. *The heat entering a system equals the sum of the external work done by the system on its surroundings and the change in internal energy of the system.* The second law of thermodynamics can take several forms and is concerned with very basic principles about the nature of our universe. The original interest in the law, however, arose from the practical concern of getting more work done when heat enters a system. The Industrial Revolution came about because people learned how to extract work from heat. Heat engines turn most of the generators of our electric power plants and also propel most of our vehicles. The question of the efficiency of these machines is obviously of prime importance.

For purposes of analysis, we can study any kind of a system that takes heat from a source at a particular temperature and uses it to perform work. The question is: Does all of the heat Q turn into useful work W? If so, the conversion efficiency is 100%. Such an ideal machine would have to be friction-free. It would also have to operate in a way that is technically called "reversible" (in thermodynamics, reversibility of a process corresponds to a friction-free situation in everyday mechanics). In a reversible process involving heat, everything happens slowly and uniformly so that the temperature and pressure of the material used in the process is uniform throughout. For instance, in a cylinder of gas with a piston, a reversible expansion is one where the piston is moving so slowly that there are no shock waves in the gas, and the temperature and the pressure of the gas at any instant are the same throughout the whole volume. Why is a process that satisfies these conditions called reversible? Suppose heat is provided to the gas in a cylinder, forcing it to expand slowly. The piston will be shoved out, performing useful work. Now if the process can be truly reversed, it should be possible to shove the piston back, using the same amount of work, compressing the gas, and driving heat out of the cylinder. The process is indeed possible in principle, providing that there is no friction between piston and cylinder and that shock waves in the gas do not spread the energy through the walls of the cylinder instead of doing work on the piston. An explosion of gasoline and air in the cylinder of a car engine is not a reversible process, since, of course, when the piston moves back in the cylinder, it cannot restore the exhaust gases to their original form of gasoline and air.

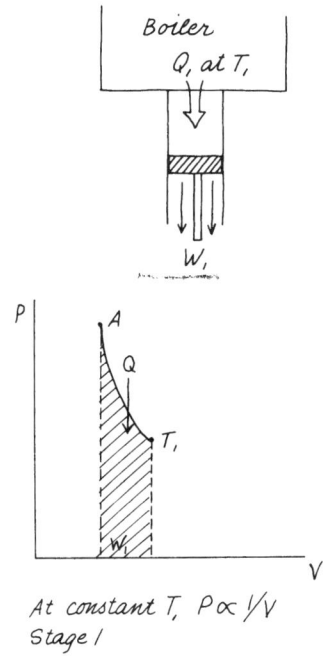

At constant T, $P \propto 1/V$
Stage 1

Figure 11.1. In an ideal isothermal expansion, $Q = W$.

The simplest ideal heat engine that we can study consists of an ideal gas in a cylinder with a piston. We could use other materials that expand when heat is provided but for an ideal gas we already know the equations. At first our task seems very simple. Let heat flow into the gas in the cylinder and the gas will expand. The piston will be shoved out, exerting a force through a distance, thus doing a known amount of work. We can make the problem even simpler by sending the heat in at constant temperature. We showed how to do this on p. 236. A metal wall of the cylinder could be fastened to the boiler. The piston is then allowed to move so slowly that the gas is maintained at the temperature of the boiler. The process is called *isothermal*. Since the temperature of the system remains constant, $\Delta U = 0$, and $Q_1 = W_1$.

It might seem that with this process we have extracted order out of chaos with an efficiency of 100%. The heat that left the boiler, Q_1, turned into the work done by the piston, which corresponds to the area under the curve on the $P-V$ graph of Fig. 11.1. No energy was absorbed by the gas itself since its temperature did not change. Unfortunately, we cannot repeat the process using the same cylinder because the piston is now shoved partly out. We could get a new cylinder and piston, but that would not be a practical way to manufacture a heat engine. The thing to do is to shove the piston back to its original position. Then we would have a cyclic process, which could be repeated over and over again. The problem is, how can we shove the piston back without using up the same work that we just gained? The isothermal process that we used was by definition reversible. If the same amount of work is done on the piston, then the gas can be compressed at constant temperature back to its original small volume and high pressure. In the process, heat would be driven out and would be absorbed back by the boiler. We would be right back where we started from with no net heat extracted and no net work done.

Here is a way to produce a cyclic process that will restore the cylinder, the piston, and the gas to their original conditions and yet not retrace the original steps. After Q_1 has been absorbed from the boiler in the isothermal process, insulate the cylinder from the boiler so that no heat can enter or leave the system. Then let the gas continue to expand, producing *more* outside work. The energy will be drawn from the internal energy of the gas as it drops in pressure and also in temperature.

$$Q = 0 = W_2 + \Delta U \rightarrow W_2 = -\Delta U.$$

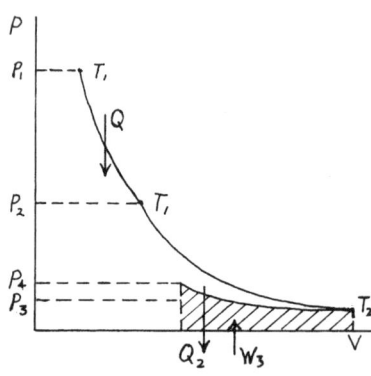

Figure 11.2. To restore the heat engine after the expansion, we must do work. In the third stage the compression is isothermal, driving out Q_2 at temperature T_2. $W_3 = Q_2$.

This second part of the cycle, which takes place without heat entering or leaving, is called an adiabatic process. Note that there is nothing sacred about our choice of using first an isothermal process followed by a completely adiabatic process. Each of the steps could have been partially isothermal and partially adiabatic or neither. The only requirement is that the expansions take place slowly enough so that they are reversible. The choice of pure isothermal and adiabatic processes does more than merely make the calculations easier. As we shall see, the efficiency of the heat engine depends on the temperatures of the heat source and sink. By starting out with an isothermal expansion we can provide Q_1 at as high a temperature as possible.

Notice, incidentally, that at the conclusion of the first two steps of expansion, the system has produced work, $W_1 + W_2$, but has used only Q_1. Since $W_1 = Q_1$, it appears that the efficiency of our heat engine so far is

$$(W_1 + W_2)/Q_1 = (W_1 + W_2)/W_1 = 1 + W_2/W_1 > 1.$$

It appears that we have obtained something from nothing! However, to repeat the process we would have to buy a new cylinder, unless we can restore the original conditions.

After the first two stages, our heat engine consists of a piston shoved out as far as it can go and the gas in the cylinder at low pressure and low temperature. If we are going to keep on using the heat engine, we must get the gas and the cylinder back to their original condition. Once again, for the sake of easy calculation, the easiest steps to consider are isothermal and then adiabatic. For the third step *we* must do the work by shoving the piston back, compressing the gas from pressure P_3 to P_4, as shown in Fig. 11.2. However, if this compression is to take place at constant temperature, the heat must be *expelled* to some reservoir at that temperature. We could do this by connecting one end of the cylinder through a metal wall to a low-temperature bath or radiator. The amount of heat expelled, Q_2, must be just equal to the work that we have to do to shove the piston in. On the $P-V$ graph, that work corresponds to the area under the compression curve.

The fourth stage of the cycle is once again adiabatic with the cylinder walls insulated so that no heat can enter or leave. We must do an amount of work, W_4, to shove the piston back to its original position. All of this work shows up as the internal energy of the gas. Its temperature rises. With this fourth stage of the cycle, we have restored the piston to its original position and the gas to its original pressure, volume, and temperature. The energy balance is as follows: an amount of heat, Q_1, was extracted from the boiler at temperature T_1. The work done *by* the piston was $W_1 + W_2$. In restoring the heat engine to its original condition, we had to do an amount of work, $W_3 + W_4$, *on* the piston. It was also necessary to expel Q_2 at the lower temperature, T_2. The net work done by the heat engine was $W = W_1 + W_2 - W_3 - W_4$. As required by the first law of thermodynamics, this net work done must be just equal to the net heat used: $Q_1 - Q_2 = W$. No energy is really lost, but, on the other hand, we were not able to convert all the original heat energy to useful work. Furthermore, the expelled heat, Q_2, is only available at a lower temperature.

Typically, in an electric power plant, this heat is expelled at the temperature of the local cooling water provided by a river, lake, or evaporation tower. In most cases the heat is wasted, although in some plants it is sent out at temperatures high enough to heat buildings. Nevertheless, as far as being useful for producing mechanical energy, the heat has been degraded below a useful level. The conversion efficiency of our ideal heat engine is equal to the net work done divided by the original heat energy used, Q_1. Since the mechanical work done is equal to Q_1-Q_2, the efficiency is equal to $(Q_1-Q_2)/Q_1$.

Most students react to this analysis with the uneasy feeling that something is being put over on them. A very particular sequence of operations with a particular heat engine has produced a particular result. How can you draw general conclusions from this particular case? Perhaps this heat engine is inefficient because we have chosen a bad sequence of expansions and compressions. Perhaps there is some other way to restore the cylinder and piston to the original condition without doing any work. For instance, why not just let the gas expand adiabatically to atmospheric pressure? Then open a valve and shove the piston back without compressing the gas and thus without doing any work. When the cylinder is at its original volume, the valve could be closed and heat could be admitted to raise the temperature and pressure of the gas to the original condition. Wouldn't this cycle be 100% efficient in changing heat into work? The cycle is shown in Fig. 11.3.

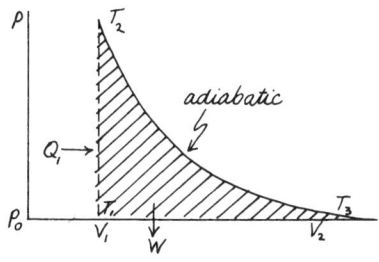

Figure 11.3. Possible Heat-Work Cycle, letting the gas escape during the change of volume from V_2 to V_1.

There are some problems with this proposed cycle. Note that as long as we keep the same quantity of gas in the cylinder, the gas must satisfy the equation $PV=nRT$. After the expansion, the pressure of the gas is back to P_0 but the volume it occupies is V_2, which is larger than the original volume, V_1. Therefore, the temperature of the gas must be greater than it was at the beginning: $T_3 > T_1$. In restoring the cylinder to its original condition we have to cool the hot gas. If we simply open a hole in the cylinder and shove the piston back, we will shove out hot gas and lose some of its internal energy that way. Note also that the work performed W must be equal to or less than the heat Q_1 injected in the first place. If the internal energy of the gas at the beginning is U_1, then after the heat is admitted in stage 1, the internal energy of the gas must be equal to U_1+Q_1 since no external work was done. After the adiabatic expansion (during which no heat is admitted or provided), the internal energy U_3 must be equal to U_1+Q_1-W. However, since $T_3 > T_1$, then $U_3 > U_1$. But to satisfy this requirement, $Q_1 > W$. Once again we have a heat–work cycle in which it is necessary to throw away some heat during the restoration part of the cycle and in which the work done is less than the heat admitted.

■ THE CARNOT CYCLE

There must be an infinite number of different cycles to turn heat into work. Back in 1824 a young Frenchman named Sadi Carnot solved the general problem of finding the conversion efficiency for *any* heat engine, using any cycle. Note that he did this about 25 years before the general acceptance by scientists of the proposition that heat is a flow of energy. The particular cycle of steps analyzed by Carnot is the one that we have just described, a sequence of isothermal and adiabatic steps. It is called the *Carnot cycle*. What Carnot demonstrated was that any *reversible* cycle has certain universal features that apply to all other *reversible* cycles. Let us take a look at the special features of the Carnot cycle and then see why Carnot argued that these features are characteristic of all ideal heat engines.

In Fig. 11.4 the details of pressure and volume of a gas in a cylinder are stripped away. The basic operation of the heat engine is to extract an amount of heat Q_1 from a boiler at temperature T_1 and produce an amount of useful work W. However, to complete the cycle it is necessary to expel a small amount of heat, Q_2, at a lower temperature T_2. Since each step of the cycle is completely reversible, the whole cycle could be run backwards to create a refrigerator. As shown in

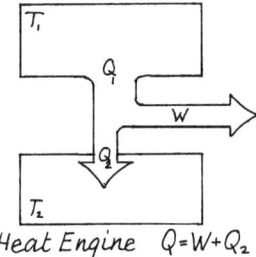
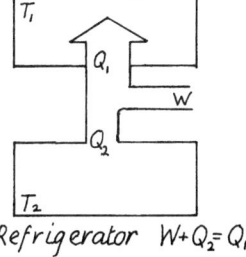

Figure 11.4. The basic operations of heat engines.

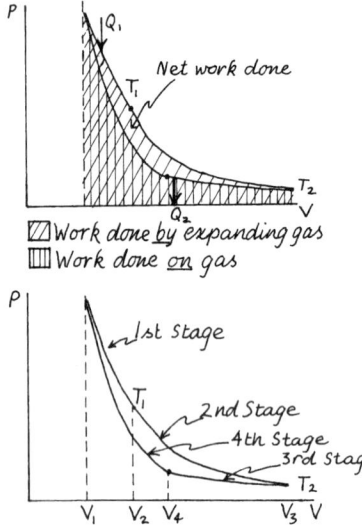

Figure 11.5. A complete Carnot cycle. The derivation in Table 1 shows that $Q_1/Q_2 = T_1/T_2$.

the schematic, the refrigerator would require an amount of work W to extract a small amount of heat Q_2 at the low temperature T_2 and expel the sum of the work energy and the low temperature energy as the heat Q_1 at the higher temperature T_1. In our home refrigerators, for instance, W is provided by the electric motor and Q_2 is extracted from the freezing compartment at the low temperature T_2. The total energy Q_1 is then expelled out the back of the refrigerator at temperature T_1. The room air takes away this heat from the radiator coils.

On the $P-V$ graph for the Carnot cycle, the work done corresponds to the area enclosed by the cycle. Graphically, it is obvious why the return path of the cycle must be different from the expansion part. Otherwise there would not be an enclosed area and hence no net work would be done. Regardless of the detailed steps of the conversion cycle, the area inside the graph corresponds to the useful work obtained.

We claimed that one reason for choosing a sequence of isothermal and then adiabatic steps was that it made calculations easier. This is indeed the case if we want to calculate the amount of work done in a cycle. The work done for a change of volume is $W = \int P \, dV$. To find the total work done as the volume of the gas expands and its pressure drops, we have to know how the pressure depends on volume. Those formulas are well known and relatively easy to use for isothermal and adiabatic expansions of ideal gases. The derivations are shown in Table 1 and the sequence is given in Fig. 11.5. The final result is very simple and very important for an ideal gas used in this particular cycle: $Q_1/Q_2 = T_1/T_2$. Consequently, we can express the efficiency of the machine in a new way:

$$\text{(efficiency)} = \frac{W}{Q_1} = \frac{Q_1 - Q_2}{Q_1} = 1 - \frac{Q_2}{Q_1} = 1 - \frac{T_2}{T_1} = \frac{T_1 - T_2}{T_1}$$

In the special case of Carnot's cycle it appears that the efficiency of conversion of heat into work depends only on the temperature of the source and of the sink (the reservoirs at T_1 and T_2). Note that although we have assumed that the gas in the cylinder is an ideal gas, apparently the process does not depend on how much gas there is or on the starting pressure or volume. The efficiency is a function only of the two temperatures.

Let us consider a practical application of this formula. The boiler of a modern steam plant operates at about 550 °C. The exhaust heat might be released into a lake or river at about 20 °C. The *ideal* operating efficiency, which the real machine cannot obtain, is

$$\frac{823 \text{ K} - 293 \text{ K}}{823 \text{ K}} = 0.64 = 64\%.$$

Now that we have seen some of the features of the Carnot cycle, we return to the question of whether any other heat engine can be more efficient. First, let us give the competing machine the best possible chance, and assume that it uses only reversible processes. Remember that these correspond to friction-free situations in mechanics. No energy is wasted in shock waves or lost as heat out the side walls. As long as the processes are reversible, our arguments will apply to any kind of machine using any kind of material. For instance, the machine might use real gas instead of an ideal gas or might even be made up of rubber bands. (Heat engines using the stretch and contraction of rubber bands can indeed be made.) Suppose that we have such a heat machine and that it is more efficient than a Carnot cycle

Second Law of Thermodynamics

TABLE 1. *Proof that $Q_1/Q_2 = T_1/T_2$. The sequence of steps is shown in Fig. 11.5.*

1st stage	Isothermal	$W_1 = nRT_1 \int_{V_1}^{V_2} \frac{dV}{V} = nRT_1 \ln \frac{V_2}{V_1}$
2nd stage	Adiabatic	$W_2 = \Delta U = nc_V(T_1 - T_2)$
3rd stage	Isothermal	$-W_3 = -nRT_2 \ln \frac{V_3}{V_4}$
4th stage	Adiabatic	$-W_4 = \Delta U = nc_V(T_2 - T_1)$

Therefore $W_2 = W_4$

$$\frac{Q_1}{Q_2} = \frac{nRT_1 \ln(V_2/V_1)}{nRT_2 \ln(V_3/V_4)} = \frac{T_1 \ln(V_2/V_1)}{T_2 \ln(V_3/V_4)}$$

Adiabatic	Isothermal
$P_2 V_2^\gamma = P_3 V_3^\gamma$	$P_1 V_1 = P_2 V_2 \rightarrow \frac{P_2}{P_1} = \left(\frac{V_2}{V_1}\right)^{-1}$
$P_1 V_1^\gamma = P_4 V_4^\gamma$	$P_3 V_3 = P_4 V_4 \rightarrow \frac{P_3}{P_4} = \left(\frac{V_3}{V_4}\right)^{-1}$
$\frac{P_2}{P_1}\left(\frac{V_2}{V_1}\right)^\gamma = \frac{P_3}{P_4}\left(\frac{V_3}{V_4}\right)^\gamma$	

Combining adiabatic and isothermal relationships

$$\left(\frac{V_2}{V_1}\right)^{\gamma-1} = \left(\frac{V_3}{V_4}\right)^{\gamma-1}$$

Therefore $\left(\frac{V_2}{V_1}\right) = \left(\frac{V_3}{V_4}\right)$ and $\frac{Q_1}{Q_2} = \frac{T_1}{T_2}$

---

in taking heat from a boiler at T_1, performing work and then emitting heat at a lower temperature T_2. For a given amount of heat extracted, the more efficient machine would produce a larger amount of work and thus expel a smaller amount of heat at the lower temperature.

Now let us use the more efficient one as a refrigerator. We can operate it with the work produced by the Carnot machine, as shown in Fig. 11.6. With that much work, another Carnot machine would just be able to take the expelled heat Q_2 and move it up the temperature hill to the boiler at T_1. The original conditions would be completely restored. However, if that same amount of work is used on a more efficient engine acting as a refrigerator, a larger amount of heat from the low temperature T_2 can be taken up to the boiler temperature of T_1. The result would be to take more and more energy from the lower temperature and pump it to the higher temperature. No external work would have to be applied to the pair of engines since the Carnot machine gets its energy from the boiler and supplies it to the more efficient machine.

So what is wrong with that? Are we creating more energy? It might seem so because the average energy of the gas molecules is larger at the higher temperature. However, their extra energy comes from the molecules at the lower tempera-

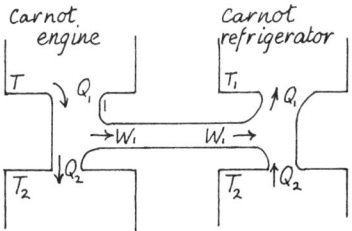

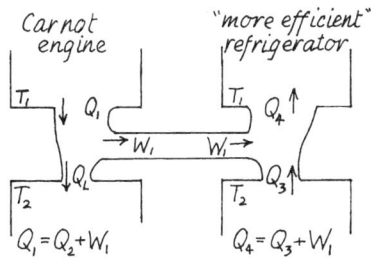

Figure 11.6. *Driving a carnot refrigerator and a "more efficient" refrigerator with a carnot engine.*

ture, which get colder. No energy is being created or destroyed; we are merely moving energy from the cold reservoir to the hot boiler.

■ THE MANY ASPECTS OF THE SECOND LAW OF THERMODYNAMICS

The first law of thermodynamics is an assertion that heat is energy in transit between two systems and that energy is conserved. Now we have come to the second law of thermodynamics, which can be expressed in many different ways. One of the simplest statements of the second law is that *heat cannot flow, without energy input, from a low-temperature region to a higher temperature*. The second law is not something that can be proved. It is one of the laws of impotence—a statement that something cannot happen. Another such law is that no energy or information or object can be transmitted faster than the speed of light. Although laws of impotence cannot be proved, they do have specific consequences. If all the consequences prove to be valid experimentally, then we have faith in the original hypothesis. In the case of the second law, we will see a number of such consequences. Furthermore, it is a matter of common experience that heat always flows from hotter things to colder things. Notice that there would be no loss of energy if it went the other way. The first law of thermodynamics would be satisfied if we could run our industrial machinery with heat energy taken from the ocean. Such a process would simply cool down the ocean a little. But the heat would have to flow (without energy being supplied) from the lower temperature of the ocean to the higher temperature of a boiler. It just does not happen. (Of course, in principle, though so far not in practice, you could create a heat engine by exploiting the temperature difference between surface water and cold water from a deep ocean trench.)

Now note one of the consequences of the second law of thermodynamics. If heat cannot flow from cold to hot without work being done, then the combination heat machine that we just analyzed will not work. Therefore, *there cannot be any heat machine that is more efficient than the Carnot cycle*. Furthermore, all machines operating with reversible processes between temperatures T_1 and T_2 must have the same efficiency. We do not have to analyze each and every possible cycle. The Carnot cycle or any cycle using reversible processes yields the maximum efficiency possible. A machine using irreversible processes would necessarily be less efficient since it would be wasting energy in friction like processes.

Since the efficiency of the Carnot cycle depends only on the operating temperatures and not at all on the mechanism of the machine, then we have a new definition of temperature. Remember that in Chap. 10 when we first defined temperature, we worried about the fact that a temperature scale using one material is different from the temperature scale using another material. We defined our absolute temperature scale based on the properties of an ideal gas. Experimentally, an ideal gas is approximately the same as any gas operating at low pressure and a temperature well above its condensation point. Now it appears that the temperatures defined in that way also define the efficiency of heat engines *that do not depend on the properties of the materials at all*. In principle we could set up an absolute temperature scale in terms of some ideal heat engine working between two temperatures. If a heat engine operated between any temperature and 0° absolute, then its efficiency would be 100%:

$$\text{(efficiency)} = \frac{T_1 - T_2}{T_1} = \frac{T_1 - 0}{T_1} = 100\%.$$

If we then arbitrarily choose a particular temperature as 100°, a temperature of 200° is defined to be such that a heat engine operating between that temperature and the defined 100° has an efficiency of 50%:

$$\text{(efficiency)} = \frac{200° - 100°}{200°} = 50\%.$$

In practice, temperature scales are not determined experimentally by measuring the efficiencies of heat engines. However, *in principle*, temperature can be defined in a way that is independent of the materials of the thermometer.

The temperatures defined in this way (called Kelvin temperatures) must be the same as the absolute temperatures that we have already deduced in terms of the properties of ideal gases. Note that we used the ideal-gas equations to show on p. 269 that $Q_1/Q_2 = T_1/T_2$. Therefore, the efficiency of a reversible heat engine using ideal gas is $(T_1 - T_2)/T_1$, where T is the absolute temperature defined in terms of ideal gases. However, since *all* reversible engines working between T_1 and T_2, regardless of working material, must have the same efficiency, the absolute temperature must be the same as the Kelvin temperature defined by the operation of heat engines. Incidentally, note that the symbol designating absolute or Kelvin temperature is simply K, not °K (e.g., helium boils at 4.2 K).

Since a large share of the energy that we use in this world comes from heat engines, there is obvious practical importance in making the process as efficient as possible. Not only is it wasteful to throw away part of the heat energy at a lower temperature, it is also a serious nuisance. The discarded heat pollutes our rivers and lakes, changing the natural ecology of many places. Carnot attacked this problem long ago because of its practical implications. Real machines are always less efficient than ideal ones. But at least we can use the ideal case as a goal and as a guide in determining the important factors in building heat machines. Large electric power plants use very-high-temperature, high-pressure steam as the operating gas. The expanding steam does not push a piston but instead turns a turbine. The pressure and temperature of the steam decrease as it flows through the turbine until finally the steam is condensed, producing a vacuum at the far end. Some of the steam is led off at earlier stages, reheated and sent back to the original boiler. To keep the final stage cold so that the final pressure will be low, cooling water is circulated through radiators. The expelled heat is then carried out into the stream or lake at the temperature of the source of water, typically around 20 °C. In order to maximize the efficiency of such a machine, the operating temperature of the boiler, T_1, should be as high as possible. Why not make it as high as the fires can reach—perhaps 2000 °C? The trouble is, steam at such a temperature would have an extremely high pressure and also be very corrosive because of the breakdown of water molecules into hydrogen and oxygen. No practical boiler could stand it. An engineering compromise is reached so that the steam temperature is as high as can be reached without excessive damage to the boiler or early stages of the turbine. Most large power plants operate at a boiler temperature of about 1100 °F, which is about 600 °C. With a cooling temperature of 20 °C, the ideal efficiency of such a machine is

$$\frac{873 \text{ K} - 293 \text{ K}}{873 \text{ K}} = 66\%.$$

The actual efficiency of such a power plant is less—usually under 40%. This means that for every joule of electricity produced, $2\frac{1}{2}$ J of energy has to be supplied by the oil or coal that is burned. Furthermore, the extra $1\frac{1}{2}$ J of energy is discarded into the cooling water, raising the temperature of the river or lake downstream.

Thermal pollution is a serious concern as we build more and more power plants. Nuclear power plants currently being used are worse offenders than those powered by fossil fuels. The reason is that the fuel elements in nuclear plants cannot be allowed to go to very high temperatures lest they suffer corrosion and spill their radioactivity. Nuclear powered boilers usually operate below 600 °F, which corresponds to about 330 °C. The Carnot efficiency of such a machine is about 50%. The practical efficiency for many nuclear power plants is only about 30%. For every joule of electricity produced, 3 J have to be supplied and 2 J have to be drained away in the cooling water.

It appears that there is a large discrepancy between the Carnot efficiency and the actual efficiency of real power plants. The difference is due not just to friction, but to the fact that the boiler and sink temperatures are appreciably different from the actual operating temperatures of the gas. The boiler must be much hotter than the gas into which it feeds energy or the energy flow will be very slow. Similarly, the sink must be colder than the gas during the return cycle. A better approximation to the optimum efficiency of heat engines is

$$\mathcal{E} = \frac{\sqrt{T_1} - \sqrt{T_2}}{\sqrt{T_1}}.$$

This is called the Curzon–Ahlborn efficiency.

Using this criterion, the efficiencies calculated earlier become

$$\mathcal{E}_{\text{oil}} = \frac{\sqrt{873} - \sqrt{293}}{\sqrt{873}} = 42\%, \quad \mathcal{E}_{\text{nuclear}} = \frac{\sqrt{603} - \sqrt{293}}{\sqrt{603}} = 30\%.$$

As you can see, these values are much closer to the ones actually achieved.

■ THE REFRIGERATOR

Most of us have heat engines in our homes—refrigerators or air conditioners. By supplying outside energy, usually in the form of electricity, these machines pump the heat from a cold region to a hot region. Let us calculate the efficiency of such a machine, assuming that it is an ideal heat engine being run backward. When converting *heat* into *work*, the efficiency is

$$\mathcal{E} = \frac{Q_1 - Q_2}{Q_1} = \frac{W}{Q_1} = \frac{T_1 - T_2}{T_1}.$$

When we use the machine as a refrigerator, however, we want to pump heat Q_2 for as little work W as possible. The *quality factor* for a refrigerator is thus

$$\frac{Q_2}{W} = \frac{Q_1 - W}{W} = \frac{Q_1}{W} - 1 = \frac{1}{\eta} - 1 = \frac{T_1}{T_1 - T_2} - 1 = \frac{T_2}{T_1 - T_2}.$$

According to this analysis, the quality factor is proportional to the temperature from which we are pumping the heat, and inversely proportional to the temperature difference through which we are pumping it. Reasonably enough, it would be expensive to pump heat from a very low temperature to a very high temperature. It is also slightly cheaper to pump heat from 20 to 30 °C with an air conditioner than it is to pump it from −10 to 0 °C with a freezer.

A typical air conditioner for a bedroom is rated at 5000 BTU's per hour with a consumption of 750 W. Let us see how the quality factor for this machine compares with that for an ideal heat engine:

$$\frac{Q_2}{W} = \left(\frac{5000 \text{ BTU}}{1 \text{ h}}\right)\left(\frac{1 \text{ s}}{750 \text{ J}}\right)\left(\frac{1 \text{ h}}{3600 \text{ s}}\right)\left(\frac{1055 \text{ J}}{1 \text{ BTU}}\right) = 1.95.$$

The *actual* quality factor for this machine is thus equal to about 2. If the room temperature is to be kept at 20 °C, and the outside temperature is 30 °C, then the ideal quality factor for a machine working reversibly is

$$\frac{Q_2}{W} = \frac{293 \text{ K}}{303 \text{ K} - 293 \text{ K}} \approx 29.$$

As a practical matter, even if the temperature difference between room and outside is only 10 °C, the actual difference between refrigeration coil and radiator exhaust will be more like 30 or 40 °C. That would bring the ideal quality factor down to less than 10. Even so, it is clear that an actual machine is using several times more energy than an ideal reversible one. The extra electrical energy is, of course, turning into heat that also has to be blown out into the surroundings, preferably outdoors.

■ ENTROPY

A complex system of interacting parts, such as a gas of many molecules, has a certain amount of internal energy. We rarely are able to measure that energy directly and, indeed, are usually concerned only with changes in the energy. Even these *changes* in the internal energy are actually measured in terms of such quantities as pressure, volume, and temperature. For a particular number of molecules of a gas, the measurement of any two of these quantities determines the third and also characterizes the internal energy. No matter what changes the gas undergoes, if it ends up with the same pressure, volume, and temperature that it has to begin with, then its internal energy is also the same.

There is another way to characterize the state of a system. Suppose that the gas in a box were arranged with all the high-energy molecules on one side and all of those with lower energy on the other side. The internal energy of the system would be the same as if the molecules were mixed in the regular fashion. There is no conservation law that forbids such a momentary arrangement, but the *probability* of it happening is very small. We ought to have some way to characterize the probability of having a particular arrangement of the interacting particles of the system. The measure of this probability is called *entropy*.

Qualitatively, entropy must also be a measure of disorder. An ordered system (all the gas molecules on one side of a box or all the children's toys neatly on the shelves) is an unlikely situation, one with low probability. With all the components of a system free to mill around and interact with each other, there are many more ways to arrange a disordered system than an ordered one. If a system starts out being highly ordered, with low entropy, then as random interactions proceed the system will degenerate into a more probable, disordered arrangement.

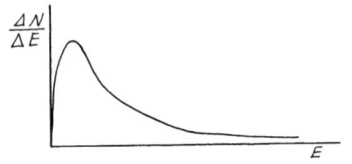

*Maxwell-Boltzmann distribution
maximum probability—maximum entropy*

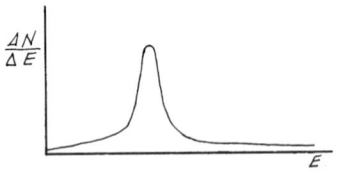

Highly unlikely distribution—low entropy

Figure 11.7. The distribution graphs show the number of molecules that have a particular kinetic energy, as a function of energy.

Suppose, for instance, that you keep casting four coins. There is only one arrangement that will produce four heads. This must be an arrangement with low probability and hence low entropy. There are six ways in which the coins can produce two heads. This is a more probable outcome, with greater entropy. [The number of combinations of n things taken r at a time is $_nC_r = n!/r!(n-r)!$. In this case $_4C_2 = 4!/2!(4-2)! = 6$, or HHTT, HTHT, HTTH, THHT, THTH, and TTHH.]

We usually cannot measure the entropy of a system directly any more than we can measure internal energy. We are mainly concerned with *changes* in entropy. As we will see, for an ideal gas these can be calculated in terms of such variables as pressure, volume, and temperature. However, the formal definition of the entropy of a system in terms of probability is $S = k \ln W$. The entropy S is equal to the product of Boltzmann's constant k and the natural logarithm of the number of possible ways of having that particular arrangement. Remember that the probability of a particular arrangement is proportional to the number of ways the arrangement can be made. (The probability of getting two heads by tossing four coins is six times greater than the probability of getting four heads—because there are six times as many arrangements of the four coins that will yield two heads.) With the molecules of a gas, each possible combination of the positions and velocities of the molecules is subject to the constraint that the total internal energy must be constant. Each possible arrangement that satisfies that condition has equal probability but there are many more arrangements where the molecules are uniformly distributed in space and have velocities corresponding to the standard distribution than there are arrangements where the molecules are all off on one side or the other. The equilibrium distribution shown in Fig. 11.7(a) is the most likely one because it can be obtained with the greatest number of molecular arrangements. In fact, for a very large number of molecules, the number of ways in which any other appreciably different distribution could be arranged is insignificantly small. If the number of arrangements of velocity and position that yield the equilibrium distribution is larger than that for any other distribution, then the entropy as we have defined it is maximum when the system is in its normal equilibrium.

We have emphasized that the probability of a particular arrangement is proportional to the number of ways that the arrangement can be formed. There is the possibility of misunderstanding the nature of W in the definition of entropy. There are six ways in which four tossed coins can yield two heads. The number of arrangements and the probability of getting two heads is six times larger than getting four heads. However, if you ask for the probability of getting two heads on a single toss of four coins, the answer is

$$_nC_r p^r(1-p)^{n-r} = \frac{n!}{r!(n-r)!} p^r(1-p)^{n-r} = \frac{r!}{2!(4-2)!}\left(\frac{1}{2}\right)^2\left(1-\frac{1}{2}\right)^{4-2} = \frac{3}{8}.$$

To find the probability of getting four heads, use $n=4$ and $r=4$. The number of combinations is 1 ($0! = 1$), and the probability is thus 1/16, or 1/6 of the probability of getting two heads.

In finding the entropy of an arrangement of coins, W represents the number of combinations, not the probability of that arrangement happening. Otherwise we would be taking the logarithm of a fraction, yielding a negative entropy! In this case, the actual value of the entropy for the coin arrangement is extremely small.

$S = k \ln W = 1.4 \times 10^{-23}$ J/K $\ln 6 = 2.5 \times 10^{-23}$ J/K.

The entropy, or entropy change, for the arrangements of a few coins is indeed very small. But as we shall soon see, the number of arrangements of a mole of gas molecules is enormous, and balances out the small size of Boltzmann's constant to yield sizeable values of the entropy.

So far, this definition of entropy seems very arbitrary. Why, for instance, define entropy as proportional to the *logarithm* of the probability, rather than proportional to the probability itself? To answer this, suppose you have two boxes of gas, A and B. The entropy of the gas in the first box is $S_A = k \ln W_A$ and the entropy of the second box is $S_B = k \ln W_B$. If we do not mix the gases, the combined entropies must be simply the sum of the two: $S_{A+B} = k \ln W_A + k \ln W_B = k \ln(W_A W_B)$. This is indeed the way probabilities combine. The probability of obtaining first one arrangement of a system that is controlled by chance and then obtaining a second particular arrangement is equal to the *product* of the probabilities of each. For instance, the probability of obtaining any particular number when casting a die is equal to 1/6. The probability of casting a die twice and obtaining first a 1 and then a 2 is equal to $(1/6)(1/6) = 1/36$.

Since it is not practical to calculate the detailed probabilities of arrangements of complicated systems, we should find some other way to characterize changes of entropy. We can find a clue about how to do this by reexamining the Carnot cycle. In any reversible heat cycle, the working substance ends up in its original condition, even though during two of the steps heat was added or taken away. As we have seen, if the original conditions in the system are to be restored, the heats and temperatures must be related by

$$\frac{Q_1}{T_1} = \frac{Q_2}{T_2}.$$

We now give another definition for the change of entropy and will then have to make it plausible that this definition has the same characteristics as the first one. If a system has an amount of heat, Q, added to it at temperature T *in a reversible process*, then its entropy is increased by $\Delta S = Q/T$.

According to this definition, the gas in the cylinder of our reversible heat engine will have an increase of entropy during the first isothermal expansion. That increase will be equal to $\Delta S_1 = Q_1/T_1$. Meanwhile, the boiler will have lost that much entropy. The total system undergoes no entropy change during the adiabatic expansion or compression, since there is no heat loss or gain. During the third stage of isothermal compression, the gas in the cylinder loses entropy equal to $\Delta S_2 = Q_2/T_2$. The cold heat absorber must gain that much entropy. At the end of the complete cycle, the gas in the cylinder has its original entropy since $Q_1/T_1 = Q_2/T_2$. The *total* entropy of the system has also not changed, although the boiler lost entropy Q_1/T_1 while the cold absorber gained entropy Q_2/T_2. Entropy has been transferred from the hot source to the cold sink but for the entire system there has been neither gain nor loss. However, for a *nonreversible* heat engine, the amount of heat absorbed from the boiler must be large enough to furnish the energy for frictional and other losses. In that case, Q_1/T_1 is greater than Q_2/T_2. At the end of the cycle, the cooling water ends up with extra heat having been added to it and the overall entropy of the entire system has increased. In fact, a reversible process can be defined by the fact that it creates no change in entropy in an isolated system. An irreversible process always creates an *increase* in entropy. This feature can be used as another statement of the second law of thermodynamics.

In any process within an isolated system, the total entropy will either increase or stay the same. This statement of the second law is equivalent to the first one, which says that heat cannot by itself flow from a low-temperature region to a higher temperature. For instance, if a pair of heat engines working together could pump a quantity of heat Q from a low-temperature sink to a high-temperature boiler with no other work being done, then the sink would lose entropy equal to Q/T_2. The boiler would gain entropy equal to Q/T_1. But since $T_1 > T_2$, $Q/T_2 > Q/T_1$. The loss of entropy by the sink would be greater than the gain of entropy by the boiler. Hence the entropy of the whole system would have been reduced. Since heat cannot flow to higher temperature without work being done, it must be that the entropy of an isolated system cannot be reduced. A corollary to this argument is that in any process involving an irreversible reaction, the overall entropy of an isolated system increases. It is only during reversible processes that the entropy of an isolated system stays the same.

■ EQUIVALENCE OF THE TWO DEFINITIONS OF ENTROPY

We have introduced two apparently very different definitions of entropy: $S = k \ln W$ and $\Delta S = Q/T$. Here is a proof of their equivalence in the case of heat exchanges in an ideal gas.

$$\text{(work done during an isothermal expansion)} = \int_{V_1}^{V_2} P \, dV = NkT \int_{V_1}^{V_2} \frac{dV}{V}$$

$$= NkT \ln\left(\frac{V_2}{V_1}\right) = Q.$$

Then, derived from the gas laws,

$$\Delta S_{\text{gas}} = \frac{Q}{T} = Nk \ln\left(\frac{V_2}{V_1}\right).$$

In terms of probability, $S = k \ln W$. The probability of any one molecule being confined in V is proportional to V. With N molecules confined in V, $W \propto V^N$. Therefore, according to probability considerations,

$$\Delta S = k \ln CV_2^N - k \ln CV_1^N = Nk \ln\left(\frac{V_2}{V_1}\right).$$

The two expressions for ΔS are the same.

We have offered several equivalent statements of the second law of thermodynamics. This law has implications from the homely to the profound. For instance, every parent is familiar with an example of the second law. In a home with small children, the disorder (the entropy) increases throughout the day. After bedtime, the toys can be filed and the living room straightened up. The entropy of the toys has been reduced but at the expense of work being done by the parents, which has increased their entropy.

One way to remember the first three laws of thermodynamics is this:

Zeroth: If $T_A = T_B$ and $T_B = T_C$, then $T_A = T_C$.
First: You cannot get something for nothing.
Second: At best, you can only break even.

There is a fourth law (called the third law), which says that you cannot reach absolute zero, and therefore no heat engine can have 100% efficiency. This might be characterized as: Third law—you will always lose.

Since most of the natural processes in this world are irreversible, it follows that the entropy of the universe, or at least our local region in it, must be constantly increasing. That means that the energy available to do useful work is continually decreasing because the energy is being degraded into low-temperature forms. In terms of the probability definition of entropy, the second law implies that the distribution of matter and energy in our current local universe is very unlikely, with a high concentration of hot matter clumped in stars and galaxies. As time continues we must be heading toward a more probable equilibrium distribution that, unfortunately, will be at a very low temperature. In effect, the universe is running down, with random chaos as the ultimate end. Of course, we have a few more years to go and during that time may find that the universe has other surprises in store for us. Perhaps the present expansion will cease and contraction will begin. No one knows the implications for that in terms of time and entropy.

As we have seen, a random distribution of the parts of a dynamic, interacting system would seem to be most likely. A particular well-ordered distribution should be highly unlikely. Nevertheless, all around us in nature we see very complex systems with an amazing degree of orderliness. Not the least among these is ourselves. A living object is able to take molecules from the earth and air and arrange them in highly organized combinations of cells and larger structures. In effect, living creatures are refrigerators of entropy. They are able to create order out of chaos. Of course, we do not escape the requirements of the second law. We have to "eat energy." As the entropy of living systems decreases, the entropy of our surroundings necessarily increases. For the whole system of living objects plus surroundings, the entropy relentlessly increases.

■ REFERENCE

Curson, F. L., and Ahlborn, B., "Efficiency of a Carnot Engine at Maximum Power Output," Am. J. Phys. **43**, 22 (1975).

12
Fluids

In everyday life we are familiar with only three phases or states of matter: solids, liquids, and gases. Our topic in this chapter is not *liquids*, one of those three phases, but *fluids*. A fluid is something that can flow. Obviously, the category includes liquids, but it also includes gases and even, under some circumstances, solids. Evidently, the definition is not so much concerned with a type of material as it is with a particular response of material to forces. A fluid changes shape *continuously* under the action of a steady shear force, giving rise to the phenomenon of flow.

Liquid water is the very prototype of a fluid—"water seeks it own level." Air is also a fluid. To a first approximation, the atmosphere acts like a sea of air (and has tides). A major difference between the fluid behavior of a gas and a liquid is that a gas is easily compressed, resulting in variable density, while most liquids are almost incompressible. The density of the atmosphere is much greater at the surface of the earth than it is at the top of Mount Everest (see p. 227 and 243). In contrast, the density of water in the deepest trench of the ocean is only about 5% greater than it is on the surface (see p. 282).

At low velocities, fluid behavior is relatively simple. The flow follows "stream lines" that remain in fixed positions. To a first approximation that has limited usefulness, the *viscosity* of a fluid can be ignored. Viscosity is the friction property of a fluid. Most of the interesting effects of real fluids involve viscosity and *turbulence*. With these added factors, the simple subject of fluids suddenly becomes fearsomely complex. Although fluids have been the subject of scientific study since at least the days of Archimedes 2200 years ago, the analysis of fluid behavior is still one of the most difficult branches of applied science. The equations are known, but until the advent of supercomputers in the 1980s, the solutions of those equations were usually too complex for calculation or even approximation.

In this chapter we will take a look at just some of the simple and easy features of fluid behavior. Even an introductory treatment reveals important and fascinating phenomena. The topic of fluids is not part of most introductory-physics curricula. The assumption is that students in these courses have already learned about siphons and Archimedes' principle in some earlier course. As we all know, this is seldom the case. Even physics graduate students are usually unfamiliar with the actual phenomena of fluid behavior, to say nothing of the theoretical explanations.

■ SUGGESTED DEMONSTRATIONS

Here are some demonstrations that can be done with simple apparatus. Even though the subject of fluids may not be in your syllabus, you can make each of these demonstrations available for a few days. Let the students play with the apparatus and then challenge them to explain what they observe. After all, it would be shameful if a student graduated from an introductory-physics course and could not explain why a boat floats!

1. Solid objects are buoyed up when submerged in a liquid. You experience this effect every time you jump into water. It is the classic Archimedes discovery. Get a rock or brick heavy enough so that it exerts a strain on the muscles that you use to lift it. (Do not use iron. Its density is too large to yield a dramatic differential when it is submerged. The density of most concrete or surface rocks is about 2500 kg/m^3.) Lower the rock into a pail of water and feel the decrease in the force needed to support it. If you want to make the experiment more quantitative, suspend a rock or stone from a rubber band. The stone should be heavy enough so that the rubber band is stretched a considerable amount. Once again, lower the stone into the water and observe what happens to the length of the rubber band.

2. This next observation requires some kind of platform weighing scale and a spring suspension scale. Put a container partly filled with water on the platform scale. Next, suspend an object by a string from the *spring scale*. Note its weight. Now lower the object into the water so that it is submerged but does not touch the bottom. The size of the container and the weight of the object depend on the size of the scales you have available. The object should be heavy enough so that its weight can be easily read on the scale. Compare the weights read by the two scales when the object is above the water, when the object is suspended in the water, and when the object is lying on the bottom of the container. Using the same apparatus, float a sizable piece of wood on the water. Compare the weight read on the scale when the wood is floating and when it is just lying beside the container.

3. For this observation you will need a graduated cylinder or a kitchen measuring cup. Float a rectangular piece of wood (or anything else that will float) in water in the cylinder or cup. Use a needle or toothpick to hold the floating object under water and then fill the container up to its top mark. Next, let the object float freely and measure the level of the water. Finally, remove the floating object and record the new level of the water. Find the volume change for each of these two cases. Compare these volumes with the measured volume (cross-sectional area times height) of the piece of wood.

4. Anyone who swims has experienced the increased pressure on the ears while swimming or diving to greater depths. If your students have access to a pool or lake, have them take a small rubber balloon filled with air down as deep as they can. There are two effects to be noticed. First, the volume of the balloon will noticeably decrease as you take it down. The second effect is a consequence of the first. The buoyancy of the balloon will decrease the further down it goes. Can you notice this effect in a pail of water? Recall that atmospheric pressure will support a column of mercury about 76 cm high or of water that is 76 cm$\times$13.5 (specific gravity)$\rightarrow$10.3 m$\approx$33 ft high. The difference in pressure under a foot of water is only 1/33 of an atmosphere, but at the bottom of an 8-ft pool the increase is about 1/4 atm.

5. Liquids act as if their surfaces consist of elastic skin. This property is called surface tension. One way to observe it is to blow soap bubbles or to observe

bubbles that are clustered together in any kind of foam. First, note that a soap bubble that is free and drifting in the air is spherical. On the other hand, soap bubbles that meet in a cluster have very sharp and specific angles where their walls meet.

You can produce soap bubbles as large as basketballs by using almost any good detergent (not the kind for dishwashers!) mixed with the usual amount of water. Roll a piece of ordinary typing paper into the shape of a cone, fastening it together with a single piece of transparent tape. Soak the large end in the soap water until the paper is sopping wet. If you draw the paper cone slowly out of the solution and blow gently into the small end of the cone, you can produce and shake loose a very large bubble. Such bubbles will last for 30 s or so and you can even hold them or bounce them with your hand if your wear a woolen mitten.

You can also observe the surface skin of a liquid by floating a needle or razor blade on water. There is an easy way to rest the metal on the liquid surface without penetrating the surface skin. Simply put the needle or the razor blade on a tissue, and then lower the tissue onto the surface of the water. With your fingers or a pencil, you can poke the paper gently down into the water and away from the floating metal. Examine closely the surface of the liquid at the edges and underneath the floating metal. Then add a small amount of soap or detergent to the water!

Sprinkle water droplets of various sizes on a piece of wax paper and examine them closely. The large drops will be flat on the bottom and then bulge up, but very fine drops will be almost spherical.

6. Very few students will ever have observed a siphon. It is particularly instructive if you can demonstrate the effect with a transparent tube so that you can see what the liquid and bubbles inside are doing. One easy way to make such a transparent siphon is to use two plastic straws that have corrugated elbows for sipping at an angle. Crease the end of one straw, insert it into the other, and you can produce an airtight U-tube. Siphon water up over the edge of a raised bowl and down into a lower bowl. What are the conditions necessary to start and continue such a flow? For instance, what happens if the bottom tube is below the *surface* of the upper liquid but higher than the bottom of the upper liquid (and thus higher than the opening in the upper tube)? What happens if there is a bubble in the upper end of the U tube?

7. The motion of air sometimes produces pressure changes that are surprising. An atomizer (a terrible misnomer) draws the liquid up a tube when you force compressed air across the open end of the tube. The same effect is used in a carburetor. You can show the effect easily by using two straws and a cup of coffee. Place one straw vertically in the coffee and use the other straw to blow directly across the open end of the first straw. Watch what happens to the level of coffee in the vertical straw.

8. There is a lot of physics, most of it impossible to solve, involved in the way water comes out of a faucet. Turn a faucet on slowly and watch the way the water flows and twists and even breaks up into droplets if it falls far enough. Notice particularly that the cross section of the stream decreases as it moves away from the immediate region of the outlet.

9. Dangle the convex side of a spoon close to the edge of a smooth stream of water coming from a faucet. You might expect that when the water hits the back of the spoon, it would bounce off in one direction while knocking the spoon in the other. What actually happens is different and surprising.

   Take a close look at the way fluid moves in a coffee cup. If you use instant coffee, you can observe the phenomena on the surface shortly after you pour in the hot water and before the foam disappears. You can also see these effects by adding one drop of milk. Notice particularly how eddies persist and move around. Notice also what happens to their rotational velocity.

10. One way to remove dust from a surface is to blow on it—or so one might think. Why is it, then, that the outside of a car remains dusty even when it has traveled at high speed? For an even more dramatic example of this paradox, look at the blades of a room fan after it has been in operation for some time. A lot of dust-laden air has blown through the fan and quite a bit of the dust remains on the blades. It is not because the dust is greasy; the dust can easily be wiped off.

    Another way to observe this effect is to sprinkle some very fine powder on a smooth tabletop. Use talcum powder or a sugar–cinnamon mixture. Now blow hard but steadily along the surface of the table. The larger clumps or grains will be blown away but a film of fine powder remains. We must conclude that either the fine powder is particularly sticky or else the wind velocity very close to the surface is too small to remove the dust.

11. You can observe the pressure as a function of height in a liquid by letting water run out of holes in the side of a waxed cardboard or plastic milk carton. Punch three holes with a nail, one about one-third of the way from the top, another near the bottom, and the third midway between them. Stagger the holes slightly so that the three streams of water will not run into each other. Now fill the carton with water and observe the trajectories of the three streams. There is a tricky point here as pointed out by Paldy. Depending on how far the streams fall before they hit the ground (or sink), the top stream may actually have a longer range even though its horizontal velocity is clearly less than that of the bottom stream.

■ FLUID PRESSURE

Pressure is exerted by gases, liquids, and solids in very different ways. Most of us have become so familiar with the model of gas molecules colliding with walls that we think fluid pressure is transmitted in the same way. With gases the pressure depends on the density of molecules, their mass, and their average speed. The temperature also depends on the mass and average speed of the molecules, and so temperature and pressure are strongly connected. Consider the paradox: if water is nearly incompressible, and therefore the density at the bottom of a pail of water is nearly the same as at the top, and if the greater pressure at the bottom is caused by harder bombardment of the molecules, why isn't the temperature greater at the bottom? With air, incidentally, the primary reason that the atmospheric pressure is greater at sea level than it is at mountaintop is that the density of molecules is greater at sea level.

In the case of liquids and solids, pressure is not transmitted by molecular

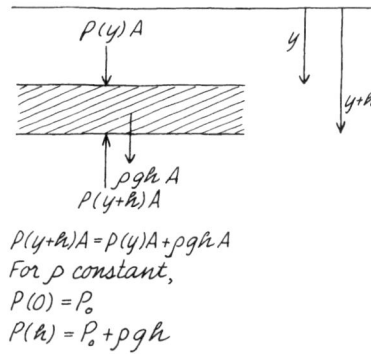

$P(y+h)A = P(y)A + \rho g h A$
For ρ constant,
$P(0) = P_0$
$P(h) = P_0 + \rho g h$

Figure 12.1. Forces on a slab of liquid.

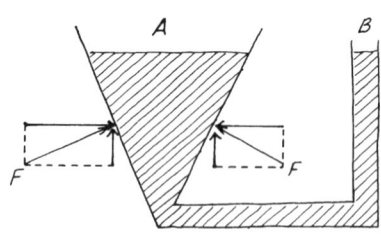

Figure 12.2. Pascal's vases and the hydrostatic paradox.

bombardment. The atoms or molecules vibrate within potential wells created by their neighbors. As molecules press closer together, exerting force on each other, the equilibrium position between molecules decreases. It is as if the molecules were being shoved against a hard spring. We know that the equilibrium position does not decrease much because the compressibility of solids and liquids is very small. The bulk modulus β is the ratio of the change in pressure, ΔP, on an object to the fractional change of its volume, $\Delta V/V$: $\beta = -\Delta P/[(\Delta V)/V]$. The minus sign means that an increase in pressure produces a decrease in volume. The bulk modulus for water is 2×10^9 N/m^2. An increase of 1 atm pressure (1×10^5 N/m^2), for instance at the bottom of a water tower 33 ft high, yields a fractional increase of density of only 2×10^{-4}. The equilibrium distance between neighboring water molecules would decrease by only $\frac{2}{3} \times 10^{-2}$%. Meanwhile, if the temperature of the water is the same at the bottom as it is at the top, the average vibration amplitude of the molecules is the same at top and bottom. Each molecule vibrates within a temporary cage formed by its neighbors, much as if it were in a solid crystal. The energy of the molecule associated with its meandering motion, caused by occasional jumps from temporary cage to cage, is usually negligible compared with the vibrational energy.

The resulting pressure in a liquid does not have a direction of its own but is exerted perpendicular to any surface placed in the liquid. At a given point in the liquid, there is the same force on a surface up as there is sideways or down. Otherwise, as you can see in Fig. 12.1, the net forces on a slab of the liquid would not be 0 and so there would be movement of the liquid. Notice that consequently there is a difference of pressure between the top and the bottom of a slab of the liquid. The total difference in the force between top and bottom is just sufficient to support the weight of the slab:

$$P(y+h)A - P(y)A = \rho g h A \quad (\rho = \text{density}).$$

Since ρ is approximately constant for most liquids and most depths,

$$P(h) = P_0 + \rho g h.$$

The pressure in a liquid at a depth h is equal to the pressure at the surface, which is usually the atmospheric pressure, plus a term proportional to the depth. Notice how different this is from the case that we calculated on p. 000 for atmospheric pressure. In that example the density of the gas is proportional to the pressure. Consequently, the pressure and density of the atmosphere decreased exponentially with increasing height.

Solids transmit pressure more like liquids than like gases. In solids, however, the atoms or molecules are locked in place. In a crystalline structure, the repulsive spring-type constants between atoms may be different in one direction from what they are in another (because of the geometrical arrangement of the different atoms in the array). Consequently, the pressure and pressure response are not the same in all directions. There can be a very complicated relationship between stress and strain. In a liquid, molecules can move around each other, and so static pressure effects are uniform in direction.

The fact that pressure in a liquid depends only on the height gives rise to the so-called hydrostatic paradox illustrated in Fig. 12.2. The containers, called Pascal's vases, are named for Blaise Pascal (1623–1662). The liquid stands at the same level in each of the containers. One might naively be surprised that the heavy weight of all the liquid in container A could be balanced by the small weight of the water in cylinder B. Of course, part of the weight of the liquid in container A is being supported by the walls of the container as shown in the diagram. As an extension of this "paradox," you might ask why all water towers are about the same height. Water distribution and plumbing systems require about the same "head" of pressure. Then why not use a high slender tower instead of going to the expense of supporting a huge tank at the standard height, or why not forgo the tower and use the pumps to maintain line pressure, since you need the pumps to raise the water into the tower in the first place? You need a large-volume reservoir to even out the fluctuating pressure of pump input and sudden output demand.

The pressures at various points of a barometer are shown in Fig. 12.3. An extension of this analysis explains the operation of a siphon. A siphon is an upside-down U tube that can provide drainage from an upper level of fluid to a lower level, as shown in Fig. 12.4. Of course, a hole in the upper container could provide the drainage as well, but a siphon has the apparently peculiar ability to raise a liquid uphill over a container wall. The liquid does indeed rise in one leg of the tube but only so long as it falls a longer distance in the other leg. To make a siphon work, you first have to fill the tube with the liquid and keep it filled until the short end is in the liquid.

Suppose that you put your finger over the end of the longer tube, stopping the flow. The pressures in the liquid column are shown in Fig. 12.5. At the surface of the upper container there is atmospheric pressure P_0. At the top of the U the pressure is reduced to $P_0 - \rho g h_0$. In the long leg of the tube of the tube at the level of the liquid surface in the upper container, the pressure is again atmospheric, P_0. The pressure on your finger is equal to the atmospheric pressure plus the pressure due to the difference in liquid heights in the two legs: $P_0 + \rho g h$. If you take your finger off the tube, the liquid that is under pressure will shoot out. The cohesiveness of the liquid keeps the column from separating and so drags water into the short end to keep the flow going. Note that the opening in the open leg does not have to be below the submerged end. All that is required is that the lower end be below the upper surface of the liquid in the upper container.

We emphasized that the siphon must be filled before its action can start. Suppose that you do not fill it or end up with an air bubble in the U section as shown in Fig. 12.6. As you can see from the analysis in the diagram, the siphon will not work unless $h_1 = h_2 > h_0$.

There is a limit to the height that a siphon can raise a liquid, even as there is a limit to the height that a liquid can rise in a barometer or when pulled by a vacuum pump. The pressure in the upper leg of the siphon decreases steadily from the atmospheric pressure at the surface: $P = P_0 - \rho g h$. The siphon will not raise water higher than the point where the pressure equals 0. For mercury this height is 76 cm [at standard temperature and pressure (STP)]. For water the height is $10\frac{1}{3}$ m, or about 33 ft. Actually under the right circumstances, such as those existing in vegetative capillaries, liquids can withstand negative tension and so can rise to even greater height.

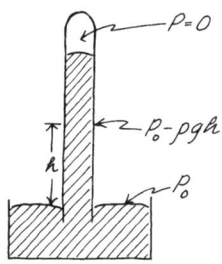

Figure 12.3. Pressures in a barometer.

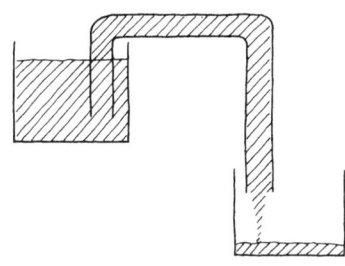

Figure 12.4. Siphon draining liquid over a barrier down to a lower level.

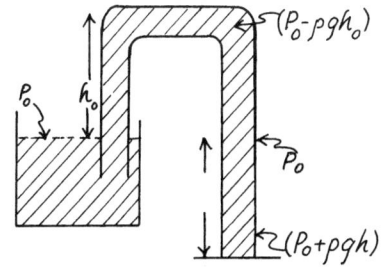

Figure 12.5. Pressures in a non-flowing siphon.

■ ARCHIMEDES' PRINCIPLE

Archimedes (287–212 B.C.) was a Greek mathematician and scientist who lived in Syracuse, Sicily. He considered his primary work to be in mathematics, but he was famous in his own day and in subsequent legends because of his

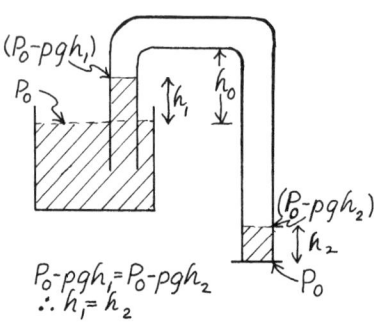

Figure 12.6. Pressures in a siphon with an air bubble.

inventions. One of these involved the measurement of the density of an object that had a complex geometry. According to legend, King Hiero of Syracuse had a crown made of gold. He suspected that silver had been substituted for part of the gold, and asked Archimedes to find out without destroying the crown. Supposedly Archimedes discovered the method while in his bath and ran naked through the streets crying, "Eureka, eureka!" ("I have discovered it; I have discovered it!") Succeeding generations have argued endlessly over exactly what it was that he had discovered. One possibility is that he noticed the overflow of the water as he got into the tub and realized that he could measure the volume of an irregularly shaped object by means of water displacement. If he could find the volume of the crown in this way, he could weigh it and then calculate the density. He could then compare the density with that of pure gold or with gold–silver alloys of various concentrations. The scheme is suitable in principle, but is not capable of yielding great precision. Because of surface tension and the requirement that the water container be at least as large as the object to be submerged, the volume of the displaced water is hard to measure with a precision of more than two significant figures. There is, however, another phenomenon that Archimedes might have noticed when he got into the tub. He got lighter. A fluid exerts a buoyant force on a submerged object. As we will now see, this effect provides another—and far more precise—method of measuring the density of an object.

Archimedes' principle is that an object in a fluid is buoyed up by a force equal to the *weight* of the *fluid displaced*. [There is a small quibble about this definition. Perhaps it would be more accurate to say "the weight of the fluid that is replaced by the submerged or floating object". To see the nature of the quibble, consider a 1 cm diameter cylinder inserted into a 1.1 cm inner diameter cylinder containing a small amount of water. The water will rise between the cylinders, but not much water is thus displaced. Nevertheless, the inner cylinder will be buoyed up by a force equal to the weight of the water replaced by the submerged volume of the inner cylinder.] This rule applies to both floating and submerged objects. As we have already seen, the upward force on the bottom is greater than the downward force at the top by an amount equal to the weight of the fluid slab. Consequently, the slab is in equilibrium and from this relationship we derived the formula for the pressure as a function of depth in a fluid of constant density. If the slab of fluid were to be removed and replaced with a slab of any other material of the same size, the pressure at each point of the surrounding fluid would not change. There would still be an excess upward force on the slab equal to the weight of the *fluid* that used to occupy that space. The volume of that fluid is the same as the volume of the object. When the object is submerged, that volume of fluid is shoved up to make room for the object. (Many students will have the naive misconception that the greater the amount of water under an object, the greater the buoyant force that the water will exert. They may therefore conclude that there is no buoyant force on an object at the bottom of a tank of water. Of course, if there is no water layer between the object and the bottom, this is true. However, with most objects and surfaces, there is a water layer.)

In one of the "homely" demonstrations described at the beginning of this chapter, we suggested that your students actually feel the effect of buoyancy by holding a rock in air and then under water. Suppose that the rock has a mass of 5 kg and therefore has a weight in air of about 50 N. Since the density of most silicon-based rocks is about 2500 kg/m^3, the volume of this rock must be 2×10^{-3} m^3. When you submerge this rock in water, the mass of the water displaced must be $m = \rho_{water} V = (1 \times 10^3 \text{ kg/m}^3)(2 \times 10^{-3} \text{ m}^3) = 2$ kg. The weight of 2 kg is about 20 N. While holding the rock in air, you had to support a weight of 50 N; while holding it under water, you need only exert an upward force of 30 N.

Notice that the effect of the buoyant force is to reduce the apparent density of the object by the density of the fluid. Since the mass and the weight of an object are proportional to the density, the ratio of weight in the fluid to the weight in air is

$$\frac{W_{\text{in fluid}}}{W_{\text{in air}}} = \frac{\rho_{\text{object}} - \rho_{\text{fluid}}}{\rho_{\text{object}}} = 1 - \frac{\rho_f}{\rho_0}.$$

The specific density of an object, ρ_0/ρ_{water}, can be found by weighing an object in air and then in water:

$$(\text{specific density}) = \frac{(\text{weight})_{\text{in air}}}{(\text{weight})_{\text{in air}} - (\text{weight})_{\text{in water}}}.$$

What happens if the density of the object is equal to or less than that of the fluid? Our formula apparently gives a nonsensical answer. If the densities of object and fluid are the same, then the ratio of weights is 0 and the object is floating and is in equilibrium at any depth. If the density of the object is less than that of the fluid, then the object will float on the surface and will not displace its volume.

How can a steel boat float if the density of steel is 5500 kg/m^3? A lot of high-school students will not know how to explain this problem, even after they have taken physics courses. Of course, the density of a *hollow* boat is considerably less than that of water and so the whole object floats, displacing a volume of water considerably less than the volume of the whole boat.

How Archimedes exploited the significance of the buoyant effect, we do not know. He could have determined the density of the crown by weighing it in air and then while it was submerged in water. The precision of weighing can be much greater than the precision of determining volume by liquid displacement. Eighteen hundred years after Archimedes considered the problem of the fraudulent crown (there is no historical record of what he found), another great scientist, Galileo, analyzed the legendary problem. To solve it, he devised a balance that would have been sensitive enough to have made the crucial measurement, *La Bilancetta* (The Little Balance), as translated by Laura Fermi and Gilberto Bernardini.

We can use Archimedes' principle to determine the unknown density of a liquid in which an object with known density is floating. Such an instrument, called a hydrometer, is commonly used to determine the charge condition in storage batteries, or the antifreeze component of radiator coolant. The liquid is sucked into a tube with a calibrated float which is weighted so that it remains upright, but rises or sinks to various depths depending on the liquid density. The level lines on the float are usually calibrated directly in terms of density. For instance, when a lead storage battery is fully charged, the sulfuric acid in it has a density of 1300 kg/m^3. As the sulfate combines with the lead during the discharge process, the density of the liquid falls. By the time the density is 1150 kg/m^3, the battery is essentially discharged.

Buoyant forces are exerted by gases as well as by liquids. This is why dirigibles or toy helium balloons float. In everyday life, we do not feel buoyed up by the surrounding air, but the effect is not negligible in precision weighing with a pan balance. Standard comparison weights are usually made of brass, which has a density of 8600 kg/m³. The density of air is only 1.3 kg/m³. The buoyant force exerted by the air on the standard weights is equal to the product of their volume and the weight density of air:

$$F_{\text{buoyant}} = Vg\rho_{\text{air}} = \left(\frac{m_0}{\rho_0}\right)g\rho_{\text{air}}.$$

The fractional loss of weight of an object submerged in air is

$$\frac{F_{\text{buoyant}}}{W} = \frac{(m_0/\rho_0)g\rho_a}{m_0 g} = \frac{\rho_a}{\rho_0}.$$

In the case of brass standard weights, the percentage of error due to the buoyancy in air is $(1.3 \text{ kg/m}^3)/(8600 \text{ kg/m}^3)100\% = 1.5 \times 10^{-2}\%$. It might seem that such a small effect is hardly worth bothering with. Furthermore, on a pan balance, wouldn't the same correction apply to the sample being weighed in the other pan? Of course, there would be cancellation if you were weighing a brass object. However, if you weigh material that has a density close to that of water, then the correction due to the buoyancy of that material in air is about $13 \times 10^{-2}\%$. The net effect would therefore yield an error of about 0.1% or one part per thousand. Chemical weighings frequently must be done to precision better than that.

■ SURFACE TENSION

The molecules at the surface of a liquid experience forces that are not symmetrical. Within the body of the liquid a molecule is, on the average, pulled equally in all directions by the attractive forces binding it to its neighbors. This force is called *cohesion* and is simply the short-range force between atoms and molecules that we described on p. 214 in the section on potential energy. It is electromagnetic in nature but is more complicated than the simple attraction of a positive point charge to a negative one. A molecule in the *surface* of the liquid experiences this cohesive force only in one direction—downward. Of course, it cannot accelerate downward because that force is balanced by the repulsion of the molecules directly underneath as the surface molecule presses slightly closer to them than the normal intermolecular spacing.

If the surface of the liquid expands, perhaps because one boundary is moved, then more molecules must come from the body of the liquid to take their place on the surface. The surface molecules must be pulled apart to make room for the molecules moving up. That takes work. Since the surface expands along a direction x, a force must be required in that same direction in order to do the work that increases the potential energy. Therefore, the force must lie in the surface and be in the direction of the expansion. This property of a liquid is called *surface tension*, γ. It is measured in newtons/meter since the force needed is proportional to the width of the surface being stretched. Values of surface tension for various liquids in contact with air are given in Table 1. Note, incidentally, that surface tension is not subject to Hooke's law. The force is constant, independent of the amount of surface stretch.

TABLE 1. *Surface tension of liquids at room temperature in contact with air.*

	N/m
Water	0.073
Carbon tetrachloride	0.027
Ethyl alcohol	0.022
Mercury	0.49

In Chap. 9 we used a rough model of atomic sizes and energies to get an approximate value for Young's modulus. Let us see if that same model can give us a reasonable value for the magnitude of surface tension. We must make an assumption about the energy involved when a molecule is pulled into the surface from the body of the liquid. That energy must be considerably less than one electron volt, which is the order-of-magnitude of binding energy of atoms within molecules. We are not removing the atom from a chemical bond when we move it to the surface. We might guess that the energy involved would be about half of the fusion energy per molecule as the substance changes from solid to liquid. The factor of one-half is used because the surface molecule is still held by bonds from below. For water, the fusion energy is 80 cal/g=335 J/g=6024 J/mol=1×10^{-20} J/molecule=0.06 eV. Let us assign this value to the depth of the local potential well in which a water molecule is held by its immediate neighbors. To allow a new molecule to come to the surface, we must exert a force along the surface to separate the layer by one full molecular diameter. That force must be equal to the force required to pull a molecule partly out of its potential well to the position where it is held only by bonds from below. The force is $F=\Delta U/\Delta x$, where $\Delta U=\frac{1}{2}\times 10^{-20}$ J/molecule and $\Delta x = 3\times 10^{-10}$ m, the molecular diameter of water:

$$F=\Delta U/\Delta x=(\tfrac{1}{2}\times 10^{-20} \text{ J})/(3\times 10^{-10} \text{ m})=\tfrac{1}{6}\times 10^{-10} \text{ N}.$$

In 1 m along a surface there are $1/3\times 10^{10}$ water molecules. The force per meter to separate all of these is 0.06 N/m. Compared with the table value for the surface tension of water, which is 0.073 N/m, our calculation is close, though perhaps fortuitous. Note that the surface tension for mercury is much larger. Nevertheless, the order-of-magnitude agreement is encouraging; our crude atomic model must be pretty good. (The heat of fusion for Hg is 2.8 cal/g=2400 J/mol. (Force)/$m=\gamma=\Delta U/d^2$, where d=atomic diameter).

The available energy (sometimes called the *free energy*) due to surface tension must be equal in magnitude to the work done in pulling out the surface from 0 size to its final area. Since the force required is constant (for a flat surface) the work done is equal to Fx. The force required is equal to the product of the surface tension and the length of the surface: $F=\gamma l$. Therefore, the free energy due to the surface tension is

$$U_{\text{surface}}=Fx=(\gamma l)x=\gamma S,$$

where S is the surface area.

The actual energy involved in forming a liquid surface is slightly greater than U. When any liquid surface is enlarged, the temperature of the liquid drops. The extra thermal energy is stored in the surface layer.

One way to measure surface tension is shown in Fig. 12.7(a). The force required is actually twice the product of the surface tension and the length since

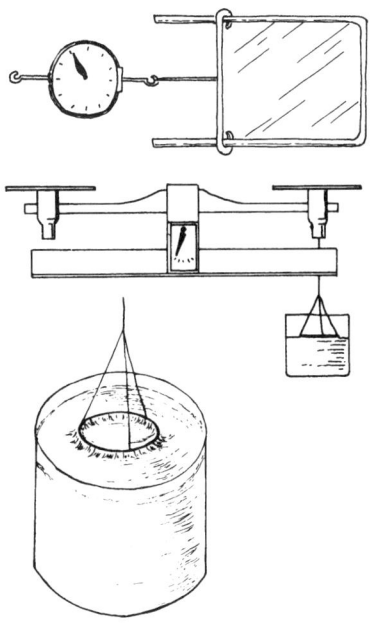

Figure 12.7. Methods of measuring surface tension.

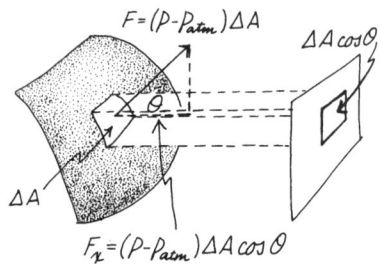

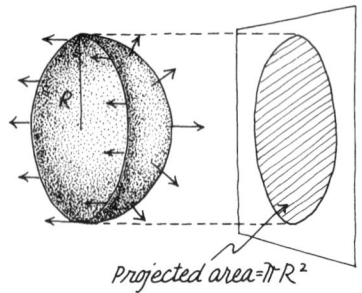

Figure 12.8. Pressure and surface tension in a liquid bubble.

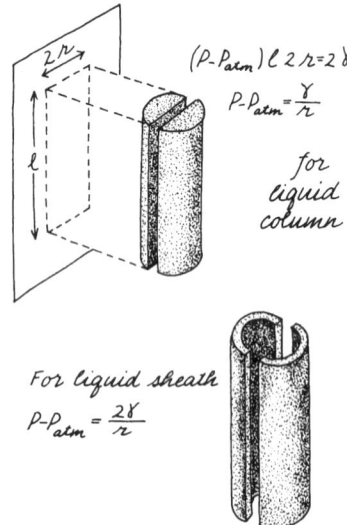

Figure 12.9. Pressure and surface tension in a liquid cylinder.

two surfaces are involved—top and bottom. A more practical way to measure surface tension is shown in Fig. 12.7(b). The force required to pull the ring out of the liquid is easy to measure with a sensitive pan balance. Consider, for example, a wire ring of diameter 6 cm. The force needed to pull up the ring (which is picking up a film along both its inner and outer edge) is

$$F = 2\gamma \text{ (circumference)} = 2(0.073 \text{ N/m})\pi(0.06 \text{ m}) = 2.75 \times 10^{-2} \text{ N}.$$

This force is the weight of 2.8 g.

Because of surface tension, any volume of liquid tends to decrease its surface area, thus decreasing its free energy. Surface tension is one of the elastic forces responsible for the motion of ripples. The water in the uplifted surface is pulled down by gravity and also by surface tension trying to make the surface smooth again. (See formula for ripple velocity on p. 328). Tiny water droplets in air are almost spherical since the sphere has a smaller ratio of surface area to volume than any other geometrical shape. As larger drops fall, they oscillate in various ellipsoidal shapes. The surface tension of mercury is so great that small drops on a surface are almost spherical. You can observe almost the same effect with water by sprinkling fine drops on waxed paper.

The tendency of a surface to achieve minimum area also explains why a needle can be floated on the surface of water. If you look closely, you can see the slight depression of the surface on which the needle rests. The restoring forces trying to make the surface flat again support the needle. When you reduce the surface tension by adding soap or detergent, the needle promptly sinks.

Because of surface tension, the surface of a liquid acts as if it were composed of an elastic skin. There is a major difference, however, between the elasticity produced by surface tension and that produced by an elastic material such as rubber. For a flat surface on a liquid, the strength of the restoring force is independent of the amount by which the surface has been stretched. In fact, the "amount of stretch" of such a surface is meaningless. You need exert only a constant force to increase the surface area by pulling more and more molecules into the surface layer. To stretch a sheet of rubber, on the other hand, you must exert a force that is roughly proportional to the stretch.

You may have noticed in blowing soap bubbles that if you take your mouth off the pipe, the bubble will contract, blowing the air back out again. Evidently, there is higher pressure inside the bubble than there is outside. The forces caused by air pressure on a hemisphere of a bubble are shown in Fig. 12.8. The forces on the spherical surface add vectorially to a resultant force tending to separate this hemisphere from the opposite one. In the diagram we show that the x component of the force on a segment of the curved surface is just equal to the product of net pressure and projected area. Since the projected area of a hemisphere of radius R is equal to πR^2, the x component of the force on the hemisphere is $(P - P_{\text{atm}})\pi R^2 = 4\pi R\gamma \rightarrow P - P_{\text{atm}} = 4\gamma/R$.

Surprisingly enough, the smaller the radius of the soap bubble, the greater the excess pressure must be. Because of this feature, whenever two bubbles come together, the smaller feeds air into the larger, and the larger grows at the expense of the smaller. Because the excess pressure in a bubble is proportional to the surface tension, we see why bubbles are more easily formed in soapy water. Bubbles in plain water (or in mercury!) would require high interior pressure.

A similar analysis of the pressure relationships and surface tension forces in a cylindrical column is shown in Fig. 12.9. Once again we find that the excess internal pressure that balances the surface tension is inversely proportional to the radius. This fact is responsible for the instability of long cylindrical columns of liquid. This is why a thin stream of water coming from a faucet breaks up into a spray of droplets. To begin with, the stream must get steadily narrower because the speed is increasing due to its fall, and there must be flow conservation (volume/second). Even so, if the column could maintain a slowly changing radius all along its path, there would be no reason why a pinch should occur in one place instead of another. However, if for any reason, such as fluctuations in the stream lines or variation in the outside pressure, one part of the stream becomes momentarily narrower than the immediately adjacent upstream or downstream sections, then the internal pressure at that point will become greater. The liquid will flow from that region to the surrounding lower-pressure regions, which will make the stream even narrower. The pressure will rise still higher at that point, making the stream yet narrower until it pinches off and forms a separate droplet.

In earning their livelihood, web-spinning spiders routinely make use of the instability of fluid cylinders. When the spider first produces the filament, the sticky fluid forms a cylindrical sheath around the strand. Since the long liquid column is unstable, it soon breaks up into tiny globules that are ideal for catching flies.

So far we have been talking about the surface between a liquid and a gas. At the edge of a container the liquid will be in contact with both a solid and a gas. A surface molecule at that boundary is affected not only by the forces of cohesion within the liquid, but also by the force of *adhesion* to the solid. The liquid surface lines itself up perpendicular to the resultant of these two forces. (Otherwise surface molecules would move along the surface.) Three cases are shown in Fig. 12.10. In the first, the force of adhesion is strong enough to produce an upward curvature of the liquid at the solid surface. This is the normal situation with water touching glass. To describe this behavior, we say that the liquid has "wet" the glass. In the second case the force of adhesion is small and the resultant force on the liquid at the junction remains directed back into the liquid. The resulting curvature of the liquid surface is convex, away from the solid container. This is the situation with mercury in a glass container. In the third case, the resultant of the forces of adhesion and cohesion is parallel to the wall and hence the liquid surface is perpendicular to the wall. This is the situation with water in a silver container or with water in some types of plastic that are not wet by water.

The curved surface of the liquid in a cylindrical tube is called a *meniscus*. In order to establish a meniscus with a curvature such that the surface is perpendicular to the resultant of the adhesion and cohesion forces, the edge of the liquid must rise or fall some distance along the solid surface. This is called *capillary* action. (A capillary is a tube of small diameter.) The surface tension pulls a column of water up a glass tube and pulls a column of mercury down a glass tube. How can this be when in all cases the resultant of the forces of cohesion and adhesion are more "down" than "up"? Remember that the surface of a liquid is always perpendicular to the net (attractive) force acting on the surface molecules. In a horizontal surface, for instance, the net force of cohesion on each surface molecule is down. (Note that the net force on a surface molecule is not an unbalanced force producing acceleration. The attractive net force of adhesion and cohesion is balanced by the molecular compressive force of the molecules underneath. Molecular forces are complex, acting to attract at ranges beyond the equilibrium distance, but to repel at shorter distances.) The *surface tension* is always

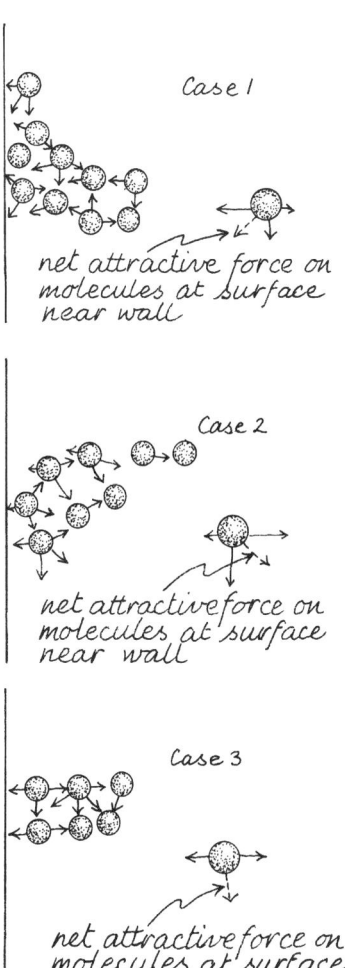

Figure 12.10. Results of adhesion and cohesion on liquid surface level.

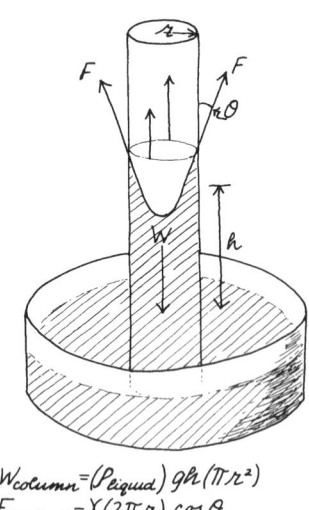

$W_{column} = (\rho_{liquid}) gh(\pi r^2)$
$F_{vertical} = \gamma(2\pi r) \cos\theta$
surface tension

Figure 12.11. Surface of water in a glass capillary tube.

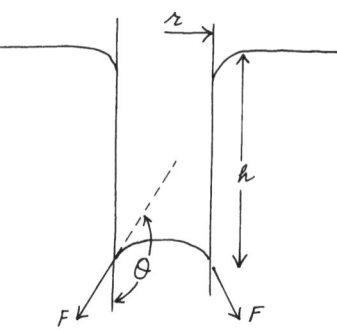

Figure 12.12. Surface of mercury in a glass capillary tube.

directed along the surface, and it is the surface tension that pulls the liquid up, or down, depending on the curvature of the surface.

The forces acting on the column of water in a capillary tube are shown in Fig. 12.11. Notice that at the upper surface where the liquid touches the glass, the geometry is very much like the hemispherical soap bubble that we analyzed before. The surface tension force lies in the liquid surface. Therefore, it is pointed upward at an angle θ from the vertical. The upward component of this force is equal to the product of the surface tension, $\cos\theta$, and the circumference of the contact ring:

$$F_{vertical} = \gamma \cos\theta (2\pi r).$$

This upward force is balanced by the weight of the liquid column, which is

$$W = \rho_{liquid} \, gh(\pi r^2).$$

When the upward capillary force balances the downward force of the weight of the column, we have

$$h = \frac{2\gamma \cos\theta}{\rho_L g r}.$$

As shown in Fig. 12.12, the depression of the column of mercury in a glass capillary is essentially the same effect and is governed by the same formula. Notice that the smaller the radius of the capillary tube, the higher the liquid can rise (or sink in the case of mercury). Capillary action is responsible for the absorption of liquids in paper towels and is partially responsible, along with osmotic pressure, for the rise of sap in plants and trees.

We have already seen that there is excess pressure inside a spherical surface of liquid. The smaller the radius of curvature, the greater the excess pressure. The same arguments show that if the curvature is negative, the pressure in the liquid is less than on the outside. Both these situations occur in capillary tubes. In the case of water in a tube, the radius of curvature of the meniscus is negative and the pressure of the liquid inside the surface is less than than the pressure on the outside. Note that the pressure *must* be less than atmospheric since the column of water is higher than that of the reservoir from which it comes. With the meniscus formed by mercury, however, the radius of curvature is positive. The pressure of the mercury just under the surface must be greater than atmospheric. Once again, this situation is reasonable since the level of the mercury is depressed and the pressure must be equal to that of the mercury at that depth.

Let us calculate two cases of capillary rise. In the case of mercury, the contact angle between the surface of the mercury and the lower wall is 140°. The surface tension of mercury is 0.47 N/m and the density of mercury is 13.4×10^3 kg/m³.

$$h = \frac{2(0.47 \text{ N/m})(\cos 140°)}{(13.4 \times 10^3 \text{ kg/m}^3)(9.8 \text{ N/kg})r} = \frac{-5.4 \times 10^{-6}}{r}.$$

If the capillary tube has an inside radius of 1 mm, then the mercury in it will be depressed by 5.4×10^{-3} m = 5.4 mm.

Water completely wets clean glass so that the angle of contact between the surface of the water and the glass wall is 0°. Therefore the height that water will rise in a glass capillary is

$$h = \frac{2(0.073 \text{ N/m})}{(1 \times 10^3 \text{ kg/m}^3)(9.8 \text{ N/kg})r} = \frac{1.5 \times 10^{-5}}{r}.$$

For a glass capillary tube with a radius of 1 mm, the water will rise to a height of 1.5 cm.

Some of the cells just inside the bark of trees are long and hollow with interior dimensions less than 1 μm. At first thought, one might expect that capillary action could explain how sap rises in trees even to heights of several hundred feet. Simply make the cellular channels thin enough. Note, however, that by the time the water has risen to a height of about 34 ft, the internal absolute pressure is approximately 0. Standard atmospheric pressure can support a column of mercury that is 76 cm high. The comparable height of a free-standing column of water is 10.3 m. The pressure in such a column starts out at atmospheric pressure at the bottom and decreases steadily to 0 at the top: $P = P_0 - \rho g h$.

Can liquids sustain negative pressures? If so, the liquid would be under tension! Apparently this situation can and does occur, although the complete explanation of the rise of sap is still a matter of controversy. To be sure, the pressure in the roots, P_0, is greater than atmospheric pressure because of osmosis. The osmotic effect is small, however, in comparison with the large pressure differential needed to raise water to the top of tall trees.

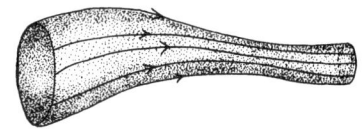

Figure 12.13. Fluid stream lines.

■ "IDEAL" FLUID FLOW—A POOR APPROXIMATION

The motion of fluids—fluid dynamics—is one of the most complicated fields of physics. Our standard method of dealing with complex phenomena is to make simplifying assumptions so that we can calculate a first approximation. For instance, in the study of mechanics we start out by assuming that there is no friction. The equivalent assumption in the study of fluids is that the *viscosity* is 0. We will define viscosity later in a formal way but for now keep in mind that the viscosity of ketchup at room temperature is high, the viscosity of water is low, and the viscosity of air is even lower. We cannot arbitrarily say, however, that the effects of viscosity can usually be ignored when objects move through air. The motion of dust particles in air is completely dominated by the viscous effects.

In spite of the fact that we must be very careful in applying our first approximation, let us analyze fluid flow under the special conditions of 0 viscosity and incompressibility. Most *liquids* are incompressible, and in regions where the pressure does not change rapidly, the volume changes of gases can be ignored. The fluid motion must be smooth and without turbulence. We will also ignore flow that is rotational with small whirlpools. Does fluid ever flow in such a simple way? Yes, the approximation is good when considering the movement of water in slow-moving streams or large pipes, as long as one stays away from the banks or walls. The approximation also helps to explain some of the phenomena that occur when air moves over surfaces at low speeds. We can characterize this flow in terms of *stream lines* that remain constant in time. Stream lines are tangent to the fluid velocity at each point. If we form a bundle of stream lines as shown in Fig. 12.13, we enclose a quantity of fluid that moves within the bundle and never passes in or out of the walls formed by the stream lines. Note that stream lines never cross each other and can never just end.

There is an apparent paradox. If the fluid is not compressible, and yet must stay inside the tube formed by the bundle of stream lines, how can the tube have

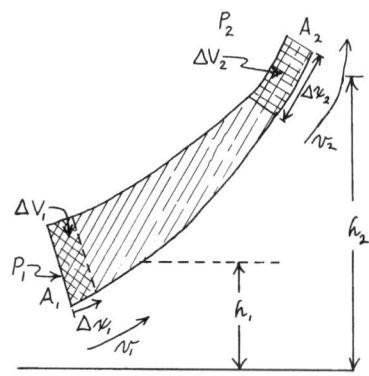

Figure 12.14. Pressures and speeds in a slug of fluid contained within stream lines—derivation of Bernoulli's equation.

different cross-sectional areas? As the tube gets larger, wouldn't the fluid have to expand? No, instead the velocity can change. Where the cross section is large, the velocity is small and vice versa.

A chunk of fluid passing within a bundle of stream lines has kinetic energy due to its motion. If in following the stream lines the fluid rises or falls, its gravitational potential energy will change. If it moves from a region of low pressure to high pressure, some work must be done on it. Each of these three forms of energy can change, but the work done on the fluid must equal the change in its kinetic and potential energies.

Let us analyze the three types of energy change. In Fig. 12.14 we show a slug of fluid (indicated by solid lines) enclosed by stream lines. During a short interval of time the fluid moves up and to the right to the position shown as dotted lines. The cross-sectional area on the left side is A_1; on the right side it is A_2. The left end moves a distance Δx_1 while the right end moves a greater distance Δx_2. Since the fluid is incompressible, the volume change on the left must equal the volume change on the right. Therefore

$$\Delta V_1 = \Delta V_2 = \Delta V \rightarrow A_1 \Delta x_1 = A_2 \Delta x_2.$$

The end volumes that are under consideration have a mass m. Since the density of the fluid is $\rho = m/\Delta V$, the volume of the small end element is $\Delta V = m/\rho$. The change of kinetic energy of the entire slug of fluid is simply the difference in kinetic energy between the two end elements:

$$\Delta(\tfrac{1}{2}mv^2) = \tfrac{1}{2}m(v_2^2 - v_1^2).$$

The change in gravitational potential energy of the whole slug of fluid is just the change of potential energy of the two end elements:

$$\Delta E_{\text{grav}} = mg(h_2 - h_1).$$

As far as the slug of liquid is concerned, P_1 is pushing it from the left and meanwhile the slug has to exert a pressure P_2 on the right. The work done *on* the slug is $P_1 A_1 \Delta x_1$. The work done by the slug is $P_2 A_2 \Delta x_2$. The net work done is $W = (P_1 - P_2) \Delta V = (P_1 - P_2) m/\rho$. The work done on the slug of fluid must be equal to its change of kinetic energy plus the change in potential energy:

$$(P_1 - P_2)m/\rho = \tfrac{1}{2}m(v_2^2 - v_1^2) + mg(h_2 - h_1),$$

$$P_1 + \tfrac{1}{2}\rho v_1^2 + \rho g h_1 = P_2 + \tfrac{1}{2}\rho v_2^2 + \rho g h_2.$$

This is Bernoulli's equation, named for Daniel Bernoulli (1700–1782), a Swiss mathematician and scientist. Let us examine the consequences that it predicts and compare these with experiments.

First, each term in the equation has the dimension of energy density, *but pressure is not an energy density*. For instance, it is not the case that the extra kinetic energy of a fluid as it enters a narrower tube is provided by the energy stored in the higher pressure in the wider tube. The work accounted for in the derivation is provided by the source of the fluid, perhaps in a remote standpipe. This work cannot be $\int P \, dV$ because we have built into the derivation that the fluid is incompressible. Let us calculate the energies and pressures involved in a particular case of a real fluid. The change in kinetic energy as a volume of water goes from a large diameter tube to a smaller one is $\Delta E_{\text{kin}} = \tfrac{1}{2}\rho V(v_2^2 - v_1^2) = V(P_1 - P_2)$. The expansion of the water can be found in terms of the bulk modulus: $\beta = \Delta P/[(\Delta V)/V]$.

The work done by the internal energy during expansion is

$$P\Delta V \approx -\left(\frac{P_2+P_1}{2}\right)\frac{V}{\beta}(P_2-P_1) = \frac{1}{2}\frac{V}{\beta}(P_1^2-P_2^2).$$

The ratio of gain in kinetic energy of the fluid to work done by the expansion is

$$\frac{\Delta E_{\text{kin}}}{P\Delta V} = \frac{(P_1-P_2)V}{\frac{1}{2}\frac{V}{\beta}(P_1^2-P_2^2)} = \frac{\beta}{\frac{1}{2}(P_1+P_2)} = \frac{\beta}{P_{\text{av}}}.$$

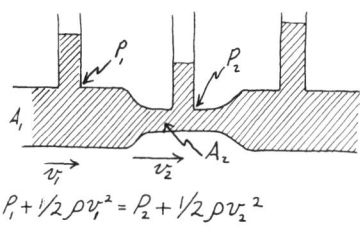

Figure 12.15. Pressure and speed relationships of fluid moving horizontally.

For water, $\beta = 2 \times 10^9$ N/m². When the average pressure of the water is around atmospheric pressure (1×10^5 N/m²), this ratio is 2×10^4. Evidently, the energy to provide the increased kinetic energy of the fluid cannot come from the very small energy of compression.

A second point about Bernoulli's equation: If the fluid is not moving, we should get the pressure–depth relationship for a static fluid:

$$P_1 + \rho g h_1 = P_1 + \rho g h_2.$$

If P_2 is the pressure at the top of a liquid and $(h_2 - h_1)$ is the depth from that top surface, h, then we have the same formula that we used previously:

$$P = P_0 + \rho g h.$$

If we leave out the potential-energy term, we get a relationship between the pressure and speed along the stream lines of a fluid moving horizontally. Where the speed is high, the pressure is low. An example of this effect is shown in Fig. 12.15. The speed in the narrow region is necessarily greater than it is in the larger region, since there must be the same volume transported per second across any cross section. If $\Delta V_1/\Delta t = \Delta V_2/\Delta t$, then

$$A_1(\Delta x_1/\Delta t) = A_2(\Delta x_2/\Delta t) \rightarrow A_1 v_1 = A_2 v_2.$$

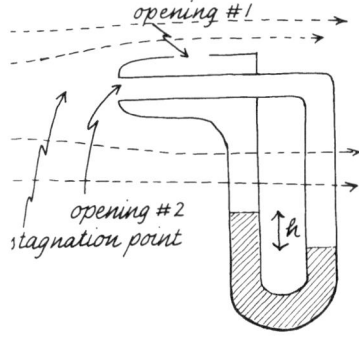

Figure 12.16. Prandtl or Pitot tube.

The pressure in the fluid is measured by the height of the static fluid in the standpipe. A tube like this, with a narrow section and standpipe, is called a Venturi tube. When it is inserted into a pipe in which fluid is flowing, it can be used to measure v_1, the velocity of flow in the main pipe. Bernoulli's equation gives one relationship between v_1 and v_2 in terms of the pressures P_1 and P_2. The other relationship needed to eliminate the variable v_2 comes from the equation for continuity of flow: $A_1 v_1 = A_2 v_2$.

A type of air-speed indicator called the Prandtl tube or sometimes a Pitot tube is shown in Fig. 12.16. The trick of its operation is that it measures both the pressure in the moving stream and the pressure at 0 velocity and takes the difference automatically. Note first of all that opening 1 is adjacent to the main flow. The pressure will be less than it would be if the fluid were standing still. Opening 2 samples the pressure in a region called the stagnation point. Observe the way the stream lines diverge and avoid the leading edge of the probe. At that leading edge the speed of the fluid relative to the probe is 0. The pressure at the stagnation point is read by the right-hand part of the U tube, while the pressure in the moving stream is read by the left-hand part of the tube. The difference in levels of the liquid in the two legs measures the difference in pressure at the two points. Bernoulli's equation requires $P_0 + \frac{1}{2}\rho v_0^2 = P + \frac{1}{2}\rho v^2$. Since $v_0 = 0$ and the difference in pressures is $P_0 - P = \rho_L g h$, we obtain $\frac{1}{2}\rho v^2 = \rho_L g h$. (The density of the

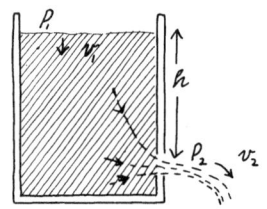

Figure 12.17. Geometry causing Vena Contracta.

moving fluid is ρ; the density of the indicator liquid in the U tube is ρ_L.) The instrument can be calibrated to read speed directly in terms of the height difference h.

Suppose that such an air-speed indicator uses oil with a density of 1×10^3 kg/m³ for the indicating fluid. What is the air speed if $h=10$ cm?

$$(1\times 10^3 \text{ kg/m}^3)(9.8 \text{ N/kg})(1\times 10^{-1} \text{ m}) = \tfrac{1}{2}(1.2 \text{ kg/m}^3)v^2 \rightarrow v=40 \text{ m/s}.$$

With a choice of indicating fluid with the proper density, the instrument can be made sensitive in any of several different velocity ranges.

When liquid flows out of a hole near the bottom of a tank, the stream lines converge as shown in Fig. 12.17. Note that they are not parallel to each other as they leave the tank. The cross-sectional area of the stream ends up smaller than the area of the hole. (This region of smallest cross section is called the *Vena Contracta*, the contracted vein.) Let us apply Bernoulli's equation to find the velocity of the stream as it leaves the hole. The top surface of the liquid is at a height h with respect to the hole:

$$P_1+\tfrac{1}{2}\rho v_1^2+\rho g h = P_2+\tfrac{1}{2}\rho v_2^2.$$

For a small container, $P_1 \approx P_2$. Both pressures are atmospheric. (Note that P_2 is the pressure of the stream out in the air.) Furthermore, $v_1 \ll v_2$. The top level of the liquid does not fall fast compared with the velocity in the stream. If we say that v_1 is approximately 0, then we are left with the equation $\rho g h = \tfrac{1}{2}\rho v_2^2$ and $v_2 = \sqrt{2gh}$.

When you observed the three streams coming from the holes in the milk carton (p. 281) you probably could not compare speeds just by the appearance of the trajectories. If the depth of the water over the bottom hole were four times that over the top hole, the horizontal velocity of the bottom stream would be twice that of the top stream. The horizontal *range* of the two streams, however, may not have been very different since the top stream had further to fall. In order to compare the velocities quantitatively, it would be necessary to express the horizontal range in terms of the horizontal velocity and the height.

Take another look at the formula for the velocity of the stream coming out of the hole. If the liquid is sent through a tube with a 90° turn and shot straight upward with the same speed, how high would it rise (assuming zero viscosity and therefore no friction losses in going through the 90° turn)? If you try this experimentally, you will find that because of frictional losses the stream will not reach anywhere near its original height.

In using Bernoulli's theorem to find the velocity of the stream, we were of course applying the law of conservation of energy. The velocity did not depend on the area of the hole and, furthermore, did not depend on the fact that the cross-sectional area of the stream is actually less than the area of the hole. Now let us take into account that the stream is also carrying momentum. The time rate of delivery of momentum by the stream must be equal to a driving force within the tank. Presumably, the driving force results from the pressure of the liquid and ultimately must be provided by the opposite wall of the tank. There is an interesting problem in connection with this calculation, however. The momentum per unit volume in the stream is equal to the density times the velocity: (momentum)/(volume)$=\rho v$. The volume of the stream that is emitted in unit time is equal to the product of the cross-sectional area A_1 and the velocity v_1 as shown in Fig. 12.18. Therefore the momentum carried away in the stream per unit time is equal to:

(momentum/unit volume)(volume emitted per unit time) = $(\rho v)(A_1 v)$

$= \rho A_1 v^2$.

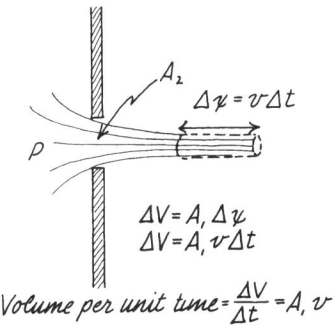

The momentum carried away per unit time must be equal to the force provided by the container: (momentum)/$\Delta t = F = PA_2$. The area of the hole is A_2, which we know is greater than A_1. In general, however, we do not know the ratio A_2/A_1 nor, surprisingly enough, do we in general know P. At first thought we would expect that $P = \rho g h$ since the pressure in a static liquid depends only on the density and depth of the liquid. The catch is that this liquid is not static. Near the hole inside the container the liquid is beginning to move fairly rapidly and therefore the pressure in the immediate vicinity of the hole must be less than it would be at that depth in a static situation.

Note the trap that we would fall into if we were not aware that $A_1 < A_2$. If we said $A_1 = A_2$, and $P = \rho g h$, then Newton's second law concerning the time rate of change of momentum would give us

$$\rho v^2 = P = \rho g h \rightarrow v = \sqrt{gh}.$$

Compare this incorrect result with that which we derived from the conservation of energy using Bernoulli's formula: $v = \sqrt{2gh}$.

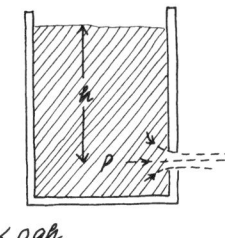

In one special case we can choose a geometry so that we can be sure of the value of P and therefore the ratio A_1/A_2. Instead of a hole in the container, we will use a reentrant tube such as that shown in Fig. 12.19. If the hole is small, the liquid in the container will be moving only in the immediate vicinity of the hole that is not near any of the walls. Therefore the walls will sustain only the static pressure. In particular, the section of the wall directly opposite the hole in the reentrant tube will have only the static pressure and that must be the source of the force that provides the momentum change of the liquid. Our momentum formula therefore becomes

$$P = \rho g h = \rho v^2 (A_1/A_2).$$

Figure 12.18. Momentum analysis of fluid flowing through circular hole.

Since we already know that $v = \sqrt{2gh}$, it must be that $A_1/A_2 = 1/2$. In this particular geometry the cross-sectional area of the *vena contracta* can be calculated in a relatively simple manner. For an ordinary hole in the bottom of a tank, the ratio is between 1/2 and 1, but the calculation is very difficult. For a round hole with a sharp edge, $A_1/A_2 = 0.62$.

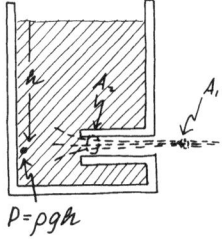

Figure 12.19. Special case of fluid exiting through reentrant tube.

■ LIFT PRODUCED BY FLUID FLOW

When the stream lines passing around an obstacle in their path are not symmetric, the speed of the flow on one side of an object may be different from the speed on the other side. If the flow satisfies Bernoulli's law, then there will be lower pressure on the side with higher speed. The resulting force on the object will have a major component perpendicular to the flow.

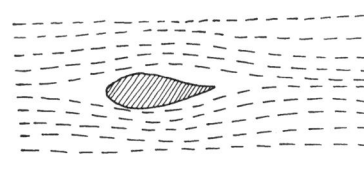

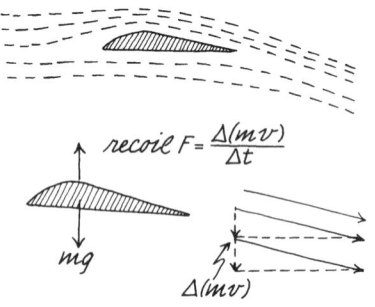

Figure 12.20. Streamlines around air foils under several conditions.

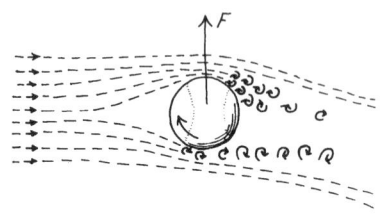

Figure 12.21. Streamlines and eddies around spinning baseball.

In Fig. 12.20 we show the stream lines around air foils under several conditions. The first air foil is symmetric and so are the stream lines around it. Note that a higher concentration of stream lines through a perpendicular area implies a higher velocity. Near the top and bottom of the air foil, the stream lines are closer together indicating that the fluid velocity in those regions is greater than it is far away from the foil. In the second case, the foil is asymmetric and so are the stream lines. There are two important characteristics of these lines. First, they crowd together above the humped section of the foil, indicating that the velocity on top is greater than it is on the bottom or in the rest of the stream. According to Bernoulli's law, the higher velocity must be accompanied by lower pressure. Consequently, we would expect a net lift force on the foil. Below the foil there is a buildup of pressure, almost as if a pillow of compressed air were supporting the plane. This result would provide a very primitive explanation of how airplanes stay up in the air. However, all these models apply only to the friction-free case of a foil that has no turbulence at its tips (an infinite wing), 0 viscosity fluid following the stream lines, and the equilibrium pressures and velocities already established. Under such circumstances, we could have perpetual lift without drag. For real foils we must take into account that energy is continually being lost due to viscosity and turbulence, and that the pattern of stream lines must change at the foil tips. As a result, the stream lines coming from both above and below the foil are deflected downward. The presence of the foil in the airstream generates a downward momentum in the air. This situation is absolutely necessary if the foil is to remain aloft and not fall. The upward force on the foil is equal to the time rate of change of momentum of the air. That change of momentum is necessarily downward. The down-wash from a wing can be felt in the immediate vicinity of the wing and produces particularly dramatic effects in the case of a hovering helicopter. The reason that you do not feel the down-wash or the excess pressure from a high-flying plane is simply because the effect is spread over a large area.

Let us stress that the foil is not supported by the bombardment of air molecules. Lift must be analyzed in terms of fluid flow, not as an application of the kinetic theory of gases. Indeed, most of the oncoming molecules do not get near the wing. They follow the stream lines that start diverging well ahead of the foil. You can feel the fluid nature of air when driving a car that is being passed by a truck. The high pressure of the air being shoved forward extends to a considerable distance beyond the truck. Furthermore, note that a simple application of Bernoulli's law is suspect in fluid foil analysis. Bernoulli's law is based on the conservation of energy of a closed system in a gravitational field. A real foil in a real fluid is energy dissipative. The fluid flow in the vicinity of the main part of a foil is best described as a combination of flow past and flow around the foil (rotational flow). The latter is caused by the flow of high-pressure air from under the wing to the low-pressure region on top. Out at the foil tips the fluid comes spiraling out trailing a vortex from each end. The real world of lift and drag is very complicated.

There is a way to create asymmetry of streamlines with a symmetrical obstacle, as shown in Fig. 12.21. The stream lines going around a stationary baseball (or a baseball that is moving slowly without spin through still air) are symmetrical. Actually, for any reasonable velocity of throw, the stream lines behind a moving baseball break down into eddy currents or turbulence. This turbulence, dissipating the ball's energy, drastically affects the motion. Nevertheless, the stream lines can persist partway past the ball and to the extent that they remain stream lines, are subject to Bernoulli's law.

In Fig. 12.21 an asymmetry has been introduced by spinning the ball. The effect is legendary. The path of the spinning ball is curved. Baseball pitchers can throw curves, but the effect is even more noticeable with Ping-Pong balls or golf balls. As you can see in the diagram, before the stream lines break up into turbulence, they crowd together more on one side of the ball than the other. A spinning surface of the ball increases the velocity of the air on one side and decreases it on the other. Since the pressure is lower in the region with higher velocity, the ball moves sideways in that direction. Necessarily, in order to conserve momentum, air is swirled sideways in the opposite direction.

A standard demonstration of Bernoull's principle is to blow between two small pieces of paper that are lightly held together along one edge. The naive expectation is that the pieces of paper will be blown apart. Instead, they are pulled together. Once again, the explanation is that the region with higher velocity (between the pieces) also has lower pressure. Remember, however, that in this case, too, the air must end up changing its momentum in the opposite direction from the thrust exerted on the obstacle. In this case, if the two pieces of paper are drawn together, the air escaping out the edges must spread out in the two directions perpendicular to the paper as shown in Fig. 12.22. The recoil action of the stream that exerts a low-pressure pull on an obstacle is easy to see when you hold a spoon in a faucet stream as shown in the diagram. In this case the spoon is snatched into the stream because of surface tension as well as the higher velocity and lower pressure on the convex side of the spoon. The stream does not "bombard" the spoon away. Notice what happens to the direction of the stream after it leaves the spoon. A force to the left on the spoon must be balanced by a force to the right on the water. The same demonstration can be performed with a stream of air. This is the basis for the demonstration of supporting a ball in a blast of air (which need not be vertical) from the pressure output of a vacuum cleaner.

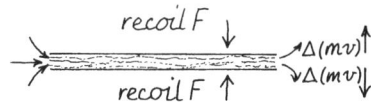

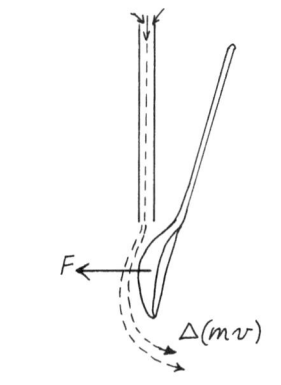

Figure 12.22. Forces on fluid foils caused by fluid thrust.

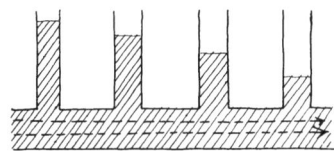

Figure 12.23. Pressure gradient in a pipe with liquid flow.

■ THE REAL WORLD OF VISCOSITY

So far we have assumed that there is no friction in fluids. The great physicist-mathematician John von Neumann characterized this model as the study of dry water. Aside from the fact that turbulence is associated with most fluid motion, even in stream-line flow there is energy lost because of friction. Consider, for example, the flow of water through a long horizontal garden hose. According to Bernoulli's law, the pressure in the tube is constant since the cross section and height remain constant. We all know very well, however, that the pressure along such a tube drops steadily as shown in Fig. 12.23. Indeed, *since there is friction,* if there is not a higher pressure at the beginning of one section than there is at the end of that section, why should the fluid in that section flow?

We have also assumed that all the fluid in a particular cross section of a pipe is moving with the same velocity. The velocity at the center of the pipe is supposed to be the same as it is at the walls. The real situation, however, is radically different and is shown in Fig. 12.24. *The velocity of a fluid at the wall of a container is always zero.* That is why dust remains on fan blades and car bodies. A leaf or stone will be blown off the top of a car, but a fine grain of dust does not stick high enough above the surface to be affected by the moving air.

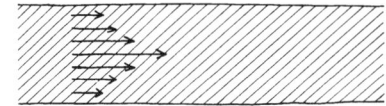

Figure 12.24. Velocity distribution in a cross section of liquid flowing in a circular pipe.

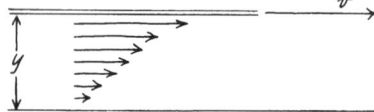

Figure 12.25. Fluid speeds between two flat plates—upper plate moving with speed v.

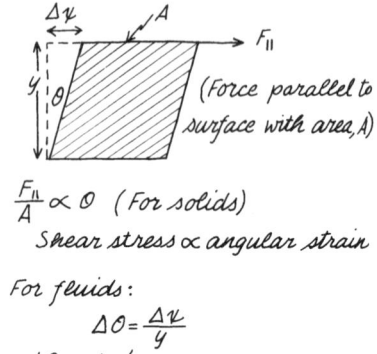

Figure 12.26. Derivation of relationships between shear stress and angular strain.

Suppose that we have a liquid between two flat horizontal plates. Let the top plate move with constant velocity to the right with respect to the bottom stationary one. Then there will be a layer of liquid attached to the bottom plate that has zero velocity and a layer of liquid attached near the surface of the top plate that has the velocity, v, of the top plate. Each layer of water in turn affects the layers next to it so that there is a continual change of velocity from top to bottom as shown in Fig. 12.25. This situation is very much like that produced by shear forces (see p. 75), shown in Fig. 12.26. A shear stress on a *solid* produces a *finite* strain measured by an angular deformation. For a given material, the strain is proportional to the stress, at least over a small range. In a *fluid*, any shear stress can produce an *infinite* shear strain. The fluid does not provide spring-type resistance that could eventually buck out the shear stress. However, not all fluids yield to the shear stress at the same rate. If you apply a shear stress to tar, for instance, the shape of the tar will change continuously but slowly. If you apply the same shear stress to water, the change in shape will be very rapid. The *time rate of change of shear strain* in the case of fluids is equal to v/y where v is the velocity of the upper plate and y is the separation distance between plates, as shown in Fig. 12.26. The shear stress in this case is the same as it would be for a solid: shear stress $= F/A$. In the case of simple fluids, the *time rate of change* of shear strain is proportional to the shear stress:

$$\text{shear stress} = F/A = \eta \times (\text{time rate of shear strain}) = \eta\,(v/y)$$

$$\eta = \frac{F/A}{v/y}$$

The proportionality constant η (eta) is called the *viscosity* and for simple fluids is independent of the velocity. Table 2 lists the values of viscosity for a number of fluids. The S.I. units for viscosity are determined by its definition:

$$\eta = \frac{N/m^2}{(m/s)/m} = N \cdot s \cdot m^{-2}$$

This unit is not yet in common use. Instead, we use the old unit based on the centimeter-gram-second system. The old unit is called the *poise*, named after a nineteenth-century French physician and scientist, Jean Poiseuille.

$$1 \text{ poise} = \tfrac{1}{10} N \cdot s \cdot m^{-2}$$

TABLE 2. Viscosities in centipoise (cP). To transform viscosity in cP to N s m^{-2}, multiply by 10^{-3}, e.g., $\eta_{water} = 1 \times 10^{-3}$ N s m^{-2} at room temperature.

Temperature (°C)	Water	Air	Mercury	SAE 10 oil	SAE 30 oil	Glycerin	Honey
0	1.79	0.017	1.68				
20	1.01	0.018	1.55	70	300	500	1500
40	0.66	0.019	1.44				
60	0.47	0.020					
80	0.36	0.021					
100	0.28	0.022	1.21				

Notice the enormous difference in viscosities between air, water, and viscous liquids: (Also notice the temperature dependence of η for water. As the temperature rises, the hydrogen-bonded clusters break up.) Notice also that the viscosity of air does not change much with temperature but that the change is one of an *increase* in viscosity as the temperature increases. On the other hand, for most liquids viscosity *decreases* with an increase in temperature. The viscosity of an ordinary motor oil is high at low temperatures, making it hard to start the engine. When the engine is hot, the viscosity of the oil decreases, thus becoming too "thin" to protect the engine just when it needs it most. Hence the fabricated oils now available are designed with several components so that the combination effect will be just the opposite.

Newton derived the viscosity relationship that we have shown. If a fluid has a viscosity that is independent of velocity, it is called Newtonian. Not all fluids act so simply. Fluids composed of suspensions such as colloids frequently have different viscosities at different velocities. For instance, as the velocity of blood increases, the cells orient themselves for minimum resistance to the flow and the viscosity decreases.

The motion of fluids is frequently characterized by the *kinematic viscosity*, which is equal to the viscosity divided by the density:

$$(\text{kinematic viscosity}) = \eta/\rho.$$

The viscosity of air is much less than that of water and yet if a whirlpool is set up in a pail of air and a similar whirlpool in a pail of water, the air will come to rest sooner than the water. The viscosity is a measure of the internal friction in the fluid. The internal energy loss as the fluid moves is proportional to the viscosity. On the other hand, the kinetic energy that a moving fluid can lose is proportional to its density. The friction in moving air is very small but moving air has little energy to lose in the first place.

As we might expect from our atomic model, the viscosity of liquids decreases with increasing temperature. The hotter the liquid, the greater the amplitude of oscillation of the molecules in their temporary cages of neighbors, and the more easily the molecules can jump from cage to cage—thus moving past each other. With gases, however, the molecules are actually moving freely between collisions. As the temperature increases, the average velocity of the molecules and hence the average momentum transfer increases as the square root of the temperature. Since the internal friction, or the viscosity, depends on momentum transfer, we might expect that $\eta \propto \sqrt{T}$. Experimentally, the viscosity increases with temperature slightly faster than $\sqrt{T}$, because of the increasing penetration into each other of the molecules at higher speeds.

Surprisingly enough, the viscosity of gases is almost independent of the pressure. You might think that as the pressure decreases, the number of collisions would decrease, which would reduce the number of molecules coming from one region to transfer momentum to the adjacent plane. On the other hand, as the pressure decreases, the mean free path of molecules increases and so molecules transferring momentum from one moving plane can affect a region further away. These two effects just cancel each other so that the viscosity in gases is independent of the pressure over a large range. When motion through a gas is dominated by viscosity effects, the terminal velocity of an object is independent of the air pressure. Although it is true that at 1% atmospheric pressure, a feather will fall with about the same acceleration as a coin, the gradual fall of a tiny dust particle

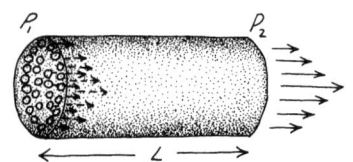

Figure 12.27. Velocity pattern of liquid flowing through cylindrical tube.

would be no faster at this low pressure that it would be at normal atmospheric pressure. In the next section we will take a closer look at the conditions that determine whether motion through a fluid is dominated by viscosity or by inertial effects.

Poiseuille, for whom the poise is named, investigated the flow of liquids through thin tubes. Being a physician, he was primarily interested in the circulation of blood. As we have pointed out, the flow of a real liquid does not follow Bernoulli's law because of the loss of energy due to viscosity. Since the velocity of a real liquid is 0 adjacent to the walls of the container, the velocity pattern of a liquid moving through a tube follows the pattern shown in Fig. 12.27. The effective size of the tube decreases drastically with radius since, for small radii, the low velocity near the wall affects most of the liquid in the tube. Poiseuille demonstrated that the volume of the liquid passed in unit time through a round tube with radius r and length L is given by the formula

$$\frac{V}{\Delta t} = \frac{\pi r^4 (P_1 - P_2)}{8 \eta L}.$$

The effective pressure across the tube is $P_1 - P_2$. We might expect that the volume per unit time would be proportional to the net pressure per unit length along the tube and that the volume of flow should be inversely proportional to the viscosity. If the velocity of flow were uniform across the entire cross section of the tube, then we might expect that the volume of flow would be proportional to the area of the cross section: πr^2. Instead, because of the wall effect, the volume of flow is proportional to r^4 (see the derivation in the Appendix).

There is such a thing as a superfluid. Below 4.2 K, at atmospheric pressure, helium becomes a liquid. As the temperature is reduced below 2.19 K, the viscosity of part of the liquid drops to 0. Without internal friction, the superfluid can penetrate tiny cracks much more rapidly than gaseous helium can. Because the atoms in the superfluid can slide over each other without energy loss, they can rapidly transmit disturbances throughout the liquid. Consequently, superfluid helium can conduct heat better than any other substance (about 100 times better than copper).

The difference between a liquid and a solid is determined by their response to shear forces. A crystalline solid will respond to a shear force with an angular distortion, the strain, which is proportional to the stress. Eventually the restoring force caused by the strain equals the shear force and there is no further strain. With a fluid, however, the distortion continues indefinitely, though in some cases slowly, so that the liquid eventually assumes the shape of the container that it is in. Many solids, however, are not crystalline and actually flow. Glass is a good example. If you support a glass rod horizontally at the two ends, it will eventually sag permanently in the middle. When glass is hot enough, it is obviously a fluid. For instance, at the temperature at which it is blown, the viscosity is about 10^7 poise.

Even crystalline solids can flow. If the strain exceeds the elastic properties of the material, the crystalline layers start sliding over each other. This effect is evident when you stretch a thin wire to the breaking point. At the point where the break begins, the wire not only lengthens, but becomes much narrower. Under sufficient pressure, such as those that exist deep in the earth, any material responds like a fluid.

■ THE FLOW OF REAL FLUIDS

The motion of real fluids or the motion of objects through fluids can be strongly influenced by the viscosity of the fluid and the adhesive forces between the molecules of the fluid and the object. The existence of a boundary layer around an object where there is a transition from zero to full velocity may have a major effect on the shape of the surrounding stream lines. Furthermore, in many situations the fluid does not follow stream lines but changes paths in a chaotic way called turbulence. Turbulent flow not only mixes up the mass elements of the fluid but also averages out velocity and momentum throughout the turbulent region. The turbulent region contains extra kinetic energy that must have been provided by the source of energy of the flow.

One way to characterize the transition between stream-line and turbulent behavior is in terms of the Reynolds number. This dimensionless quantity is derived and discussed on p. 74 in the chapter on Units, Dimensions, and Scaling. To form the Reynolds number, take the ratio of the two forces acting on a small chunk of moving fluid—the "inertial reaction" and the viscous drag:

$$\text{Re} = \frac{\text{(inertial reaction)}}{\text{(viscous drag)}} = \frac{\rho L^2 v^2}{\eta v L} = \frac{\rho L v}{\eta}.$$

The density of the fluid is ρ, its viscosity is η, the velocity of the object is v, and L is its "characteristic" length. This length is approximately the maximum size of the object that is disturbing the fluid. For instance, for a dust particle or a raindrop falling in air, the diameter of the particle or drop is about the size of the disturbed chunk of air under consideration. An airplane will disturb air all around it in a region with a characteristic dimension of the length of the wing (actually, the disturbed air in front of and behind the plane extends over much greater distances, but that is due to more complicated effects.) As you can see, the assignment of the proper size for L is inexact by a factor of 2 or more. Nevertheless, the order of magnitude of the Reynolds number is useful in characterizing the type of motion that the fluid undergoes. For instance, if Re is between 0.1 and 10, both the inertial reaction of the fluid and its viscous drag play important roles in the motion. However, if Re<0.001, we can ignore the inertial reaction. Viscous drag dominates the motion and there is essentially no acceleration of the fluid. If Re>1000, the viscous drag is negligible although the viscosity may still produce changes in the boundary region close to the object, thus influencing the flow pattern which may control the phenomenon. Usually, with Re>3000, the flow is turbulent.

For Re≪1, we are in the regime of small dust particles in air or microbe-sized creatures in air or water. In this regime, Stoke's law holds and drag is proportional to $\eta v L$ (see p. 97). Surface effects also become very important for tiny creatures because of surface tension and because the boundary layer for transition between 0 and full velocity becomes as thick as the object itself. Note that drag is proportional to the first power of velocity and the length or diameter of the object and is proportional to the viscosity of the fluid. The drag is independent of the density of the fluid because the fluid is not being accelerated.

For Re≫10, the inertial reaction of the fluid as it is shoved out of the way of the moving object becomes more important than the viscous drag. In this regime the

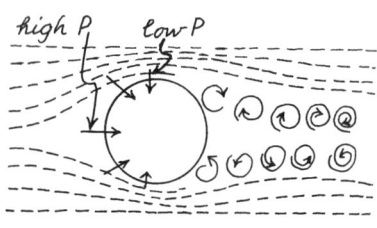

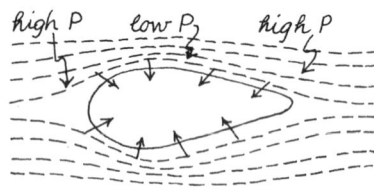

Figure 12.28. Stream lines and drag around spherical and "streamlined" fails.

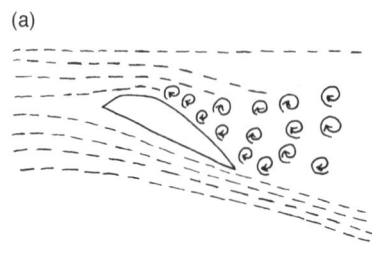

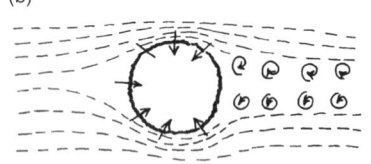

Figure 12.29. (a) Stream lined foil at steep angle leads to stall. (b) Flow lines around sphere with rough surface.

drag is proportional to the square of the velocity, the cross-sectional area, and the density of the fluid. Most human actions take place at large Reynolds numbers. Let us calculate the Reynolds number for a thrown baseball. Let $L=0.1$ m and $v=30$ m/s:

$$\text{Re} = \frac{(1.3 \text{ kg/m}^3)(1\times 10^{-1} \text{ m})(30 \text{ m/s})}{(1.8\times 10^{-5} \text{ N s m}^{-2})} = 2\times 10^5.$$

Evidently, the drag on a baseball is proportional to its cross-sectional area and to the square of its velocity. In this case the Reynolds number is also well into the regime of turbulent flow, and indeed, a turbulent region exists in the wake of a baseball.

Even at large Reynolds numbers, viscosity can affect the boundary layer where the velocity is very small. The viscous region of the boundary layer itself usually does not add much drag to the motion of the object. It does, however, affect the stream lines in the vicinity of the boundary and thus influences the way turbulence builds up in the wake of the object. In Fig. 12.28 we show the stream lines around a sphere and a "streamlined" foil. The drag on either shape is caused by two effects. There is resistance to the motion due to the pressure changes in the vicinity of the object and there is resistance caused by the viscous effects as the fluid passes the boundary. In the case of the streamlined foil, a chunk of the fluid passing from in front of the foil up to the region of the shoulder would go from a high-pressure region to a low-pressure region. In doing so, it exerts a backward shove on the foil as shown in the diagram. However, as the chunk of air moves down the stream line following the long tail, it moves from low pressure back to high pressure and in doing so exerts a thrust on the foil, sending it forward. As long as all the fluid follows stream lines the backward pressure on the foil in front is matched by the forward thrust along the tail. The net pressure resistance would be zero. The only drag would be caused by the viscous friction along the boundary layer. If the viscosity were 0 or negligible, a foil would move through a fluid with 0 resistance.

Look what happens, however, in the case of fluid moving past a sphere. The transition between 0 velocity at the surface and high velocity a short distance away becomes too abrupt and flow becomes unstable. Turbulence sets in. Along the surface where there is turbulence, there is no net thrust on the ball. Consequently, there is resistance to the motion due to the pressure drop from in front of the ball up to the point where turbulence sets in and there is no forward thrust to make up for this loss. The resistance to motion is thus very large. The energy lost to this fluid friction becomes the kinetic energy of the turbulent region.

Although the viscosity drag on a baseball is small compared with the pressure drag, it is the viscosity that creates the conditions leading to turbulence. As a chunk of fluid moves from in front of the ball up the curved surface, it loses velocity and hence energy because of viscosity at the boundary layer. Once the chunk of air gets up to the top of the curve, it does not have enough momentum to work its way down from the region of low pressure to high pressure at the tail, and so it stalls. This gives rise to the instability of turbulence. Even a streamlined foil can suffer stall at the wrong angle as shown in Fig. 12.29a. Once the turbulence begins too high up on the convex surface of the wing, the lift drops abruptly and the plane will drop.

From the diagrams that we have just shown, it is evident that a ball moving through air must meet a lot of wind resistance. Golf balls have a dimpled surface.

It might seem that this roughness would produce extra viscous drag and it does. At low speed, a dimpled golf ball has greater air resistance than a smooth one. At high velocities, however, the pressure drag is more important and the dimpled golf ball has much less pressure drag than a smooth one. The boundary layer of a rough surface is turbulent and clings to the ball far down toward the tail as shown in Fig. 12.29b. The turbulent wake is thus relatively small allowing a considerable region on the back surface of the ball where a forward thrust is exerted by the pressure difference. Since the boundary layer on a moving object may be very thin, and since the action in the boundary layer determines where turbulence begins, small changes in the smoothness of a surface may produce large effects in the drag forces. Notice that on airplane wings the rivet heads are flush with the rest of the surface.

Although the actual value of the Reynolds number for a particular situation can only indicate flow behavior within an order of magnitude, the comparison of Reynolds numbers for two different situations can be made very precise and useful. The application of this technique to analyzing scale models is discussed in Chap. 3 on p. 83.

■ REFERENCE

Fermi, Laura, and Bernardini, Gilberto, *Galileo and the Scientific Revolution* (Basic Books, New York, 1961), Appendix.

Paldy, Lester, "The Water Can Paradox," Phys. Teach. **1**, 126 (1963).

■ POISEUILLES'S EQUATION FOR CYLINDRICAL FLOW

In the flow of a viscous fluid, the velocity of the fluid is zero at the walls. Therefore, the velocity profile of fluid in a cylindrical tube must look something like that in Fig. 12.30(a). Our problem is to find $v(r)$.

Consider a cylinder of the fluid at the center of the tube, with radius r, as shown in Fig. 12.30(b). The force driving this cylinder is:

$$F_{\text{driving}} = (P_1 - P_2)\pi r^2$$

Viscosity, η, provides the retarding force that leads to constant velocity. This is a shear force acting along the curved surface of the cylinder because of the difference in speed between the cylinder walls and the adjacent fluid.

$$F_{\text{retarding}} = -\eta(2\pi rL)\frac{dv}{dr}$$

For steady flow: $F_{\text{driving}} = F_{\text{retarding}}$

$$(P_1 - P_2)\pi r^2 = -\eta(2\pi rL)\frac{dv}{dr}$$

$$\frac{dv}{dr} = -\frac{(P_1 - P_2)r}{2\eta L}$$

Integrate to find $v(r)$. Note that $v(R) = 0$

$$\int_0^v dv = -\frac{(P_1 - P_2)}{2\eta L}\int_R^r r\, dr$$

$$v = \frac{(P_1 - P_2)}{2\eta L}\frac{r^2}{2}\bigg|_R^r = \frac{(P_1 - P_2)}{4\eta L}(R^2 - r^2)$$

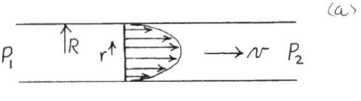

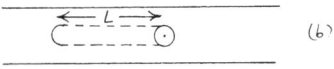

Figure 12.30. Geometry for derivation of Poiseuilles's equation.

The total flow through the tube is $Q = dV/dt$, where V is the volume passing a point. Choose a thin tube of fluid at radius r with thickness dr. The flow for this tube is

$$dQ = vdA = v2\pi r dr = \frac{(P_1 - P_2)}{4\eta L}(R^2 - r^2)2\pi r dr$$

Integrate over r to find the total flow for all the thin tubes from $r=0$ to $r=R$.

$$Q = \frac{\pi(P_1 - P_2)}{2\eta L}\int_0^R (R^2 r - r^3)dr = \frac{\pi(P_1 - P_2)}{2\eta L}\left|\frac{R^2 r^2}{2} - \frac{r^4}{4}\right|_0^R = \frac{\pi(P_1 - P_2)}{8\eta L}R^4$$

13
Vibrations

Almost everything twangs! Nuclei, atoms, molecules, crystals, bells, violin strings, pot covers, electrical circuits, bridges, the sea, yea, the great globe itself. Disturb any one of these and it will oscillate around some equilibrium position, gradually dissipating the energy that disturbed it. If you drive the system at its resonance frequency, the oscillations may grow to alarming size.

We present three approaches to describing simple oscillatory motion. The first exploits the consequences of simple observations about the position of the bob as a function of time. The second makes use of similarities between vibrations and circular motion. The third appeals to the force laws and requires solution of a differential equation. We will work out the mathematics in detail, both for the sake of reference for those who need it, and as preparation for the more complicated problems of forced vibrations.

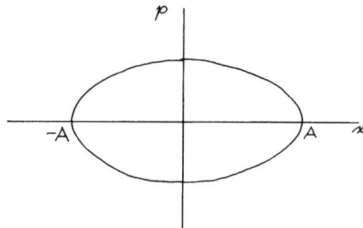

Figure 13.1. $p(x)$ (phase space) graph for SHO. The formula of the ellipse is $(p^2/2m + \frac{1}{2}kx^2 = E)$.

1. A naive observer of a pendulum or of a bob oscillating on a spring might note the way *velocity* depends on *position*: $v(x)$. It is indeed interesting and frequently useful to plot $v(x)$, or more generally, $p(x)$, where p is the momentum. These are called phase space diagrams. An example is shown in Fig. 13.1. However, since the days of Galileo, all right-thinking people have learned that the proper observation of motion consists of finding *position* as a function of *time*: $x(t)$. Thus, the brainwashed observer would note that the bob oscillates symmetrically between two maxima and passes through the equilibrium point at the half-times. These end-point observations are plotted in the $x(t)$ graph of Fig. 13.2.

To find the in-between points without making detailed measurements, we can make some plausibility arguments about the general shape of $x(t)$. This is because we are constrained by the velocity and acceleration relationships. Since $\bar{v} = \Delta x/\Delta t$, or $v = dx/dt$, the slope of the $x(t)$ curve at any time is the velocity at that particular time. Consider the consequences. Suppose that we guess that the simplest graph of $x(t)$ is created by joining the end points with straight lines. [The "end points" occur at the quarter period times, shown by the dashed vertical lines on the graph in Fig. 13.2(a).] Then the velocity between end points would be constant, and the acceleration at the turnaround points would be infinite. Since this prediction is contrary to observation, we might next guess that the end points of $x(t)$ should be connected with semi-circles as shown in Fig. 13.2(b). (After all, if the oscillator is a pendulum bob, it is traveling along a circular arc.) This curved function is an improvement over the straight lines since it calls for a changing slope (and hence a changing speed) as the turnaround points are approached, with horizontal

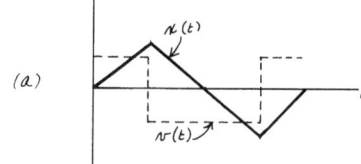

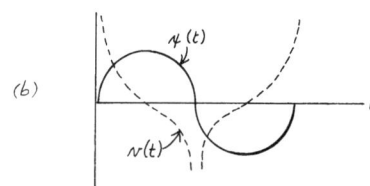

Figure 13.2. Constraints on connecting "end-points" of $x(t)$ of pendulum motion—the slope of $x(t)$ must equal $v(t)$. Straight line or semi-circular connections yield non-physical velocity values.

305

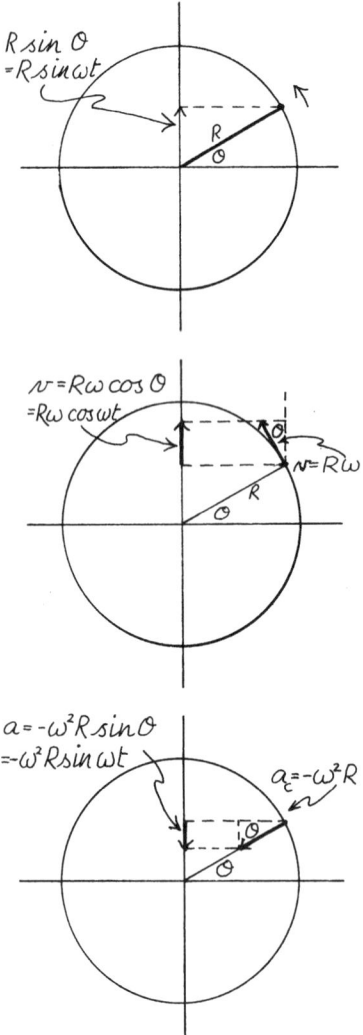

Figure 13.3. Huygens' relationship between circular motion and Simple Harmonic Motion.

slope (zero speed) at the turnaround points themselves. However, we would also have infinite slope and speed at the equilibrium position ($x=0$), which seems unlikely.

What we need is an $x(t)$ curve that has 0 slope at the turnaround points and a large but finite slope where $x=0$. A sine curve has such a shape and might plausibly be a good model for $x(t)$. Of course, these plausibility arguments cannot guarantee that $x(t)$ must be a sine curve. Nevertheless, the exercise of manufacturing a math model from the physical constraints is a useful one for introductory students. Such an exercise can also serve to reinforce understanding of the $x(t)$, $v(t)$, and $a(t)$ relationships.

2. A standard demonstration device shows the relationship between oscillatory and circular motion. Fasten a Ping-Pong ball to the side of the rim of a mounted wheel and use a projector light to cast a shadow of the ball and the edge of the wheel as it rotates at constant angular speed. The projected shadow of the ball will oscillate up and down as if it were a bob on a spring. Indeed, you can hang a bob on a long spring next to the wheel and match the oscillation period to the rotation period. Then the two shadows will move up and down in synchronism.

The geometrical relationship between the circular position of the ball and its projection is shown in Fig. 13.3. The projection on the y axis is $y = R \sin \theta = R \sin \omega t$. Since the magnitude of the tangential velocity is $V = R\omega$, the projection of velocity on the y axis is $v = \omega R \cos \omega t$. The rotating ball is subject to centripetal acceleration: $-\omega^2 R$. The y component of that acceleration is $\omega^2 R \sin \omega t$.

Thus we appear to have proved that oscillatory motion of a spring can be described by a sine function. The method is only as good as our observations that the motion of the spring bob matches that of the shadow ball. The analogy lacks generality or insight into other oscillations, but does serve to relate circular motion and projections of circular motion, which are oscillatory, such as length of daylight or height of tides. Whether a first-year physics student can usefully follow such a complicated argument is another matter. The method appears to derive the necessary formulas without using calculus, but we suspect that few students really understand the derivation or concepts. Still, the method is hoary with distinction. Huygens used it to solve oscillatory motion in 1673 in his study of clockwork—*Horologium Oscillatorium!*

3. The analysis with the greatest generality involves the mathematical consequences of linking two physics laws. An object disturbed from its equilibrium position is subject to a restoring force. To find the result, we specify that force and insert it into Newton's second law.

Within their elastic limits, many springs obey Hooke's law. The restoring force is proportional and opposite to the displacement: $F = -kx$. (Or, as Hooke wrote, "ceiiinosssttuv," a cryptogram standing for "ut tensio sic vis." Fearing that his discovery might be usurped, and rushing to ensure priority, Hooke announced his law in this form to the Royal Society, with explanation and details later.) Some springs, however, such as those in seat belts and measuring tapes, are designed to yield approximately *constant* restoring force, independent of stretch. Otherwise, seat belts would strangle us and measuring tapes would slash us.

It seems peculiarly appropriate to let the F in Hooke's law provide the F in Newton's law. Newton and Hooke were contemporaries and also bitter antagonists. In this case, we require them to join forces. If the restoring force on an object is $F = -kx$, then $ma = -kx$ and $a = -(k/m)x$. The required motion is such that the acceleration must be proportional to the negative of the displacement at all times. For 0 displacement, there must be 0 acceleration. For maximum displacement there must be maximum acceleration *and* in the opposite direction (because of the negative sign). A sine function yields this behavior. If $x = A \sin \omega t$, then $v = \omega A \cos \omega t$, and $a = -\omega^2 A \sin \omega t$. The acceleration is 180° out of phase with the displacement; the velocity is 90° out of phase.

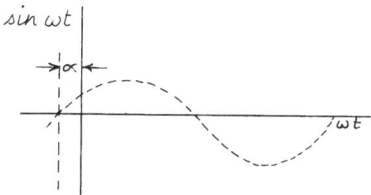

Figure 13.4a. The phase angle, α, in $\sin(\omega t + \alpha)$ determines the starting time.

In terms of differential equations, $d^2x/dt^2 = -(k/m)x$. To solve this equation we ask: What function, $x(t)$, has a second derivative that is proportional to the negative of itself? A complete solution that has this property is $x = A \sin(\omega t + \alpha)$. When we differentiate once, we get $dx/dt = v = \omega A \cos(\omega t + \alpha)$. When we differentiate again, we get $d^2x/dt^2 = -A\omega^2 \sin(\omega t + \alpha)$. This satisfies the original differential equation if $\omega^2 = k/m$. The angular frequency is therefore $\omega = \sqrt{k/m}$. The frequency is $f = (1/2\pi)\sqrt{k/m}$ Hz, and the period is $T = 2\pi\sqrt{m/k}$. It is plausible that the larger the mass, the longer the period; and the stronger the spring constant, the shorter the period. Our second-order differential equation allows two arbitrary constants in the solution: the amplitude A and a phase angle α, which arises from our choice of starting time, as shown in Fig. 13.4(a).

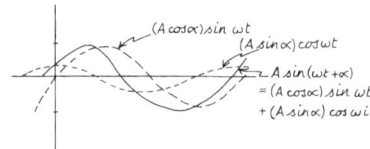

Figure 13.4b. The role of the phase, α, can also be represented by $\chi = B \sin \omega t + C \cos \omega t$ where $B = A \cos \alpha$ and $C = A \sin \alpha$.

This same motion can also be described by $x = B \sin \omega t + C \cos \omega t$, since $\sin(\theta + \alpha) = (\cos \alpha)\sin \theta + (\sin \alpha)\cos \theta$. Therefore, $B = A \cos \alpha$ and $C = A \sin \alpha$. This construction is shown in Fig. 13.4(b). (Note that we still have two arbitrary constants, B and C, but the phase angle is no longer explicit.)

For a Hooke's law restoring force, the resulting oscillation is called simple harmonic motion (SHM). "Simple" in this case means that the motion is pure sinusoidal. "Harmonic" refers to the fact that any oscillation can be described by a Fourier series of sinusoidal motions, such as in the musical case of a fundamental and its harmonics. (We describe the use of Fourier series on p. 353.)

For completeness, and for more complicated cases, note that one other mathematical function has the property that its second derivative is proportional to the negative of itself—the imaginary exponential, where $i = \sqrt{-1}$:

$$x = Ae^{i\omega t}, \quad \frac{dx}{dt} = i\omega A e^{i\omega t}, \quad \frac{d^2x}{dt^2} = -\omega^2 A e^{i\omega t}.$$

With this function it is much easier to solve the more complicated differential equations that arise with damped or driven oscillations, or with coupled oscillations.

The physical meaning of an imaginary exponential solution is inherent in an identity first described by Leonhard Euler in 1748:

$$e^{i\theta} = \cos \theta + i \sin \theta.$$

Consequently, if the mathematical solution to the oscillation equation is

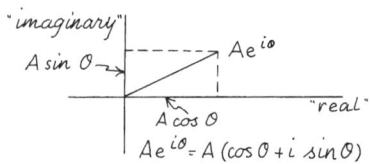

Figure 13.5. Graphical representation of $Ae^{i\theta}$.

$x = Ae^{i(\omega t + \alpha)}$, then the solution can also be written as $A\cos(\omega t + \alpha) + iA\sin(\omega t + \alpha)$. Since the differential equation is linear, each part of the complex function is also a solution:

$$x = A\cos(\omega t + \alpha) \quad \text{or} \quad x = A\sin(\omega t + \alpha).$$

You can justify that each function is separately a solution by substituting it back in the original differential equation.

Geometrically, $Ae^{i\theta}$ can be represented as a vector in the complex plane, as shown in Fig. 13.5. The magnitude of the vector is A; its component along the "real" axis is $A\cos\theta$ and its component along the "imaginary" axis is $A\sin\theta$. This representation provides a revealing model of oscillation as a function of time. If $\theta = \omega t$, then $Ae^{i\omega t}$ is a vector of constant amplitude rotating with constant angular velocity, ω. Its projection on either x or y axis represents the vector changing in phase, corresponding to simple harmonic motion.

■ ENERGY AND SHM

We can derive another property of SHM by noting that the total energy of the oscillating system must be conserved. If the system has only kinetic and potential energy (no dissipation), $E_{\text{kin}} = \frac{1}{2}mv^2$, and for $F = -kx$, $E_{\text{pot}} = \frac{1}{2}kx^2$. Therefore, $E = \frac{1}{2}mv^2 + \frac{1}{2}kx^2$, at any and all times. It is always fascinating to see examples of energy transformation. Students are particularly impressed with the exchange of kinetic and potential energy of a pendulum, or a bob on a spring. The transformations should be described to them step by step while moving the actual apparatus in slow motion.

For the mathematically inclined, the energy-conservation requirement can be derived by taking the first integral of the defining differential equation:

$$F = ma = -kx,$$

$$\int m\frac{d^2x}{dt^2}\,dx = \int m\frac{d^2x}{dt^2}\frac{dx}{dt}\,dt = -\int kx\,dx$$

$$\tfrac{1}{2}m\left(\frac{dx}{dt}\right)^2 = -\tfrac{1}{2}kx^2 + \text{const.}$$

The integration constant is the total energy of the system:

$$\tfrac{1}{2}mv^2 + \tfrac{1}{2}kx^2 = E.$$

Remember that $k = m\omega^2$. Then, if $x = A\sin\omega t$, $v = \omega A\cos\omega t$, and $E = \frac{1}{2}m\omega^2 A^2 \cos^2\omega t + \frac{1}{2}kA^2 \sin^2\omega t = \frac{1}{2}kA^2(\sin^2\omega t + \cos^2\omega t) = \frac{1}{2}kA^2$. At the equilibrium position, where $x = 0$, the speed and kinetic energy are maximum. When displacement is maximum, and $x = A$, the speed is 0 and the potential energy is maximum and equal to the total energy. Note that the energy in SHM is proportional to the square of the amplitude.

The energy equation allows us to derive a simple relationship between the speed of the oscillating bob and its position. The largest value of x is the amplitude A. At any time and position, $\frac{1}{2}mv^2 + \frac{1}{2}kx^2 = E = \frac{1}{2}kA^2$. Therefore

$$v = \sqrt{\frac{k}{m}(A^2-x^2)} = \omega\sqrt{A^2-x^2}.$$

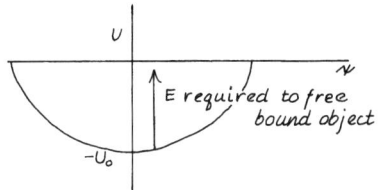

We get SHM from springs that obey Hooke's law, but we also get vibrations from almost everything else in the world. A tectonic plate slips; the earth rings. A string is plucked; the violin sings. An atom is disturbed; it produces an oscillating electromagnetic wave. Wherever there is a bound system, some object can be described as being held in a potential-energy well, such as the one shown in Fig. 13.6(a). Because the object is trapped, we describe the potential energy as being negative; it would take positive energy to free it so that its total energy would be 0. For instance, a human on the surface of the earth has negative gravitational potential energy $= -GmM/R$, as shown in Fig. 13.6(b).

Not every potential energy well can be described by a term containing $\frac{1}{2}kx^2$ (and thus not every well is parabolic in shape). Nevertheless, if there is a vibration there must be an equilibrium position, x_0, at some lowest negative energy, U_0. If the object moves from that position, there will be a restoring force shoving it back. Regardless of the exact shape of the well, it can be described as a power-series function: $U(x) = -U_0 + \frac{1}{2}k(x-x_0)^2 + a_3(x-x_0)^3 + \cdots$. For small excursions around x_0, the potential-energy well is approximately parabolic; $U \approx -U_0 + \frac{1}{2}k(\Delta x)^2$. The restoring force approximately obeys Hooke's law: $F = -dU/dx \approx -k(\Delta x)$. These are the conditions for SHM.

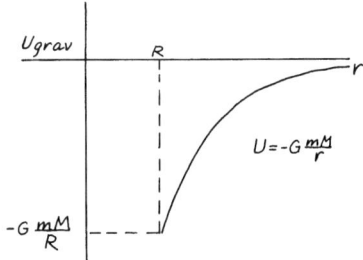

Figure 13.6a and 6b. Graphs of energy of objects trapped in potential wells.

In Fig. 13.6(c), we show the special case of a bob on a spring hanging vertically in a gravitational field. Since $U_{\text{total}} = -mgx + \frac{1}{2}kx^2 = -U_0 + \frac{1}{2}k(x-x_0)^2 = -U_0 + \frac{1}{2}kx^2 + \frac{1}{2}kx_o^2 - kxx_0$ the linear term merely moves the equilibrium point: $x_0 = mg/k$ and $U_0 = \frac{1}{2}kx_o^2$. The conclusions of this graphical and algebraic analysis are not at all obvious to most introductory students. It almost seems mysterious that a constant gravitational force and a linear spring force combine to form the linear restoring force needed for SHM.

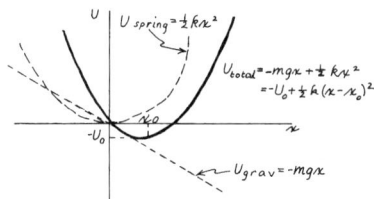

Figure 13.6c. Potential well for bob hanging from vertical spring The resulting well is displaced but is parabolic.

■ SPRINGS AND SHM

If an object with mass m is hung on a spring obeying Hooke's law, the extension will be $x_0 = F/k = mg/k$. If the object is disturbed it will oscillate with SHM with a frequency,

$$f = \frac{1}{2\pi}\sqrt{\frac{k}{m}} = \frac{1}{2\pi}\sqrt{\frac{mg/x_0}{m}} = \frac{1}{2\pi}\sqrt{\frac{g}{x_0}}.$$

Note how the frequency can be calculated by knowing the extension of the spring when the weight is attached. The effect of mass and spring constant seem to disappear. The frequency is the same as that of a simple pendulum with length equal to the extension.

We have been assuming that the spring is massless. A real spring has mass and would oscillate on its own. Let us find the contribution of this extra mass for the simple case where $m_{\text{spring}} \ll M_{\text{bob}}$, and the mass of the spring is uniformly distributed along its length. This latter condition is clearly not satisfied by a slinky, especially when it is holding only its own weight.

If we ignore the reality and if the spring has an unstretched length L, then an element along the spring has mass $dm = (m/L)dl$. (see Fig. 13.7.) The displacement of the element is a fraction, l/L, of the total spring displacement x. The kinetic energy of this element is

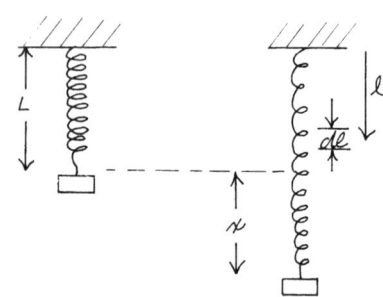

Figure 13.7. Role of mass of spring in oscillation frequency.

$$dE_k = \tfrac{1}{2}\left(\frac{m}{L}dl\right)\left(\frac{l}{L}\frac{dx}{dt}\right)^2 = \frac{m}{2L^3}\left(\frac{dx}{dt}\right)^2 l^2 dl.$$

The total kinetic energy of the spring at any given instant (and at the bob velocity dx/dt) is

$$E_{\text{kin spring}} = \frac{m}{2L^3}\left(\frac{dx}{dt}\right)^2 \int_0^L l^2 dl = \tfrac{1}{2}\frac{m}{3}\left(\frac{dx}{dt}\right)^2.$$

For the whole system, including the bob with mass M the total energy is

$$E = \tfrac{1}{2}(M+\tfrac{1}{3}m)\left(\frac{dx}{dt}\right)^2 + \tfrac{1}{2}kx^2.$$

The angular frequency of this system becomes

$$\omega = \sqrt{\frac{k}{M+\tfrac{1}{3}m}}.$$

Testing this formula makes an interesting laboratory excursion. Plot ω^2 versus $1/M$ and find the intercept on the $1/M$ axis. It should be particularly challenging for students to use a Slinky and see what happens when the approximation conditions are not met and $M \to 0$.

When dealing with springs it is useful to keep in mind the roles of the material and the length of the spring. For a wire, the stretch $\Delta l = -(1/Y)L(F/A)$, where Y is Young's modulus, which depends on the material. The restoring force is F, the cross section is A, and L is the original length. For a coil spring the stretch is $\Delta l = -(L/C)F$, where C is a constant depending on the material and its cross section and shape. Transforming this formula to Hooke's law, we obtain $F = -k\Delta l = -(C/L)\Delta l$. The spring constant k is inversely proportional to the length.

When adding two springs in series we have

$$-(\Delta l_1 + \Delta l_2) = F\left(\frac{L_1}{C_1} + \frac{L_2}{C_2}\right) = F\left(\frac{1}{k_1} + \frac{1}{k_2}\right) = \frac{1}{k}F,$$

$$\frac{1}{k} = \frac{1}{k_1} + \frac{1}{k_2}.$$

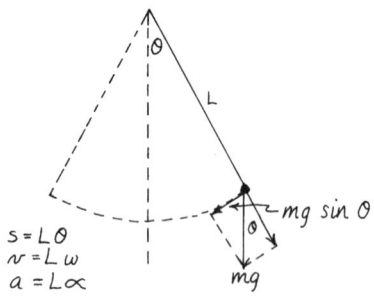

Figure 13.8. Force diagram for pendulum bob. For a "simple pendulum," the string is massless and swings from a point. The mass of the bob is concentrated at a point. For a good approximation, use thread or thin string, swinging from a small knot (not slithering around a support), and a small, dense, spherical bob.

Doubling the length of a spring reduces k by a factor of 2. Cutting a spring in half doubles k.

Springs in parallel share the load. Therefore

$$-k\Delta l = F = F_1 + F_2 = -k_1 \Delta l - k_2 \Delta l,$$

$$k = k_1 + k_2.$$

■ PENDULUMS AND OTHER OSCILLATORS

The most used example of oscillation in introductory physics is simple pendulum motion. The motion is SHM only at small angular amplitudes. As shown in Fig. 13.8, the restoring force on the bob is $mg \sin \theta$.

$$-mg \sin \theta = ma = mL\alpha,$$

$$\alpha = \frac{d^2\theta}{dt^2} = -\frac{g}{L}\sin\theta \approx -\frac{g}{L}\theta \quad \text{for small } \theta,$$

$$\omega = \sqrt{\frac{g}{L}}$$

The question is, how good is the approximation? The exact solution requires math that will not be familiar to most introductory-physics students, though some college students could follow it. Here are some steps that reduce the equation to an elliptic integral. Multiply both sides by $2(d\theta/dt)$:

$$2\left(\frac{d\theta}{dt}\right)\frac{d^2\theta}{dt^2} = 2\omega\frac{d\omega}{dt} = -2\frac{g}{L}\left(\frac{d\theta}{dt}\right)\sin\theta.$$

Integrate with respect to time and obtain

$$\left(\frac{d\theta}{dt}\right)^2 = 2\frac{g}{L}\cos\theta + C.$$

(Note that if you take the derivative of this equation with respect to time, you get the previous equation.) If $d\theta/dt = 0$ when $\theta = \theta_M$,

$$0 = 2\frac{g}{L}\cos\theta_M + C.$$

Therefore,

$$\left(\frac{d\theta}{dt}\right)^2 = 2\frac{g}{L}(\cos\theta - \cos\theta_M),$$

$$\frac{dt}{d\theta} = \sqrt{\frac{L}{2g}}\frac{1}{\sqrt{\cos\theta - \cos\theta_M}}.$$

The integral of this is

$$\frac{T}{4} = \sqrt{\frac{L}{2g}}\int_0^{\theta_M}\frac{d\theta}{\sqrt{\cos\theta - \cos\theta_M}}.$$

A series approximation to this integral yields the period

$$T = 2\pi\sqrt{\frac{L}{g}}\left(1 + \tfrac{1}{4}\sin^2\frac{\theta_M}{2} + \tfrac{9}{64}\sin^4\frac{\theta_M}{2} + \cdots\right).$$

If $\theta_M = 30°$, the first term correction is $\tfrac{1}{4}\sin^2 15° = 0.017$, or 1.7%. The period of a simple pendulum is independent of amplitude within 1% for angles up to 23°.

Many systems yield a restoring force proportional to the negative of the displacement. In all such cases, the period can be expressed in a form equivalent to

$$2\pi\sqrt{\frac{\text{(inertia term)}}{\text{(restoring term)}}}.$$

Here are some examples.

Physical Pendulum

$$T = 2\pi\sqrt{\frac{I}{Mgd}},$$

where I is the moment of inertia about the fulcrum, and d is the distance between the center of mass and the fulcrum. (The derivation is on p. 191.) For instance, the period of a meter stick swinging from one end is

$$T = 2\pi\sqrt{\frac{\frac{1}{3}ML^2}{Mg\frac{L}{2}}} = 2\pi\sqrt{\frac{2L}{3g}} = 1.6 \text{ s}.$$

(The period of a simple pendulum 1 m long is 2.0 s.)

Torsion Pendulum

$$T = 2\pi\sqrt{\frac{I}{\kappa}},$$

where I is the moment of inertia of the object about the axis, and κ is the torsion constant: $\tau = -\kappa\theta$.

Solid Wire Suspension

$$T = 2\pi\sqrt{\frac{M}{YA/L}},$$

where M is the mass of the suspended object (assuming $M \gg m$, where m is the mass of the wire), Y is Young's modulus, A is the cross section, and L is the unstretched length of the wire. This provides an interesting way to measure Y. Hang a weight on a long wire in a stair well, and time the oscillation.

Gas Piston

$$T = 2\pi\sqrt{\frac{L^2\rho}{\gamma P}},$$

where L is the length of the gas column, ρ is the density of the gas, P is the pressure, and γ is the ratio of specific heats (1.67 for ideal monatomic gases, 1.40 for diatomic gases, and less than 1.4 for all others.) This formula is for adiabatic conditions, where no heat flows to keep the gas at constant temperature as it compresses and rarifies, and therefore best satisfied for frequencies equal to or higher than audio.

Floating Object

$$T = 2\pi\sqrt{\frac{M}{g\rho A}},$$

where M is the mass of the object, A is its (uniform) cross section, and ρ is the density of the fluid. A hydrometer makes a convenient floating object for such a measurement.

LC Circuit

$$T = 2\pi\sqrt{LC},$$

where L is the inductance and C is the capacitance. Note that here inductance has the characteristics of an inertia term, and $1/C$ represents the restoring function.

■ DAMPED OSCILLATIONS

In our everyday world, motion always slows down and stops. The kinetic and potential energy turns into internal energy of the material and surroundings. In oscillatory motion, the amplitude decreases as the energy dissipates. The friction may also affect the frequency.

One of the anomalies of our practice in introductory physics is that we concentrate on dry sliding friction that is independent of velocity, except in the analysis of oscillatory motion. There we assume that the friction is proportional to the velocity. Actually, the drag in fluids of most human-size objects at human-size speeds is proportional to v^2. The internal friction in springs is essentially independent of velocity. If this internal stretching and contraction is the only source of friction, then the energy loss in vibrating springs will be proportional to the distance traveled. More energy would be lost during a cycle with large amplitude than with small.

The reason that we assume a velocity-proportional friction term is that we can easily write and solve the associated linear, homogeneous differential equation. An ordinary differential equation cannot be written for either the case of velocity-independent friction or friction proportional to v^2. The problem is that the friction term must always have the opposite sign from the velocity, but in an oscillator the velocity keeps changing sign. A constant friction term, or one proportional to v^2, would not change sign. Of course, the problem can be solved by numerical integration, tracking the effect up to the point where the velocity is zero, and then changing the sign of the friction term for the next part of the oscillation.

Here is the equation for the case where the friction, $b\,dx/dt$, is proportional to the velocity. You might consider the proportionality constant b to be the "streamlining" factor:

$$m\frac{d^2x}{dt^2} + b\frac{dx}{dt} + kx = 0.$$

This same equation describes the damped oscillations of electric charge in an LCR circuit:

$$L\frac{d^2q}{dt^2} + R\frac{dq}{dt} + \frac{q}{C} = 0.$$

In this case the friction term is exactly proportional to the velocity of the charge, which is proportional to the current.

This type of equation is always satisfied by an exponential solution:

$$x = Ae^{\alpha t}$$

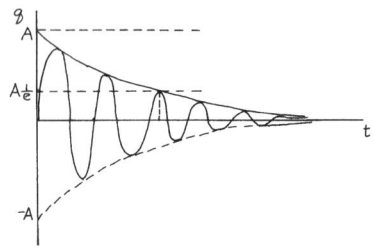

Figure 13.9. Graph of damped oscillation.

Substituting in the equation: $m\alpha^2 A e^{\alpha t} + b\alpha A e^{\alpha t} + kA e^{\alpha t} = 0$,

$$m\alpha^2 + b\alpha + k = 0.$$

The differential equation is satisfied if

$$\alpha = \frac{-b \pm \sqrt{b^2 - 4mk}}{2m}.$$

Here is the general solution with two arbitrary constants, A and B, allowing for arbitrary amplitude and phase. Recall the proof in Fig. 13.4(b): $A\sin(\omega t + \beta) = (A\cos\beta)\sin\omega t + (A\sin\beta)\cos\omega t$.

$$x = A e^{-(b/2m)t + \sqrt{b^2/4m^2 - (k/m)}\,t} + B e^{-(b/2m)t - \sqrt{b^2/4m^2 - (k/m)}\,t}$$

For the electrical case:

$$q = A e^{-(R/2L)t + \sqrt{R^2/4L^2 - (1/LC)}\,t} + B e^{-(R/2L)t - \sqrt{R^2/4L^2 - (1/LC)}\,t}$$

Each term contains a damping factor, $e^{-(b/2m)t}$, and another exponential that may be real, or imaginary and thus sinusoidal. If $k/m \leq b^2/4m^2$, the bob returns to its equilibrium position exponentially, the stored energy being dissipated without any oscillations. If $k/m > b^2/4m^2$, the bob oscillates, exchanging energy between kinetic and potential forms. If $1/LC > R^2/4L^2$, the charge flows back and forth around the circuit, storing energy alternately between the capacitor and the inductor.

A graph of damped oscillation is shown in Fig. 13.9. The time constant for decay to $1/e$ of the starting amplitude is $t = 2m/b$ or $t = 2L/R$. If the friction term were 0, the solutions would be

$$x = A e^{i\sqrt{k/m}\,t} + B e^{-i\sqrt{k/m}\,t},$$

$$x = A e^{i\omega_0 t} + B e^{-i\omega_0 t} \quad \text{or} \quad q = A e^{i\omega_0 t} + B e^{-i\omega_0 t},$$

where $\omega_0 = \sqrt{k/m}$ or $\omega_0 = \sqrt{1/LC}$. Transforming the imaginary exponentials into their sinusoidal equivalents,

$$x = x_0 e^{-(b/2m)t} \sin\left(\sqrt{\frac{k}{m} - \frac{b^2}{4m^2}}\,t + \alpha\right)$$

or

$$q = q_0 e^{-(R/2L)t} \sin\left(\sqrt{\frac{1}{LC} - \frac{R^2}{4L^2}}\,t + \alpha\right).$$

The angular frequency of oscillation is given by the square-root terms. Without friction, $\omega = \sqrt{k/m}$ or $\omega = \sqrt{1/LC}$. Note that the presence of friction decreases the frequency.

For ease of analysis, many texts (as well as common practice) define the following quantities:

The "natural" resonant frequency:

$$\omega_0 = \sqrt{\frac{1}{LC}} \quad \text{or} \quad \omega_0 = \sqrt{\frac{k}{m}}.$$

The damping factor:

$$\gamma = \frac{R}{2L} \quad \text{or} \quad \gamma = \frac{b}{2m}.$$

The "quality" factor:

$$Q = \frac{\omega_0}{2\gamma} = \frac{\sqrt{L/C}}{R} = \frac{\omega_0 L}{R} \quad \text{or} \quad Q = \frac{\sqrt{mk}}{b} = \frac{\omega_0 m}{b}.$$

In addition,

$$\omega_a^2 = \omega_0^2 - \frac{R^2}{4L^2} = \omega_0^2 - \gamma^2 \quad \text{or} \quad \omega_a^2 = \omega_0^2 - \frac{b^2}{4m^2} = \omega_0^2 - \gamma^2,$$

where ω_a is the actual resonant frequency.

The equations for damped harmonic oscillation then become

$$\frac{d^2x}{dt^2} + 2\gamma\frac{dx}{dt} + \omega_0^2 x = 0 \quad \text{or} \quad \frac{d^2x}{dt^2} + \frac{\omega_0 dx}{Q\, dt} + \omega_0^2 x = 0.$$

Let us investigate the nature of the Q of a system. This factor is commonly used in many branches of physics and engineering to characterize the properties of oscillating systems. Evidently, the larger the resistance or friction, the lower the Q. Using the sinusoidal expression:

$$x = x_0 e^{-\gamma t} \sin(\sqrt{\omega_0^2 - \gamma^2}\, t + \alpha) = x_0 e^{-\gamma t} \sin(\omega_a t + \alpha).$$

If friction is low so that $\omega_a \approx \omega_0$

$$x = x_0 e^{(-\omega_0/2Q)t} \sin(\omega_0 t + \alpha).$$

In terms of number of cycles of oscillation, n,

$$t = \frac{2\pi n}{\omega_0} = nT_0.$$

Therefore

$$x = x_0 e^{-n\pi/Q} \sin(\omega_0 t + \alpha).$$

The amplitude falls by e in Q/π cycles, and the energy falls by e in $Q/2\pi$ cycles (since the energy is proportional to the square of the amplitude). If A is the amplitude and E the energy, and for $\omega \approx \omega_0$ so that to first approximation the exponential is linear: $(\Delta A)/A = \pi/Q$ and $(\Delta E)/E = 2\pi/Q$ in one cycle.

For a Q of 600, about 1% of the energy is lost in each cycle. A piano or violin string sings for a second or so after it has been plucked. With a fundamental frequency of several hundred, the Q of such a string must be of the order of 10^3. Seismic vibrations lose intensity very slowly and have Q values of several hundred. An atomic transition producing visible light has a duration of about 10^{-8} s. The period of visible light is about 10^{-15} s, and so the Q must be about 10^7. A γ ray from the nucleus of Fe^{57} (when bound in a crystal—the Mössbauer effect) has a Q of over 10^{12}. As we will see, the Q value determines the sharpness of response when an oscillator is driven by an imposed frequency. For instance, a typical Q value for a portable amplitude-modulated radio tuner is 10^2. If the receiver is tuned to 1.000 MHz (the center of the AM band), the response curve would have a full width (at half maximum) of 10 000 Hz. This bandwidth is

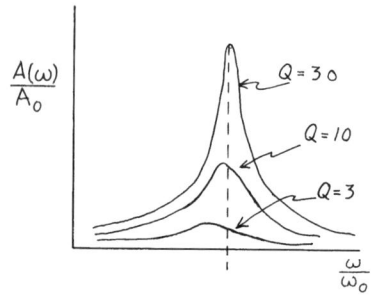

Figure 13.10a. Resonant response of amplitude of a driven system for different Q values.

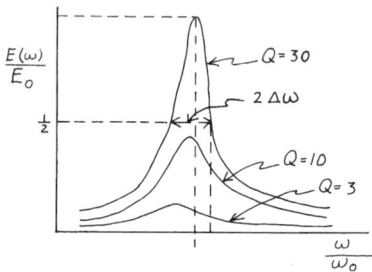

Figure 13.10b. Resonant response of energy of a driven system for different Q values.

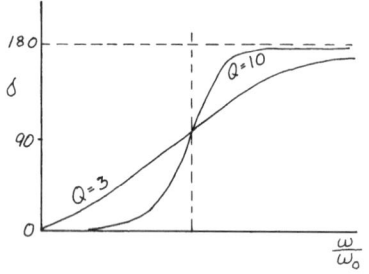

Figure 13.10c. Phase relationships of a driven system for different Q values.

necessary in order to carry the range of audio frequencies covered, which for AM radio is only 5000 Hz.

■ DRIVEN OSCILLATIONS

Simple Demonstrations

When an external force drives an oscillator with a frequency close to its natural frequency, oscillations of large amplitude build up. This phenomenon is important in both natural and human-made devices. The common behavior of any such system is characterized in terms of the ratio of driving frequency to natural frequency, ω/ω_0. When ω is much smaller or much larger than ω_0, the amplitude of the driven system is small. When ω is close to or equal to ω_0, the amplitude becomes large. The smaller the internal resistance or friction of the system (the larger the Q), the greater will be the resonant response when $\omega \approx \omega_0$ and the faster the amplitude will fall off as ω differs from ω_0. This behavior is shown in Fig. 13.10(a). The response in terms of energy absorption from the source is shown in Fig. 13.10(b).

The driven system lags in phase from the driver. When the driving frequency is low ($\omega/\omega_0 < 1$) the phase lag is small. When $\omega/\omega_0 = 1$, the phase lag is 90°. When $\omega/\omega_0 > 1$, the phase lag grows toward 180°. Once again, the Q of the circuit determines the sharpness of these phase changes, as shown in Fig. 13.10(c).

Several commercial devices are available to demonstrate the resonance and phase phenomena of driven oscillation. It is also possible to demonstrate the qualitative effects with a very simple apparatus. Hang two pendulums as shown in Fig. 13.11. The horizontal string should be stout and its tautness adjustable. If the two bobs have the same mass, the system will display complete interchange of energy between the two pendulums, and the existence of normal modes. For our demonstration of driven oscillation, however, make one of the bobs at least 10 times heavier than the other. Then the heavy one will be the driver, and you can vary ω by changing the length of its string. Both amplitude and phase of response as a function of ω/ω_0 are easily demonstrated and observed.

Another fascinating way to demonstrate and experiment with forced oscillations is to measure and observe your walking behavior. We all swing our arms as we walk, and, of course, we have to swing our legs. Prigo and Bachman described the process. By relaxing and letting your arm swing freely, and then by standing on a step and letting the other leg swing freely, you can measure the natural periods of your limbs. These turn out to be remarkably close to the periods you might calculate assuming your limbs to be uniform cylinders. Next, measure the period of your step when walking at a comfortable gait. For most people the walking period is equal to their arm-swinging period, and slightly shorter than their leg-swinging period. It takes extra effort to walk either faster or slower than your natural period. You can observe the phase lag of your arms below, at resonance, and above. At your natural pace—at resonance—your arms are 90° out of phase with your legs. For another experiment, try holding two heavy weights, such as suitcases, thus increasing the moments of inertia of your arms and their natural periods. Your natural frequency will be reduced and it will be very tiring to keep up with someone walking at what was your natural pace.

Quantitative Study

Let us study the case of an oscillator driven by a sinusoidal voltage. The resulting formulas will look more familiar when expressed in terms of the electrical parameters. Our defining equation is

$$L\frac{d^2q}{dt^2} + R\frac{dq}{dt} + \frac{q}{C} = Ve^{i\omega t}.$$

If we can find a particular solution that satisfies the complete equation, we can add it to the general solution for the free equation—since the general solution makes the left-hand side 0. For a particular solution, note that the final function of q must oscillate with the same frequency as the driving function, though not necessarily in phase. Let $q(t) = Ae^{i\omega t}$, and substitute into the equation

$$-LA\omega^2 e^{i\omega t} + iRA\omega e^{i\omega t} + \frac{1}{C} Ae^{i\omega t} = Ve^{i\omega t}.$$

Cancel $e^{i\omega t}$. We are left with real and imaginary terms. Solve for the complex amplitude A:

$$A = \frac{V}{1/C - L\omega^2 + i\omega R} = \frac{V/L}{1/LC - \omega^2 + i\omega(R/L)} = \frac{V/L}{(\omega_0^2 - \omega^2) + i\omega(R/L)}.$$

Multiply the numerator and denominator by the complex conjugate of the denominator:

$$A = \frac{(V/L)(\omega_0^2 - \omega^2)}{(\omega_0^2 - \omega^2)^2 + \omega^2(R^2/L^2)} - i\frac{(V/L)(\omega R/L)}{(\omega_0^2 - \omega^2)^2 + \omega^2(R^2/L^2)}.$$

Picture this complex amplitude in terms of the real and imaginary components of a vector in the complex plane. The magnitude of the vector is the square root of the sum of the squares of the components. For details of the analysis, see Fig. 13.12.

The particular solution is

$$q = \frac{(V/L)e^{i(\omega t - \delta)}}{\sqrt{(\omega_0^2 - \omega^2)^2 + \omega^2(R^2/L^2)}}.$$

The complete solution is the sum of the general and particular solutions:

$$q = q_0 e^{-(R/2L)t} \sin\left(\sqrt{\frac{1}{LC} - \frac{R^2}{4L^2}}\, t + \alpha\right) + \frac{V/L}{\sqrt{(\omega_0^2 - \omega^2) + \omega^2(R^2/L^2)}}\, e^{i(\omega t - \delta)}.$$

The complete solution satisfies the equation because, when substituted in, the general solution yields 0 and the particular solution yields the driving term. The general solution is a *transient* that eventually damps out with a time constant $\gamma = 2L/R$. After that the particular or *steady-state* solution dominates. Note first that although the system oscillates at the driving frequency ω, it is not in phase with it. The phase angle δ and the amplitude depend on the relationship of the driving frequency ω and the natural system frequency, ω_0. When $\omega = \omega_0$, $\tan \delta = \infty$

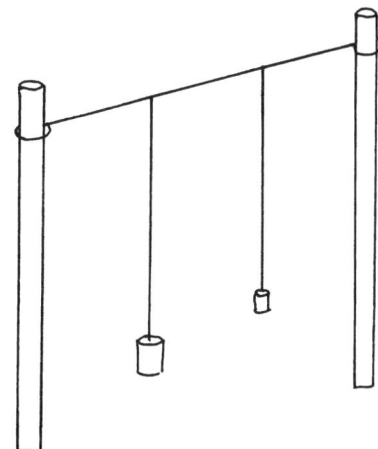

Figure 13.11. Inexpensive way to produce driven oscillations. The mass of one bob is about 10 times that of the other.

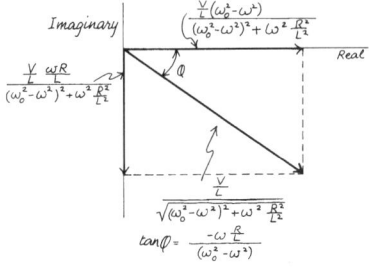

Figure 13.12. Graphical representation of the steady state solution of the equation for a driven LRC circuit.

and $\delta=90°$. For ω smaller than ω_0, $\delta<90°$. For ω greater than ω_0, $\delta>90°$. If the driving frequency is close to the natural frequency, the amplitude rises. The maximum amplitude occurs when the denominator is a minimum. If ω_0 is varied, the resonance is reached when $\omega_0=\omega$. Usually the imposed frequency is swept to obtain resonance. In that case, both terms of the denominator are varying. To find the resonance point, differentiate the quantity inside the square root by ω:

$$-4\omega(\omega_0^2-\omega^2)+2\omega\frac{R^2}{L^2}=0,$$

$$2(\omega_0^2-\omega^2)=\frac{R^2}{L^2},$$

$$\omega_{resonance}=\sqrt{\omega_0^2-\frac{R^2}{2L^2}}=\sqrt{\omega_0^2-2\gamma^2}.$$

If the damping constant $R/2L$ is small compared with ω_0, the resonance frequency is approximately equal to ω_0.

The Q of a system is also a criterion of the sharpness of the resonance curve when the system is being driven. For $\omega_{resonance}\approx\omega_0$, the amplitude at resonance is

$$\frac{V/L}{\omega_0(R/L)}=\frac{V}{\omega_0 R}=(VC)Q$$

(because $Q=\omega_0 L/R$, $R=\omega_0 L/Q$, and $\omega_0^2=1/LC$). Since the maximum charge across a capacitor would normally be VC, where V is the maximum voltage across the whole circuit, it appears that the resonant condition increases the charge by a factor of Q.

Now consider the frequency $\omega_{1/2}$, for which the two terms in the amplitude denominator are equal. When $\omega_0^2-\omega_{1/2}^2=\omega^2(R^2/L^2)$, the amplitude is $1/\sqrt{2}$ the resonant amplitude (if $\omega\approx\omega_0$). For this condition

$$\omega_0^2-\omega_{1/2}^2=(\omega_0+\omega_{1/2})(\omega_0-\omega_{1/2})\approx 2\omega_0\Delta\omega\approx\omega_0\frac{R}{L}.$$

Then

$$\frac{\omega_0}{2\Delta\omega}=\frac{\omega_0 L}{R}=Q.$$

For large Q, and hence when $\omega_{1/2}\approx\omega_{resonance}\approx\omega_0$, Q is the ratio of the resonance frequency and the full width, $2\Delta\omega$, of the amplitude curve at the height where $A=(1/\sqrt{2})A_{resonance}$. Since the energy in the oscillating system is proportional to the square of the amplitude, Q is the ratio of the resonance frequency and the *full* width of the energy curve where $E=\frac{1}{2}E_{resonance}$.

Figure 13.10(a) shows the dependence of amplitude on the ratio of ω/ω_0 for three different values of Q. Fig. 13.10(b) shows the energy dependence for the same three values.

The magnitude of the (complex) amplitude of the particular solution can also be written

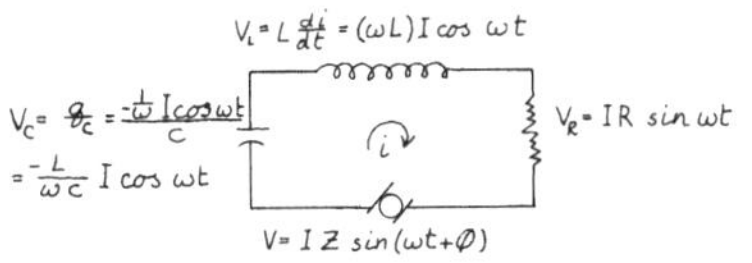

$$|A| = \frac{V/L}{\sqrt{(\omega_0^2 - \omega^2)^2 + \left(\frac{\omega\omega_0}{Q}\right)^2}} = \frac{V/L}{\omega_0^2} \frac{\omega/\omega_0}{\sqrt{\left(\frac{\omega_0}{\omega} - \frac{\omega}{\omega_0}\right)^2 + \frac{1}{Q^2}}}$$

$$= VC \frac{\omega_0/\omega}{\sqrt{\left(\frac{\omega_0}{\omega} - \frac{\omega}{\omega_0}\right)^2 + \frac{1}{Q^2}}}.$$

Figure 13.13. Voltages in a driven LRC circuit.

In terms of γ and Q:

$$\tan\delta = \frac{2\gamma\omega}{\omega_0^2 - \omega^2} = \frac{\omega\omega_0/Q}{\omega_0^2 - \omega^2} = \frac{1/Q}{\frac{\omega_0}{\omega} - \frac{\omega}{\omega_0}}.$$

As $\omega/\omega_0 \to 0$ (for frequencies $\ll \omega_0$), $|A| \to VC = A_0$ and $\delta \to 0$.
As $\omega/\omega_0 \to 1$ (at resonance), $|A| \to A_0 Q$ and $\delta \to +90°$.
As $\omega/\omega_0 \to \infty$ (for frequencies $\gg \omega_0$), $|A| \to 0$ and $\delta \to +180°$.

The charge q always lags the driving voltage (or, the displacement always lags the force). At resonance the voltage (force) is in phase with the current (velocity), since the current (velocity) of the system leads the charge (displacement) by 90°. In this way the oscillator absorbs maximum power from the driving force at resonance, since power equals the product of voltage and current (force and velocity). Fig. 13.10(c) shows the variation of δ as a function of ω/ω_0.

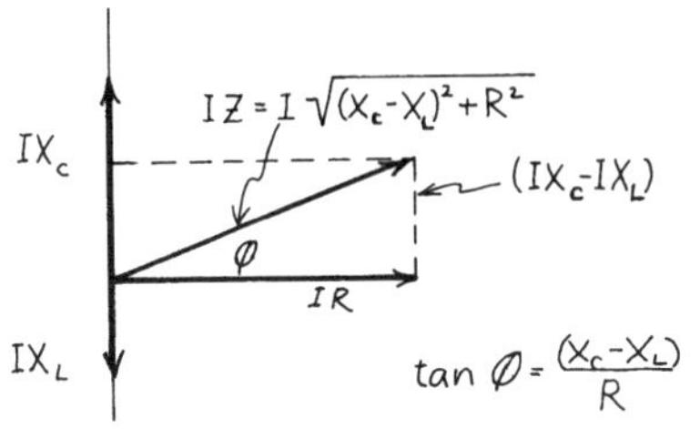

Figure 13.14. Phasor diagram for driven LRC circuit.

■ AC PHASOR NOTATION

In the early years of this century, Charles Steinmetz popularized a graphical way of describing steady-state oscillations without using differential equations. This is now the standard way to describe electric ac power systems. We illustrate this method in terms of the circuit shown in Fig. 13.13. Notice that we are now dealing with current (dq/dt) rather than with charge q. The current is the same in each of the series components and is equal to $I \sin \omega t$. The voltage across the resistor is in phase with the current: $V_R = IR \sin \omega t$. The voltage across the inductor *leads* the current by 90°:

$$V_L = L\frac{di}{dt} = (\omega L)I \cos \omega t.$$

The voltage across the capacitor *lags* the current by 90°:

$$V_C = \frac{q}{C} = \frac{-(1/\omega)I \cos \omega t}{C} = -\frac{1}{\omega C}I \cos \omega t.$$

Now, define ωL as *inductive reactance*, X_L; $1/\omega C$ as *capacitive reactance*, X_C; and R as resistance. Since the voltages are out of phase, they can be represented by "vectors" on a phase diagram as shown in Fig. 13.14. The horizontal (or real) axis is the reference phase of the current, and therefore of the voltage across the resistor. IX_L is plotted on the positive vertical (or imaginary) axis, and IX_C on the negative vertical axis. The resulting series voltage is derived from the geometry: $V_{\text{circuit}} = I\sqrt{(X_L - X_C)^2 + R^2}$. In analogy with Ohm's law for dc, $I = V/Z$, where Z is the *impedance*; $Z = \sqrt{(X_L - X_C)^2 + R^2}$.

Note how this formula compares with the formula for q on p. 317. Taking the steady-state (particular) solution:

$$i = \frac{dq}{dt} = \frac{i\omega(V/L)e^{i\omega t - \delta}}{\sqrt{(\omega_0^2 - \omega^2)^2 + \omega^2 R^2/L^2}}.$$

The factor of i brought down by the differentiation tells us that the current leads the charge in phase by 90°. Consequently, $ie^{i(\omega t - \delta)} = e^{i(\omega t - \phi)}$, where $\phi = \delta + 90°$. Now divide numerator and denominator by ω/L:

$$i = \frac{Ve^{i(\omega t - \phi)}}{\sqrt{(\omega_0^2 - \omega^2)^2 L^2/\omega^2 + R^2}} = \frac{Ve^{i(\omega t - \phi)}}{[(L/\omega)\omega_0^2 - L\omega]^2 + R^2}.$$

In this electrical engineering system, it is assumed that resonance is obtained by varying the reactances or ω until the reactances are equal in magnitude. Then $\omega_0 L = 1/\omega_0 C$ and $\omega_0^2 = 1/LC$. And

$$i = \frac{Ve^{i(\omega t - \phi)}}{\sqrt{(1/\omega C - \omega L)^2 + R^2}}.$$

We have arrived at the phasor formula.

The phase angle between voltage and current is given by $\tan \phi = (X_L - X_C)/R$. [Note that for these same conditions, $\tan \delta = R/(X_L - X_C)$].

Suppose you want a series circuit that contains a 10-H choke to be resonant at 60 Hz. $X_L = 2\pi(60 \text{ Hz})(10 \text{ H}) = 3770 \text{ }\Omega$. $X_C = 3770 \text{ }\Omega = 1/(377 \text{ C}) \rightarrow C = 0.70$ mF. The dc resistance of the choke might be 10 Ω. If there is a 10-V peak across the circuit, $I_{max} = 1.0$ A. The peak voltage across the inductor and across the capacitor (180° out of phase) would be $IX_L = 3770$ V. That, of course, would probably break down the coil insulation and destroy the capacitor. Notice, incidentally, that the circuit $Q = \omega_0 L/R = 337$. Here is another use of Q. It gives the voltage amplification across the reactances in a series circuit.

One of the convenient features of the phasor technique is that complicated circuits can be analyzed in terms of reactances just as if they were dc resistances. For instance, the parallel impedance of an inductor and a capacitor is found from

$$\frac{1}{Z} = \frac{1}{1/\omega C} - \frac{1}{\omega L} = \omega C - \frac{1}{\omega L} = \frac{\omega^2 LC - 1}{\omega L} \rightarrow Z = \frac{\omega L}{\omega^2 LC - 1}.$$

When $\omega^2 LC = 1$, that is, when $\omega = \sqrt{1/LC}$, the parallel impedance is infinite.

In doing the algebra for circuits that contain R as well as L and C, the full complex expression for reactances must be used: $X_L = i\omega L$ and $X_C = -i(1/\omega C)$. For instance, the impedance for the parallel inductor and capacitor is found by

$$\frac{1}{Z} = \frac{1}{i\omega L} + \frac{1}{-i(1/\omega)C} = \frac{-i}{\omega L} + i\omega C = i\left(\omega C - \frac{1}{\omega L}\right),$$

$$Z = -i\frac{\omega L}{\omega^2 LC - 1}.$$

Vibrations

The $-i$ in the impedance indicates that the impedance is a pure reactance, inductive if $\omega^2 LC > 1$ and capacitive if $\omega^2 LC < 1$.

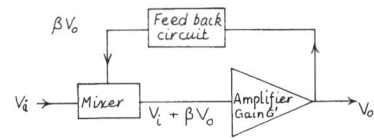

Figure 13.15. Schematic of feedback oscillator and/or amplifier.

■ OSCILLATIONS WITHOUT INDUCTORS

Most electronic oscillators do not contain inductors. Clocks, watches, and many calculators contain a piezoelectric crystal, either quartz or manufactured ceramic. These crystals have a natural oscillation frequency of expansion–contraction, which depends on the size of the crystal. As strain develops in the swelling crystal, electric charge is produced on the surfaces, creating a voltage across the crystal. Similarly, if an oscillating voltage is imposed across the crystal, the crystal will mechanically vibrate. The crystal is incorporated into an electric circuit which feeds part of its output energy back to the circuit that imposes the voltage across the crystal. Because of the high Q of the crystal (10^5 for a cheap watch that is precise to 1 second/day $= 1$ s/86 400 s), the frequency of the oscillator is locked to that of the crystal.

Another way to produce oscillations is to create feedback from the output of an amplifying device to the input. Examine the schematic circuit shown in Fig. 13.15. The gain of the amplifier by itself is G'. A fraction, β, of the output is fed back and added to the input. The input to the *system* is V_i but the input to the amplifier is $V_i + \beta V_o$, where V_o is the output voltage. The gain for the amplifier alone is $G' = V_o/(V_i + \beta V_o)$. The gain for the whole system is $G = V_o/V_i$. Solving for G in terms of G', we get $G = G'/(1 - \beta G')$. Depending on the feedback ratio *and phase*, the system can have much greater gain than G'. For positive feedback, the feedback voltage must be in phase with the input signal. If $\beta G' \geq 1$, the system will oscillate. With negative feedback and large gain, the system's gain depends mostly on the feedback ratio and so is approximately constant regardless of amplifier gain: $G = G'/(1 + \beta G') \rightarrow 1/\beta$ for $\beta G' \gg 1$.

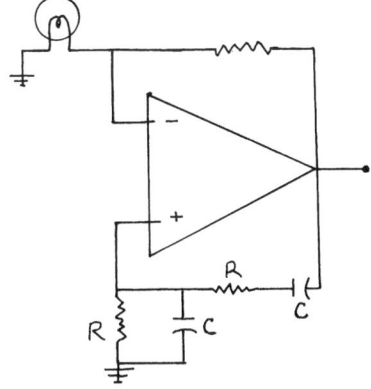

Figure 13.16. Wien bridge low distortion oscillator phase shift through amp. $= 0$.

We have all heard the results of this feedback condition when an open microphone picks up an amplified signal from the loudspeaker it is driving. The fractional feedback, the system gain, and most important, the phase delay, are all frequency dependent, and so the result is usually a high-frequency squeal. Out-of-phase feedback is used to produce ordinary audio oscillators. In Fig. 13.16 we show the classic Wien circuit, the basis for the Hewlett–Packard audio oscillators. The slow-response gain (with a time constant longer than that of any output frequencies) is moderated by the nonlinear resistance of a glowing filament so that the amplifier stays well below saturation.

■ SUMMARY OF DEFINITIONS OF Q

The "quality factor" or Q of an oscillating system characterizes the rate of decay of an initial oscillation and the response of the system to a driving force. We group together here four definitions and uses of Q.

1. For a series LRC circuit:

$$Q = \frac{\omega_0}{2\gamma} = \frac{\omega_0 L}{R} = \frac{1}{\omega_0 CR},$$

where $\gamma = R/2L = \omega_0/2Q$. The amplitude falls by e in Q/π cycles. The energy falls by e in $Q/2\pi$ cycles.

$$\frac{\Delta A}{A} = \frac{\pi}{Q}, \quad \frac{\Delta E}{E} = \frac{2\pi}{Q}, \quad Q = 2\pi \frac{\text{(energy stored)}}{\text{(energy lost in 1 cycle)}}.$$

2. For a driven LRC circuit, $Q = \omega_{\text{resonance}}/2\Delta\omega$ where $2\Delta\omega$ is the *full* width of the *amplitude* curve at the height where $A = (1/\sqrt{2})A_{\text{resonance}}$, or where $2\Delta\omega$ is the *full* width of the *energy* curve, where $E = \frac{1}{2}E_{\text{resonance}}$.

3. For a driven LRC circuit, Q determines the amplification factor at resonance of q on the capacitor or of the voltage across the capacitor. For $\omega_{\text{resonance}} \approx \omega_0$, $q \to (VC)Q$ and

$$IX_L = I\omega_0 L = IX_C = I\frac{1}{\omega_0 C} = Q(IR) = QV.$$

4. For a series LRC circuit, the actual resonant frequency is less than ω_0:

$$\omega_{\text{resonant}}^2 = \omega_0^2 - \frac{R^2}{4L^2} = \omega_0^2 - \gamma^2 = \omega_0^2 - \frac{\omega_0^2}{4Q^2} = \omega_0^2\left(1 - \frac{1}{4Q^2}\right).$$

In this case, Q is a measure of the displacement of the actual resonant frequency from ω_0. (Note that even if $Q = 1$, $\omega_{\text{resonant}} = 0.87\omega_0$.)

There is a problem of usage with these circuit parameters, γ and Q. Some books define $\gamma = R/L$. In that case, the solution to the LRC circuit is $q = q_0 e^{-\gamma t/2} \sin\sqrt{\omega_0^2 - \gamma^2/4}$. The value of Q is not affected, however. In the notation we have used, $Q = \omega_0/2\gamma = \omega_0 L/R$. In the other convention, $Q = \omega_0/\gamma = \omega_0 L/R$.

■ REFERENCE

Prigo, Robert, and Bachman, C. H., "Some Observations on the Process of Walking," Phys. Teach. **14**, 360 (1976).

14

Wave Transmission

Most introductory texts have only two or three chapters exclusively committed to the study of wave motion. As we all know, however, the subject permeates every field of physics. We hear waves, we see waves, we feel waves, and when we describe atoms we imagine waves. A disturbance in a medium propagates, transporting energy, momentum, and information without transporting the medium.

One remarkable feature of wave motion is that under the right circumstances wave energy can be sent over vast distances with very little loss. We are used to the fact that the earth is heated by electromagnetic radiation coming from the sun. We know that a sudden shift of the earth in one location can produce vibrations that destroy houses many miles away. But it is startling to realize that a *tsunami* can roll across the whole Pacific ocean at almost constant amplitude. On April 1, 1946, an earthquake near the Aleutian Islands made the Pacific ocean slosh with such an amplitude that villages on shores 4000 miles away were destroyed.

In teaching wave motion there is some temptation to avoid demonstrations of the phenomena, substituting pictures or videos. In 1987 we discovered that among a group of senior physics majors, only 10% had ever seen a ripple tank. Somehow they had missed it in high school and then again in their college introductory course. Such a failure to display the actual phenomena so that students can play and experiment with it does tremendous disservice to our common mission. The subject of waves is one of the most important in our entire curriculum. In the 1960s many devices for studying waves were invented or made readily available. These are not old-fashioned. Students should still send pulses down ropes and slinkies. They can and should create ripples in ponds, puddles, and sinks. They should learn how to use ripple tanks for demonstrating both transmission and interference of waves. (We provide some essential information about using ripple tanks on p. 329.)

■ DIMENSIONAL ANALYSIS OF WAVE VELOCITY

A disturbance in one region can propagate energy and momentum to adjoining regions without transmitting any of the original mass or substance of the source. (Of course, mass *is* transported since energy is transported!) For each kind of disturbance there must be a special analysis of the way a change at one point causes a change in the surrounding points. However, in all cases there must be some force or forcelike property that tends to restore a disturbed region to equilibrium. If a water wave goes up, gravity and surface tension will try to flatten it. If a loop forms in a rope, the tension in the rope will try to straighten it. The velocity of propagation of the disturbance should be proportional to some *positive* power of this restoring force.

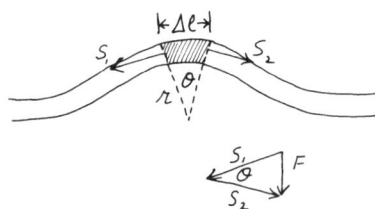

Figure 14.1. Idealized pulse in stretched rope.

In all cases there must be some inertia or inertialike property that resists change in the medium. Waves travel more slowly along a hawser than along a string, providing the tension is the same. The velocity of propagation of a disturbance should be proportional to some *negative* power of the inertia term. In the case of disturbances along a line, this inertia would be expressed in terms of linear mass density, e.g., kg/m.

Although the detailed analysis of propagation must depend on the special features of the medium, we can use a dimensional argument to find a general feature of the velocity. Let us express the conclusions of the two preceding paragraphs in terms of symbols:

$$v \propto F_r^a \mu^b,$$

where F_r is the restoring force and μ is the inertia term. The dimensions of this proportionality are

$$L^1T^{-1} = (L^1T^{-2}M^1)^a (M^1L^-)^b,$$

$$L^1T^{-1} = L^{a-b}T^{-2a}M^{a+b}.$$

Therefore

$$a-b=1, \quad -2a=-1, \quad a+b=0.$$

And

$$a=1/2, \quad b=-1/2.$$

Therefore

$$v = K\sqrt{\frac{F_r}{\mu}}.$$

For an example of the plausibility of our result, let us choose reasonable units to analyze the velocity of a pulse in a rope. The restoring force must be proportional to the tension in the rope in newtons, and μ would be the linear mass density in kg/m. Then

$$v = K\sqrt{\frac{N}{kg/m}} = K\sqrt{\frac{kg\,m\,s^{-2}}{kg/m}} = K\sqrt{\frac{m^2}{s^2}} = K \text{ m/s}.$$

By satisfying dimensionality, it appears that we have deduced the functional dependence of the velocity of wave propagation on two properties of the medium. However, there are hidden assumptions. We implicitly assumed that the velocity does not depend on the amplitude or shape of the disturbance. Our simple argument is good only for linear disturbances (the amplitude of two disturbances acting together is equal to the sum of the two independent amplitudes), without dispersion (independent of frequency).

■ WAVE VELOCITY IN A ROPE

In Fig. 14.1, we show an idealized pulse in a stretched rope that has tension S. We assume that the shape of the pulse is smooth enough so that at the top it can be approximated by the arc of a circle. The small section of rope in this arc has length δl and mass $\mu \, \delta l$. The small section is being pulled in both directions by the tension of the rope, but these two forces are not quite in line. Since the tension is always in line with the rope, the force acting at each end of the arc is tangent to the arc and therefore perpendicular to the radius. The net force acting on the small section is radial downward and has the value $F_r = 2S \sin(\theta/2) \approx S\theta$ (for small θ). The angle θ is equal to the arc it subtends divided by the radius. Therefore

$$F_r = S \frac{\delta l}{r}.$$

In our reference frame, where the rope is stationary, any piece of the rope just goes up and down. Let us change reference frame and travel to the right with the pulse. Then any particular piece of the rope shoots along to the left, and swings up, around, and down the pulse. In order for that piece to travel around the circular arc, it must be subject to centripetal force, which can be provided only by the restoring force in the rope. Therefore

$$\frac{(\mu \delta l)v^2}{r} = S \frac{\delta l}{r}, \qquad v = \sqrt{\frac{S}{\mu}}.$$

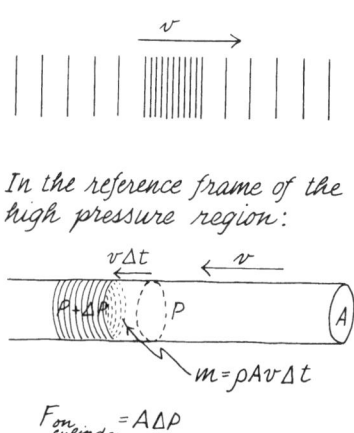

Figure 14.2. Longitudinal (or compression) pulse traveling in a fluid.

Notice how this formula agrees with the one that we derived on the basis of dimensional analysis. In this case we have replaced the restoring-force term F_r with the actual tension in the rope. We have assumed that the restoring force is proportional to that tension. With this change, the proportionality constant K is equal to unity. We have obtained a formula for the velocity of a pulse that is independent of the shape of the pulse—but notice how the restrictions enter. The amplitude of the pulse must be small and the shape must be smooth so that $\sin \theta$ can be replaced by θ, and so that we can assume that the tension is constant throughout the rope, especially in the region of the pulse.

Let us try the formula by calculating the speed of a pulse in a thick rope with a linear mass density, $\mu = 1$ kg/m. For a tension of 500 N (about 110 pounds), $v = \sqrt{(500)/(1)} \approx 22$ m/s ≈ 50 miles/h.

■ VELOCITY OF A LONGITUDINAL PULSE

A solid can transmit either transverse or longitudinal waves; an ideal gas can support only longitudinal waves of compression. Waves in most fluids are usually longitudinal except at interfaces, such as between air and water. For a simple derivation of the velocity of a longitudinal wave, let us assume that a square pulse of uniform pressure is moving to the right as shown in Fig. 14.2. Once again we will change reference frame and travel along with the pulse. In our frame it appears that the fluid is traveling to the left with velocity v. When a region of this fluid strikes the high-pressure zone, its velocity will be reduced because it will be meeting a force directed to the right. Let us choose a particular volume of fluid and set up Newton's second law for it just as it enters the high-pressure zone. We choose a cylinder of fluid, as shown in the diagram, with a cross section A and a length $v \, \delta t$. The short interval δt is just the time that it takes this cylinder to completely enter the high-pressure zone. The mass of the fluid cylinder is $\rho v A \, \delta t$ (ρ is the volume mass density in kg/m^3). During the time that the fluid cylinder is entering the high-pressure zone, the pressure on its left end is $P + \Delta P$ while the pressure on its right end is simply P. Therefore, the net force on the cylinder exerted toward the right is $A \Delta P$. Because of this net force to the right, the fluid cylinder undergoes a deceleration equal to $-\delta v/\delta t$. Consequently, $A\Delta P = (\rho v A \delta t)(-\delta v/\delta t)$ and $\rho v^2 = -\Delta P/(\delta v/v)$.

The relative change of velocity, $\delta v/v$, is related to the relative change of volume of the fluid cylinder:

$$\frac{\Delta V}{V} = \frac{A \delta v \delta t}{A v \delta t} = \frac{\delta v}{v}.$$

The volume must change as it enters the high-pressure zone because the volume of a fluid is always less at higher pressure, even though for liquids the volume change is small. The cross section of the sample cancels out of the equation, but therefore the fractional decrease in volume must equal the fractional decrease in velocity.

We can replace the relative change of velocity with the relative change of volume:

$$\rho v^2 = \frac{-\Delta P}{\Delta V/V}$$

The right-hand side of this equation is the bulk modulus B. It is the ratio of the excess pressure on a chunk of material to the fractional change of volume caused by that excess pressure. It is negative because a positive increase in pressure produces a decrease in volume. In terms of the bulk modulus, we now have a simple formula for the velocity of a longitudinal pulse:

$$v = \sqrt{\frac{B}{\rho}}.$$

Notice that once again we have a formula for velocity that is proportional to the square root of a force term divided by a density. In this case the force term has the units of pressure, and the density is a volume density, providing the correct units and dimensions for velocity.

For pulse transmission along a solid rod, the same analysis would lead to a velocity formula with Young's modulus in the numerator. Although we assumed a particularly simple form of high-pressure pulse for the derivation, the formula is good for a pulse of any shape as long as the bulk modulus remains constant. For a pulse that starts at one point in a solid and then spreads throughout the volume, a more complicated expression must be used since the solid is not only compressed but also suffers tangential or shearing forces.

The bulk modulus of a gas (for adiabatic changes) is equal to γP, where γ is the ratio of specific heats. The value of γ depends on the structure and number of atoms in the gas molecule. For monatomic gases, $\gamma = 5/3$. For diatomic gases, including air, $\gamma = 7/5$. (Newton derived the formula for the velocity of sound, but assumed that the process was isothermal and that therefore the bulk modulus was equal to P. His value for v was too small.)

The formula for the velocity of sound in a gas can be transformed so that it is in terms of the more usual gas constants. From the gas law, we have $P = NkT/V$. The density is equal to $\rho = nM/V$, where n is the number of moles and M is the molar mass. The ratio of the number of molecules in a sample to the mass of the molecules is equal to the ratio of the number of molecules in a mole (Avogadro's number) to the molar mass: $N/nM = N_A/M$. Substitute these values into the formula for the velocity of sound in a gas:

TABLE 1. *Speed of sound in various materials.*

Material	Speed (m/s)
Air (20 °C)	343
Helium (20 °C)	981
Hydrogen (20 °C)	1328
Carbon dioxide (20 °C)	267
Water	1490
Seawater	1530
Mercury	1450
Iron	~5000
Glass	~4500
Lucite	~1800
Aluminum	~5000

$$v = \sqrt{\frac{\gamma P}{\rho}} = \sqrt{\frac{\gamma(NkT/V)}{nM/V}} = \sqrt{\gamma kT(N/nM)} = \sqrt{\gamma kT(N_A/M)}.$$

Notice that the speed of a longitudinal pulse in a gas, which is the speed of sound in the gas, is proportional to the square root of the absolute temperature and to a number of fixed constants characteristic of that type of gas. Let us substitute the values for air and see what the formula gives us for the velocity of sound in everyday life: $\gamma=1.4$ and $N_A=6\times10^{23}$ molecules per mole. The Boltzmann constant $k=1.38\times10^{-23}$ J/(molecule K). The mass of one mole of an oxygen–nitrogen mixture of air is equal to 0.0288 kg. Substituting these values into the equation, we find that the velocity of sound in air $=20.1\sqrt{T}$ m/s. At room temperature, where $T=293$ K, the speed of sound is 344 m/s. A convenient approximation of this formula is to take the velocity at 0 °C and add a small amount for each degree above 0 °C. This approximate formula is $v=(331+0.6t)$ m/s. In this formula, t is the temperature in degrees Celsius.

It is revealing to compare the speed of sound in air with the root-mean-square speed of the molecules. If we solve for v in $\frac{1}{2}mv^2=\frac{3}{2}kT$, using the values in the paragraph above, we get $v=503$ m/s. It is plausible that the molecular speed is about the same as the speed of sound but slightly larger.

The speed of sound in various materials is shown in Table 1. Notice how much faster sound travels in hydrogen and helium than it does in air. The large difference in atomic mass accounts for the difference in speed. Also notice that the speed of sound in metals and water is much greater than that in air. In this case the density of metal is much greater than that of air, but the relative bulk modulus is even greater.

■ VELOCITY OF ELECTROMAGNETIC WAVE IN VACUUM

We derive the velocity of light from Maxwell's equations on p. 482. His astonishing result, derived just a little over a century ago, was that the velocity of light is determined by the electrostatic and magnetostatic properties of space. We write the equation here in order to point out its relationship to the general form of velocity that we have obtained:

$$c = \sqrt{\frac{1}{\epsilon_0\mu_0}},$$

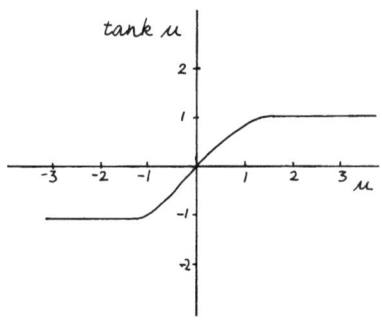

Figure 14.3. Hyperbolic tangent of μ as function of μ.

where $\epsilon_0 = 8.85 \times 10^{-12}$ is the electrical permittivity and $\mu_0 = 4\pi \times 10^{-7}$ is the magnetic permeability. The permeability plays the role of the inertia term and $1/\epsilon_0$ is the restoring-force term. Recall the analogous formulas for resonant frequency. For mechanical systems,

$$f = \frac{1}{2\pi}\sqrt{\frac{k}{m}}.$$

For electrical,

$$f = \frac{1}{2\pi}\sqrt{\frac{1}{LC}}.$$

In these formulas, the magnetic inductance L corresponds to inertia, and the reciprocal of electrical capacitance plays the role of the restoring spring constant.

■ VELOCITY OF WATER RIPPLES

Wind-driven water waves are usually very complicated. The motion of any particular chunk of water is not simply up and down, but rather is roughly ellipsoidal with some to-and-fro movement. Breaking waves are obviously chaotic. For ripples, however, where the amplitude is small compared with the wavelength and with the depth, analysis is feasible. It is a good thing! We like to use ripple tanks in our physics classes, and it helps to have reasonably simple relationships between velocity and depth and frequency. Actually, the derivation of the formula for velocity is complicated and so is the algebraic expression. A good treatment is given by Kuwabara, Hasegawa, and Kono.

Here is the formula, where g is the gravitational field strength (9.8 N/kg), $\lambdabar = \lambda/2\pi$, γ is the surface tension (73×10^{-3} N/m for water), ρ is the volume density (1×10^3 kg/m³ for water), D is the water depth in meters, and tanh is the hyperbolic tangent:

$$v = \sqrt{\left(g\lambdabar + \frac{\gamma}{\rho\lambdabar}\right)\tanh\left(\frac{D}{\lambdabar}\right)}.$$

To interpret this formula, keep in mind the graph of the hyperbolic tangent shown in Fig. 14.3. For values of the argument, u, greater than 1, tanh $u = 1$. Where u is between 0 and 1, tanh $u \approx u$. Consider first wavelengths very short compared with the depth. Then $D/\lambdabar > 1$, and tanh $u = 1$. Of the two remaining terms, the first provides the restoring force of gravity and the second depends on surface tension. A small wavelength means large curvature of the surface, which increases the effect of surface tension. For example, suppose $\lambda = 1$ cm $= 0.01$ m. Then

$$v = \sqrt{(9.8)(0.01/2\pi) + (73 \times 10^{-3})/(1 \times 10^3)(0.01/2\pi)}$$
$$= \sqrt{1.6 \times 10^{-2} + 4.6 \times 10^{-2}} = 0.25 \text{ m/s}.$$

The surface tension effect is 3 times greater than the gravitational. However, if $\lambda = 3$ cm, with $D/\lambdabar > 1$, the surface tension effect decreases by 3 and the gravitational increases by 3. Notice that the velocity is independent of water depth, but depends strongly on wavelength. The medium is *dispersive*, an effect that we did not have with pulses in ropes or air.

Now consider cases where $D/\lambdabar < 1$, and therefore $\tanh(D/\lambdabar) \approx D/\lambdabar$. The surface-tension effect will be small for these long wavelengths, and v

Wave Transmission

$\approx \sqrt{(g\bar{\lambda})(D/\bar{\lambda})} = \sqrt{gD}$. Under these conditions the medium is not dispersive; the velocity is independent of wavelength (or frequency). The dependence on water depth is very simple. This is the equation for the velocity of the actual tidal waves (which have a period of $12\frac{1}{2}$ h) or of tsunamis in deep oceans. For instance, in the Pacific a tsunami travels with speed $v = \sqrt{(9.8)(4\times 10^3)}$ m/s = 200 m/s ≈ 450 miles/h!

What happens to our dimensional argument that velocity must depend on the square root of the ratio of a restoring force and an inertial term? For the last approximation considered, create a fraction by multiplying and dividing by m/D:

$$v = \sqrt{\frac{mg}{m/D}}.$$

The numerator is the weight of a column of water displaced upward in a wave, and the denominator is the mass per unit depth. Once again we have the square root of a restoring force divided by the mass density of material being moved vertically by the wave.

For purposes of using student laboratory ripple tanks, the algebraic equation for ripple velocity is opaque. In a ripple tank, wavelengths (divided by 2π) are about the same size as the depth, and both gravity and surface tension contribute. Under such circumstances, the formula is best understood in terms of a family of curves. The graph shown in Fig. 14.4, reproduced from the article by Kuwabara, Hasegawa, and Kono (and also available in the original PSSC Teachers' Guide) is indispensible for successful operation of a ripple tank. Note first of all the range of values on the axes. Ripples travel at about 30 cm/s. At frequencies greater than about 15 Hz there is almost no dispersion. Velocity is independent of depth and frequency. If you want conditions for no dispersion as a function of frequency, choose a depth of about half a centimeter. On the other hand, if you want to observe refraction, create one region with a depth of only a couple of millimeters and the other region with a depth of at least one centimeter. Then (and this is the hardest part), make the vibrator oscillate under 5 Hz. Most of the commercial vibrators tend to stall when operated under 5 Hz. Of course, if you operate in the region below 5 Hz, the ripple velocity in the shallow region depends strongly on the depth. Consequently, the tray must be very flat and horizontal.

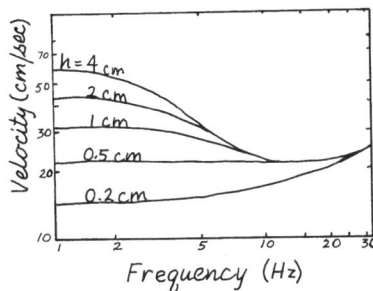

Figure 14.4. Dispersion relation of water ripples. [Reprinted from Kuwabara, Hasegawa, and Kono, Am. J. Phys. **54**, 1002 (1986)].

■ WAVE FUNCTIONS

In deriving formulas for wave velocity, we talked in terms of "pulses" of disturbance. We assumed that at any given instant a horizontal disturbance in a rope, or a pressure pattern in air, or a vertical displacement of a water surface can be described as $f(x)$, a function of distance along the line of propagation. The function might be $e^{-(x/a)^2}$, modeling a Gaussian bump, or a sine wave, or a square wave, all of which are shown in Fig. 14.5. If these disturbances move along the x axis, the mathematical function remains the same with a change of argument: $f(x)$ becomes $f(x \pm vt)$. In effect, this is a (Galilean) relativistic change of origin of the axis. For $x \rightarrow x - vt$, the disturbance moves to the right with speed v. (From the reference frame of the disturbance, the origin is moving to the left.) As time goes on, the argument in the parentheses will maintain constant value *if* x increases so that $\Delta x = \Delta(vt) = v\Delta t$. For $x + vt$, the same value is maintained for $\Delta x = -v\Delta t$, and the disturbance moves to the left with negative v.

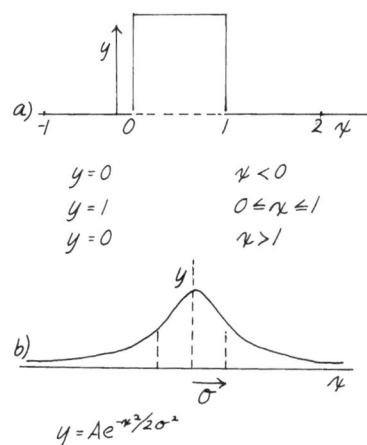

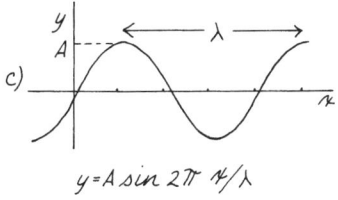

Figure 14.5. The shapes of possible disturbances in a medium.

The argument of any mathematical function must be dimensionless. Notice that in writing the expression for the Gaussian disturbance, the argument in the exponential was x/a, where a is a unit length along the axis. Similarly, for a sinusoidal disturbance we must write $\sin(360°x/\lambda)$ or $\sin(2\pi x/\lambda)$. The distance along the axis is measured in terms of fractions or multiples of the wavelength, and hence in terms of fractions or multiples of a complete sinusoidal cycle of $360°$ or 2π radians. In order to get the expression for a traveling sine wave we transform x to $x - vt$. For a rope with maximum excursions in the y axis of amplitude A:

$$y = A \sin \frac{2\pi}{\lambda}(x - vt) = A \sin 2\pi \left(\frac{x}{\lambda} - \frac{v}{\lambda}t\right) = A \sin 2\pi \left(\frac{x}{\lambda} - \frac{t}{T}\right).$$

This last form makes apparent the dimensionless nature of the argument and shows the dependence on the x/λ or t/T fraction of the full cycle.

The argument of the sinusoidal expression can be changed further by using other variables: $\omega = 2\pi f$, where ω is in radians and f is in cycles/s or Hz. The "reduced" wavelength is $\lambdabar = \lambda/2\pi$. The angular wave number is $k = 1/\lambdabar$. Then $y = A \sin(kx - \omega t)$. This is a common form, particularly in advanced texts and references. It gets rid of the 2π's and the reciprocals. Unfortunately, there are very few branches of physics where anyone has a quantitative feel for the size of an angular wave number. Spectroscopists sometimes report results in terms of the wave number $= 1/\lambda$, which is the number of wavelengths per meter. The angular wave number is evidently the number of *radians* per meter, or the number of wavelengths in 2π meters. For instance, the value of k for the 440-Hz sound wave to which the orchestra tunes is 8.04. The value of k for 1 MHz in the middle of the AM radio band is 0.02.

■ WAVE EQUATIONS

As is customary in introductory physics, we derived and discussed formulas for waves and for wave speeds before setting up the wave equation itself. This is because any wave equation is a partial differential equation, involving derivatives of the function (pressure, height excursion, electric field) with respect to both space and time. Usually this level of math is considered beyond any introductory course. Note, however, that for students who have already studied calculus, we almost had the wave equations in hand when we derived the speed for waves in a rope. For the pulse in a rope, the net force acting on a small section was given by $F_r \approx S\theta$ (for small θ), where S is the tension in the rope. For this small piece,

$$F_r = (\mu \delta l)a = (\mu \delta l)\frac{d^2 y}{dt^2}.$$

The angle θ is equal to the ratio of the arc it subtends and the radius of curvature: $\theta = \delta l/r$. From analytical geometry, we learn that the radius of curvature of a function is

$$r = \frac{[1 + (dy/dx)^2]^{3/2}}{d^2 y/dx^2}.$$

For a slope that is small compared with 1,

$$r \approx \frac{1}{d^2 y/dx^2}.$$

The larger the spatial second derivative, the smaller the radius of curvature, and hence the larger the "curvature." Our equation of motion for the small section of rope now becomes

$$(\mu \delta l)\frac{d^2 y}{dt^2} \approx S\theta \approx S\delta l \frac{d^2 y}{dx^2}.$$

Since we are now using y as a function of both x and t, we must write partial differentials. The wave equation in its usual form is

$$\frac{\partial^2 y}{\partial x^2} = \frac{\mu}{S}\frac{\partial^2 y}{\partial t^2} = \frac{1}{v^2}\frac{\partial^2 y}{\partial t^2}, \quad v = \sqrt{\frac{S}{\mu}}.$$

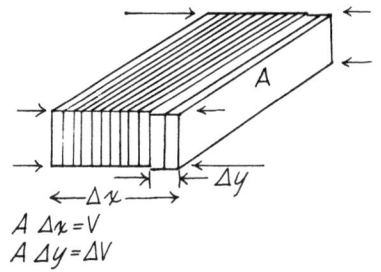

Figure 14.6. Variables for a longitudinal disturbance.

The wave equation for longitudinal disturbances can be derived with a short extension of our previous calculation for the velocity. In Fig. 14.6 the infinitesimal column has length Δx and is shortened by Δy due to the difference of pressure across it. We derived previously the relationship

$$\Delta P = -B \frac{\Delta V}{V} = -B \frac{\delta v}{v}.$$

But this quantity must be equal to $B\Delta y / \Delta x$. The net force acting on the column is

$$(\Delta P_1 - \Delta P_2)A = ma = (\rho A v \Delta t)a = (\rho A \Delta x)\frac{\partial^2 y}{\partial t^2}.$$

Rearranging the terms and substituting for the ΔP's yields

$$B\frac{\left(\frac{\partial y}{\partial x}\right)_1 - \left(\frac{\partial y}{\partial x}\right)_2}{\Delta x} = \rho \frac{\partial^2 y}{\partial t^2} \rightarrow \frac{\partial^2 y}{\partial x^2} = \frac{\rho}{B}\frac{\partial^2 y}{\partial t^2} = \frac{1}{v^2}\frac{\partial^2 y}{\partial t^2}, \quad v = \sqrt{\frac{B}{\rho}}.$$

These wave equations are satisfied by any function whose second spatial derivative is proportional to its second time derivative. Sinusoidal (or imaginary exponential) functions of $x \pm vt$ have this property:

$$\frac{\partial^2}{\partial x^2}\left(\sin \frac{2\pi}{\lambda}(x-vt)\right) = -\frac{4\pi^2}{\lambda^2}\sin \frac{2\pi}{\lambda}(x-vt),$$

$$\frac{\partial^2}{\partial t^2}\left(\sin \frac{2\pi}{\lambda}(x-vt)\right) = -v^2 \frac{4\pi^2}{\lambda^2}\sin \frac{2\pi}{\lambda}(x-vt).$$

Substitute these expressions into the wave equation and cancel the common terms:

$$1 = \frac{\mu}{S}v^2, \quad v = \sqrt{\frac{S}{\mu}}.$$

We will see examples in the next chapter of how we can combine, or superimpose, sinusoidal functions of different frequencies to produce pulses or waves of almost any shape. In Chap. 6 we show how time-dependent oscillations are transformed into spatial oscillations by a Galilean transformation of reference frame. The same technique will then be used with traveling waves to derive the Doppler shift.

■ THE SCHRODINGER WAVE EQUATION

One of the most famous wave equations is the one invented by Erwin Schrödinger in 1926. It mathematicizes the proposal by de Broglie that particles have wavelengths given by $\lambda = h/p$, where h is Planck's constant and p is the particle momentum. Schrödinger's equation is the foundation stone of quantum mechanics. For a free particle the equation is

$$\frac{\partial^2 \Psi(x,t)}{\partial x^2} = -i \frac{2m}{\hbar} \frac{\partial \Psi(x,t)}{\partial t}.$$

Note that although this expression is called a wave equation, it relates the second spatial derivative to the *first* temporal derivative. Furthermore, the equation contains i, $\sqrt{-1}$. Nevertheless, the equation has wavelike solutions. Let $\Psi(x,t) = Ae^{i(kx-\omega t)}$ and substitute into Schrödinger's equation:

$$-k^2 = -\frac{2m}{\hbar} \omega = -\frac{2m}{\hbar^2} E,$$

where

$$E = \hbar \omega = h\nu, \quad E = \frac{k^2 \hbar^2}{2m} = \frac{h^2}{\lambda^2} \frac{1}{2m} = \frac{p^2}{2m}.$$

We get the right energy relationships for a particle with momentum p and mass m, which is nevertheless associated with a wave of frequency ν. To be sure, the wave equation contains the imaginary i, but the crucial distinction between this and the ordinary wave equation is that the solution—the wave function—is necessarily complex (in the sense of being a mixture of real and imaginary values in the complex plane). With waves on a rope, or for electromagnetic waves, a solution can also be written as $Ae^{i(kx-\omega t)} = A\cos(kx-\omega t) + iA\sin(kx-\omega t)$. However, for these waves either the "real" cosine term or the "imaginary" sine term is also a solution. This is not the case for Schrödinger's equation. Only the full complex function satisfies the equation.

With Schrödinger's equation there is also the problem of interpreting the meaning of Ψ. With the other wave equations, the solution function represents the excursion of a rope segment or the strength of an electric field. It turns out that Ψ is a probability function, yielding the probability P of finding the particle in the region dx when manipulated as follows:

$$P(x,t)dx = \Psi^*(x,t)\Psi(x,t)dx,$$

where Ψ^* is the complex conjugate of Ψ, e.g., for $\Psi = Ae^{i(kx-\omega t)}$, $P\,dx = Ae^{-(kx-\omega t)} A e^{+i(kx-\omega t)} dx = A^2 dx$.

Note the analogy with other waves where the energy density is proportional to the square of the wave function.

■ TRANSMISSION OF PULSES

A static picture of a traveling pulse on a rope gives no hint of which way the pulse is traveling or even that it is in motion. How does the pulse know which way to go? Note that the rope is under tension, and any particular section is subject to the opposing forces on either side of it. As long as the rope is straight, the forces on any section cancel out. Where the rope is curved, however, the forces on either side of the section are equal in magnitude but are not in line. Therefore, there will be a net transverse force. In Fig. 14.7(a), we show the force pairs operating on several sections of a rope as a pulse sweeps past. In part (b) we show the resulting net transverse forces. At section A, the pulse has not yet arrived, and the opposing forces cancel. At B, there is a net upward force tending to raise that section and give it an upward velocity. At C, where the shape of the pulse is straight, the opposing forces cancel, and the resulting force is 0. Section C must still be moving upward, however, because of the previous acceleration. As soon as the pulse has a curvature downward, such as in section D, there is a transverse force down, acting to stop the upward motion of the rope in that region. In section E, the rope is traveling downward but the net force is 0. From that point on to section F there is a net upward force that slows the downward movement of the rope, bringing it to rest. This same sequence is repeated for each section of the rope over which the pulse passes. In Fig. 14.7(c), we show how the resulting transverse *velocities* of the rope produce a pulse that travels to the right.

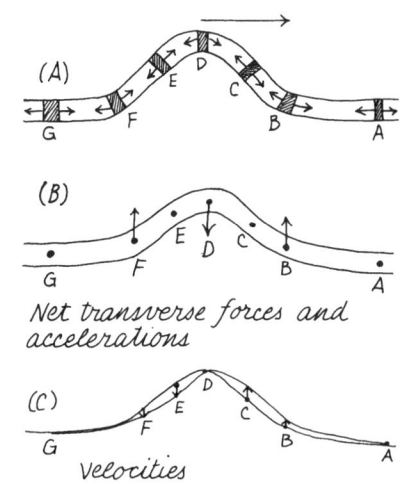

Figure 14.7. Dynamics of a pulse traveling to the right on a rope.

Note that we can describe the same sequence using the algebraic relationships of the partial differential wave equation. The second derivative of y with respect to time is the transverse acceleration, which is proportional to the net transverse force. According to the wave equation, that acceleration is proportional to the second derivative of y with respect to x, which is proportional to the curvature. At A the curvature is 0; the pulse has not arrived yet. At B the curvature is concave upward, calling for an upward acceleration. At C and E the rope [and $y(x)$] is straight, but at D the rope is concave downward, yielding a downward acceleration.

If a rope is plucked in the middle, so that a pulse momentarily exists there, the shape of the rope will at first look as if a pulse is traveling one way or the other. Furthermore the net forces on the rope segments will be just the same as we showed in Fig. 14.7. However, since no segment has any initial momentum the forces will produce two pulses traveling symmetrically in opposite directions.

Similar diagrams could be drawn for a longitudinal pulse. Instead of showing the transverse displacement, the graph could show the displacement of segments in the line of travel of the pulse. The same form of graph might also portray the pressure or density variation along the direction of travel. These graphs are usually harder for students to understand, and care must be taken to distinguish between pressure graphs and longitudinal displacement graphs. The two variables are out of phase! In Fig. 14.8 we show the two graphs of $y(x)$ and $P(x)$, along with a pictorial representation of the corresponding displacements of regions. At point B the displacement of the region is 0, but the regions to the left have positive displacements (to the right) and the regions to the right of B have negative displacements (to the left). Consequently they gang up at point B and produce high density and high pressure. Similarly at point D there is 0 displacement but the regions on either side have been displaced away from D producing low density. At A and C where there is maximum displacement, adjoining regions have all been displaced in the same direction, leading to no buildup and therefore no density or pressure increase or decrease.

The phase relationship between pressure and longitudinal displacement can be

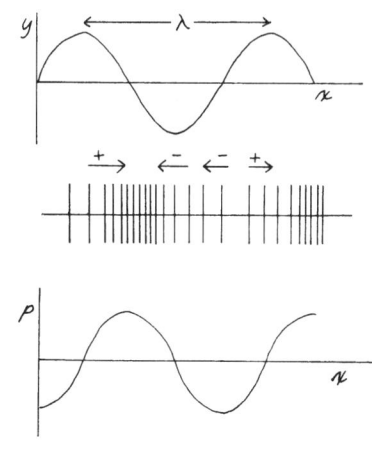

Figure 14.8. Dynamics of a longitudinal traveling pulse.

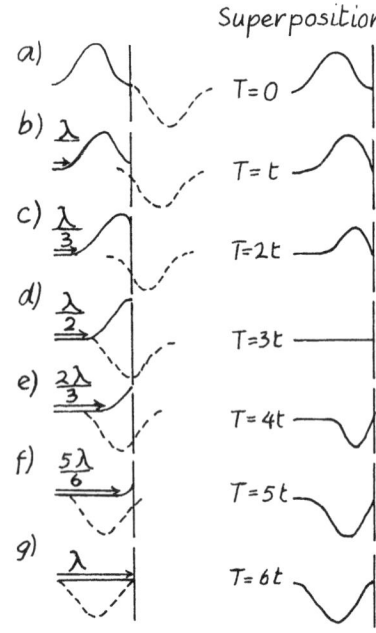

Figure 14.9. Behavior of pulse on rope reflected from fixed end. On the left side, the sequence shows the original pulse (solid line) traveling to the right. The reflected pulse (dotted line) is coming out of the boundary, traveling to the left. The right hand sequence shows the superposition of these two moving pulses. Note that at T=3t, the rope is flat. A similar model may be used to analyze pulse reflection from an open end, but in this case the pulse coming through the boundary (dotted line) should not be inverted. Diagram from: McGervey, John and Schluchter, Clay, "Pulses upon Reflection" Phys. Teach. **34**, 4 (1996).

derived from the algebra that we have already used to find the longitudinal wave velocity. The (excess) pressure is given by

$$p = \Delta P = -B \frac{\Delta V}{V} = -B \frac{\Delta y}{\Delta x} \rightarrow -B \frac{dy}{dx}.$$

If $y = A \sin(kx - \omega t)$, then $p = -kAB \cos(kx - \omega t)$ and is 90° out of phase with y.

■ REFLECTION OF PULSES AND WAVES IN LINES

Every student should have the experience of sending pulses down a rope or Slinky-type spring and observing the reflection and transmission of the pulses as they hit a fixed or loose end. The fixed end is obtained by having someone hold it; the loose end is obtained by tying a long string (at least several times longer than the pulse) at the end, and having someone hold the string. With the fixed end the pulse is reflected with a 180° phase change (on the opposite side of the rope from the original pulse). With a loose end the reflection is in phase *and* a small pulse continues on in the string. We will discuss the relative transmission and reflection in the later section on impedance. In the meantime, note that the detailed action during the reflection usually happens rapidly and is hard to examine. In Fig. 14.9 we show sequences of the rope shapes for the fixed end reflections. If you know what to look for, you may be able to confirm the details through observation. The examination of this simple phenomenon by photography would make a great student project, surely of more value than painting a model rocket or designing a poster about black holes.

Consider why the reflection from the fixed end produces a pulse on the opposite side from the original one. As the pulse arrives at the fixed boundary, the leading edge, which is accelerating upward, pulls on the boundary which does not move but does pull back. In effect, the spring constant of the rope at the leading edge is very great since the next section of the rope cannot move. Consequently, the rope at the leading edge springs downward as if it had been pulled down by a force at the boundary. The higher the advancing pulse tries to pull the material near the boundary, the greater the restoring force on that material to snap it down in the opposite direction. The result is that the same pulse is formed, now traveling in the opposite direction and on the opposite side of the rope.

At the open-ended boundary, the situation is just the opposite. The light string has very low mass density and so, when the leading edge of the pulse reaches it, the upward force produces a large acceleration and a high velocity upward. The junction point whips upward to an amplitude almost twice that of the original pulse before being pulled back down again. The effect is the same as if someone had raised and lowered a free end of a rope, sending a pulse back down the line on the original side.

■ REFLECTION AND REFRACTION IN TWO DIMENSIONS

Some of the phenomena of reflection and refraction are best described in terms of Huygens's principle. We will postpone the discussion of these topics until the next chapter when we consider wave interactions. Here are some points about wave paths that can be taken up in terms of "rays."

Fermat's Principle

Over 2000 years ago, Hero of Alexandria (the same Hero who invented a rotary steam engine) proved mathematically that a light ray reflected from a plane mirror takes a path shorter than any other path (only paths leading from source to detector by way of the mirror are considered). The proof is clever and, of course, classic. The geometry is shown in Fig. 14.10. It is essential that the image is as far behind the mirror as the object is in front of it, which can be shown with the usual geometry. Of all the rays leaving the object and striking the mirror, the only one to reach the observer obeys the law of reflection. The reflected angle is equal to the incident angle. The diagram shows two hypothetical paths that do not obey this law.

Hero's argument is that the actual path taken by the light is shorter than any hypothetical path. Certainly the shortest distance between the *image* and the observer is along the straight-line path. Because of the symmetrical geometry, however, that distance is equal to the path length from the object to the observer. The hypothetical paths from the *image* to the observer are longer than the straight-line path, and each of these equals the corresponding hypothetical path length from the *object* to the observer.

The distance from object to mirror and then to observer can be described in terms of the general variable x along the mirror, as shown in the top part of Fig. 14.11. The path length is plotted as a function of x in the bottom part. When the observer is also at a distance a from the mirror, the path is a minimum for $x = L/2$. The length L is the distance along the mirror between perpendiculars dropped from the object and from the observer.

In the middle of the seventeenth century, Pierre de Fermat (1608–1665) extended Hero's principle to light rays passing through various media, such as glass and air. Fermat transformed the rule from "shortest path" to "least time," raising the possibility of some grand economy of nature. For some reason, light would take a path between two points that would require less time than any nearby path. This principle can serve as the basis for the analysis and design of optical instruments. As we will soon see, however, the path taken by light actually produces an *extremum* of time—either a maximum or a minimum.

As an example of minimum path time, consider the path of the refracted ray going from a point in air to a point in water. The actual path is shown in Fig. 14.12, along with several hypothetical paths. Richard Feynman in his *Lectures on Physics* pointed out that the problem is similar to that faced by the lifeguard on a beach who wants to reach a drowning person in the water in the shortest possible time. The lifeguard can run faster on the beach than he can swim in the water. Should he run immediately to the water and then swim in a straight line, should he run in a straight line on the beach to a point opposite the person and then swim the shortest distance out, or should he run and then swim, always moving in a straight line between his original position and the drowning person? None of these paths yield as short a time as the path given by Snell's law: $(\sin\theta_1)/(\sin\theta_2) = v_1/v_2$. The law holds for light and for the lifeguard.

There is no simple purely geometric proof that Snell's law produces minimum time in this case, but an algebraic proof can be given. Once again, it is possible to

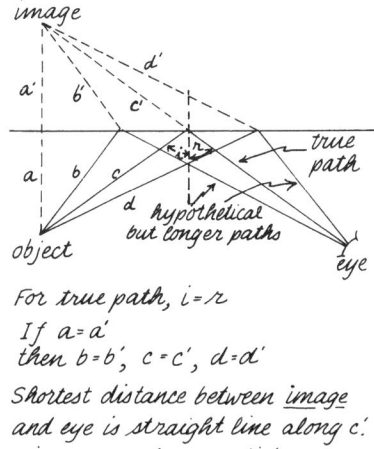

For true path, $i = r$
If $a = a'$
then $b = b'$, $c = c'$, $d = d'$
Shortest distance between *image* and eye is straight line along c'.
Since $c = c'$, shortest distance between *object* and eye is along c.

Figure 14.10. Hero's geometric proof that a light ray reflected from a plant mirror takes the shortest path between source and image.

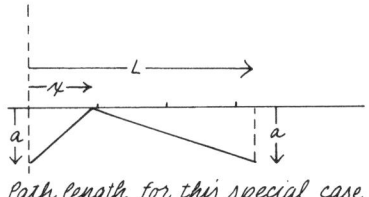

Path length for this special case where both object and eye are at distance a from mirror:

$$\sqrt{a^2 + x^2} + \sqrt{a^2 + (L-x)^2}$$

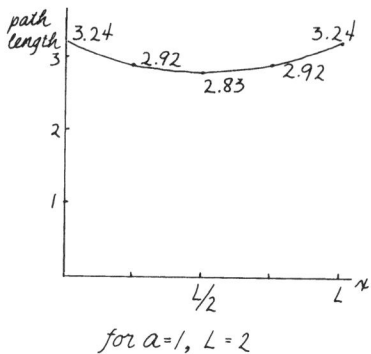

for $a = 1$, $L = 2$

Figure 14.11. Algebraic proof of Fermat's principle for reflection.

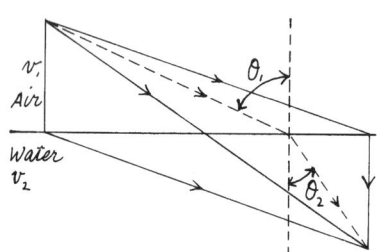

Figure 14.12. Snell's law for refraction satisfies Fermat's principle.

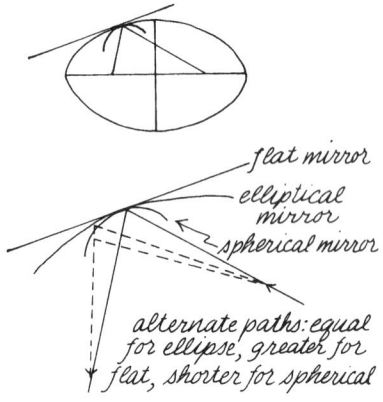

Figure 14.13. Fermat's principle is not for minimum time, but for an extremum.

describe the time in terms of the variable x where the path intercepts the boundary. Take the first derivative of the path *time* with respect to x, and set it equal to 0. The value of x that gives minimum time also yields the angles corresponding to Snell's law.

The path that light takes in going from one point to another is not necessarily a minimum. Consider the geometry of an ellipse. The distance from one focus to any point on the ellipse and back to the other focus is a constant. If one focus is a point source of light, all of the light will reflect from the ellipse and pass through the second focus (in three dimensions, the ellipse would be an ellipsoid). Since the distance from the source to any point on the ellipse and back to the other focus is a constant, so is the time taken. In this case, therefore, rays of light can and do leave the source and go to any point of the ellipse and then reflect back to the other focus, always taking the same time regardless of where they strike the surface of the ellipse.

There is another important point about the equal-time paths for light going from one focus of an ellipse to the other. Since the flight time is the same regardless of the path, the light always arrives in phase at the other focus no matter which way it started out. Therefore, the various waves will all reinforce each other at the second focus. The arrival in phase and the mutual reinforcement is a requirement of having waves come to a focus. As we will see, this is a requirement in analyzing the focusing properties of mirrors and lenses.

In Fig. 14.13 we show one particular ray being reflected from an ellipse. At the point where the ray reflects, we have drawn a straight line, representing a flat mirror touching the ellipse at that point. We have also drawn an interior curve, which represents a mirror with greater curvature. The particular angle of reflection at that point yields a flight time that is independent of position for the ellipse, is a *minimum* for the flat mirror, and evidently is a *maximum* for the mirror of greater curvature. Here we have an example of the generalization of Fermat's principle. The time taken by light in going from one point to another is a minimum, a maximum, or a constant. In all three cases, the time does not change appreciably for small excursions around the actual path.

In Fig. 14.14 we show the geometry for the reflection of light that arrives parallel to the axis of a paraboloid. A paraboloid is the locus of all points such that the perpendicular distance to a plane is equal to the distance to a focal point. If all the rays coming toward the paraboloid are parallel to each other, they represent a plane wave that would arrive at the plane (if the mirror were not in the way) with all points in phase. For each ray striking the mirror, the distance to the plane is the same as it is to the focal point. Consequently, all of the rays will arrive at the focal point in phase and will reinforce each other there. This phase relationship is the requirement for a focus. A detailed drawing of the ray geometry shows that each ray satisfies the law that the angle of reflection must equal the angle of incidence.

A converging lens with spherical surfaces produces a focus only for paraxial rays—close to the axis and at small angles. It is possible to create a lens that deviates all parallel rays entering it to the same focus. However, spherical surfaces are easier to produce, and the special lens would not be similarly perfect for rays entering at other angles. Note the qualitative application of Fermat's principle to the operation of a lens. A ray passing through the top of the lens clearly must travel further to get to the focal point than a ray passing through the center of the lens. How can the two arrive at the focal point in phase? The ray passing through the center travels a longer distance through a slower medium. Flight time is the same for both rays. Quantitatively, this requirement serves as a basis for some computer programs for complicated lens designs.

A glass lens that focuses parallel light must be thicker in the middle than at the edges. In Fig. 14.15 we show a lens that brings parallel microwaves to a focal point. The lens consists of metal channels or wave guides. Notice that the lens is *thinner* in the center than at the edges. The *phase* velocity of microwaves in such a channel is greater than c. Therefore, the microwaves going through the channels at the edge of the lends spend less time in the lens and arrive at the focal point in phase with the rays on the axis, even though they have traveled a longer distance. (We describe the difference between phase and group velocity on p. 365)

We usually deal with wave refraction in terms of light. Sound waves also refract. Remember that the velocity of sound in air is proportional to the square root of the absolute temperature. Under normal conditions, the temperature of the air decreases with increasing height above the ground. Consequently, the velocity of sound in air is faster near the ground. If a plane wave starts out moving parallel to the ground, the part close to the ground will move faster than the upper part. The face of the planewave will tilt upward so that at some distance away the sound can be heard better at a height than on the ground itself.

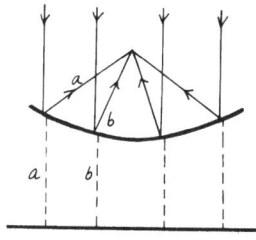

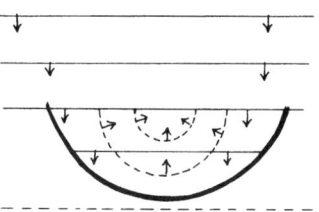

Figure 14.14. Parallel rays reflecting to a focus from a parabolic surface; Parallel wave fronts reflect from parabolic surfaces in such a way that they arrive in phase with each other at the focus.

The refraction of sound has caused many strange and unfortunate occurrences in connection with warning signals, such as from lighthouses. Joseph Henry, the great American physicist in the first half of the nineteenth century, investigated and solved several such mysteries. Ships would report that they had heard a warning bell but that as they advanced toward it the sound disappeared, only to reappear too late. The sound had been refracted upward and then, due either to a temperature inversion or air movement had been refracted downward, leaving a region of silence in between. A similar phenomenon occurs if you try to shout upwind. Since the velocity of sound in air is about 340 m/s, you would not expect that a wind of 10 m/s or so would make much difference in the transmission of sound. A wind in the opposite direction would indeed reduce the speed of sound slightly, but that is not why a person upwind has a hard time hearing you. The explanation is that the velocity of wind near the ground is almost always less than it is higher up. Therefore, the opposing velocity of the wind is less near the ground than it is higher up and so the velocity of your sound is greater near the ground. Once again, refraction is produced and the wave front of your voice bends upward, passing over the person who is trying to hear you. If the person upwind is on a balcony or in a tree, she may hear you very well. People living near water frequently hear sounds from boats that are far away. If there is a temperature inversion so that the air is warmer higher up than it is near the surface of the water, then sound will be refracted downward and will travel long distances along the surface because the energy is confined to a thin two-dimensional sheet.

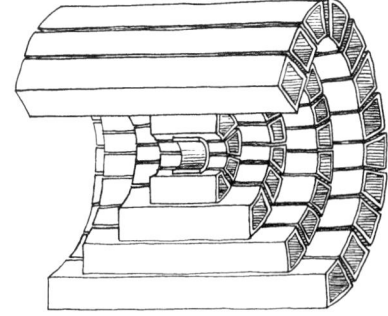

Figure 14.15. Microwave focusing lens, thinner in the middle. In the channels, the phase velocity is greater than c.

Mirages are caused by the refraction of light passing through regions with slightly varying index of refraction. The higher the temperature of air, the less the density, and thus the lower the optical index of refraction. If we are traveling along a hot road, we may see blue water ponds on the road ahead. Light from the blue sky coming down toward the road is refracted upward in a gradual bounce by the hot air above the road. That light reaches us without ever actually having touched the ground; we interpret it as water on the road. Similar effects are common in deserts, where the hot air close to the ground inverts the usual gradual decrease of density with height. The shortest time path for light rays traveling from tall trees in an oasis to a distant observer may be a swooping path down close

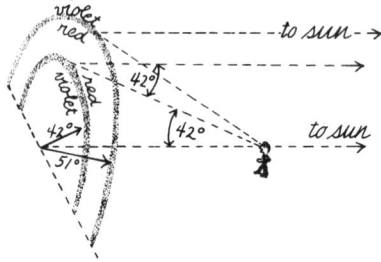

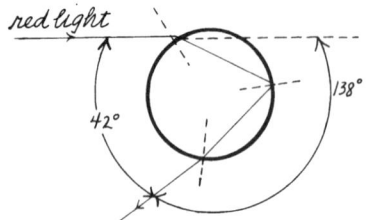

red: $180° - 138° = 42°$
blue: $180° - 140° = 40°$

Angular relationships for the primary bow.

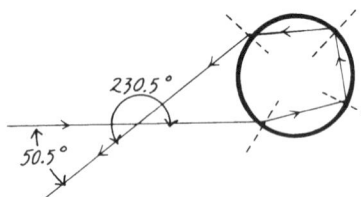

red: $230.5° - 180° = 50.5°$
blue: $234° - 180° = 54°$

Angular relationships for the secondary bow.

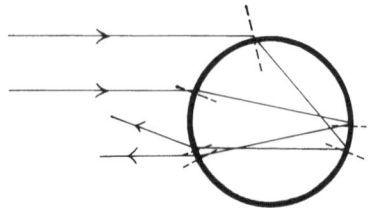

Rays striking above or below the special region of a droplet suffer greater deviation than the rays that form the bow.

Figure 14.16. *Rainbow geometries.*

to the ground. The observer will see the trees upside down in the middle of a distant lake.

The best known example of the refraction and dispersion of light is first described in the book of Genesis (9:13), which tells how God set his bow in the cloud as token of a covenant between himself and the creatures on the earth. He would not destroy them again by flood (the fire next time).

The rainbow and many other types of light dispersion seen in the sky are caused by light interacting with water droplets or ice crystals in the air. Some of the effects are very complicated. The most common rainbow is shown in Fig. 14.16. It is best seen after a rain storm when the sun is low but is shining brightly through a section of the sky that is clear. To see the rainbow you must turn your back to the sun and look toward a region that still has rain clouds (and is therefore dark). Under good conditions you can see a colored arc, with blue-violet on the inside and red on the outside. Above the primary bow there may be a fainter secondary bow with the colors reversed—red on the inside and blue-violet on the outside.

The light that forms the bows has been reflected and refracted in raindrops, as shown in the diagram. The light from the primary bow enters the top of a droplet and is reflected just once. The light in the secondary bow enters the droplets near the bottom and reflects twice inside the drop. Since the light entering and leaving the droplet undergoes refraction, it also undergoes dispersion, so that each color leaves at a different angle. It might seem at first that this effect is enough to explain the existence of the rainbows. The trouble with the simple explanation is that the total angle of deviation depends not only on the wavelength but also on the point of entry of the original ray into the droplet. As you can see in the diagram, a ray entering near the top of the droplet undergoes a larger deviation; most of the light entering near the middle of the droplet either goes all the way through or reflects through almost 180°. The light that reaches our eyes from all these regions will be a mixture of colors and hence white, since the sun's rays strike the droplets over their whole surface.

There is a special condition, however, that makes all the difference. There is a particular region near the top of the droplet that produces *minimum* deviation compared with the regions above or below. This minimum angle is 138° for red light and 140° for violet. As often happens for such an extremum condition, the rays that strike the droplet just above or just below the critical point are deviated almost through the same angle. Therefore, there is a slight concentration of light deviated at these particular angles.

The geometry of the sun, the droplets, and the observer is shown in the first part of the diagram. The observer sees the concentration of colors in the rays that have been deviated from droplets in a conical shell. The opening angle of this cone as seen by the observer is 42° (180° − 138°) for red light and 40° for violet light. For the secondary bow, the angles are 50.5° for red and 54° for violet.

The apex of the light cone is at the eye of the observer (so that each person sees his or her own individual bow). The center line of the cone extends through the observer to the sun. Consequently, if the sun is on the horizon, the observer will see a full 180° of the rainbow. If the sun is higher in the sky, the light cone is tilted down into the horizon, so that only a small section of the upper arc can be seen. If the sun is higher than 42° above the horizon, the rainbow disappears completely. Of course, if you are looking down into a canyon where there is a waterfall or down at the mist produced by a law sprinkler, you can see rainbows

Wave Transmission

even though the sun is high in the sky. Under the right conditions, particularly from an airplane, you may be able to see the whole 360° of the bow.

■ TRANSMISSION OF ENERGY

Transverse Waves

The important consideration about the energy of wave motion is not how much energy is *in* the wave but, instead, how much energy is *delivered* per second by the wave. In other words, we want to know the *power* delivered by the wave. In the case of a wave spreading out in three-dimensional space, such as a sound wave or an electromagnetic wave, we should calculate the power delivered per square meter.

First, let us calculate the power delivered in one dimension by a continuous sine wave in a rope. We can almost see the energy in this case since the rope is actually vibrating back and forth. Any particular point on the rope oscillates with a displacement perpendicular to the velocity of the wave and with a value

$$y = A \sin \omega t.$$

The velocity of that segment of rope is

$$u = \frac{dy}{dt} = \omega A \cos \omega t.$$

The velocity of the segment is 90° out of phase with its displacement and is proportional to the frequency. The higher the frequency of the oscillating segment, the faster it must travel to go between $+A$ and $-A$ during half a period.

Averaging over time, any segment of rope in the wave has the same amount of energy as any other segment of the same length, and the average kinetic energy must be equal to the average potential energy. Both kinetic and potential energy go hurtling down the rope together; the two forms of energy are in phase! When a particular segment has 0 displacement, $y=0$, its speed u is maximum. Therefore its kinetic energy is maximum. The potential energy of that segment is also maximum because at that moment there is the greatest stretch on the segment. The segment has twisted its maximum amount from the equilibrium position. (At all other times the direction of the segment is more parallel to that of the undisturbed rope.) Half a period later the same segment is momentarily at rest at $y=A$ and, thus, has 0 kinetic energy. Its potential energy is also 0 because the segment itself is not stretched perpendicular to the rope. The net perpendicular force on the segment is maximum, but this force produces a perpendicular acceleration, not a strain representing potential energy. (It may help to see this situation, by considering the transverse motion to be horizontal, not vertical. That avoids the problem of gravitational potential energy.)

Let us calculate the kinetic energy of a segment of the rope. The mass of the segment is m.

$$E_{\text{kin}} = \tfrac{1}{2} m u^2 = \tfrac{1}{2} m \omega^2 A^2 \cos^2 \omega t.$$

The average value of $\cos^2 \omega t$ over one period is exactly 1/2. Therefore, the average kinetic energy of a rope segment is

$$E_{\text{kin av}} = \tfrac{1}{4} m \omega^2 A^2.$$

The average potential energy has the same value. Therefore, the average energy of the segment is

$$E_{\text{total}} = \tfrac{1}{2} m \omega^2 A^2.$$

The total energy in one wavelength is

$$E_{\text{total}/\lambda} = \tfrac{1}{2}(\rho \lambda) \omega^2 A^2,$$

where ρ is the mass per unit length. The power delivered is the energy contained in one wavelength divided by the period of the oscillation:

$$P = (\tfrac{1}{2} \rho \lambda \omega^2 A^2)/T.$$

Since $\lambda/T = v$, the velocity of the wave, the power is also

$$P = \tfrac{1}{2} \rho v \omega^2 A^2.$$

The velocity of a wave in a stretched string with tension S is $v = \sqrt{S/\rho}$. Therefore the power is

$$P = \tfrac{1}{2} \rho \sqrt{\tfrac{S}{\rho}}\, \omega^2 A^2 = \tfrac{1}{2} \sqrt{S\rho}\, \omega^2 A^2.$$

Look at the various terms in this expression. The power delivered by a wave in a rope is proportional to the square of the amplitude of the wave and to the square of the frequency. The squares of these quantities enter because the average energy of a segment is proportional to u^2 (the transverse velocity squared) which is proportional to $\omega^2 A^2$.

In our expression for the power delivered by an oscillating rope, the frequency and amplitude terms are characteristic of the waves that we generate. The first term, $\sqrt{S\rho}$, is a characteristic of the rope itself. Note that it is made up of the two properties that control the velocity of the waves—the tension S and the mass per unit length ρ. This time, however, we are not dealing with their ratio but with their product. This combination is given a special name. It is the *impedance* of the rope, usually given the symbol Z. The power can thus be expressed in terms of the impedance:

$$P = \tfrac{1}{2} Z \omega^2 A^2.$$

The impedance of a medium is not only a factor in the power delivered by a wave, but is also the major consideration in determining the extent to which a wave will be reflected from a boundary with another medium or will be transmitted through the boundary. We will describe impedances for other types of waves in other media. In each case they will play a similar role, but the impedance is not always simply related to a force term and a mass-density term. In the case of the rope, the impedance has the units of kilograms per second. It is reasonable that the greater the tension in the rope, and the greater the mass density, the greater should be the energy contained in a segment and thus the power transmitted. By manipulating the units in the expression for Z, you can show that Z can also be expressed in terms of N/(m/s). As we will see, impedance can be defined as the ratio of the force to the velocity of a particle being affected by a wave.

Longitudinal Waves

In longitudinal pressure waves such as sound, each particle oscillates back and forth in line with the direction of the wave. It is much harder to represent this action in simple drawings than it is to show transverse waves. The problem is partly pedagogical; even though the diagram is correct, what does the naive reader see? In Fig. 14.8 we show several graphs and sketches. The pressure graph, $P(t)$, looks much like the displacement graph, $y(t)$, though out of phase, and neither looks much like the diagram that attempts to show the densities of particles displaced back and forth along the line of travel.

We use y for the displacement of each particle from its equilibrium position (even though the particle is moving back and forth along the x axis). For the particle at $x=0$, $y=-A \sin \omega t$. As we have seen (on p. 334), the excess pressure at that point is equal to

$$p = -BkA \cos \omega t = -B\frac{2\pi}{\lambda} A \cos \omega t.$$

This excess pressure, which is the sound pressure, is very small compared with atmospheric pressure. The sound pressure at which humans begin to feel pain is about 30 N/m^2, whereas atmospheric pressure is 1×10^5 N/m^2. The pressure amplitude of the faintest sound that we can hear (depending on frequency) is about 3×10^{-5} N/m^2.

The longitudinal velocity of particles in a sound wave is

$$u_{\text{particles}} = \frac{dy}{dt} = -\omega A \cos \omega t.$$

Notice that the longitudinal velocity of the particles is in phase with the pressure, both of which are 90° out of phase with the displacement. Both the velocity and the pressure are proportional to the frequency.

The power transferred across the unit area is equal to the product of the pressure differential and the velocity. (Pressure is equal to the force per unit area, and the product of force and velocity is work per unit time.)

$$pu = \left(-B\frac{2\pi}{\lambda} A \cos \omega t\right)(-\omega A \cos \omega t) = \omega B \frac{2\pi}{\lambda} A^2 \cos^2 \omega t.$$

The average value of $\cos^2 \omega t$ over a full cycle is 1/2. Therefore, the average intensity or power per unit area is

$$I = \tfrac{1}{2} \omega B \frac{2\pi}{\lambda} A^2.$$

This expression can be developed into something more familiar by using simple substitutions. Call the velocity of the wave v to distinguish it from the oscillatory velocity u of the individual particles. Then

$$\frac{\omega}{2\pi}\lambda = v, \quad \frac{2\pi}{\lambda} = \frac{\omega}{v}, \quad v = \sqrt{\frac{B}{\rho}}, \quad B = v^2 \rho,$$

$$I = \tfrac{1}{2}\frac{\omega^2}{v} BA^2 = \tfrac{1}{2}\rho v \omega^2 A^2 = \tfrac{1}{2}\sqrt{B\rho}\,\omega^2 A^2 = \tfrac{1}{2} Z \omega^2 A^2,$$

where $Z = \sqrt{B\rho}$. We end up basically with the same formula for the power transmitted in a longitudinal pressure wave that we had for the power transmitted in a rope. The power is proportional to the square of the amplitude of the wave and to the square of the frequency. It is also proportional to the impedance of the medium. In this case, the impedance Z is equal to the square root of the product of the bulk modulus and the mass density. The units for Z are kilograms per second per square meter, or newtons per square meter per meter-per-second (pressure over velocity).

Our formula for the power transmitted per unit area in a longitudinal wave in a gas is derived in terms of the amplitude of oscillation of the particles. Actually it would be very difficult if not impossible to measure that amplitude. Let us calculate the amplitude for the weakest sound that we can hear. The excess pressure is only 3×10^{-5} N/m^2.

$$p = B \frac{2\pi}{\lambda} A.$$

For a frequency of 1000 Hz, to which the ear is very sensitive, $\lambda \approx 0.3$ m. Therefore, $A \approx 1 \times 10^{-11}$ m. ($B_{air} = \gamma P_{atm}$, where γ is the ratio of specific heats, $c_p/c_v = 1.4$.) Compare this amplitude with the diameter of an atom which is 1×10^{-10} m. Nevertheless, our eardrums can respond to repetitive signals of that amplitude.

Let us find the intensity of sound in terms of the pressure amplitude:

$$I = \tfrac{1}{2} Z\omega^2 A^2 = \tfrac{1}{2} Z\omega^2 \frac{p^2}{B^2(2\pi/\lambda)^2} = \tfrac{1}{2} Z \frac{f^2\lambda^2}{B^2} p^2 = \tfrac{1}{2} Z \frac{v^2}{B^2} p^2 = \tfrac{1}{2} Z \frac{1}{B\rho} p^2 = \tfrac{1}{2} \frac{p^2}{Z}.$$

The intensity is proportional to the square of the pressure amplitude and inversely proportional to the impedance. Notice that in this form, the frequency dependence has disappeared. However, the pressure amplitude is related to the displacement amplitude in terms of the wavelength.

Let us calculate the intensity of a sound wave that is produced by normal conversation. The excess pressure amplitude is about 1×10^{-3} N/m^2. The impedance of air is equal to $Z = \sqrt{B\rho} = \rho v = (1.3 \text{ kg/m}^3)(343 \text{ m/s}) = 4.5 \times 10^2$ kg/m^2 s. Then $I = p^2/2Z = (1 \times 10^{-3} \text{ N/m}^2)^2/2(4.5 \times 10^2 \text{ kg/m}^2 \text{ s}) = 1 \times 10^{-9}$ W/m^2. If you assume that your voice has this intensity over an area of about 10 m^2, then the total power that you deliver is about 1×10^{-8} W. If you can deliver this intensity to the back of a lecture hall with an area of about 100 m^2, then you would be producing about 10^{-7} W. Compare this power with that nominally produced by high-fidelity sets that are rated from 10 to 100 W from each speaker! (Presumably, such claims are only for peak power.)

■ IMPEDANCE

Qualitative Impedance

We started the chapter by marveling at the way waves can transport energy through large distances. Now we must be concerned with how wave energy can be transferred from one medium to another. Usually when waves in one region enter another material there is reflection and only part of the energy is transmitted across the boundary. Furthermore, many vibrating systems radiate away very little energy because they do not couple well with the surrounding medium.

For a startling example of coupling a vibrator to a wave medium, make a coat hanger chime. Unwind a wire coat hanger and bend it into a U shape. Tie each end of a loop of string to each end of the wire and hang the string over your head so that the wire loop is dangling in front of you. Tap the wire with a pen and observe that you can hear only a faint tinkling sound. The wire is vibrating but only high frequencies are coupled to the air and hence to your ears. Now press the string into your ears with your index fingers so that the string and wire hang freely without touching anything except your fingers and ears. Have someone tap the wire again. This time you will hear the awesome tones of a great bell. The low-frequency vibrations in the wire can travel through the string directly to your ears, but cannot get out into the air.

Stringed instruments are prime examples of ways of coupling small vibrating systems to the air. A vibrating string cannot push air very well. However, if the string is fastened to a sounding board (such as the body of a violin) that has a large surface area, the vibrating board moves the air and transfers the vibration energy very efficiently. You can demonstrate this with a rubber band and a drinking glass. Stretch the rubber band between your fingers and twang it. Very little sound comes out. Then stretch the band across the glass held in your hand. Now you get a louder sound. You can enhance the effect still more by placing the glass on a table top.

With brass musical instruments, the flaring bell determines what fraction of the vibration energy in the pipe gets out into the air and how much reflects back into the pipe to trigger the next pulse. Evidently the effectiveness depends on wavelength. Note that efficient radiation of low-frequency sound from a high-fidelity set requires the large surface area of a "woofer," while the high-frequency sounds are emitted by a small "tweeter."

Wave reflections are determined not by the relative indices of refraction of two media, but by their impedances. For instance, part of the sound coming out of a tube into the air is reflected even though the speed of sound is the same inside the tube and out. To enhance the transfer of energy across a boundary, the two impedances must be "matched." For instance, a TV antenna must be matched to the impedance of free space in order to absorb the signal efficiently. The transmission cable down to the TV set must match this impedance, and the cable must be properly "terminated" at the set to prevent reflections back up the cable. (The impedance of free space for electromagnetic waves is 377 ohms. As a practical matter this impedance is hard to match for AM radio but is approximated for TV and FM sets. The common twin-lead TV antenna wire has an impedance of 300 ohms. Interconnecting cables usually have an impedance of 75 ohms and must be connected together through "impedance match junctions.")

> *The speed of an E-M wave in free space is given by:* $c = \sqrt{\frac{1}{\mu_0 \varepsilon_0}}$, *where* $\frac{1}{\varepsilon_0}$ *is the "forcing" term, and* μ_0 *is the "inertial" term. (See p. 327 and p. 482.) The impedance of free space is given by the square root of the product of these two terms. (See p. 340.)*
>
> $$Z = \sqrt{\mu_0 \frac{1}{\varepsilon_0}}$$
> $$= \sqrt{(4\pi \times 10^{-7})\left(\frac{1}{8.85 \times 10^{-12}}\right)}$$
> $$= 377 \, \Omega$$

There is a classic example in physics demonstrations of an impedance mismatch that is used incorrectly to try to prove something else: the good old alarm-clock-in-the-bell-jar trick. As you pump the air out of the bell jar, the sound of the ringing alarm gets fainter and fainter, presumably proving that sound cannot travel in a vacuum. Actually, sound travels just fine in any vacuum we normally produce in the classroom. (The speed of sound in a gas depends only on the temperature and general constants, and does not depend on the pressure. See p. 327). However, the impedance of the gas does depend on pressure and density. At low pressures the impedance match between vibrating bell and gas is so poor that little energy is transferred.

For an exquisite example of impedance matching by nature, consider how sound energy is transferred from the air into our ear. The large surface area of our outer eardrum is connected to the small inner-ear diaphragm by a lever system of three bones. The relative area of the outer and inner diaphragms is 25. The combination of levers and areas produces pressure changes in the inner liquid about 35 times greater than the pressure changes in the outside air. Otherwise the low-density air could not transfer much wave energy to the high-density liquid (See p. 371 in Chap. 15, Complex Waves.)

Quantitative Impedance

Now we must define and examine the property of impedance more formally. We saw one set of definitions in connection with power transmission. For a plane wave, or a wave along a line, impedance is the square root of the *product* of a tension or pressure term and the density of the material. (The speed of such a wave is equal to the square root of the *ratio* of these two terms.) More generally, impedance characterizes the response of a disturbed medium and is a function not only of the material of the medium but also of its geometry.

Let us determine that function. In the case of a rope with transverse waves, the restoring force at a point is equal to

$$-S \tan \theta = -S \frac{\partial y}{\partial x}.$$

Similarly for longitudinal waves in a column of gas, the excess pressure at a point $= p = -B \partial y/\partial x$. If we assume that the wave motion is described by $y = A \sin(\omega t - kx)$ for waves traveling to the right, then the restoring force is

$$-S \frac{\partial y}{\partial x} = +SkA \cos(\omega t - kx) = S \frac{k}{\omega} \frac{\partial y}{\partial t}.$$

For waves traveling to the left. The restoring force is

$$-S \frac{\partial y}{\partial x} = -S \frac{k}{\omega} \frac{\partial y}{\partial t}.$$

Note that $\partial y/\partial t$ equals u, the transverse velocity of the rope segment, and $k/\omega = 1/v$, where v is the velocity of the wave along the rope. Therefore, the restoring force at a point on the rope is proportional to its transverse velocity!

$$F_r = \frac{S}{v} u.$$

The proportionality constant is

$$\frac{S}{v} = \frac{S}{\sqrt{S/\rho}} = \sqrt{S\rho} = Z,$$

the impedance. A similar argument for the case of the longitudinal wave would yield a similar result. In general, the impedance of a medium is equal to the ratio of the restoring force (or pressure) at a point to the velocity of the *medium* at that point. (This is the velocity u, not the velocity of the wave, v.)

$$Z = \frac{F_{\text{restoring}}}{u}.$$

1. Note several features of this definition of impedance. First, a wave sweeping along a medium must exert a force at a given point that will be equal and opposite to the resultant restoring force at that point. The power delivered through that point will be the product of the applied force and the velocity:

$P = Fu = Zu^2 = Z\omega^2 A^2 \cos^2(kx - \omega t)$. The average power delivered in a cycle will be $\frac{1}{2}Z\omega^2 A^2$, the same expression that we obtained earlier from energy-storage considerations.

2. The second feature about this definition of impedance is that the physical situation is not obvious. Why should the restoring force be proportional to the *velocity* instead of the acceleration? This relationship has the characteristic of a drag force. On the other hand, it is not a dissipative force. The energy delivered is not turned into heat by this drag force, but is propagated along the line. This restoring force is the force at a point and is not the net force on a segment of the rope. The latter is the difference between the restoring forces at the two ends of the segment. The acceleration of a segment is given by

$$\frac{\partial^2 y}{\partial t^2} = a = \frac{F_{net}}{m} = \frac{S\left(\frac{\partial y}{\partial x}\right)_2 - S\left(\frac{\partial y}{\partial x}\right)_1}{\rho \Delta x} \rightarrow \frac{S}{\rho}\frac{\partial^2 y}{\partial x^2} = v^2 \frac{\partial^2 y}{\partial x^2}.$$

This is another way to derive the wave equation for a rope. (Compare p. 331.)

3. The third feature about impedance is that the same definition and relationships apply to any kind of wave motion. We are already familiar with the notion of impedance with alternating electric current. In that case, the "force" is the voltage across a circuit and the "velocity" is the current: $Z = V/I$. Here we are not surprised by the proportionality between V and I because we can appeal to a model of charge carriers in a conductor and picture the charges having a drift velocity. However, we also note that electrical impedance is a complex quantity, with the "real" part representing resistance and the "imaginary" part representing capacitive and inductive reactance. Even though the current is proportional to the voltage, it may not be in phase. The same situation applies to other wave motions. With sound waves in air, for instance, the right geometry can produce "acoustic mass," analogous to electrical inductance. In an open-ended tube for instance, some of the air is displaced outward by a pressure pulse, instead of being compressed. The magnitude of this acoustic mass is $\rho l/A$, where ρ is the linear mass density of the gas (mass per unit length), l is the length, and A the cross-sectional area of the tube. In the inverse situation, the gas in a closed container can be compressed but cannot be displaced. The magnitude of its "acoustic compliance," analogous to electrical capacitance, is $V/\rho v^2$, where V is the volume of the container, ρ is the linear mass density, and v is the velocity of the acoustic wave. The reactance of the acoustic mass or compliance has the same form as the analogous electrical reactance: $X_m = \omega \rho l/A$ and $X_C = \rho v^2/\omega V$. When a sound wave passes through a gas in these geometries, the longitudinal particle velocity is still proportional to the wave pressure, but is not in phase with it.

4. The fourth, and perhaps most important, point about the concept of impedance is that it is the criterion determining the reflection and transmission of waves from one medium to another. The *angle of refraction* is determined by the relative indices of refraction, i.e., the relative *velocities* in the two media. But wave reflections can occur at geometrical boundaries within the same medium where there is no change of velocity! (e.g., when sound comes out of a pipe). At any boundary the wave function and its first time derivative (the

velocity) must be continuous, and so must the transverse force. (Otherwise, a particle in the medium at the boundary would have two different accelerations.) Let us define the incident wave, coming from the left, as $I \sin(\omega t - k_1 x)$; the wave reflected from the boundary, going back to the left, as $R \sin(\omega t + k_1 x)$; and the transmitted wave to the right of the boundary as $T \sin(\omega t - k_2 x)$. We will describe the continuity of the equations for transverse velocity and force at a boundary where x is chosen to be zero.

Continuity of v_y: $\quad I\omega \cos \omega t + R\omega \cos \omega t = T\omega \cos \omega t \quad \rightarrow \quad I + R = T$

Continuity of $F_y = Zv_y = Z\dfrac{\partial y}{\partial t}$: $\quad \rightarrow \quad Z_1(I - R) = Z_2 T$

For waves traveling to the right, $F_y = +Z \partial y/\partial t$, and, for waves traveling to the left, $F_y = -Z \partial y/\partial t$. Solve for the reflection and transmission coefficients:

$$\frac{R}{I} = \frac{T-I}{I} = \frac{Z_1(I-R)/Z_2 - I}{I} = \frac{Z_1}{Z_2}\left(1 - \frac{R}{I}\right) - 1,$$

$$\frac{R}{I}\left(\frac{Z_1}{Z_2} + 1\right) = \frac{Z_1}{Z_2} - 1,$$

$$\frac{R}{I} = \frac{Z_1 - Z_2}{Z_1 + Z_2} \quad \text{and} \quad \frac{T}{I} = \frac{I + R}{I} = 1 + \frac{R}{I} = \frac{2Z_1}{Z_1 + Z_2}.$$

Consider the physical implications of these formulas in several extreme cases. If the second medium has infinite impedance—a closed end or a solid wall—then $R/I = -1$ and $T/I = 0$. The wave is entirely reflected, but with opposite phase. For an open-end boundary, with $Z_2 = 0$, $R/I = 1$. There is a reflection at an open end, in phase with the incoming wave. It might appear that in this case there is also a transmitted wave with twice the incident amplitude, but remember that Z_2 is 0 and therefore there is no energy transmitted into the second medium. If $Z_1 = Z_2$, $R/I = 0$ and $T/I = 1$. In this case there really is no boundary, hence no reflection and full transmission. One other requirement is that the incident energy equals the reflected plus transmitted energy: $\tfrac{1}{2}Z_1\omega^2 I^2 = \tfrac{1}{2}Z_1\omega^2 R^2 + \tfrac{1}{2}Z_2\omega^2 T$. If you substitute the values for R and T in terms of I, you will arrive at an identity.

5. There is another feature of the importance of impedance. Note that it is not necessary to have the same values of restoring force in two media in order to eliminate reflection at their boundary. Nor must the two media be the same. The criterion is that the ratios of $F_{\text{restoring}}$ to u must be the same. For instance, in waves on a string, there might be a junction between two strings with different densities and different tensions. It is possible to have

$$Z_1 = \sqrt{S_1 \rho_1} = Z_2 = \sqrt{S_2 \rho_2},$$

and yet

$$v_1 = \sqrt{\frac{S_1}{\rho_1}} \neq v_2 = \sqrt{\frac{S_2}{\rho_2}}.$$

(To be sure, usually when two strings are fastened together, their tensions are the same. However, you can arrange the circumstance above by connecting the strings to either side of a metal ring floating on a vertical rod. It would make a good student project to create and explore such an arrangement.) Another way to achieve zero reflection of a wave on a string is to anchor the string to a dashpot immersed in oil. To first approximation, the drag force is proportional to velocity. If the value of F_{drag}/u is equal to the impedance of the string, there will be no reflection; in electrical terms, the line has been "terminated."

6. Another way to get zero reflection is to create a thin layer of a third medium at the boundary between the first two. There will then be two reflecting boundaries. You can arrange the thickness of the thin layer and its impedance such that the reflection from the second boundary will cancel the reflection from the first boundary. The usual geometry calls for the middle layer to have a thickness of $\frac{1}{4}\lambda$ and $Z_2 = \sqrt{Z_1 Z_3}$. This arrangements is commonly used in optical coatings of lenses to reduce reflected light. Usually the formula is seen in terms of indices of refraction. For electromagnetic waves, the speed is $\sqrt{1/\mu\epsilon}$ and the impedance is $Z = \sqrt{\mu/\epsilon}$. The magnetic permeability μ has about the same value for most transparent materials. Thus Z is proportional to $\sqrt{1/\epsilon}$, which is proportional to $1/n$. Substitute $Z_1 = k/n_1$ into the quarter-wave zero-reflection formula to obtain $n_2 = \sqrt{n_1 n_3}$.

7. Yet another way to eliminate reflections is to change the boundary conditions slowly compared with the wavelength. For instance, a tapered horn with gradually decreasing impedance will lead waves out from a tube with very little reflection. One geometry that can be easily calculated is to assume that you can produce a whole series of quarter-wavelength plates. At every quarter wavelength, change the impedance from Z to $Z+\Delta Z$. The reflection from that step is

$$\Delta R = \frac{Z_1 - Z_2}{Z_1 + Z_2} \approx \frac{\Delta Z}{2Z} \approx \frac{1}{2Z} \frac{dZ(x)}{dx} (\tfrac{1}{4}\lambda).$$

In order to have the reflection from each boundary cancel the one from the earlier boundary, ΔR must be a constant independent of x, call it α. Then

$$\frac{dZ}{Z} = -\frac{8\alpha}{\lambda} dx.$$

If the wavelength does not change with Z, as is the case with sound in a tapered horn, then $Z = Z_0 e^{-8(\alpha/\lambda)x}$. The horn should be tapered exponentially.

8. Note one other important feature about impedance. The formula for "characteristic impedance" is $Z = \sqrt{S\rho}$ for a rope or $Z = \sqrt{B\rho}$ for sound in air. The *actual* impedance depends on the geometry and is always given by the ratio of restoring force to velocity. When a wave leaves a narrow tube and abruptly enters a large open region, its velocity does not change, but the impedance does. In this case the restoring force becomes very small and so does the impedance.

■ LOSS AND SPREAD OF WAVE ENERGY

So far we have been dealing with waves that travel along a line or whose wave fronts are plane and parallel to each other and all move in one direction. We have assumed that the amplitude of these waves stays constant and that no energy is lost. Now let us examine situations where wave energy is dissipated or spread, in either case changing the amplitude A. Waves can lose energy if they set into oscillation some medium that has friction. For instance, radio waves do not penetrate steel buildings because they set up electric currents in the metal surface that both reflect the beam and also absorb part of it. The absorbed energy turns into thermal energy of the metal. If you stretch a rope along a floor and send a pulse down it, the amplitude and hence the transmitted energy will steadily decrease as the wave progresses. If sound waves strike a smooth, solid wall, most of the energy will reflect since the solid mass is barely set into vibration. If sound is directed through low-density material, however, the waves penetrate and may be damped. Energy loss of sound in air is proportional to the square of the frequency. Although the absorption effect is small, the effect can be noticed if you listen to an orchestra at a great distance; the high notes are lost.

When there is an absorption loss, such as with light going through fog or sound through low-density walls, the fractional energy loss per meter is frequently constant: $\Delta E = -kE\Delta x$. In such cases, the energy loss is exponential: $E = E_0 e^{-kx}$. The *amplitude* of the sinusoidal field is therefore $A = A_0 e^{-kx/2}$.

The intensity of waves (the power per square meter) can decrease even when no wave energy is lost. You can observe this effect by watching circular spreading waves on the surface of a pond. The frictional energy loss for ripples is very small and approximately the same amount of power is spreading out from each circle. The circumference of the circle, however, is increasing. Consequently the intensity (the power per meter in this two-dimensional case) is decreasing. Since the power per meter is inversely proportional to the distance from the center, the amplitude of the two-dimensional wave must be inversely proportional to the square root of the distance: $A \propto \sqrt{1/r}$. You can produce the inverse of this effect in a pool or pond by dropping a large ring into the water. The amplitude of the outer wave will decrease as $\sqrt{1/r}$ but a circular pulse will also rush into the center. As $r \to 0$, the pulse builds up to a peak at the center.

In the case of waves spreading out uniformly in three dimensions, the wave front must be spherical. Since the spherical surface of the wave front has an area proportional to r^2, the intensity must be proportional to $1/r^2$. Consequently the amplitude is proportional to $1/r$.

There are cases where waves seem to travel long distances without attenuation. In some cases, the intensity even increases at particular points. We mentioned cases of focusing or refraction of waves due to changes in the wave velocity in media. At the top of the atmosphere there is a region of ionization where the temperature begins to rise again with height. This region provides a natural sound channel, since the temperature and hence the wave velocity is higher at both the upper and lower edges of the channel. Sound entering this region, for instance from large explosions on earth, is refracted back and forth so that the energy spreads out in a thin shell around the earth rather than spreading in three-dimensional space. Because of this channeling, sounds from large explosions can be detected around the world if the detecting instruments are at the right height. There is a similar narrow channel in the deep oceans. A combination of water temperature and salinity creates a thin shell where the wave velocity is less than it is above or below. This channel has been used for both detection purposes and for signaling across oceans.

1. Kuwabara, Hasegawa, and Kono, "Water Waves in a Ripple Tank," Am. J. Phys. **54**, 1002 (1986).

15
Complex Waves and Wave Interactions

How distressing it would be to listen to music produced by an orchestra of turning forks! Musical sounds not only have pitch or frequency but also timbre. The waves are periodic but not sinusoidal. The irregular shapes and the jagged edges of the harmonics produce tonal richness. They also produce complications in how we analyze and teach wave phenomena.

Consider some of the complications. (1) Many waves are dispersive; their velocity depends on frequency. (2) The observed frequency of a wave depends on the relative velocity between source and observer. (3) Although most waves can pass right through each other without disturbing their progress, the disturbance at each point where they overlap is the sum of the separate disturbances, producing interference patterns. (4) With the right geometry, this effect gives rise to standing waves where certain points appear to experience no disturbance at all, although both waves are passing through. (5) A plane wave passing through a slit or around an obstacle will diffract, with the deflection of the path depending on the ratio of wavelength to width of slit or obstacle. For all of these complications, we have to be able to describe how sine waves add.

Why sine waves? (By *sine* we mean sinusoidal; a cosine is just a sine displaced by 90°.) First, they are prevalent. Most vibrating mechanical systems obey Hooke's law to first approximation and therefore produce simple harmonic motion. Electromagnetic oscillations, whether from LC circuits or from atomic transitions, produce sine waves. Second, more complicated periodic wave forms, or even pulses, can be described as Fourier series of sinusoidal waves of varying frequencies.

■ ADDITION OF SINE WAVES

For introductory physics there are three levels of understanding how sine waves combine. All students, whether in high school or college, should be familiar with the graphical or pictorial addition of sine curves. Looking is not enough (not even at a computer!); the students should actually do the arithmetic and graph at least one example of each case. The second level involves trigonometric identities. These should be accompanied by illustrative graphs and simple numerical examples. The third level appeals to phasors, or the rotating vector representation of sine waves, described mathematically by $e^{i\theta} = e^{i(kx-\omega t)}$. The notation may seem abstract at first but once learned it is extremely powerful and actually simpler than ordinary trig.

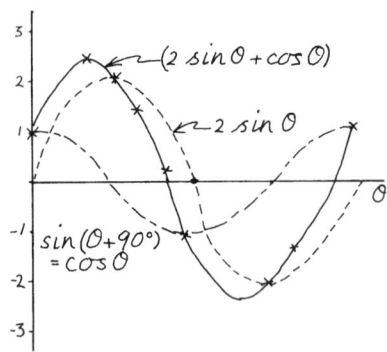

Figure 15.1. Graphical addition of two sine waves of the same frequency yields another sine wave.

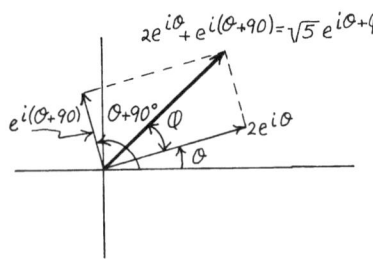

Figure 15.2. Complex vector representation of addition of sinusoidal functions.

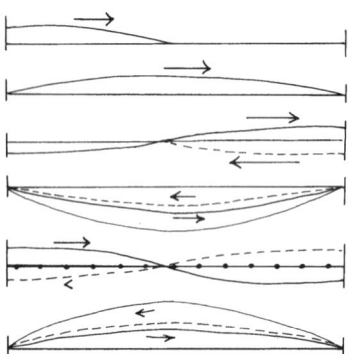

Figure 15.3. Build up of a standing wave by reflection.

1. The first thing to be learned about the addition of sine waves is that the addition of any two sine waves *of the same frequency* always produces another sine wave. This is not obvious. In Fig. 15.1 we show the addition of $2 \sin \theta$ and $\sin(\theta + 90°)$. As you can see, it yields a sine wave with amplitude of $\sqrt{5}$.

   The usual trigonometric identity tables provide combinations only for functions of equal amplitude. For instance: $\sin \theta + \sin \phi = 2 \sin \frac{1}{2}(\theta + \phi) \cos \frac{1}{2}(\theta - \phi)$. For the case where $\phi = \theta + 90°$, $\sin \theta + \sin(\theta + 90°) = 2 \sin(\theta + 45°) \cos 45° = 1.4 \sin(\theta + 45°)$. The cosine term in the identity contributes only to the amplitude. The cosine term remains constant as a function of θ because $\theta - \phi$ is a constant phase difference. If the sum of any two sine functions (of the same frequency) is a sine function, then the sum of any number of such sine functions is also a sine function (since any two can be added, and their sum then added to a third, etc).

   We can use the vector notation for sine functions to demonstrate the points made above as well as for an easy proof of the usual identity. In Fig. 15.2 we show the graph of $2e^{i\theta} + e^{i(\theta + 90°)}$. The imaginary exponential simply describes a vector in the complex plane. (Also see p. 307.) The amplitude of the exponential is the amplitude of the vector, and the phase angle θ is the angle that the vector makes with the real (x) axis. The projection on the x axis is the resultant sine function, varying in length as θ varies. Note that in this representation it is immediately obvious that the sum of any two sine functions *of the same frequency* is another sine function. Any two vectors simply form another vector. As θ changes, with the same angular frequency for each, all the vectors rotate together. The two extreme cases of sine-wave interference are constructive and destructive. In the first case the amplitude of the resulting vector is twice that of one, and in the second case the two vectors (180° apart) cancel each other.

2. A *standing wave* is an oxymoron—a contradiction of terms. How can a wave stand still? Actually, it does not. A standing wave is a sum of two waves of equal amplitude and frequency, but traveling in opposite directions. Usually the second wave is created by the reflection of the first. We show an example of the buildup of such a pattern in Fig. 15.3.

   The algebraic description of the situation is derived from the same trigonometric identity that we used before: $\sin(kx - \omega t) + \sin(kx + \omega t) = (2 \sin kx) \cos \omega t$. We have a sinusoidal term (the cosine) varying with the same temporal frequency as the individual waves but with an amplitude that has a sinusoidal dependence on position along the x axis. Although the resultant was built up out of traveling waves, there is no indication in the final appearance that there is any transfer of energy along the line. Instead, each point on the line appears to be oscillating up and down. The amplitude of oscillation varies along the line and there are nodal points where there appears to be no disturbance at all. One might well wonder how energy can be fed along the line to maintain this standing wave if these are true nodal points. But in the ideal case, no energy is lost or transferred once the reflections occur and the oscillations are established. If no energy is lost, no energy has to be fed into the system, and therefore no energy need cross a nodal point. In a real situation, energy is continually lost along the line or in the reflection. To make up for the loss, energy must be fed in, the incident wave is larger than the reflected wave, and there are only approximate nodal points.

In the vector representation of standing waves, the two vectors are rotating in opposite directions, as shown in Fig. 15.4. The projection of their sum on the x axis is the resulting signal. If at a point (for instance a boundary) the vectors lie on the x axis and are 180° out of phase at a given time, then the sum of the x projections of the counter-rotating vectors will remain 0 at all times. At a point (perhaps in the middle of a tube) where the vectors lie on the x axis and are in phase at a given time, the sum of their x projections will vary from maximum positive to maximum negative. Any particular point will oscillate in time with angular frequency ω and with amplitude given by $2 \sin kx = 2 \sin 2\pi x/\lambda$.

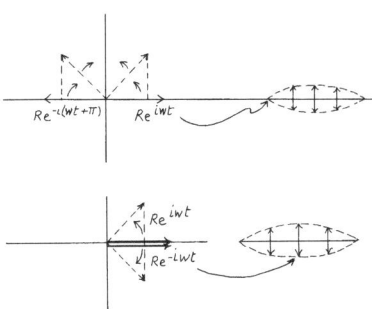

Figure 15.4. Complex vector representation of standing waves.

3. One of the simplest cases of interference between two waves occurs when the waves have equal amplitude and are moving in the same direction with the same velocity but have slightly different frequencies. This situation produces the phenomenon of "beats." The situation is shown graphically in Fig. 15.5. The upper part of the diagram shows the two separate sine waves moving along the same line. In the lower part of the diagram, the waves have been superimposed by adding the displacements of the two waves at every point. In examining the resultant pattern, the first thing to observe (and it is not obvious that it should be true) is that the sum looks like a sine wave with changing amplitude. The second thing to note is that the envelope of the amplitude is itself sinusoidal. Suppose that the two signals are 8 and 10 Hz. If they start out in phase, they will be back in phase half a second later. In that time one of them will have been through five periods and the other will have been through four periods. At 1/4 s the 10-Hz note will have gone through $2\frac{1}{2}$ periods and the 8-Hz note will have gone through two periods. At that time they are 180° out of phase. During 1 s the resultant amplitude goes through two minima and two maxima. The frequency of these "beats" is 2 Hz, which is the difference in frequency between the two separate signals.

We can describe the general situation algebraically by using our standard identity for adding two sine waves. In this case, the sum is

$$\sin \omega_1 t + \sin \omega_2 t = 2 \sin \tfrac{1}{2}(\omega_1 + \omega_2)t \cos \tfrac{1}{2}(\omega_1 - \omega_2)t$$

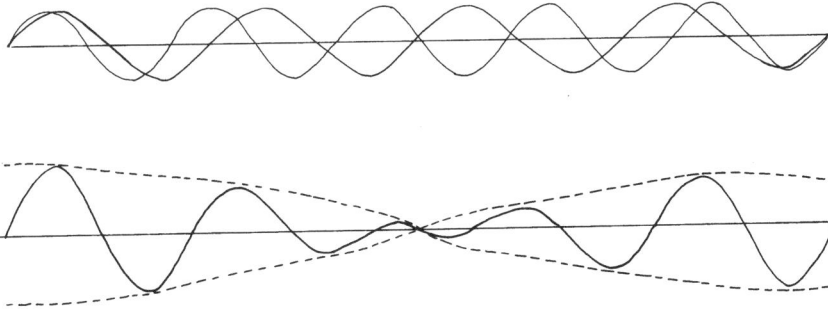

8 Hz and 10 Hz signals combined.

Figure 15.5. Addition of two sine waves of slightly different frequencies to produce beats.

$$= \left(2\cos 2\pi \frac{f_1-f_2}{2} t\right)\left(\sin 2\pi \frac{f_1+f_2}{2} t\right)$$

$$= A(t)\sin 2\pi ft.$$

Here we have the algebraic proof that the superposition of the two sine waves produces another sinusoidal wave. The fine structure of the oscillation is described by the sine curve. The frequency is $\frac{1}{2}(f_1+f_2)$, which is just the average of the two separate frequencies. For instance, in our example the resultant average frequency is 9. The amplitude of this sinusoidal high-frequency wave is given by the cosine term, which has a frequency equal to $\frac{1}{2}(f_1-f_2)$. In our example, the frequency of this slowly varying amplitude is 1. It appears at first that the beat frequency given by the algebraic derivation is different from the one we derived graphically. However, the frequency $\frac{1}{2}(f_1-f_2)$ is the frequency for the amplitude to go through a complete cycle including a maximum positive amplitude, and a maximum negative amplitude. Each of these, however, is heard as a beat and therefore the frequency of the beats is twice that of the varying amplitude term. The *beat* frequency is indeed f_1-f_2.

In the phasor description of beats, the two vectors have equal amplitude and rotate in the same direction but with different velocities. The period for the first vector is $1/f_1$. It takes n/f_1 seconds to go through n periods. If it is then back in phase with the second vector which has gone through $n+1$ rotations, the time taken for the two of them has been the same: $n/f_1=(n+1)/f_2$. Consequently, $n=f_1/(f_2-f_1)$. Using our numerical example above, if $f_1=8$ Hz and $f_2=10$ Hz, $n=4$. If the two vectors start out in phase, then in four rotations of the first vector the second one will complete five rotations and they will be back in phase. The frequency of this occurring is $f_1/n=f_2-f_1$.

Skilled musicians or piano tuners use the beat phenomenon for tuning purposes. When two tones have almost the same frequency, you can hear a definite pulsation. You can even determine the difference in frequencies by counting the number of beats per second, providing the number is under about five. The tuning proceeds by adjusting the frequency of one instrument until the beat disappears.

The presence of beats between two notes is a factor in our pleasure in listening to the notes. Each string in the middle and upper registers of a piano is really a triplet of strings, each of which can be tuned separately by adjusting the tension. If the frequencies of the separate strings are slightly different, beats will be produced. The effect is not particularly noticeable unless the beat frequency is more than several per second. If they are out of tune by that much, the strings produce a jangling sound typical of the sound supposedly produced by barroom pianos.

Combining sinusoidal functions to form Fourier series yields both astonishing and delightful results. Although a power series, if enough terms are used, can represent a periodic function—such as a sine or cosine—it makes more sense to approximate periodic functions with a series of terms that are also periodic. Any periodic function can be described as a sum of sines and cosines with appropriately chosen amplitudes and with frequencies equal to integral multiples of the fundamental frequency. Most nonperiodic functions (such as a pulse) can be described by an *integral* of sine waves with continually varying frequencies. This property of sine waves is not at all obvious. Probably the best pedagogical appeal to prior experience is to remind students (for many, that means informing them for the first time) about fundamentals and harmonics in music. When seen on an oscilloscope, sounds from instruments that produce brilliant overtones are ragged, not at all like smooth sine waves.

In Fig. 15.6 we show graphically the addition of a fundamental sine wave and its second harmonic (the fundamental is the first harmonic): $f(\omega t) = b_1 \sin(\omega t) + b_2 \sin 2(\omega t)$. In the first curve, $b_1 = b_2$. In the second curve, $b_1 = 2b_2$. Note that the resulting curves are not sine waves. In the third graph, we have plotted the sum of a fundamental and its *third* harmonic, first with $b_1 = b_2$, and then with $b_1 = 2b_2$, $f(\omega t) = b_1 \sin(\omega t) + b_2 \sin 3(\omega t)$. Notice how the symmetry of the second group differs from that of the first. The addition of higher *even* harmonics to the first group would make the sum look more like a sawtooth. The addition of higher *odd* harmonics to the second group would make the sum look more like a square wave.

Jean Fourier (1768–1830) published a treatise in 1822 showing that if $f(x)$ is a finite continuous function that is periodic in 2π, then

$$f(x) = \tfrac{1}{2}a_0 + a_1 \cos x + a_2 \cos 2x + a_3 \cos 3x + \cdots + b_1 \sin x + b_2 \sin 2x + b_3 \cos 3x + \cdots .$$

In Fig. 15.7 we show a square wave and a sawtooth wave and the first few terms of the Fourier series that approximate them. For complicated wave shapes, many sinusoidal terms must be added to get a good approximation. Sharp corners on the original function are particularly hard to produce and require high frequency terms (high harmonics). If a function is even [$f(x) = f(-x)$], then only cosine terms are required in the expansion. If a function is odd [$f(x) = -f(-x)$], then only sine terms are needed.

The amplitudes are given by the following formulas:

$$a_0 = \frac{1}{\pi}\int_{-\pi}^{\pi} f(x)dx, \quad a_n = \frac{1}{\pi}\int_{-\pi}^{\pi} f(x)\cos nx\, dx, \quad b_n = \frac{1}{\pi}\int_{-\pi}^{\pi} f(x)\sin nx\, dx.$$

As an example of how to use these formulas, suppose $f(x)$ is a square wave: $f(x) = -1$ from $-\pi$ to 0, $+1$ from 0 to π, and -1 from π to 2π, as shown in Fig. 15.7. The general shape of this square wave looks more like a sine wave than a cosine wave. The function is odd: $f(x) = -f(-x)$.

$$a_0 = \frac{1}{\pi}\int_{\pi}^{-\pi} f(x)dx = \frac{1}{\pi}(-1\pi + 1\pi) = 0$$

$$a_n = \frac{1}{\pi}\left(-\int_{-\pi}^{0} \cos nx\, dx + \int_{0}^{\pi} \cos nx\, dx\right) = 0.$$

All the a_n are 0, since $\cos nx$ is an even function and the two terms in the bracket will always cancel:

$$b_n = \frac{1}{\pi}\left(-\int_{-\pi}^{0} \sin nx\, dx + \int_{0}^{\pi} \sin nx\, dx\right) = \frac{1}{n\pi}(\cos nx\big|_{-\pi}^{0} - \cos nx\big|_{0}^{\pi})$$

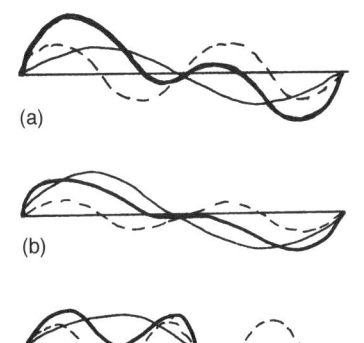

Figure 15.6. Addition of a fundamental frequency sine wave and its second and third harmonies. (a) Fundamental and 2nd harmonic of same amplitude. (b) Fundamental and 2nd harmonic with half amplitude. (c) Fundamental and 3rd harmonic of same amplitude. (d) Fundamental and 3rd harmonic with half amplitude.

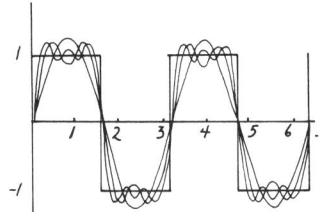

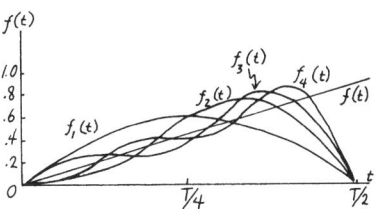

Successive Fourier approximations for a sawtooth.

$$f(t) = \frac{2}{\pi}\sin\frac{2\pi}{T}t - \frac{2}{2\pi}\sin\frac{2\pi}{T}2t + \frac{2}{3\pi}\sin\frac{2\pi}{T}3t - \frac{2}{4\pi}\sin\frac{2\pi}{T}4t + \ldots$$

Figure 15.7. Fourier series approximations to a square wave and a sawtooth wave.

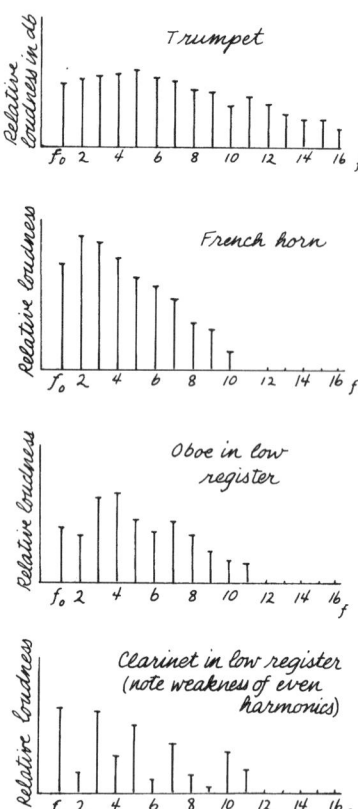

Figure 15.8. Relative loudness in db of fundamental and harmonies for instrument tones after the initial attack.

$$= \begin{cases} \dfrac{1}{n\pi}(2+2) & \text{for odd } n \\ \dfrac{1}{n\pi}(0+0) & \text{for even } n. \end{cases}$$

Therefore, the series for this square wave is

$$f(x) = \frac{4}{\pi}\sin x + \frac{4}{3\pi}\sin 3x + \frac{4}{5\pi}\sin 5x + \cdots .$$

If a trumpet and oboe both sound middle A, each is producing 440 repetitive pulses per second. However, on an oscilloscope the oboe pulses look different from the trumpet pulses. Each musical instrument produces periodic pulses with characteristic shapes, at least during the initial attack of the tone. In the Fourier series for these pulses, the difference between one pattern and another is determined by the distribution of amplitudes of the harmonics. Each instrument has its own recipe for these amplitudes. A graphical representation of these recipes for several instruments is shown in Fig. 15.8.

A good rule of thumb for the reproduction of musical notes is that harmonics up to the tenth are required. For a 2000-Hz note, the peculiarities of shape may require harmonics up to 20 000 Hz for faithful reproduction. Of course, this rule of thumb depends on the nature of the instrument. Any note produced by plucking or striking is apt to have pulses with sharp edges involving harmonics of high frequency. One test for a high-fidelity system is to reproduce the sounds from bells or triangles, or even pianos, so that they sound natural.

If a reproducing system is not linear, the output of a sine wave will be distorted. A typical example of this would be the saturation of the output caused by too high gain or too large an input. The distorted sine wave can be represented by a Fourier series of higher harmonic frequencies. The output will contain higher frequencies than existed in the original signal.

■ HUYGENS'S PRINCIPLE

Back in 1678, Christiaan Huygens (1629–1695) in Holland proposed a way to predict the motion of waves in various geometries. His idea is simple in principle, but can become very complicated if you calculate all the details. Huygens proposed that each point of a disturbed medium (such as water, air, or space) must be the origin of a new wave front coming from that particular point.

At first thought, Huygens's idea seems reasonable enough. If for any reason you disturb a medium at a point, then that oscillating point will radiate a wave. On second thought, the wave front from such a point would be an expanding circle on a two-dimensional surface, and an expanding sphere in three dimensions. If you have a long straight pulse moving along the surface of water, how can this generate a whole line full of circular pulses, expanding in every direction? A complicated analysis shows that in every direction, except one, the disturbance from any point along the line is canceled out by the corresponding disturbances from all other points. The surface motion at any point ahead of the advancing waves will be determined by contributions from the *entire* wave front. It turns out that the resulting wave front one period later is located along a line, tangent to the circular wave fronts that originate from each of the points in the original crest. The line tangent to all of these circles is parallel to the original crest and represents a front that has moved one wavelength to the right in one period.

An easy way to observe this effect experimentally is to dip a coarse comb in shallow water. Each tooth of the comb will be the source of a circular wave front. The resultant disturbance, however, will be a plane wave, the same as if the comb had been a solid ruler. Actually, the resultant disturbance creates two plane waves, moving in opposite directions. Why shouldn't Huygens's construction always produce a backward wave? The reason basically is the same as we saw on p. 333 when we analyzed the velocities and forces on a rope while a pulse was passing through. The particles in the medium directly behind the crest of the pulse have downward velocities. The pulse moving away exerts upward forces on those particles, leaving them with 0 velocity as the pulse passes.

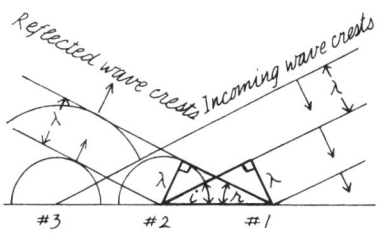

Figure 15.9. Huygens's wave construction for reflection.

Let us use Huygens's construction to see what happens when a plane wave is reflected at an angle from a boundary. In Fig. 15.9 we show the parallel wave fronts advancing at an angle i toward a boundary. We construct the reflected wave fronts by drawing tangents to the reflected circular wave fronts generated at each point. However, we must connect wave crests of the same phase. One period before the positions shown in the diagram, crest 1 was in the position now occupied by crest 2. Therefore we must connect point 1 with the crest reflected earlier from point 2.

The resulting reflected waves are also plane and parallel, and furthermore they maintain a special angular relationship with the boundary and the incoming waves. The two right triangles showing this relationship are emphasized in the diagram. They have a common hypotenuse, and the short side of each has the same length 1. Therefore, the triangles are identical and $i = r$.

Next we will use Huygens's construction to see what happens with refraction. The geometry is shown in Fig. 15.10. Because the velocity in the second medium is less than in the first, the radii of the circular wave fronts are smaller in the second medium. The tangent lines to these smaller circles are parallel to each other but not to the incident wave fronts. The angular relationship between the incident angle and the refracted angle can be obtained from the geometry of the two adjacent triangles. They have a common hypotenuse. Each of them is a right triangle with one leg being the wavelength in that particular medium. Therefore

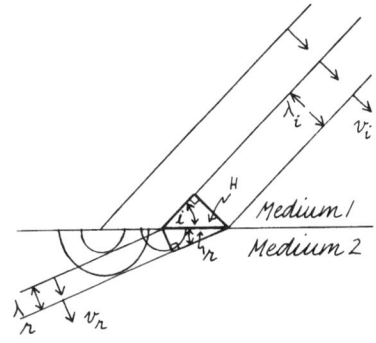

Figure 15.10. Huygens's wave construction for refraction.

$$H \sin i = \lambda_i = v_i T_i,$$

$$H \sin r = \lambda_r = v_r T_r,$$

since $T_i = T_r$

$$v_r \sin i = v_i \sin r.$$

■ COMBINING FERMAT'S AND HUYGENS'S PRINCIPLES

As we have seen, the simple rule for ordinary reflection, $i = r$, satisfies both Huygens's wave front geometry and also Fermat's principle of minimum time (or, rather, extremum of time). (See p. 335) We can combine the arguments of these two methods, setting the stage for a different and very useful analysis of wave phenomena. In Fig. 15.11 we show the two-dimensional geometry of a point wave source and a detector above a boundary that acts as a mirror. According to

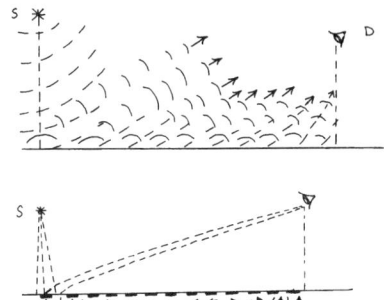

Figure 15.11. Huygens's wave construction for reflection from a plane mirror, taking into account phase differences for different paths.

Huygens's principle, all points of the mirror will respond by oscillating and thus creating new circular ripples. It appears that the detector will receive waves from every point on the mirror. In the lower part of the diagram we have divided the mirror into many segments, each of which is a source of reflection. However, instead of drawing spreading circles from each one, we have drawn the phasor representations of the reflected signals that will reach the detector. Remember that a phasor is an arrow whose length represents the magnitude of a sine wave, and whose angle with the direction of the x axis represents the phase of the wave. The magnitude of a wave traveling from S to D in our two-dimensional model will be inversely proportional to the square root of the distance from S down to the intersection with the mirror and back to D. (Thus the *intensity* will be inversely proportional to this distance.) The paths we show in the diagram differ in length but for first approximation we assume that the amplitudes of the different waves, and hence the lengths of their phasor arrows, are about equal. However, the different path lengths have a major effect on the *phases* of the waves from different segments. For instance, we could choose the segment widths such that for waves traveling almost straight down to the mirror, adjacent paths would be different in length by 1/2 a wavelength. The waves reflecting from this region—and the symmetrical one underneath D—would all cancel each other. Waves reflecting from the middle region of the mirror all travel about the same distance and thus arrive at D with about the same phase. They reinforce each other. The diagram shows a graph of travel times versus position on the mirror. The simple law of reflection holds because the geometry produces an extremum condition: for small variations about the usual reflection point, the path lengths are about the same.

We can take away the end pieces of the mirror and still get the same reflection. Our simple law of reflection would claim the same thing. However, we cannot reduce the mirror to too small a region around the optimum point. The reflection takes place, and the waves add constructively, for a sizable region on either side of the point for which $i = r$. If we reduce that region so that it becomes comparable to a few wavelengths, then the mirror will act like a point source and reflect the waves over a circular spread. Yet more remarkable, we can indeed get reflection from the end pieces. If we cover over every other segment of the mirror out at the ends, we will be left with reflected waves that are in phase when they reach the detector. The mirror will have become a reflection grating, designed to produce constructive interference for a particular wavelength at a particular angle.

In *QED*, Feynman's profoundly simple book on quantum electrodynamics, Feynman introduces phasors as arrows that follow certain combination rules. For photons, the arrows are probability functions, the solutions of Maxwell's equations. The absolute square of the function at any point is the probability of the electromagnetic event at that point. In this model the arrow is not the amplitude and phase of an actual wave at a point, but describes the probability of a photon occurring at that point. The "net" or total arrow at a point is the vector sum of all the arrows at that point that have been contributed by photons traveling all possible paths. As an example of the subtle simplicity of this concept, consider the proof that in free space light (photons) travel in straight lines. Once again we appeal to a Fermat-type extremum argument, shown in Fig. 15.12. All the paths in the immediate vicinity of a straight line—those in sort of a bundle around the line—differ only slightly in distance and thus in phase. The arrows (phasors) from all these paths add constructively. Paths that deviate through larger excursions will

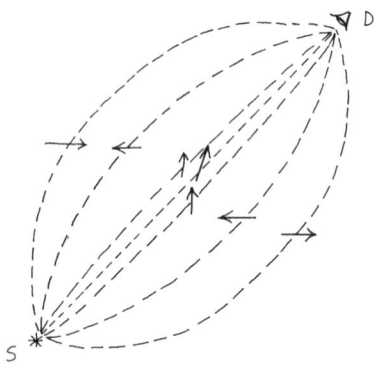

Figure 15.12. Feynman's argument that light travels in a straight line because other paths cancel due to phase differences.

yield larger phase differences. The phasors or arrows will have greater rotations. For every longer path there will be one yet longer by $\frac{1}{2}\lambda$, yielding cancellation. It appears that light travels in straight lines only if some of the photons can travel through slightly curving paths on either side of the straight line. If this seems like a fanciful argument, try blocking those curved paths. Make the straight line and the light go through a pinhole. As we all know, the light then spreads out into a diffraction pattern. These arguments are not peculiar just to light and photons; they apply to any kind of wave motion.

■ DIFFRACTION

Following Huygens's model, we claim that each point on an advancing wave front can be considered a source of a new set of waves. If the wave front is plane to begin with, then it continues on as a plane, and in the same direction. Interference cancels out any wave effects except along the tangent of crests. For complete cancellation in any direction except forward, contributions of waves are needed from *every* point along the original plane wave front. If no contributions arrive from some part of the original plane wave, then there may be reinforcement of the waves at one or more angles besides the forward direction. If an advancing two-dimensional plane wave goes through a narrow slit, then the wavelet contributions on either side of the slit are eliminated. What comes through the slit will not be simply part of the original plane wave. In the limit of a slit smaller than a wavelength, for example, the wave getting through the slit will act as if it came from a point source and will spread radially in two dimensions.

For certain simple geometries it is easy to add up the effects of the Huygens's wavelets and so predict the interference patterns. One of our simplifying factors will be to consider that it is indeed a plane wave that advances on the narrow opening. The situation would be more complicated if the waves had originated from a point source only a small distance from the opening. In that case, the original wave would be spreading out radially and consequently the wave that enters the opening would not be all in phase. If we use plane waves to begin with and look at the interference pattern far away from the barrier, the phenomenon is called Fraunhofer diffraction. If the original wave front is not plane and if we study the interference pattern just past the barrier, the phenomenon is called Fresnel diffraction. Frequently, Fraunhofer diffraction, which is easier to calculate, makes a good approximation to what happens in a real system of Fresnel diffraction.

With light waves there is an additional feature that makes the Fraunhofer approximation a good one. Because the wavelength of light is so small and the slits producing diffractions are so narrow, the diffraction pattern is usually observed at a large distance compared with the slit opening. Furthermore, the light from the slit frequently passes through a lens before it interacts to form the diffraction pattern. If we form a slit with our fingers and observe a distant small light source we have all the conditions necessary for using the Fraunhofer approximation. Since the source light is distant, the wave fronts striking the slit between our fingers are plane. After the light passes between our fingers, it enters the lens of our eye and is focused on our retina at the back of the eye ball. Since a convex lens always focuses a plane wave front to a point (but not necessarily on the axis!), we can use the very simple geometry of plane waves, or parallel rays, in our analysis. There is a pedagogical problem here. Compare the two diagrams

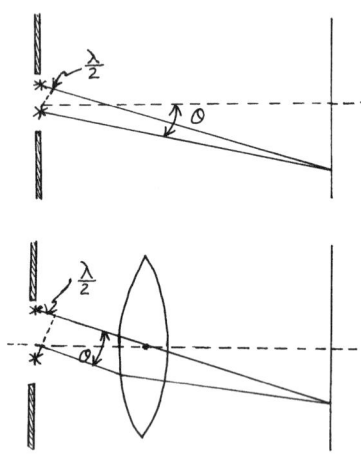

Figure 15.13. Two views of the geometry of single slit diffraction.

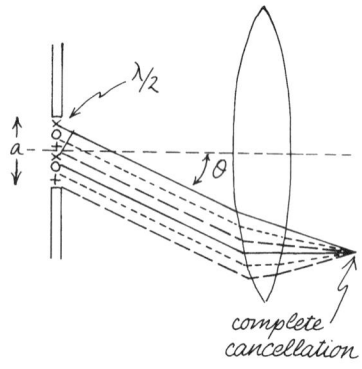

Figure 15.14. Geometry for first minimum in single slit diffraction.

of Fig. 15.13. The lower one uses parallel rays and no approximations, but assumes some knowledge about the behavior of a lens. The upper one eliminates the lens but requires small angle approximations (which are generally valid). There is yet another pedagogical problem in explaining what it is we see when looking at a distant light through a crack between two fingers. With only a little bit of good will and determination it is possible to see very good diffraction patterns this way. The light is not only spread out, but shows the nodal bands. While most students can see the effect, it is not clear that many of them can combine what they see with what they know about the optics of the eye and with the text diagram of the geometry.

Now let us analyze the interference of the waves coming from a slit opening in a barrier. In two-dimensional geometry, we can divide the opening into an arbitrary number of small points, each of which can be the source for circular wave fronts. All of these point sources are being driven in phase because we assume that they are points on the original plane wave front. In the original forward direction, the wave fronts remain in phase and will be brought to a central focus point by the lens. For small angles on either side of the forward direction, the wave fronts are almost in phase and these wave fronts will be focused to points on either side of this central image with only a small amount of cancellation. However, at larger angles from the forward, the wave crests are not all in phase with each other and there will be some destructive interference reducing the intensity of the light when the lens brings it to a focus. At certain particular angles, this destructive interference is complete.

In Fig. 15.14 we show the geometry that produces complete destructive interference. We have divided the narrow slit into two sections; a top and a bottom. Within each half there are numerous point sources. A line has been drawn perpendicular to the direction θ. At that line the light coming from the top point of the top half has had to travel $\frac{1}{2}\lambda$ farther than the light from the top point of the bottom half. Therefore the light from the upper point is 180° out of phase with the light from the bottom point. This same geometrical relationship holds for each pair of points in the bottom and top half that are separated by $\frac{1}{2}a$. For every point source in the bottom half, there is a matching point source in the top half. At the particular angle θ, the contributions from these matching pairs cancel each other. Consequently, there is no light intensity at all in the interference pattern at the point corresponding to the angle θ.

The critical angle θ is determined by a relationship between the wave length of the light and the width of the slit:

$$\sin\theta = \frac{\frac{1}{2}\lambda}{\frac{1}{2}a} = \frac{\lambda}{a}.$$

As long as θ is small (less than 30° or so), we can use the approximation that

$$\theta \approx \frac{\lambda}{a}.$$

The same geometrical argument can be used if the slit is divided into four or six or eight equal parts. In each case there will be an angle such that pairs of points will produce waves that cancel. In general, there will be complete cancellation for any angle that satisfies the condition:

$$\sin\theta = \frac{n\lambda}{a}, \quad n=1,2,3\ldots .$$

We would expect to see bands of light between the lines of complete cancellation. It is not quite so easy to calculate exactly the intensity within these bands or even the position of maximum brightness. Notice, for example, that the intensity must fall off on either side of the central maximum, since for any angle different from the forward direction, there will be partial cancellation. One way to estimate the amount of light entering the first bright fringe (between the first and second nodes) is to divide the slit mentally into three sections as shown in Fig. 15.15. If we draw a line such that $\sin\theta = (\frac{1}{2}\lambda)/(\frac{1}{3}a)$, then matching points in two of the sections will produce wave contributions that cancel each other. However, the contributions from the remaining third do not all add constructively. Note that the light from the top part of the remaining section must travel half a wavelength further than the light from the bottom part of that section. So there is partial cancellation of the contributions within this remaining section. As a first approximation we might guess that about half the points cancel each other. If we add the amplitudes of the wavelets that survive, we have only about one-half of the one-third of the total sources in the slit.

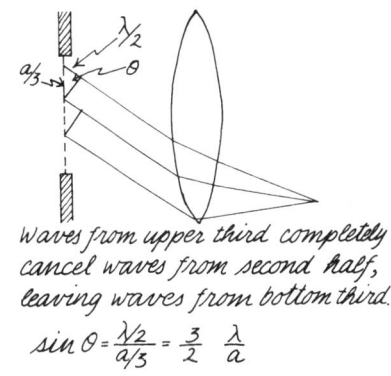

Figure 15.15. Geometry for first bright sideband in single slit diffraction.

The crude argument that we have used predicts that the *amplitude* of the resulting wave in the first bright fringe will be only 1/6 of the total amplitude of the original wave. That does not mean, however, that the *intensity* of the light in the first bright fringe as 1/6 of the original intensity. The intensity of a wave is proportional to the square of its amplitude. Therefore the intensity of the light in the first bright fringe is 1/36 of the original intensity, or about 3%. The actual intensity, which can be derived from a more complete analysis, is 4.7%. To analyze the interference in the second bright band, we would have to divide the slit into five equal parts. The contributions from two of these would cancel the contributions from two others and the contributions from the fifth section would cancel about half the remaining amplitude. The resulting amplitude would be only 1/10 of the original, and the resulting intensity only about 1%. The actual intensity in this band is 1.7%. Since there are bright fringes on either side of the central maximum, 9.4% of the energy goes into the first side band pair, and 3.4% into the second side band pair. Altogether, the side bands get about 15% of the energy. The remaining 85% is in the central maximum.

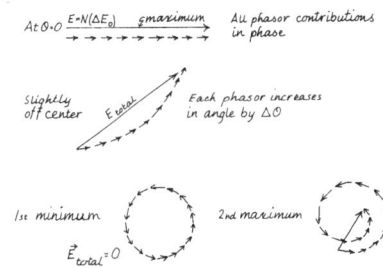

Figure 15.16. Phasor representation of single slit diffraction.

The analysis of single-slit diffraction in terms of phasors is now presented in all calculus-based introductory texts. It is unlikely that many students comprehend the derivation, which used to be included in optics courses at the sophomore college level. We illustrate the procedure and geometry in Fig. 15.16 as an example of the usefulness of phasors. Each small arrow represents the wave contribution at the image from one small segment of the slit. All the arrows have the same length, since each segment is the same size and contributes the same amount (for small θ), and is about the same distance from the image point. However, the small differences in path length between source and image produce small phase differences between neighboring contributions. These are shown by the different angles of the phasors. The total contribution is the vector sum of all the phasors, shown graphically by stringing them together head to toe. In Fig. 15.16 we show summations for several different angles, θ, including the one for the first minimum.

When the amplitudes from sources in the slit are summed in the limit as they become infinitesimal, the resultant amplitude is

$$A(r,\theta,t)=A(r,0)\left(\frac{\sin \frac{1}{2}\Phi}{\frac{1}{2}\Phi}\right)\cos(kr-\omega t).$$

The time-averaged intensity is

$$I(r,\theta)=I_{max}\frac{\sin^2 \frac{1}{2}\Phi}{(\frac{1}{2}\Phi)^2}.$$

Φ is the total phase shift difference between sources at top and bottom edges of the slit for light at the angle θ: $\Phi=2\pi(D\sin\theta)/\lambda$. Notice that these formulas predict null points at the same angles given by the simple analysis.

Other patterns produced by diffraction can be analyzed with the same method we used for the single-slit pattern. In each case, the diffraction is caused by the fact that some of the contributions from the original plane wave are blocked. The contributions from the remaining sections interfere constructively or destructively both inside and outside their geometrical shadow region. Along straight edges there will be parallel light and dark fringes. Around circular holes or obstacles there will be circular fringes. One of the most spectacular examples of diffraction is the bright region in the center of a circular shadow, called the Poisson spot (derived by Poisson in 1818 as an example to show the silliness of the wave theory of light—and then immediately observed by Arago). At other points of the shadow, the wave contributions being received from the surrounding regions cancel to 0. At the center, however, the symmetric contributions from the ring closest to the disk dominates and the contributions with varying phase from the surrounding concentric regions are not enough for complete cancellation. Creating this effect is a splendid special project for students. Have them aim a laser at a ball bearing and examine the shadow of the ball bearing for the bright spot at the center. It is necessary to expand the cross section of the beam so that it is larger than the obstacle. Blacken the ball so that the reflection of the laser does not hurt anyone. The effect can also be produced with sound waves in appropriate geometry.

Any time that waves pass through an opening or over an obstacle, the angular location of the first minimum of the fringe pattern depends on the ratio λ/a, where a is the width of the slit or the diameter of the hole or obstacle. For the circular fringes surrounding a pinhole, or any hole or obstacle with circular geometry, the angle for the first minimum is given by $\sin\theta \approx \theta=1.2\lambda/a$. You can demonstrate this effect with sound waves by observing the shadow cast by a human or an upholstered chair as a function of pitch or frequency. For frequencies higher than middle A (440 Hz→$\lambda=\frac{3}{4}$ m), the shadow effect is easily detected. Similarly, you can detect the diffraction effect of sound waves going through doorways.

Although we usually think of diffraction patterns produced by pinholes, an opening of any size produces diffraction. For example, the largest diameter of your eye pupil is about 8 mm. The image on your retina of a point source will not be a point but a circular diffraction pattern. The angular half-width has a size $\theta=1.2\lambda/a$. For visible light, $\theta=[1.2(5\times10^{-7}\text{ m})]/(0.8\times10^{-2}\text{ m})=7.5\times10^{-5}$ rad$=4.3\times10^{-3}$ deg.

It might seem that such a small angle is hardly worth bothering about. Suppose, however, that you are looking at two point light sources very far away, such as the headlights of a distant car. If the car is close enough, the pattern on your retina will look like part (a) of Fig 15.17. If the car is far away, the pattern on your retina may

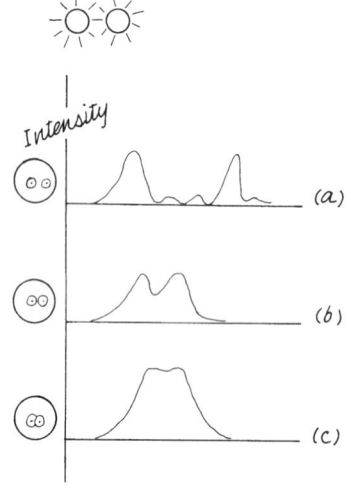

Figure 15.17. Images on the retina from two distant light sources. In a the sources can be resolved. In c the sources are so far away that their diffraction images merge.

look like part (b) or even (c). In this last case, the maximum of one diffraction pattern falls on a minimum of the diffraction pattern of the other light. (This is known as Rayleigh's criterion.) If the lights were any further away the two images would be so completely merged that your eye could not detect whether it is seeing two sources or one. It could not *resolve* the two sources. The angular resolution limit of a lens or a circular opening is given by $\theta = 1.2\lambda/a$. In our example, if the car were at a distance of 17 km, the human eye could just resolve the two headlights. Actually, glare or mist would probably make the lights appear to fuse at a shorter distance.

The resolving power of a microscope or a telescope is a crucial measure of its performance. Since the objective lens of a microscope is about the same size as a human eye, its angular resolution limit is approximately the same as that of the eye. If the focal length of the lens is 1 cm, then the separation of two objects that can just be resolved is

$$\Delta x = (7.5 \times 10^{-5} \text{ rad})(1 \times 10^{-2} \text{ m}) = 7.5 \times 10^{-7} \text{ m}.$$

A microscope with these characteristics could just resolve two point objects if they were 0.75 μm apart. The human eye has a focal length of about 1.8 cm, but cannot focus on an object close to the lens. With only one lens (instead of the compound system of the microscope), the magnification of the eye is relatively small. The resolution limit cannot be seen when looking at objects that are close. The resolving power of the 200-in. reflecting telescope on Mount Palomar is much greater. The resolving angle for the big mirror is approximately 1×10^{-7} rad, which is equal to about 0.02 seconds of arc.

Diffraction limits the parallelism of a laser beam. For most lasers, the controlling aperture is not the diameter of the opening or of the beam as it leaves, but rather a narrower cross section in the middle of the lasing column. In a typical helium–neon laser, the minimum diameter of the beam within the tube is only $\frac{1}{4}$ mm. We might therefore expect the diffraction angle to be $\theta = 1.2(5 \times 10^{-7}$ m$)/(0.25 \times 10^{-3}$ m$) = 2.4$ mrad. The actual diffraction angle for a laser beam is only about half that which we just calculated. The intensity of the beam is not uniform across its diameter but is Gaussian. In effect, the limiting aperture is "softer" and produces less diffraction than an aperture with sharp edges.

The spread of a laser beam is limited by diffraction, but the spread of most searchlight beams with ordinary incandescent sources is limited by the finite size of the filament or light source. There is a conservation requirement imposed on any kind of beam. If the beam *cross section* is made smaller, with lenses or mirrors for example, then the *angular divergence* must increase. Conversely, in the case of the narrow laser beam, the beam can be expanded to have a larger cross section and consequently will have a smaller divergence.

■ INTERFERENCE

Historically, the interference of waves coming from the *same* source or through the *same* opening was called diffraction. The interactions between the waves coming from *separate* sources or through *separate* channels was called interference. There is no physical difference between these processes. In one sense the wave interference produced by two sources is easier to understand than diffraction. To calculate a diffraction pattern from a single source, we have to divide the source into many small regions and then calculate the phase and intensity arriving from each of these sources. If we have only two sources or slits through which the waves may come, we can assume that each one acts like a point source and that at the sources the waves start out in phase with each other.

This last point is an important requirement. We do not get interference patterns from sources that are *incoherent*. If the two sources are two light bulbs, the phase relationship between their light at any particular point is completely random. Actually, *coherence* is a function of the frequency response of the observer. On the human time scale the light from two separate bulbs in incoherent, but at a point some distance from the bulbs the phase between the waves coming from the two sources might stay constant for times of the order of 10^{-8} s.

If two slits close together are illuminated by the light from a single source far enough away so that it is essentially a point source, then the light reaching the slits is in phase. The wave train reaching the slits may start and stop abruptly and shift phase every 10^{-8} s, but at every instant the phase in one slit is the same as the phase in the other.

The light in a slide projector beam is not coherent from one point to another in its cross section. The light has come from the surface of a filament that is not even approximately a point. The light from one point on a filament is not in phase with that coming from any other point. A laser beam, however, consists of light waves that are in phase throughout a cross section of the beam. Even a laser, however, does not maintain a constant phase for very long. The ordinary helium–neon laboratory laser keeps constant phase for only a few nanoseconds. You cannot see interference between two laser beams (of this type). Perhaps the greatest long-term coherency of signals produced by humans is that of the nation wide electric power grids. If the 60-cycle power from one station is only slightly out of phase with that from another station, power will flow in unwanted ways between them. Actually, they are locked in phase *because* power flows if one falls behind.

One way to produce a plane wave to illuminate a double slit is to send the light first through a single slit. Then the central maximum from the diffraction of light from the first slit, which is all in phase, can cover the double slit downstream.

One feature about double slit interference is usually not obvious to students. At a large distance from the double source, the basic pattern in essentially that resulting from the diffraction of a *single* source. You can see this in a pond by throwing two stones into the water close together. The spreading ripples eventually merge into one large circular pattern. In the case of light, the central patterns of the two slits overlap if the screen is far away, or fall exactly on each other if a lens is used. *The interference pattern between the two slits then exists within the diffraction envelope produced by each slit.*

■ INTERFERENCE PRODUCED BY SPLITTING AND RECOMBINING A SINGLE BEAM

Since it is hard to get long-term coherent light sources, the best way to produce interference is to use just one beam, split it, and then recombine the separate beams. Nature does this for us with reflections from thin films or natural gratings. One of the most dramatic devices that humans have created to accomplish this is the Michelson interferometer. (See p. 530.) The way to consider the geometry of that device is that if you look back from the final image position you will be seeing two images of the source, straight ahead of you, one slightly closer than the other, and perhaps slightly tipped. The two sources are necessarily in phase, since they are the same source, but the different paths will produce phase differences where the beams interact.

In the usual analysis of thin-film interference, we assume that part of the light reflects at each surface and part goes through. Following the rules for waves reflecting from boundaries between media of different impedance, we claim that light reflected off a film back into air suffers a phase change of 180°. However, if we try to analyze the phenomenon in terms of the interactions of photons, we must realize that there is really nothing special about the surface of a transparent medium. The photons just go through the material with only an occasional interaction per layer of atoms. When a photon does interact, it excites an atom in the material which then deexcites, emitting another photon. (For visible light in transparent material, these interactions are just Rayleigh scattering, where the wavelength is much larger than the scatterer. This is essentially a classical effect where an incident electromagnetic field drives the atom as an oscillating dipole, which then radiates. The emitted photon has the same wavelength as the incident photon, and the process takes place with a delay of one-fourth period.) These interactions take place at all depths of the transparent medium. Photons always travel at the speed of light c, but the photons produced by the interactions are delayed in phase by 90°. The resultant phasor representing the light at any particular depth is the sum of the phasor for the light that has not interacted plus the sum of the phasors of the reemitted light. The direction of this latter phasor is perpendicular to the original phasor, and hence the resultant is delayed in phase. The effect is the same as if the light had been traveling more slowly through the medium. (That is why $n = c/v$ is greater than 1.)

The photons that are emitted back out into the air from the film will be concentrated at the appropriate reflection angle for the reason that we have already seen—the probability for any other angle is small because of phase cancellation. To find the magnitude of the reflected beam we must add up the phasors from contributions of all depths of the film. We show the geometry and the phasor addition in Fig. 15.18. Note that the resultant *could* be constructed of two components of equal length, one representing reflection from the front surface of the film, and the other from the back surface. If we stick to the photon model, what happens to the 180° phase change at the front surface and the 0° phase change at the back surface? After all, we appeal to those phase changes to explain why a film is black if it is thinner than a wavelength. But "black" just means there is no reflection. For such thin films there is little probability of a photon interacting before passing out the other side. What about the case where the film thickness equals $\frac{1}{4}\lambda/n$ and there *is* reflection? The diagram also shows this phasor geometry; it produces a maximum resultant. The first phasor in the chain represents the probability of a reflected photon being emitted near the front surface. We take its phase as 0. The last phasor in the chain represents the probability of a reflected photon being emitted near the back surface of the film. Because of the time taken for the incident photon to go a quarter wavelength and for the reflected photon to return to the front surface, the relative phase is delayed by 180°. To be sure, the photons from the upper and lower surfaces are in opposite phase, but the phasor chain of all the photons reradiated from the layer add up to a maximum amplitude. As the film increases in thickness to half a wavelength, the phasor representing a

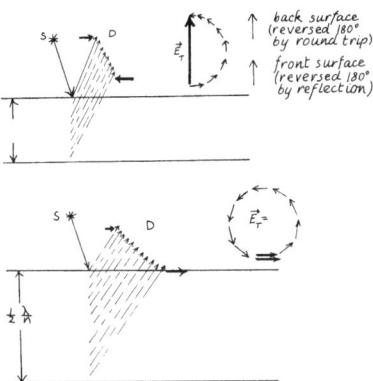

Figure 15.18. Phasor explanation of thin film interference.

photon emitted from the back surface will be delayed by 360°. In the phasor addition chain, that will produce a closed circle of the arrows with 0 resultant—no reflection.

■ DISPERSION OF LIGHT

The simple derivations of wave velocity assumed no dependence of velocity on frequency (or amplitude). We have seen that water waves are more complicated and that in certain regimes of wavelength and depth the velocity depends strongly on frequency. This formula was presented and analyzed in Chap. 14. We also make common use of the dispersion of light in transparent materials. The qualitative features of this dispersion can be explained by a simple model of electromagnetic waves setting atomic dipoles into oscillation and these in turn radiating waves that merge with the original wave.

For wavelengths that are long compared with an atom, it is a good approximation for many purposes to consider light an electromagnetic field and the atom a bound collection of positive and negative charge with various natural frequencies of oscillation. The oscillating electric field in the wave provides a force on an electron and sets it into oscillation. If the electron is bound like a particle on a spring, then the amplitude and phase of the forced oscillation depends on the relationship between the frequency of the incoming wave and the natural frequency of the bound electron. If the driving frequency is small compared with the natural frequency, then the electron will follow the driving oscillation with small amplitude, but will be slightly lagging in phase. If the driving frequency is close to the natural frequency, a resonance condition will be set up where the electron's amplitude of oscillation is large and the phase difference is 90°. If the driving frequency is larger than the natural frequency, the electron's amplitude of oscillation will be smaller again, and its phase will be between 90° and 180° behind that of the driving electromagnetic wave. Since these atomic oscillators are high-Q systems, the lag is close to 0° or close to 180° except very close to resonance.

An accelerated electric charge radiates electromagnetic waves. Each electron driven into oscillation by the original wave becomes the source of Huygens's wavelets. The resultant wave is produced by the addition of the original wave and the new ones that have been produced. What happens next is very hard to calculate quantitatively, although it has been done and the calculations agree very well with experiments. The problem is that each electron is affected not only by the original radiation, but also by the contributions from all its neighbors. The phase of each contribution depends on the distance of the neighbor—and its oscillation in turn depends on its neighbors.

It turns out that the resultant radiation at a distance from a whole plane of oscillators, being driven in phase, is 90° out of phase with the oscillators themselves. (The radiation from such a group is proportional to the velocity of the oscillators, not the acceleration.) The advancing electromagnetic wave is made up of the original wave plus the combined contributions from all the local oscillators. This contributed wave is 90° out of phase with the original because of the dependence on velocity of the oscillators, *plus* the phase lag caused by the mechanical response of the electron oscillators themselves. The resultant wave lags between 90° and 180° if the driving radiation is below resonance of the local oscillators, and between 180° and 270° if the radiation is above resonance. However, a lag of

$180° + \theta$ is the same as a *lead* of $360° - \theta$. For visible light in glass, the driving frequency is below the natural frequencies of the bound electrons. Because the contributed wave from these oscillators is slightly delayed in phase, and the effect accumulates as the wave proceeds, light travels more slowly in glass than it does in vacuum.

With some materials the oscillation of the bound electrons can produce other motions of the microstructure that dissipate the original energy. This effect may be frequency-dependent. For instance, oscillation at the frequency of red light may convert the energy to random thermal motion. In that case, red light is absorbed, even though blue light may be transmitted without loss.

The closer the frequency of the light to the resonant frequency of the bound electrons in a material, the greater the phase lag of the driven electrons. In clear transparent materials like glass, the natural frequencies of oscillation of the electrons are greater than the frequencies of visible light. Therefore, blue light with its higher frequency will be closer to the natural frequency of oscillation of the bound electrons and therefore will suffer a larger phase lag than will red light. Furthermore, the amplitude of scattering of the blue light will be greater because it is nearer resonance. Both of these effects cause blue light to travel more slowly in glass and hence it has a larger index of refraction. However, at frequencies higher than the resonant frequency the phase lag, which is greater than 180°, is equivalent to a phase lead. This can occur at certain frequencies in the ultraviolet or soft x rays. In these regions, the *phase* velocity is actually greater than c. As we will see in the next section, however, the *group* velocity is less than c.

When light shines on a smooth metal surface, most of the light is reflected. A little bit is absorbed, and none is transmitted far into the metal. The electrical conduction electrons in the metal are not bound to their parent atoms and therefore have no natural frequency of oscillation. At low frequencies, and even up to the frequency of light for some metals, the electrons are free to respond instantly to the driving electromagnetic wave. The field that they radiate in the forward direction is 180° out of phase and completely cancels the original field that is advancing into the metal. This same field radiated in the backward direction becomes the reflected wave.

■ PHASE VELOCITY AND GROUP VELOCITY

Dispersion causes blue light to travel more slowly in glass than red light and hence refract at a greater angle. Dispersion will also affect the shape of nonsinusoidal waves of any kind. Consider that most musical notes are made up of a fundamental plus many harmonics whose frequencies are integral multiples of the fundamental. If these higher frequencies traveled at different speeds in air, music would be very different. Fortunately, air is not dispersive for sound waves.

When a medium is dispersive, a wave form composed of waves of many frequencies will have a velocity different from the velocities of its individual components. A pulse travels at a *group velocity*; a wave of a particular frequency has a *phase velocity*. You can study this effect by watching a pulse travel along the surface of water. Use a depth greater than 10 cm so that $D/\lambda > 1$. According to our formula on p. 328, under those conditions $v = \sqrt{g\lambda}$. (For long wavelengths, the surface tension term is small.) These are the conditions for large smooth waves (called *swell*) on deep bodies of water, such as the ocean. Notice that this is a different approximation than for the case of tsunamis or the tides where $D/\lambda < 1$.

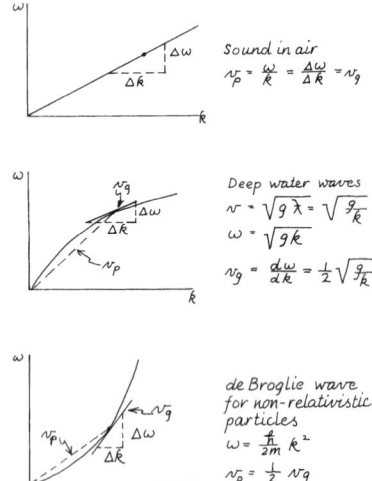

Figure 15.19. Relationship between group velocity and phase velocity for three cases of $\omega(k)$.

In the latter case velocity is independent of frequency. For ordinary deep-water waves the phase (single-frequency) velocity is strongly dependent on frequency. By watching individual pulses under these conditions you can see small ripples advance out of the pulse. These long-wavelength components evidently have higher velocity than the main group. One might think that the group would rapidly dwindle away—thus disperse! Nevertheless, individual *swells* or groups of waves can travel hundreds of miles across oceans, without losing appreciable energy or shape. It appears that the high-velocity lead wave provides the energy to set into motion the water ahead of the main group and then is regenerated at the rear as the group passes by and the water settles down.

We have already dealt with an oscillation that has two component frequencies close to each other. The combination produces beats, which can be described as an oscillation with a frequency that is the average of the original two, modulated by a sine wave that has a frequency equal to half the difference between the two. Let us rewrite these two wave equations and then when we combine them take into account the possibility that the two *velocities* may be different:

$$y_1 = A \sin(k_1 x - \omega_1 t), \quad y_2 = A \sin(k_2 x - \omega_2 t),$$

$$v_1 = \frac{\omega_1}{k_1} = f_1 \lambda_1, \quad v_2 = \frac{\omega_2}{k_2} = f_2 \lambda_2.$$

Adding the two sines, we get

$$y = y_1 + y_2 = 2A \cos \tfrac{1}{2}[(k_1 - k_2)x - (\omega_1 - \omega_2)t] \sin \tfrac{1}{2}[(k_1 + k_2)x - (\omega_1 + \omega_2)t].$$

Define $\Delta k = k_1 - k_2$, $\Delta \omega = \omega_1 - \omega_2$, $k = \tfrac{1}{2}(k_1 + k_2)$, and $\omega = \tfrac{1}{2}(\omega_1 + \omega_2)$:

$$y = 2A \cos \tfrac{1}{2}(\Delta k x - \Delta \omega t) \sin(kx - \omega t).$$

The sine term describes the (average) high-frequency component and the cosine term describes the modulating envelope. In this case of just two frequency components, the "beat" is the pulse or the "group." The velocity of the high-frequency component, the average phase velocity, is $v_{\text{phase}} = \omega/k$. The velocity of the envelope, or the group, is $v_{\text{group}} = \Delta \omega / \Delta k$. This relationship holds even for groups that contain an infinite number of frequencies. In general, $v_g = d\omega/dk = v_p + k \, dv_p/dk$ (since $\omega = k v_p$).

Let us apply this formula to the case of deep ocean waves where $v_p = \sqrt{g \lambda} = \sqrt{g/k}$. Then $\omega = \sqrt{gk}$ and $v_g = d\omega/dk = \tfrac{1}{2}\sqrt{g/k} = \tfrac{1}{2} v_p$. The group velocity is only half the average phase velocity. Ocean swells usually consist of trains of almost sinusoidal waves. The individual waves travel at the phase velocity given by the formula for deep water. However, the lead wave keeps disappearing, expending its energy in lifting up the water, and then being brought back into existence at the tail where it recovers its energy as the water returns to its normal level. Meanwhile the group moves at half the average phase velocity.

The frequency-wavelength characteristics of waves in a particular medium can be displayed by plotting ω versus k, which is the same as plotting f versus $1/\lambda$. We show three such plots in Fig. 15.19. The first is for sound in air. The graph is a straight line whose slope is the velocity of sound. In this case velocity is not a function of frequency, and so the phase velocity is equal to the group velocity: $d\omega/dk = \omega/k$. In the second case, for deep-water waves, ω/k is clearly larger than the slope, $d\omega/dk$, confirming what we have already seen. The individual waves travel faster than the group.

The third case involves a seeming paradox in quantum mechanics. According to de Broglie, the wavelength associated with a particle with momentum p is $\lambda = h/p$, and the energy $E = hf$. Since $f\lambda = v$, $(E/h)(h/p) = E/p = v_p$. For nonrelativistic particles, we get $\frac{1}{2}mv^2/mv = v_p$. Consequently, $\frac{1}{2}v_{\text{particle}} = v_{\text{phase}}$. The particle is going twice as fast as the probability waves that form the packet representing it! Actually, the particle is represented by a packet of such waves. We must find the group velocity. From de Broglie's relationship we get $p^2 = \hbar^2 k^2$. Since the energy of the particle is $p^2/2m = hf = \hbar\omega$, we have $\omega = (\hbar/2m)k^2$. Therefore, $d\omega/dk = (\hbar/m)k$ and $\omega/k = (\hbar/2m)k$. In this case, $d\omega/dk > \omega/k$, and the group velocity, which is the velocity of the particle, is greater than the phase velocity! Although this may seem peculiar, note that since $\omega = E/\hbar$, $d\omega = dE/\hbar$, and since $k = 1/\lambda = p/\hbar$, $dk = dp/\hbar$. Then

$$\frac{d\omega}{dk} = \frac{dE}{dp} = \frac{d(p^2/2m)}{dp} = v.$$

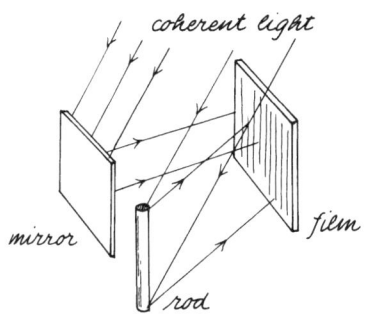

Figure 15.20. Hologram production of a one-dimensional object.

So the group velocity is equal to the velocity of the particle, which is reassuring! The slow moving phase velocities of the probability function inside the wave packet leak out the back and reform at the front.

■ HOLOGRAMS

An ordinary photograph records the amount of light reflected from each point of an object to the camera lens and hence to the photographic film. Each point of the object corresponds to a point on the film. Another type of photograph, called a *hologram*, records the *interference* between coherent light striking the film directly and reflections of that same coherent light from the object being photographed. The geometry involved is shown in Fig. 15.20. Notice that no lens is needed in front of the film. Every section of the film is recording information from the entire object. The coherent light is provided by a laser that must be intense enough to expose the film in a time short enough so that mechanical vibrations do not move the relative position of light source, object, and film. Since an interference pattern is being recorded, the relative positions must remain stationary to within a small fraction of a wavelength of light. The temporal coherence of the beam must be good enough so that the length of the coherent wave packet is longer than the difference in path lengths of the direct and reflected beams. The ordinary He–Ne laser usually has coherence for nanosecond times, yielding coherent wave packets of 10 cm or so. (The He–Ne beam is made up of a number of modes, each with a slightly different frequency.)

The resulting photograph, called a hologram, does not contain a picture that can be seen with ordinary light. What has been recorded is an interference pattern containing information about the phase differences between light reflected from an object and light that reached the film directly. If the hologram is viewed with transmitted coherent light from a laser or even with nearly coherent light from a strong point source, the original scene is reconstructed and can be seen as a real image on the near side of the film or a virtual reflected image on the far side. The light reaching the eye has been influenced by the interference pattern, passing through where there was a node and being blocked where there was an antinode. The resulting light forms an interference pattern (of the hologram—which is itself an interference pattern) in your eye—a pattern that is exactly the one that would have been formed had you been looking directly at the object in the first place.

Furthermore, this pattern is different in the left eye from what it is in the right eye just as it would have been if you had viewed the object directly. Hence you see the object in three dimensions. By moving your head with respect to the film, you can look around the object to the same extent that you could if you had been in that position looking at the object directly.

One way to understand why an interference pattern allows you to see the original object is to analyze the interference pattern produced by a very simple geometry. Suppose that the object being photographed is a long rod as shown in the diagram. The coherent light reflected from the rod interferes with the light striking the film directly. The resulting interference pattern is a series of dark and light bands covering the entire film. If a positive transparency is made of the film and it is then illuminated with a plane wave of coherent light, the cylindrical wave fronts originating from the transparent bands on the film will produce a reinforcement maximum at the position P. The waves diverging from that point will look as if they had come from the original rod. Another set of wave crests diverge from the film as if they were coming from an illuminated rod on the other side of the film.

Note that the interference pattern corresponding to the shape of the rod exists in every part of the film. The position of the rod is determined by the spacing of the light and dark fringes. If the rod is close to the film, the spacing is large at first and then decreases rapidly. If the rod is far away from the film, the spacing is almost uniform. If the object consists of two rods perpendicular to each other forming a cross, the resultant image on the film will consist of perpendicular fringes. If the object were a point, the interference pattern would be a series of concentric circles. Of course, a complicated object produces a complicated interference pattern, but when coherent light shines through that pattern and produces its own interference pattern in our eyes, we see exactly the pattern of light and darkness that we interpret as the image of the original object.

■ SOUNDS AND THE EAR

Whenever possible, physics principles and formulas should be linked with familiar phenomena, preferably quantitatively. Wave motion is a natural for this treatment, since we are surrounded by sound waves, frequently in the form of music. The wavelength of visible light is on the threshold of ordinary comprehension, but the length of sound waves is human scale. Let us calculate the wavelength of the musical note, A_4, to which the orchestra tunes. The frequency is 440 Hz, and the velocity of sound in air (at room temperature) is 344 m/s.

$$\lambda(440 \text{ Hz}) = 344 \text{ m/s}, \quad \lambda = 0.78 \text{ m} = 78 \text{ cm}.$$

As you can see, sound waves have human dimensions. In Table 1 we list various frequencies and their corresponding wavelengths in the audio spectrum. As far as humans are concerned, the audio spectrum extends from 20 to 20 000 Hz. Actually, those limits are only for very young humans. Men begin losing the upper range of hearing sooner than women do. Most middle-aged men cannot hear notes higher than about 12 000 Hz. After menopause, women rapidly lose their ability to hear high frequencies and their hearing falls to that of men of the same age.

Notice that our response range is much greater than the small range of frequencies that we use for speaking or singing, which is usually between 100 and 1000

TABLE 1. *Frequency and wavelength range for sound.*

	f in Hz	λ	
Lower limit of audio	16	21.5	m
Lowest piano note	27.5	12.4	m
"60 cycle hum"	60	5.7	m
Lower limit of tabletop radios	100	3.4	m
Middle A for orchestra tuning	440	78	cm
"High C"	1048	33	cm
Highest piano note	4186	8.2	cm
Upper limit for cassette tape	8000	4.3	cm
Upper limit of audio	20,000	1.7	cm
Ultrasonic used by bats	100,000	0.34	cm
Ultrasonic medical scanning	2 MHz	0.17	mm in air
		0.74	mm in water (or flesh)

Hz. Cheap sound systems such as those in small radios or tape recorders usually reproduce sounds only in the range from 100 or 200 Hz to about 6000 Hz. Although speech and tunes can be reproduced with this limited range, the sound seems to lack "quality," "presence," "richness." These qualitative words describe the effect that you feel. What is missing, of course, are the overtones. When these high frequencies are cut off, the notes lack sharpness or brilliance. When the low frequencies are cut off, our ears may interpret the pitch correctly but the notes will sound muffled or dull. When older people lose their high-frequency sensitivity, they have a hard time understanding words even when the sound is loud because the sibilants and hard-edged consonants are missing.

Other creatures can make and hear sounds far beyond the human range. Whales and elephants sing (and apparently communicate) at frequencies below 20 Hz, while bats and porpoises use sound as a form of radar called sonar to locate objects. It is surprising at first to learn that both bats and porpoises use 100 000 Hz for sonar. The wavelength in air of 100-kHz audio is 3.4 mm. That's about the size of an insect that the bat would find interesting. In the case of the porpoise, the velocity of sound in water is about 1500 m/s, leading to a wavelength of 1.5 cm. That provides good resolution for hunting small fish.

In Fig. 15.21 (p. 370) we show a piano keyboard with frequencies and musical notation for many of the keys. You will also see the approximate range of other instruments and voices.

What happens when those sound waves hit our ears? We say that we hear sound or music, but how do changes in air pressure turn into music in our brains? The first parts of that process are mechanical and are fairly well understood. To be sure, the mechanical construction of the ear is extremely clever. Consider the problem—air pressure changes of a millionth or even a billionth of an atmosphere have to produce some sort of effect on nerves inside the ear. Remember that it is hard for a vibrating object to set air into vibration unless there is a large surface area pushing back and forth against the air. In the same way it is hard for the air to be very effective producing vibrations in something solid like flesh.

Look at the cross-sectional view of the ear in Fig. 15.22. Note first that the outer ear helps direct the air pressure changes down into the ear canal. Without your outer ears you could not hear so well, and you would not have any place to hang your glasses. About 2 cm inside the opening the ear canal ends in a diaphragm, or

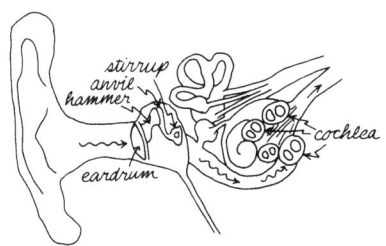

Figure 15.22. Cross-section of human ear.

eardrum. This thin membrane of flesh vibrates back and forth as the air pressure changes. The distance it moves back and forth is extremely small. As we calculated on p. 342, we can hear sound when the vibration amplitude is no larger than the diameter of an atom. The motion of the eardrum is not itself felt by any nerves. Instead, the eardrum is fastened to three bones that connect to yet another

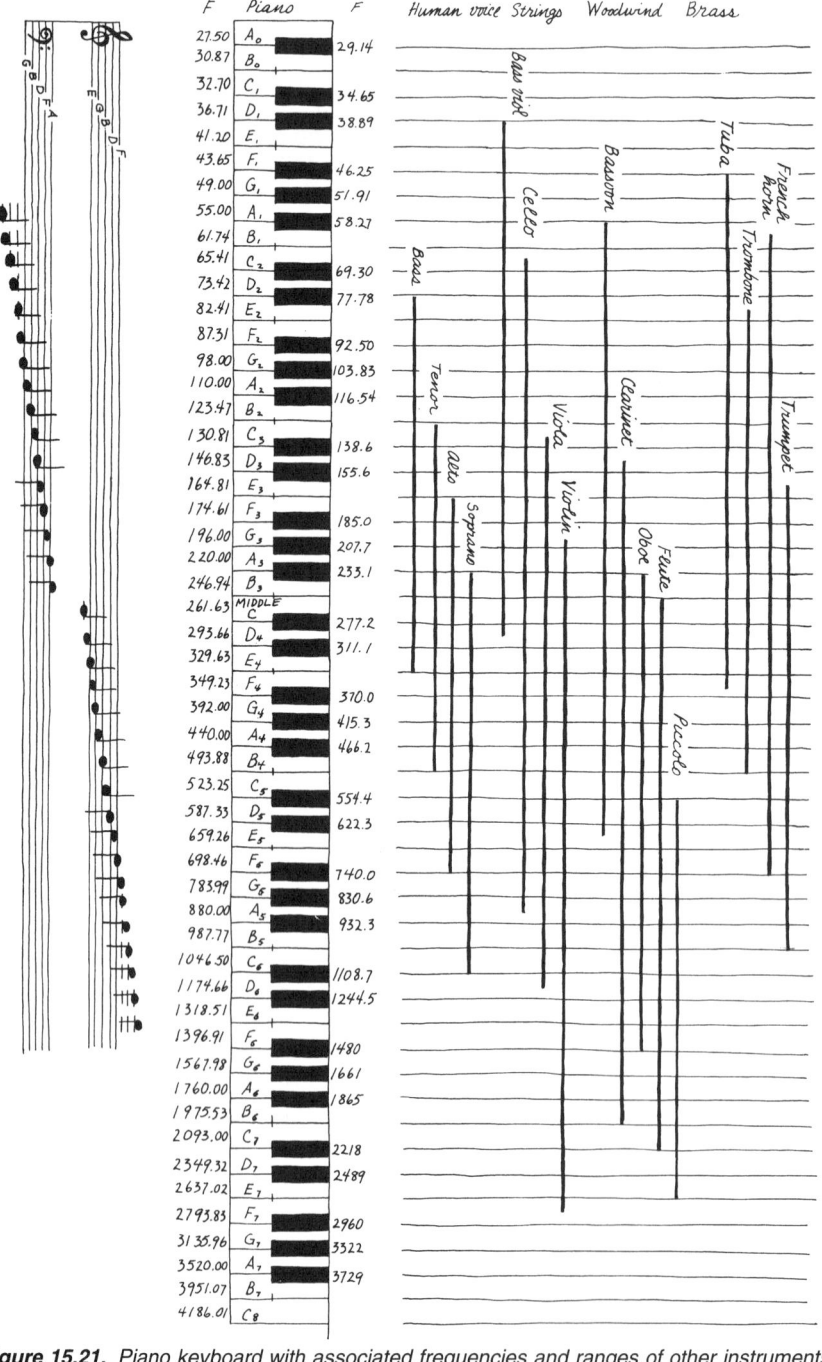

Figure 15.21. Piano keyboard with associated frequencies and ranges of other instruments.

diaphragm in the inner ear. Inside this third part of the ear the pressure waves travel in a jellylike material. The function of the bones and diaphragms in the middle part of the ear is to turn air-pressure changes into liquid-pressure changes. The surface area of the eardrum is about 25 times that of the inner diaphragm and the connection between them is made with bones acting like levers. The combination produces pressure changes on the inner liquid about 35 times greater than the changes in air pressure.

On p. 346 there is a discussion of impedance and its role in reflections. As derived there, the impedance for linear geometry is proportional to the square root of the density of the wave medium. The ratio of densities of water to air is about 10^3. Therefore the ratio of their impedances is about 35, which is just the mechanical transformation provided by the ear diaphragm and lever system. Evidently the ear is a well-designed impedance transformer for vibrations leaving air and entering liquid.

The vibrations in the liquid pass along the tube called the *cochlea*, which is shown in Fig. 15.23. Note that the cochlea is separated along its length into two main parts, one of which is filled with nerves. Some of the action of this organ is still unknown, but it is certain that different parts of the cochlea are sensitive to different frequencies. Although the pressure waves of all different frequencies travel the whole length of the cochlea, resonance for a particular frequency occurs at a particular region of the long canal. At that particular point, the dividing wall between the two parts of the cochlea expands and contracts more than at any other point along the tube. The nerves that are attached to that region are therefore more strongly triggered to send messages on to the brain. High frequencies are detected near the beginning of the cochlea and low frequencies toward the far end. Apparently, therefore, different pitches excite different nerves. A louder sound makes the same set of nerves send more signals to the brain. Just how the brain sorts out all this information is still a mystery.

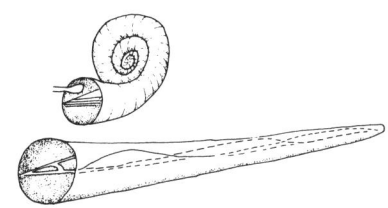

Figure 15.23. Cochlea of ear shown in normal coiled position and the way it would look if uncoiled.

■ PRODUCTION OF SOUNDS BY MUSICAL INSTRUMENTS

To set the stage for describing the operation of musical instruments, we should review the simple rules for wave transmission and reflection in lines and in pipes, and the conditions for creating standing waves. These rules are usually presented in introductory texts, but, as we shall see, the simple explanations are poor approximations to what really happens in musical instruments. First, for pulses traveling down a line, a pulse will reflect at a fixed end with a phase change of 180°. At an open end, the pulse will reflect with no phase change. For a sinusoidal wave reflected from a fixed end, the reflected wave will create a node at the end because of the phase change and under certain conditions can create a standing wave. These conditions are satisfied if $\lambda = 2L/n$, where L is the length of the line (fixed at both ends) and n is an integer. For wave velocity, v, the allowed frequencies are $f = nv/2L$. These conditions are shown in Fig. 15.24.

For pressure pulses of air in pipes, the situation is a little more complicated. As we showed on p. 341, the pressure and displacement in a longitudinal wave are 90° out of phase. In Fig. 15.25 we show the pressure and displacement diagrams for standing waves in open and closed pipes. At the closed end of a pipe, there must be a displacement node; the molecules cannot move back and forth into the solid end of the pipe. At that end, however, there will be a pressure antinode, since the pressure oscillates from positive maximum to negative maximum as the

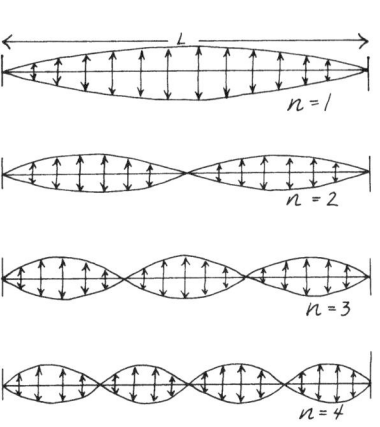

Figure 15.24. Standing waves for a line closed at both ends $\lambda = 2L/n$, $n = 1, 2, 3$.

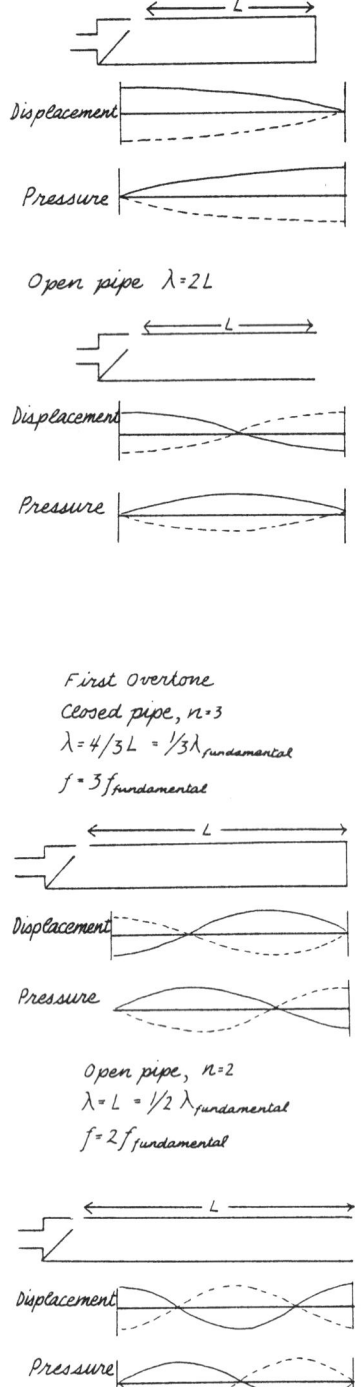

Figure 15.25. Displacement and pressure graphs for standing waves in pipes driven at one end.

molecules pile up and bounce away from the fixed boundary. A pressure pulse reflected from a fixed end does *not* change phase. A pressure pulse reflected from an open end *does* change phase. For a pipe open at one end and closed at the other (for instance, the trumpet), the wavelength of the fundamental is four times the length of the pipe. The other allowed wavelengths, as shown in Fig. 15.25 are $\lambda = 4L/(2n+1)$, where n is an integer, starting with 0. The corresponding frequencies are $f = (2n+1)v/4L$. Only the odd harmonics are allowed.

These diagrams of standing sinusoidal waves on strings and in pipes might lead a student to think that musical instruments work that way. Reality is quite different. For one thing most sounds have many overtones or harmonics, so that the actual shape of the pressure wave reaching the ear is a Fourier sum of all those harmonics. Second, the shape of a string that is plucked, or struck, or bowed cannot look like a sine way, at least at the beginning. Instead, the string will have a *kink* after it has been plucked, or released from the bow, and will consist of two straight-line segments, as shown in Fig. 15.26. As for the wind instruments, some source emits a series of pressure pulses, which are not at all sinusoidal. Each pulse travels toward the open end of the pipe where it is partly emitted and partly reflected. The reflection, after one or two round-trips, triggers its own replenishment. At any given time there may be many such pulses in the pipe, with the frequency of the tone equal to the number of emitted pulses per second. Each pulse can be represented by a Fourier group of harmonics. In Fig. 15.27 we show the shape of typical pulses from several different instruments.

Now let us consider the mechanism of musical instruments, rather than just pipes and strings.

Every musical instrument consists of some source of vibrations and some arrangement for transmitting the energy to the air with reasonable efficiency. The basic vibration is rarely sinusoidal but instead consists of a series of pulses of various shapes. The vibrating system is part of or is connected to a resonating system with its own pattern of preferred frequencies of oscillation. The resonating system responds to the fundamental and overtones of the driving pulse and radiates these into the air with varying efficiency depending on the frequency of each and the shape of the resonator. The initial vibrator may be a stretched string, or vibrating lips, or vocal cords, or turbulent air in a constricted channel. The resonating system may be a carefully shaped wooden box that can vibrate in response to a whole range of frequencies or it may be a column of air enclosed in a pipe that will respond only to certain harmonic frequencies.

There is a large range of relationships between the vibrator and the resonant system to which it is attached. In a mechanical siren, for example, compressed air blows through holes in a rapidly moving disk. The frequency is strictly determined by the number of holes per second passing the air blast. Even if there is a horn attached in front of the air blast to concentrate the sound in one direction, reflections of pressure from the end of the horn exercise no control over the frequency of the source.

The frequencies of the stringed instruments are also independent of the resonator. The basic frequency and the shape of the individual pulses on the string depend only on the mass, tension, and length of the string and how it is disturbed in the first place. Music that we hear from a stretched string, however, depends on the response of the resonant system to the various harmonics present in the pulse shape and to the efficiency with which the resonator can send these frequencies into the air.

At the other extreme of frequency control, the source vibrations of an organ pipe are caused by the self-triggered release of pulses of compressed air into one end of a pipe containing traveling pulses or pressure waves. The reflected pulses trigger their own replenishment; thus the vibration frequency is determined by the shape of the resonating system. However, even in this case the source has some privileges. If the air pressure, or the direction in which the air is admitted, is arranged correctly, the turbulent air at the opening may admit an extra pulse into the pipe when the first one is up at the other end. There will then be two pulses in the tube at the same time passing through each other, each triggering its own replenishment when it gets back to the opening. The pipe will then sound its second harmonic, which in the case of an open pipe produces a note one octave higher.

Organ pipes are designed to play only the pitch corresponding to the fundamental frequency. Consequently, there is only one pressure pulse at a time in the pipe. With many other wind instruments, however, the driving vibrator can be controlled to send in one, or two, or three, or many pulses into the pipe before the first one has returned. The pitch of the horn is determined by how many pulses per second that it contains. Thus, with a flute, for a given setting of the holes that determine a given length of pipe, the player can change the position of her mouth to send from one to seven pulses at a time into the pipe. Seven different pitches can thus be produced (though not necessarily seven different octaves). Similarly, a bugler can sound five (or more) different pitches with an instrument that contains no keys. In the case of the bugle (and most other brass instruments), the lowest *musical* pitch is actually the second harmonic containing two pulses of pressure in the tube at a time. (A note close to the fundamental, which is the first harmonic, can sometimes be sounded. It sounds like a growl, and is called the "pedal" note.)

With the reed instruments such as the clarinet or the oboe, the combination of players' lips and reeds merely takes the place of the turbulent air gate of the organ pipe. The frequency at which the reed opens to admit another pulse of air is controlled by the reflection of the previous pulse as it returns to the reed. The player chooses different frequencies by opening or closing holes in the pipe. These actions change the effective length of the pipe and therefore the travel time for the pulse to make the round trip. For a particular setting of the holes, the player can raise the pitch to a harmonic by opening one of a set of holes that throws the pipe into the two-pulse-at-a-time mode.

Production of sound by the human voice and by the brass instruments is intermediate between control of the frequency by the source or by the resonator. The vocal cords in the throat vibrate as air rushes past them. Men usually have more massive muscles in these cords than do women. Consequently, you might expect that male vocal cords would vibrate at a lower frequency. It is indeed true that some small men can have deep bass voices while some large women can be lyric sopranos.

In the case of brass instruments played with the lips pressed against a mouthpiece, the lips vibrate at the particular frequency that they wish to sound. In fact, a bugler can play recognizeable bugle calls with just a mouthpiece and no bugle. On the other hand, the bugler's lips are not completely free agents in determining their frequency of vibration. The motion of the lips depends to some extent on the arrival of reflected pulses to trigger and maintain the proper frequency of vibration. As evidence for this, try playing a trumpet or bugle that is filled with helium.

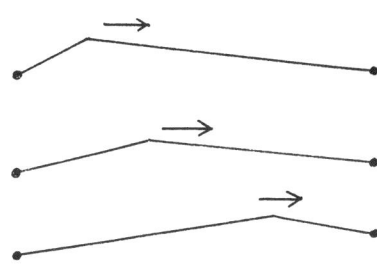

Figure 15.26. *Traveling kink* in bowed violin string.

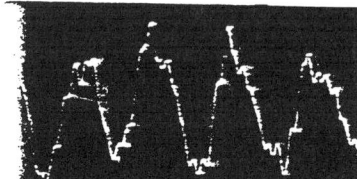

Middle A (440 cycles per second) on piano. This is the note for orchestra tuning. This particular picture was taken right after the note was struck. As the tone died away, it became smoother.

Girl's voice at middle A-440 cycles per second. The extra wiggles on the trace are called harmonics.

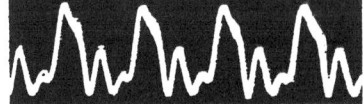

Middle A on the flute. The extra harmonic here is so strong that the note one octave higher is almost as loud as the fundamental.

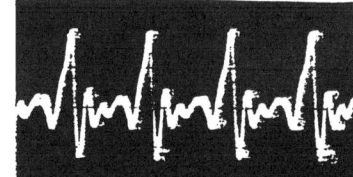

Same note (A-440 cycles per second) on trumpet. Notice the sharp spikes that give the trumpet its brilliant tone.

Figure 15.27. Shapes of typical sound waves from several instruments.

The pitch of the note will be much higher than expected, but then will return to normal as the helium drains out of the instrument. Because of the low atomic mass of helium, the speed of sound in helium is several times that in air. A pressure pulse can travel to the end of the instrument and back in a much shorter time and will trigger the lips into a higher vibration rate.

There is a standard lecture demonstration apparently showing the same effect with the human voice. If a lecturer inhales one good breath of helium, the next few sentences will sound as if they were spoken by Donald Duck. While this effect is a stunt for a lecturer, it is a serious problem for deep-sea divers who breathe a mixture of helium and oxygen instead of nitrogen and oxygen. Their words cannot be understood or easily decoded by surface listeners. The frequency shift is a complicated one because the mechanism is more complicated than it is for wind instruments. Apparently the vocal cords are not much affected by any feedback. Instead, they produce a large number of overtones that resonate in various cavities in the throat, nose, and mouth. The basic frequency of the vocal cords does not change when the throat is filled with helium, but the resonating frequencies of the cavities are shifted upward. The higher overtones are thus amplified more than the lower ones. (If you ever have occasion to try the helium stunt, inhale no more than one deep breath of helium. Helium is not poisonous but it is not oxygen, either. It takes only about three breaths to fill the lungs. If you deplete your lungs of that much oxygen, you will pass out immediately.) Instead of helium, you can now readily obtain and use sulfur hexafluoride from laboratory supply houses. This gas is nontoxic. Because of the very large molecular mass, the gas has a low speed of sound and that will lower your pitch.

Any of the instruments that help control their own vibration rates also determine the shape of the pressure pulses that are in them. Instead of a smoothly varying sinusoidal type of pulse, the pulses are usually abrupt, containing many harmonics. The harmonic recipes for several instruments were shown in Fig. 15.8. Remember, however, that the pulse shapes, and thus the harmonic pattern, change from the beginning to the end of the note. The beginning of each note, depending on how the note is attacked, usually has many more high-frequency harmonics with larger amplitude. These create the characteristic timbre of that particular instrument. Experiments have been done where the initial attack of each note is electronically removed from a recording. Even skilled musicians cannot then tell what kind of instrument was being used.

When an instrumental string is hit or plucked, the kink hurtles back and forth along the string, gradually reducing in amplitude and also gradually losing the sharp point of the kink and turning into something resembling half a sine wave. At the beginning, the shape of the kink represents many harmonics. If the string is not attached to a resonating system, the sound transmitted to the air is weak and nonmusical. The higher harmonic frequencies disturb the air more efficiently than does the fundamental, making the sound seem thin. The basic frequency of the note is determined by the time that it takes the kink to make one round trip.

When you bow a violin string, you are merely plucking it in a very special way. The rosin on the bow helps to grab the violin string and pull it along with the moving bow. When the kink gets too sharp, the violin string is freed from the bow and snaps back. The disturbance travels in both directions along the line—toward the bridge and toward the far end—and reflects at both ends. Meanwhile the bow has grabbed the string again and has pulled it up to form a new kink. The crucial part of the sequence now arises. The part of the previous pulse that started toward

Complex Waves and Wave Interactions

the bridge, after a complete round-trip, triggers the release of the next kink and the whole sequence repeats itself. The fundamental frequency is therefore determined by the time it takes the kink to make a round-trip along the string. There is only one such traveling kink in the string at a time. Under exceptional and unfortunate circumstances, however, two kinks can be introduced into the string as the result of improper bowing. This gives rise to a harmonic screech and happens particularly frequently with the cello or other bass stringed instruments. It is one of the reasons why children should never be allowed to play stringed instruments in public.

With the strings attached to a resonating box, such as a violin body, the kink-shaped pulse drives the wood and the air in the cavity of the instrument. The entire system now resonates to all of the harmonics that make up the kink-shaped pulse. The larger surface area of the instrument helps to transmit the lower harmonics including the fundamental. Consequently, the tone that we hear is mellow and rich. A struck string, such as in a piano, rapidly transmits and radiates its high-frequency components, leaving half a sine wave. To reproduce the full tone of a piano requires a sound system with very wide frequency response to capture the many overtones of the beginning of each note.

So far we have described two apparently different responses of strings and pipes to repetitive disturbances. First, we showed how standing waves can be set up by reflections within a system. These standing waves are sinusoidal with frequencies that are simple multiples of each other, and with wavelengths that are integral fractions of the length of the line. Although we explained standing waves in terms of the interference of two equal waves traveling in opposite directions, a standing wave itself does not appear to be traveling. Second, we claimed that in musical instruments the sound is produced by a pulse of some sort that hurtles the length of the instrument, reflects at the end, and then triggers the next pulse when it gets back to the origin. Which of these two views is correct? They both are.

The traveling pulse, which may be a pluck in a string or a high-pressure region in a pipe, can be represented mathematically by the Fourier sum of many sinusoidal waves. Each of these harmonic waves has one of the special wavelengths required to form standing waves in the line or pipe.

The pitch that we hear in a sound usually (but not always) corresponds to the frequency of the fundamental. If a sustained tone in an instrument gradually loses its overtones, it is because the higher harmonics have been absorbed or radiated away more efficiently. What remains is a standing wave of the fundamental. From the viewpoint of the traveling pulse model, the sharp edges of the pressure pulse or the pluck shape have been lost, leaving a sinusoidal pulse interfering with its own reflection.

The size and shape of instruments is obviously associated with the particular sounds that they produce. In general, the small instruments produce soprano notes and the large instruments produce bass notes. In all cases the frequency of the fundamental is determined by the time it takes a pulse to make a round-trip within the instrument. The longer the instrument, and the slower the speed of the pulse (such as in a massive string), the longer that time, and therefore the lower the frequency.

With wind instruments there is a direct relationship between the length of the pipe and the wavelength of the sound in air. Since most instruments act like pipes that are closed at one end and open at the other (with the driving vibration occurring at the closed end), the fundamental wavelength is approximately four times the length of the pipe. As we have already mentioned, this fundamental or "pedal" tone cannot be sounded easily in most brass instruments and is nonmusical. The frequency of the lowest note that can be used is usually about four times that of the pedal note. For that note four different pulses are introduced into the instrument during the time for one round-trip. The length of the pipe can be changed by pressing down valves and adding extra lengths in the case of the trumpet, by sliding out a length of pipe in the case of the trombone, or by opening and closing holes in the side of the tube, which effectively changes its length, in the case of the flute or clarinet.

Unfortunately for introductory students, and those who teach them, our simple explanations of brass instruments in terms of uniform pipes are not valid. Note that a pipe with length L, closed at one end, should produce standing waves with $\lambda = 4L, \frac{4}{3}L, \frac{4}{5}L, \frac{4}{7}L, \ldots$. The resulting frequencies form a series of odd harmonics! This is, indeed, the behavior of a pipe with uniform bore. We can explain this nicely in terms of pulses traveling back and forth. A pressure pulse reflects from the open end as a rarefaction. Then after one round trip it reflects from the closed (lips) end as a rarefaction, without changing phase. At the open end it reflects with a change of phase and returns after two round-trips as a pressure pulse, triggering the opening of the lips for its own replenishment. The time taken for two round-trips should be the period of the fundamental, and it is. We might think that if there were two evenly spaced pulses in the horn at the same time, we would get the second harmonic. Not so. At the end of the second return trip, the first pulse, which is again high pressure, coincides with the first round-trip of the second pulse, which at that point is a rarefaction. They cancel and do not trigger the lips, and the same is true for all even number of pulses. However, if an odd number of pulses are evenly spaced in the horn, each will trigger a replenishment and the odd harmonics are produced.

Yet brass musical instruments, without changing the length, can sound all the harmonics, both even and odd. It turns out that the mouthpiece and bell play a vital role in the performance of the instrument. Of course, one function of the bell is to serve as an impedance match between the air in the small bore tube and the comparatively infinite cross section available outside the horn. Enough signal must get out to create music, but enough must also reflect to trigger the timely replenishment of the signal from the lips. But the bell also drastically affects the frequencies that can be produced. In Fig. 15.28 we show how a proper bell for a bugle affects the low harmonics, raising them, and how a proper mouthpiece affects the high harmonics, lowering them. By art and experience (largely) the skilled instrument maker designs the combination so that a complete harmonic series can be played.

For a standard bugle, with length (depending on where you measure toward the end of the bell) of 1.75 m, the pulse travel time, one way, is 1.75 m/(340 m/s)≈5 ms. In the simple model, we would expect that the fundamental would correspond to 2 round-trips, taking 20 ms. The frequency of the fundamental would thus be 50 Hz. However, the actual series has a fundamental of 92 Hz, which cannot even be sounded. Even the pedal tone (which can be sounded by using a trombone mouthpiece) has an indistinct frequency somewhere between 50 and 92 Hz. The first note that can be played with the bugle is $F^\sharp$ at 185 Hz. This corresponds to four pulses at a time in the tube, each triggering its own replenishment after two complete round-trips. The travel time is longer, however. Evidently the effective length of the horn is longer than its physical length; it is 1.86 m, larger by about

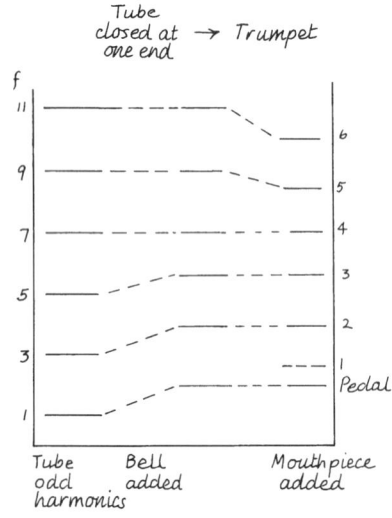

Figure 15.28. Effect on resonant frequencies of adding bell and mouthpiece to tube closed at one end.

Complex Waves and Wave Interactions

7%. The effective length is actually a function of frequency. Shorter wavelengths (or short pulses) reflect from further out in the bell.

The wavelength *in the string* of the fundamental note produced by a violin string is twice the length of the string, but this is not true of the wavelength in air. The frequency of the oscillation is the same for both air and string, but of course the velocity is different. Note that all four strings of a violin are the same length, yet each produces a different fundamental note.

The shape and size of the instruments are related in yet another way to the frequency range in which they work. Not only is the oboe shorter than the bassoon, and the trumpet shorter than the sousaphone, and the violin strings are shorter than those of the bass viol, but the soprano instruments are also thinner or smaller in diameter. This feature is connected with the problem of transferring energy from the internal vibrating system to the external air. The lower the frequency, the larger the area of the surface boundary that is needed for effective transfer. As a rule of thumb, an instrument does not produce or at least send into the air a note that has a wavelength in air very much longer than the size of the instrument itself. In the case of a flute, the lowest note produced has a wavelength about twice that of the length of the flute. In the case of the trumpet, the wavelength of the lowest note that can be sounded is about the length of the instrument itself (about 5 ft). With the violin the lowest note produced by the G string is about four times the length of the case.

■ HARMONIC RELATIONSHIPS OF MUSIC

"All God's chillun got rhythm," but not all of us hear harmony in the same way. Indian and Chinese music often sounds jangling and confused to people brought up in Europe or the United States. Cultural experiences make a large difference in how we respond to combinations of notes. To people accustomed to western classical music, certain notes on the piano seem to go together while certain other combinations sound harsh or even wrong. Some instruments such as bells produce sounds that just cannot go together no matter what the basic pitch is. Sometimes people who play a chime try to produce chords by striking two bells at once. The result is usually a clanging noise instead of a pleasant harmony.

In Fig. 15.29 we show the musical notations and the respective frequencies for several different tonal clusters. The first three correspond to classical chords. The next three would have been considered discords in classical music although modern composers frequently make use of such chords. What is there about certain notes that make them sound good together? Why does this work with only certain instruments? It turns out that the shapes of our familiar music instruments are carefully designed so that they will produce harmony when played with other instruments.

There is something very special about the overtones or harmonics of musical instruments. They do not consist of just any combination of higher frequencies. If you have a fundamental note of 500 Hz, then the musical harmonics are 1000, 1500, 2000 Hz, and so on. In other words, if the overtones are musically harmonic, they must be at twice the fundamental frequency or three time the fundamental and so on. Vibrating strings automatically produce higher-frequency harmonics that are integral multiples of the fundamental, as do most, but not all, musical pipes. Indeed, for some of the horns the higher harmonics are only approximately integral multiples of the fundamental. Bells produce lots of over-

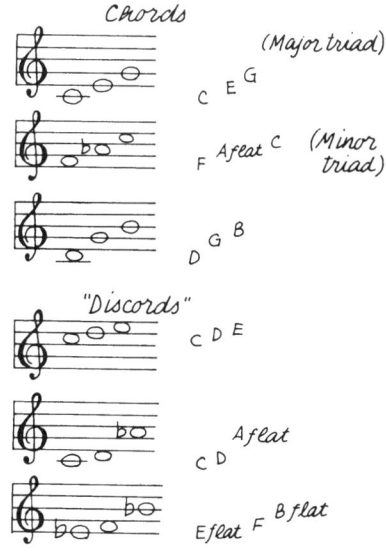

Figure 15.29. Musical chords and discords.

tones but not *harmonic* ones. The resulting sound is interesting in itself but when you try to mix sounds coming from two different bells, their overtones do not match at all.

When a note has harmonic overtones, the note sounds rich and pleasant. What happens if you start out with two notes played at the same time? If they are pure sine waves and differ by only a few vibrations per second, then you will hear beats. The combination tone will seem to pulsate in a rather unpleasant way. If we want to produce a chord of music with two or more notes in it, we must make sure that the notes are far enough apart in pitch so that the beat frequency is higher than 20 Hz or so. But we must also face the fact that real musical tones are not sinusoidal but contain many harmonics. In producing a chord, we must make sure that none of the *harmonics* produce loud beats. In many cases the amplitude of the second or third harmonic is almost as great as that of the fundamental. As an example, suppose we play together three notes with the fundamental frequencies of 311, 392, and 523 Hz. Those frequencies correspond to the notes $D^\#$ ($3\frac{1}{2}$ notes below middle A), G (one note below middle A), and C (two notes above middle A). In Fig. 15.30 we show the frequencies of the various harmonics of these notes. If the notes were played with tuning forks there would be no discord. But if the notes are played by instruments that produce lots of harmonics, then these frequencies will produce beats with each other. Try it on a piano and hear for yourself.

Since harmony and the quality of music create such subjective experiences, there is considerable discord among musicians over whether certain tonal clusters are pleasant to listen to, and there is also controversy about the physical reasons for this situation. Yet another feature of the ear's response to tonal clusters is that most people find it pleasant to hear sounds that are multiple frequencies of each other—in other words, harmonics. That is why the A produced by a musical instrument such as the violin sounds richer and more pleasant than the rather mechanical sound of a pure A from a tuning fork or signal generator. When two or more notes are sounded together from different instruments, the ear is pleased if their various overtones have a harmonic relationship. The more combinations there are of overtones that are the same as each other, or two or three or four times the frequency of each other, the better the ear likes it.

Certain combinations of notes produce the same overtones and so reinforce each other. The most important of these is produced by two notes, one octave

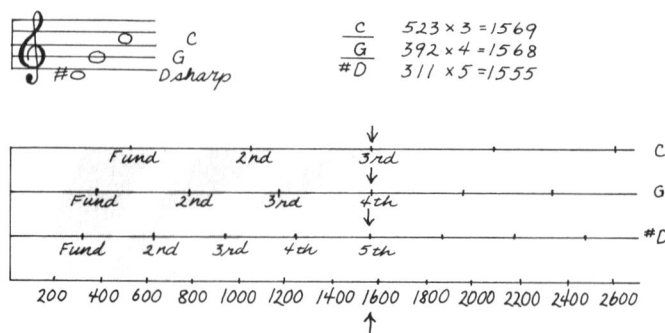

Figure 15.30. Beats between third harmonic of C, fourth harmonic of G, and fifth harmonic of $D^\#$.

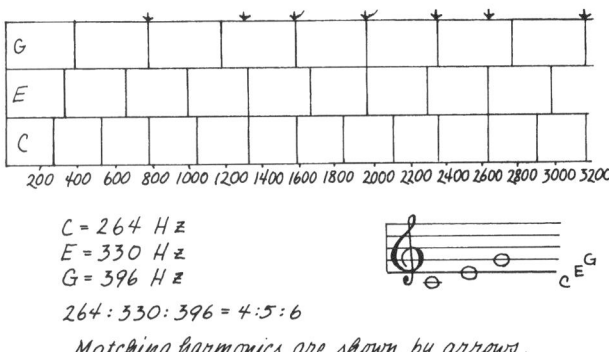

Figure 15.31. Harmonic matches of Major Triad.

apart. The upper note has twice the frequency of the lower note and many of their harmonics are the same. For instance, if one note has a frequency of 400 Hz, then the note an octave higher will have a frequency of 800 Hz. If these notes are produced by musical instruments, each of them will have harmonics at 2, 3, 4, and 5 times the fundamental. Every harmonic of the upper notes matches one of the harmonics of the lower note. There are similar matches of harmonic overtones if two notes have frequencies with a ratio of 3:2; for instance, 600 and 400 Hz. Other ratios that have matching overtones are 4:3, 5:3, 5:4, and 6:5.

One combination of three notes is very important as the basis of chords and also for forming the standard scale of the notes we know as "do, re, mi, fa, sol, la, ti, do." This basic trio of notes must have frequency ratios of 4:5:6. It is called a major triad. Notice in Fig. 15.31 how many of the harmonic overtones of these three notes match each other. A fourth note can always be fit into such a triad if its frequency is just twice that of the first note (the fourth note would thus be the octave of the first note).

The scale we sing with "do, re, mi" is based on this major triad. In Fig. 15.32 we show the notes and frequencies of the C scale. Notice that the middle A to which the orchestra tunes has the standard frequency of 440 Hz. It is "la" in the "do, re, mi" scale, where "do" is the middle C. Notice the way in which the whole scale is made up out of three major triads. If it were not for the preference of our ears, it might seem logical to divide an octave into eight equally spaced notes. However, this is not the way the standard scale works. As you go up the scale, you sometimes advance a whole tone and sometimes only a half tone. For instance, the ratio of frequencies between D and C (between "re" and "do") is 9:8. The "re" note has a frequency about 12% larger than the "do" note. The next step up to "mi" is almost as large—in this case a ratio of 10:9 or a frequency that is higher by about 11%. As you go on to "fa," however, you go up only half a tone. The frequency ratio is 16:15 or an increase of about $6\frac{1}{2}$%. As you go on to "sol," "la," and "ti" you go up whole tones; but between "ti" and "do" there is only a half tone again. That is the way it sounds good. If you think that it would sound better to go up a whole tone every time, try it on a piano. If you start with middle C and use only the white keys, you will be playing the standard scale we have just described. To go up a whole note each time, you will have to use some of the black keys in a way that you can figure out by looking at the diagram and the piano scale shown on p. 370.

Figure 15.32. Major scale in C.

Note	C	C#	D	D#	E	F	F#	G	G#	A	A#	B	C
Name	do		re		me	fa		sol		la		ti	do
Frequency of equally tempered scale	264	279.7	296.3	314	332.6	352.4	373.4	395.6	419.1	444	470.4	498.4	528
Frequency of perfect key of C	264		297		330	352		396		440		495	528

Figure 15.34. *Comparison of frequencies of equally tempered scale with those of perfect key of C.*

Note	D	E	F#	G	A	B	C#'	D'	E'
Triads	4	5	4	6	5	6			
				4	5	6			
Name	do	re	me	fa	sol	la	ti	do	re
Frequency	297	333	371.3	396	445.5	495	556.9	594	668.3
Interval	9/8	10/9	16/15	9/8	10/9	9/8	16/15	9/8	

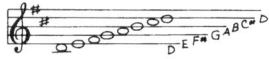

Figure 15.33. *Major scale in D.*

If you play a musical instrument, you may have had to spend time learning how to play scales. The easy one is the scale in C. There are no sharps or flats. With the piano, you need use only the white keys. Try to do that, however, starting one note higher in the key of D. If you go marching up from D using no sharps or flats, you will get a scale that does not at all sound like the standard "do," "re," "mi." That's because you haven't maintained those same frequency ratios that are determined by the major triad frequencies of 4:5:6. In the key of D, for instance, the scale can start with D, go to E, but then must go to F♯. This scale and frequency ratio are shown in Fig. 15.33.

There is another important combination of frequencies called the minor triad. The frequency ratios of this triad are 10:12:15. This combination of notes has matching harmonics that also sound pleasant to the ear but western culture interprets the result as a sad sound.

As a matter of fact, the exact frequency ratio that we have described cannot work for more than one scale without producing all sorts of complications. For instance, suppose that we say we would like to build a scale starting with D, based on the same magic triad: 4:5:6. We must start out with a frequency for D that we got from the scale in C. That is 297 Hz. The note that has a frequency of 6/4 that of D should be A at 440 Hz. Instead, 6/4 of 297 is 445.5. Most of the other notes would also be slightly off. If a person is playing an instrument with continuous tuning like a slide trombone or a violin, and that person is very skillful, he or she can make the slight adjustments so that the frequency ratios of the notes are exactly right. With valve instruments, however, that is harder to do and with a piano it is impossible. If we had to have a different set of keys and strings on the piano for every key that might be played, the piano could not fit into a living room and no human would be able to play it. This problem was recognized long ago and various piano makers have used various compromises so that a reasonable number of notes will be almost right for a large number of scales. The most common system is called an equal-tempered scale. The chart in Fig. 15.34 compares an octave of standard notes of the equally tempered scale of a piano with the frequencies of the notes in the perfect key of C. Using the equal-tempered system, only 12 notes are needed to play all the tones and half tone of a full octave. As you can see, none of the notes are off frequency by very much. The progression of frequencies in the equal tempered scale is geometric. The frequency of each of the 12 notes in an octave is higher than the preceding note by $2^{1/12} = 1.059\,46$.

16
Electrostatics

The electromagnetic interaction is responsible for most of the phenomena of everyday life. Gravity is important to us here on earth, of course, but the effects of the strong and weak nuclear interactions are so hidden within atomic nuclei that we seldom experience any evidence of them. (The weak interaction, which is one aspect of the combined electroweak force, is responsible for β decay, among other things.) The electromagnetic interaction, on the other hand, is responsible for holding the electrons and nuclei together to form atoms and is also responsible for all the various forms of molecular bonding. Polar bonding, hydrogen bonding, van der Waals's forces, and covalent bonding are all electromagnetic. Only the geometry and nature of the atoms produce the apparently different effects. But if molecular forces are electromagnetic, then almost certainly all of the biological phenomena are controlled by electromagnetism. That means you and me. The interactions within biological cells as well as the signals they send to each other are electromagnetic. The forces in everyday life are electromagnetic—whether we classify them as muscular, air pressure, hydrostatic, spring, or thermoexpansion. Furthermore, electromagnetic energy can be radiated. We are familiar with it in the form of light, radiant heat, x ray, radio, and television. Even sound can be analyzed in terms of electromagnetism since pressure waves in air or solids are caused by density variations between atoms or molecules and those atomic and molecular forces are electromagnetic.

We will divide the subject of electromagnetism into three realms. In the first, the electric charges that are the source of the interaction will be at rest with respect to each other and to us, the observers. In Chap. 17 we will study the second realm and learn how to deal with electric charges that are moving with constant average speed in electrical circuitry. This is the subject that we normally think of as "electricity." In Chap. 18 we will see another aspect of electric charges moving with constant speed—magnetic fields are produced. In Chaps. 19 and 20 we will review what happens when the electric charges are moving, but not with constant velocity. In this third realm of accelerated charges, energy is radiated into space.

First we consider how to produce these electric charges and see what effect they have on each other when they are standing still.

■ HOMELY DEMONSTRATIONS

In the case of electrostatics, it is hard not to demonstrate the effects. The situation is shocking, particularly in dry weather. If you pull off a sweater, throw back the bed blankets, or walk across a rug, you become a minor-league Thor. Tiny streaks of lightning flash and thunder crackles. On a really dry day with a thick rug you can be a hazard to yourself and friends.

Here are some ways to demonstrate electrostatic effects on a more organized basis, but using only homely materials.

1. Blow up a small party balloon and you will have an excellent source of negative electric charge. Rub the balloon on wool or fur, or easiest of all, your own hair, and you will find that the balloon will then stick to the thing that it was rubbed on or to almost anything else, including the nearest wall. In this particular case, as in all others that we will describe, it is not necessary to rub the material vigorously or more than once. It is contact between two different materials that produces the separated electric charges. Rubbing two objects together merely increases the area of contact. Not all balloons work well in electrostatic demonstrations. Apparently the dye in some colored rubber lowers the resistance. Always try a sample from the balloon package before using them before an audience.

2. Blow up two balloons, preferably both spherical. Suspend each with a thread to a common point on the ceiling or some high suspension. At first the balloons will just hang together. Then rub each of them with wool or your own hair and watch them repel each other. In a later section we will show how to use this demonstration for an order-of-magnitude calculation of the magnitude of the charges on the balloons.

3. Cut several small pieces of paper and aluminum foil. Each should have an area of less than 1 cm^2, but make some of them long and narrow. Run a comb through your hair and then pass the comb near the pieces of paper and foil.

4. Wrap a tiny piece of aluminum foil around the end of a nylon or polyester thread. Smush the foil into a little ball and then hang it by the thread so that it is some distance away from any object. Rub a balloon or your comb and bring it near the aluminum foil ball. What happens next is complicated. Watch the whole sequence. Repeat it, but each time start out by "grounding" the aluminum ball with your fingers.

■ CHARGE SEPARATION

Over 200 years ago, Benjamin Franklin arbitrarily assigned the name "negative" to the charge that appears on hard rubber when it is rubbed with wool or hair. Hard rubber when rubbed with almost any other material will grab electrons; thus, according to the convention started by Franklin, electrons have a negative electric charge. Since, as we now know, electrons are the charge carriers in metal wires, there is sometimes confusion as to whether electric current is in the direction of electron movement or in the direction of positive charge motion. The worldwide convention is that *current is in the direction of positive charge flow*. No need to blame Benjamin Franklin, however. In transistors, fluorescent tubes, and electrolytic cells, the charge carriers are as apt to be positive as negative. Attempts to change the convention in hopes of making understanding easier for students are misguided. Student forced to learn the rules of negative charge flow will be confused throughout their career. (This is the case in New York State schools where the state Department of Education decrees that current is in the direction of *negative* charge flow.)

The outer electrons of atoms are frequently only loosely tied to the parent nucleus. If two atoms come together, the outer electrons may rearrange themselves to bond the two atoms together in a molecule, or one atom may simply steal an electron from the other. Oxygen molecules, for instance, have a tendency to pick up spare electrons forming negatively charged oxygen ions. All materials have this varying power to hang onto electrons whether they are conductors or insulators. However, the effects are most commonly seen with good insulators such as rubber or hair or plastic. The reason for this is that when electrons are added to insulators, they remain fixed in place; if they are taken from a region, it becomes and remains positive. If a piece of metal is rubbed with fur, it too will gain electrons but they apread across the whole surface. If you are holding the metal with your hands, the electrons will spread throughout your body. Unless you are holding the metal with an insulator, any electrical effects will be diluted. However, note that although your body is relatively large and a fairly good conductor of electricity, if you shuffle across a rug you may pick up enough electrons so that your whole body becomes charged to high voltage. You can estimate the voltage by using the rule of thumb that in dry weather the breakdown voltage of air is about 12 000 V/cm. The exact voltage depends on whether the spark jumps from a point or a smooth rounded surface. But if you can produce an inch-long spark from your finger to a doorknob, or a friend's neck, you are probably at a potential difference of at least 25 000 V.

When you wipe the charges off a charged insulator onto an electroscope, the actual contact area is very small. Negligible charge would be deposited if only the charge in immediate physical contact transferred. In a darkened room you can see and hear what actually happens. When the charged insulator is brought near the electroscope knob, tiny sparks jump from the surface charges of the insulator to the metal of the electroscope. These discharges ionize the air, allowing the flow of charges from the surrounding area of the insulator.

■ LIGHTNING

The separation of electric charges can produce dramatic effects in nature. Almost any material that slides or is blown past some other object will gain or lose electrons. When this happens with rising or falling droplets of water in clouds, one part of the cloud can become negative or positive with respect to another part, or to the earth. Usually the bottom and top of a thundercloud are positive and the middle negative. A typical value of separated charge is 40 C. The potential between the bottom of the cloud and the ground is typically 10^8 V. The amount of energy contained in a lightning bolt is in the range of 10^9 J or 300 kW h. When the concentration of charge in one region is sufficient, part of the charge is driven away, forming a conducting path to ground or to another part of the cloud. Breakdown occurs rapidly, raising the temperature of the path to the point of luminescence and creating a column of high pressure that radiates outward to be heard as thunder.

The main lightning bolt follows a path that has been created by a "leader." The leader works its way down to earth by following the easiest breakdown path, sometimes temporarily ending up in dead-ends, giving the appearance of "forked lightning." By the time the branching leaders approach the ground, they may contain about 5 C. Usually when the first leader is within 50 m or so from the ground (or a tall object), an upward discharge rises from the ground. Once the conducting path is completed, the main charge surges down in a succession of strokes that may last from 10^{-2} to 1 s. Peak currents within each surge are from

10 000 to 20 000 A, lasting about 100 μs, and with about 40 ms between surges. A typical average current for the whole stroke is 3000 A. Copper wire with a diameter of about 1/8 in. (#9 gauge) can conduct such a transient surge without danger—as long as there is not a break or a sudden bend in the wire on the way to ground.

A lightning rod on a building does not prevent lightning from striking, but provides a safe path to earth for any lightning bolt that arrives close to the rod. A lightning rod provides a cone of protection for everything underneath it out to about 60°. However, moist air is a good conductor of electricity, and so the cone of protection will be influenced by air currents, which, of course, are prevalent during thunderstorms. The rod does not prevent lightning by discharging the cloud, which is usually far above it. Nor does it make any difference whether the lightning rod ends in a point or a sphere, although a really sharp point might have the effect of creating a small localized cloud of charged air around it, thus slightly increasing its region of protection. The small-scale demonstrations of lightning and lightning rods that employ model houses between big capacitor plates have nothing to do with the real thing. For this complicated phenomenon, scaling is almost impossible.

For safety in a thunderstorm, avoid being the tallest thing around. Also, do not stand under the tallest thing around. The insides of cars or planes are usually safe, because there you are surrounded by a conductor which keeps the current on the outside. If you are caught on a golf course, huddle as low as possible and crouch so that your feet are together and the rest of you is not touching ground. Sometimes cattle are electrocuted by lightning that strikes the ground and travels along a thin topsoil layer. A four-legged creature acts like a parallel circuit in such a case, allowing current to go up one leg, through the heart, and down another leg. Perhaps that is why mountain goats are always seen standing on crags with all four feet together.

Electrostatic effects can be annoying in industrial processes as well as in the atmosphere. Any process that involves moving sheets of material or moving grains or liquids will produce charge separation. If you throw off wool blankets at night, you may see the whole surface glow. One of the hazards of blowing grain into storage elevators is that the separated charges will produce a spark in an atmosphere filled with flammable dust. On the other hand, in the making of sandpaper, the electrostatic effect is put to good use. Instead of sifting the abrasive grain onto glued paper, the grain is kept underneath the paper and jumps up into the glue because of an electrostatic attraction that is created in the region. In the process the abrasive grains line themselves up lengthwise, producing much sharper sandpaper. Electrostatic attraction is also employed in copying machines. A uniform layer of charge is wiped onto a sheet of photoconductive semiconductor that is bonded to a conducting metal, such as selenium on aluminum. Light striking the photoconductor removes the charge. Ink powder is then spread on the surface, adhering to the remaining charges. The powder is transferred and baked onto paper, forming a printed image.

■ VAN DE GRAAFF GENERATORS

In the Van de Graaff machines, electric charges are sprayed onto a moving belt and literally carried up to an insulated dome where they accumulate. In the research machines the charges are sprayed onto the belt from needles that are raised to high voltage. Electrons can either be sprayed on or taken off, thus charging the belt and the dome negative or positive. (Most research machines use a positive dome.) In most of the tabletop machines, a rubber belt is driven at the base by an axle that delivers electrons to the *inside* of the rubber belt as the axle makes rolling contact with the rubber. The inside of the belt is thus made negative but the electrons cannot migrate through the rubber to the outside of the belt. Meanwhile a wire attached to the base of the machine on the outside of the belt, and close to the spinning axle, sees a large *positive* charge on the *axle*. (All the electrons on the inside of the whole belt have been taken from that one region of the axle.) Electrons jump off the wire toward the positive axle but of course land on the outside of the belt. There they are carried up until they are inside the metal dome; at that point they hop over to a wire connected to the inside of the dome and immediately spread out over the outer surface of the dome. This sequence of events makes the outside of the dome negative. The electrons are forced up into this negative region by the physical motion of the belt. However, once the dome is negative, why should the electrons jump off the belt onto the collecting wire inside the dome? That's because the *inside* of the dome is not charged negative. All the electrons have gone to the outside surface of the dome. See Fig. 16.1.

The toy Van de Graaff generators can raise the voltage of the dome to 20 000 V or so, producing sparks several centimeters long. Demonstration Van de Graaff generators (about 50 cm high) can go up to 100 000 V and should be used with caution. As for the quantity of charge, consider that a dome with a 10-cm radius has a capacitance of only 10 pF, and at 10^5 V would hold $(10^{-11}$ F$)(10^5$ V$)=1$ μC. Since it takes at least 1 s for one of these demonstration devices to attain full charge, the charging current must be about 1 μA. Since it takes at least 1 mA through the heart to cause any damaging effects, you cannot be wounded even if you let the current pass from hand to hand and thus right through the heart. Besides, only a fraction of the hand-to-hand current would pass through the heart itself. Of course, the shock from the fully charged dome hurts, and the startled victim might fall backwards and suffer head injuries. Also, if your demonstration helper has a pacemaker, why take a chance? Remember that while Franklin survived the kite experiment, neither he nor any of his assistants are alive today.

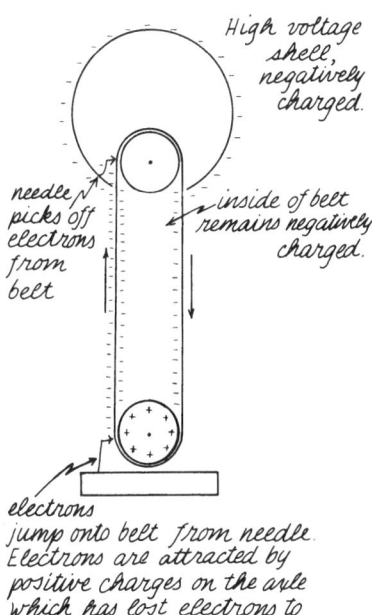

Figure 16.1. Schematic diagram of table top Van de Graaff generator. For best operation, swab the outside of the insulating cylinder with alcohol. This should be done for any insulating handle used in an electrostatics demonstration, especially in humid weather.

■ THE SIZE OF THE COULOMB

Coulomb's law for the force between charged *point* sources in free space is

$$F = k\frac{q_1 q_2}{r^2} = \frac{1}{4\pi\epsilon_0}\frac{q_1 q_2}{r^2}.$$

In Système International (SI) units, F is in newtons, r is in meters, q is in coulombs (C), $k=9\times10^9$, and ϵ_0 is the permittivity of free space, which equals 8.85×10^{-12} C^2/N m^2. If we start out using only the constant k, the 4π will show up in other formulas. This problem is an inescapable consequence of geometry. (There are 4π steradians, unit solid angles, in a sphere.) The magnitude of the proportionality constant depends on the size of the electric charge chosen as unit charge. It might seem reasonable to choose the charge on the electron as our unit. That much charge is very small, however, and partly for historical purposes a much larger amount of charge has been chosen as the unit. The coulomb is defined in terms of the unit of electric current, since a current is a flow of electric charge.

When 1 C of electric charge moves past a point in 1 s, the current is 1 A: $I = \Delta q/\Delta t$.

A coulomb of separated charge is enormous. For instance, the force between two 1 C charges, 1 m apart, is

$$F = 9\times 10^9 \frac{(1\ \text{C})(1\ \text{C})}{(1\ \text{m})^2} = 9\times 10^9\ \text{N} \approx 10^5\ \text{tons}.$$

The potential produced by a point charge is $V = kq/r$. The potential 1 m away from 1 C would be

$$V = 9\times 10^9 \frac{1\ \text{C}}{1\ \text{m}} = 9\times 10^9\ \text{V}.$$

Evidently, except for thunderstorms, we do not usually deal with isolated charges in the coulomb range. However, since the late 1980s, there are inexpensive capacitors available in the farad range that can store coulombs of charge.

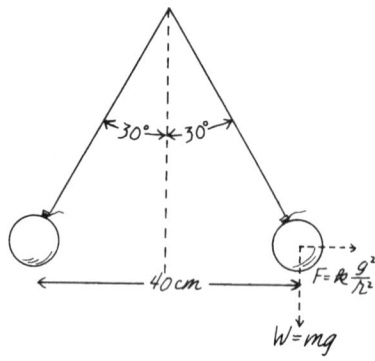

Figure 16.2. Geometry for calculating the approximate charge on two balloons repelling each other.

Let us figure out how much separated charge exists on hanging balloons that repel each other. We assume that each balloon has a radius of 10 cm and that each has approximately the same amount of negative charge on it, uniformly distributed over the surface. We will also assume that the balloons are spherical and that the force between two such spheres depends on the distance between their centers. We show the geometry in Fig. 16.2. For each balloon, its weight is acting down and the electrostatic repulsion is acting horizontally. For an assumed separation angle of 60° (each balloon hanging at 30° to the vertical), the two forces form the legs of a 30°–60° right triangle. Therefore, the electrostatic force is about half the balloon's weight. The mass of a typical small balloon is a few grams. Let us assume that its weight is equal to 0.04 N. The electrostatic repulsion between the two balloons must therefore be 0.02 N. With the arrangement shown in the diagram, the center-to-center distance between balloons is about 40 cm.

$$F = (0.02\ \text{N}) = 9\times 10^9 \frac{q^2}{(0.4\ \text{m})^2}, \quad q = \tfrac{1}{2}\times 10^{-6}\ \text{C} = \tfrac{1}{2}\ \mu\text{C}.$$

Evidently the microcoulomb is the more natural unit of separated charge in ordinary electrostatic demonstrations.

The potential *of* the spherical balloon is

$$V = k\frac{q}{R} = 9\times 10^9 \frac{\tfrac{1}{2}\times 10^{-6}\ \text{C}}{0.1\ \text{m}} \approx 50\ \text{kV}.$$

A useful approximation is that the capacitance of a sphere in picofarads (10^{-12} farads, pF), is equal to the radius in centimeters. For this balloon, the capacitance (assuming uniform surface distribution) would be ≈ 10 pF. Using this approach, we find a charge of $Q = CV = (10^{-11}\ \text{F})(5\times 10^4\ \text{V}) = 5\times 10^{-7}\ \text{C}$, the same magnitude that we first calculated. You can use these same approximations to calculate your own charge as you shuffle across a rug on a dry day. If you can generate a 1-in. spark, your potential is about 25 kV. To the extent that you can model yourself in terms of a sphere, your radius might be 30 cm and your capacitance 30 pF. The charge on your body is $Q = (3\times 10^{-11}\ \text{F})(2.5\times 10^4\ \text{V}) \approx 10^{-6}\ \text{C}$.

In terms of the coulomb, the natural unit of electric charge is $e = 1.6\times 10^{-19}$ C. Consequently, there are 6×10^{18} electrons per coulomb. Consider a particular case where electrons are collected to produce a sizable signal. The central wire in a Geiger tube and the associated input leads to the amplifier have a capacitance of

10^{-11} F, a typical capacitance for several tens of centimeters of circuit wire. For a pulse of 1 V, the collection wire must get 10^{-11} C $\approx 10^8$ electrons. In order to detect a subatomic particle that goes through the Geiger tube and that liberates about 100 ion pairs, the tube must provide intrinsic gain of the order of 10^6.

■ ELECTROSTATIC FIELD

Coulomb's law in its simple form applies only to the force between charge points or uniformly charged spheres. The dependence on the inverse square of the distance is primarily a geometrical property of space, just as it is for gravitational attraction between spheres. If an influence spreads out from a point, or uniformly from the surface of a sphere, without decaying and without being absorbed, then the amount of influence *per unit area* is inversely proportional to the square of the distance from the center of the source. The reason for this is that the surface area of the spreading sphere of influence is equal to $4\pi r^2$. Since the quantity of the influence going out through the surface of each concentric sphere is always the same and since the area of the sphere is proportional to r^2, then the influence *per area* must be proportional to $1/r^2$.

The question is, how does one object influence another? Is something shot out of one object toward the other (virtual photons, perhaps)? If so, does the influence take some time to travel between objects? A very useful model to explain *static* effects is to picture a "field" of influence. This field model was ingeniously developed by Michael Faraday in the early part of the nineteenth century. Instead of asking how one charged particle affects another, Faraday described the effect that one or more charges has at each point in space. He imagined that space is electrically warped by the presence of electric charges. We find out what this total influence or warping of space is at any particular point by bringing a unit test charge to that point, and measuring the force on it. The *field strength* at that point is then defined to be the force per test charge: $\mathbf{E} = \mathbf{F}/q$. The units are newtons per coulomb.

The field model describes the interaction between charged particles in two steps. First, you calculate the effect of one of the charges on the point in space where the other charge exists. The first charge creates a particular field strength at that point. The second charge then experiences that field strength as a force equal to the field strength times its own charge. What advantage does such a system have over Coulomb's simple law? The advantage is that a different mental image can lead to new concepts and easier mathematical models. If there are many charges in a region, or if they are spread out over some unusual geometric shape, then we can start out by calculating the combination effect that all the charges have at a particular point in space. Each small segment of charge produces its own particular electric field at that point. These individual contributions must then be added *vectorially* to produce the net field at that point. The field at any point is thus a function of the *geometry* of the sources of charge. Coulomb's law, with its dependence on $1/r^2$, applies only to point sources or to uniformly charged spheres.

■ ELECTROSTATIC POTENTIAL

In everyday life you expect to be shocked if you touch objects that are at high *voltage* or high *potential*. Let us define these terms and make them quantitative. Take a positive test charge and move it in a constant electric field so that constant

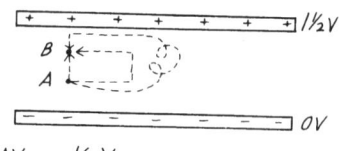

$\Delta V_{A-B} = \frac{1}{2} V$
(A-B is 1/3 distance between plates)

Potential difference, or work done per charge in going from A to B is the same for all three paths shown.

Figure 16.3. Three possible routes to take a test charge from A to B. The work done in one way must be equal to that done in any other way.

force is required. You can arrange these circumstances by moving the charge from bottom to top in a parallel-plate capacitor. Except near the edges of the plates, the electric-field lines in such a capacitor are perpendicular to the plates, parallel to each other, and uniformly spaced. The electric field E is constant. If you move the positive test charge against the direction of the field (from negative plate to positive plate), you must exert a force $\mathbf{F}=\mathbf{E}q$. The work you do in moving the charge a distance d is $W = \mathbf{F} \cdot \mathbf{d} = q\mathbf{E} \cdot \mathbf{d}$. The *work done per coulomb* in moving a charge from one point to another in an electric field is called the *electric potential difference*: $\Delta V = W/q$. The unit of potential difference is the joule per coulomb, which is called the *volt*.

If you connect a $1\frac{1}{2}$-V cell across the parallel-plate capacitor, the potential difference across the plates is equal to $1\frac{1}{2}$ V. But we have not defined the actual electric potential of either plate. The situation is the same as it is with gravitational potential energy. We are free to choose the zero point. In this case we might say that the negative or bottom plate of the capacitor is zero potential, in which case the top plate is at $+1\frac{1}{2}$ V. Or we could define the top plate to be zero potential, in which case the bottom plate is at $-1\frac{1}{2}$ V. In either case it would take $1\frac{1}{2}$ J to force 1 positive coulomb from the bottom plate to the top plate. (Actually, as we have seen, it would be impossible with this particular geometry to separate a coulomb of charge. However, if 1×10^{-9} C were separated in this case, it would require $1\frac{1}{2} \times 10^{-9}$ J.)

It is very convenient to be able to assign a potential to each point in space. In describing the *electric field* at any point, we have to give both its magnitude and direction. A *potential*, however, has only magnitude; it is a scalar quantity. That means that we can find the work done in going from one point to another in an electric field simply by finding the potential difference between the two points. The route we take with our test charge makes no difference. In Fig. 16.3 we show three ways to go from point A to point B in the electric field between parallel plates. In all three cases the work done is equal to $\frac{1}{2}$ J/C. (More realistically, if we used 1×10^{-12} C as our test charge, the work done would be $\frac{1}{2} \times 10^{-12}$ J). Rather than prove this analytically, we appeal to the energy-conservation law. Suppose that it takes only two units of energy to go from A to B by the straight path but three units to go by the long path. Then the smart thing to do would be to escort the charge upward on the left-hand path using up two units of energy and then let it slide back down on the right-hand path, gaining three units of energy. We could make money that way if the system worked, but, of course, it does not in the case of static surroundings.

It is easy to find the relationship between the electric field in a parallel-plate capacitor and the potential difference across the plates:

$$\Delta V = \frac{W}{q} = \frac{\mathbf{F}}{q} \cdot \mathbf{d} = \mathbf{E} \cdot \mathbf{d}.$$

At any point in the region between the parallel plates, the electric field is equal to the potential difference divided by the spacing between the plates:

$$|E| = \frac{\Delta V}{d}.$$

Note that we now have another way of calculating E and expressing its units. The units are volts per meter. These are equivalent to our previous expression:

Electrostatics

$$1 \text{ N/C} = 1 \text{ V/m}.$$

Actually, electric fields are most commonly measured in terms of volts per meter. For instance, a radio or TV signal with an electric field strength of 1 μV/m is just at the limit of detectability by a good receiver.

The relationship between E and ΔV in the parallel-plate capacitor can be generalized. Even in regions where the electric field is not constant, it is approximately true that $E_x = -\Delta V/\Delta x$ for small displacements. For instance, suppose that in moving a distance of Δx equal to 1 cm, the potential increases 1 V. Then the component of the electric field *in that direction* is $E_x = -1 \text{ V}/0.01 \text{ m} = -100$ V/m. The sign is negative because if we take a positive test charge *up* a potential hill of 1 V, the electric field must be pointing down. Notice that the relationship between E and ΔV allows us to find a vector quantity, **E**, if we know the distribution of a scalar quantity, V. Actually, we find the vector *component* in the direction of the displacement.

The inverse relationship also holds approximately. For small displacements, $\Delta V = E_x \Delta x$. The electric field in the region of a point charge is $E = kq/r^2$. For a charge of 1×10^{-9} C, the field at $r = 10$ cm is

$$E = (9 \times 10^9) \frac{1 \times 10^{-9} \text{ C}}{(0.1 \text{ m})^2} = 900 \text{ V/m}.$$

The potential difference between $r = 10$ and 11 cm is *approximately* equal to $\Delta V = -(900 \text{ V/m})(0.01 \text{ m}) = -9$ V.

More generally, the relationships between the scalar V and the vector **E** are

$$E_x = -\frac{dV}{dx} \quad \text{and} \quad V_b - V_a = -\int_a^b \mathbf{E} \cdot d\mathbf{x}.$$

Notice the sign convention in the dot product and in the negative sign in front of the integral. If the electric field is in the $-x$ direction, then you must exert a force and do work in shoving a positive test charge from a to b in the positive x direction. Therefore, V_b must be more positive than V_a, since you are shoving a positive charge up a potential hill. Sure enough, the dot product yields a negative sign since E and dx are in opposite directions, and the negative sign in front of the integral makes the right-hand side positive.

■ FIELDS AND POTENTIALS DUE TO POINT CHARGES

1. Point charge. Since the force between the source charge and a test charge q_t is equal to $F = kqq_t/r^2$ and since the electric field produced by q is F/q_t, then $E = kq/r^2$.

   To find the potential due to a point charge, we must first determine the location of 0 potential. The convention is that 0 potential is at a region beyond the influence of the source charge. Surely, that would be at infinity. With electric charges there is another possibility. The potential of a very large conducting object is not appreciably affected by small changes of small nearby charges. The largest thing around is the earth. When we separate charges by rubbing balloons against our hair we do not affect the earth's potential. Consequently, it makes sense to define "ground" as 0 potential. To "ground" one side of a circuit, you must make sure that there is a good conducting path to conducting earth. Usually that requires pipe connections to

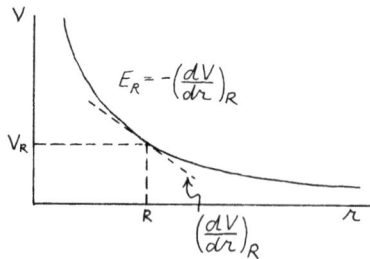

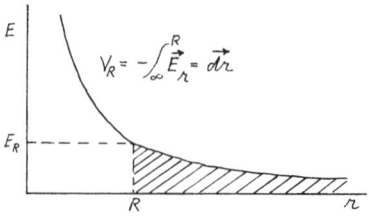

Figure 16.4. Relationship between E_r and V_r for point charge.

moist soil. A metal rod stuck into dry sand underneath the eaves of a house is not a good "ground."

To find the potential at a particular distance R from a point charge that is isolated from other objects, we calculate the potential difference between infinity and R:

$$V_R - V_\infty = -\int_\infty^R \mathbf{E}_r \cdot d\mathbf{r} = \int_\infty^R k \frac{q}{r^2} dr = -k \frac{q}{r}\bigg|_R^\infty = k \frac{q}{R}.$$

Notice the way the signs and dot product work in this case. There is a positive source charge at the origin. To determine the potential at $r = R$, we pull a unit test charge from infinity to R. At any point the force exerted on the unit test charge by the field is $+E$, but the displacement is $-dr$. The dot product therefore yields a sign of -1. The solution to the integral still assumes, however, that dr is heading in the positive direction. Since it is not, one more change of sign is required, which is effected by changing $\int_\infty^R$ to $\int_R^\infty$. Alternatively, change the problem to finding $V_\infty - V_R$. Then the dot product yields a sign of $+1$, and dr is in the direction of positive r. Then $\int_\infty^R k(q/r^2) dr = -kq/R$. This is the value for $V_\infty - V_R$, which is the negative of the desired potential difference $V_R - V_\infty$. The integral is the area under the $E(r)$ curve, from $r = \infty$ to $r = R$, as shown in Fig. 16.4.

2. *Two point charges separated by a short distance.* The two point charges might both be positive (or negative), or one positive and the other negative. In Fig. 16.5 we have drawn field lines produced by a positive and negative charge of equal magnitude, placed close together—a *dipole*. The field lines indicate the direction of the force on a positive test charge placed at any point in the vicinity of the dipole. The lines are not, however, the paths that a test charge would follow, since the direction an object moves depends both on the force that it experiences at any given instant and also its previous velocity.

Along the axis:

$$E = kq\left(\frac{1}{(r+a)^2} - \frac{1}{(r-a)^2}\right) = kq \frac{r^2 + a^2 - 2ar - r^2 - a^2 - 2ar}{(r+a)^2(r-a)^2}.$$

When $r \gg a$, $(r+a)^2 \to r^2$ and $(r-a)^2 \to r^2$.

$$E \to -4k \frac{qa}{r^3} = -2k \frac{p}{r^3},$$

where the *dipole moment* is $p = 2aq$.

Perpendicular to the axis:

$$E = 2k \frac{q}{r^2 + a^2} \cos\theta = 2k \frac{q}{r^2 + a^2} \frac{a}{\sqrt{r^2 + a^2}} \to k \frac{p}{r^3}.$$

Note the dependence of dipole field on the inverse *third* power of distance from the center. The dipole field is proportional to the strength of the dipole, defined as $p = 2aq$. The dipole is a vector with direction from the negative to the positive charge. The equations for the fields along the axis and perpendicular to the axis differ only by a factor of two. The complete equation for the *magnitude* of the dipole field is

Electrostatics

$$|E| = k\frac{p}{r^3}\sqrt{1+3\cos^2\theta}.$$

For $\theta=0°$, along the axis, $\sqrt{1+3\cos^2\theta} = 2$. For $\theta=90°$, perpendicular to the axis, the square root factor $=1$.

The calculation of the potential due to a dipole is much simpler than the calculation of the field because the potentials add algebraically. The geometry of the calculation is shown in Fig. 16.6.

$$V = V_1 + V_2 = k\left(\frac{+q}{r_1}+\frac{-q}{r_2}\right) = kq\frac{r_2-r_1}{r_1 r_2}.$$

For $r \gg d$, $r_2 - r_1 = d\cos\theta$ and $r_1 r_2 \approx r^2$:

$$V \to kq\frac{d\cos\theta}{r^2} = k\frac{p\cos\theta}{r^2}.$$

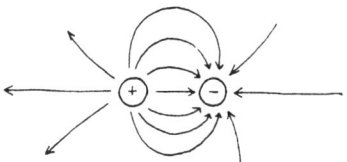

Field lines produced by two charges, equal in magnitude. Upper drawing - same polarity. Bottom drawing - opposite polarity.

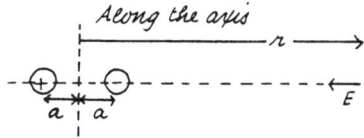

The potential due to a point charge is proportional to r^{-1}; the potential due to a dipole is proportional to r^{-2}. The two charges in a dipole partially cancel each other. At any point on the perpendicular bisector of the dipole, where $\theta=90°$, the potential is 0. This makes sense because if you started the test charge on its journey far away from the dipole it would start out at 0 potential by definition. As you move the test charge along the perpendicular bisector, you are always moving it perpendicular to the electric-field lines. Therefore, no work is done. Since you can move the test charge from infinity right to the center of the dipole without doing any work, all of those points must be at 0 potential.

We found the *field* due to a dipole along two particular directions, but did not derive the general expression. Here is an example of how it is sometimes easier to calculate the potential first and then deduce the field. For the polar components, r and θ of the dipole field:

$$E_r = -\frac{\partial V(r,\theta)}{\partial r} = 2k\frac{p\cos\theta}{r^3}$$

(this is the field we derived along the axis) and

$$E_\theta = -\frac{\partial V(r,\theta)}{r\,\partial\theta} = k\frac{p\sin\theta}{r^3}$$

(this is the field we derived perpendicular to the axis).

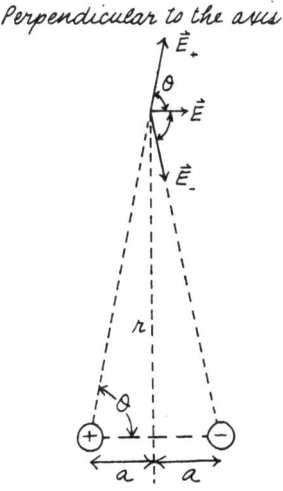

Figure 16.5. Field produced by a dipole.

■ FIELD LINES AND EXTENDED CHARGE SOURCES

If we used the vector-addition method to find the net field produced at a point by all the charges spread over a sphere, we would have a difficult calculation on our hands. Of course, it is possible to do the calculation by dividing the sphere into appropriate symmetrical zones and integrating their effects. An easier method is available for certain geometries. To make use of this method we have to add more details to our model of electric field lines.

Faraday developed a very useful model out of his mental image of electric field lines. Not only does the direction of the lines indicate the direction of the force that a positive test charge would experience at that point, but the magnitude of the

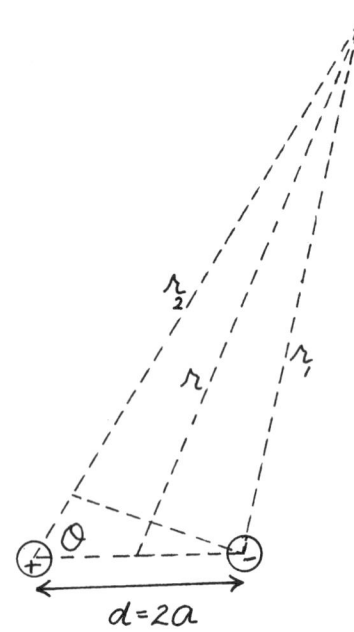

Figure 16.6. Geometry for calculating the potential at a point due to a dipole source.

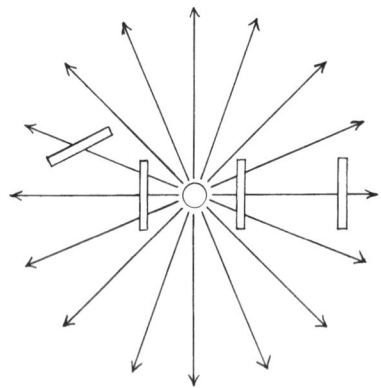

Figure 16.7. Electric field is proportional to the area density of field lines.

force can be represented by the *density of the lines* near the point. Such a scheme makes qualitative sense. In Fig. 16.7 the field lines are most crowded close to the source charge. Farther away, where the field is weak, there are fewer field lines per unit area. We can make this idea quantitative by defining the magnitude of the electric field in any region to be the number of field lines per unit area in that region. Of course, when we map out the unit area in order to count the number of lines, we must make sure that the surface of the area is perpendicular to the direction of the electric field lines:

$$|E| = \frac{\text{(number of field lines going through area)}}{\text{(area)}}.$$

Therefore, the number of field lines going through area is $\mathbf{E} \cdot \mathbf{A}$, where $\mathbf{A}$ is the area. The dot product accounts for the requirement that the effective surface area in question is perpendicular to the field lines. (The direction of an area is the normal to the surface.) The reaction of most students to this model of field lines is that it may be a useful qualititative description but that it seems too abstract for quantitative work. Faraday thought of these lines as if they were stretched elastics in warped space. They cannot cross each other (if they did, a test charge would experience a force in two different directions at the same point), but they tend to contract to exert an attractive force between positive and negative, and they tend to repel each other to exert a repulsive force between like charges. Naturally, the "lines" represent the average situation in their neighborhood—not just on the lines themselves.

To make the description a quantitative tool, we define *one electric field line per square meter* to represent a unit field of *one newton per coulomb*. Let us see how many lines are coming from each positive coulomb of charge. If we put the coulomb of charge at the center of a sphere with a 1-m radius, then the lines of force extend uniformly as radii, piercing the sphere perpendicular to its surface. See Fig. 16.8(a). The field produced by 1 C at a distance of 1 m is equal to

$$E = k\frac{q}{r^2} = 9 \times 10^9 \frac{1 \text{ C}}{(1 \text{ m})^2} = 9 \times 10^9 \frac{\text{N}}{\text{C}}.$$

The surface of a sphere has an area of $4\pi r^2$. Therefore, when $r = 1$ m, the number of lines piercing the sphere must be $\mathbf{E} \cdot \mathbf{A} = E(4\pi r^2) = (9 \times 10^9 \text{ N/C})4\pi (1 \text{ m})^2 = 4\pi \times 9 \times 10^9$ lines. That may seem like a strange number and a lot of lines to be coming from a point charge. Of course, a coulomb is a lot of charge, and besides this is just a mathematical model and the actual number of lines in the model depends on the units we have chosen. The number of lines, $4\pi \times 9 \times 10^9$, is equal to $4\pi k$. In other systems of units (cm instead of m, for instance), k would have a different value. At any rate, if we use this number, we will end up with the electric field in our standard units.

The next feature of the model was described by Karl Gauss (1777–1855). Gauss pointed out the simple fact that the total number of field lines leaving the surface of a three-dimensional volume must be related to the number of source charges inside the volume. The crucial point is that these field lines can start only on positive charges and can stop only on negative charges. They do not just start or stop in midspace. Therefore, if there is no electric charge inside a volume, the net number of lines entering or leaving the volume must be 0.

If more lines leave than enter, then there must be net positive charge inside; if more lines enter than leave, then there must be net negative charge inside. Of course, field lines can enter a volume that has no net electric charge but then they will pass right through and leave again. This simple-minded theorem does not depend on the location of the electric charge inside the volume. The charge can be at the center or near one surface. Furthermore, there can be many charges inside the volume, both positive and negative, located at any point. But only the *net* charge will contribute to the *net* number of lines leaving or entering. The net number of lines leaving the volume is called the *flux*, Φ. The formal statement of Gauss's law is

$$\Phi = \iint_{\text{surface}} \mathbf{E} \cdot d\mathbf{A} = 4\pi k q_{\text{in volume}} = \frac{q_{\text{in volume}}}{\epsilon_0}.$$

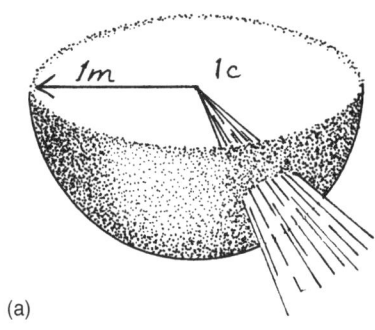

(a)

The permittivity in free space is $\epsilon_0 = 1/4\pi k$. The net charge in the volume could also be expressed as an integral of all the infinitesimal distributed charge inside the surface boundary: $q_{\text{in volume}} = \iiint_{\text{volume}} \rho \, dV$, where r is the charge density in the volume element, dV. Gauss's law relates a volume integral to a surface integral, and it applies to any closed surface, regardless of shape or symmetry. At each patch of area on the surface, take the dot product of the electric field and the area. Sum these products (perhaps some are positive and some negative) over the whole surface and the result is equal to the net charge enclosed by the surface. As a practical matter, we apply Gauss's law only to geometries with symmetry such that the surface integration is easy.

1. *E produced by a uniformly charged sphere.* If electric charge is uniformly distributed on the surface of an isolated sphere (or uniformly distributed throughout the volume of the sphere), then symmetry requires that the field lines be uniform and radial. After all, there is no reason why there should be more field lines on one side of the sphere than on the other. If the lines were to leave the surface in some direction other than radial, then the spherical symmetry would not hold.

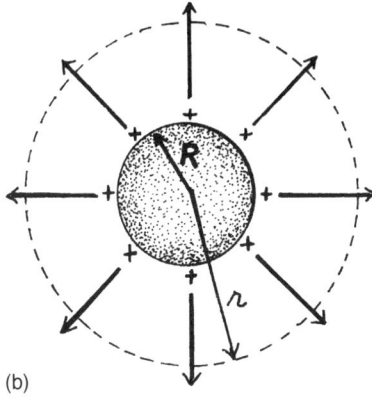

(b)

Figure 16.8. Electric field lines from uniformly charged sphere of radius R.

Under these symmetrical circumstances, the field lines at a distance from the sphere all have the same direction and spacing as if they had come from the center of the sphere itself. You cannot tell whether the lines are coming from uniformly distributed charges on the surface or from that same amount of charge concentrated at the center. Let's apply Gauss's law to find the electric field at a distance r from the center of the sphere, but outside the surface of the sphere itself which has a radius R. Construct an imaginary sphere with radius r around and concentric to the charged sphere. According to Gauss's law, the total number of field lines passing through the outer sphere is proportional to the net charge inside a volume of that size:

$$(\text{number of field lines}) = \Phi = \iint_{\text{surface}} \mathbf{E} \cdot d\mathbf{A} = 4\pi k q.$$

[See Fig. 16.8(b).] The electric field is radial, hence in the same direction as the normal to the elements of surface area, $d\mathbf{A}$. Since the electric field has the same value at every point on the surface, the dot product is everywhere the same and the surface integral is equal to $EA_{\text{sphere}} = 4\pi r^2 E$:

$$4\pi r^2 E = 4\pi k q \rightarrow E = k \frac{q}{r^2}.$$

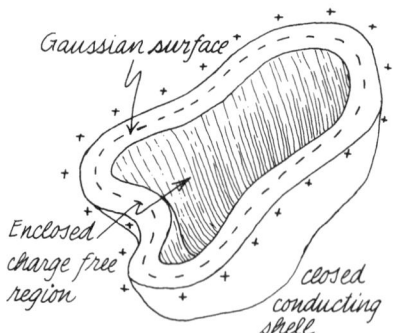

E in conductor = 0 for static conditions

∴ *since* $E = \frac{\text{number of lines}}{\text{unit area}} = 0$

no lines cross Gaussian surface

∴ *the net charge in the interior is zero, and any surplus charges must reside on the outer surface of the conductor.*

Figure 16.9. The interior of a conducting shell is free of electric field (produced by outside charges) and contains no net charge.

The same reasoning holds if the charged sphere is a point. In that case, Gauss's law yields Coulomb's law; the two are equivalent.

Gauss's law confirms that the electric field produced outside a uniformly charged sphere is just the same as if all the charge were concentrated at the center of the sphere. The result is not at all obvious. Newton faced the problem long ago in trying to calculate the force of gravitational attraction between the earth and the moon. He assumed that this force was determined by the distance between their centers but he had to invent the calculus to prove it. With Gauss's law, the proof tumbles out easily because of spherical symmetry. Actually, the mass of the earth is not uniformly distributed throughout its volume. The core is much more dense than the surface layer. Nevertheless, the spherical symmetry remains because at any given radius the density is almost the same regardless of the angular latitude or longitude. (There is a small nonspherical component because of the equatorial bulge.) In Chap. 5 on gravity, we appeal to Gauss's law to find what happens to the gravitational field *inside* a uniformly charged sphere.

There is yet another corollary to Gauss's law applied to regions enclosed by insulated, closed, *conducting* shells. In such regions, no matter how much charge is placed on the outer shell, there can be no net charge in the interior. For this demonstration, imagine a Gaussian surface the same shape as that of the conducting shell, but lying *in the metal* of the conducting surface as shown in Fig. 16.9. There can be no static electric fields in a conductor that has no power source. If there were, charges would move until static equilibrium were reestablished. Therefore, the electric field is 0 everywhere on the Gaussian surface; the number of field lines coming through the Gaussian surface must be 0, which means that the enclosed net charge must be 0. This is the proof of the situation with Van de Graaff generators that we described earlier. Although there is a strong repelling force on the incoming charges as they are hauled up on the belt, once inside the sphere the charges are shielded from the charges already on the sphere. Therefore, they can easily hop off the belt onto the collection wire without being repelled.

A closed conducting shell is called a "Faraday cage." Even before Faraday's experiments that demonstrated this effect, Franklin had shown experimentally that no external charge penetrates to the interior of a conducting shell. Such a device provides shielding against strong electric fields. This is why a car is a safe place to be during a lightning storm. The shielding property of a closed conducting surface depends crucially on the validity of Gauss's law. Gauss's law holds only if Coulomb's law holds for point sources. Both depend on the Euclidean nature of three-dimensional space (the fact that the surface of a sphere is $4\pi r^2$) and on the fact that the electrical influence does not decay and is not absorbed in free space. Consequently, a shielding experiment is a very sensitive test of the inverse-square law. Faraday himself got into a metal cage to see if he could detect any interior charges when the outside of the cage was highly charged. Similar experiments, done in recent years, demonstrate that the exponent, 2, in Coulomb's law is exact to one part in 10^{17}!

2. *E at a distance from a uniformly charged cylinder.* Once again we apply Gauss's law by picturing the field lines and appealing to symmetry. All the

field lines from a charged infinite wire or uniformly charged, infinite cylinder must be radial. They are perpendicular to any concentric cylindrical surface. Suppose that there is constant charge per unit length along a wire or cylinder that is very long. This charge density is usually called λ, (the Greek letter for *l*, for length) and is equal to the total charge divided by the length: $\lambda = Q/L$. To find the electric field at a distance *r* away from the center of the wire or cylinder, apply Gauss's law to a concentric (imaginary) cylinder of radius *r* and height *h*. We must assume that the distance *r* is very small compared to the length of the wire so that we do not have to worry about the asymmetry of the field lines near the ends. The net charge in the Gaussian cylinder is equal to the charge density λ times the height of the cylinder: net charge = λh. According to Gauss's law, the total number of field lines leaving the Gaussian cylindrical volume of radius *r* and height *h* is proportional to the amount of charge in the volume:

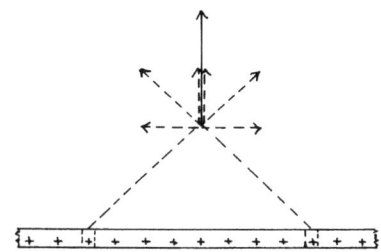

Figure 16.10. The field at a point distant from a long, uniformly charged wire is radial. The components parallel to the wire cancel.

$$(\text{number of lines}) = \Phi = \int\int \mathbf{E} \cdot d\mathbf{A} = 4\pi k(\lambda h).$$

The field lines pierce only the curved surface of the Gaussian cylinder. They do not come out through the end caps. Since the electric field is radial, in the same direction as $d\mathbf{A}$, and constant at *r*, then

$$E(2\pi rh) = 4\pi k(\lambda h) \rightarrow E = 2k\frac{\lambda}{r}.$$

Notice that *h*, the arbitrary height of the Gaussian cylinder, cancels out of the equation.

Once again Gauss's theorem produces something that is not obvious. The electric field outside a uniformly charged wire or cylinder decreases, but only with the inverse *first* power of the distance, not with the inverse second power of the distance. The electrostatic attraction or repulsion does not necessarily obey an inverse-square law. The dependence of the field on distance depends on the geometry of the source.

Students might well wonder why the field lines from a row of point charges should not go off in all directions instead of just radially. In Fig. 16.10 we show how the field at a remote point is made up of the contributions from all the point sources on the cylinder. The horizontal component of the field produced by one segment of the cylinder is exactly canceled by the horizontal component of the field produced by its symmetrically opposite segment. Only the outward or radial components are not canceled. This diagram explains why the argument holds only for points close enough to the cylinder so that the cylinder appears approximately infinite and thus symmetrical.

3. *Potential near a charged cylinder.* To find the potential difference between two points at radii *a* and *b*, we use the defining relationship between *V* and *E*:

$$\Delta V_{b \rightarrow a} = -\int_b^a \mathbf{E}_r \cdot d\mathbf{r} = +2k\lambda \int_a^b \frac{dr}{r} = 2k\lambda \ln\frac{b}{a}.$$

If the charge density is positive, and $b > a$, then ΔV is positive in going from *b* to *a*. Conversely, there is a drop in potential as you move to larger radii. However, we cannot define 0 potential to be at the point where $b \rightarrow \infty$. As $b \rightarrow \infty$, $\ln(a/b) \rightarrow -\infty$. Since the formula for the field assumes that the wire is

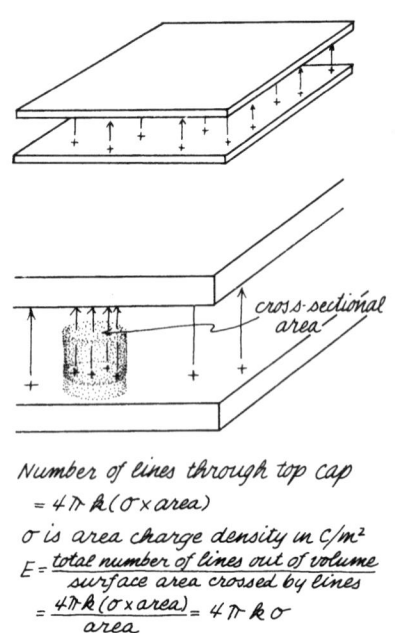

Number of lines through top cap
$= 4\pi k (\sigma \times area)$
σ is area charge density in C/m²
$E = \dfrac{\text{total number of lines out of volume}}{\text{surface area crossed by lines}}$
$= \dfrac{4\pi k (\sigma \times area)}{area} = 4\pi k \sigma$

Figure 16.11. Gaussian construction to find field between parallel plates.

infinite, there would be charge at infinity and therefore the potential there could not be 0. For the case of charged cylinders, we cannot specify the potential *at a point*; all we can do is to measure the potential *difference* between two points.

4. *Electric field produced by uniformly charged parallel conducting plates.* The parallel-plate geometry shown in Fig. 16.11 is the basic form of an electric capacitor. If the bottom plate is uniformly charged positive and the upper plate negative, then the field lines will go from bottom to top, parallel to each other, and uniform. Except near the edges, the electric field will be constant and uniform within the volume. If there is a total charge Q on each plate, then the surface charge density σ will be equal to Q/A, where A is the area of the plate. (The presence of the oppositely charged plate holds the total charge on the inner surface of each conducting plate.) To find the field at any point between the plates, construct an imaginary volume as shown in the diagram. Note that the bottom surface of the cylinder is embedded in the metal of the bottom plate, and the top surface is between the plates at the point where we wish to calculate the electric field. The total electric charge inside this volume is equal to the surface density of charge on the bottom plate times the cross-sectional area of the cylinder. The number of field lines coming from this much charge is equal to $4\pi k[\sigma \times (\text{area})]$. All those field lines leave the upper flat area of the cylinder. None pass through the bottom because that is buried in the metal and there are no field lines inside the conductor. No field lines pass through the curved surface of the cylinder because that surface is parallel to the field lines. Therefore the electric field in the region at the top of the Gaussian cylinder is equal to

$$E = \dfrac{(\text{number of field lines})}{(\text{area})} = \dfrac{4\pi k \sigma \times (\text{area})}{(\text{area})} = 4\pi k \sigma.$$

In this case the electric field does not at all depend on distance from the source. As we should expect from the field model, the field is constant between the plates from top to bottom. (If this were not the case, the electric field lines would not be parallel to each other and uniformly spaced.)

The relationship that we have just developed, between the electric field and the surface density of charge ($E = 4\pi k \sigma$), also applies to the field close to the surface of *any* charged conductor, regardless of its shape and regardless of whether there is another plate nearby. The reason for this generalization is that *in a static situation* electric field lines near the surface of a conductor must be perpendicular to the surface. In a very small region near the surface of a conductor we can apply the same Gaussian cylinder that we did between parallel plates. The same arguments apply and we find that E (near surface of a conductor) $= 4\pi k \sigma$. Of course, the surface may be curved and so the field lines are not necessarily parallel to each other. Furthermore, the surface charge density depends on the surface curvature (as we show on p. 401), and therefore the electric field is not constant over the surface.

5. *Potential between parallel plates.* With constant field between the plates, the relationship between E and the voltage across the plates, ΔV is $\Delta V = Ed$, where d is the distance between plates. If we define the 0 of potential to be at the bottom plate, then the potential rises linearly from 0 to V as you go

from the bottom plate to the top. The geometry is just like that for the gravitational potential close to the surface of the earth. In that case, the gravitational potential is gh, and the gravitational potential energy of an object with mass m is mgh. Between parallel plates the electrostatic potential is Ey, and the electrostatic potential energy of an object with charge q is qEy. In both these cases the field lines are uniform and parallel to each other; the force per charge is constant in the region. Therefore, the work done per charge in moving along the field lines is proportional to the distance moved: $V = Ey = 4\pi k\sigma y = (\sigma/\epsilon_0)y$.

■ SUMMARY OF THE DEPENDENCE OF E AND V ON SOURCE GEOMETRY

We have developed formulas for E and V produced by points, dipoles, spheres, cylinders, and planes. What about all the other shapes that may be sources of electric influence? The calculations would be very complicated in general, but for many cases we can approximate unusual shapes with the ones we already have. Suppose, for example, that you want to know how the gravitational field of the earth depends on the distance from its center. True enough, the gravitational field is proportional to the inverse square of the distance from the center of the earth. However, for small distances above the surface of the earth, the gravitational field is almost constant. This is the flat-earth approximation, the one that we use all the time in everyday life. For vertical distances of the order of a few hundred meters, the gravitational field lines are almost parallel to each other and are constant in density—just like the electric field lines in a parallel plate capacitor:

$$F = m\left(\frac{GM}{R^2}\right) = m(g),$$

$$(\text{work})_{R \to R+h} = mGM \int_R^{R+h} \frac{dr}{r^2} = -mGM \left.\frac{1}{r}\right|_R^{R+h} = mGM\left(\frac{1}{R} - \frac{1}{R+h}\right)$$

$$\approx m\frac{GM}{R^2} h = mgh.$$

The approximation is good for $h \ll R$.

For another example, take the case of the light produced by a long fluorescent tube. For distances of a few millimeters away from the surface of the tube, the light intensity is almost constant. The surface of the lamp looks flat. Farther away, where the distance to the center of the tube is more than about $1\frac{1}{2}$ times the radius of the tube, the geometry should be considered cylindrical. In that case the intensity of the light is inversely proportional to the distance from the center of the tube. That approximation holds only so long as the distance from the tube is small compared with the distance out to the ends of the tube. When you are at a distance from the tube many times its own length, the geometry is approximated by that of a point source. The light intensity will be approximately proportional to the inverse square of the distance from the source.

In the table below we summarize the formulas for E and V for several geometries. We use Coulomb's constant k but frequently it is convenient to use the permittivity $\epsilon_0 = 1/4\pi k = 8.5 \times 10^{-12}$.

Geometry of source	E	V
Point	$E = k\dfrac{q}{r^2}$	$V = k\dfrac{q}{r}$
Dipole	$\lvert E\rvert = k\dfrac{p}{r^3}\sqrt{1+3\cos^2\theta}$	$V = k\dfrac{p\cos\theta}{r^2}$
Uniformly charged sphere, $r \geq R$	$E = k\dfrac{Q}{r^2}$	$V = k\dfrac{Q}{r}$
Uniformly charged sphere, $r \leq R$	$E = k\dfrac{Q}{R^3}r$	$V = \tfrac{3}{2}k\dfrac{Q}{R} - \tfrac{1}{2}k\dfrac{Q}{R^3}r^2$
Uniformly charged spherical shell, $r < R$	$E = 0$	$V = k\dfrac{Q}{R}$
Uniformly charged cylinder, $r \geq R$, $r \ll L$	$E = 2k\dfrac{\lambda}{r}$	$\Delta V_{b\to a} = 2k\lambda\ln\dfrac{b}{a}$
Uniformly charged plates, or near surface of conductor	$E = 4\pi k\sigma$	$\Delta V = 4\pi k\sigma y$

There is an interesting relationship between electric-field lines and lines of equipotential. As long as you move a test charge always in a direction perpendicular to the electric field lines, no work is done and every point on the line must be at the same potential. In Fig. 16.12 we have sketched the field lines and equipotential surfaces around our standard geometries of a charged sphere, cylinder, parallel plates, and dipole.

■ ELECTRICAL CAPACITANCE

Every time objects touch and produce charge separation, the positive charge gets stored on one object and the negative on the other. The stored charge creates

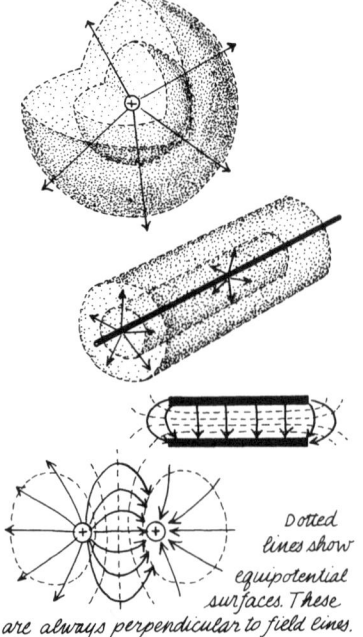

Source Geometry	Potential
Point or sphere $r \geq R$	$\Delta V = kq[1/r_1 - 1/r_2]$ $V = kq/r$ if $V=0$ as $r_2 \to \infty$
cylinder	$\Delta V = 2k\lambda \ln(r_1/r_2)$
parallel plates	$\Delta V = Ed$ $V = Ey$ if $V=0$ for $y=0$ and $0 \leq y \leq d$ also $\Delta V = \dfrac{Qd}{\varepsilon_0 A}$
Dipole	$V \propto \dfrac{p}{r^2}$ on axis $V = 0$ everywhere on $\perp$ bisector

Dotted lines show equipotential surfaces. These are always perpendicular to field lines.

Figure 16.12. Field lines and equipotential surfaces around four standard geometries.

an electric field around each object and raises or lowers its potential. In all of the formulas for potential, the potential is proportional to the charge on the source. The potential also depends on the geometry of the object and on Coulomb's constant, or the permittivity.

The electrical capacitance of an object is defined as $C=Q/V$. If an object has a large electrical capacitance, then it can hold a large electric charge at a relatively small potential. The unit of capacitance is called the farad, with the symbol F. A capacitor with a capacitance of 1 F could hold 1 C with a potential difference of 1 V. Until recent years no device had a capacitance anywhere near that size, but now they exist and are very useful for class demonstrations of long-term electrostatic effects. In electrical circuits it is more customary to have capacitances in the range of microfarads (μF) or smaller. One picofarad=1 pF=1×10^{-12} F.

1. *Capacitance of a sphere.* For the capacitance of a sphere, simply substitute the formula for the potential of a sphere into the formula for capacitance using the radius of the sphere, R, as the particular point for the potential:

$$C = \frac{Q}{V} = \frac{Q}{kQ/R} = \frac{R}{k}.$$

The capacitance of a sphere is simply proportional to its radius. Here is the basis for the approximation we used on p. 386.

$$C = \frac{(R \text{ in m})}{9\times 10^9} \approx 1\times 10^{-10}(R \text{ in m}) = 1\times 10^{-12}(R \text{ in cm}) = (R \text{ in cm}) \text{ pF}.$$

The capacitance of a sphere in picofarads is about equal to its radius in centimeters. As a matter of fact, it is a good rule of thumb that the approximate capacitance (in pF) of any isolated conductor is equal to the radius (in cm) of a sphere that has about the same size as that object.

It is amusing to calculate the capacitance of the largest sphere that we have available. The radius of the earth is 6.4×10^6 m. Therefore, the electrical capacitance of the earth is $C=(6.4\times 10^6)/(9\times 10^9)=700$ μF. Even the earth does not have an electrical capacitance of 1 F!

2. *Capacitance of parallel plates.* We can make a device with much more capacitance than a sphere out of a parallel plate arrangement. The parallel-plate potential difference given previously in the table on p. 398 can be transformed:

$$\Delta V = 4\pi k \sigma y \rightarrow V_{\text{across plates}} = 4\pi k \frac{Q}{A} d = \frac{Qd}{\epsilon_0 A}.$$

The capacitance of the parallel plates with free space between them (or approximately for air) is: $C=Q/V=\epsilon_0 A/d$. The permittivity of other materials is greater than ϵ_0: $\epsilon=\kappa\epsilon_0$. The factor κ is called the dielectric constant of the material. Certain ceramics such as barium titanate can increase the capacitance by a factor of 10 000. The complete formula for the capacitance of a

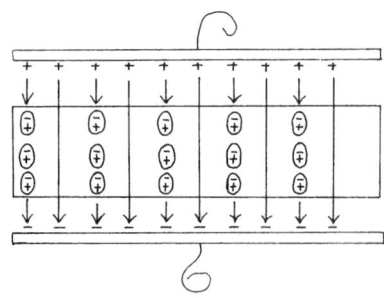

Figure 16.13. The electric field (the number of lines per area) in the dielectric is less than it is in the open region. Therefore, the work done per charge (the voltage) moving a test charge from negative to positive is reduced by the presence of the dielectric. For the same amount of charge on the plates, the voltage between the plates is less. Thus the capacitance is larger.

parallel-plate arrangement is:

$$C_{\text{parallel plate}} = \frac{\kappa \epsilon_0 A}{d}.$$

■ PRACTICAL CAPACITORS

Capacitors used in circuits or for energy storage usually exploit one or two of the factors controlling the capacitance. For instance, a *paper and foil capacitor* consists of a spool of two layers of foil and one very thin sheet of insulator sandwiched together. The total surface area is very large and the spacing between foils is just the thickness of the thin insulating paper or plastic. Students can make such a capacitor out of aluminum foil and plastic wrap, rolling the sandwich in such a way that they have access to the two sheets of aluminum and yet do not short them out. The making of such a capacitor is a good homely lab project since it is necessary to do some calculating of sizes in advance.

In *electrolytic* capacitors the surface area of the foil is not very large but the spacing between foils is produced by a chemical layer that may be only 10 to 100 atoms thick. Since d is so small, the capacitance can be very high. However, in this case the thin chemical layer is maintained only as long as the electric field is in one direction. An electrolytic capacitor must be polarized with one electrode maintained always positive with respect to the other one.

In the case of *ceramic* capacitors, the area of the plates is not large, nor is the spacing particularly small. The material between the plates, however, may have a dielectric constant of the order of 10^4, thereby producing a large capacitance.

■ THE MICROSTRUCTURE OF A DIELECTRIC

Why should the material between the plates of a capacitor increase the capacitance? Figure 16.13 shows a model of how the charges in the material arrange themselves. The original electric field between the plates polarizes the material so that every atomic or molecular cluster is warped. The effect is to produce an inner layer of induced negative electric charges near those on the surface of the positive plate and a similar layer of induced positive charges close to the negative charges on the other plate. The field between the plates in the material is thus greatly reduced. That means, however, that the potential between the plates is reduced. For a given amount of stored charge, if the potential difference between the plates is reduced, then the capacitance is increased.

■ ENERGY STORAGE IN A CAPACITOR

Capacitors are made for many purposes. In some circuits capacitors allow rapid changes in potentials to go through easily while blocking slow voltage changes. In other applications capacitors are used to store charge or electric energy for short periods. Some capacitors that are made for energy storage at high voltage are large physically even though their electrical capacitance is not large. These are filled with oil that provides a dielectric constant of less than 10 but prevents breakdown sparks between the plates.

The energy stored in a capacitor by the addition of charge dq is $V\,dq$. The total energy in a capacitor is

$$U = \int_0^Q V\, dq = \int_0^Q \frac{q\, dq}{C} = \frac{1}{2}\frac{Q^2}{C} = \tfrac{1}{2}QV = \tfrac{1}{2}CV^2.$$

A typical oil-filled, high-voltage capacitor might be rated at 1 μF with $V_{max} = 2000$ V. The maximum energy that could be stored would be $U = \tfrac{1}{2}CV^2 = \tfrac{1}{2}(1 \times 10^{-6}$ F$)(2000$ V$)^2 = 2$ J. Two joules do not seem like much energy. The capacitor has a mass of about 500 g and thus weighs about 5 N. If it were raised only 40 cm in the air it would have a *gravitational* energy increase of 2 J. The 2 J of *electrical* energy are stored in a very convenient way, however. With the right electrical circuit, that energy can be extracted from the capacitor in a time of less than 1 μs. Such a burst would provide momentary power of 2×10^6 W.

When a dielectric is between the plates of a capacitor, the capacitance is greater. Consequently, for a given potential difference across the plates, the stored energy is greater. The extra energy is stored in the distortion of the dielectric molecules. When some types of capacitors are short-circuited, not all the charge drains off the plates immediately. The dielectric remains partially distorted, preventing some of the charge on the plates from leaving. If you touch such a capacitor some time after you have discharged it for the first time, you may get a nasty shock.

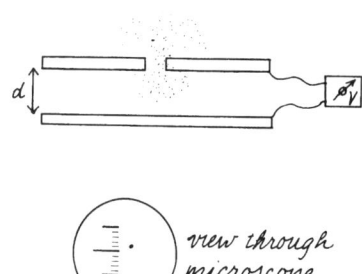

Figure 16.14. Schematic drawing of Millikan's oil drop apparatus.

■ THE EFFECT OF CURVATURE OF CONDUCTOR SURFACE ON CHARGE DENSITY

The charge density and consequently the electric field are highest in regions of a conductor surface where the curvature is highest. At a sharp point on a conductor, for instance, the charge density is very high and the surrounding field is very strong. Using the idea of capacitance, we can make a simple argument about why this effect should occur. Suppose that a very small sphere whose radius is 1 lies on the surface of a much larger sphere whose radius is 100. They are at the same potential but the big sphere has a capacitance 100 times the charge of the little sphere, and therefore carries 100 times as much charge. However, let us calculate the respective charge *densities* on the surfaces of the little sphere and the big sphere. The big sphere has a surface 10 000 times that of the little sphere. Therefore, the charge density of the little sphere is greater by a factor of 100. The electric field in the immediate vicinity of the tiny sphere will be 100 times stronger than that near the surface of the big one. Because of the high electric fields produced in the vicinity of points on charged conductors, the air there may become ionized and provide a leakage path for the charge. Any device meant to store charges in air should have a smooth rounded surface.

■ MILLIKAN'S MEASUREMENT OF THE ELECTRON CHARGE

The charge of the electron was first measured accurately by Robert Millikan in 1909. His apparatus is shown schematically in Fig. 16.14. Actually, Millikan measured the electric charge not on electrons but on tiny oil droplets. He sprayed a mist of oil over the top of a parallel-plate capacitor. The droplets drifted through a hole in the top plate and gradually fell, pulled down by their weight, mg. As seen through a low-powered microscope (about magnification of 50), the droplets appear to be bright points of light.

When droplets come out of an "atomizer," they are usually electrostatically charged. Suppose a droplet has picked up a charge, $-q$. If there is an electric field in the space between the plates, the droplet will be subject to a force, $-Eq$. The electric field, E, is equal to V/d, where V is the potential difference between the plates, and d is the separation distance. We can measure and control the electric field by adjusting a known voltage across the plates. With a small tabletop device and convenient voltage, it turns out to be easy to balance the downward force of gravity with an upward electrical pull. As you watch the pinpoint of light coming from some particular droplet, you can turn the voltage control and see the droplet slow down, stop, and even start drifting up if you choose. When the droplet is standing still:

$$qE(\uparrow) = q(V/d) = mg(\downarrow).$$

Note that q is the charge on the oil droplet and m is the mass of the oil droplet—not the mass of the extra electrons, which is negligible.

We have no way of knowing how many surplus electrons are on the droplet. However, we can knock some off, or at least change the size of q, by bringing a weak radioactive source nearby. All of a sudden, the droplet will start rising or falling again. The voltage source must be readjusted to achieve a new equilibrium. This sequence of operations is done many times with the same droplet, and then with succeeding droplets.

If we knew the mass of the droplet, we could calculate the size of the charge on it. In most school laboratories these days, the "droplets" are actually tiny spheres of plastic of uniform size manufactured for the calibration of electron microscopes. The diameter of a typical sphere is one micron (micrometer, μm), and they are produced with remarkable precision and uniformity. If you know the diameter and density of a sphere, then you can calculate its volume and mass. Poor Millikan (and generations of students since) did not have spheres of known and uniform size. He found the mass of each droplet by measuring its drift speed as it fell. That speed is related by Stoke's law to the diameter and weight of the droplet and the viscosity of air.

It is useful to ask students why Millikan didn't just measure the diameter of the droplets by observing them against a reticle in the microscope. The constraints of the system in terms of number of charges per droplet and reasonable voltages dictates that the droplets be no larger than about 1 μm. Consider (or calculate!) then, the precision with which you could determine a diameter of 1×10^{-6} m using visible light which has a wavelength about half that size. Even if you could measure the diameter to 10%, you would still have an error of 30% in volume and mass.

TABLE 1. *Illustrative values of quantized voltages in Millikan experiment.*

V in volts	q in coulombs	n
319	1.6×10^{-19}	1
160	3.2×10^{-19}	2
106	4.8×10^{-19}	3
80	6.4×10^{-19}	4
64	8.0×10^{-19}	5

Let us put in some typical numbers for such an experiment. Assume that the plastic spheres have a density of 1×10^3 kg/m³. The mass of each one will be 5.2×10^{-16} kg. Assume a separation distance between the plates of exactly 1 cm. Then, when the droplet is in equilibrium:

$$q = \frac{mg}{V/d} = \frac{(5.2 \times 10^{-16} \text{ kg})(9.8 \text{ N/kg})(1.0 \times 10^{-2} \text{ m})}{V} = \frac{5.1 \times 10^{-17} \text{ N m}}{V}.$$

If the experiment were done under these conditions, the values of V would be those shown in Table 1, though they would not be obtained in that particular sequence. Note, first, that only certain values of V produce the condition of balanced force. The mass of the droplet does not change; only the quantity of charge changes as we bombard the droplet from time to time with radioactivity. The data can be explained only if the electric charge is quantized; that is, if the charge can exist only in integral multiples of some basic charge: $q = ne$. The minimum charge that can exist on the droplet corresponds to the maximum voltage that is needed. That charge, which is the charge on one electron, is -1.6×10^{-19} C.

17
Electric Current

One of the most primitive laws of electricity is that every wire has two ends, usually covered with insulation. Although fifth graders are old enough to play successfully with batteries and bulbs, few get the chance. Many teenagers still have trouble wiring a simple circuit. It is the continuity requirement that seems to cause confusion. All the "electricity" that goes in one end of a wire must come out the other end. It does not get used up. Just in case they did not have the experience in fifth grade, we should make sure that each of our students can use real wires, bulbs, and batteries to wire combinations in series and parallel.

Students can make their own electrolytic cell with a lemon (or any other citrus fruit), a copper penny, and a zinc-coated nail. Roofing nails or ones that appear gray and nubbly are usually coated with zinc. Insert the penny and the nail into the lemon, close together but not touching. There is not enough surface contact in this kind of cell to provide sufficient energy to light a bulb but you can measure the voltage with a high resistance voltmeter. If you do not have a voltmeter, put your tongue across the nail and penny. "Tasting" the potential difference will not hurt, but it will provide a tingling sensation similar to a sour taste.

Most of our everyday use of electricity is with alternating current. Unfortunately, ordinary ac household voltage is potentially too dangerous for introductory laboratory exercises, so we use direct current. There are some ac phenomena that students can observe around the house. After they learn about kilowatts and kilowatt-hours, they should examine their home *power* meter and see that it is recording energy, not power. If they watch the meter while someone turns on a major appliance, such as a stove or washer, they can see the dials spin faster. It is also easy and instructive to see line voltage drop. House wires are supposed to be thick enough so that they do not get hot or cause much of a voltage drop when large amounts of current are drawn. Most such house circuits are limited to 15 A. An ordinary toaster draws about 10 A. Plug both a lamp and a toaster into the same wall plug and watch what happens to the lamp when you turn on the toaster. Better yet, start up a heavy motor (such as a power drill or pump) while watching the lamp and see the light dim during the initial current surge. If you have an electric shaver, plug that in instead of the lamp and you can hear the effect when the toaster is turned on. The shaver, incidentally, draws very little current compared with the lamp.

■ CURRENT AND POWER IN FAMILIAR CIRCUITS

The unit of current is the ampere, a flow of 1 coulomb per second. The ampere is the fundamental unit from which the coulomb is derived. The defining calibration of the ampere is in terms of the magnetic attraction between two parallel wires. (See Chap. 18.) The unit of current was named for André Ampère (1775–1836).

Most of our familiar household appliances are rated in terms of the power they use rather than in terms of the current. The two, of course, are related. If we maintain a potential difference across a wire or a lamp or a motor, then the energy used in shoving a small charge through the device is equal to $\Delta U = V \Delta Q$. The power required is equal to the time rate of using the energy: $P = \Delta U / \Delta t$. But $\Delta Q = I \Delta t$. Therefore, $\Delta U = V I \Delta t$, and $P = VI$. The power is measured in joules per second. The unit of power is the watt, named for James Watt (1736–1819). (As you can see, many of these early scientists have achieved a certain nominal immortality.)

A typical flashlight bulb is rated at 1 W. The potential difference is provided by two $1\frac{1}{2}$-V cells in series. Since $P = VI$, then 1 W $= (3$ V$)I$. The current in such a bulb is, therefore, $\frac{1}{3}$ A. Consider next the current in a 60-W house light; to be sure, it uses alternating current, but the Ohm's law relationships hold for average power, current, and voltage. The equivalent potential difference in this case is 118 V (with a permissable short-term variation of ±5%). If we assume that the value is 120 V, then $I = P/V = \frac{1}{2}$ A. Most toasters are rated at about 1000 W. The current drawn by such a toaster would be about 8 A. A car's starting motor operates on direct current. A typical motor requires about 1 horsepower=746 W. With a 12-V battery, the current drawn by such a motor is about 60 A. This is why the wires between the battery and starting motor are so large. In some generators and research magnets, currents of 10 000 A are common. The "wires" to carry such large currents are usually copper bars with cross sections of several square centimeters.

■ WHAT'S IN A NAME?

Purists like to point out that individual electrolytic cells are not batteries. Batteries are combinations of cells. As a practical matter of long usage, any source of dc can be called a battery. (Try asking your local hardware salesman for a "flashlight cell.") Similarly, in studying electrostatics, there are pedagogical reasons for referring to "potential differences" between two points. However, for practical applications of electrical circuitry, the word "voltage" causes no confusion and is perfectly acceptable. As for whether you can have current "in" a wire, or "through" a wire, it is the fact that charges move, currents exist. Perhaps in this case, since a mental image is at stake, teachers should refer to currents "in," but not overly worry if students misspeak.

■ MODEL OF CHARGE MOVEMENT IN A WIRE

The atoms of a metal are bound in a crystal lattice. The one or two outer electrons, which would be considered valence electrons if the atom were free, are not tied to their parent atom. Instead, they are free to roam throughout the crystal. We will consider two models of their behavior. The first was analyzed by Paul Drude in 1900. It assumes that the electrons behave very much like the molecules of a gas, moving about at random, and bumping into the massive ions that are vibrating in locked positions. The second model recognizes quantum requirements that the electrons must occupy potential energy levels, which they fill up in pairs.

Only the electrons at the top of the occupied levels are free to accept small amounts of energy and move to unoccupied states. These few electrons can accept thermal energy and so contribute a small amount to the specific heat of the metal. However, when an electric field is imposed on the conductor, there is a shift in the velocity distribution of *all* the conduction electrons so that they all take part in the drift velocity.

The classical model of electrons in metal behaving like gas molecules has the same limitations that occur in the Bohr model of the atom. Electrons with the energy of an electron volt or less cannot be localized to regions smaller than an atom. Therefore, we cannot accurately describe their motion in terms of orbits or of particle–particle collisions. Instead, quantum mechanics allows us to predict the probability of finding an electron in a particular region and with a particular energy. The probability functions are in terms of waves that contain all the familiar wave properties, such as reflection, refraction, and, most important, interference. In this more sophisticated theory, the motion of electrons through a lattice is described in terms of a wave moving through a periodic lattice and being scattered by imperfections in it. Furthermore, there are quantum conditions on the energy levels of the wandering electron. These conditions are similar to the ones that exist for atomic orbits. For instance, no two electrons of the same spin can have the same energy, and only two values of the spin are possible. The electrons fill the energy levels in pairs with opposing spins. In conductors there are more available levels than there are electrons. The electrons occupy all the positions up to the "Fermi level," which has values from 3 to 8 eV.

Surprisingly enough, the crude model of an electron gas in a metal explains many of the ordinary phenomena of electric current. That in itself will require explanation. In the meantime we can use the model as a good first approximation, and borrow additional facts from the second model whenever we need them. Note first the difference between conductors and insulators. If the outer atomic electrons of a substance are being used for molecular bonding, then they are not free to conduct current. It is not the case, however, that materials are necessarily either insulators or conductors. There is an infinite range of possibilities between, including some pairs of materials that permit electrons to move one way across their boundary but not the other. Some materials, such as diamond, are ordinarily very good insulators; but if a subatomic particle passes through and releases electrons from their bound positions, the crystal can momentarily become a good conductor.

Assuming that we have a long copper wire filled with free electrons, how can we impose a continuous electric field that will make the electrons move? As a practical matter, we just connect the ends of the wire to the terminals of a battery. Note that before the circuit is complete, the electric field around the battery and wires is electrostatic. The lines of force around the battery electrodes act as if they were coming from a dipole, distorted by the local geometry of wires and battery casing. As soon as the circuit is completed, the field lines must change so that the field *in* the wire is constant in magnitude and follows every turn of the wire.

If the electric field at one point of the wire is stronger than it is at another, then the electrons in the first part will be shoved harder and will accumulate in the second part. But, on a continuous basis, that cannot be. The current must be the same at every point in a wire or else the charges will pile up. Therefore, the electric field must be uniform everywhere along the wire (assuming that the wire is uniform and has no divisions). How can the wire have a uniform electric field when it may extend a long way from the battery and may even be tied in knots? The answer must lie in the fact that if at first the electric field is not uniform, charges will pile up and produce forces that tend to slow the motion of charges in one part and speed them up in others. When equilibrium is established, there may be small surpluses of charges along the outer boundaries of the wires or at corners. This distribution of surplus charges all along the wire must guide the electric field, causing it to be uniform.

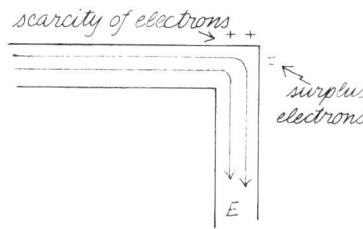

Figure 17.1. Surface charges on a wire needed to guide and maintain constant electric field in the wire.

Let us take a particular case with simple geometry to see how much surplus charge is required on the outer wall of the wire in order to lead an electric field around a sharp corner. In Fig. 17.1 there is a right angle bend in the wire. In order for the electric field to be uniform in such a geometry, the field lines coming from the left must look as if they were going to terminate in a negative charge at the corner. The field lines starting down must have a pattern as if they had originated in a positive charge on the other side of the corner. For a typical circuit using hookup wire and flashlight batteries, there might be 1 A in a wire of 1 mm^2 cross section. In order to produce this current, there would have to be a potential difference of 17 mV across 1 m of the wire. (The resistivity of copper is 1.7×10^{-8} Ω m.) Therefore, the field in the wire is $E = -(17 \times 10^{-3} \text{ V})/(1 \text{ m}) = -17 \times 10^{-3}$ V/m. Note that in calculating the electric field in the wire, we assumed that it was constant in magnitude throughout the wire and that, therefore, the relationship between field and potential difference was the same as it is in a capacitor where the field is also constant: $E = -\Delta V/\Delta x$.

Now we can find how much charge must be at the corner of the wire in order to terminate this much electric field. Remember that when field lines end on the surface of a conductor the relationship between electric field and charge density is $E = \sigma/\epsilon_0$. In this case, 17×10^{-3} V/m $= \sigma/(8.9 \times 10^{-12})$. Therefore, the surface charge density is $\sigma = 1.5 \times 10^{-13}$ C/m^2. Since the total charge is equal to the charge density times the area, $Q = \sigma A = (1.5 \times 10^{-13} \text{ C/m}^2)(1 \times 10^{-6} \text{ m}^2) = 1.5 \times 10^{-19}$ C. This is approximately the charge on one electron! For these particular conditions of electric field and cross-sectional area of the wire, the presence of one extra electron on one side of the corner and the absence of an electron on the other are all that is needed to swing the constant electric field lines through 90°.

When a circuit is first closed, the electric field cannot be uniform all the way along the wire. During the initial surge, however, electrons travel more rapidly where the electric field is strong and pile up in such a way that they reduce the strong field. The surplus charges rapidly establish a dynamic equilibrium so that the electric field becomes constant all along the wire, creating a constant current at every point. The time taken for the equilibrium to be established is similar to the time taken to fill up a capacitor. To reach $1 - 1/e$ of the final voltage takes $t = RC$ seconds. For a meter length of wire with $C = 10^{-10}$ F, and $R = 10^{-2}$ Ω, the settling-down time is 10^{-12} s.

The current must be constant and the electric field uniform only for the particular case where the circuit consists of just one wire with constant cross section and characteristics. If there are branch points in the wire, the current will divide, some of the charges going one way and some the other. If two different kinds of wire are fastened together to make one long wire, the current will be constant at every point; but in order to produce this condition, the electric fields in the two wires must be different. In order to understand these complications, we must develop a relationship between electric field and current.

■ VELOCITY OF CHARGE FLOW IN A WIRE

According to our simple model of an electron gas, a wire is like a pipe filled with electrons. Although the electrons are dashing about randomly because of their thermal energy, they must also be drifting down the pipe because of the imposed electric field. Let us calculate this drift velocity.

Fig. 17.2 shows the geometry of our calculation. All the electrons in the cylinder that has length $v_d \Delta t$ and cross-sectional area A will pass the measuring point in the time Δt. The density of the free electrons in the metal is n, and since each electron carries a charge e, the charge density is ne. The amount of charge in the cylinder is, therefore, $\Delta q = neAv_d \Delta t$. The current is $i = \Delta q/\Delta t = neAv_d$. If we know the density of the free electrons in a metal, then we can calculate the velocity with which they travel in order to produce a particular current. How many electrons in the wire are free to move?

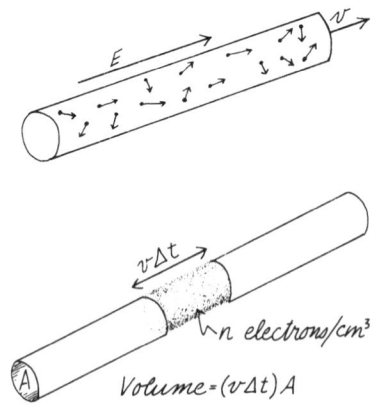

Figure 17.2. Model of charge flow in a wire.

Most of the electrons in an atom are bound to the nucleus. In compounds, the outer, or valence, electrons are locked in the molecular bonding. In metals, however, one or two of the outer electrons are free to roam throughout the whole crystal, while the heavy ions remain fixed in the crystalline lattice. Copper provides one free electron per atom. Since we know the number of atoms per mole and the number of moles per gram and the number of grams per cubic centimeter, then we can find the number of atoms per cubic centimeter. For copper,

$$n = \left(\frac{1 \text{ charge carrier}}{\text{atom}}\right)\left(\frac{\text{atoms}}{\text{mole}}\right)\left(\frac{\text{mole}}{\text{gram}}\right)\left(\frac{\text{grams}}{\text{cm}^3}\right)$$

$$= (1)(6 \times 10^{23})\left(\frac{1}{64}\right)(9) = 8.4 \times 10^{22} \text{ charge carriers/cm}^3.$$

Right here we can see that our model ought to be in trouble. The density of charge carriers is the same as the density of atoms in a solid, which is larger than the density of molecules in a gas by a factor of more than 10^3. If the charge carriers were molecules, we would not expect to be able to apply the laws of ideal gases to such a dense mixture. With electrons the situation ought to be worse because of the long-range nature of the electric force between them.

Clinging to the model a little longer, we can calculate the drift velocity of the electrons: $v_d = i/neA$. For a current of 1 A in a copper wire with a cross section of 1 mm²:

$$v_d = \frac{1 \text{ A}}{(8.4 \times 10^{22} \text{ carriers/cm}^3)(1.6 \times 10^{-19} \text{ C/carrier})(1 \times 10^{-2} \text{ cm}^2)}$$

$$= 7 \times 10^{-3} \text{ cm/s}.$$

The rms *thermal* velocity of electrons at room temperature is 1.2×10^5 m/s—much higher than the velocity produced by the electric field.

$$\tfrac{1}{2}mv^2 = \tfrac{3}{2}kT \rightarrow v = \sqrt{\frac{3kT}{m}} = \sqrt{\frac{3 \times 1.38 \times 10^{-23} \times 293}{9.1 \times 10^{-31}}} = 1.2 \times 10^5 \text{ m/s.}$$

Evidently, the electron drift velocity created by the electric field is just a small perturbation imposed on the basic random motion. The actual ratio of velocities is even greater than we calculate from our first model. Because of quantum requirements, the electrons do not share the thermal energy distribution of the ions of the lattice. Instead, they are organized in closely spaced energy levels up to the Fermi-level energy. The speed corresponding to the Fermi energy of several eV is about 2×10^6 m/s.

A "drift" velocity implies that the charge carriers are undergoing continual small collisions. The electrons are subject to a constant force, which, in our first model, might be expected to accelerate them between collisions. Let us calculate the time and distance between collisions. For the field and drift velocity that we proposed before, $E = 0.017$ V/m and $v_{\text{drift}} = 7 \times 10^{-5}$ m/s. If an electron starts out at zero drift velocity after a collision with an ion, how long does it take to accelerate to drift velocity?

$$\Delta v = v_{\text{drift}} = 7 \times 10^{-5} = at = \frac{eE}{m}t = \frac{(1.6 \times 10^{-19})(0.017)}{9 \times 10^{-31}}t,$$

$$t = 2.3 \times 10^{-14} \text{ s.}$$

This is the average time interval between collisions. To find the distance l traveled in this time, we must use the random, thermal speed, not the drift velocity. $l = (1.2 \times 10^5 \text{ m/s})(2.3 \times 10^{-14} \text{ s}) = 2.8 \times 10^{-9}$ m $= 28$ Å. This is a surprisingly long distance. The copper atoms are shoulder to shoulder, with their centers separated by only 2.6 Å. Note that if we use the *actual* Fermi speed of the electrons, the time between collisions (determined by the drift velocity) remains the same, but the average distance between collisions increases by a factor of 10.

There is another problem with our first model. Once we determine the average distance traveled between collisions, that value should not change with temperature. However, as the random thermal speed of the electrons increases, there will be a shorter time interval between collisions and so the drift velocity will decrease. That will reduce the electrical conductivity. Since the random thermal speed is proportional to the square root of the temperature, Drude concluded that the conductivity of a metal should be proportional to $T^{-1/2}$. Instead, except at very low temperatures the conductivity of most metals is proportional to T^{-1}.

Even though the drift velocity is very small, there are many electrons traveling at that velocity, and therefore, the current is appreciable. Since we have been assuming that the current is proportional to the electric field, we must conclude that the electron drift velocity is proportional to the field. If the electric field is constant throughout the wire, $E = V/L$, where V is the potential difference across the wire of length L. According to our assumptions, it appears that if the current in a wire is proportional to the electric field in the wire, then it is also proportional to the voltage across the wire: $I \propto E \rightarrow I \propto V$. It is usually easier to measure the potential across a wire or a circuit than to measure the electric fields in it. In many cases we can analyze circuits assuming that current is proportional to voltage, but as a matter of fact, the assumption is not always valid. There are important

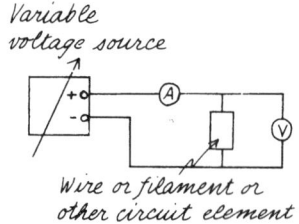

Figure 17.3. Circuit diagram for measuring $I(V)$ of various circuit elements.

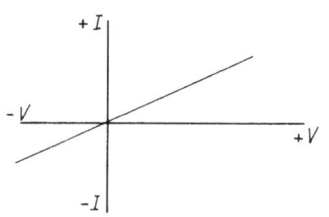

Figure 17.4a. $I(V)$ For a copper wire or a carbon resistor $I(V)$ for a diode.

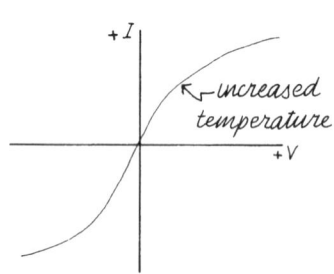

Figure 17.4b. Current as a function of voltage for a tungsten filament. The shape of the curve depends on the temperature of the filament at a particular voltage. Current must be measured after temperature equilibrium is attained.

exceptions, and as often happens in science, the exceptions reveal more about the theory than does the rule.

■ CURRENT AS A FUNCTION OF V

We now have a formula for current as a function of the velocity of the charge carriers, and a relationship between that velocity and the potential difference across the circuit. Let us combine these two expressions:

$$I = neAv, \quad v = KE = K\frac{V}{L}$$

The proportionality constant between the drift velocity and the electric field is K. (It is called the *mobility*.) The constant field in the wire is equal to the potential difference across the wire divided by its length:

$$I = \left(ne\frac{A}{L}K\right)V, \quad I \propto V.$$

The current in a wire is proportional to the voltage across it, although the proportionality factor may change with varying conditions. For instance, a change of temperature may change n, the density of charge carriers; it may also change K, which relates the electric field to the drift velocity. Furthermore, the proportionality constant between current and voltage depends on the length and the cross-sectional area of the wire.

Let us examine the current–voltage relationship in three cases. It is easy in the laboratory to vary the voltage across a wire or a filament or some other circuit element and to measure the current that is produced. The operation is shown schematically in Fig. 17.3, where the circle containing A represents an ammeter, and the circle with V represents a voltmeter. Note that the ammeter is reading the current *in* the circuit, whereas the voltmeter is reading the potential difference *across* the circuit element.

1. *$I(V)$ for a metal wire at constant temperature.* For an ordinary metal wire at constant temperature, the current is proportional to the voltage. The number of charge carriers remains constant and is independent of the direction of the current. The graph of $I(V)$ is shown in Fig. 17.4(a). The positive or negative signs of the current and voltage merely represent the directions. In this case, if you reverse the battery, you reverse the current, and the proportionality constant stays the same.

2. *$I(V)$ of a light-bulb filament.* The temperature of a tungsten filament in a light bulb rises to about 2000 K when it is fully lighted. The density of charge carriers in the wire does not change appreciably with this much of a temperature change. But the drift velocity does. Remember that $v_{\text{drift}} = KE = K\,V/L$. Considering our model of electrons in the metal, we might expect that as the temperature increases the drift velocity would be reduced by the shorter times between collisions and by the increased thermal oscillations of the lattice ions. At a higher temperature and a larger amplitude of oscillation, the ions present larger targets to the drifting electrons, thus reducing their mobility. Notice the effect that a change in K has in our basic equation: $I = ne(A/L)KV$. If K decreases with temperature, then there will be less current for a particular voltage. As the voltage across a filament increases, so

does its temperature. This negative feedback produces a self-limiting situation so that there is a terminal current in a tungsten lamp filament. The resulting curve of $I(V)$ looks like the one shown in Fig. 17.4(b).

When Thomas Alva Edison started making electric light bulbs in the latter part of the nineteenth century, he used carbon filaments in vacuum tubes. (Our present light bulbs are made with tungsten filaments enclosed in a bulb that contains an inert gas at atmospheric pressure.) The carbon filaments were very fragile and also had an electrical characteristic quite different from that of tungsten. The $I(V)$ graph for a carbon filament looks like the one in Fig. 17.4(c). Since an increase of voltage makes the light glow brighter, implying an increase of temperature, it must be that the proportionality constant between I and V *increases* with temperature. Nevertheless, it is still true that the drift velocity constant K decreases as the temperature rises. With carbon, however, the density of charge carriers, n, increases strongly with increased temperature. The reason for this is that most of the valence electrons in solid carbon are only loosely tied up in covalent bonds. The number of electrons per atom that are free to roam depends on the temperature (and on the type of carbon crystal). If the voltage increases across a carbon filament, the current increases and the temperature rises. But as the temperature rises, more electrons are freed from the covalent bonds, increasing the current. In turn, the larger current makes the temperature rise still higher. These processes would compound until the carbon filament burned up except for one other factor. The radiation given off by the glowing filament is proportional to the fourth power of the temperature. Before the filament boils away, equilibrium is established between the energy supplied by the voltage source and the energy being radiated. The temperature stops rising and the current stops increasing.

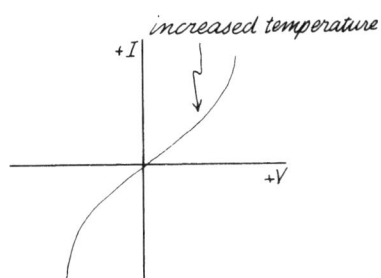

Figure 17.4c. $I=ne(A/L)KV$. With carbon, as temperature increases, K decreases, but n increases.

When a voltage is applied across a tungsten filament, the current is large at first because the metal is cold. In about a hundredth of a second, the temperature rises and the current decreases to its equilibrium value. With a carbon filament, the sequence is just the opposite. The initial current is small but then increases to its equilibrium value as the temperature of the filament increases. It is instructive for students to measure the resistance of a 60-W bulb with an ordinary volt-ohm meter. The meter measures resistance using a $1\frac{1}{2}$-V internal battery. The reading will be the low-temperature resistance of the filament and will be much lower than the resistance calculated from $P=V^2/R$.

3. *I(V) of a diode*. The junction between a copper wire and a silver wire allows electron flow in either direction. This is not true of the junctions of certain materials, such as copper and copper oxide or specially prepared combinations of germanium or silicon. Such junctions that favor flow of charge in one direction are called diodes. A typical $I(V)$ curve for a diode is shown in Fig. 17.4(d). In one direction, shown as positive in the graph, electrons can move across the junction as if it were an ordinary wire. In the other direction, however, there is great resistance to electron flow. For most diodes, the curve does not sharply change direction at 0 voltage. Instead, the forward characteristic slurps gradually into the backward characteristic over a region of several tenths of a volt. Also in most cases, the changeover takes place on the positive side of the voltage graph. It is necessary to get beyond a threshold

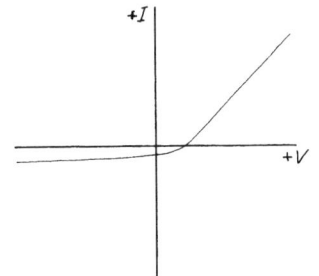

Figure 17.4d. I(V) for a diode.

■ THE SPECIAL—BUT IMPORTANT—CASE OF OHM'S LAW

In many cases the current in a circuit element is *not* proportional to the voltage across it. In many other circuit arrangements, however, materials are used so that the current *is* proportional to the voltage. The conditions must be such that the number of charge carriers does not depend on the voltage, and such that the proportionality constant K between v_{drift} and E must remain at least approximately constant. Usually these restrictions mean that the temperature of the circuit must not change much when the circuit is activated. In that case, we can lump all the constants relating current and voltage into one new constant, the resistance R:

$$I = ne\left(\frac{A}{L}\right)KV = \frac{V}{R}.$$

The misanthropic Englishman, Cavendish, measured the proportionality of current and voltage in the 1770s. Characteristically, he did not bother to tell anyone about it at the time. The law was announced for the first time by Georg Ohm (1787–1854) in about 1820, but the law was not generally recognized for another 10 years. Now it is known as Ohm's law: $I = V/R$.

Notice the relationship between the resistance R and the other factors determining current: $1/R = ne(A/L)K$. It may seem unreasonable to group those factors together and then define them as being the *reciprocal* of a new constant. A different arrangement might have been chosen with the product of those factors equal to a constant called the *conductivity*. However, history and custom have ordained otherwise.

The resistance can be expressed in terms of the other factors in another way:

$$R = \left(\frac{1}{neK}\right)\left(\frac{L}{A}\right) = \rho\frac{L}{A}.$$

By making this rearrangement, we have separated the geometrical aspects of the conductor from those concerned with its microstructure. The new constant ρ is called the *resistivity*. It is a property of the type of material in the conductor and depends on the density of charge carriers and on the relationship between electric field and drift velocity. The geometrical factor L/A depends on the thickness of the wire and its length. If we know the size of a wire and the resistivity of the metal from which it is made, we can calculate its resistance. The units of resistivity are ohm-meters (Ω m). When you multiply resistivity by the length of the wire and divide by its cross-sectional area, the resulting unit of the resistance is the ohm. Note that the slope (derivative) of the $I(V)$ graph is the *reciprocal* of the resistance.

The resistivity of various materials is shown in Table 1. There is an enormous range of values. As you can see, there are both good and poor conductors and also good and poor insulators, with a very large difference between the extremes of the two groups.

Let us calculate the resistance of a copper wire 1 m long with a cross section of 1 mm^2. This is about the size of a thin wire (AWG gauge #18) that we assumed we would use earlier in several of the calculations with batteries and bulbs.

Electric Current

TABLE 1. *Resistivities (ρ) and temperature coefficients (α) (definition of α on p. 415).*

	ρ (Ω m)	α (°C^{-1})
Aluminum	2.7×10^{-8}	0.0043
Boron	2×10^4	
Carbon (amorphous)	3.5×10^{-5}	-0.0005
Copper	1.7×10^{-8}	0.0068
Germanium	0.46	-0.05
Gold	2.4×10^{-8}	0.004
Iron	9.7×10^{-8}	0.0065
Lead	20.7×10^{-8}	0.0034
Mercury	98×10^{-8}	0.0009
Platinum	10.6×10^{-8}	0.0039
Silicon	100–1000	-0.075
Silver	1.6×10^{-8}	0.0041
Sulfur	2×10^{15}	
Tungsten	5.6×10^{-8}	0.0045
Nichrome (alloy of Fe, Ni, and Cr)	100×10^{-8}	0.0004
Fused quartz	10^{18}	
Glass	10^{10}–10^{14}	
Hard rubber	10^{13}–10^{18}	
Hardwood	10^8–10^{11}	

$$R = (1.7 \times 10^{-8} \ \Omega \ \text{m}) \frac{1 \ \text{m}}{1 \times 10^{-6} \ \text{m}^2} = 1.7 \times 10^{-2} \ \Omega.$$

Let us calculate the resistance of a flashlight circuit. A typical bulb used with a typical double-cell flashlight has a power of 1 W and uses about $\frac{1}{3}$ A. Therefore, $R = (3 \ \text{V})/(\frac{1}{3} \ \text{A}) = 9 \ \Omega$. This resistance is large compared with that of a few meters of copper wire. It is provided by the hot filament, which would have a much lower resistance at room temperature (by about a factor of 10!).

The drift speed proportionality constant K and hence the resistivity ρ depend not only on the temperature of the metal but also on the crystalline structure. The values of resistivity given in Table 1 for the conductors are for very pure metals. Small amounts of impurities or imperfections in the lattice produced during crystallization can greatly increase the resistivity. In the probability wave model of electron conduction, the electron wave merely diffracts around stationary atoms. Resistance is caused by the scattering of the wave as it travels down the wire. Any deviation of regularity of the crystal structure, such as would be caused by a gap or an impurity atom, produces an increase in resistivity. Here we have the quantum explanation of how the high-density electrons ($n = 10^{23}$ charge carriers/cm^3) can act like a low-pressure gas. [For a gas at standard temperature and pressure (STP), $n \approx 3 \times 10^{19}$ molecules/cm^3.] Although copper atoms in a solid are only about 3×10^{-10} m apart, electrons in copper at room temperature typically travel 300×10^{-10} m between collisions. In effect, their wave behavior reduces their effective density by a factor of $100^3 = 10^6$.

■ MENTAL MODELS OF RESISTANCE

Frequently students think of electrical resistance as representing some sort of obstacle course that electrons must pass through. Such a notion leads to apparent

414   Electric Current

paradoxes that are unnecessary. Remember that resistance is just a proportionality constant between current in a circuit and the potential difference across it. Better yet, think of Ohm's law as a relationship between the current and the electric field.

$$I = \frac{V}{R} = \frac{V}{\rho(L/A)} = \frac{V/L}{\rho/A} = \frac{E}{\rho/A}.$$

For an example of this altered point of view, consider the two wires in parallel, as shown in Fig. 17.5(a). They are made of the same material and have the same cross section A and the same resistivity ρ. However, $L_2 = 2L_1$. The voltage across the two wires is the same, but their electric fields are different: $E_1 = V/L_1$ and $E_2 = V/L_2 = V/2L_1$. Therefore, $E_1 = 2E_2$. Because the electric field in the shorter wire is twice that in the longer one, the drift velocity of the electrons in the first will be twice that of the second. Hence the current in the first will be twice that in the second. Of course, you can get the same result by calculating that $R_2 = 2R_1$. Since $I = V/R$, it follows that $I_1 = 2I_2$. If you ask how much work you must do to escort a unit charge through the two resistances, you might be tempted to think that you must do more work to shove the charge through the larger resistance, but there is the same potential drop across both of them; the work done per charge must be the same by either route. To be sure the second route is twice as long as the first, but the force per charge required in the second route is half that in the first route.

Let us look at the meaning of resistance in another case, which is shown in Fig. 17.5(b). Once again we have two wires in parallel with the same potential difference across them, but this time they have the same length. Therefore, there is the same electric field in each of them. They can have the same cross-sectional area A and the same charge carrier density n, but because they are different metals or because they are at different temperatures, the drift velocity constant K will be different. Let us assume that $K_1 = 2K_2$; then since $v_{\text{drift}} = KE$, it must be that $v_1 = 2v_2$. The drift velocity in the first wire is twice the drift velocity in the second. Therefore, the current in the first one will be twice that in the second: $I_1 = 2I_2$. Since $\rho \propto 1/K$ it must be that the resistance of the first one is only half that of the second: $R_1 = \frac{1}{2}R_2$. Once again, we face the question of how much work is done in escorting a charge through each of the two routes. Of course, that work must be the same, since the two routes are in parallel. But route 1 has only half the resistance of route 2. Nevertheless, the fields in the two routes are the same and so are the distances through which the charges must be taken. Since the work per charge is equal to the force per charge times the distance, the work done is the same in both routes in spite of the fact that the resistances are different.

Let us examine one more case where electric fields and resistances are different in two parts of the circuit. Fig. 17.5(c) shows a junction between a thin wire and a thick one. Let us assume that the wires are made of the same material at the same temperature but that $A_2 = 2A_1$. Since the wires are in series, there must be the same current in each. Since $I = neAv$, the drift velocity in the larger wire must be just half that in the small wire: $v_2 = \frac{1}{2}v_1$. But the drift velocity is proportional to the electric field: $v_{\text{drift}} = KE$. Therefore, the field in the big wire must be only half that in the small wire. The field lines in the small wire must spread out as they enter the larger wire. The number of lines is the same in the two sections, but the number of lines *per unit area* is reduced by a factor of 2. Around the surface of the junction, there must be a distribution of pseudostatic charge, warping the

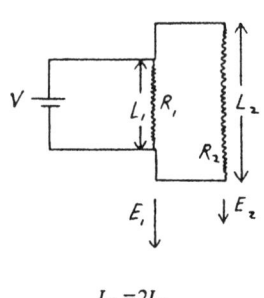

(a)
$L_2 = 2L_1$
$V_2 = 2V_1$
$E_1 = 2E_2$

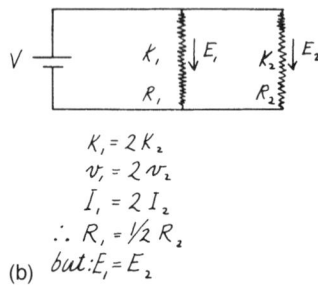

(b)
$K_1 = 2K_2$
$v_1 = 2v_2$
$I_1 = 2I_2$
$\therefore R_1 = \frac{1}{2}R_2$
but: $E_1 = E_2$

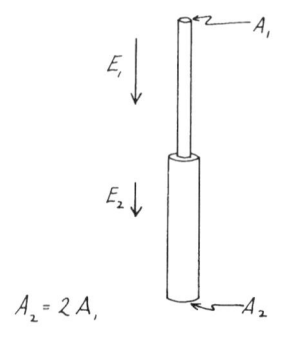

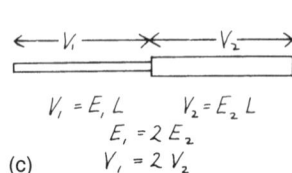

(c)
$V_1 = E_1 L \quad V_2 = E_2 L$
$E_1 = 2E_2$
$V_1 = 2V_2$

Figures 17.5a, b, and c. Relationships among E, V, L, K, A, and v.

electric field lines into the required pattern.

If the small wire and the large one in series each have the same length, then the potential difference across the thin wire is twice that across the thick wire, because $V = EL$. In this case, the resistance of the thin wire is twice that of the thick wire, and the work done in escorting a charge through the thin wire is twice that of escorting that charge through the same length of the thick one. The reason, however, that twice as much work is done is not because there is more resistance in the wire but because there is a stronger electric field in the wire.

■ TEMPERATURE DEPENDENCE OF ρ

The resistivity as a function of temperature is shown for a "pure" and an "impure" metal in Fig. 17.6. Over a considerable range, in the region of room temperature, the curve can be described by $\rho = \rho_0(1 + \alpha t)$. In this formula, t is the number of degrees above room temperature, 20 °C. Values of the temperature coefficient α are given for a number of materials in Table 1 on p. 413.

On the scale of the graph in Fig. 17.6, it appears that the resistivity of metals goes to 0 at absolute zero. In general, this is not true. A residual resistance remains. On the basis of both the electron-gas model by Drude and the probability-wave model, we should expect resistance to decrease as the temperature drops. In both models, the vibration of the bound ions decrease, presenting smaller collision targets or a more regular lattice. In the electron-gas model, the thermal velocities of the electrons decrease with decreasing temperature. During the longer interval between collisions, the electric field can have a greater effect on the electrons, thus increasing their drift velocity and decreasing the resistance. In this model, $\rho \propto v_{\text{thermal}} \propto \sqrt{E_{\text{thermal}}} \propto \sqrt{T}$. In the probability-wave model, the electron energy and random speed do not change appreciably with temperature, since the electrons free to move are at the top of the energy distribution with a Fermi energy of about 10 eV. However, the probability of scattering with the thermally oscillating lattice ions is proportional to the square of their oscillation amplitude and hence to their thermal energy which is proportional to T. Therefore $\rho \propto T$, which agrees with experiment for very pure metals, except at temperatures close to 0. However, the lattice vibrations do not stop at absolute zero. Even in a perfect crystal we should expect particle collisions on the one hand, or wave scattering on the other, yielding nonzero resistance. For a perfect crystal, without lattice imperfections, the resistance is caused by the thermal oscillations of the bound ions. For real metals, any impurity atoms or breaks in the crystal lattice produce a residual resistance that does not depend on temperature.

Zero resistance is displayed by a few compounds below 90 K and by a large number of elements below 10 K. The phenomenon of *superconductivity* was discovered by Kamerlingh Onnes in 1911. A superconductor does not have just small resistance; it has 0 resistance. Widespread use of superconductivity started in the 1970s with the successful fabrication of superconducting wires from various combinations of molybdenum, niobium, tin, vanadium, germanium, indium, gallium, and aluminum. The mixtures are part compound, part solid solutions, and become superconducting in the 18 K region. The major use of the new materials has been in research magnets where strong magnetic fields are required over large regions. Although such magnets do not require power supplies to maintain the current, they do require expensive cryogenic containers and a supply of liquid helium. In 1986–1987, a new family of superconductors was discovered. These

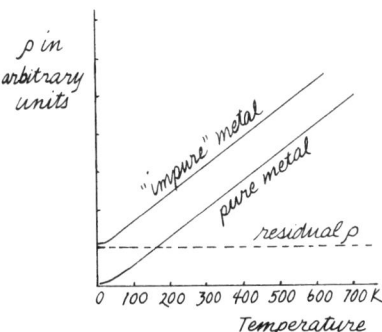

$\rho(T)$ for a "pure" and "impure" metal. The resistivity for a pure metal, without lattice imperfections, is caused by thermal vibrations of the atoms. For real metals, any impurity atoms or breaks in the crystal lattice produce a residual resistivity that does not depend on temperature.

Figure 17.6. Resistivity as a function of temperature.

Figure 17.7. *"Voltaic pile."*

have threshold temperatures of 90 K or above. These materials become superconducting when kept in an insulated container of liquid nitrogen, which liquefies at 77 K. If the new materials can sustain high current densities and strong magnetic fields, and if they can be fabricated in coils suitable for producing magnets, they will open up a whole new technology. It is much easier and cheaper to produce refrigeration that can liquefy nitrogen than it is to go down to liquid-helium temperatures.

The theoretical explanation of superconductivity was formulated only within the last three decades. It is strictly a quantum effect without any classical analogy. It turns out that pairs of electrons interact with each other and with the lattice in such a way that the two electrons are loosely held together [by about $(1-2) \times 10^{-3}$ eV]. Although the binding is weak, it is sufficiently strong so that one electron cannot undergo a collision (its probability wave cannot scatter) from an imperfection or vibration in the lattice without disturbing the other electron. Below a critical temperature, the lattice ions cannot provide sufficient energy in such collisions to break up the electron pair. The electrons cannot exchange any smaller amount of energy, and so they move through the crystalline lattice without any resistance at all. This explanation (the BCS theory—Bardeen, Cooper, Schrieffer) describes the "classical" superconductivity. It is not clear that it explains the effect observed with the new family of high-temperature superconductors.

■ VOLTAIC CELLS AND BATTERIES

When Benjamin Franklin and other eighteenth-century experimenters worked with electricity, they had to use electrostatic generators. It was not until 1800 that Alessandro Volta (1745–1827) created the first battery. Some years earlier, Volta's friend, Luigi Galvani, had discovered that when a wire made of iron on one end and copper on the other was touched to various nerve and muscle combinations of dissected animals, the muscles would twitch. Galvani thought that he had demonstrated "animal" electricity. Volta found that as long as two different metals were used, you could substitute a salt solution and do without the animal.

A diagram of a typical "voltaic pile" is shown in Fig. 17.7. Alternate layers of silver and zinc are separated by paper soaked in a salt solution. Each layer forms an electrolytic cell producing an emf of about 1 V. By arranging the cells in a pile, Volta could produce a considerable voltage. With batteries available that could produce direct current for long lengths of time, the nature of experiments in electricity changed drastically. Joseph Henry in America and Michael Faraday in England were able to power large magnets and also to do electrolysis experiments in which various compounds form or dissociate in conducting liquids.

Regardless of the details of the particular materials used in a cell, "oxidation" takes place at one electrode and "reduction" at the other. When an atom is *oxidized* it loses electrons; when it is *reduced* it gains them. If one of the electrodes dissolves in the electrolyte, its atoms go off as positive ions, leaving electrons behind on the electrode. At the other electrode where reduction is taking place, positive ions are gaining electrons. In the reduction in some voltaic cells, metallic ions are plated as solids on the electrode. In many other cells, the reducing electrode turns hydrogen ions into atoms, which then combine and bubble off as a gas.

The oxidation and reduction can continue only if electrons released at one electrode can get over to the other one. An external wire provides such a path.

Each electron freed by oxidation circulates through the external circuit (or shoves other electrons ahead of it) and ends up on the other electrode where it reduces a positive ion.

In a lemon cell, zinc dissolves more readily than copper and goes into solution as an ion with a double positive charge. Each zinc ion leaves behind two electrons that can circulate through an external connecting wire. If the external wire is connected to the copper electrode, the electrons are available to reduce hydrogen ions in the (citric) acid electrolyte and to turn them into hydrogen atoms.

A voltaic cell turns chemical energy into electrical energy. In the process, some chemical compound breaks down into a more stable arrangement of atoms or ions, thus providing the energy to force an electron into a region of the circuit where there are already surplus electrons. Of course, the chemical energy is really electromagnetic energy in the first place. The amount of energy released depends on the particular chemical reactions (the kind of electrodes and electrolyte), on the concentration of the electrolyte, and on the temperature and pressure. In all cases, the emf produced is of the order of a volt (in practical cells, from 1 to 2 V). Therefore, the energy provided to each electron emitted by such a cell is of the order of an electron volt.

The standard dry cell that is most commonly used in flashlights was first invented by a man named Leclanché in 1868. The outer casing is zinc, which forms the negative electrode; the top button, which is the positive electrode, is fastened to a carbon rod that goes down the axis of the cylinder. The carbon just serves as a chemically inert connection to a surrounding layer made of a paste of manganese dioxide and powdered graphite. Between the manganese dioxide and the zinc is a wet paste of ammonium chloride and zinc chloride, which serves as the electrolyte. When the zinc dissolves and goes into the electrolyte, it leaves electrons behind on the bottom case. If there is an external circuit connecting the bottom case to the top positive button, electrons can pass down the carbon rod and reduce the manganese dioxide from an oxidation state of 4+ to 3+. This particular reaction produces an emf of about 1.6 volts. Commercial batteries may have many variations on this basic scheme, including the use of sandwiched sheets of zinc between the other layers and also small amount of other chemicals in the electrolyte to control some of the chemical interactions.

Several other kinds of batteries are available on the market. The "alkaline battery" is basically the Leclanché cell, with a high concentration of sodium hydroxide in the electrolyte. It can temporarily withstand a heavier drain of current than the standard dry cell. A "mercury battery" has electrodes made of zinc and mercuric oxide with an alkaline electrolyte. This combination produces an emf of 1.34 V, with the advantage that the internal resistance does not change appreciably as the battery ages. A "nickel–cadmium" cell is rechargeable. That is, if an exterior voltage is applied backward across the cell, the chemical interactions will reverse themselves, storing up energy for future use. The nickel–cadmium cell has been known for many years but has become available only recently in a practical sealed container. The problem that had to be solved was how to contain the gases that are generated during the charging process.

An amusing and instructive problem to give students is to find the cost of electricity from a flashlight battery. They must find the cost of some particular battery—perhaps a D cell—and then measure the length of time that it can provide a particular current. For instance, a battery that costs 50 cents can provide 2/3 A for about 5 h. If 5 W h costs half a dollar, 1 kW h costs $100. Evidently, the power coming from the plug in the wall is relatively cheap.

The workhorse of batteries is the lead storage battery used in all cars. One electrode is lead and the other is lead dioxide, with the electrolyte being sulfuric acid. The lead would tend to go into solution as a doubly charged positive ion, releasing two electrons at that electrode. However, the positive lead ion immediately joins with a negative sulfate ion from the sulfuric acid, forming a deposit of lead sulfate on the plate. When the external circuit is completed and electrons arrive at the lead oxide electrode, the lead oxide is reduced. The oxygen is stripped off and combined with hydrogen ions from the acid to form water, leaving a doubly charged positive lead ion on the plate. This immediately joins with a negative sulfate ion and forms a deposit of lead sulfate. Thus, both electrodes become plated with lead sulfate, and the sulfuric acid slowly turns into water. During the charging process, the chemical reactions are reversed, building up the concentration of sulfuric acid again. The specific gravity of the sulfuric acid is a good indicator of the extent of the battery charge. When the battery is fully charged, the specific gravity of the sulfuric acid is 1.28. At 50% charge, the specific gravity is 1.18, and at zero charge it is 1.08. The emf produced by each cell is about 2 V. Six cells are connected in series to form the standard 12-V car battery.

The amount of charge stored in a typical car battery is 50 A h. Since 1 A s is a coulomb, such a battery could supply 180 000 C. Let us calculate how many electrons are transferred in a complete discharge of the battery and the mass of the chemicals that undergo transformation. The charge on one mole of electrons is called a faraday:

$$1 \text{ faraday} = (6 \times 10^{23} \text{ electrons/mol})(1.6 \times 10^{-19} \text{ C}) = 96\,500 \text{ C}.$$

As you can see, a 50-A h battery can supply about 2 faradays. However, the process does not involve two moles of lead, since each lead atom provides two electrons as it transforms into a double charged positive ion. Consequently, 1 mol or 207 g of lead atoms are transformed into lead sulfate at the negative electrode. Two moles of sulfuric acid are used up, one at the positive plate and one at the negative plate. Therefore, 196 g of sulfuric acid are transformed during the discharge.

■ CIRCUITRY

1. *Parallel resistances.* Sometimes it is revealing to make use of the algebraic solution of the formula of reciprocals to find directly the equivalent resistance of two resistors in parallel.

$$\frac{1}{R} = \frac{1}{R_1} + \frac{1}{R_2} \rightarrow R_{\text{equivalent}} = \frac{R_1 R_2}{R_1 + R_2}.$$

In this form it is apparent that the equivalent parallel resistance must always be less than either of the separate resistances, since the formula could also be written

$$R_{\text{equivalent}} = R_1 \frac{R_2}{R_1 + R_2} = R_2 \frac{R_1}{R_1 + R_2}.$$

In either case, the individual resistance would be multiplied by a fraction less than 1. If one of the parallel resistors is less than one-tenth of the other one, then usually the *large* one can be ignored. For instance, the equivalent resistance of a 10-Ω resistor in parallel with a 100-Ω resistor is (10×100)/(10 +100)=9.1 Ω. There is a 10% error in ignoring the 100-Ω resistor and assuming that the parallel resistance is just 10 Ω.

There are other common situations that depend on reciprocal relationships like this. For instance, if you have leaks in a container, it is inefficient to make any effort to seal small leaks as long as there is still one large leak. With home insulation it does no good to increase the thickness of high R material in 50% of the house if there is still 50% of the surface area with low R. (See page 258.)

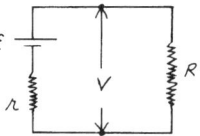

Figure 17.8. *Circuit diagram showing internal resistance, r, of power supply.*

It is often useful to know that the resistance of two identical resistors in parallel is half the individual resistance; the resistance of three identical resistors in parallel is one third the individual resistance, etc.

2. *Internal resistance of the power supply.* A battery *or any other power source* must contain an internal resistance. For instance, if you short-circuit a flashlight battery, you may get 5–10 A, sufficient to raise the temperature of the cell and connecting wire so that you can feel them get hot. The resistance of the short piece of wire would not be more than 0.01 Ω. If there were no internal resistance in the battery, we might expect a current of $I=1.5$ V/0.01 Ω=150 A. Consequently, the battery itself must contain the limiting resistance.

In drawing circuit diagrams of any power supply, the internal resistance should be included as shown in Fig. 17.8. The potential difference available is called the terminal voltage V. For historical reasons, the actual potential difference created by the battery is called the electromotive force $\mathcal{E}$. In spite of its name, an electromagnetic force or emf is not a force; it is a potential difference—the energy provided per coulomb. The current in the circuit shown in the diagram is $I=\mathcal{E}/(r+R)$. The actual voltage available at the terminal of the battery is

$$V=\mathcal{E}-Ir=\mathcal{E}-\frac{\mathcal{E}r}{r+R}=\frac{\mathcal{E}}{r+R}R.$$

If the internal resistance r is very small compared to the load resistance R, then the terminal voltage is almost equal to the emf.

The terminal voltage depends on the amount of current drawn from the source. A new $1\frac{1}{2}$-V D cell normally has an internal resistance of about 0.1 Ω. Even if the cell provides 1 A, its terminal voltage is not much reduced from its rated emf:

$$V=\mathcal{E}-Ir=1.5 \text{ V}-(1 \text{ A})(0.1 \text{ Ω})=1.4 \text{ V}.$$

As a battery ages, its internal resistance increases. If you measure the terminal voltage of a dead flashlight cell, your voltmeter might read close to 1.5 V, but this is because the voltmeter draws very little current and, therefore, the Ir drop inside the battery is small. An old dry cell may well have an internal resistance of 5 Ω.

With household wiring you may notice the dimming of light or the slowing

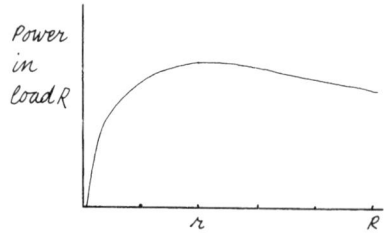

Figure 17.9. Power in load as function of external load resistance, R. Internal resistance, r, is fixed.

down of a shaver when a toaster draws current from the same outlet. In this case, the internal resistance of the power supply may be considered to be the resistance of the wires leading from the wall socket to the main fuse box. Depending on the length of the run of the wire in the walls and on the age of the house, the internal resistance may easily be 1 Ω. If the toaster draws about 10 A, the Ir drop in the wires is 10 V. A 10-V drop in line voltage is enough to make itself seen in a lamp or heard in a shaver.

Any power supply has an internal resistance. The size of that resistance in a large city generator is small, but the current provided is large. The resulting internal energy loss is wasteful, though inevitable, and the heat produced must be carried away by cooling water.

One of the standard text exercises is to derive the conditions for extracting maximum power from a generator *with a fixed internal resistance*. Let us combine the expressions for current and terminal voltage and get the formula for the power supplied to the load resistance, R:

$$P = IV = \frac{\mathcal{E}}{r+R} \frac{\mathcal{E}}{r+R} R = \frac{\mathcal{E}^2}{(r+R)^2} R.$$

Now vary the load resistance to get maximum power. If R is 0, then the current is maximum, but the voltage and external power are 0. On the other hand, if we make R much larger than r, the terminal voltage is nearly equal to the emf. However, the power supplied to the load is small:

$$P \to \frac{\mathcal{E}^2}{R} \quad \text{for } R \geqslant r.$$

The graph of power in the load as a function of R is shown in Fig. 17.9. Note the two extreme conditions that we just analyzed. The curve rises to a maximum at the point where $R=r$. This condition can be derived analytically by setting $dP/dR=0$:

$$\frac{dP}{dR} = \frac{\mathcal{E}^2}{(r+R)^2} + \mathcal{E}^2 R[-2(r+R)^{-3}] = 0 \to 1 = \frac{2R}{(r+R)} \to r = R.$$

When the external load is made to equal the internal resistance, maximum power is drawn from the supply and consumed in the external load. But the same amount of power must also be dissipated inside the supply. Of course, these are not the conditions under which large power stations operate; they try to minimize the internal resistance of their generators. For maximum power extraction with certain electronic circuits, however, the load is made equal to the fixed internal resistance. For alternating current, these conditions of matched impedance also eliminate reflections at the boundary between source and load.

3. *Ammeters and voltmeters.* An ammeter must always be connected in series, since it measures the current *in* a circuit. A voltmeter, on the other hand, is connected in parallel with a circuit component, since it measures the potential difference *across* the component.

Ammeters and voltmeters of the standard type (as opposed to electronic versions) contain the same basic measuring device, called a galvanometer. It consists of a coil of wire that can rotate in the gap of a permanent magnet.

When there is current in the coil, it becomes a magnet with a strength proportional to the current. The angular rotation of the coil, which is opposed by a spring, indicates the amount of current. The sensitivity of the galvanometer depends on its construction, which in turn affects its price. A typical galvanometer movement costing several dollars has a coil that will rotate to a full-scale reading with a current of 1 mA. The resistance of the coil winding in this case might be 50 Ω.

A galvanometer is basically a *current*-measuring device. How can we make it into a *voltmeter*? Suppose we want a meter to read full scale for a potential difference of 10 V. A resistance can be added in series to the galvanometer coil so that there is only 1 mA in the circuit with 10 V across it. The total resistance of the series resistor and the galvanometer coil must be 1×10^4 Ω. The 50 Ω of the galvanometer coil itself is only $\frac{1}{2}$% of the total resistance needed. Such meters cannot be read closer than about 2%, and they are seldom calibrated to better than 5%. The 50 Ω could be ignored. With 10 000 Ω in series with this galvanometer, it will read full scale when placed across a potential difference of 10 V.

Let us make an *ammeter* out of the galvanometer. By itself, the galvanometer *is* an ammeter, reading full scale with a current of 1 mA. To make the ammeter read full scale with a current of 5 mA, wire a "shunt" resistor in parallel with the moving coil. If 5 mA is in the parallel network and the galvanometer is reading full scale, then 1 mA must be in the coil and 4 mA in the shunt. The potential drop across both coil and shunt must be the same if they are in parallel:

$I_c R_c = I_S R_S \rightarrow (1\times 10^{-3}$ A$)(50$ $\Omega) = (4\times 10^{-3}$ A$)R_S \rightarrow R_{shunt} = 12.5$ Ω.

We could use the same procedure to find the shunt resistance needed for any other full-scale reading. For large currents, the resistance of the coil can be ignored in comparison with the very small resistance needed for the shunt. For instance, suppose that we want to use this galvanometer for an ammeter that reads full scale with 1 A. The coil requires 1/1000 of the total current, and the shunt will carry 999/1000 of the current. Since the parallel shunt carries approximately 1000 times the current of the coil, its resistance must be smaller by a factor of 1000. Therefore, $R_{shunt} = 0.05$ Ω

For ammeters and voltmeters of this type, there are two characteristics that have to be known when making measurements. First, there is always a potential drop across the coil when the instrument is being used. At full scale reading, that potential is $IR = (1\times 10^{-3}$ A$)(50$ $\Omega) = 50$ mV. The second feature that must be known if the device is used as a voltmeter is the size of the resistor in series. That value is usually given in small print someplace on the face plate. A typical reading would be 10 000 Ω/V. If the voltmeter has multiple ranges, there is a different series resistor for each range. If one range is 1.5 V full scale, then the series resistor has a value of 15 000 Ω. If you switch to the 10-V range, you have switched to a series resistance of 100 000 Ω. Even if the voltmeter reads 5 V, the series resistance is still 100 000 Ω.

It is hard to make a measurement without interfering with the thing that you are measuring. When putting voltmeters or ammeters into circuits, you may alter the circuit appreciably. Since the inner galvanometer of such a

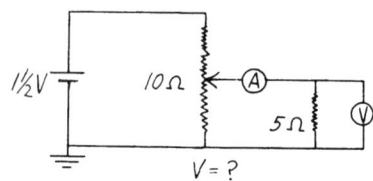

Figure 17.10. Circuit for continuously variable voltage divider.

meter has a resistance and requires a current, it evidently absorbs energy that has to be supplied by the circuit.

Suppose that you want to measure the current in a 0.1-Ω resistor. Let us use the ammeter whose characteristics we have calculated. In the 1-A range, the internal resistance of the ammeter is 0.05 Ω. If you insert the ammeter into the circuit, you are putting a resistance of 0.05 Ω in series with a resistance of 0.1 Ω. The ammeter will faithfully read the current in this circuit, but the ammeter is now a major part of the circuit since it provides one-third of the total resistance.

Suppose that you have two 10 000-Ω resistors in series and want to measure a potential drop across one of them. If the expected range is slightly less than 1 V and you use the voltmeter whose characteristics we have calculated, the voltmeter itself will have an internal resistance of only 10 000 Ω. When you place the device in parallel with the 10 000 Ω in the circuit, you are reducing the parallel resistance to 5000 Ω. The voltmeter will give the correct reading for the potential drop of this particular circuit, but the circuit is quite different from what it would be if the voltmeter were removed.

As a rule of thumb, if meters are to be used as test instruments and then removed from the circuit, their internal resistance should not cause more than a 10% change in the circuit. For an ammeter, this means that its internal resistance must be less than 10% of the series resistance of the circuit into which it is being inserted. In the case of a voltmeter, the internal resistance should be at least 10 times that of the circuit component across which it is placed.

The internal resistance of a lemon cell might be anywhere from 1000 to 10 000 Ω, depending on how far apart you place the penny and the nail. The emf of the copper–zinc pair is about 1 V. If you try to measure this potential with a cheap voltmeter having a resistance of only 1000 Ω/V, then you may detect very little terminal voltage. To measure the voltage of lemon cells, you should have a voltmeter with an internal resistance of at least 10 000 Ω/V. In many electronic circuits, particularly if electronic vacuum tubes are involved instead of transistors, critical control voltages often exist across resistors of 1 MΩ (10^6 Ω) or more. If you use a galvanometer-coil voltmeter to measure such voltages, you will completely distort the circuit, and the readings will be meaningless. For such measurements, vacuum tube voltmeters must be used. These usually have input resistance of 10 MΩ or higher.

4. *The voltage divider.* It is frequently necessary or convenient to produce continuously variable potential differences. By putting cells in series, all you can get are quantized voltages $-1\frac{1}{2}$, 3, $4\frac{1}{2}$ V, and so on. The circuit shown in Fig. 17.10 is a continuously variable voltage divider. Its output could be used in the measurement of $I(V)$ of circuit components. As a fixed voltage divider, it is also a basic element in all circuits. In the variable form, the divided resistance (R_1+R_2) is usually a long coil with a tap connection that can slide along the coil. Such a device is called a *rheostat*. The voltage between the tap and one end of the rheostat is

$$V_2 = IR_2 = \frac{V}{R_1+R_2}R_2 = V\frac{L_2}{L_1+L_2}.$$

In Fig. 17.10 the voltage divider is used as a power supply for the load resistor, R_3. Both R_3 and the voltmeter are in parallel with R_2. The equation that we derived is valid only if $R_2 \ll R_3$ and the internal resistance of the voltmeter. If this is not the case, then R_2 must be replaced in the formula with the equivalent resistance of the whole parallel network.

In the circuit diagram we have used the standard symbol to indicate the "ground" or 0 potential point. In most circuits the choice of this point is quite arbitrary. The voltage-divider circuit would work just as well if we chose the positive terminal of the battery to be at 0 potential. However, the usual convention is to choose the negative side of the battery as 0 potential, thus making all the other potentials in the circuit positive.

When various circuits are fastened together, it is usually important to make sure that the ground connections are consistent with each other. In particular, if circuits are used with test instruments powered by alternating current, such as oscilloscopes or vacuum-tube voltmeters, then the ground connections may already be determined by the construction of the ac instrument. The case is usually connected to water pipe grounds through the ground wire of the three-prong ac power plug. When wiring a complicated circuit, it is good practice to connect all the ground lines first. It is also standard practice not to connect the final lead to any power supply until all the rest of the circuit is wired and checked.

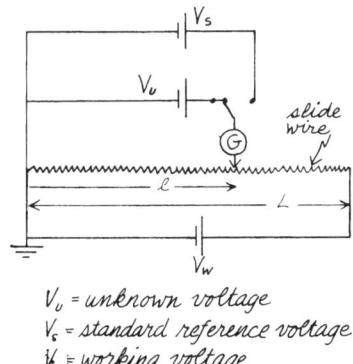

V_u = unknown voltage
V_s = standard reference voltage
V_w = working voltage

Figure 17.11. Potentiometer circuit—a null reading device.

5. *The potentiometer.* Precision instruments often involve a null reading. The chemical pan balance is an example of such an instrument. The weight to be measured is balanced against a known standard, with the equality of unknown and standard determined by the level of the balance. The circuit of a null-reading instrument for voltage measurements is shown in Fig. 17.11. It is called a potentiometer. Basically, it is a voltage divider whose tap voltage is compared with that of the unknown source. The galvanometer, which determines the equality of the tap voltage and the unknown, can be very sensitive because when the two voltages are nearly equal, there is very little current. You can usually measure zero current more precisely than you can measure the value of some larger current.

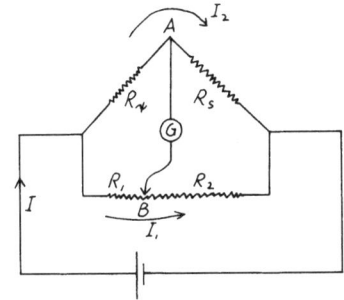

Figure 17.12. Wheatstone bridge.

In the simple circuit shown in the diagram, the voltage of the unknown is $V_U = (l/L) V_W$. Laboratory potentiometers are considerably more elaborate than the simple version shown here. For instance, in a good instrument the galvanometer would be provided with several shunts to protect it and make it less sensitive until the 0 point region is nearly reached. In our simple circuit, the unknown voltage is found in terms of a working voltage V_W. In the real instruments, another cell is used as a standard. As shown in the diagram, the standard cell can be interchanged with the unknown voltage. In a preliminary procedure, it is used to calibrate V_W, which is then used in the actual measurement. In that way, the standard cell does not have to supply current.

6. *Bridge circuits.* There are many "bridge" circuits used in electrical instruments. They are null-reading devices that compare an unknown with a known fraction of a standard value. One of the simplest of these circuits was designed by Charles Wheatstone over a century ago. It is shown in Fig. 17.12. The null condition is obtained when there is no current in the galvanometer.

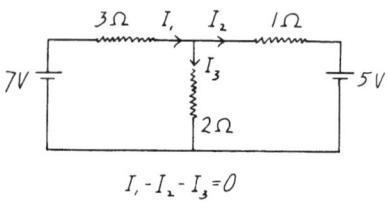

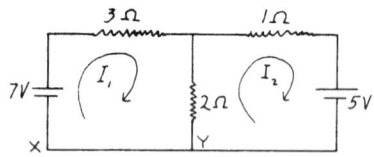

This circuit cannot be solved by using series and parallel equivalent rules.

Figure 17.13. Three currents can be identified related by: $I_1 - I_2 - I_3 = 0$. In the bottom version of the circuit only two loop currents are specified. The actual current in the 2Ω resistor is $(I_1 - I_2)$.

Since a very sensitive galvanometer can be used to measure a very small current, the null condition can be determined very precisely. When there is no current in the galvanometer, the circuit consists of just two parallel branches. The current provided by the battery divides into the upper current, I_2, and the lower current, I_1. The voltage drop across the unknown resistor R_x must be the same as the voltage drop across R_1; and the voltage drop across the standard resistor R_s must be the same as the voltage drop across R_2. Equate the potential drops and take the ratio of the two equations:

$$\frac{I_1 R_1 = I_2 R_x}{I_1 R_2 = I_2 R_s} \rightarrow R_x = \frac{R_1}{R_2} R_s.$$

Note that the value of the unknown resistance depends on the accuracy with which the standard resistance R_s is known and on the uniformity of the slide wire or rheostat that determines the ratio R_1/R_2. The voltage rating of the battery and the characteristics of the galvanometer do not appear in the equation. However, those characteristics do affect the sensitivity of determining the null condition.

7. *Circuit analysis with Kirchhoff's rules.* It is not always possible to reduce a circuit to equivalent series and parallel resistance. A simple example is shown in Fig. 17.13. All circuits, including this one, can be solved at least in principle by an application of two rules that are known by Gustav Kirchhoff's name. The first is that at any branch point, the current entering must equal the current leaving. This requirement is just common sense but, in applying it, you must be careful to be consistent about the directions of the currents. In the example shown in the diagram, the current entering a junction point is labeled I_1 and the currents leaving are I_2 and I_3. Kirchhoff's law requires that $I_1 - I_2 - I_3 = 0$. The algebraic signs of the currents are determined by the fact that I_1 is entering the junction and I_2 and I_3 are leaving it.

The second Kirchhoff rule is that if you trace a closed-loop path around any part of the circuit, the potential increases and decreases must add up to 0. Again, this requirement is just common sense. If you could go around a closed loop and end up at a higher potential, you could get something for nothing (and make a fortune). The electrostatic potential, like the gravitational potential, is conservative. In the case of lines of electric force produced by a changing magnetic field, a charged particle can gain energy in going around a closed loop, but in that case, the changing magnetic field furnishes the energy.

Let us apply these laws to the simple circuit in Fig. 17.13. Notice that we have labeled the currents in two different ways. In the first part of the diagram, there are three unknowns: I_1, I_2, and I_3. To solve for three unknowns we need three equations. One equation is given by Kirchhoff's first rule: $I_1 - I_2 - I_3 = 0$. The other two equations can be derived by writing down the potential increases and decreases for two different loops. We can simplify the arithmetic by using the notation shown in the second part of the diagram. In this case, there are only two unknown currents: I_1 and I_2. Therefore, only two equations are needed. However, it appears that there are two currents in the parallel 2-Ω resistor! The actual current is the algebraic sum of the two.

Now let us apply Kirchhoff's second rule to the two loops shown in the

Electric Current

bottom diagram. As we go around the left-hand loop starting from the *x* mark, the potential first rises by the battery potential of 7 V. Then there is an *IR* drop as the I_1 current goes through the 3-Ω resistor. That potential difference must be listed as $-3I_1$. As our route goes down through the 2-Ω resistor, there is a voltage drop due to the current, I_1 equal to $-2I_1$. However, this route takes *us* in the opposite direction from the current I_2, which is also going through the 2-Ω resistor. There is a voltage *rise* due to that current, equal to $+2I_2$. The final equation for the first loop is

$$+7-5I_1+2I_2=0.$$

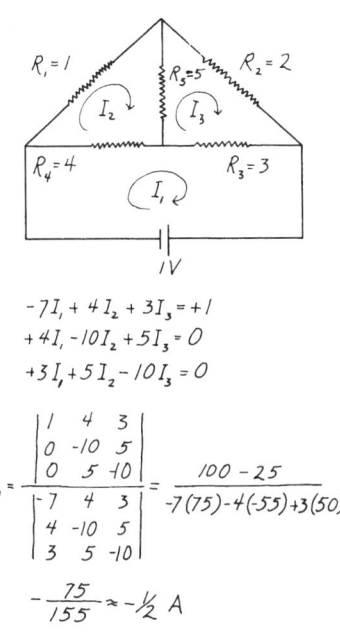

Figure 17.14. Kirchhoff's solution to unbalanced Wheatstone Bridge.

Each term in the equation represents a voltage. If we take the right-hand loop, the arguments are much the same. Starting at the *y* mark, we travel up through the parallel 2-Ω resistor. We are traveling downstream with I_2 and, therefore, there is a potential drop equal to $-2I_2$. Because we are bucking the current I_1, there is a voltage rise of $+2I_1$. There is a voltage drop going through the 1-Ω resistor equal to $-1I_2$. Traveling through the 5-V battery, from positive to negative, there is a voltage fall of -5 V. The complete equation for this loop is

$$-5+2I_1-3I_2=0.$$

We have two simultaneous equations that can be solved for two unknown currents. By multiplying the first equation by 2, the second equation by 5, and adding the two resulting equations, I_1 can be eliminated. The solutions are $I_2=-1$ A, and $I_1=+1$ A. Notice that I_2 is negative. This merely means that the current is going in the opposite direction from which we assumed. Since the current in the 2-Ω resistor is I_1-I_2, the current is equal to $+1-(-1)=2$ A.

The *balanced* Wheatstone bridge circuit is easy to solve because no current exists in the central branch. In Fig. 17.14 we show a circuit with the same configuration, but now there must be current in the central branch. As you can see, there is no way to arrange these branches into equivalent parallel and series resistances. In the diagram we have shown three possible loop paths along which we can apply Kirchhoff's second rule. Each loop is characterized by a current. In this way there are only three unknowns, but the net current in three of the resistors is the difference between two loop currents. The three equations are written underneath the diagram, along with the determinant method of solving a triple simultaneous equation. The equations could also be solved by using two equations at a time to reduce one unknown, but the determinant method is more elegant.

In the two examples that illustrated applications of Kirchhoff's rules, we built equations for just enough loops to solve for the unknowns. In the first case, we had two unknown currents and needed two equations, and so we chose two particular loops. In the second case, we had three unknowns and so we wrote equations for three loops. In each case it is possible to devise other closed loops around the circuit, thus providing more equations. In the first case, for example, we could have followed a path around the outside of the circuit. It turns out that these extra equations are not independent, and so in practice we choose the most convenient loops.

18
Magnetism

In Chap. 16 we described the first realm of electromagnetism in which the charges and the observers are standing still. In Chap. 17 we studied the second realm where charges move with constant velocity in circuits. In this second realm another phenomenon occurs. There is a velocity-dependent force between moving charges, producing an interaction that occurs between two currents. It is called magnetism. Until the early 1800s magnetism and electricity were considered separate phenomena. One of the great discoveries of the nineteenth century was that the change of a magnetic field produces an electric field and vice versa. Indeed, the special theory of relativity requires that electric and magnetic fields are just different manifestations of a single electromagnetic interaction.

The unification of electricity and magnetism, a process culminating in Maxwell's equations, presents teachers with a strange pedagogical problem. The natural introduction to electricity is through the study of electrostatics. The existence of an electric field is made plausible in terms of "warped space," described by force lines emanating from stationary sources. Now that we know that the electromagnetic interaction between charges is really velocity dependent, perhaps we should start immediately by exploring the forces between electric currents. The currents are the "sources." Such an approach would complicate the concept of magnetic field. Indeed, note how strange the model is for a magnetic field produced by a current. Instead of radial lines, we must hypothesize circular lines of force around the current. On the other hand, we might start instruction with the old-fashioned picture of magnetic "poles." Radial lines of force would stretch out from these static poles in complete correspondence with the electrostatic model.

The trouble is, no one has ever discovered an isolated magnetic pole. It is not for lack of trying; the search for the monopole is as old as the history of electricity and has been pursued in various ways by every generation of physicists. In 1928 Paul Dirac, one of the founders of our modern quantum theory and the man who successfully predicted the existence of antiparticles, devised a simple but rigorous argument concerning the existence of monopoles. *If even one magnetic monopole exists anywhere in the universe, then electric charge must be quantized.* The inverse of the theorem does not necessarily hold, but it is certainly provocative to note that electric charge is indeed quantized. Recent searches for the monopole have been in the field of cosmic rays and very-high-energy subatomic particles.

However, we know very well that in everyday life there are no isolated magnetic poles. Whether with currents or with permanent magnets we are stuck with dipole sources and dipole probes. For a qualitative introduction to magnetism, we could adopt the old-fashioned method of using a long magnetized knitting needle as a probe. The poles at the ends are sufficiently isolated to serve as useful single-pole probes of magnetic fields. Then the procedure of definitions and thought experiments could follow that of electrostatics. The difficulty is that there is no fundamental way to define the magnitude of the unit pole.

Alternatively, we can accept the dipole nature of the sources and use very small dipoles as probes. Small compasses arranged around a current-carrying wire line up tangentially to indicate the circular lines of B. Iron filings scattered on a plane above a magnetic configuration organize themselves into patterns representing field lines, each tiny iron needle becoming an induced dipole.

If we start out by defining magnetic fields in terms of the responses of small permanent dipole magnets, or with the approximation of a single pole at the end of a knitting needle, we will obtain the standard geometric rules for field dependence. From a "point pole" the field falls off as $1/r^2$; from a dipole the field falls off as $1/r^3$. We can measure the field patterns produced by a C magnet, plot the magnetic field of the earth, and observe how magnetic field lines crowd together in iron. Without a fundamental definition of pole, however, we cannot make quantitative links between magnetic and electric fields. Note that magnetic lines of force do not really "end" on magnetic poles or "start" from them. The poles are merely regions at the ends (and sometimes other places) of magnets. The lines of force leave a north pole region, enter a south pole region, and continue right on around. Magnetic lines of force are continuous loops.

■ THE FORCE BETWEEN PARALLEL CURRENTS

Since magnetism really is a current–current interaction, let us analyze what happens when two current-carrying wires are near each other. If the current is in the same direction in two parallel wires, there is an attractive force between the wires. When the currents are antiparallel, the wires are repelled from each other. This is strictly a velocity-dependent effect of the charges. Although the wires are filled with electric charges, no region has a surplus of negative or positive charge. The negative electrons are simply drifting past the positive ions.

The actual magnitude of this force between parallel currents and its dependence on the distance between the wires can be measured with a simple pan balance arrangement. This is a fundamental experiment in determining the standard unit of current—the ampere. That definition, in turn, determines the coulomb. Experiments show that the force between parallel wires is inversely proportional to the distance between the centers of the wires: $F \propto 1/r$. Since the force must depend on an influence spreading from one wire to the other, it is not surprising that the *cylindrical* geometry produces a force dependent on the inverse first power of the distance. The electrostatic field also spreads out from a charged wire with a $1/r$ dependence on distance.

Experiments also show that the force of interaction between the wires depends on the product of their currents. From symmetry we would expect that, if the force were proportional to I_1, then it would also be proportional to I_2. That the force is directly proportional to each current is simply an experimental fact, although at the end of this chapter we will show another way of justifying the fact.

By inserting a proportionality constant, we get our basic formula for the force per meter between two parallel currents:

$$\frac{F}{l} = \frac{\mu_0}{2\pi} \frac{I_1 I_2}{r}.$$

The force per meter, F/l, is measured in newtons per meter. The value of μ_0 is

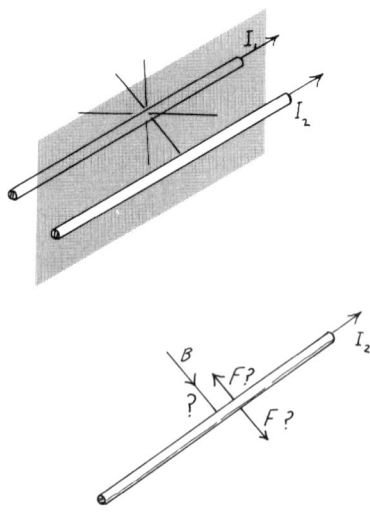

Figure 18.1. Parallel currents produce forces that attract the wires. But, if the magnetic field lines were radial, how would current #2 know whether current #1 is parallel or antiparallel?

chosen to be exactly $4\pi \times 10^{-7}$, thus defining the size of the ampere. The constant μ_0 is called the *permeability of free space*. It is analogous to ϵ_0, the electrostatic permittivity of free space. There is an operational difference, however, in assigning values to these two constants. We can choose one or the other to be any value we like, but then the other must be determined experimentally, since the coulomb and ampere are linked. In the SI (Système International) we choose μ_0, then measure ϵ_0. As we will see at the end of this chapter μ_0 and ϵ_0 are also related by the speed of light.

Our basic equation gives the attractive force per meter between two parallel currents. The same formula gives the repulsive force between two antiparallel currents. If the wires are perpendicular to each other, there is of course only a very small region of influence where the wires are close together, and so we might expect that the force between the wires would be small. As a matter of fact, the force is 0. Since the force could be described as positive if the currents are parallel, and negative if the currents are antiparallel, it is plausible that, when the wires are perpendicular, the force should be 0, which is right in the middle—between positive and negative. Note that we cannot write any simple formula for the force between two wires with an arbitrary angle between their directions. Evidently some vector relationship exists, but as we change the angle we also change the proximity relationships. The solution to this problem lies in defining a magnetic field produced by one current, and then finding the force experienced by the other current in that field.

We propose that *moving* charges produce *magnetic* fields; this is analogous to our approach in defining the electric field, except then we used stationary charges. The strength and direction of the magnetic field can be calculated in terms of the force on a unit test *current*. In the parallel-wire interaction, one of the wires could serve as the testing device to measure the magnetic field produced by the other wire. The magnetic field at a point, which is given the symbol B, would then be measured as the *force per meter on a unit current* at that point. The units of B are N/A m.

Now we have a problem. If we describe magnetic field in terms of field lines, what is the *direction* of the magnetic field produced by a wire? (Remember that the field is a human-made model, which we invent for our convenience.) If we were to make the field lines radial, like those from an electrostatically charged wire, then they would be perpendicular to the parallel-unit test current. Remember that the force on that test current can be positive or negative or 0, depending on its orientation with respect to the source wire. In all cases, however, the direction of the force on the test wire must be along the radial line that joins the two wires. As shown in Fig. 18.1, there would be no way, by just using the orientation of the radial field lines and test current, to predict the direction or size of the force. Such a situation is not good. We must devise a way of portraying the magnetic field so that its interaction with a test current will yield a force in the proper, *unique* direction.

■ THE MAGNETIC FIELD PRODUCED BY A CURRENT— QUALITATIVE ASSESSMENT

Faraday's mental images of elastic lines of force, shoving against each other and stretching from pole to pole, are powerful guides for predicting the behavior of currents and magnets. In Fig. 18.2 we sketch several of these field patterns. In

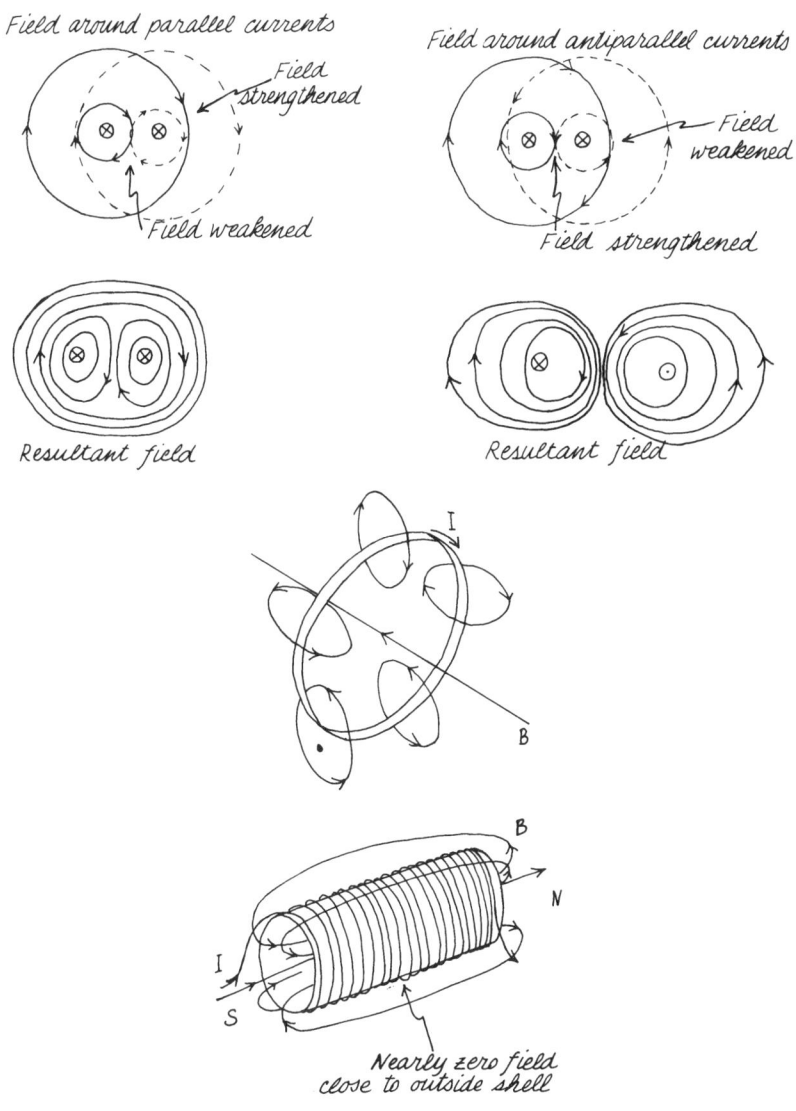

Figure 18.2. Magnetic field lines for various source current geometries.

all cases we appeal to the right-hand rule to determine the direction of field lines produced by currents. With the thumb of your right hand in the direction of the current, your fingers curl in the direction of the magnetic lines of force. To this rule we add these constraints: all lines are closed loops; no lines can cross each other; crowded lines produce higher pressure than scarce lines. *These simple rules are all you need.* They can take the place of any more complicated rule about pointing your fingers in the direction of the current, curling them into the field, and finding that your thumb is in the direction of the force.

[As for crowded lines yielding higher pressure, we should be aware (though we need not tell our students at this point) that magnetic field energy density is $u = \frac{1}{2}(B^2/\mu_0)$ J/m^3, which is also equal to the pressure in N/m^2. A solenoid with an iron core may have an internal field of 1 T (10 000 G). The pressure against the solenoid walls, tending to make it explode, is $P = \frac{1}{2}(1\text{ T})^2/(4\pi \times 10^{-7}) = 4 \times 10^5$ N/m^2 = 4 atm.]

■ QUANTITATIVE CALCULATIONS OF MAGNETIC FIELD

Ampère's Law

There are two laws relating magnetic field strength to current, even as there are two different forms of laws for the electrostatic field. (Coulomb's law gives the force between two point charges; Gauss's law relates the number of electric field lines coming out of a surface to the enclosed electric charge.) Let us start out with the magnetic equivalent of Gauss's law and relate the magnetic field lines to the current that produces them.

Ampère's law:

$$\oint \mathbf{B} \cdot d\mathbf{l} = \mu_0 i.$$

The dot product of the magnetic field strength **B** and the path distance **l** integrated along a *closed loop* is equal to the product of the permeability and the current encircled by the loop. The equation is true for any loop surrounding a current, regardless of where the current is located within the loop. If the magnetic field is not constant around the loop or changes direction with respect to the path, then the dot product must be taken segment by segment to form the integral. In such a case it may be impractical to calculate the integral, but the law still holds.

As a practical matter, Ampère's law is applied only where symmetry provides constant magnetic field values and constant orientation with the path in order to make the dot product simple. The prime case is for the field due to a long straight wire. Take the Ampèrian loop as a concentric circle around the wire. Because of symmetry the field must have the same value at any point around the circumference of the circle with radius r. Furthermore, the circumferential field is always in the same direction as the circumferential path elements, so the dot product of each segment is simply $B\,dl$. Adding up the path lengths around the circle yields $2\pi r$. Hence Ampère's law gives us the formula for the field due to a long straight wire:

$$B 2\pi r = \mu_0 i \quad \text{or} \quad B = \frac{\mu_0 i}{2\pi r} = 2 \times 10^{-7} \frac{i}{r}.$$

Note that Ampère's law would also apply if we were to take the integration path around the wire in the form of a square. However, the field would be different point by point, and the angle for the dot product would continuously change.

The units for magnetic field are those that we deduced from the interaction between two parallel currents.

$$B = (\text{force})/[(\text{current})(\text{length})] = \text{N/A m}.$$

The newton per ampere-meter has a proper name; it is called a *tesla* in honor of Nikola Tesla (1856–1943). A tesla (T) is a large unit of magnetic field. A smaller unit is the gauss (G). 1 T = 10 000 G.

Let's calculate the magnetic field strength 1 cm away from a wire carrying 10 A:

TABLE 1. *Range of measurable magnetic fields.*

Source	Strength (T)
Brain's α rhythm currents	10^{-15}
Currents triggering heartbeat	10^{-14}
Typical TV signal	10^{-11}
Light from 100-W bulb at 3 m	10^{-8}
1 m from long wire carrying 1 A	2×10^{-7}
Earth's surface	10^{-4}
Between jaws of toy permanent magnet	10^{-2}
Beam-focusing research magnet, or electric motor	1
Superconducting research magnets	10^1
At atomic nucleus from valence electrons	10^2
Laboratory implosion of trapped fields	10^3
Surface of neutron star	10^8

$$B = 2 \times 10^{-7} \frac{10 \text{ A}}{1 \times 10^{-2} \text{ m}} = 2 \times 10^{-4} \text{ T} = 2 \text{ G}.$$

The strength of the earth's magnetic field varies, depending on the location on earth, but in most of the United States it is about $\frac{1}{2}$ G. The further north you go, the more steeply the earth's field points downward. The *horizontal* component of the earth's field in the northern United States is only about $\frac{1}{3}$ G. In Table 1 we give values of magnetic field for various natural and technological conditions.

The Biot–Savart Law

Ampère's law is concerned with the magnetic field along a closed loop surrounding a current. The other law for magnetic field describes the field at just one point. It is the counterpoint of Coulomb's law for electrostatics. It is called the Biot–Savart law, named after Jean Biot (1774–1862) and Felix Savart (1791–1841). There are two slightly different forms of the law, one for the field produced by a small segment of current in a wire and the other for the field caused by an isolated moving charge.

$$dB = \frac{\mu_0}{4\pi} \frac{i dl \sin \varphi}{r^2} \quad \text{(field from current segment } dl),$$

$$B = \frac{\mu_0}{4\pi} q \frac{v \sin \varphi}{r^2} \quad \text{(field from moving charge } q).$$

In the first form of the law, the current in the small segment, dl, provides a contribution to the field, dB at a particular point that is at a distance r from the segment. Each segment of the wire yields its own contribution, and the net field at the point is the vector sum of all these contributions. In the second form of the law, there is only one charge q and so it provides the only field there is at the distant point. The geometry on which these laws are based is shown in Fig. 18.3. Note that the field depends on the inverse *square* of the distance between the point and the source. The field from a long straight wire depends on the inverse *first* power of the distance to the wire. In the Biot–Savart geometry, however, the source of the field is a pointlike segment on the wire.

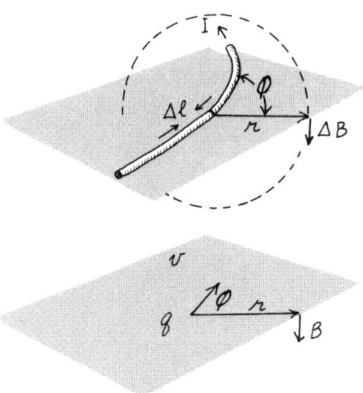

Figure 18.3. *Geometry of the magnetic field due to a current or a moving electric charge.*

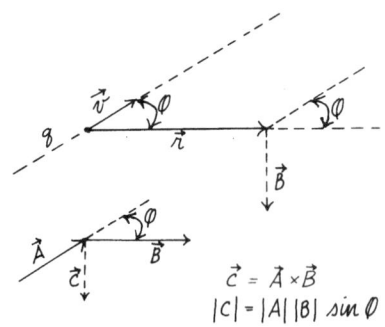

Figure 18.4. Geometry for Biot-Savart Formula.

In both forms of the law, the strength of the field is proportional to the current, since a moving charge is a current. The field also depends on the sine of the angle φ between the velocity of the charge and the line between the point and the charge. The field produced at the point is a maximum when the line from point to current is perpendicular to the direction of the current or the charge velocity. The magnetic field produced at the point is a vector and is perpendicular to the plane defined by the current direction and the line from point to moving charge.

Another way to represent the Biot–Savart law is in terms of a vector product. In Fig. 18.4 we show the geometry for this product:

$$\mathbf{B} = \frac{\mu_0}{4\pi} q \, \frac{\mathbf{v} \times \mathbf{r}}{r^3}.$$

At first you might think that we have turned the Biot–Savart law into one that depends on the inverse *cube* of the distance between point and charge. However, this is not the case. There is an extra r in the numerator, representing the radius vector from the measuring point to the moving charge. The r is there to represent the vector direction, but, since it also has magnitude, we must cancel it out by putting one more power of r in the denominator. Some texts finesse this problem by using unit-vector notation. In our own experience, unit-vector notation is unnecessary in any development normally tackled in introductory physics at any level, and most students find it forbidding.

Suppose that you could measure the magnetic field from a single proton as it went past you. Consider how its strength would change as a function of its distance from you. As shown in Fig. 18.5, there are three reasons why the magnetic field would be extremely weak when the proton is far away. First, the charge on the proton is very small. Second, the inverse square of a large distance makes the field small, and third, the sine of the angle between the velocity of the proton and the line toward you would also be very small until the proton is almost opposite you. If the proton heads straight for you, that angle remains 0. If the proton is passing by, however, there will be an instant when the radius vector from you to the proton is perpendicular to its velocity. At that moment, $\sin \varphi$ has its maximum value, and the separation distance has its minimum value. Hence the field would be a maximum. But whether strong or weak, the *direction* of the magnetic field at your position does not change. It is perpendicular to the horizontal plane that contains the radius vector to the proton and its velocity vector. In the case shown in the diagram, the magnetic field at your position would be directed downward.

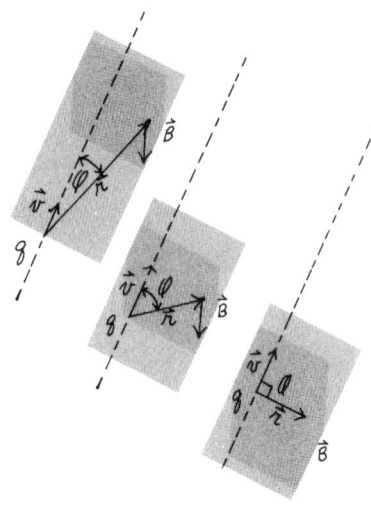

Figure 18.5. Change in strength of B as charged particle passes by.

Let us see whether it is possible to detect the magnetic field that is produced by a single proton as it goes past you at a very high velocity. We will make the measurement when φ equals $90°$ and r equals 1×10^{-3} m. The proton might be coming down a very small pipe, for instance, and we might try to sense its magnetic presence with a small loop around the pipe. If the proton is traveling with one-third the velocity of light, then the magnitude of the magnetic field that it produces is approximately (using our nonrelativistic formula):

$$B = \frac{4\pi \times 10^{-7}}{4\pi} (1.6 \times 10^{-19} \text{ C}) \frac{(1 \times 10^8 \text{ m/s})(1)}{(1 \times 10^{-3} \text{ m})^2} = 1.6 \times 10^{-15} \text{ T} = 1.6 \times 10^{-11} \text{ G}.$$

The magnetic field produced by a single proton under these conditions evidently is very small. It is on the borderline of detectability. But magnetic detection methods have been used for *beams* of protons, where there may be as few as 10^6

protons passing through the pickup ring at any one time.

We have two laws—Ampère's and the Biot–Savart law—describing the same phenomenon. Do they always predict the same results? Can we, for instance, use the Biot–Savart law to calculate the magnetic field near a current in a long straight wire? We already know the result given by Ampère's law. In Fig. 18.6 we show how such a calculation can be made by integrating the magnetic-field contributions of each little segment of the long wire. It is a standard calculus derivation found in most college-level physics texts. Its main usefulness is as a math integration exercise, or as a case study to contrast the calculus derivation with the simple power of Ampère's law when symmetry exists.

The Biot–Savart law is easy to use with the geometry of a current loop in order to find the resulting field on its axis. Each segment of current in the loop yields the same contribution to the total field at the center of the loop. Furthermore, at every point along the circumference, the current is perpendicular to the radius vector, making the angular contribution of the vector product always unity. The field contributions, dB are all in the same direction along the axis, and so they simply add up. The contribution to the field by any one segment is

$$dB = \frac{\mu_0}{4\pi} \frac{dl \sin\varphi}{r^2} = \frac{\mu_0}{4\pi} \frac{i\,dl}{r^2} \quad \text{since } \varphi = 90°.$$

The sum of all the dl segments is equal to the circumference. Therefore the resultant field is

$$B = \frac{\mu_0 i}{2r}.$$

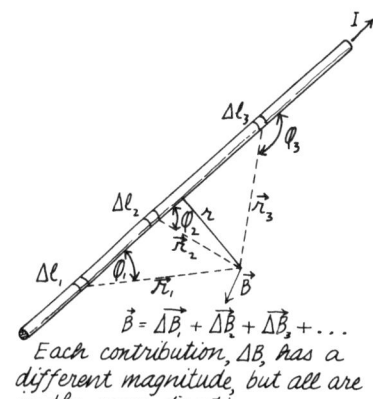

Figure 18.6. Calculating the magnetic field at a distance r from a long straight current, using the Biot-Savart law.

Notice that in this case the geometry and different symmetry dictate the use of Biot–Savart law rather than Ampère law. While an Ampèrian loop could be formed around the circular current, threading back down through the center of the current loop, the magnitude and direction of the field on the loop would be continually changing, making the evaluation of the integral impractical.

Let us see how strong a magnetic field we can get at the center of a small loop of wire. If we have a single turn with a radius of 10 cm carrying a current of 10 A, then the field at the center is equal to

$$B = \frac{4\pi \times 10^{-7}}{2} \frac{10 \text{ A}}{1 \times 10^{-1} \text{ m}} = 2\pi \times 10^{-5} \text{ T} \approx 0.6 \text{ G}$$

This is about the same strength as the earth's field.

Using the Biot–Savart law it is also easy to calculate the magnetic field on the axis of a current loop, but at a considerable distance from the loop. The simplicity of the calculation arises if the distance to the center of the loop r is much larger than the radius of the loop R. The geometry of the calculation is shown in Fig. 18.7. The contribution to the field of any one segment is

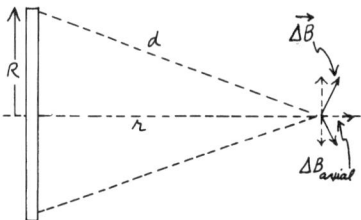

Figure 18.7. Calculation of the magnetic field on the axis of a circular current of radius R.

$$dB = \frac{\mu_0}{4\pi} \frac{i\,dl}{d^2} \approx \frac{\mu_0}{4\pi} \frac{i\,dl}{r^2} \quad \text{for } r \gg R.$$

The magnetic field of each segment is perpendicular to the line d. Only a small component of that field is axial. The large component perpendicular to the axis is exactly canceled out by the field contribution from the small segment on the opposite side of the loop. This cancellation of the perpendicular components occurs for all segments of the loop. Only the *axial* components reinforce each

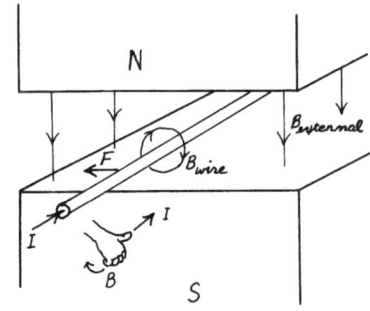

Figure 18.8. Right hand rule for finding force on a current carrying wire in a magnetic field.

other. The axial component of each field contribution dB is

$$dB_{\text{axial}} = dB\,\frac{R}{d} \approx dB\,\frac{R}{r} = \frac{\mu_0}{4\pi}\,R\,\frac{idl}{r^3}.$$

The total field due to all the individual contributions is:

$$B = \frac{\mu_0 iR}{4\pi r^3}\,2\pi R = \frac{\mu_0}{4\pi}\,\frac{2}{r^3}\,(\pi R^2 i)$$

Compare this formula with the expression for the electric field on the axis of an electrostatic dipole: $E = k2p/r^3$, where $p = (ql)$ is the electric-dipole moment. The field due to a dipole is inversely proportional to the cube of the distance from the dipole. As you can see, the magnetic field on the axis of a current loop has this same functional dependence on distance. The similarity of the formulas is complete if the magnetic dipole moment of the loop has the value $(\pi R^2 i)$. A current loop acts like a magnetic dipole with a magnetic moment (or dipole "charge" strength) equal to the product of the area of the loop (πR^2) and the current in it. This relationship is of prime importance when we study the cause of magnetism in solid materials.

■ QUANTITATIVE CALCULATION OF THE FORCE ON A CURRENT IN A MAGNETIC FIELD

When we started analyzing the magnetic force between two parallel currents, we claimed that the equation deduced from experiments is

$$\frac{F}{l} = \frac{\mu_0}{2\pi}\,\frac{i_1 i_2}{r}.$$

At that point in our development we proposed that we create a model of a magnetic field produced by one current with which the second one could interact. Now we have a formula for the magnetic field produced by a long, straight wire carrying current i. It is

$$B = \frac{\mu_0}{2\pi}\,\frac{i}{r}.$$

The field formula is evidently just part of our original formula for the force between two wires. We can now write that formula as follows: $F/l = i_2 B$. The force experienced by a current that is *perpendicular* to a magnetic field is $F = ilB$. The quantitative way to describe this geometrical dependence is to use a vector product. The force on a current is a vector, and it is proportional to the product of two other vectors, **B** and **l**. The complete formula for the force on a current in a magnetic field is

$$\mathbf{F} = i(\mathbf{l} \times \mathbf{B}).$$

The geometry of this vector product is shown in Fig. 18.8. The force must be perpendicular to both the magnetic field and the wire.

Let's calculate the force on a single wire that is 20 cm long and is carrying 6 A in a field of 1.5 T. This is a typical situation in an electric motor. The force is $F = ilB = (6\text{ A})(0.2\text{ m})(1.5\text{ T}) = 1.8\text{ N}$. That force by itself is not large, but a motor contains many such windings.

The force on a moving charge in a magnetic field can be deduced from the formula for the force on a current since the current is the time rate of passage of charge: $i = dq/dt$. The magnetic force on a moving charge is

$$\mathbf{F} = q(\mathbf{v} \times \mathbf{B}).$$

Since the magnetic force on a moving charge is always perpendicular to its velocity, it may cause the charged particle to travel in a circle. The magnetic force would then provide the necessary centripetal force:

$$qvB = \frac{mv^2}{r}$$

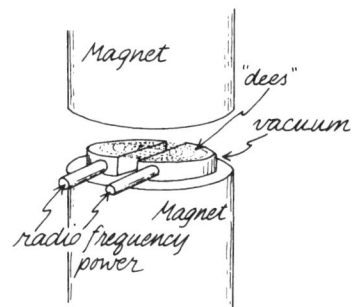

Figure 18.9. Schematic diagram of main parts of a cyclotron.

This equation is the basis for many different phenomena and instruments in the study of subatomic physics. For instance, if we cancel a velocity term on each side of the equation, we find that the momentum of the particle is equal to $mv = qrB$. If we can see or measure the track of a charged particle in a magnetic field, we can determine its momentum by measuring the radius of its track.

Let us rearrange the terms in the centripetal force equation to find the angular speed ω, which is equal to v/r:

$$\omega = \frac{v}{r} = \frac{q}{m} B.$$

A charged particle going in a circle in a magnetic field has an angular frequency that depends only on its charge-to-mass ratio (q/m) and the strength of the magnetic field. The frequency is independent of the energy of the particle or of the radius of its orbit. The orbital frequency does not depend on the velocity of the particle or the radius of curvature of its path because the velocity and radius are proportional to each other. If a particle has a large velocity, the radius of curvature of its path will also be large, and so the time that it takes to go once around its circular path is the same as if its velocity and radius were small.

The cyclotron is a particle accelerator that depends on the principle that the angular frequency of a charged particle in a magnetic field is independent of its velocity or radius. The "cyclotron frequency" is

$$f = \frac{\omega}{2\pi} = \frac{1}{2\pi} \frac{q}{m} B.$$

Let us calculate the frequency with which a proton moves in a circle when it is in a typical cyclotron magnetic field of 1.5 T:

$$f = \frac{1}{2\pi} \frac{1.6 \times 10^{-19} \text{ C}}{1.7 \times 10^{-27} \text{ kg}} (1.5 \text{ T}) = 2.3 \times 10^7 \text{ Hz}.$$

This frequency is midway between that used by AM radio (1 megacycle per second or 1×10^6 Hz) and the FM range (90×10^6 Hz).

A diagram of the working parts of a cyclotron is shown in Fig. 18.9. There is a vacuum chamber in the gap of the magnet. A very small amount of hydrogen gas is continuously leaked into the chamber at the center. The gas is bombarded by electrons and ionized so that raw protons as well as molecular ions are available. Meanwhile, the two dee's are being driven at the cyclotron frequency by what amounts to a radio station just outside the machine. When a positive proton is formed, it will be attracted toward the negative dee. In the magnetic field, its path

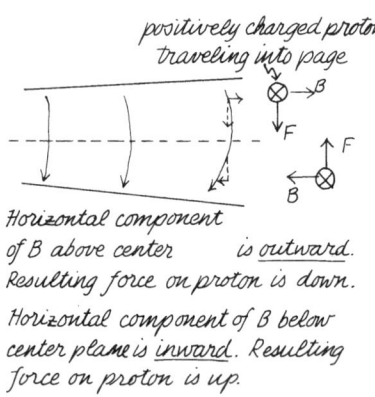

Horizontal component of B above center is outward. Resulting force on proton is down.

Horizontal component of B below center plane is inward. Resulting force on proton is up.

Figure 18.10. Focusing forces on proton in slightly non-uniform magnetic field of cyclotron.

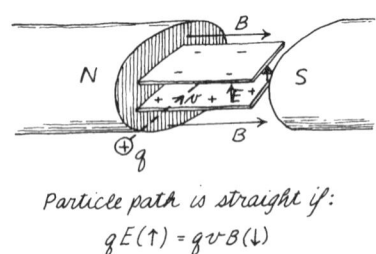

Particle path is straight if:
$qE(\uparrow) = qvB(\downarrow)$

Figure 18.11. Particle velocity selector with crossed electric and magnetic fields.

will be a circle. While it is inside a dee, it is electrostatically shielded from any electric-field lines (but not the magnetic-field lines). The proton simply coasts around a semicircle until it emerges across the gap between the two dee's. If, at that instant, the other dee is negative, the proton will be accelerated across the gap, gaining energy. The radius of its orbit will now be larger. But once inside a dee, the proton is unaffected by the changing electric field and will coast in a semicircle until it comes to the gap again. If the potential across the gap is changing at the cyclotron frequency, the proton will always arrive at the gap in time to see negative regions facing it, and so it will be accelerated. Its energy increases at each passage by an amount equal to the product of its charge and the potential difference across the gap. The radius of its orbit keeps increasing, along with its energy, until finally it reaches the outside edge of the vacuum chamber. There it either hits a target or is led toward external targets or detectors.

In a typical cyclotron each proton picks up about 10 000 eV during each revolution. After a thousand revolutions, the proton energy is 10 MeV. Cyclotrons of this size were common research instruments during the late 1930s. There is a problem about making machines of this type for yet higher energies. As the proton speed increases, its effective mass increases. That will change the cyclotron frequency. Even when the fractional change is small, the cumulative effect will ruin the resonance condition after many revolutions. A variety of solutions to this relativistic problem exists, including frequency-modulated oscillators that escort a group of protons from small to large radius with gradually decreasing frequency, and magnets that are contoured to compensate for the increase of effective mass. Several such cyclotrons have been built in recent years.

There is a crucial design problem for cyclotrons, or any particle accelerator. The original cyclotron designed by Lawrence did not work at first because the problem had not been recognized. The route of the charged particles must contain magnetic or electric fields that keep the particles from leaving the prescribed orbit. There must be "focusing." For an example of the problem that is dramatic but turns out to be trivial, consider that a proton beam takes about 1 s to be accelerated in some of our large synchrotrons. During the first second, a proton, like anything else in the earth's gravitational field, would fall 5 m. Clearly there must be some focusing force to prevent the protons from falling. Actually, there are far larger forces tending to bump the protons up, down, or sideways. The focusing forces must be strong enough to keep the beam in a narrow bath. Figure 18.10 illustrates how these focusing forces can be produced. It is essential that the magnetic field *not* be uniformly vertical and constant in strength. Instead, the magnetic field must be slightly weaker toward the larger radius, so that the field lines are slightly curved as shown. If a proton gets above the horizontal plane, it will then interact with a small horizontal component of the magnetic field that will produce a downward force on the proton. If the proton is below the equilibrium plane, this horizontal component of the field is in the opposite direction and so will produce an upward force on the proton. In any kind of particle accelerator, including the electron accelerators in your dentist's x-ray machine and the electron beam in your television tube, electric and magnetic fields are designed to focus the beam in much the same manner that glass lenses focus beams of light.

One useful combination of electric and magnetic fields provides a way to select particles with a specific velocity. Send a beam of particles through crossed electric and magnetic fields as shown in Fig. 18.11. With the orientation of the fields as shown, the electrostatic force is directed upward and the magnetic force downward. If the two forces are just equal in magnitude, the particle will continue in a straight line. The balanced forces on the particle are $qE = qvB$. Only particles with a specific velocity can satisfy these condition. Regardless of its charge or mass, a particle will continue in a straight line if $v = E/B$.

If electric and magnetic fields are crossed in a *conductor*, the charge carriers will drift in the general direction determined by the electric field, but will also experience a weak force due to the magnetic field, tending to make them concentrate along one face of the conductor. The geometry for negative charge carriers is shown in Fig. 18.12. The extra concentration of charges along the top face produces a slight difference of potential across the thickness of the conductor. The sign of this potential difference depends on whether the charge carriers are positive or negative. If negative charge carriers drifting to the right experience a magnetic force that makes them turn up so that they concentrate along the top, then positive charge carriers drifting to the *left* in the same electric field would experience a magnetic force tending to make them turn upward also. Therefore, if the charge carriers are negative as they are in metals, the top face of the bar would be *negative* with respect to the bottom face. If the charge carriers are positive as they are in some semiconductors, then the top face of the conductor would be *positive* with respect to the bottom face. The magnitude of the potential difference is also a function of the density of charge carriers and hence of the number of charge carriers contributed per atom. This phenomenon and experimental technique is called the *Hall effect*, after its discoverer, Edwin Hall (1855–1938).

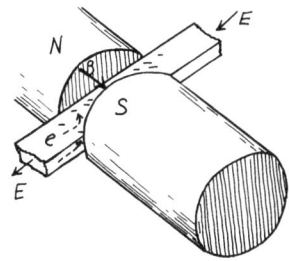

Figure 18.12. Hall effect for charges moving in a conductor perpendicular to a magnetic field.

Earlier we said that the path of a charged particle moving with constant speed in a uniform magnetic field is a circle. Actually, that is the case only if the original velocity of the particle is in the plane perpendicular to the magnetic field. If there is also a component of velocity along the magnetic field lines, then the particle will spiral in a helix, as shown in Fig. 18.13. A surprising situation arises if the magnetic field lines are curved. As you can see in Fig. 18.14, a change in the direction of the field line implies that there is a component of the magnetic field perpendicular to the original direction. The charged particle will react to that component in the way shown in the diagram. The resulting forces change the direction of the spiraling orbit so that it follows the magnetic field lines. It is this effect that makes protons and electrons spiral along the magnetic field lines of the earth far beyond the atmosphere. These regions are called the *Van Allen radiation belts*. Except for one other effect, the spiraling particles would come into the atmosphere along with the magnetic field lines near the regions of the north and south magnetic poles. As a matter of fact, many particles do come in along those lines, causing the auroras seen in northern and southern latitudes. But many other particles lie within momentum ranges such that they are reflected from the regions of higher magnetic fields near the poles before they enter the atmosphere. When a particle trapped in this way enters the region of a stronger field near the poles, the pitch of its helical path is reduced and then reversed, so that it starts spiraling back. Particles in such orbits remain trapped in the magnetic fields for many traversals. Eventually they get scattered out through collisions with other particles. But their numbers are continually replenished by protons and electrons shot out from the sun.

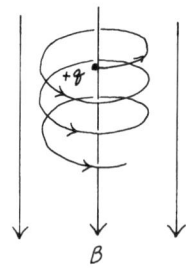

Figure 18.13. Spiral path of particle with charge $+q$ in a magnetic field and with an initial component of velocity downward.

■ THE TORQUE ON A WIRE LOOP

In a uniform magnetic field, a magnetic dipole such as the compass needle does not experience any net force. The attraction in one direction on the north pole is

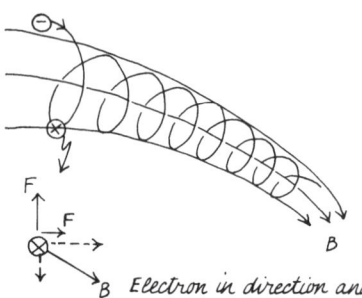

Electron in direction and position shown will experience upward centripetal force from main horizontal component of B, and also a horizontal force to the right from the small vertical component of B.

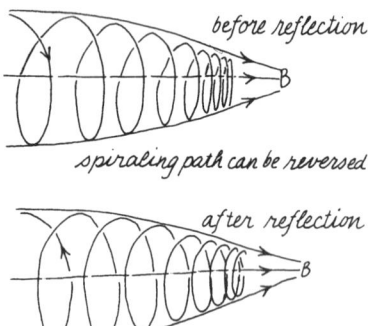

Figure 18.14. Spiraling movement of electrons trapped in magnetic field lines around Earth.

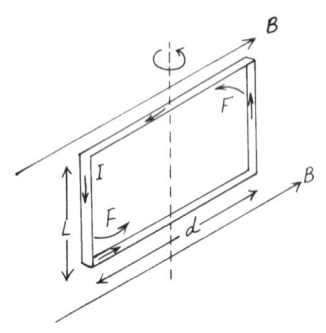

Figure 18.15a. Torque produced on a rectangular current loop in a magnetic field with the plane of the loop lying in the direction of the magnetic field.

balanced by the attraction in the other direction on the south pole. There is, however, a *torque* on the dipole. In analogy with the formula for the strength of an electric dipole, the *magnetic moment* of a compass needle could be defined as **m**=lp—if we knew the "pole strength" p and could measure the distance between poles, l. Actually, we can assign a meaning and a value to the magnetic moment by measuring the torque on a bar magnet in a known magnetic field and solving for **m**. The magnitude of such a torque is equal to $\tau = mB \sin\theta$. Note that the torque is maximum when the needle is perpendicular to the field lines and is 0 when the needle is aligned.

Although in the case of the compass needle we have defined the dipole moment in terms of the product of a pole strength and length of the needle, the experimentally measured quantity is actually just the magnetic moment itself. The poles cannot be isolated and, as we have seen, exist in a spread out region so that their separation distance is not a definite quantity. The general and most accurate expression for the torque on a magnetic dipole is:

$$\tau = \mathbf{m} \times \mathbf{B}.$$

The torque, which is a vector, is the vector product of the magnetic moment and the magnetic field. Because of the nature of the vector product, the magnitude of the torque is proportional to the sine of the angle between the magnetic moment and the field. Furthermore, the direction of the torque is along the axis of the dipole, in the sense given by the right-hand rule.

Since we know that magnetic *poles* do not exist, how do we explain the existence of magnetic *dipoles*? We have already seen that the magnetic field produced on its axis by a current loop depends on the inverse third power of the distance from the loop. The field on the axis of a dipole has that same functional dependence on the distance. In fact, the formulas for the loop and the dipole are the same if we assign the value of iA to the magnetic moment of the loop. Now let us study the other side of the interaction. Does a current loop in a uniform magnetic field behave like a magnetic dipole?

The easiest way to calculate the effect of a magnetic field on a current loop is to make the loop rectangular, as shown in Fig. 18.15(a). When the magnetic field lines lie in the plane of the rectangle, there is no force on the top current or the bottom current, since these are parallel or antiparallel to the field lines. The force on the left wire is out of the paper and the force on the right wire is into the paper, thus creating a torque:

$$\tau = 2(\text{force on each vertical wire})(\text{lever arm}) = 2(BiL)(d/2) = BiA.$$

As the rectangular coil turns, there are nonzero forces on the top and bottom wires. These forces are in opposite directions and tend to distort the coil but not to turn it. Meanwhile, the force on the side wires is always perpendicular to the current and to the field. The component of the force that is perpendicular to the lever arm is proportional to $\sin\theta$, as shown in Fig. 18.15(b). Therefore, the torque goes to 0 when θ goes to 0, and the coil is lined up with its face perpendicular to the field lines. If we assign a vector direction to area, then the torque on the rectangular current loop can be expressed as a vector product:

$$\tau = i(\mathbf{A} \times \mathbf{B}).$$

This expression for the torque on the special geometry of a rectangular current loop can be generalized if we define the magnetic moment of *any* plane loop to be equal to the product of the current and its area. This definition agrees with the one that we have already used for the magnetic field *produced* by a current loop. In Fig. 18.15(c) we show some justification for making the generalization from a rectangular loop to any current loop. The torque on any such loop is:

$$\tau = \mathbf{m} \times \mathbf{B}, \quad \text{where} \quad \mathbf{m} = i\mathbf{A}.$$

A current loop produces a magnetic field that behaves like the magnetic field of a dipole. A current loop also experiences a torque in a magnetic field as if it were a dipole. Let us consider one other feature of the behavior of dipoles in fields. When a compass needle or a current loop lines up in a magnetic field, its potential energy with respect to the field is a minimum. How much energy does it take to twist the dipole 90° to where the torque is maximum, or to 180° where the dipole is antiparallel? The work done is simply $\int_0^\theta \tau \, d\theta$, where $\tau = mB \sin\theta$. The work done to twist the dipole from 0° to 90° is $\mathbf{m}B$, and from 0° to 180° is $2\mathbf{m}B$. Under these circumstances, we can define the potential energy of a dipole in a magnetic field to be:

$$U = -\mathbf{m} \cdot \mathbf{B}$$

According to this definition, the energy of the dipole is zero when the axis of the loop is perpendicular to the magnetic field. (We can, of course, define the 0 of potential energy at any position we wish.) In the perpendicular direction, the torque on the dipole is maximum. When the dipole is lined up with the field, $\theta = 0°$, and the potential is negative. When the dipole is antiparallel to the field, $\theta = 180°$, and the potential energy is positive.

Let us find the torque and the potential energy of an actual coil in a magnetic field. Can we, for example, send current from a D cell into a coil and see it act like a compass in the earth's magnetic field? If the coil has a diameter of 4 cm, it has an area of the order of 10 cm². Make a coil of 10 turns and connect it across a D cell, obtaining a short circuit current of about 5 A. Each turn has a magnetic moment equal to iA. The magnetic moment of a coil with N turns is therefore $m = NiA = (10)(5 \text{ A})(1 \times 10^{-3} \text{ m}^2) = 5 \times 10^{-2} \text{ A m}^2$. In the horizontal magnetic field of the earth, the maximum torque on this coil would be $\tau = mB \sin 90° = (5 \times 10^{-2})(\frac{1}{5} \times 10^{-4} \text{ T})(1) = 1 \times 10^{-6}$ N m. That's a very small torque.

The deflection of such a coil in the earth's field could be detected only if the coil were suspended by a very thin and flexible wire. The potential energy difference between the in-line direction and the 90° orientation is $\Delta U = mB = (5 \times 10^{-2})(\frac{1}{5} \times 10^{-4}) = 1 \times 10^{-6}$ J. That's the amount of energy required to lift 0.1 g through 1 mm in the earth's gravitational field. The magnetic effectiveness of such a coil can be increased by a factor of 1000 or so by winding the coil on an iron core.

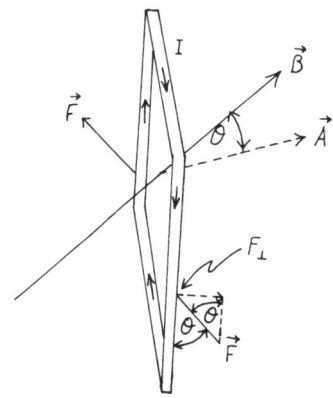

Figure 18.15b. Torque produced on a rectangular current loop in a magnetic field with the plane of the loop at an angle to the field.

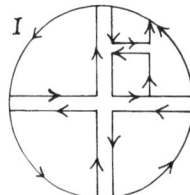

Figure 18.15c. Any current loop can be approximated as an assemblage of rectangular loops. The interior currents cancel leaving only the effects on the boundary currents. The magnetic moment of any plane loop is $m = iA$.

■ MAGNETISM IN MATERIALS

We have claimed that there are no such things as magnetic poles. Current loops or solenoids can duplicate the effects of permanent bar magnets, both in the fields they produce and in the torques that they experience. Must we therefore conclude that there are perpetual electric currents in iron magnets that circulate in such a way that we have coils or solenoids without power sources? If this is the case, why are these currents established in iron and a few other elements but not in most materials? Once again we must appeal to a model of the microstructure in order to explain a natural phenomenon.

Perpetual currents do exist in material. The electrons in each atom are in motion around the central nucleus. The old model of electrons that pictures them orbiting the nucleus, like planets around the sun, has a limited usefulness and is completely wrong in a number of ways. Nevertheless, whenever an electron motion can be characterized as having angular momentum, it seems reasonable to expect that a magnetic moment should be produced. Let us calculate the relationship between angular momentum of a charged particle and the magnetic moment that we might expect.

The magnetic moment of a current loop is $m = iA$. If an electron with charge e is going in a circular path with a period T, then the current produced is equal to e/T. The period of rotation is equal to the circumference divided by the velocity: $T = 2\pi r/v$. The magnetic moment produced must therefore be equal to

$$iA = \frac{evA}{2\pi r} = \frac{ev\pi r^2}{2\pi r} = \frac{evr}{2}.$$

The angular momentum of the electron about the axis is equal to $L = mvr$. If we multiply both numerator and denominator of the expression for the magnetic moment by the mass of the electron, we end up with a formula for the magnetic moment containing the angular momentum:

$$m = \frac{emvr}{2m} = \frac{e}{2m}L, \quad \text{where} \quad L = mvr.$$

Atomic electrons have angular momentum due to two different properties: one corresponds to the orbital motion around the central nucleus and the other is due to the apparent spin of the electron on its own axis. Both types of angular momentum are quantized; that is, the angular momentum can change only in integral multiples of a basic unit of angular momentum. The basic unit of angular momentum is Planck's constant, which has the value (in terms of rad/s) of $\hbar = 1.054 \times 10^{-34}$ kg m²/s. Although some orbital motion of an electron might produce several units of angular momentum, the *order of magnitude* of magnetic moment produced by any electron due either to its spin or its orbital motion will be:

$$m = \frac{e}{2m}L = \frac{(1.6 \times 10^{-19} \text{ C})}{2(9 \times 10^{-31} \text{ kg})}(1 \times 10^{-34} \text{ kg} \cdot \text{m}^2/\text{s}) \approx 10^{-23} \text{ A} \cdot \text{m}^2$$

Each electron evidently has only a very small magnetic moment, even if it has two or three units of orbital angular momentum plus its spin angular momentum. Still, if all those magnetic moments lined up for all electrons, every piece of material would be a very strong magnet, as we will see. Actually, most of the magnetic moments cancel out. The structure of atoms is dominated by the Pauli exclusion principle which, among other things, requires most of the electrons to exist in pairs with opposing spins. If two electrons are in the same general region, then the spin of one will be opposed to the spin of the other and, therefore, the spin angular momenta will cancel. Indeed, we might at first expect that any atom with an even number of electrons would have no magnetic moment at all. This is not the case, however; iron has 26 electrons. You might also expect that the magnetic properties of an atom would depend on whether the atoms were free in a gas or linked with other atoms in a solid. A free atom might have a valence electron, with the spin angular momentum not canceled by a paired electron. In a solid, the valence electron of such an atom would usually go into a bonding state, or into a metallic electron sea, pairing with another electron and thus canceling the magnetic moments of each. The magnetic properties do indeed depend strongly on how atoms are combined with others.

In spite of the cancellation of most atomic angular momenta, many materials do end up with a magnetic moment per atom that is equal to the basic unit that we calculated: approximately 10^{-23} A m^2. Suppose that the magnetic moments of one mole of these atoms were all lined up. A mole of most atoms in the solid state has a volume about the size of a thick pencil. The total magnetic moment of a mole of atoms, each with a magnetic moment of 10^{-23} A m^2, is equal to

$$m_{\text{total}} = (6 \times 10^{23} \text{ atoms})(10^{-23} \text{ A m}^2/\text{atom}) = 6 \text{ A m}^2.$$

Compare that magnetic moment with the one that we calculated on p. 439 for a 10-turn coil having an area of 10 cm^2 and carrying 5 A. The solid bar would have a magnetic moment over 100 times greater. Indeed, 6 A m^2 is approximately the magnetic moment of a magnetized *iron* bar of this size. But only iron, cobalt, nickel, and some of their alloys behave in this manner. Why don't the magnetic moments of other elements line up?

The atoms in a solid are continually subject to thermal agitation. Let us compare the potential energy that an atom could gain by lining up in a magnetic field with the average thermal energy that is knocking it around. If the atom had one unit of magnetic moment and swung 180° from the antiparallel to the parallel direction in a magnetic field of 1 T, then its potential energy would decrease by

$$\Delta U_{\text{max}} = -2mB = -2(10^{-23} \text{ A m}^2)(1 \text{ T}) = -2 \times 10^{-23} \text{ J}.$$

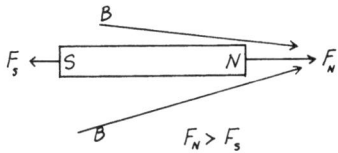

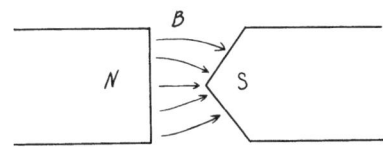

Figure 18.16. Movement of a magnet in a nonuniform magnetic field.

The average thermal energy of such an atom at room temperature is $U_{\text{thermal}} = \frac{3}{2}kT = \frac{3}{2}(1.38 \times 10^{-23} \text{ J/K})(300 \text{ K}) = 6 \times 10^{-21}$ J. Both of these energies are small compared with the binding energy of an atomic valence electron, which is of the order of 1 eV (1.6×10^{-19} J). But the average thermal energy of an atom is 300 times the maximum energy that can be gained by an atom lining itself up with a magnetic field as strong as 1 T. Even if materials are composed of atoms that have individual magnetic moments, a random thermal agitation would destroy most of the alignment. In fact, now the question should be phrased the other way. Instead of asking why aluminum is not strongly magnetic, we must figure out why iron is. First, however, let us characterize the two types of very weak magnetism that appear in most materials.

In a uniform magnetic field, a permanent magnet will rotate until it is aligned with the field. So will a rod of unmagnetized iron, and many other materials for that matter, although the torques are very weak—except for iron, cobalt, nickel, and some of their alloys. An even more interesting magnetic effect is produced in a *nonuniform* magnetic field such as that shown in Fig. 18.16. If a magnetic dipole is put in such a field, it will not only rotate but will also move along the field lines. If a piece of iron is suspended in such a field, whether the iron is a permanent magnet or not, it will move strongly toward the direction of the strong magnetic field. This effect is characteristic of *ferromagnetism*. Many other materials, such as chromium, manganese, and palladium, will also move toward the stronger field, but the force on these materials is smaller by a factor of 1000 or so than it is for iron. Materials showing this very weak magnetic effect are called *paramagnetic*.

Surprisingly enough, some materials such as arsenic, mercury, and silver will actually move toward a *weaker* field. They act as if the atomic dipoles preferred to be out of line with the external field. This effect is called *diamagnetism*. Diamagnetism is an induced effect, but in one crucial aspect it is different from electrostatic induction. Remember the way that a charged balloon clings to a wall because of electrostatic induction. The electrostatic field induces dipoles in the molecular structure of the wall. The electric charge arrangement in the molecules is warped so that the positive and negative charge centers are separated. The amount of the warping, and thus the strength of the dipoles, is proportional to the strength of the external electric field. In a similar way, when a magnetic field is imposed on any material, the electron *paths* are affected. As we have seen, there is a force exerted on a charged particle that is moving in a magnetic field. The resulting rearrangement of the electron paths produces a magnetic dipole moment proportional to the strength of the external field. However, the induced dipoles are in a direction to *oppose* the field. This diamagnetic effect occurs in *all* materials in whatever state they are in—gas, liquid, or solid. The effect is very small, however, and is washed out if the atom has a permanent magnetic moment due to the orbital or spin motion of one of its electrons.

In Table 2 we list a parameter known as the *magnetic susceptibility* χ for a number of elements at room temperature. The magnetic susceptibility times the strength of the external field (divided by μ_0) gives the magnetic dipole moment per unit volume for the material. For our purposes, this parameter serves as a comparison of the relative magnetic responses of various materials. Note that the susceptibility of iron is in the range from 1000 to 10 000. Compared to ferromagnetism, the diamagnetic and paramagnetic effects are extremely weak.

Diamagnetism is not much affected by temperature, since the dipoles are induced by the external field and are not permanently attached to the individual atoms. Their strength is independent of the way the atoms are jostled around by the thermal motion. The electron motions automatically adjust to oppose the external field.

In the case of both diamagnetism and paramagnetism, the resulting magnetic

TABLE 2. *Susceptibilities of various Materials:* χ.

Paramagnetic	
Oxygen (liquid at -183 °C)	1.5×10^{-3}
Aluminum	2×10^{-5}
Sodium	7×10^{-6}
Iron ammonium alum (used in magnetic cooling near absolute zero)	4.8×10^{-2}
Manganese	9×10^{-4}
Chromium	3×10^{-4}
Palladium	8×10^{-4}
Diamagnetic	
Bismuth	-1.7×10^{-4}
Mercury (liquid)	-3×10^{-5}
Silver	-2.6×10^{-5}
Lead	-1.8×10^{-5}
Copper	-1.0×10^{-5}
Arsenic	-2×10^{-5}
Ferromagnetic	
Iron	up to 10 000
Cobalt	depending on initial
Nickel	magnetization

moment of the material is approximately proportional to the strength of the external field. This is true in diamagnetism since the strength of the individual dipoles is proportional to the inducing field. In the case of paramagnetism, the magnetic moments are intrinsic parts of each atom, independent of the external field strength. The extent to which the individual magnetic moments line up, however, depends on the competition between the external field and the thermal motions. A typical graph of the dependence of magnetization on temperature is shown in Fig. 18.17. The graph is known as a Curie plot, in honor of Pierre Curie, who established both the experimental work on magnetism and the classical theory concerning it in the 1880s and 1890s before helping his wife to discover radium. Note that the units of the horizontal axis are teslas per degree. At room temperature in the strong field of 1 T, the horizontal coordinate of a data point would be at 1/300. To get to point 1 on the horizontal axis, we would have to have a magnetic field of 1 T at a temperature of 1 K, or 300 T at 300 K. The vertical axis is in terms of the magnetization produced in the material, divided by the maximum magnetization possible. The maximum magnetization would occur if all the dipole moments were aligned; thus the curve shows the degree of alignment of the individual atomic dipoles.

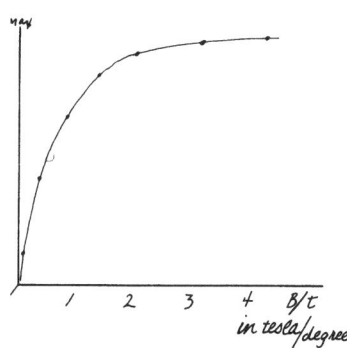

Figure 18.17. Curie plot for a paramagnetic salt. Note that in a strong field of 1 T, at room temperature, $B/T \approx 1/300$. At that point, the ratio M/M_{max} is very small. Only a small fraction of the magnetic dipoles are aligned.

■ FERROMAGNETISM

So far we have given arguments why materials cannot be strongly magnetic. Most of the intrinsic magnetic dipoles within an atom that are due to the orbital or spin motion of electrons cancel out. Even when each atom is left with a small net magnetic moment, the amount of energy that it might gain by lining up with an external magnetic field is much smaller than the thermal energy that makes the directions random. The mutual energy to be gained by lining up with immediate neighbors is even smaller. Nevertheless, iron, cobalt, nickel, and some of their alloys do respond strongly to magnetic fields and even retain the atomic alignment to become permanent magnets. Let us see what is so special about iron, cobalt, and nickel—and particularly, iron.

In Fig. 18.18 we portray the parameters of the electrons in the iron atom. There are 26 electrons, most of which go into shell-like positions in pairs. The inner shell, which is spherical and is labeled $n=1$, contains two electrons. These electrons have no orbital angular momentum, and their spins are in opposite directions so that their magnetic moments cancel. In the next region, labeled $n=2$, there can be eight electrons. Two are in a spherical shell at a larger radius than the one for $n=1$, and have 0 angular momentum. There are also three other nonspherical regions, at about the same radius and energy, each containing two electrons. Although each electron in these nonspherical regions has 1 unit of angular momentum, all of the angular momenta, including that due to the spins, cancel out because of the geometry and electron pairing. In the next major region, labeled $n=3$, 18 electrons can be accommodated; 2 with 0 angular momentum, 6 with 1 unit of angular momentum, and 10 with 2 units of angular momentum. Since iron has only 26 electrons, and the three complete shells can hold 28, it might seem that iron would have two empty places in the $n=3$ region, thus having a valence of -2. Instead, two of the electrons from iron establish themselves in the $n=4$ region, thus providing a positive valence number. Meanwhile, back in the outer subregions of $n=3$, there are six electrons that are partially shielded from the chemical interactions that are performed by the outermost electrons. We might

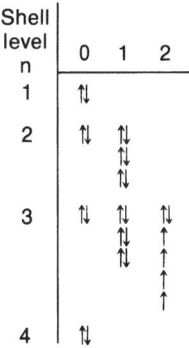

Figure 18.18. Parameters of electrons in the iron atom.

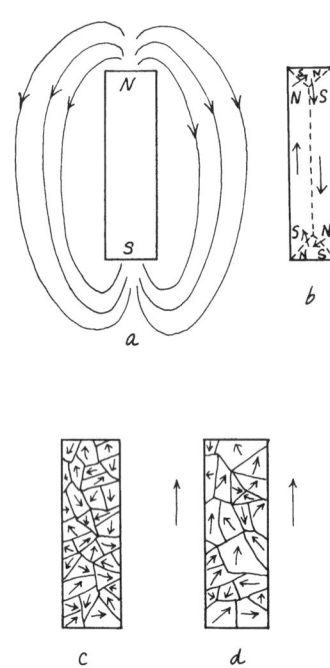

Figure 18.19. Magnetic domains in a bar magnet. a) False assumption that all atoms are aligned in same direction. b) An arrangement of large domains with lower energy than (a). c) A random arrangement of small domains representing actual situation in nonmagnetized iron. d) Domains in magnetized iron. Domains in Favored direction have grown.

expect that these six electrons would be paired so that their magnetic effects would cancel out. Instead, five of the electrons spin in one direction and one in the opposite direction. The atom is left with 4 units of angular momentum, all in line.

The detailed calculations about why the electrons arrange themselves in this way are very complicated. The atom actually has a lower energy state if two of the electrons go out to the $n=4$ shell, away from the densely populated $n=3$ region, and if the six shielded electrons in the outer subshells of $n=3$ have their spins mostly aligned, requiring them to stay far away from each other. The elementary analysis of electron positions in orbitals assumes that an atomic system will go to the lowest energy if all the electrons fall as low in the Coulomb potential well of the nucleus as they can. The electrons cannot all go to the lowest energy level, however, because of the Pauli exclusion principle. The electrons are constrained to fill the lower energy levels in pairs with opposing spins. However, in the next approximation we must take into account the repulsion of the electrons from each other. If the electrons are densely packed, the energy of the system will *increase*. This is why in iron the energy is reduced if two electrons go into the $n=4$ shell, further from the nucleus, but also further from the other electrons in $n=3$. That's also why four of the electrons in the 3 shell have their spins aligned—they are further apart from each other that way. The situations with cobalt and nickel are similar. Cobalt has 27 electrons, with three aligned, and nickel has 28. with two aligned. With cobalt there are seven electrons in the outer subshells of $n=3$, and in nickel there are eight. Just to add to the confusion, note that manganese, atomic number 25, also sends two electrons to $n=4$, leaving five in $n=3$, *all* of which are aligned! An isolated manganese atom (for instance, in a vapor) has a very strong magnetic moment.

In spite of these almost accidental geometrical arrangements that provide atoms with permanent magnetic moments, there is still no reason to expect that the individual atoms will line up with each other. The argument is still valid that the magnetic potential energy is small compared with the thermal energy. With the ferromagnetic materials, however, the quantum conditions and the Pauli exclusion principle yield lower energy states for the metallic system if the atomic magnetic moments are aligned. It is not the magnetic field from neighbors that aligns each atom; it is the quantum requirements of locating electrons to reduce the system energy. The electrons in the $n=3$ shell have wave functions extending out to other atoms; they are not completely sheltered from their neighbors. The whole system tries to arrange the electrons in the lowest energy state compatible with the requirement that no two electrons in the same energy state have the same spin. With iron, cobalt, and nickel, the electrons can be arranged far away from each other, thus having lower electrostatic energy if the atomic magnetic moments of neighbors are aligned. With manganese, the energy balance is tipped, so that the system has lower energy if the atoms alternate alignment.

Magnetism is a subject where the obvious consequences of each new addition to the theory are frustrated by yet another condition or circumstance. If the iron atoms can go to a lower energy state by aligning with each other, why isn't *every* piece of iron a strong permanent magnet? In Fig. 18.19(a) we show the field lines produced by a completely magnetized iron bar. The pattern is the familiar one expected with a bar magnet but, when an iron bar freezes out of a liquid state, the magnetic moments do not line up that way. The field, as pictured, represents a lot of energy stored in space. A lower energy state would be produced if the atoms lined themselves up as shown in Fig. 18.19(b). Now the field lines are concentrated within the volume of the iron, reducing the amount of energy. However, it turns out that it takes extra energy to produce a wall between two atomic regions. Furthermore, as the iron liquid freezes into a solid, there are geometrical constraints on the atomic locations other than just the direction of magnetic moments. Crystals of various size form with boundaries that are affected by impurities and imperfections. Within a crystal, magnetic moments are more easily aligned in one direction than in another. The end result of all these complications is shown in Fig. 18.19(c). The solid iron consists of domains that are microscopic in size. (Of course, there are still a lot of atoms lined up in each domain. Within a typical cubic domain, 10 μm on a side, there are about 10^{15} atoms.)

The magnetic moments of all the atoms within a domain are aligned; therefore, the tiny region is a strong permanent magnet. However, in an ordinary piece of iron just frozen out of the melt, the magnetic moments of the domains are completely random in direction. An iron bar ordinarily has no net magnetic moment. If such a bar is put into a strong external magnetic field, the domains that have magnetic moments in line are in a lower energy state than those with magnetic moments opposed. The boundaries between domains start moving, with the aligned domains gaining volume at the expense of those opposed. The (d) part of Fig. 18.19 shows such a transformation.

The growth of the favored domains can be detected experimentally in a dramatic way. Wind a coil of several thousand turns of fine wire around a bar of iron. Connect the coil to an audio amplifier, with the output fed into a loudspeaker. Bring a permanent magnet slowly toward the bar of iron. As the iron is magnetized by the slowly increasing external field, the loudspeaker will make a sound like static. The walls of the domains are not moving smoothly in response to the external field, but adhere to impurities or imperfections in the crystals. As the external field increases, the energy difference between one domain and its neighbor becomes sufficient to shove the wall abruptly past the imperfection, thus creating an abrupt increase in the magnetic moment of the bar as a whole. This phenomenon is called the *Barkhausen effect*. The sudden change in magnetic field induces a voltage in the pickup coil, which in turn makes the loudspeaker crackle.

The magnetic field needed to align the magnetic domains may be small compared with the resulting magnetic field in the iron. This provides us with a way to amplify greatly the magnetic fields produced by currents in wires. With tabletop solenoids or loops and currents of a few amperes, we can produce magnetic fields of only one to a hundred gauss. With an iron core in these coils, the same current can produce fields 100 to 1000 times greater.

The amplification factor of a magnetic material is determined by measuring the magnetic field in a ring solenoid containing a core of the material. Without the core, the field inside the solenoid would be approximately that of a long straight solenoid: $B_0 = \mu_0 n i$. The actual formula for a ring solenoid is slightly different, but if the ring is large compared with its cross-section diameter, the simple formula applies well. Furthermore, with ring geometry, we do not have to worry about end effects.

In principle, you could measure the actual field B inside the core by cutting a small gap in the material and inserting a small current-carrying wire. The force on the wire would be proportional to B. In practice, another "pickup" coil is wound around the primary coil. The actual measurement of B involves turning on the current in the primary to create the energizing field B_0. The induced signal in the pickup coil is proportional to B.

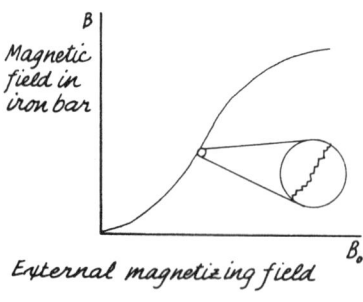

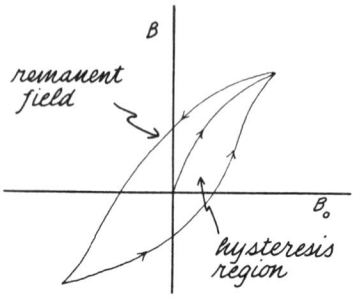

Figure 18.20. Hysteresis curves.

In Fig. 18.20, we show a B versus B_0 curve for a typical piece of iron. Consider what happens to an unmagnetized iron core as B_0 increases from 0 strength. The magnetic domains in the material that have magnetic moments in the direction of the energizing field grow at the expense of their neighbors. The resulting magnetic moment of the material produces a strong internal field. Eventually, however, the domains have grown as large as they can, and most of the atomic magnetic moments of the whole bar are aligned. The material is saturated; an increase of the energizing field can produce no more amplification. If the energizing field is then reduced in strength and finally reversed in direction, the field in the material will decrease, but not along the same curve by which it grew. Depending on the material and the way that it was fabricated, there will be a large or small remanent field left in the material, even when the energizing field is 0. If this remanent field is very large, the material may be a good permanent magnet.

The plot of B versus B_0 is called a *hysteresis* curve. When an energizing field goes through a complete cycle, energy is lost in the material if the domain walls do not move smoothly. A material with large hysteresis and remanent fields loses considerable energy during a complete cycle of the energizing field. The area within the hysteresis curve is proportional to this energy loss. Consequently, only iron with low hysteresis loss should be used in transformers where the energizing field may be changing at 60 Hz. So-called "soft" iron is used in transformers. Such iron should have few impurities and consist of small crystals, so that the domains can grow or shrink with minimum energy loss.

The ratio of B to B_0 is approximately equal to the susceptibility χ. Note that in older texts, or in advanced texts, a distinction is made between B and H, the energizing field. In some units they even have different dimensions. In modern introductory texts, H is not mentioned (nor is D, the electric displacement). Our B_0 corresponds to H. Different kinds of iron or iron alloys have values of χ ranging from 100 to 10 000. Most kinds of iron saturate with internal fields from 1 to 2 T.

There are many special kinds of ferromagnets. One alloy containing 51% of iron, and the rest a mixture of aluminum, nickel, and cobalt, is called Alnico. It is a brittle metal with a large remanent field that is not easily destroyed, thus making it a very powerful and long-lasting permanent magnet. Another iron alloy called μ-metal has a value of χ of about 10 000. It is a soft metal and does not maintain a permanent field. It is used for magnetic shielding in external fields of under 1 G. Note that above a 1-G energizing field, the μ-metal would saturate.

The growth of the favored magnetic domains is not instantaneous as the energizing field increases. If the external field is cyclical and changes directions faster than about 20 000 Hz, the growth of the domains cannot follow, and the value of χ will fall. There are materials called ferrites that get around this problem. Ferrites are mixtures of magnetic materials and ceramics. The embedded ferromagnetic materials are small and dispersed. Ferrites are electrically nonconductors and retain high values of χ up to frequencies in the television range of 10^8 Hz.

It is easy to shield a region from *electrostatic* fields just by enclosing it with metal walls. The external field lines will terminate on surface charges on the walls. As we have seen, there cannot be a static electric field inside a conductor. However, it is harder to shield a region from magnetic fields. Since there are no magnetic poles, external field lines cannot terminate on the walls of a shield. All we can do to shield a region from an external magnetic field is to surround it with a ferromagnetic wall that has a large value of χ. Most of the external field lines then concentrate in the shield walls and continue out the other side, bypassing the interior region. Sometimes it is necessary to shield sensitive instruments by placing them in a nest of concentric iron cylinders, each one reducing the strength of the field penetrating to the next cylinder. For the ultimate shielding, a region can be surrounded with a superconductor, which sets up opposing currents to block any external field.

■ THE RELATIVE NATURE OF ELECTRIC AND MAGNETIC FIELDS

In exploring the second realm of electromagnetism, we have described the actions of electric charges moving past us with constant velocity. If we have observers in two different reference frames, measuring the same phenomenon, they must be able to use the same laws—providing they make the right transformations of distances and times. Let us set up a thought experiment with electric charges and describe the results from two different reference frames moving past each other with constant velocity.

In Fig. 18.21 we show segments of two infinite rods separated by the distance y, each having a uniform, linear, static charge density of 1 C/m: $\lambda = \Delta q / \Delta x$. The electrostatic field produced by a line charge is

$$E = \frac{\lambda}{2\pi\epsilon_0 y}.$$

The force on the other wire is just equal to its total charge times the electric field produced by the first wire. Therefore, the force per unit length between the two wires is

$$\frac{F_y}{\Delta x} = \frac{\lambda^2}{2\pi\epsilon_0 y}.$$

So far we have assumed that the charges are standing still, not only with respect to each other but also with respect to us, the observers. Now, however, let us see what happens if the same charged rods are moving steadily to the right along the x axis with velocities v. We observe that the lengths in the moving system are contracted. For instance, at a speed of $0.87c$, a meter stick attached to the moving wires would appear to us to be only 50 cm long. However, there is no relativistic decrease in the actual number of charges. The charge consists of a certain number of electrons, and that number is independent of the velocity of the observer. The linear charge *density* does change, however, because $\lambda = \Delta q / \Delta x$. If we observe that the unit length has shrunk by a factor of 2, then we also observe that the charge density has *increased* by a factor of 2, since we would count the same number of charges in the contracted meter that we did when the charges were standing still.

From now on we will label all the quantities in the moving system, *as measured in the moving system* with a subscript zero. These are the quantities that would be measured by an observer riding on the rods along with the moving charges. The symbols without zero subscript indicate the magnitudes that *we* measure as we watch the rods hurtle past us. The length relationship, for instance, is

$$\Delta x = \Delta x_0 \sqrt{1 - v^2/c^2}.$$

(Note that if $v = 0.87c$, $\Delta x = \frac{1}{2}\Delta x_0$). The charge density that *we* measure is

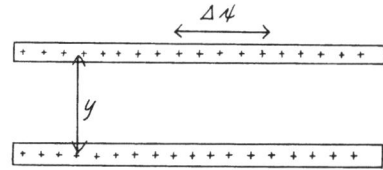

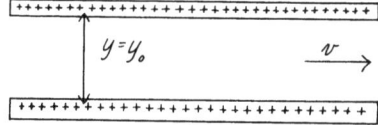

In our reference frame, the charge density on the moving rods is higher.

Figure 18.21. Relative nature of electric and magnetic fields.

$$\lambda = \frac{\Delta q}{\Delta x} = \frac{\Delta q}{\Delta x_0 \sqrt{1 - v^2/c^2}} = \frac{\lambda_0}{\sqrt{1 - v^2/c^2}}.$$

Since we measure a larger charge density than does the observer riding on the rods, let us compare the repulsive force between the wires as measured by the two observers. The force per unit length as measured by the observer riding on the wire is

$$\frac{(F_y)_0}{\Delta x_0} = \frac{\lambda_0^2}{2\pi\epsilon_0 y}.$$

The force per unit length that we measure is $F_y/\Delta x$

$$\frac{F_y}{\Delta x} = \frac{1}{2\pi\epsilon_0}\frac{\lambda^2}{y} = \frac{1}{2\pi\epsilon_0 y}\frac{\lambda_0^2}{1-v^2/c^2} = \frac{(F_y)_0}{\Delta x_0}\frac{1}{1-v^2/c^2}.$$

Evidently, we and the moving observer disagree about the magnitude of the repulsive force between the wires. This is a serious problem because it would lead to different predictions about the same event. Let us calculate the magnitude of this *difference* between the two observers.

$$\frac{F_y}{\Delta x} - \frac{(F_y)_0}{\Delta x_0} = \frac{1}{2\pi\epsilon_0}\left(\frac{\lambda^2}{y} - \frac{\lambda_0^2}{y}\right) = \frac{1}{2\pi\epsilon_0 y}\lambda^2\left[1 - \left(1 - \frac{v^2}{c^2}\right)\right] = \frac{1}{2\pi\epsilon_0}\frac{\lambda^2}{y}\frac{v^2}{c^2}.$$

At ordinary velocities the ratio v/c is very small, and so it might seem that the two observers do not disagree by much, although there is a discrepancy. Let us carry this discussion a little further. Note that although the charges are standing still with respect to the observer on the wire, they are moving with respect to us, and so they form an electric current. That current is $I = \lambda v$. Therefore, we can rewrite the size of the discrepancy in observations as

$$\left(\text{extra } \frac{F}{\Delta x} \text{ observed by us}\right) = \frac{1}{2\pi\epsilon_0}\frac{I^2}{y}\frac{1}{c^2}.$$

Now we are faced with another discrepancy. If we observe two parallel *currents*, we know that there is an *attractive* force between them. That magnetic force between two wires is

$$\left(\frac{F}{\Delta x}\right)_{\text{mag}} = \frac{\mu_0}{2\pi}\frac{I^2}{y}.$$

There are now two discrepancies between our measurements and those of the observer on the moving wires. She sees only a static repulsive force. We see a *larger*, static repulsive force because the charge density that we measure is larger. But we also see an *attractive* force because of the magnetic effect. As far as the traveling observer is concerned, there is no magnetic field.

If the force per unit length is not the same as measured by the two observers, then they would predict different events for the same phenomenon. Such a contradiction would be the downfall of our scientific methods. For instance, what if the magnetic field that we measure were sufficiently strong to cause the wires to come together? (Actually, the repulsive force would always be stronger than the attractive force for the geometry we have assumed. However, there is a situation, called the "pinch" effect, where high velocity beams do come together. For the pinch effect to occur, there must also be neutralizing charges present between the currents.)

We can escape the controversy between the two observers if we say that the extra repulsive force that we observe must be just equal to the extra attractive magnetic force. This condition is satisfied if

$$\left(\frac{F}{\Delta x}\right)_{mag} = \left(\frac{F}{\Delta x}\right)_{extra\ electrostatic}, \quad \frac{\mu_0 I^2}{2\pi y} = \frac{1}{2\pi\epsilon_0}\frac{I^2}{y}\frac{1}{c^2}.$$

These two expressions are equal if $\mu_0 = (1/\epsilon_0)(1/c^2)$, or

$$c = \sqrt{\frac{1}{\mu_0 \epsilon_0}}.$$

The permeability μ_0 and permittivity ϵ_0 are constants defined and measured in static situations. The permittivity is measured in terms of the fields and geometry of a capacitor; the value of the permeability is chosen in an experiment that defines the ampere. Let us substitute those values into the equation forced upon us by the demands of relativity:

$$c = \sqrt{\frac{1}{(4\pi \times 10^{-7})(8.85 \times 10^{-12})}} = 3 \times 10^8 \text{ m/s}.$$

Astonishingly, the properties of static electricity and magnetic fields are connected with the velocity of light!

Whether we observe an electric field or a magnetic field apparently depends on our velocity relative to the sources. One person's electric field may be another person's magnetic field. It appears that magnetism is just a relativistic effect. The magnitude of the effect depends on the square of the ratio of the relative velocity to the velocity of light. We might expect that the phenomenon could be observed only with subatomic particles that have been accelerated close to the speed of light; but we have already observed that the magnetic phenomenon can be observed with ordinary wires fastened across a D cell. The electrons in that current are drifting with a velocity of a fraction of a millimeter per second. Why is it that we can make powerful electric motors that exploit such a weak effect?

The magnetic field produced by electron drift in a wire is indeed extremely small compared with the electric field produced by that same charge. The electron is not alone in the wire, however. All the negative and positive charges are paired, so that there is very complete electrostatic neutrality. The magnetic field due to the drift velocity of each individual electron, small though it is, is not canceled out. For every ampere in the wire, 6×10^{19} electrons drift past the given point each second. The individual field is small, but the number of contributions is enormous. The result is an influence that can be detected in our human scale of things. Every time that you feel a magnetic force, you are experiencing the consequences of the special theory of relativity.

There is a provocative corollary to the arguments above. They could equally well be applied to the gravitational attraction between two rods. Is there, indeed, a gravitational interaction corresponding to magnetism? If so, when two long, massive, parallel rods hurtle pass us, there should be a slight antigravitational force between them. The trouble is, this force would be in competition with the main gravitational attraction between the rods, which in itself is very small and hard to detect. The reason that the relativistic effect can be seen in electromagnetism is that two parallel currents exert no electrostatic force between them because of the absolute neutrality of charges in matter. So the electrostatic force is essentially 0, but the relativistic effect caused by current remains. If we are to produce antigravity in this manner, we will first have to obtain material with antimass!

19
Currents and Fields that Change with Time

In our comments on teaching about electromagnetic interactions, we have now covered two realms. In the first, the electric charges are stationary with respect to each other and to us, the observers. In the second realm, the charges move with constant velocity, giving rise to the study of circuitry and to the phenomenon of magnetism. Now our study must enter the third realm where currents are not constant and where electric and magnetic fields change with time.

In this chapter we are concerned only with relatively slow changes in current. The resulting phenomena affect only the circuits themselves or their immediate neighborhood. In the next chapter we will see what happens when charges accelerate more rapidly, and where the electric and magnetic fields that are produced travel away from the source charges, carrying energy and momentum that do not come back.

In 1820 Hans Christian Oersted (1777–1851) discovered that an electric current produces a magnetic field. Many of the early researchers in electricity had sought a connection between electricity and magnetism. Oersted has been an inspiration to physics teachers ever since because he discovered the effect while giving a lecture. Besides his scientific work, Oersted was famed as a teacher and popularizer of science. The American Association of Physics Teachers chooses someone each year to receive the Oersted Medal as the Association's highest award for physics teaching. All of us show our introductory students Oersted's simple demonstration of placing a compass needle on top of a current-carrying wire.

After Oersted discovered that electric currents produce magnetic fields, many attempts were made to find the inverse effect. Can strong magnetic fields somehow produce a current? People wound wires in various ways in magnetic fields and tried with sensitive galvanometers to detect the weak currents that might be produced. Legend has it that Ampère, who had built the most sensitive current detectors in the world, kept the detectors in a laboratory separate from the magnetic fields with which he hoped to generate current. He would make a new arrangement of loops and fields and then go across the hall to see if any current existed, thus missing the crucial observation that the current was induced only while there was motion between magnetic field and wire loops. It was Michael Faraday (1791–1867) who discovered the secret. Within a year of this discovery in 1831, Faraday and others were making crude generators that turned mechanical energy into electric energy.

452 Currents and Fields that Change with Time

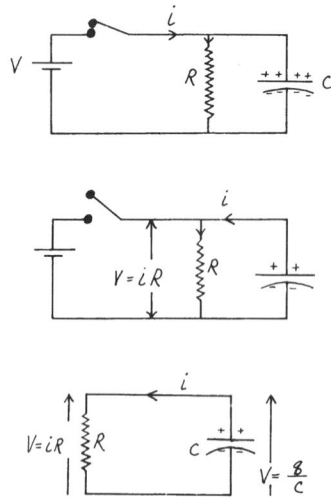

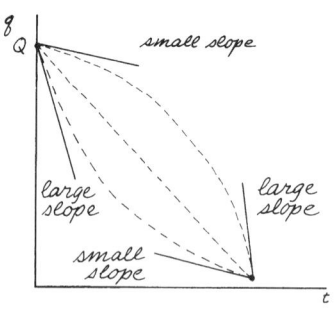

Figure 19.1. Discharge of a capacitor through a resistor.

This topic concerns both phenomena and practical techniques. The phenomena involve fundamental and profound questions about the interactions of nature and the way that we observe them. The practical techniques have become fundamental to our industrial civilization and to the conveniences of our everyday life. Motors, generators, and transformers depend on currents and fields that increase and decrease sinusoidally. The electricity available in the wall sockets of our homes is alternating current. Even though such devices are common, it is not easy to study their operation in the introductory course or laboratory. Household ac voltage is potentially lethal. Skill, knowledge, and caution are necessary when doing anything more complicated than turning on a wall switch. Nor is it easy or wise to take apart household motors or transformers that are still working. Most of these machines depend on quite sophisticated engineering details that tend to mask the basic principles involved. Of course, if you can provide old shavers, mixers, or hair dryers with motors that can be sacrificed, give them to your students and tell them to use screwdrivers (but not while the devices are plugged in!).

■ THE DECAY OF CHARGE ON A CAPACITOR

When we described capacitors before in Chap. 16, we did not worry about the charging process. Now we must face the fact that it takes time for the charges to build up on a capacitor or to leave during a discharge. No new phenomena are involved in the process, but some characteristics of this type of changing current apply to some of the more complicated cases that we must study later. Furthermore, we can illustrate a graphical method of solving a simple differential equation.

To simplify the analysis, let us see what happens when a capacitor discharges as shown in Fig. 19.1. With the switch closed, the capacitor is fully charged with a potential difference equal to that of the battery. Current exists in the resistor, producing a voltage drop of iR equal to the voltage of the dry cell. When the switch is opened, the charges on the capacitor will drain off through the resistor. The resulting current in the resistor will continue to provide a "back" voltage that opposes the field produced by the charges stored on the capacitor. This opposing field weakens, however, as the charges drain out of the capacitor, reducing the value of i, and thus of iR.

We can analyze this circuit with Kirchhoff's law: the potential rise and drop around a complete loop must equal zero. At any given time the potential difference across the capacitor is q/C. The potential drop across the resistor is iR (we use the lowercase q and i to indicate that they are variables):

$$\frac{q}{C} - iR = 0.$$

The charge q and the time rate of passage of charge, i, are related, but we have to be careful in describing the relationship. In Fig. 19.1 we show a graph of the charge on the capacitor as a function of time. When $t=0$, $q=Q$, the original charge on the capacitor. We know very well that at some later time there will be no charge left on the capacitor. Some function $q(t)$ joins the two data points on the graph. We have sketched three possible graphs that might describe the function. Note that all of them have negative slope. The slope of a $q(t)$ graph represents the current i. However, in our circuit, we know that the current consists of charges flowing out of the capacitor and down through the resistor in the direction shown by the arrow in the circuit diagram. In order to have a positive current when dq/dt is intrinsically negative, we must define the current to be

$$i = -\frac{dq}{dt}.$$

When we substitute this value for i into the Kirchhoff equation, we get

$$\frac{q}{C} - \left(-\frac{dq}{dt}\right)R = 0 \rightarrow \frac{dq}{dt} = -\frac{1}{RC}q.$$

Now we are in a position to choose between the possible functions that describe the decay of the charge in the capacitor. According to our formula, the slope of the $q(t)$ curve, dq/dt, is proportional to the magnitude of the charge itself and is negative. Sure enough, the slopes of all three curves are negative but the upper one starts out with a small slope when q is large and ends up with a large slope when q is small. That behavior is not what the equation calls for. The slope of the middle curve is constant, which does not agree with our formula. The bottom curve, however, starts out with a large slope for large q and then the slope continually decreases as q decreases. The mathematical function describing the decay of the capacitor charge must look something like the bottom curve of our graph. The mathematical function that satisfies these conditions is the exponential. The complete formula is

$$q = Qe^{-(1/RC)t}.$$

To test that this is the solution, try the extremes for t. When $t=0$, $q=Q$, and as $t \rightarrow \infty$, $q \rightarrow 0$. Furthermore, $dq/dt = -(1/RC)Qe^{-(1/RC)t} = -(1/RC)q$, which satisfies the original equation.

Since the argument of an exponential must be a pure number without any dimensions, it must be that RC has the dimensions of time. This is reasonable since $R = V/i$, $C = q/V$, and $i = dq/dt$. When R is given in ohms, C is given in farads, the time RC is in seconds.

When we open the switch in our circuit, $t = 0$. At a later time when $t = RC$, $q = Qe^{-1} = (1/2.7)Q$. The characteristic time, RC, is called the time constant of the circuit, τ. This is the time that it takes the charge in such a circuit to decay to $1/e$ of its original value. The time that it takes for the charge to decay to half its original value is called the half-life: $t_{1/2} = 0.69\tau = 0.69RC$. (Since $\frac{1}{2} = e^{-t/\tau} \rightarrow \ln\frac{1}{2} = -t/\tau \rightarrow -0.69 = -t/\tau$.)

As the stored charge decays, the capacitor loses energy. When a small amount of charge leaves, the lost energy is equal to the product of that charge and the potential difference across the capacitor at that moment:

$$\Delta U_C = \Delta q(q/C).$$

The power loss at that moment is equal to

$$P = \Delta U/\Delta t = \frac{\Delta q}{\Delta t}\frac{q}{C} = i\frac{q}{C}.$$

Since the Kirchhoff law for the circuit is $q/C - iR = 0$, the power lost by the capacitor is equal to

$$P = i\frac{q}{C} = i(iR) = i^2 R.$$

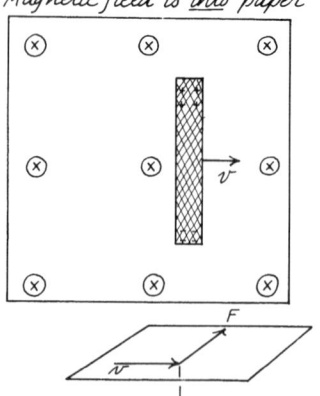

Figure 19.2a. When a conductor moves in a magnetic field there is a force on the charges in the conductor.

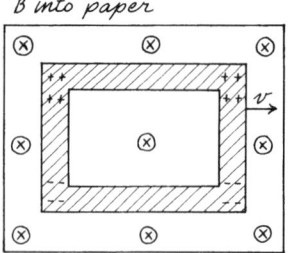

Figure 19.2b. Conducting loop moving in a magnetic field. No continuous current is created.

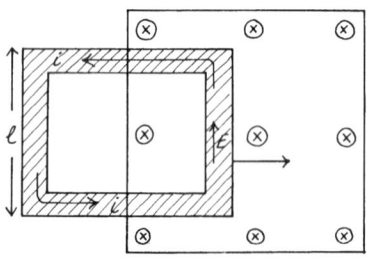

Figure 19.2c. Conducting loop moving in magnetic field, with one leg outside of field. Current is generated in loop.

The power lost by the capacitor as it discharges is just the power being dissipated in the resistor and turned into heat.

In Chap. 17 we said that when a dc circuit is closed it takes a short time for the charges to arrange themselves to produce a constant current. Now we can calculate the order of magnitude of time for a typical circuit. The time constant depends on the capacitance and resistance of the circuit. Wires and other electrical components have capacitance, even if they do not look like capacitors. The capacitance depends on the size of the object and on its proximity to ground or other components. A typical circuit of battery, wire, bulb, and switch might have a capacitance of 10^{-10} F. If the resistance of the circuit is 10 Ω, the RC time constant is 10^{-10} s. Consequently, in a nanosecond or so the charges would be distributed along the circuit in a way that would establish constant current.

■ MOVING A CONDUCTOR IN A MAGNETIC FIELD

Now we are going to analyze phenomena involving generators and transformers. The analysis starts out with a familiar thought experiment that is common in introductory physics. But look out! Before we are through we will extrapolate the argument to a conclusion that is not at all contained in the original rules and formulas, and leads to a deep mystery in our understanding of electricity and magnetism.

Starting simply, consider the results of moving a metal bar in a magnetic field as shown in Fig. 19.2(a). The bar moves perpendicular to the magnetic field and also perpendicular to its own length. The neutral conducting bar is filled with an equal number of positive and negative charges. These are being forced to move in a magnetic field, and so there is a force on each given by the Lorenz equation: $\mathbf{F} = q(\mathbf{v} \times \mathbf{B})$. In our diagram the resultant force on the positive charges is up and on the negative charges is down. Since some of the electrons are free to move, they will pile up at the bottom of the bar, leaving the top charged positive. An electrostatic field will be created by these concentrations of charges which will oppose any further motion of electrons. When the electrostatic force balances the force produced by the motion through the magnetic field, the two fields must be related in the following way:

$$qE = qvB \rightarrow E = vB.$$

An observer riding on the bar could explain the motion of the charges by saying that in her reference frame there is an electric field pointed up. She could measure the excess charges by letting them flow onto a gold-leaf electroscope and observing the deflection of the leaves.

So far we have not arranged to produce any continuous current with our thought experiment. After the initial surge, the piled-up charges repel any further motion. We have a similar situation with electric batteries. To produce a current, you must complete a circuit between the positive and negative poles. In Fig. 19.2(b), we show how to make a return path for the charges by forming a closed loop. Unfortunately, with this arrangement there is a pileup of charges on *both* vertical legs. Once again, no current is produced. We can get around this difficulty in a simple way by keeping the return path out of the magnetic field as shown in Fig. 19.2(c). Only the leading side of the loop generates the electric field, which then drives charges around the closed loop. The potential difference from top to bottom of that leading side is equal to the product of the electric field and the length of the

Currents and Fields that Change with Time

side. Such a source of potential difference is what we have previously called an electromotive force or emf, symbolized by $\mathscr{E}$:

$$-\mathscr{E} = El = (vB)l.$$

The width of the loop is l. The negative sign appears in front of the $\mathscr{E}$ because the field is pointed up in the direction of positive charge motion but there is a potential *drop* in that direction. Notice that we have substituted for the electric field E its magnitude in terms of the magnetic field and the velocity of the loop.

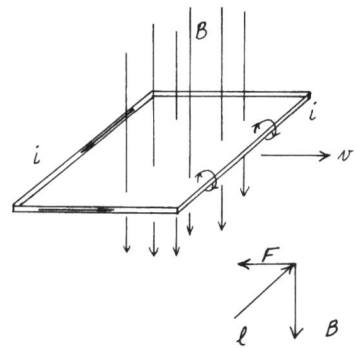

Figure 19.3. The motion of the closed loop into the magnetic field produces a current which interacts with the magnetic field, creating a force opposing the motion.

There is another way to describe the velocity of the loop through the magnetic field. In this particular case the other way comes simply from a geometric substitution, but we will then claim that the result is generally true for any geometry. Note that as the leading side of the loop enters the magnetic field, more and more lines of magnetic force thread through the loop. We can relate the strength of the field, and the size and the velocity of the loop, to the number of lines of magnetic field *entering the loop*:

$$-\mathscr{E} = lvB = l\frac{dx}{dt}B = \frac{dA}{dt}B = \frac{d\Phi}{dt},$$

where Φ is the number of lines of magnetic force. (Note that the area of *the* loop does not change with time. What is changing is the area of the loop *through which magnetic field lines go*.) Sometimes the source of emf is ascribed to the "cutting" of magnetic field lines by the conductor. I think that this term, though historical, is misleading.

The number of lines of force is called the *flux* Φ. The magnetic flux in a region is equal to the product of the magnetic field B and the area A providing that the field lines are perpendicular to the surface of the area. According to our derivation, the emf produced by the motion of this loop into the magnetic field is equal to $d\Phi/dt$, the time rate of change of the magnetic flux through the loop. (Remember that the flux can be pictured as being the total number of field lines. The emf does not depend on the *number* of field lines in the loop but rather on how *rapidly* the lines are leaving or entering the loop.)

Let us give some physical reality to our thought experiment by assigning reasonable values to the magnetic field, the velocity, and the dimensions of the loop. If the length of the leading side of the loop is 1 m, and it is moving with a velocity of 1 m/s into a magnetic field produced by a large research magnet with $B = 1$ T, then the emf produced is -1 V. In this case an emf of 1 V is available to produce a circulating current. The direction of the current is counterclockwise, which according to our usual convention agrees with the negative sign for the emf.

■ EFFECT OF INDUCED CURRENT: LENZ'S LAW

The motion of a *conducting* loop in a magnetic field produces a current in the loop. The current is said to be "induced." Consider now the fact that the induced current must interact with the magnetic field that gave rise to it in the first place. A current-carrying wire perpendicular to a magnetic field experiences a force: $\mathbf{F} = i(\mathbf{l} \times \mathbf{B})$. Let us figure out the direction of that force in the case of the loop that we used in the thought experiment.

The right-hand rule of vector multiplication, as shown in Fig. 19.3 predicts that the resulting force will be to the left in the direction opposite to the velocity. We get the same result if we use the argument about combining magnetic fields. As shown in the diagram, the induced current produces field lines that reinforce the field in front of the leading edge of the loop and reduce the field in back. According to our model, the resulting force on the conductor is therefore to the left, in the direction opposite its velocity.

There is another way to predict the direction of the induced current. Note that the field produced by the induced current tends to cancel the field lines that are sweeping into the loop. A generalization concerning the direction of the induced current is called Lenz's law. Heinrich Lenz (1804–1865) proposed that induced current is always established in a direction to oppose the change of magnetic field that produced it in the first place. A conducting loop moving into a magnetic field experiences a drag force. Furthermore, the induced current produces a magnetic field that tends to maintain the status quo of magnetic flux in the loop.

Consider the alternative! If the induced current produced fields so that the force on the loop were in the direction of motion, then the loop would move faster in that direction. The resulting higher velocity would induce a larger current. This current in turn would produce a larger force which would provide an even larger velocity. We would have perpetual motion, but, of course, it does not work that way. (Later in this chapter we will see how to make induction motors that produce an effect something like the one we have just described. The energy, however, is provided by continuously changing magnetic fields.)

If Lenz's law requires that the induced current create a drag force on the moving loop, then we will have to provide an equal force in the forward direction just to keep the loop moving with the same velocity. That force is $F_{\text{needed}} = ilB$. The induced current i satisfies Ohm's law for the loop circuit: $i = \mathcal{E}/R = Blv/R$. Therefore, the force that we must provide is

$$F_{\text{needed}} = \left(\frac{Blv}{R}\right) lB = \frac{B^2 l^2 v}{R}.$$

The power that we must provide to exert this constant force on an object moving with velocity v is

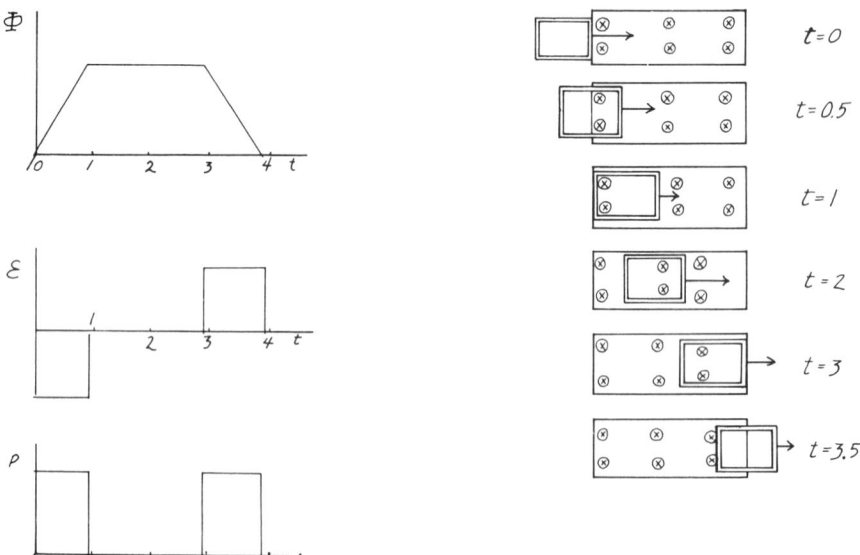

Figure 19.4. Magnetic flux (Φ) through the coil, emf($\mathcal{E}$) generated, and power (P) required as loop moves into and out of field.

Currents and Fields that Change with Time

$$P_{\text{needed}} = Fv = \frac{B^2 l^2 v^2}{R}.$$

Note that this expression is equal to $i^2 R$. The energy that we supply to keep the loop moving with constant velocity does not go into kinetic energy, but rather is dissipated in the $i^2 R$ loss of heat.

In Fig. 19.4 we show how three different quantities vary as a function of time while the loop is passing through the magnetic field. We have graphed Φ, the flux through the loop; $\mathscr{E}$, the emf produced; and P, the power required. Up until now we have considered only what happens as the loop enters the magnetic field. Once the entire loop is in the magnetic field, the induced field on the back side of the loop is in the same direction as that on the front side. The total emf is 0 and there is no current. As the leading side of the loop leaves the magnetic field, the flux in the loop decreases and the emf is generated by the back side of the loop with the front side serving as the return circuit. The emf and the current are now in the clockwise direction. Note how the direction of emf and changing flux agree with the formula. $\mathscr{E} = -d\Phi/dt$. The emf is the negative of the slope of the graph of $\Phi(t)$. When the slope of $\Phi(t)$ is positive, the emf is negative and vice versa. The power required to maintain the velocity of the loop, which, of course, is also the power dissipated in the loop, is positive, independent of the polarity of the emf or the direction of the current.

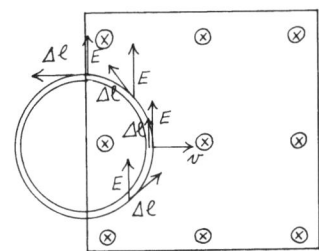

Figure 19.5a. Effect of a circular conducting loop entering a magnetic field.

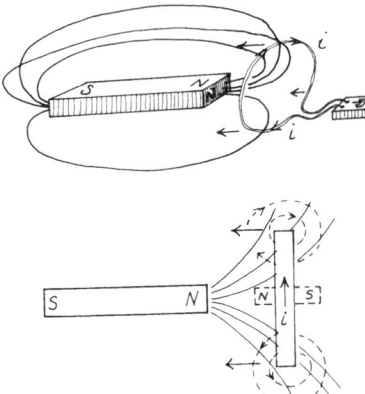

Figure 19.5b. Conducting loop passes over bar magnet, generating current.

■ FARADAY'S LAW—THE NONOBVIOUS GENERALIZATION

For the particular geometry of the rectangular loop entering a uniform magnetic field, we can express the emf induced as $\mathscr{E} = -d\Phi/dt$. In this case the emf is just equal to the product of the induced electric field and the length of the region in which the field exists: $\mathscr{E} = -El$. If electric fields are induced in other parts of the loop, the products of those fields and distances add up to produce a total emf around the loop. We now claim that regardless of the geometry of the *closed* loop, the total emf around it is equal to the time rate of change of the magnetic flux through the loop. This generalization is called Faraday's law:

$$\mathscr{E} = \int_{\text{loop}} \mathbf{E} \cdot d\mathbf{l} = -\frac{d\Phi}{dt}.$$

Let us apply Faraday's law to a number of special situations; each will be only a small departure from the geometry or the operation of the preceding example, but step by step we will arrive at phenomena and concepts very different from that with which we started.

1. In Fig. 19.5(a) we show a round loop entering a uniform magnetic field. The only difference between this situation and that of the rectangular loop is that in this case the induced electric field is not generally in the direction of the conducting path. The emf produced along a short distance is equal to the dot product of electric field and line segment: $\Delta\mathscr{E} = -\mathbf{E} \cdot d\mathbf{l}$. The integration of the dot product is possible but messy. According to Faraday's law there is no need to do the detailed integration. The total emf around the loop is just proportional to the time rate of change of magnetic flux threading through the loop. That change of flux is proportional to the change of area in the magnetic field.

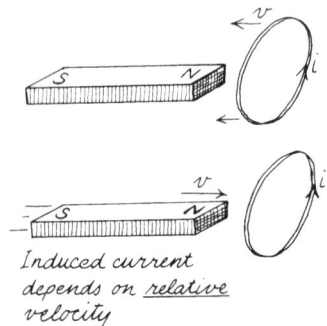

Figure 19.5c. Bar magnet passes through conducting loop, generating current, or loop passes over magnet. Only relative velocity matters.

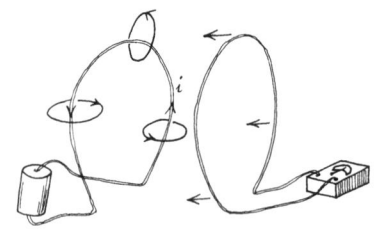

Figure 19.5d. The bar magnet can be replaced by a current-carrying loop. Current is generated in the other loop if the two loops are moving toward or away from each other, or if the current in the first loop is changed.

2. In Fig. 19.5(b) we show a loop about to pass over the end of a bar magnet that is lying on the axis of the loop. The closer the loop gets to the end of the magnet, the more field lines pass through the loop. As the flux through the loop changes, there will be an emf around the loop as required by Faraday's law. The direction of the induced current in the loop will be such as to produce a magnetic field that opposes the motion. Since a current loop acts like a magnetic dipole, the induced current will make this loop act like a small bar magnet with its north pole in the direction of the north pole to which it is heading. The induced current thus produces repulsion. When the loop is around the center region of the magnet, the number of magnetic field lines passing through the loop is maximum. However, there is very little *change* in the flux, and therefore the induced current will be small. As the loop passes beyond the center of the magnet, the number of field lines through the loop decreases, slowly at first, then rapidly, inducing a large current in the loop in the *opposite* direction. The retreating loop thus has the magnetic appearance of a dipole with its north end being pulled away from the south pole of the magnet.

   Note that so far we have been talking about loops moving into and through magnetic fields. The next step of generalization is obviously true in the case of the loop passing over a bar magnet; it makes no difference whether the loop moves over the magnet or the magnet moves through the loop in the opposite direction. Only the *relative* velocity is important. [See Fig. 19.5(c).]

3. The next case to consider [Fig. 19.5(d)] is a simple extension of the one in number 2. Instead of a bar magnet, we have an equivalent magnetic dipole produced by a current loop. It is reasonable to expect the same kind of induced currents in the primary coil as we had when there was a permanent magnet. As before, only the relative velocity between the two coils is important.

4. Once again we have two coils placed face to face, but this time we will not require relative motion between the two coils. Instead, the primary coil is equipped with a switch so that the current can be turned on or off. When the current is off there is no magnetic field produced by the primary and thus threading through the secondary. As soon as the switch is thrown, however, the primary produces a magnetic field, which links the secondary coil. This means that the flux through the secondary coil has increased with time and therefore there is an emf around the secondary coil, producing an induced current.

5. Each of these cases of induction has been only slightly different from the preceding one and yet notice how far removed we are from the original situation of a rectangular coil moving in a magnetic field. In that original situation we could explain the induced emf in terms of Lorentz's law that the force on a moving charge is caused by the cross product of **v** and **B**. In number 4 there was no physical motion yet, nevertheless, an induced current was produced. At least we can justify the effect, however, by noting that the free electrons in the secondary coil are in a region of changing magnetic field that somehow exerts a force on them.

   In Fig. 19.5(e) we show two geometries of a coil in a magnetic field. In both cases the magnetic field can be turned on, rising from 0 strength to maximum. The situation shown at the top is essentially the same as we saw in the earlier cases, and we would expect to find an induced emf. There is a peculiar feature of the case shown at the bottom, however. There is a changing magnetic flux through the center of the coil, but *nowhere is the conductor itself in a magnetic field*. Nevertheless, according to Faraday's law, there must be an induced emf in the loop equal in magnitude to the time rate of change of the magnetic flux through the loop.

Now we can see why Faraday's law is not an obvious extension of our earlier formulas concerning the force on a charge moving in a magnetic field. There are really two different laws at work in these cases, both of which can be used to predict the same result in some geometries but each applicable by itself in certain other cases. Consider, for instance, our original geometry of a conducting bar moving perpendicular to a magnetic field. We claimed that the magnetic force on the moving charges in the bar made them move as if they were subject to an electric field: $E = vB$. At that point of the argument, however, we said nothing about the conductivity of the bar. As a matter of fact, the same electric field would be produced in a broomstick shoved across the magnetic field. Even the broomstick is not necessary. Depending on how you arrange the voltmeter and the leads, the same emf produced by the electric field operating over the length l could be detected by moving the two probes of the voltmeter without any rod between them at all. As we have seen before, a magnetic field in one reference frame can turn into an electric field in a reference frame moving with constant velocity relative to the first. [See Fig. 19.6].

An example of the way in which electric and magnetic fields depend on the velocity of the observer is shown in Fig. 19.7. A long conducting bar moves along the direction of its length and perpendicular to a magnetic field. If an ammeter is arranged with skid contacts, as shown in the diagram, then it will register a continuous current as the bar moves. In a later section we will see how Faraday made a crude generator out of a similar geometry. The effect produced in the case shown here might be explained in terms of the Hall effect, which we described on p. 437. The free electrons in the bar experience a force perpendicular to the velocity of the bar and to the magnetic field. Hence, they experience an electric field, directed across the bar, that will keep them moving up through the ammeter and down the other side, forming a continuous current. However, if the ammeter and the whole detection system moves along with the bar through the magnetic field, then there will be no current. In such a case, the top part of the loop near the ammeter will also have an induced emf, which will buck the effect produced in the bar.

In the case of the moving bar, there is no closed loop in which there is a changing flux. Nevertheless, we get an induced electric field because of the relative motion. It is also possible to have a case where the flux through a closed loop changes and yet there is no induced emf. We show such a situation in Fig. 19.8. The magnetic field through the region remains constant but the area of the conducting region changes drastically as the tiny contact point slides along the gap. If Faraday's law applied to such a situation, we could get a large emf and induced current even though very little work would have to be done to move the small contact. Here is another way to produce the same effect. If a bicycle clip encircles a bar magnet and is pulled off over the end of the magnet, there will be an induced emf in the metal loop of the clip. However, if the clip is simply pulled off the magnet over its center part by letting the jaws of the clip open—even

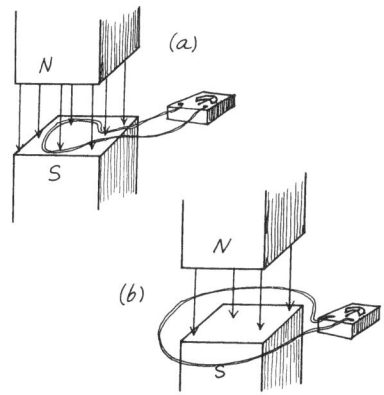

Figure 19.5e. Current is generated in the loop by a changing magnetic field, whether or not the wire is actually in the field.

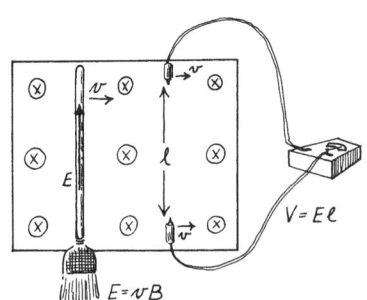

Figure 19.6. The emf between two points moving through a magnetic field is independent of the resistance between the two points.

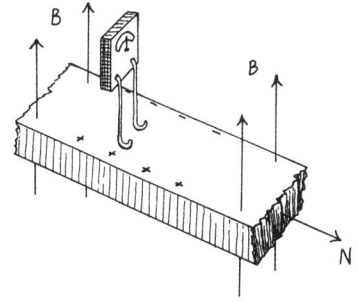

Figure 19.7. A variation of the Hull effect generates a continuous current.

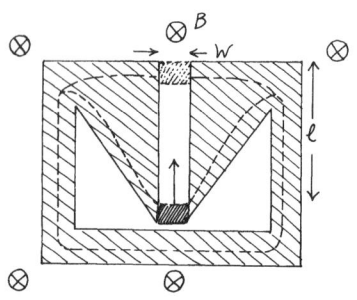

Figure 19.8. Copper sheet is perpendicular to magnetic field. As zipper is pulled, possible current path changes, increasing area of loop in B. Since there is ΔA, there is apparently ΔΦ. However, if there is ε=-(ΔΦ/Δt), current would be produced and there would be ohmic heating. The work done in pulling the zipper is W=lBiw, where l is the path length, i is the current, B the magnetic field, and w the width of the gap. Since w can be made arbitrarily small, the work done must go to 0. Therefore, there can be no ε.

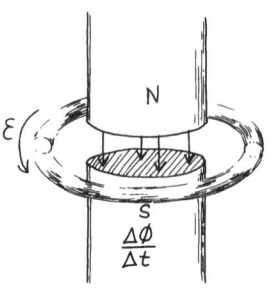

Figure 19.9. In a changing magnetic field, the electric potential between two points is a function of the route. In this case, a test charge would gain energy each time it went around the enclosed changing field.

though electrical continuity is maintained by having the jaws in contact with the iron magnet—there will be no induced emf. In both cases the flux through the clip changes to 0. Evidently, Faraday's law cannot apply to a circuit in which the material in the path changes.

Let us summarize the two quite different laws that give rise to induced emfs. The first, which is named after H. A. Lorentz (1853–1928) describes the effects of electric and magnetic fields on an electric charge. It also implies that motion through a magnetic field is equivalent to an electric field in a stationary frame:

$$\mathbf{F} = q(\mathbf{E} + \mathbf{v} \times \mathbf{B}).$$

The velocity **v**, the electric field **E**, and the magnetic field **B** may be measured with respect to *any* inertial frame, but all must be measured in the same frame. The other law of induction is Faraday's: $\mathscr{E} = -d\phi/dt$.

■ AN EMF IS NOT A CONSERVATIVE POTENTIAL

When we first defined electric potential, we were dealing with a property of electric fields around *static* charges. If you take a test charge around any closed path in such a field, the net work done is 0. Otherwise, as we pointed out on p. 388, energy would not be conserved. If you could take a test charge around a closed loop and have a gain of energy, you could provide perpetual motion.

Now we have found how to produce emfs around closed loops through which the magnetic field is changing. In the system shown in Fig. 19.9, there is a constant electric field around the circular path surrounding the constantly increasing magnetic field in the center. (Of course, the magnetic field can only increase for a limited time.) If an electric charge is allowed to travel along such a path (guided by a magnetic field of the correct, and changing, strength), it will experience a continual force and gain energy. The energy in joules gained per coulomb in one tour around the loop is just equal to the emf in volts. It is not necessary for the charge to travel in a circular path in order to pick up this extra energy. According to Faraday's law, there is the same total emf around *any* closed path that surrounds the changing flux.

Does this phenomenon violate the law of conservation of energy? Not at all. In the case of static fields, there is no source to provide energy to a charged particle going around a closed path and returning to the original point. In the case of the changing magnetic flux, the energy given to the test charge comes from the source that is providing the change of the magnetic field. When we describe the action of transformers, we will see the mechanism by which the source must provide this energy.

Although induced emf is measured in volts, it is frequently confusing to think of a potential difference between two points in a changing magnetic field. For instance, suppose that the circular ring in Fig. 19.10 has a resistance of 1 Ω and is threaded by a changing magnetic field that produces an emf around the loop of 1 V. While the field is changing at that rate, there will be a current of 1 A in the ring. What then is the potential difference between the points *A* and *B*? Since these are the end points of an arc that is 1/6 the circumference of the ring and since there is an emf of 1 V around the whole ring, you might argue that there would be a potential difference of 1/6 V between points *A* and *B*. On the other hand, *A* and *B* are also the end points of the arc that is 5/6 of the whole circumference. Therefore, the potential difference across *A* and *B* should be 5/6 V. As a matter of fact, if two voltmeters are connected to the ring as shown in the diagram, V_1 will read 1/6 V and V_2 will read 5/6 V. Yet, the two voltmeters are connected to the same two points! You can check the consistency of these readings by applying Kirchhoff's laws to the circuit. Just add up the emfs and the IR drops around the several closed loops indicated in the diagram. Finally, note that the leads going to the voltmeters are also part of the circuit through which the changing magnetic field is passing.

■ EDDY CURRENTS

A magnet has almost no effect on a piece of copper that is standing still. However, if the magnet and copper move with respect to each other, or if a magnetic field through the copper increases or decreases with time, there is an interaction. The changing magnetic field or the motion of the conductor through the field sets up emfs that produce current. The induced currents interact with the magnetic field, producing a force on the conductor. These currents frequently swirl through the conductor along the leading and trailing edges of the moving magnetic field region. Because of the similarity of their patterns to those seen around an object moving in water, these induced currents are called *eddy* currents. In some machines they are a nuisance; in others they provide the main mechanism for operation.

The damping force produced by eddy currents is proportional to the relative velocity of conductor and magnetic field. This effect is deliberately used as a brake in certain machines. Radial-arm saws, for example, are usually equipped with a horseshoe magnet that swings down over the saw blade when the power is turned off. The resulting eddy currents bring the blade rapidly to rest.

Eddy currents can also provide lift to a magnet that is moving rapidly over the surface of a conductor. The induced poles in the conductor oppose the intrusion of magnetic field and thus oppose the approach of the moving magnet. Both drag and lift are created. As Rossing pointed out in an article on Maglev devices, the drag force is proportional to v at low velocities, reaches a maximum and then decreases as $1/\sqrt{v}$. The lift force, however, is proportional to v^2 and can be used to provide practical lift for large moving objects. The qualitative explanation appeals to a model of the magnetic lines of force from the advancing magnet penetrating the conductor. At high speeds the lines do not have time to penetrate far; the opposite pole, which provides lift, is induced close to the surface. The flux that penetrates the conductor must be dragged along by the moving magnet, creating drag, but the faster the magnet moves, the smaller the amount of flux that has penetrated and so the smaller the drag.

Eddy currents must be minimized in machinery that rotates in magnetic fields or in the iron cores of transformers where magnetic fields continually change in magnitude and direction. Eddy currents in the rotors of motors or generators would produce drag. In the cores of transformers the eddy would create I^2R loss. To avoid the eddy currents, the rotors or cores are made of thin sheets of iron separated by even thinner layers of insulating material such as enamel or mica. Such laminated pieces of iron provide good paths for the magnetic fields but do not provide electric conduction paths in the perpendicular direction. The humming sound heard around some high-power transformers is caused by the cyclic compression of the laminations.

If a bar magnet is moved toward a copper sheet as shown in Fig. 19.11, the eddy

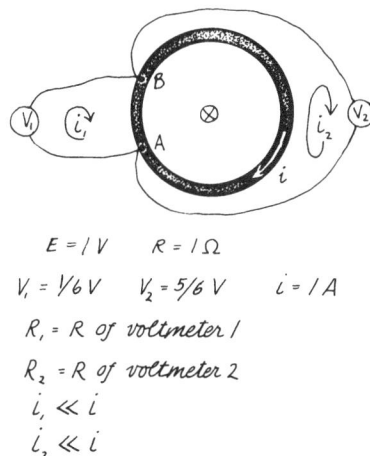

$E = 1 \text{ V} \quad R = 1 \, \Omega$
$V_1 = 1/6 \text{ V} \quad V_2 = 5/6 \text{ V} \quad i = 1 \text{ A}$
$R_1 = R \text{ of voltmeter 1}$
$R_2 = R \text{ of voltmeter 2}$
$i_1 \ll i$
$i_2 \ll i$

Figure 19.10. The "potential difference" between two points in a changing magnetic field depends on the geometry of the measuring system.

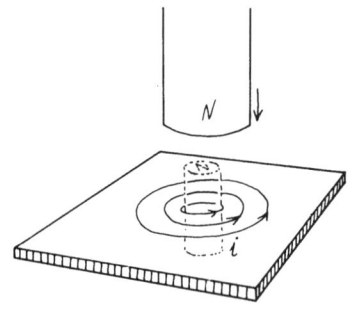

Figure 19.11. A north pole advancing toward a conducting plate induces currents that create a reflection north pole on the opposite side of the plate. The real north pole is repelled by its image.

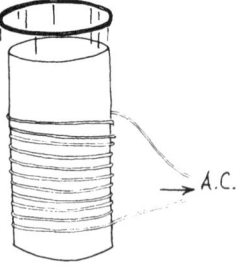

Figure 19.12. ac Propulsion Coil.

currents in the sheet will oppose the penetration into the sheet of the incoming magnetic field. The induced current loops set up a magnetic dipole with magnetic fields opposing the incoming dipole. Because of the resistance of the copper sheet, the electric currents die out and the upward force on the magnet stops. However, if the conducting sheet is a concave shaped type 1 superconductor, such as a lead bowl at a temperature close to absolute zero, then the magnet will float above the surface. The eddy currents do not die out and if the magnet tries to slip to the edge it has to climb up hill. With the warm type 2 superconductors (above the temperature of liquid nitrogen), magnets will float above surfaces that are flat. There is some penetration of field lines with these superconductors, resulting in a locking action that keeps the magnet from floating over an edge.

Eddy currents can be continuously generated with an alternating magnetic field driven by alternating current. Figure 19.12 shows a standard classroom demonstration that makes use of this effect to shoot a copper or aluminum ring off the core of a solenoid. The magnetic field in the central core changes strength and polarity at 60 Hz. The conducting ring that slips over the core has an emf generated around its circumference, which produces circumferential currents. These induced eddy currents also change direction at 60 Hz but with a phase difference. Since $\mathscr{E} = -dB/dt$, the induced emf is proportional to $-\omega \cos \omega t$ if B is proportional to $\sin \omega t$. If the induced field were exactly 270° out of phase with the primary field, there would be no net repulsion during a complete cycle. However, the loop circuit driven by the induced emf is a combination of inductance and resistance. If the load were purely resistive, the current would be in phase with the emf, and there would be no net repulsion. If the load were purely inductive, the induced current would lag the emf by 90°, and thus the field it generates would continually be 180° out of phase with the primary field. It turns out that for rings of the standard geometry used, the inductive reactance is about equal to the resistance, yielding a phase angle providing overall repulsion. During more than half of each cycle the induced current is in a direction to produce a magnetic field that opposes the primary field. The result is dramatic. The ring is subject to forces that repel it from the central core and shoot it into the air. To complete the demonstration, try a ring that has a cut through it, so that there cannot be a circumferential current. The ring does not move. Also cool the copper ring in liquid nitrogen. The lower resistance allows a larger induced current and also moves the phase angle closer to the optimum 90°. The ring jumps higher. (For a detailed analysis of this effect, see the article by Quinton.)

■ ELECTRIC MOTORS

Electric motors are so commonplace in our civilization that it is hard to realize how new they are. Although Faraday produced an electric motor 160 years ago, very few American homes contained anything with an electric motor before the 1920s. Before that time, for instance, cars were started with a crank; rugs were cleaned by sweeping; and alarm clocks were powered by springs.

There are many different kinds of electric motors, most of them designed with very sophisticated engineering techniques. The efficiency of energy conversion for large motors must be very high. If an appreciable fraction of the electrical energy supplied stayed in the motor, the high temperatures developed would ruin it. The usual classroom demonstration motor gives little indication of the engineering features required.

The basic principle is simple enough. A current-carrying wire in a magnetic field experiences a force $\mathbf{F}=i(\mathbf{l}\times\mathbf{B})$. If the wires are wound on a rotor in a stationary magnetic field, then the current in the coil must be changed in direction each half cycle so that the rotor will always experience a force in the same direction. With stationary north and south poles the rotor coil would experience variable torque. The torque could be smoothed out by winding several coils on the rotor and having the stationary magnetic fields divided into several segments.

Many other variations are possible. The current supplied to the rotor coils need not change direction if the external magnetic fields change their direction. Most large electric motors are fed by a type of alternating current called three phase. Without any moving mechanical parts, this type of power supply creates magnetic fields in the stationary part of the motor that rotate continuously. The rotor coils just follow along, rotating at the same frequency, as shown in Fig. 19.13.

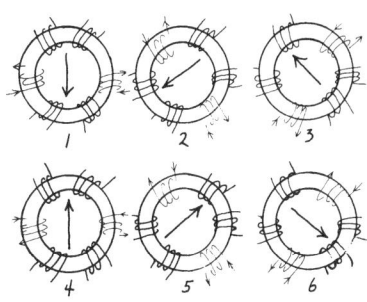

Figure 19.13. As the coils are energized in sequence, as shown, the return flux in the center of the doughnut iron ring rotates. Three phase A-C, each phase connected to an appropriate pair of coils, would smoothly energize the coils in the required sequence.

It is not even necessary to supply current to the rotor coils, at least with wires. If the surrounding magnetic field rotates, then current is induced in the coils of the rotor. The effect is basically the same as the drag produced by eddy currents when a conductor moves in a magnetic field. In this case the rotor coils are dragged along following the rotating magnetic field. They cannot rotate at the same frequency, however. If they did, there would be no relative motion and hence no induced currents. Therefore there must be a constant slip. If the stationary coils produce a magnetic field that rotates at 60 Hz, the rotor may follow along at 58 Hz, providing that the motor does not deliver much power. As the external load on the motor increases, the induced current in the rotor coils must increase. This can happen if the relative velocity between rotor coils and rotating magnetic field increases. Therefore, the rotor slips a little more, perhaps decreasing in frequency to 56 Hz.

Some motors, such as those in clocks, must be locked to a specific frequency determined by the alternating current frequency provided by the power station. A primitive rotating magnetic field is produced in the clock with single phase alternating current by "shading" alternate legs of the stationary magnet. A conducting ring around one leg of an ac magnet (or sometimes an aluminum plate over the end) experiences induced currents that slow down the change of the magnetic field in that region. The result is that the magnetic field in that leg lags in phase behind the field in the other leg. The rotor in a clock motor is a steel disk. The rotating magnetic field induces a magnetic field in the disk, which because of hysteresis in the steel creates a temporary magnet that tags after the rotating field. As the rotor magnet sweeps around, it feels a pull from the unshaded leg and as that force dies down, the rotor is moving into the delayed pull of the shaded leg.

Notice that the induction or synchronous motors do not have commutators to change the direction of current in the armature. A dc motor must have commutators or slip rings. The design of large motors and the choice of induction, dc, single-phase ac, or three-phase ac involves complicated engineering and cost considerations. These topics used to be part of introductory-physics courses up until about 60 years ago. There is still a lot of good physics to be learned from the analysis of electrical machinery, and such studies might form special projects for advanced students. Let us mention just one particular phenomenon that is frequently observed with medium-size household motors, such as those in air conditioners or pumps. When these motors are turned on, they usually draw much more current than when they are up to speed. The resultant drop in supply voltage is enough to dim lights or shrink TV patterns for units drawing power from the same line. The armature (rotor) windings of the motor have low resistance and thus will draw large current. As the rotor starts turning, however, the circuit loops move through the magnetic field of the stationary coil, producing a back-emf proportional to the speed. The final current through the armature is proportional to the imposed voltage minus the back-emf. Very large motors must have special turn-on switches to limit the initial current.

■ ELECTRIC GENERATORS

When Franklin studied electricity, he had to use electrostatic devices to separate charges and store them. Experimenters of that era were able to produce high-voltage effects but currents were small and brief. The discovery of electrochemical cells in the early 1800s led to rapid development of electromagnets and electrochemistry, which require large currents. Meanwhile Faraday had discovered how to turn mechanical energy into electrical energy through the induction effect. It took about another 50 years for the widespread development and installation of large electrical generators.

One of Faraday's first versions of a generator was the so-called homopolar generator shown schematically in Fig. 19.14. The device uses the same induction principle that we have seen before. A radius of the moving disk sweeps through the magnetic field and hence has an emf generated across its (radial) length. Electric charge of one polarity builds up on the perimeter, and charge of the other polarity on the axis. If the circuit is completed by sliding contacts, then there will be a steady current down the axis, out the radius, through the external circuit, and back again. Such a generator works well but is not particularly efficient in its use of space or materials. There are also problems in extracting high currents from sliding contacts.

Generators and motors can have the same basic design. If the rotor windings are provided with current, then the rotor will turn and provide mechanical energy. On the other hand, if some external source of mechanical energy turns the rotor, the induced electric current in the windings can be extracted and used elsewhere. Some motor-generators are designed to be used either way. For instance, electric motors for subways have been designed so that when the trains are braked, the wheels drive the motors as generators sending electrical power back into the system. The resulting drag furnishes a major share of the braking power needed.

In very large generators, such as in a central power station, a small current is supplied to the rotor, which then creates a rotating magnetic field. The large induced current is taken from the stationary surrounding coils, thus eliminating some of the problems caused when large currents have to pass through sliding contacts.

■ INDUCTANCE

A current in a coil produces an axial magnetic field. If the current changes, the magnetic field changes. On the other hand, if the magnetic field through the coil changes, an emf will be set up in the coil. The current thus induced will set up a magnetic field in a direction to resist the changing magnetic field that produced the effect in the first place. Consider the consequences of this complicated circular interaction. If you try to change the current in a coil and thus change the magnetic field through it, an extra current will be induced in the coil to prevent what you are

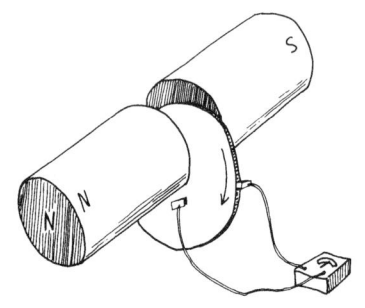

Figure 19.14. The generator disk is off-center from the axis of the magnetic field to provide a way to mount the axis of the disk.

trying to do. If you throw a switch to impose a potential difference across a coil, there will be an induced back-emf to slow down the rise of the current and the magnetic field. On the other hand, if a current and a magnetic field exist in the coil and you try to turn them off, an emf will be generated that will try to keep the current going.

Regardless of whether the magnetic field is generated by the coil itself or is imposed from the outside, the emf of the coil is equal to

$$\mathscr{E} = -\frac{d(N\phi)}{dt}.$$

The number of turns in the coil is N. Since these turns are wound in series, the total emf is simply the product of N and the emf induced in a single turn.

If the magnetic field in the coil is generated by the coil itself, then that magnetic field is proportional to the current in the coil: $B \propto I$. The total flux threading the coil is equal to the product of the area of the coil and the average magnetic field through the plane of the coil: $\Phi = AB_{av} \propto AI$. In all cases, the magnetic flux depends on the geometrical factors involved and is proportional to the current. We can separate out the geometrical factors and describe the self-induced emf in the following way:

$$\mathscr{E} = -L\frac{di}{dt}.$$

The new constant L, which describes the geometrical factors of the coil, is called the *inductance*. The unit of inductance is the henry (H), named for Joseph Henry (1797–1878). Henry was the American contemporary of Faraday. He discovered many of the same electromagnetic effects that Faraday did and was particularly knowledgeable about the construction of electromagnets.

Let us calculate the inductance of a solenoid. Since by definition $N\Phi = Li$, the inductance itself is equal to $L = N\Phi/i = N(BA)/i$. In equating Φ to BA, we are assuming that the magnetic field in a solenoid is uniform throughout its cross section A. This is approximately the case for a solenoid that is much longer than the diameter of its core. The field of such a solenoid is $B = \mu_0 ni$. Therefore, the inductance is equal to

$$L = (NA/i)(\mu_0 ni) = N\mu_0 nA = \mu_0(N/l)Al = \mu_0 n^2 V.$$

In this derivation we first found that the inductance of a solenoid is proportional to the total number of turns, N, and also to the number of turns per meter, n. To avoid having two kinds of n in the same formula, we then divided the total number of turns by the length l and compensated by multiplying the cross-sectional area by l. The final result is that the inductance of a solenoid is proportional to the square of the number of turns per meter and to the volume enclosed by the solenoid. Notice that the inductance is proportional to the *square* of n because the greater the value of n the stronger the field B, and also the greater the value of n the more turns of emf are connected in series.

A standard school laboratory solenoid has a length of 10 cm and the cross-sectional area of its core is about 75 cm^2. It has 3400 turns piled into several layers. Let us use the formula that we have developed to calculate the inductance of this solenoid, even though our formula will provide only a rough approximation:

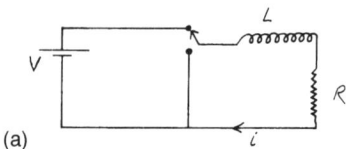

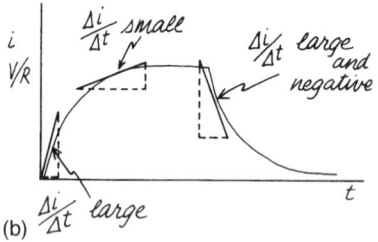

Figure 19.15. (a) A circuit for charging and discharging an inductor. (b) Current as a function of time in the circuit.

$$L=\mu_0 n^2 V=(4\pi\times 10^{-7})(34\,000 \text{ turns/m})^2(0.1\times 75\times 10^{-4} \text{ m}^2)=1.1 \text{ H}.$$

A coil designed to introduce inductance into a circuit is called an *inductor*. The inductance of an inductor can be greatly increased if the core is filled with iron instead of air. In that case the magnetic field is increased by the χ factor for the iron and the resulting inductance may be larger by a factor of 100 to 1000. In electronic circuitry these devices are called chokes. We will soon see why.

■ CHARGING AN INDUCTOR

It takes time for current to rise and fall in an inductor. The circuit shown in Fig. 19.15(a) is a way to connect a battery across a resistance and inductor in series and then to bypass the battery and let the current in the circuit decay. In Fig. 19.15(b) we show the current as a function of time for the charging and discharging. When the battery is connected, the current rises to a final value given by V/R. When the switch is thrown to ground, the current eventually dies to 0.

Consider what happens when the switch is first connected to the battery. Current suddenly starts to rise in the inductor but that means that di/dt is large. The back-emf produced by the coil has the value $L(di/dt)$. The potential difference across the resistor is given by the difference between the battery voltage and the back-emf across the inductor: $i=[V-L(di/dt)]/R$. The current is not blocked completely and as it increases, the *slope* of $i(t)$ decreases. Therefore the back-emf decreases allowing yet more current in the circuit. During discharge the same phenomenon takes place. Suddenly, $V=0$. The current starts to decrease rapidly but now the slope of $i(t)$ is negative. The emf is opposite in sign to what it was before and tends to keep the current going *in the same direction* that it had been.

Let us analyze the circuit during the decay process when the battery is not connected. According to Kirchhoff's law, the potential changes are

$$-L\frac{di}{dt}-iR=0 \rightarrow \frac{di}{dt}=-\frac{R}{L}i.$$

According to this formula, the slope of the $i(t)$ curve is proportional to the current at each instant and is negative. We had the same equation and the same graph on p. 452 to describe the decay of charge on a capacitor. The curve is that of an exponential function:

$$i=Ie^{-(R/L)t}, \quad \text{where } I=V/R.$$

When $t=0$ (when we throw the switch to the ground connection) the equation says that $i=Ie^0=I$. For very long times, as $t\rightarrow\infty$, the current goes to 0. At the particular time when $t=L/R$, the current is equal to $i=Ie^{-(R/L)(L/R)}=Ie^{-1}=I/e$.

Let us find the time constant for the school solenoid described above. The inductance is $L=1.1$ H and the resistance is $R=50$ Ω. Therefore the time constant is $\tau=L/R=(1.1 \text{ H})/(50 \text{ Ω})=0.022$ s. The time that it would take the current to fall to half of its original value is slightly less than this: $t_{1/2}=0.69\tau=15$ ms.

The back-emf produced by trying to stop the current in a large inductor can be large and dangerous. The inductance of a large research magnet, for instance, might be 10 H. The current in the coil might be 100 A. If you try to interrupt the current by throwing a large switch or if there is an accidental break in the circuit, then the back-emf that would appear across the switch or the break would be equal to $\mathscr{E}=-L(di/dt)=-(10 \text{ H})(100 \text{ A})/\Delta t$. Even if Δt were 1 s, the emf would still be 1000 V. An actual break in the line or the operation of a switch would take much less time than 1 s and so the back-emf would be much greater. This large potential difference would appear across the switch or the break and create an arc that would melt the leads—and anything else that got in the way. Special provisions must be made for handling large currents that exist in large inductors, even though the original driving voltages may be low.

■ THE ENERGY IN AN INDUCTOR

If current is suddenly interrupted in an inductive circuit, where does the energy come from to fuel the arc that melts the leads at the break point? An inductor stores magnetic field. It takes energy to create such a field and the energy becomes available again as the magnetic field decreases.

The power used during the creation of the magnetic field in an inductor is equal to the average value of the product of current and potential difference across the coil. P=average of $i[L(di/dt)] = dU/dt$. The stored energy is U_L. The final stored energy when the equilibrium current I has been reached is $\int dU = L \int i\, di = \frac{1}{2}LI^2$. Compare this expression for stored magnetic energy with the similar one for stored electrostatic energy in a capacitor: $U_C = \frac{1}{2}CV^2$.

Let us find out how much energy can be conveniently stored in a solenoid like the one described on p. 465. The inductance for that solenoid is 1.1 H and its resistance is 50 Ω. If we use a 6-V battery the equilibrium current is equal to I =(6 V)/(50 Ω)=0.12 A. The stored magnetic energy is equal to $U_L = \frac{1}{2}LI^2 = \frac{1}{2}(1.1\text{ H})(0.12\text{ A})^2 = 7.9 \times 10^{-3}$ J. This solenoid has a mass of about 2 kg. The stored energy would barely be enough to lift the solenoid 0.4 mm. There would seem to be no danger in handling the amount of energy that we have just calculated. Note, however, that the stored energy depends on the square of the current. Let us find a way to calculate the energy *density* of a magnetic field and then we will have a better way of finding the stored energy in a large research magnet.

In a solenoid the energy density is equal to the stored energy divided by the volume of the solenoid:

$$u_B = \frac{U_B}{V} = \frac{\frac{1}{2}LI^2}{V} = \frac{\frac{1}{2}\mu_0 n^2 V I^2}{V} = \frac{1}{2}\mu_0 n^2 I^2.$$

The magnetic field in the solenoid is $B = \mu_0 n I$. Therefore the energy density can be expressed in terms of the magnetic field:

$$u_0 = \tfrac{1}{2}B^2/\mu_0.$$

The question is: Where is this energy stored? Since the energy is a function of the whole system, it is not obvious that it is actually stored in any particular place. However, we will find it convenient and consistent to assign the stored energy to the space in which the magnetic field lines exist. According to this model, the energy of the solenoid is mostly stored in the core. It is as if space itself were warped with the magnetic field.

Now let us calculate the energy in a typical research magnet used for beam transport around the high-energy accelerators. At full current the magnetic field in such a magnet, and hence in the gap, is about 1.5 T. The iron part of the magnet has a much larger volume than the gap alone does but the energy density is inversely proportional to the magnetic permeability μ. In iron, the permeability is increased approximately by the factor χ: $\mu = \chi\mu_0$. Hence, the value of μ for iron

Currents and Fields that Change with Time

is larger than μ_0 by a factor of from 100 to 1000. Therefore, most of the energy in the magnet is located in the gap between the pole pieces. If the volume of the gap is about 1/5 m³, then the stored energy is

$$U = \tfrac{1}{2} B^2 V/\mu_0 = \tfrac{1}{2}(1.5\ \text{T})^2 (0.2\ \text{m}^3)/(4\pi \times 10^{-7}) = 1.8 \times 10^5\ \text{J}.$$

If you try to get rid of that much energy in a fraction of a second without special precautions, something will melt. If $\Delta t = 0.1$ s, for instance, the momentary power is 1.8 MW.

■ INDUCTANCE AND CAPACITANCE IN SERIES

We have seen that it takes time for charge to drain off a capacitor through a resistance. It also takes time for the current in an inductor to rise to its full value after a potential difference has been imposed across the inductor and the resistance in series. In the first case, electrostatic energy that had been stored in the capacitor must dissipate in the resistance. In the second case, the magnetic field energy in the inductor must build up as the current rises.

Let us combine the two actions. In Fig. 19.16(a), we show a circuit that allows you to charge the capacitor from a battery and then discharge it through an inductor. Although any real wires and real inductors must contain resistance (unless they are superconducting), let us ignore that resistance to begin with. As soon as the switch is thrown to the right, the charge on the capacitor will start to drain through the inductor. The resulting change of current in the inductor will create a back-emf that will slow the decay of charge from the capacitor. Meanwhile the stored electrostatic energy in the capacitor is decreasing and the magnetic energy in the inductor is beginning to increase. Eventually all the electrostatic energy will be turned into magnetic energy. At that time there will be no charge and hence no potential difference across the capacitor. At that same time, however, the current in the inductor is a maximum. Without a potential difference to maintain the current in the inductor, the current and field will start to decrease but the current will continue in the same direction, piling positive charges on the bottom plate of the capacitor. The potential difference across the capacitor rises again but with the opposite polarity. Soon the capacitor is fully charged again but in the opposite sense and the current in the inductor is 0. The energy of the system is once again in the electrostatic form. Now the action repeats but this time the positive charges flow in the opposite direction causing the current and the magnetic field in the inductor to increase but in the opposite sense from the original one. The energy surges back and forth between capacitor and inductor. If there is no dissipation in a resistor, the action will continue indefinitely.

In Fig. 19.16(b) we have plotted graphs of the current, the charge on the capacitor, the magnetic field energy, the electrostatic field energy, and the total energy of the system. The current and the charge vary sinusoidally but 90° out of phase with each other. The potential difference across the system is in phase with the charge on the capacitor since $V = q/C$.

Following the voltage drops of the circuit clockwise, Kirchhoff's law is $q/C - L\,di/dt = 0$. When the slope of the $q(t)$ curve is negative, the current is positive; therefore $dq/dt = -i$. The differential equation is $q/C + L\,d^2q/dt^2 = 0$, which becomes

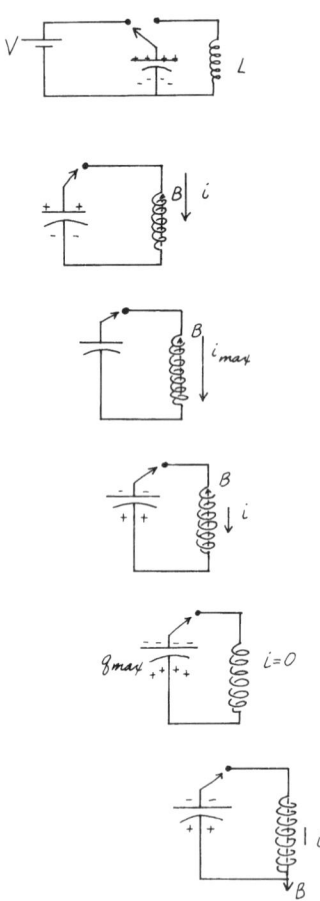

Figure 19.16a. Circuit to charge capacitor from a battery and then discharge the capacitor through an inductor.

Currents and Fields that Change with Time

$$\frac{d^2q}{dt^2} = -\frac{1}{LC}q.$$

To solve this equation, we look for a function $q(t)$ that has the property that its second derivative is proportional to the negative of itself. The sine and cosine have this property. A solution could be $q = q_0 \cos \omega t$. At time $t=0$, $q = q_0$. Take the second derivative of q and substitute it in the equation. The equation is satisfied if $\omega^2 = 1/LC$. Therefore, the frequency of oscillation of the charge will be

$$f = \frac{\omega}{2\pi} = \frac{1}{2\pi}\sqrt{\frac{1}{LC}}$$

If an inductor of 1.1 H and a capacitor of 1 μF are used, the oscillation frequency will be 151 Hz. The period of oscillation is $T = 1/f = 6.6$ ms.

If you actually wire the circuit shown in the diagram and observe the potential difference across the inductor with an oscilloscope, you will not see a continual sine wave with constant amplitude. Instead, you will see a damped sine wave because there is a continuous I^2R loss in the resistor. The differential equation with a resistor becomes

$$L\frac{d^2q}{dt^2} + R\frac{dq}{dt} + \frac{q}{C} = 0.$$

which is satisfied by the function

$$q = q_0 e^{-(R/2L)t} \cos \omega t.$$

This equation is discussed on p. 313 in the chapter on vibrations. The function represents a sinusoidal curve with an exponentially damped amplitude. The damping time constant is $\tau = 2L/R$. In that time the maximum charge on the capacitor, or current in the inductor, has decreased to $1/e$ of the original value.

The energy in the whole system is proportional to the square of the maximum charge in any given oscillation or to the square of the maximum current: $U = \frac{1}{2}q^2/C$ or $U = \frac{1}{2}Li^2$. Since the maximum charge is equal to $q_0 e^{-(R/2L)t}$, then $q_{max}^2 = q_0^2 e^{-(R/L)t}$. The energy in the system is dissipated with a time constant of L/R.

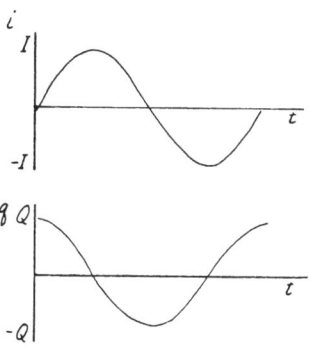

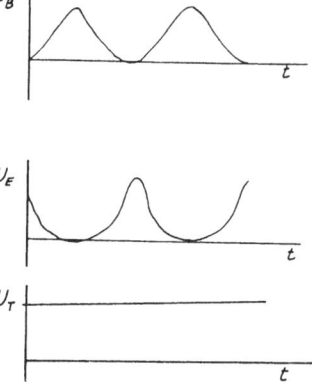

Figure 19.16b. The resulting graphs of current (i), charge on capacitor (q_c), the magnetic field energy (U_B), the electrostatic field energy (U_E), and the total energy (U_T), as a function of time.

■ ALTERNATING CURRENT

Thomas Edison thought that electricity provided to homes should be direct current (dc). The electricity coming out of a rotating generator will always be in the same direction if the current is taken off by commutator rings and brushes. Depending on the number of groups of windings, the resulting current fluctuates in magnitude but is, at least, always in the same direction. George Westinghouse and others championed the cause of alternating current (ac). (Westinghouse won the bid to make the first electric chair, thus, according to Edison, justifying Edison's claim that ac was dangerous.) In the standard form of ac, the voltage and current are sinusoidal. Nowadays in the United States, the standard frequency for ordinary ac is 60 cycles per second, or 60 Hz. In most of the rest of the world the standard is 50 Hz. In the early years of this century, many public utilities produced alternating current at 25 Hz. As Edison kept pointing out, this low frequency produced a slight fluctuation in the intensity of light from ordinary filament bulbs which was annoying to some people.

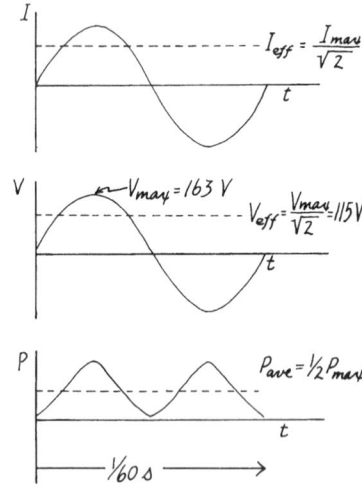

Figure 19.17. Current, voltage, and power for alternating current in a resistor.

[The standard frequency for motion pictures is 24 frames per second. For television in the United States, the frequency is 30 frames per second. The eye is actually sensitive to flickering *intensity* of light up to frequencies of about 50 Hz, especially if the light goes from complete darkness to medium brightness. Although the current in a filament bulb goes through zero twice in every ac period, the emitted light does not go to zero because of the thermal lag of the filament. Nevertheless, filament bulbs driven by 25 Hz ac used to bother some people (including one of the authors, C.S.). To avoid the flicker problem with motion pictures, each frame is actually blanked out by the shutter in the middle of its presentation, thus raising the flicker rate to 48 Hz. A similar but neater trick is used in television. Each frame is composed of two sets of interlaced horizontal lines. One set is made in 1/60 s and then during the next 1/60 s the other half of the lines are created.]

One of Edison's worries about alternating current concerned the higher peak voltage required in order to deliver power equivalent to that produced with a given dc voltage. The power in a dc circuit is given by IV or I^2R. These formulas still hold for ac but the power delivered changes from instant to instant as the current and voltage change. In Fig. 19.17 we show current, voltage, and power for ac in a resistor. Even though the current and voltage go both positive and negative, the power production is always positive. However, it is pulsating, going from 0 to maximum at twice the frequency of the ac. The average power delivered is clearly less than the maximum; as a matter of fact, it is exactly one-half the maximum. (The average of $\sin^2 \theta$ or $\cos^2 \theta$ over a complete cycle is just one-half. This must be the case, since for any angle, $\sin^2 \theta + \cos^2 \theta = 1$, and over a whole cycle the average of $\sin^2 \theta$ must be just equal to the average of $\cos^2 \theta$.)

$$P = \tfrac{1}{2} I_{max} V_{max} \quad \text{(for sinusoidal ac in a resistor)}.$$

Let us define *effective current* and *effective voltage* for ac to be equal to:

$$I_{eff} = \frac{I_{max}}{\sqrt{2}}, \quad V_{eff} = \frac{V_{max}}{\sqrt{2}}.$$

The average power produced in a resistor in an ac circuit is then equal to

$$P_{av} = \tfrac{1}{2} I_{max} V_{max} = I_{eff} V_{eff} = I_{eff}^2 R.$$

Now let us calculate the peak voltage that Edison was worried about. The standard potential difference across the wall outlets in our homes is 120 *effective* volts. The maximum voltage plus and minus from the wall socket is larger than 120 V by a factor $\sqrt{2}$, making the power source more dangerous. The maximum voltage is 170 V, which is much more effective than simply a factor of 1.4 in breaking down the insulating barrier of the skin.

In Chap. 17 we calculated the drift velocity for electrons in a typical wire and circuit. That velocity is a fractional millimeter per second. With alternating current, the charges never get a chance to drift anywhere. They are pulled in one direction for 1/120 s, and then are shoved in the other direction for 1/120 s. During that time, however, they are delivering thermal energy to the conductor just like dc and also are producing and responding to magnetic fields.

Alternating current is usually generated and transmitted with a four-wire, three-phase system. The voltage between any two wires varies sinusoidally, but the phases of the sine waves in the three "hot" wires differ as shown in Fig. 19.18.

Each phase is driven by a particular pair of coils in the generator. Induction motors are most efficiently driven by a three-phase alternating current that can produce a rotating magnetic field in the motor. Although three-phase alternating current is transmitted throughout neighborhoods, usually only single phase from two wires and the neutral is brought into each house, providing either 120 or 240 V.

Let us see what happens to the current and voltage in an inductor and capacitor in a series ac circuit. The current in each component will be the same at any given instant. Any charge that is piling up in the capacitor must be coming out of the inductor. The alternating current is described by $I = I_m \sin \omega t = I_m \sin 2\pi f t$. The standard frequency f is 60 Hz. The angular frequency ω is equal to $2\pi f = 377$ rad/s.

The voltage across the inductor is $V_L = L\, di/dt$. The slope of the $i(t)$ curve is equal to $di/dt = \omega I_m \cos \omega t$. Note that the values for the slope of a sine curve form a cosine curve. The slope is also proportional to ω. The higher the frequency of the sine function, the greater the slope. The result of this argument is that the voltage across the inductor is equal to $V_L = L\, di/dt = (\omega L) I_m \cos \omega t$. The voltage across the inductor is sinusoidal, just like the current, but *leads* the current by 90°, as shown in Fig. 19.19(a). The magnitude of the voltage can be related to the magnitude of the current by an expression similar to that of Ohm's law: $(V_L)_{max} = (\omega L) I_{max}$. The expression ωL is called *inductive reactance* and is given the symbol, X_L. The unit of reactance is the ohm; it is the equivalent of a resistance in a dc circuit. Note, however, that the reactance is frequency dependent. For a given current in the inductor, the back voltage is proportional to the frequency because the higher the frequency, the faster the current is changing, and the greater the slope di/dt.

Now let us calculate the potential difference across the series capacitor as the current changes sinusoidally. At all times the voltage across the capacitor is equal to q/C. As we have seen before on p. 468, when the sinusoidal current in a capacitor is maximum, the charge on the capacitor is 0. When the charge is maximum, the current is 0. If the current is described by a sine function, the charge and hence the voltage must be proportional to a negative cosine function as shown in Fig. 19.18(b). In this case, the voltage across the capacitor *lags* the current by 90°. The faster the current changes, the less time the charge has to pile up on the capacitor and so the smaller the accumulated charge: $q = (1/\omega) I_{max} \cos \omega t$. The voltage across the capacitor equals

$$V_C = \frac{q}{C} = \frac{-(1/\omega) I_m \cos \omega t}{C} = -\frac{1}{\omega C} I_m \cos \omega t.$$

The maximum voltage across the capacitor is related to the maximum current by an Ohm's law type of equation: $(V_C)_m = (1/\omega C) I_m$. The term $1/\omega C$ plays a role similar to that of a resistance. This term is called *capacitive reactance* X_C: $X_C = 1/\omega C$. Note that its frequency dependence is just opposite that of inductive reactance. As the frequency ω becomes large, the capacitive reactance becomes small. On the other hand, as the frequency ω goes to 0, toward dc, the capacitive reactance goes to infinity—an open circuit.

Let us calculate some typical values of reactances. For instance, what is the inductive reactance of a 1-H coil at 60 Hz?

$$X_L = 2\pi(60 \text{ Hz})(1 \text{ H}) = 377 \text{ }\Omega.$$

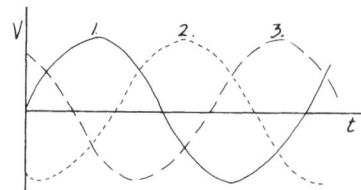

Figure 19.18. Voltage of each wire with respect to neutral wire as function of time. At neighborhood transformers, the houses are divided into 3 groups—each line of houses fed from one phase. At the cylindrical transformers on local poles, the voltage is reduced, usually from 400–450 V to 115 V. Each house, however, can be fed with three lines. One is neutral, and the other two are 115 V with respect to the neutral, but 180° out of phase with each other. Thus 230 V is available for stoves but it is only single phase.

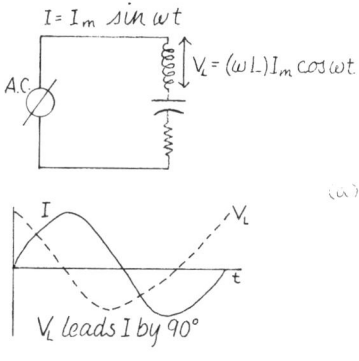

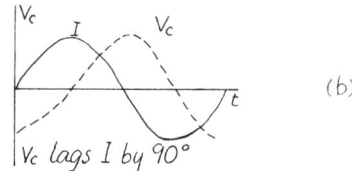

Figure 19.19. Phase relationships between current and voltage in an LRC circuit.

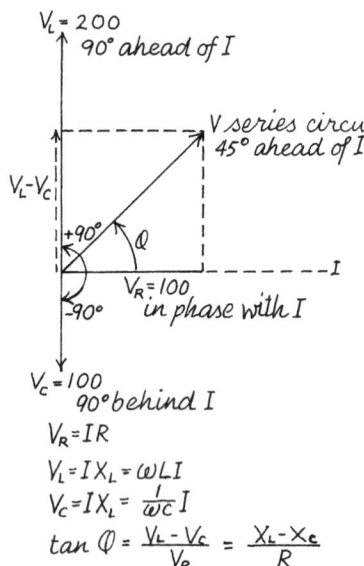

Figure 19.20. Phasor diagram of voltages in an LRC circuit.

At the AM radio frequency of 600 kHz (6×10^5 Hz), the reactance of 1 H would be equal to 3.77×10^6 Ω—greater by a factor of 10^4.

A 1-μF capacitor at a frequency of 60 Hz has a capacitive reactance of

$$X_C=\frac{1}{\omega C}=\frac{1}{2\pi f C}=\frac{1}{2\pi(60\text{ Hz})(1\times10^{-6}\text{ F})}=2.65\times10^3\text{ }\Omega.$$

At 600 kHz, the reactance of 1 μF would be only 0.265 Ω.

When resistors are in series with a dc circuit, the effective resistance is simply the sum of the individual resistances. The derivation of that formula depends, however, on the fact that the total potential difference across the series circuit is equal to the sum of the individual potential differences of the component resistors. With alternating current we have a more complicated situation. The *current* in a series circuit consisting of an inductor, a capacitor, and a resistor is the same in each component at any given moment. The current is all in phase, but the voltages are not. The voltage across the capacitor is 90° behind the current and hence 90° behind the voltage across the series resistor. Meanwhile, the voltage across the inductor is ahead of the current by 90° and hence is 180° out of phase with the voltage across the capacitor.

■ PHASOR REPRESENTATION OF AC

On p. 319 in the chapter on vibrations, we showed how to represent phasors graphically. Now in Fig. 19.20 we show a particular example for ac. Each voltage is represented by a "phase vector" or *phasor* on a type of graph used to show the real and imaginary components of complex numbers. Any voltage in phase with the current lies on the real (horizontal) axis. A voltage leading the current in phase by 90° lies on the positive imaginary (vertical) axis. A voltage lagging by 90° lies on the negative side of the imaginary axis.

In the case shown, $X_C=100$ Ω, $X_L=200$ Ω, and $R=100$ Ω. If there is a current of 1 A maximum in the series circuit, then the maximum voltage across the inductor is 200 V, the maximum voltage across the capacitor is 100 V, and the maximum voltage across the resistor is 100 V. Since the potential differences across the inductor and capacitor are 180° out of phase, the potential difference across both of them is simply the difference between their two voltages. In this case that difference is 100 V, and it is leading the voltage across the resistor by 90°. The total potential across the whole circuit is the sum of these two sinusoidally varying voltages. That voltage is also sinusoidal with a magnitude equal to the vector sum of the magnitudes of the reactance potential and the resistance potential. In this case that vector sum is $V_{\max}=\sqrt{(100\text{ V})^2+(100\text{ V})^2}=141$ V. Compare this net voltage across the series circuit with the individual voltages across the components. In particular, note that the voltage across a component can be much greater than the voltage imposed on the whole series circuit, sometimes leading to breakdown of a capacitor or inductor even though the circuit voltage is less than the rated breakdown voltage of the component.

The power consumed in an ac circuit is equal to the product of the effective current and effective voltage only if the voltage and current are in phase. Otherwise, the power consumed is equal to the product of the effective current and the component of the effective voltage that is in phase:

$$P=I_{\text{eff}}V_{\text{eff}}\cos\Phi\quad\text{(where }\Phi\text{ is the phase angle between }I\text{ and }V\text{)}.$$

In an inductor with 0 resistance or a capacitor, the voltage is 90° out of phase with the current and therefore no power is consumed. (Of course, an inductor always has resistance because it is composed of many turns of wire. Power is dissipated in the resistance.) In most household ac circuits the net reactance is small compared with the resistance and so the current supplied to the house is usually almost in phase with the voltage. Motors usually have some inductive reactance and fluorescent lights usually have some capacitive reactance. In large office buildings or in certain industries, the load may be strongly inductive or capacitive. In that case the power company might have to furnish a large current at the standard voltage and yet the product of $I_{eff}V_{eff}\cos\Phi$ might be considerably smaller than just the product $I_{eff}V_{eff}$. Even though the building is using less power because the current and voltage are out of phase, the power company still has to supply a large current which costs money because of the I^2R losses along the transmission lines. In such a case, the power company charges extra, depending on the magnitude of the phase angle.

■ TRANSFORMERS

The main advantage of alternating current over direct current is the ease and efficiency with which ac voltages can be raised or lowered. When electrical power is produced at a central station and then sent over long distances, there is I^2R loss in the transmission lines. For the same amount of power transmitted, this loss can be reduced by increasing the voltage and thus lowering the current. Transmission lines between cities usually operate at potentials over 100 000 V. Within cities the standard potential is 2300 V. Within neighborhoods, this potential is reduced to 230 or 115 V for the lines going to individual homes. The voltages are changed up or down by passive devices called transformers, which usually operate with an efficiency close to 99%.

A basic transformer consists of two coils, a primary with N_P turns and a secondary with N_S turns, wound on the same iron core. The sinusoidal current in the primary produces a sinusoidally changing magnetic field in the core. The core is usually made of iron laminations separated by thin insulating sheets in order to reduce power loss due to eddy currents. If no current is being drawn by the secondary coil, then the primary coil acts very much like an inductor. In an ideal transformer, without losses, I_P and V_P are 90° out of phase with each other, and so no power is consumed. For that matter, the current drawn under these conditions is usually very small. The inductance of a typical small household transformer is about 1 H. At 60 Hz its reactance is $X_L = 2\pi(60 \text{ Hz})(1 \text{ H}) = 377 \ \Omega$. For $V_{eff} = 115$ V, $I_{eff} \approx \frac{1}{3}$ A.

The primary coil is driven by a generator with a sinusoidal emf. Kirchhoffs law requires that $\mathscr{E} + V_P = 0$, where $V_P = -N_P(d\phi/dt)$. V_P is the induced back-voltage in the primary. It might appear that the induced voltage in the primary completely bucks out the emf of the generator. How can there be any current in the transformer to produce the flux? Remember that the current and voltage in the transformer are 90° out of phase. The induced voltage does indeed prevent the emf of the generator from providing more current, but the current in the primary that was built up during the turn-on time remains. In a real transformer or inductor, there is some loss of energy due to the resistance and therefore the generator must continually supply some energy. It can do this if the primary current is not quite 90° out of phase with the emf.

Since the same magnetic flux exists in both the primary and the secondary coils, the induced voltage in the secondary is equal to $V_S = -N_S(d\phi/dt)$. The ratio of secondary voltage to primary voltage is therefore

$$V_S/V_P = N_S/N_P \quad \text{or} \quad V_S = (N_S/N_P)V_P.$$

In a step-up transformer, $N_S/N_P > 1$. An extreme example of such a transformer is the common spark coil found in many cars. It is used to transform the 12 V from the battery to the 20 000 V needed to produce the ignition spark in the cylinders. To produce the necessary ac, the direct current from the battery is interrupted at the crucial moment by the distributor breakers.

If current is drawn from the secondary coil, the secondary current tends to produce a change in the magnetic flux in the common core. However, the magnitude of that flux is controlled by the driving emf of the generator, and must remain essentially unchanged. To counteract the effect of the secondary current, the primary must draw current from the source, 180° out of phase with the secondary current. Each current produces an additional flux proportional to the number of turns in its coil and the magnitude of the current, but these fluxes must cancel. Hence, the primary and secondary currents are related by

$$N_P I_P = -N_S I_S \quad \text{or} \quad I_P = -(N_S/N_P) I_S.$$

The negative sign shows that the primary and secondary currents are 180° out of phase with each other. Note than in a step-up transformer, there are more turns in the secondary coil than in the primary: $N_S/N_P > 1$. Consequently, the primary current is greater than the secondary current. Here is another case where you cannot get something for nothing. In exchange for an amplification of voltage, the primary must draw more current.

If the current in the secondary is in a resistive load, it is in phase with the secondary emf. The extra current drawn by the primary is also in phase with the primary emf, and hence power is drawn from the generator. For the ideal transformer, without core losses, primary power drawn = secondary power consumed: $I_P V_P = I_S V_S$.

A transformer not only transforms current and voltage magnitudes, but also the effective resistance or impedance of a circuit. For a resistive load, R_S in the secondary, the primary coil furnishes current, I_P, at a voltage, V_P. Therefore, the effective resistance in the primary is

$$R_P = \frac{V_P}{I_P} = \frac{(N_P/N_S)V_S}{(N_S/N_P)I_S} = \frac{1}{(N_S/N_P)^2} R_S.$$

A step-up transformer changes a load resistance in the secondary into a smaller effective resistance in the primary.

If the secondary load is inductive or capacitive, we must analyze more carefully the phase relationships of currents and voltages in primary and secondary. Since the primary and secondary currents must be 180° out of phase with each other, the transformed effective impedances are also changed in sign. A capacitor in the secondary circuit is seen as an inductor by the primary; an inductor is seen as a capacitor. This transformation of impedances, both in sign and magnitude, is a vital function of transformers in circuits transmitting signals. As we saw when studying wave motion, a signal is transmitted without reflection when passing from one medium to another if the two media have the same impedance. For efficient transfer of power from one circuit to another, the "output impedance" must be equal to the "input impedance." For instance, the output resistance of vacuum tubes driving audio loudspeakers is usually higher than the resistance of the loudspeaker coils. The final stage of an audio amplifier using tubes, therefore, is not the tube output, but instead a step-down transformer.

1. Quinton, Arthur, Phys. Teach. Jan, 40 (1979).
2. Rossing, Thomas, Phys. Teach. Dec. (1991).

20
Electromagnetic Radiation

In Chap. 19 we saw what happens *in circuits* when currents change. A changing current produces a magnetic field, which, as it changes, sets up an emf. The effects that we described were all localized within the circuit or the immediate neighborhood. Now we must take a look at a whole new phenomenon that arises when electric charges accelerate. Electric and magnetic fields are produced that radiate from the source, carrying away energy, momentum, and angular momentum. This electromagnetic radiation comprises all the effects of radio, TV, visible light, and x rays.

We start our exploration of electromagnetic radiation by reviewing the basic laws of electricity that we have seen so far. As they stand, these laws are incomplete. When we add one more term, we will see that the equations predict a radiation phenomenon. The prediction even extends to the speed of the radiation, which turns out to be the speed of light.

Once we have seen how electromagnetic radiation arises out of the basic laws, we will explore the various ranges of electromagnetic radiation. Although radio waves and x rays are basically the same phenomenon, their manifestations are very different.

■ DEMONSTRATING THE PHENOMENA

1. You can transmit and detect radio and TV signals without a license. All you need is a source of current that can be turned on and off rapidly. Electric motors that have commutators (not induction motors) are unfortunately usually good generators of radio signals. Such motors are usually found in electric shavers or kitchen mixers or sewing machines. A television set or an AM (amplitude-modulated) radio is a good detector for locally produced static. TV *sound* is FM (frequency-modulated), which is not much influenced by static. The video part of TV, however, is AM and subject to local interference from currents that turn on or off abruptly.

   Run a motor at various distances from the radio or TV. The closer the commutator-interrupted current to the detector, the greater the effect. Another way to produce static is to short-circuit a battery momentarily with a loop of wire. To see or hear this effect you must usually have the battery and wire close to the receiver.

   Notice that the effect on the TV picture depends to some extent on the orientation of the motor and also on small sideways displacements from the TV set. The motor is both generator and antenna and so the transmission of the electromagnetic radiation depends on its position and orientation with respect to the receiver. If you are using a TV set as your receiver, you should disconnect the input from the cable (if you have cable) and give the static a chance to be picked up by the unshielded antenna leads.

2. Take a close look at the antennas used for radio, TV, or radar. A radio station for ordinary AM radio must have a tall vertical tower for its antenna. We will see later how the proper height of the antenna depends on the radio frequency or wavelength. Many AM radio receivers do not have a visible antenna, but if they do it will also be a vertical rod, such as the one used for car radios. Of course, for practical reasons, the receiver antenna is not as long as the radio station transmission antenna. However, the taller the receiver antenna, the better the reception, at least up to the same height as the transmitting antenna. If you have a telescoping car antenna, you can observe the effect by tuning to a weak station and raising or lowering the antenna.

Portable AM radios usually have antennas that are sensitive to the magnetic part of the radiation. If you have such a radio, tune it to a station some distance away, whose compass direction you know, and observe the reception as you turn the radio first around a vertical axis and then around the horizontal axis in the direction of the station. Usually, good reception depends on whether the front or the side of the radio is facing the distant transmitter. If your radio has a telescoping antenna, which is responsive to the electric part of the signal, observe what happens if the antenna is horizontal instead of vertical. The same kind of observation can be made with a portable phone, where the transmitting phone is in your house with a vertical antenna.

FM and TV antennas are very different from those used for AM. Both transmitting and receiving antennas for FM and TV are about the same size and may even be the same shape. The shapes of TV antennas on rooftops can vary but note that they are all about the same size. Furthermore, in a given neighborhood, they are probably all pointed in the same direction. Note that the long cross bars are perpendicular to the direction toward the station.

Radar antennas can be considerably smaller than those used for TV. Consider, for instance, that police radar can be generated and received by a telescopelike device that may have a diameter of only 10 cm or so. The shape of radar antennas is something like that of reflectors for optical telescopes or searchlights.

3. If you have tested the sensitivity of a portable AM radio to various orientations with respect to the station, you have already seen the effect of *polarization* of electromagnetic radiation. You can see some of the same effects with visible light if you have a pair of polarizing sunglasses. Look at the blue sky away from the sun through the polarizer and observe the effect of rotating the glass through 90°. Now look at glare or light reflected from a road or tabletop and see how the appearance changes as you rotate the polarizer. Evidently, electromagnetic radiation, whether radio or TV or visible light, is transmitted by some mechanism that depends on the orientation with respect to the direction of transmission.

4. A stream of water from a hose makes a very precise analogy of electric field lines. If you move a garden hose very slowly (compared to the stream velocity) in a direction perpendicular to the stream, you will have a model of the field lines coming from a slowly moving electric charge. (Ignore the fact that a stream of water also follows a parabolic path in the vertical plane because of gravity. Observe only the horizontal component of the stream.) The stream from the hose will form a straight line leading back to the nozzle, perpendicular to the direction of motion. However, if you give the nozzle a sudden acceleration in the direction of motion, you will generate a *kink* in the stream. This kink has a pronounced component *perpendicular* to the main direction of the stream.

A particularly dramatic effect can be produced by moving the nozzle back and forth perpendicular to the direction of the stream in a simple harmonic oscillation. Loops of wave motion appear in the stream. Notice particularly that the magnitude of the loops *increases* with the distance from the nozzle. To understand the radiation from an accelerated charge you should really perform this simple demonstration for yourself.

■ REVIEW OF THE BASIC EQUATIONS OF ELECTROMAGNETISM

In Fig. 20.1 we show the five basic laws of electromagnetism that we have presented so far. We illustrate these with drawings of field lines, with integral

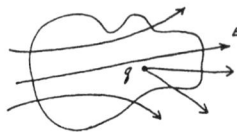

1. Gauss' Law for E

Net number of lines of E leaving closed surface $= \Phi_E = 4\pi k q = \dfrac{q}{\varepsilon_0}$

q is net charge enclosed by surface

$E = \dfrac{\text{number of lines}}{\text{unit area}} = \dfrac{\Delta \Phi_E}{\Delta \text{area}}$

where unit area is taken so that lines are perpendicular to surface.

In integral form: $\varepsilon_0 \oint_A \mathbf{E} \cdot d\mathbf{A} = q$

In differential form: $\nabla \cdot \mathbf{E} = \dfrac{q}{\varepsilon_0}$

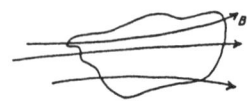

2. Gauss' Law for B

Net number of lines of B leaving closed surface $= \Phi_B = 0$

$B = \dfrac{\text{number of lines}}{\text{unit area}} = \dfrac{\Delta \Phi_B}{\Delta \text{area}}$

Lines of B are continuous, since there are no monopoles.

In integral form: $\dfrac{1}{\mu_0} \oint_A \mathbf{B} \cdot d\mathbf{A} = 0$

In differential form: $\nabla \cdot \mathbf{B} = 0$

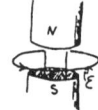

3. Faraday's Law

$\mathcal{E} = \sum_{\text{loop}} \mathbf{E} \cdot \Delta \ell = -\dfrac{\Delta \Phi_B}{\Delta t}$

In integral form: $\mathcal{E} = \oint_{\text{loop}} \mathbf{E} \cdot d\ell = -\dfrac{d\Phi_B}{dt}$

In differential form: $\nabla \times \mathbf{E} = -\dfrac{\partial \mathbf{B}}{\partial t}$

4. Ampère's Law

$\sum_{\text{loop}} \mathbf{B} \cdot d\ell = \mu_0 i$

Integral form: $\dfrac{1}{\mu_0} \oint_{\text{loop}} \mathbf{B} \cdot d\ell = i$

Differential form: $\nabla \times \mathbf{B} = \mu_0 \mathbf{j}$

5. Lorentz Equation

$\mathbf{F} = q\mathbf{E} + q\mathbf{v} \times \mathbf{B}$

$\nabla \cdot \mathbf{E} \equiv$ divergence of $\mathbf{E}$

$\nabla \times \mathbf{E} \equiv$ curl of $\mathbf{E}$

$\mathbf{j} \equiv$ current density

Figure 20.1. Representations of the basic laws of electromagnetism.

equations which are included in university introductory-physics courses, and with the symbols of differential operators that are used in more advanced work. All three forms are equivalent and each has particular usefulness in understanding the phenomena. The first two laws are concerned with static situations. In terms of field lines, they link electric and magnetic fields to static sources. Electric-field lines start on positive charges and end on negative charges. Since there are no magnetic monopoles, magnetic field lines must be continuous.

The third law, due to Faraday, describes a dynamic way to produce an electric field. A time-varying magnetic flux produces closed lines of electric field. In the fourth equation, due to Ampère, closed lines of magnetic field are produced by the motion of electric charges. The fifth equation is due to Lorentz and links electric and magnetic fields to forces on charged particles.

The first four equations were gathered together by James Clerk Maxwell (1831–1879). He was impressed by the elements of symmetry between electric and magnetic fields in the equations, but on the other hand he was distressed by some of the obvious lacks of such symmetry. First, as we have already pointed out, there is no magnetic charge and hence no magnetic current. But there is another gap in the equations. Although Eq. (3) in Fig. 20.1 describes what happens when magnetic fields vary with time, there is no corresponding term in Eq. (4) to describe what happens if *electric* fields change. In Fig. 20.2 we show one of the problems caused by this lack. When current is charging a capacitor, there must be magnetic-field lines around the lead-in wires. However, a closed loop on the circumference of a circle cutting between the plates of the capacitor would contain no current. According to Ampère's law, as it stands, there should be no magnetic-field lines around the gap in the capacitor. Such an abrupt change in magnetic field would be an unnatural phenomenon. Since the electric field in the capacitor is changing during the charging process, Maxwell proposed that magnetic field lines must be produced around the time-varying electric field. With the additional term, the complete fourth equation looks like this:

$$\int \mathbf{B} \cdot d\mathbf{l} = \mu_0 i + \mu_0 \left(\epsilon_0 \frac{d\Phi_E}{dt} \right), \quad \nabla \times \mathbf{B} = \mu_0 \mathbf{j} + \mu_0 \epsilon_0 \frac{\partial \mathbf{E}}{\partial t}.$$

Notice that the term in parentheses must have the units and dimensions of a current. Since according to Gauss's law, the net electric flux leaving a closed region is $\Phi_E = q/\epsilon_0$, then $\epsilon_0 \Phi_E$ has the units and dimensions of an electric charge. Dividing this quantity by time yields a current.

Let us compare the current and changing electric field in a capacitor that is discharging. In Fig. 20.3 we show such a capacitor. It has been charged to 100 V and now is discharging through a 1-Ω resistor to ground. As you can see from the calculations that accompany the figure, the electric field at the beginning is 1×10^5 V/m and the electric flux is 10^3 lines. When the discharge begins, the charge on the capacitor decays exponentially with a time constant equal to RC. With the dimensions given, that time constant in this case is equal to $\tau = 8.9 \times 10^{-11}$ s. The voltage, the electric field, and the flux also decay with this same time constant. Meanwhile, the charge flowing out of the capacitor creates a current in the lead-in wires. For an approximate calculation, let us assume that the discharge is linear during a short period, instead of exponential. Then at the beginning of the discharge, $I = V/R = (100 \text{ V})/(1 \Omega) = 100$ A. This current of 100 A in the wires will produce circular lines of magnetic field around the wires. The equivalent current

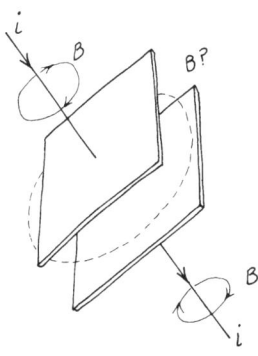

Figure 20.2. A magnetic field is produced around current-carrying wires. Similarly, there must be a magnetic field around a region where the electric field is changing.

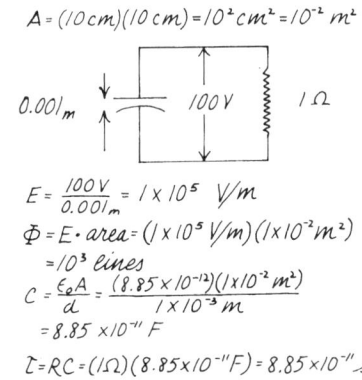

Figure 20.3. The "displacement" current produced by the change of electric field in the capacitor yields approximately the same value of current that exists in the resistor.

caused by the change of electric flux in the capacitor is equal to

$$\epsilon_0 \frac{d\phi_E}{dt} = (8.9 \times 10^{-12}) \frac{10^3 \text{ lines}}{8.9 \times 10^{-11} \text{ s}} = 10^2 \text{ A}.$$

It appears that the changing electric field in the capacitor produces a result equal to that of the current in the wire. The magnetic-field lines around the gap in the capacitor will be equal to those around the lead-in wires.

■ THE IMPLICATION OF MAXWELL'S EQUATIONS

The four equations that describe electric and magnetic fields and their sources are called Maxwell's equations, in spite of the fact that some of the individual equations and phenomena were discovered by other people. Thus, Maxwell's equations include those named for Gauss, Ampère, and Faraday. It was Maxwell, however, who gathered the separate equations together and proposed the crucial extra term that describes the effect of time-varying electric flux. Maxwell then proceeded to combine the four equations and predict a whole new range of phenomena.

Note that with the electromagnetic effects that we have seen so far, we would have a hard time transmitting signals through space. We could, of course, start current in a wire or coil to produce a magnetic field, or we could separate electric charge to produce an electric field. In most such geometries we would be producing a magnetic or electric dipole whose fields fall off as $1/r^3$. At best we would have a point source whose field would fall off as $1/r^2$. We could send signals this way, of course. The detector could be a compass needle for magnetic fields or a small charged ball on a thread for electric fields. You might at first think that the response of these detectors to the new fields created by the distant sources would be proportional to $1/r^3$ or $1/r^2$. The sensitivity of the detector response, however, depends not on the force exerted, which is proportional to the field strength, but on the energy that can be extracted from the field. The energy density of a magnetic field is proportional to B^2 as we saw in Chap. 19. Consequently if we turn on electric or magnetic fields and hope to detect these effects at a distance, we will find that our detection efficiency falls off as $1/r^4$ or $1/r^6$, depending on whether the source is an isolated charge or a dipole. Evidently, we would have a hard time transmitting electric or magnetic signals if detectability faded so rapidly with distance.

(Why should a detector depend on the energy density rather than on the force? Any kind of a detector must have some indicator, such as a pointer on a dial, which will return to its original position if there is no signal. The indicator may be fastened to a spring or might be in a potential well. In either case, a force must move the indicator through a distance if the signal is sensed. Work is required and the energy must be provided by the signal).

Maxwell's analysis of the four equations for electromagnetism showed that there is another way for electric and magnetic fields to propagate through space. In this new mode, the fields fall off only as $1/r$. Let us see how this can happen. Here are the last two Maxwell equations written for free space where there are no electric or magnetic charges or currents:

$$\oint_{\text{loop}} \mathbf{E} \cdot d\mathbf{l} = -\frac{d\Phi_B}{dt}, \quad \oint_{\text{loop}} \mathbf{B} \cdot d\mathbf{l} = \mu_0 \epsilon_0 \frac{d\Phi_E}{dt}.$$

A new and remarkable effect is predicted by these two equations. If a magnetic field somehow enters a region of free space there will be a time-varying magnetic flux there. That will produce an electric field. Since the electric field did not exist in that region previously, its arrival is equivalent to a time-varying electric flux. But a change of electric flux produces a magnetic field. This circular sequence of events generates electric fields from changing magnetic fields and magnetic fields from the subsequent change of electric fields. It is a self-generating system, which, as we will now see, is also self-propagating.

Suppose that a wall of magnetic field oriented in the z direction advances along the x direction as shown in Fig. 20.4. (Later we will ask how such a wall of magnetic field could have been generated in the first place.) We now have a problem and a geometry very much like the one with which we introduced the subject of electric induction. If we take a rectangular loop in the x-y plane, the magnetic field will sweep across it. The change of magnetic field inside the loop provides an emf around the loop. The only contribution to the emf comes from the electric field along the side of the loop that is in the advancing magnetic field. That emf or voltage is equal to $-Ew$. The width of the loop, w, is arbitrary and will cancel out. The electric field in that leg of the loop, E, is constant because the magnetic field is sweeping over it at a constant rate. The induced electric field is in the positive y direction, which makes the emf negative as we go around the loop in the proper x-y direction. The time rate of change of magnetic flux in the loop is equal to

$$\frac{d\Phi_E}{dt} = \frac{d(BA)}{dt} = B\frac{dA}{dt} = Bw\frac{dx}{dt} = Bwv.$$

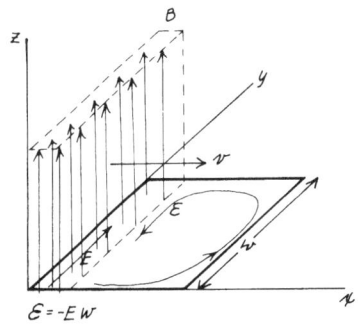

Figure 20.4. A wall of magnetic field oriented in the z direction advances toward $+x$. This changing magnetic field generates an electric field.

The change of magnetic flux in the loop is proportional to the velocity v with which the wall of magnetic field is advancing.

How would we go about controlling the velocity of the advancing wall of magnetic field? You might at first think that we could move a magnet past the observing plane, even as in Chap. 19 we described an experiment where we moved a loop through a magnetic field with a velocity v. In this case, however, we have launched a magnetic pulse by some method of turning on a magnet far away. The magnetic pulse is traveling on its own with an arbitrary velocity; at least it is arbitrary as far as we know at this stage. In the next section we will see that the arbitrary velocity of the *magnetic* field is linked with the velocity of the induced *electric* field.

According to the third Maxwell equation for free space,

$$\oint_{\text{loop}} \mathbf{E} \cdot d\mathbf{l} = -\frac{d\Phi_E}{dt}.$$

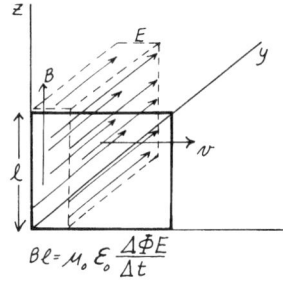

Figure 20.5. The electric field induced by the advancing wall of magnetic field induces the formation of a moving wall of electric field.

Therefore $-Ew = -Bwv$ and $E = Bv$. Here is a simple relationship between the strength of the advancing magnetic field and the electric field that is induced. Note that qualitatively the electric-field lines exist only in the region of the advancing wall of magnetic field. There is no electric field ahead of or behind the advancing wall. But this means that we also have a wall of electric field sweeping along the x direction with the same velocity v. We can now use Maxwell's addition to the fourth equation to see how this moving electric field generates a magnetic field. The same geometry is shown again in Fig. 20.5, but this time the wall of electric field is shown with the field lines beginning to penetrate a loop in the x-z plane. The induced magnetic field is in the positive z direction but exists only within the region of the electric-field wall. Therefore, only the front leg of the loop contributes to the sum of the magnetic field and distance products around the loop. The time rate of change of electric flux in the loop is equal to

$$\frac{d\Phi_E}{dt} = E\frac{dA}{dt} = El\frac{dx}{dt} = Elv.$$

Substituting into the fourth Maxwell equation:

$$\oint_{\text{loop}} \mathbf{B} \cdot d\mathbf{l} = \mu_0 \epsilon_0 \frac{d\Phi_E}{dt},$$

$$Bl = \mu_0 \epsilon_0 Elv \rightarrow B = \mu_0 \epsilon_0 v E.$$

We have now derived another relationship between the magnetic and electric fields that are generating each other. Once again the relationship involves the velocity with which the fields are moving. When E is generated by a changing B, $E = Bv$. When B is generated by a changing E, $E = B(1/\mu_0 \epsilon_0 v)$. These two equations cannot both be true at the same time unless there is a special relationship between the velocity v and the constants of permeability and permittivity. Let us solve for that relationship:

$$v = \frac{1}{\mu_0 \epsilon_0 v} \rightarrow v^2 = \frac{1}{\mu_0 \epsilon_0} \rightarrow v = \sqrt{\frac{1}{\mu_0 \epsilon_0}}.$$

The permittivity ϵ_0 is determined from measurements with electrostatics, and the numerical value depends on the assigned value for permeability μ_0. We are not free to choose arbitrary values for ϵ_0 and μ_0. Now it appears that the velocity with which the electric and magnetic fields sweep along the x axis cannot be arbitrary either. Only one value is possible and that is determined by the electric and magnetic properties of free space. Let us put the numbers in and find the magnitude of this special velocity:

$$v = \sqrt{\frac{1}{(4\pi \times 10^{-7})(8.85 \times 10^{-12})}} \rightarrow v = 3.0 \times 10^8 \text{ m/s}.$$

Note first that this unique speed of propagation of the electromagnetic waves is very fast. Second, as you know, it is the speed of light. It must be that light itself is an electromagnetic wave. Consider what a triumph of theory is represented by these simple arguments. Maxwell's equations contain only a few terms and these require relatively simple relationships between electric and magnetic fields and their rates of change. When the equations are combined they not only suggest that time-varying electric and magnetic fields will continually generate each other, but also predict that the varying fields will propagate through space with a unique speed. That speed is determined by constants that are measured in static experiments. The permittivity ϵ_0 is measured with a capacitor experiment. The permeability μ_0 has an assigned value that determines the size of the ampere and coulomb. Yet when these values are combined, they lead to the speed of light. Imagine the thrill of discovering the nature of light! In Maxwell's words, "I made out the equations in the country before I had any suspicion of the nearness between the two values of the velocity of the propagation of magnetic effects and that of light, so that I think I had reason to believe that the magnetic and luminiferous media are identical".

Maxwell did not have at his command our modern system of differential operators—nor do students in our introductory courses. In these terms, however, Maxwell's equations can be expressed compactly and a wave equation for electric and magnetic fields can be derived in a few steps. For those readers familiar with this notation, here are Maxwell's equations using the del operator ∇, where ∇V is the gradient of the scalar V, $\nabla \cdot \mathbf{E}$ is the divergence of the vector $\mathbf{E}$, and $\nabla \times \mathbf{B}$ is the curl of the vector $\mathbf{B}$. For free space:

$$\nabla \cdot \mathbf{E} = 0, \tag{1}$$

$$\nabla \cdot \mathbf{B} = 0, \tag{2}$$

$$\nabla \times \mathbf{E} = -\frac{\partial \mathbf{B}}{\partial t}, \tag{3}$$

$$\nabla \times \mathbf{B} = \mu_0 \epsilon_0 \frac{\partial \mathbf{E}}{\partial t}. \tag{4}$$

Take the curl of Eq. (3) and the time derivative of Eq. (4):

$$\nabla \times (\nabla \times \mathbf{E}) = -\nabla \times \frac{\partial \mathbf{B}}{\partial t}, \quad \frac{\partial}{\partial t}(\nabla \times \mathbf{B}) = \mu_0 \epsilon_0 \frac{\partial^2 \mathbf{E}}{\partial t^2}$$

Since

$$\nabla \times \frac{\partial \mathbf{B}}{\partial t} = \frac{\partial}{\partial t}(\nabla \times \mathbf{B})$$

and $\nabla \times (\nabla \times \mathbf{E}) = \nabla(\nabla \cdot \mathbf{E}) - \nabla^2 \mathbf{E}$,

$$\nabla(\nabla \cdot \mathbf{E}) - \nabla^2 \mathbf{E} = -\mu_0 \epsilon_0 \frac{\partial^2 \mathbf{E}}{\partial t^2}.$$

Since $\nabla \cdot \mathbf{E} = 0$,

$$\nabla^2 \mathbf{E} - \mu_0 \epsilon_0 \frac{\partial^2 \mathbf{E}}{\partial t^2} = \mathbf{0}.$$

The standard wave equation in three dimensions is

$$\nabla^2 \mathbf{E} - \frac{1}{v^2} \frac{\partial^2 \mathbf{E}}{\partial t^2} = \mathbf{0}.$$

Evidently, the velocity is $v = \sqrt{1/\mu_0 \epsilon_0} = c$. In one dimension, the wave equation would be

$$\frac{\partial^2 E}{\partial x^2} = \frac{1}{c^2} \frac{\partial^2 E}{\partial t^2},$$

which has a general solution,

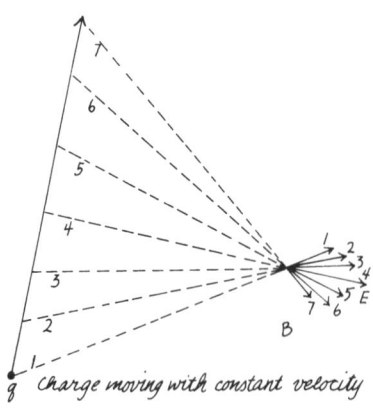

Figure 20.6. The field at a distant point is radial to the position of the charge at that time (not the earlier time when the information was sent).

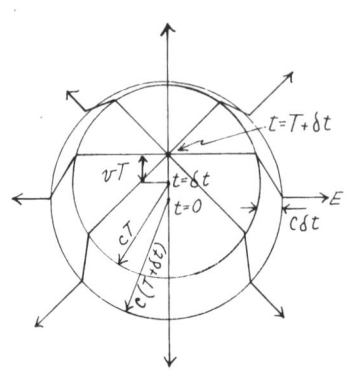

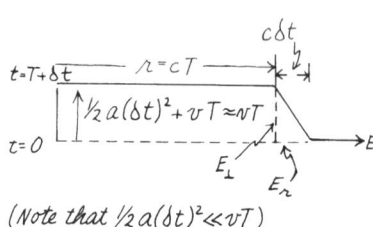

Figure 20.7. Model of the electric field lines produced by an accelerated charged particle.

$$E = E_0 \sin \frac{2\pi}{\lambda}(x - ct + \alpha),$$

where α is a phase constant.

■ GENERATION OF ELECTROMAGNETIC FIELDS FROM ACCELERATED CHARGES

Maxwell himself did not solve his equations to demonstrate that accelerated charged particles shake off electromagnetic energy. The detailed calculations are complicated but we can understand the approximate situation through a field-line model. The fields experienced at a distance from a moving electric charge were produced by that charge at an earlier time. If a charged particle is moving at constant velocity, but with speed considerably less than that of light, the electric field at a point far away will gradually change in strength and direction as shown in Fig. 20.6. The magnetic field at that point will remain in the same direction as the charge moves, but will gradually change strength. It takes no energy to keep the charged particle moving at constant velocity nor does the charged particle lose energy as it trails the electric and magnetic field lines with it.

To *accelerate* a charged particle along its direction of motion, a force must be applied to it and hence work must be done. Part of this energy is radiated away in the form of electric and magnetic fields that carry the message of the abrupt change in velocity of the source.

Let us analyze the field produced at a distant point by an electric charge that has been at rest and then is suddenly accelerated upward for a short time δt. (Fig. 20.7) Its speed at the end of that acceleration period is $v = a\,\delta t$. It continues to drift upward with the velocity v. Consider the electric-field line that extends outward horizontally from the charge. Even after the charge has started moving, the electric field at a distant point remains unchanged and horizontal. No information has reached this point that the charge has started to move. At the time T, which is equal to the separation distance divided by the speed of light, the remote field starts changing its direction. At the time $T + \delta t$, the electric field has been changed in direction so that it is in line *with the position of the charge at $T + \delta t$*. But this is unexpected! Why shouldn't the field at that time be in line with the position of the charge at δt, the time when the signal started out? One way of considering this question is to think of the field lines as being generated by photons thrown off from the electric charge. For such a model the mathematical analogy with the motion of a water stream from a hose becomes exact. At time δt, the charged particle has attained a velocity of v. For a photon to be thrown off in a direction perpendicular to the particle velocity (in order to reach the observer), it must start with a *backward* component equal to v (since it is coming from a source with *forward* velocity, v). When the photon reaches the observer, it will appear to have arrived from the line extending to the source charge at time $T + \delta t$.

During the time interval from T to $T + \delta t$, there is an abrupt change in the direction of the electric field at the remote measuring point. Let us look down on the scene and picture the lines of force as shown in Fig. 20.7. Beyond a circle with radius $c(T + \delta t)$, the lines of force all point radially back to the charge's original location at $t = 0$. No information about the charge's movement has arrived beyond this circle. Within a smaller circle of radius cT, the lines of force point radially to the position of the charge at $t = T + \delta t$. Both circles sweep outward radially at the speed of light. Between them there is a transition shell where the field lines can have a large component perpendicular to the radial direction. (Note that the field lines must be continuous, and so those in the inner sphere must connect with the corresponding ones in the outer sphere.) The "kink" in the field line is not in the same direction that the field line will assume after the pulse has passed by. That direction will change very little, since the source charge moves only a short distance, vT, while the light signal has traveled a distance of cT. Therefore, the field line will end up changing direction abruptly only by the small angle, v/c:

$$\frac{E_\perp}{E_r} = \frac{vT}{c\delta t} = \frac{a\delta t T}{c\delta t} = \frac{aT}{c}.$$

The *radial* field is just that due to a point charge, and is given by Coulomb's law: $E_r = kq/r^2$. Consequently, the perpendicular component of the electric field is equal to $E_\perp = kqaT/r^2c$. The distance between the moving charge and the remote measuring point is equal to r, which is equal to cT. Substituting this value for T yields the following value for the perpendicular component of electric field:

$$E_\perp = \frac{kqa}{r^2 c} \frac{r}{c} = \frac{kqa}{c^2 r} \quad (r=cT).$$

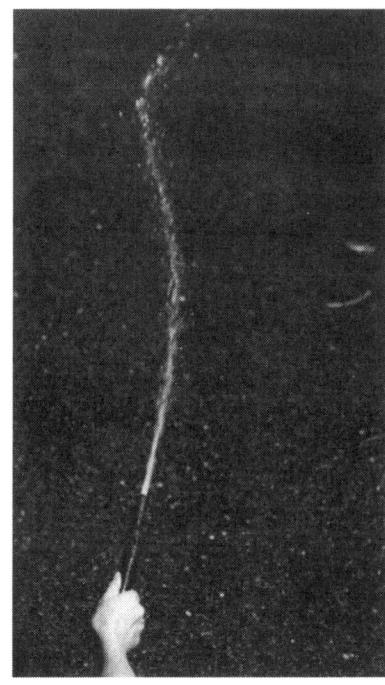

Figure 20.8. A moving kink in the stream from a water hose. The kink was produced by an acceleration of the nozzle, perpendicular to the stream.

Remember that this perpendicular component of electric field carries the information that the charge has been accelerated. Its magnitude is proportional to the acceleration but, more remarkably, is proportional only to the inverse *first* power of the distance. Why should a *part* of an electric field that depends on the inverse *square* of the distance depend on the inverse first power of the distance? If you watch closely the behavior of the water stream from a hose as you move the nozzle back and forth sideways, you will see that the perpendicular component increases with distance from the nozzle. A picture of the effect is shown in Fig. 20.8. From an analytic point of view, note that the perpendicular component is proportional to the time taken for the message to go from source to observing point. This is caused directly by the fact that the abruptly shifted field at the observing point lines up with the position of the drifting charge *at that time*. The greater the transmission time T, the further the charge has drifted upward and therefore the greater the shift in the electric field and the greater the perpendicular component.

The perpendicular component of the electric field—the kink in the field line—also depends on the angle between the field line and the direction of the acceleration. We examined the case where the field line is perpendicular to the acceleration, where the effect is maximum. If we had observed the behavior of the field line in the same direction as the acceleration, we would have found no effect. Regardless of the acceleration, the electric field line would maintain its same direction and therefore there would be no perpendicular component. The size of the perpendicular component depends on the sine of the angle between the line of acceleration and the line of the observer, as shown in Fig. 20.9. The complete formula for the strength of the perpendicular component is

$$E_\perp = \frac{kqa \sin \theta}{c^2 r}.$$

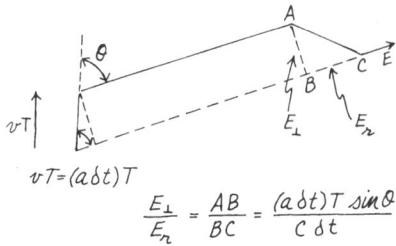

Figure 20.9. Geometry of the electric field lines at an angle to the direction of the moving particle.

Since the ordinary electric and magnetic fields caused by isolated charges or dipoles or electric currents fall off as the square or cube of the distance, the component of the field created by the acceleration, which depends on the inverse first power of the distance, will be the only field detectable at large distances from the source. Note first that this part of the field is perpendicular to the line back to the source. Furthermore, this perpendicular kink is hurtling outward at the speed of light. As we have already seen, such a pulse of electric field must be accompanied by a magnetic-field pulse that is perpendicular to the electric field and also perpendicular to the direction of propagation. The energy density in such a pulse is proportional to the square of the electric or magnetic fields. Therefore, the energy radiated from an accelerated charge must be proportional to the square of the charge, the square of the acceleration, and, inversely, to the square of the distance from the source. The light reaching us from a distant flashlight or a distant star is proportional to $1/r^2$ because by "light" we mean the energy that we can detect from that source. Because the energy produced from a charge accelerated in a straight line is proportional to $\sin^2 \theta$, the radiation density bulges in the equatorial plane perpendicular to the acceleration.

In a radio-transmitting antenna, electrons are sent up and down a vertical line, forming a resonant circuit. The line has length L on either side of the power input. The electric charge undergoes simple harmonic motion, with the position of the center of charge given by $y = L \sin \omega t$. The velocity of the center of charge is equal to $v_q = dy/dt = \omega L \cos \omega t$. The acceleration of the charge is equal to $a_q = dv/dt = -\omega^2 L \sin \omega t$. Therefore, the magnitude of radiated energy is proportional to the fourth power of the frequency! Consequently, for equivalent radiated power, low-frequency sources require much higher currents in the antenna than do high-frequency sources.

■ THE ENERGY RADIATED IN ELECTROMAGNETIC FIELDS

When we calculate energy density of static electric and magnetic fields, we usually make assumptions about the location of the fields. In the case of a capacitor, we say that the electric field and hence the stored energy exist in the volume between the plates. The magnetic energy calculation is usually made in terms of a solenoid in the form of a torus or a hollow doughnut. In this case we assign the location of the energy to the interior volume. While it is plausible and useful to picture these interior spaces as filled with energy, we might alternatively assign the energy to the stored charges on the capacitor plates or to the currents in the solenoid.

In the case of radiated energy, the concept of energy actually existing and traveling in space becomes much more physical. A receiver antenna can extract power from apparently empty space. Let us compare the electric and magnetic energy densities in radiation and see another way of describing its properties.

The static electric and magnetic energy densities are given by

$$u_E = \tfrac{1}{2}\epsilon_0 E^2 \text{ J/m}^3,$$

$$u_B = \tfrac{1}{2}\frac{1}{\mu_0} B^2 \text{ J/m}^3.$$

(Note that the symbol for energy density is u, not μ.) For a plane wave of electromagnetic radiation, there is a simple relationship between the strength of the electric and magnetic fields: $B = E/c$. Substituting this value for B into the formula for the magnetic field energy density,

$$u_B = \tfrac{1}{2}\frac{1}{\mu_0}B^2 = \tfrac{1}{2}\frac{1}{\mu_0}\frac{E^2}{c^2} = \frac{\epsilon_0\mu_0}{2\mu_0}E^2 = \tfrac{1}{2}\epsilon_0 E^2.$$

Evidently, the magnetic-field energy density is equal to the electric-field energy density in a plane wave of radiation. The total energy density in the wave is equal to

$$u_{\text{total}} = u_E + u_B = \epsilon_0 E^2.$$

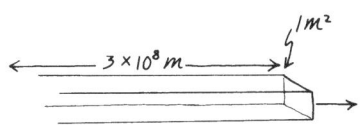

Figure 20.10. Energy in this box will pass through end during one second.

The *intensity* or radiation, or *irradiance*, is defined to be the amount of energy passing through unit area in unit time:

$$I = \frac{\text{(energy)/(time)}}{\text{(area)}} = \frac{\text{(power)}}{A}.$$

In Fig. 20.10 we show how to calculate the amount of energy passing through unit area in unit time. The long box has unit area cross section and a length equal to the product of unit time and the speed of light. In our standard units, the cross-sectional area of the box would be 1 m² and its length would be 3×10^8 m. During 1 s all the energy in this box would pass through the 1 m² of unit area at the end. The total energy in the box is equal to the product of the energy density and the volume of the box. Consequently, the intensity is equal to

$$I = \frac{u\times(\text{volume})}{(\text{area})\times(\text{time})} = \frac{u(1\ \text{m}^2)(c\Delta t)}{(1\ \text{m}^2)(\Delta t)} = uc = \epsilon_0 E^2 c.$$

Note that the units work out. Since $\epsilon_0 E^2$ is an energy density with units J/m³, the product of this quantity and a velocity must yield the units $(J/m^3)(m/s)=(J/s)(1/m^2)$. These are the units of watts per square meter.

There is another way of describing the intensity of electromagnetic radiation. What we have described so far is the *magnitude* of the intensity. Since the radiation has direction, it would be convenient to represent the intensity as a vector. Let us transform the expression that we have so that it contains both electric and magnetic fields:

$$I = \epsilon_0 E^2 c = \epsilon_0 E(Bc)c = \epsilon_0 EBc^2 = \frac{\epsilon_0 EB}{\epsilon_0\mu_0} = \frac{1}{\mu_0}EB.$$

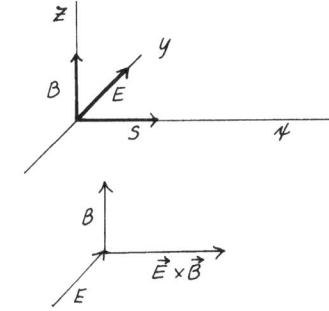

Figure 20.11. Geometry of energy flow in an electromagnetic wave, and the Poynting vector.

We now assert a new definition that was first derived by John Poynting (1852–1914). The *Poynting vector*, a particularly descriptive term, is defined as

$$\mathbf{S} = \frac{1}{\mu_0}\mathbf{E}\times\mathbf{B}.$$

In Fig. 20.11 we show the cross-product relationship between **E** and **B**, and the Poynting vector. Note that the magnitude of **S** is just the intensity that we have already derived and the units are in watts per square meter. The additional feature of the cross product gives us the direction of the power flow. Note also that if the electric and magnetic fields are not perpendicular to each other, perhaps because the electric or magnetic field is twisted in passing through boundaries, the power passing through that region will be less.

Let us calculate the electric and magnetic fields produced by the radiation coming from a 100-W bulb. In the first place, most of the 100 W consumed by the bulb is not converted into visible radiation. The efficiency is only about 2.5%. Let us assume that the bulb is a point source and is emitting uniformly in all directions. At a distance of 3 m from the bulb, the surface area of the surrounding sphere is $A = 4\pi r^2 = 4\pi (3\text{ m})^2 = 113\text{ m}^2$. The watts per square meter for the visible radiation at the distance of 3 m is therefore $I = (2.5\text{ W})/(113\text{ m}^2) = 0.022\text{ W/m}^2$. Half of this power is provided by the electric field and half by the magnetic field. For the electric field

$$\tfrac{1}{2}\epsilon_0 E^2 c = \tfrac{1}{2}I = \tfrac{1}{2}(0.022\text{ W/m}^2),$$

$$E = \sqrt{\frac{0.022}{(8.85 \times 10^{-12})(3 \times 10^8)}} = 2.9\text{ V/m}.$$

The value of E that we have calculated is for a steady-state field. Since the electric field in a light beam is sinusoidal, the peak electric field is $\sqrt{2}\,E = 4.1$ V/m. As you can see, the electric-field strength of the light that you use for reading is fairly large. (The electric-field strength of TV or FM for fringe-area reception is only a few microvolts per meter.) Now let us calculate the strength of the *magnetic* field in the visible radiation from a 100-W bulb 3 m away:

$$\tfrac{1}{2}\frac{1}{\mu_0}B^2 c = \tfrac{1}{2}I = \tfrac{1}{2}(0.022\text{ W/m}^2),$$

$$B = \sqrt{\frac{(0.022)(4\pi \times 10^{-7})}{3 \times 10^8}} = 9.6 \times 10^{-9}\text{ T} = 9.6 \times 10^{-5}\text{ G}.$$

Once again, because the field in the light beam is sinusoidal, the peak magnetic field is $\sqrt{2}\,B = 1.4 \times 10^{-8}$ T. Although the power in the magnetic field is equal to the power in the electric field, the magnetic-field strength is evidently very weak.

The question of the localization of the electromagnetic energy has some strange but revealing consequences if we ask for the direction of energy flow. Let us take the familiar case of a constant current in a wire or resistor. Clearly there is energy flow; the wire or resistor gets hotter. Our intuition would tell us that the energy must in some way flow along the wire. Nevertheless, let us calculate the Poynting vector and find its direction in the region of the wire. The electric field in the wire and at its surface is $E = V/l$. The electric field is in the direction along the wire. The magnetic field at the surface of the wire is circumferential. Its magnitude is $B = \mu_0 I / 2\pi r$. Notice that the electric and magnetic fields are perpendicular to each other. The magnitude of the Poynting vector at the surface is

$$|S| = \frac{EB}{\mu_0} = \frac{1}{\mu_0}\frac{V}{l}\frac{\mu_0 I}{2\pi r} = \frac{IV}{(2\pi r)l} = \frac{IV}{(\text{surface area of wire})}.$$

The total rate of flow of electromagnetic energy into the wire must be equal to the product of the Poynting vector (watts per square meter) and the surface area of the wire. That product is equal to the thermal power dissipation, IV, in the wire. Furthermore, as you can see in Fig. 20.12, the direction of the energy flow as given by the Poynting vector is radially in toward the wire everywhere on the surface. According to our model, the thermal energy dissipated in the wire is continuously supplied by the electric and magnetic fields at the surface of the wire. But aren't these quasistatic fields? How can they supply a continuous flow of energy? The fields do indeed remain constant but only because there is a steady supply of energy from a battery or generator. One way to interpret the localization of the energy is that the chemical or mechanical energy of the source continuously supplies energy for the arrangement of the field lines in order to maintain the

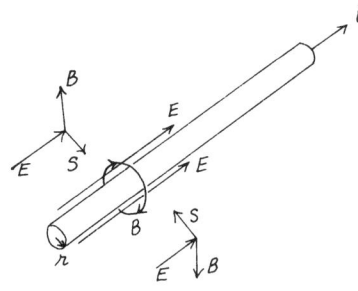

l is length of wire
r is radius
IV is thermal energy provided

Figure 20.12. Model of energy dissipation and flow in a current carrying wire, making use of the Poynting vector.

constant flow of charges and the constant feed of thermal energy to the wires. Remember that the electric-field lines in the wire are maintained by pseudostatic charges set up on the surface of the wire to direct the field lines along the wire.

■ MOMENTUM CARRIED BY ELECTROMAGNETIC RADIATION

If electromagnetic radiation transports energy, does it also carry momentum? Can it exert a force on an object? Certainly we do not experience such a force in everyday life. Furthermore, it is not immediately obvious how a plane electromagnetic wave could produce a force on a charged particle in the direction of the wave propagation. If the electric and magnetic fields are varying sinusoidally in the wave, then the principal reaction of a charged particle would be to oscillate up and down in response to the electric field. This motion is perpendicular to the direction of propagation, and the time average motion is 0. However, as the electric charge is driven by the electric field, it moves in the presence of the accompanying magnetic field. Therefore the charge experiences a $\mathbf{v} \times \mathbf{B}$ force. In Fig. 20.13 we show the relationship among the directions of $\mathbf{E}$, $\mathbf{B}$, and the resultant force on the moving electric charge. Half a cycle later when the charge is moving in the opposite direction in response to the electric field in the opposite direction, the magnetic field is also reversed. Once again, the force on the electric charge is in the direction of propagation of the wave. This is true for both negative and positive charges.

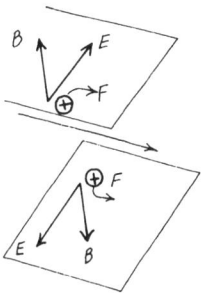

Figure 20.13. Model for momentum delivered to a surface by an impinging electromagnetic wave.

The magnitude of the radiation force on the charged particle is equal to

$$F_r = evB = ev(E/c).$$

The electric force on the charge is $F_{\text{electric}} = eE$. Therefore the radiation force on the charge is

$$F_r = \frac{vF_{\text{electric}}}{c} = \frac{\text{power}}{c}.$$

The power absorbed by the charge is equal to Fv (since the work on the charge is $W = F\Delta x$ and power $= W/\Delta t$). Since $\Delta(\text{momentum})/\Delta t = F$, and $\Delta(\text{energy})/\Delta t =$ power, the relationship between momentum and energy carried by the electromagnetic wave must be

$$\frac{\Delta(\text{momentum})}{\Delta t} = \frac{1}{c} \frac{\Delta(\text{energy})}{\Delta t},$$

$$(\text{momentum}) = \frac{1}{c}(\text{energy}).$$

There is a catch to this neat derivation! The E and B sinusoidal fields are in phase with each other, but the velocity of the moving charge will be 90° out of phase (since Eq is the sinusoidal force, which is 180° out of phase with the displacement, and 90° out of phase with the velocity). The radiation force is proportional to vE, but the product of two sinusoidal terms 90° out of phase over a full cycle will be equal to zero. The resolution to this apparent paradox is to note that the electrons in a solid lag in phase because they are not free. Even "free" electrons (perhaps in a plasma) experience a phase lag because of radiation damping. If $E = E_m \sin \omega t$, then $y = y_m \sin(\omega t - \alpha)$, and $v = v_m \cos(\omega t - \alpha)$

$= \omega y_m \sin(\omega t + \pi/2 - \alpha)$. Over one full cycle, $\langle(\sin \omega t)\sin(\omega t + \pi/2 - \alpha)\rangle > 0$. Therefore, F_r and the power are positive.

Of course, the energy of electromagnetic radiation is not contained in a continuous smooth wave, but, at least during interactions, appears to be quantized. Energy is emitted and absorbed in chunks. The corresponding model for light is that it travels in a bundle called a photon whose energy is proportional to the frequency of the light. For a photon, (energy)$=h\nu$. The momentum carried by such a photon must therefore be

$$(1/c)h\nu = h/\lambda$$

(since $\lambda = c/\nu$). The constant of proportionality h is Planck's constant. Its value is 6.6×10^{-34} J s. The energy and momentum carried by an individual photon of visible light are very small and yet sufficient to be detected.

Let us do an order-of-magnitude calculation of the pressure of sunlight on the earth. The radiation force is equal to the power delivered divided by the velocity of light: $F_r = $(power)$/c$. The power per unit area delivered to the earth in sunlight is approximately one kilowatt per square meter. Therefore, pressure of sunlight on earth $= F_r/A = [$(power)$/c]/A = [$(power)$/A]/c = (1 \times 10^3$ W/m$^2)/(3 \times 10^8$ m/s$)$ $= \frac{1}{3} \times 10^{-5}$ N/m^2. Evidently sunlight exerts only negligible pressure on the earth. (Atmospheric pressure is 1×10^5 N/m^2.) On the other hand, the earth presents a target with large area. For a rough calculation of the total force exerted on the earth by sunlight, assume a flat area equal to the cross-section disk of the earth. The area is $\pi r^2 = \pi(6.4 \times 10^6$ m$^2) = 1.3 \times 10^{14}$ m^2. The total force of sunlight on the earth is, therefore, $(1.3 \times 10^{14}$ m$^2)(\frac{1}{3} \times 10^{-5}$ N/m$^2) = 4.3 \times 10^8$ N. Since 10^4 N is about a ton, the force of sunlight on the earth is about 40 000 tons.

Comet's tails are blown away from the sun partly by electromagnetic radiation and partly by other particles that are shot out from the sun (mostly protons and electrons in the "solar wind"). Radiation pressure of visible light was demonstrated in the laboratory by E. F. Nichols and G. F. Hull in 1903. The standard technique for such a demonstration is to shine an intense beam of light on a small mirror that is part of a torsion suspension in a very good vacuum chamber. The momentum delivered by light to a mirror is twice that which occurs if the light is absorbed. The recoil of the mirror not only accounts for stopping the momentum of the light in one direction but also in sending it back again in the opposite direction. The intensity of light that can be focused on a tiny mirror in such a suspension is too small to make a sizable deflection of the system. A standard technique is to pulse the light with a frequency equal to the natural frequency of the mirror suspension. In this way, like repeatedly pushing a swing, the incident light sets the mirror into oscillation. The magnitude of the momentum delivered agrees well with the theoretical prediction.

A toy called "Crooke's radiometer" is sometimes presented as a demonstration of light pressure. It is not, although the four-vaned propeller does whirl around in sunlight, or even in the proximity of a light bulb. Look closely at the vanes and observe that each has one side silvered and the other blackened. Contrary to the explanation above, these vanes rotate with the blackened sides trailing. Whatever is pushing the vanes pushes harder on the black side than on the mirror. There is not a good vacuum in these bulbs; lots of gas molecules are left. A century ago it was assumed that the black side of each vane got hotter and that therefore the gas molecules would bounce off the black side with greater speed, thus requiring a greater momentum recoil from the black side. The explanation is much more complicated and has been given by Woodruff.

■ POLARIZATION OF ELECTROMAGNETIC RADIATION

When we first considered how electric and magnetic fields would generate each other and so propagate through space, we assumed that we could create a moving wall of electric field with the field lines all in the same direction perpendicular to the direction of propagation. A radio signal coming from the antenna of an AM radio station is a practical example. Because the source electrons accelerate vertically up and down the tall antenna, the electric field of the radiated signal is also vertical. The corresponding magnetic field is horizontal.

In the first section of this chapter, "Demonstrating the Phenomena," we suggested that you turn a portable AM radio in various directions with respect to a radio station. If your radio has a vertical antenna like a radio in a car, then it is designed to respond to the electric field. If you turn the antenna horizontal, the signal will be greatly diminished. Some portable radios have horizontal antennas. They consist of a coil of many turns of wire wound around a rod of *ferrite*. This ceramic material has a high magnetic χ value but, unlike iron, it can change its magnetization many millions of times per second. A radio that contains such an antenna must not only remain horizontal in order to respond to the horizontal magnetic field, but must also be lined up perpendicular to the direction to the radio station. You may find strange exceptions to these rules. Radio waves may be reflected from hills or buildings so that the optimum signal may not come from the straight line direction to the station. Furthermore, in these reflections the vertical and horizontal orientation of the field may be twisted.

Visible light is not generated by the organized motion of electrons up and down an antenna. The frequency of visible light is so high and the wavelength so short that it is usually generated by the transition of an atom or molecule from an excited state to a lower energy level. Each atom or molecule produces one photon at a time with a phase and polarization that is usually unrelated to those of any of the other photons being produced. (There are certain exceptions such as with lasers or with synchrotron radiation.) With ordinary visible light, the electric field vector at a given point at a given instant may be vertical, and a billionth of a second later may be horizontal or at any angle in between. The electric and magnetic-field vectors remain perpendicular to each other and to the direction of propagation, but the fields are oriented at random, with continually changing directions. Light from a student laboratory He–Ne laser (1 mW) maintains temporal coherence for times of a few nanoseconds (which is about 10^6 periods of visible light).

The efficiency of scattering or reflection of light depends on the orientation of the electric-field vector with respect to the interacting surface. Most light that is reflected from surfaces with a large angle of incidence and reflection appears as glare and contains little information about the surface. The glare part is light that has been reflected as from a mirror. Such *specular* reflection only contains information about the source. In effect, the surface of the mirror has not been probed. When we observe glare images from a road ahead of us, we are actually seeing the reflection of the sky.

Glare reflection from a horizontal surface is highly polarized in the horizontal direction. The vertical component acts as if it dug into the surface and was

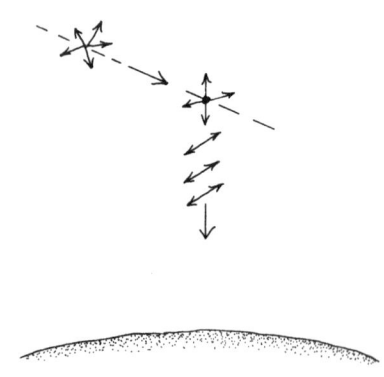

Figure 20.14. Sun light scattered by 90° is preferentially blue and is polarized.

absorbed. We can reduce glare by wearing polarizing sunglasses that accept only light polarized in the vertical direction. If you have such glasses available, notice how they eliminate glare from horizontal surfaces and how the glare returns if you turn the polarizers 90°.

Blue sky is caused by the preferential scattering of blue light from the molecules of the upper atmosphere. The scattering probability (for wavelengths much longer than the size of the scatterer) is proportional to the fourth power of the frequency. The scattered light is partially polarized as you can tell by examining a section of the sky through polarizing glasses. A polarizing filter on a camera is sometimes used to darken the background sky and enhance the whiteness of clouds. In Fig. 20.14 we show the geometry of light scattered down to earth from the sun in afternoon. Each molecule acts as a tiny dipole set into oscillation by the incident light. The reradiated energy has its maximum amplitude along the perpendicular to the dipole. Therefore the scattered light in the case shown will be polarized in the north–south horizontal direction.

The long-chain molecules embedded in plastic-sheet polarizers are oriented during manufacture so that they are all in one direction. If the electric vector of the light is perpendicular to the preferred direction, then the polarizer appears almost black (actually, the effect depends to some extent on wavelength, and blue light can still get through). If light passes through two polarizers, the light will be plane polarized by the first and then "analyzed" by the second. If the orientations of the two pieces are in the same direction, then light will pass through. But if the orientations are at 90°, most of the light will be blocked. If the electric field of the incident light is at an angle θ to the preferred orientation, then only the component of $E \cos \theta$ will get through. Since the intensity of the light is proportional to E^2, the intensity of plane polarized light that is passed by a polarizer is $I = I_m \cos^2 \theta$.

If you sandwich three polarizing sheets together, with the outer ones crossed by 90° and the middle one at 45°, you might think that no light could get through because the first one yields plane polarized light only in the direction forbidden by the final one. Keep in mind, however, that light goes through transparent materials by being repeatedly absorbed and reradiated. The light coming through the first filter has been largely generated *in* that filter. Half of the original intensity will leave the first filter. Filter number 2 will treat that light as coming at 45° to its generating direction and will reradiate light with an electric field of $E \cos 45°$. The intensity leaving the middle filter is $\frac{1}{2} I_0 \cos^2 45° = \frac{1}{4} I_0$. There is a component of that light in the direction allowed by the third filter. Because the angle is once again 45°, there will be another factor of $\frac{1}{2}$ multiplying the intensity. The fraction of the original intensity getting through the polarizing sandwich is thus $\frac{1}{8}$.

■ THE EXPERIMENTAL CONFIRMATION OF MAXWELL'S PREDICTION

Maxwell developed his equations and their consequences on the basis of Faraday's model of electric and magnetic fields. The mental models that his mathematics described were more complicated than those we use today. Maxwell and others at that time thought that the fields and the wave motions were physical properties of an actual fluid permeating all matter and called the aether. Nevertheless, in 1862 he proposed that "light consists in the transverse undulations of the same medium which is the cause of electric and magnetic phenomena." By that time he had calculated the speed of electromagnetic waves from his equations and knew that the speed was approximately the same as that which had recently been measured for light.

The experiments that confirmed Maxwell's predictions were first made by Heinrich Hertz (1857–1894), between 1885 and 1889, about 25 years after Maxwell's predictions and only shortly after Maxwell's death. In a brilliant series of experiments, Hertz first learned how to make oscillating circuits with frequencies in the range from 1×10^8 to 1×10^9 Hz. He did this by interrupting currents in induction coils to impose sudden high voltages across spark gaps. The spark that jumps the gap consists of electrons and ions from the air, which oscillate back and forth. The frequency is determined by the inductance and capacitance of the coils or rods that form the gap.

Figure 20.15 is a schematic drawing of the apparatus that Hertz used. The receiver consisted of a "split dipole," a rod with a length of half a wavelength, separated in the middle by a small spark gap. The *transmitter* spark jumped between small polished spheres; the resulting spark was fat and carried a large current. However, the radiation energy available at the receiver is relatively weak. Even with the receiving antenna tuned to the radiated frequency, the voltage buildup across the receiver gap was small. The *receiver* gap was formed by sharp points separated by only a fraction of a millimeter. Hertz detected the radiation by looking for the tiny sparks jumping between the points.

At a frequency of 1×10^9 Hz, the wavelength is only 30 cm. Hertz measured this wavelength by reflecting the radiation from metal plates and looking for the nodes in the standing waves. (At a node, there would be no spark in his receiver.) He also demonstrated that the radiation was polarized and that it was subject to refraction (in a prism of pitch) and diffraction.

One of the ironic features of the experiment was Hertz's observation that the gap in the receiver broke down to produce a spark only if the receiver was in a position to be illuminated by the ordinary light from the spark. We now know that the ultraviolet radiation from the original spark helped to release electrons from the receiver electrodes, enabling the detection spark to form. This phenomenon, of course, is the photoelectric effect, which is frequently thought of as a decisive demonstration of the photon nature of light. Hertz's experiment, which demonstrated and seems to depend on the wave nature of electromagnetic radiation, also depended critically on this other aspect of radiation. Hertz was not able to measure directly the speed of the radiation produced by his apparatus but that experiment was done a few years later. The value was indeed equal to the speed of light.

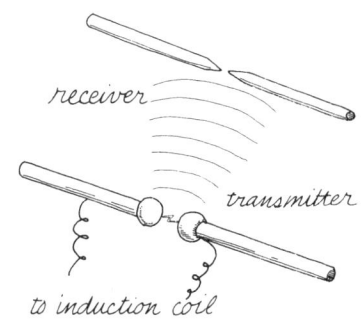

Figure 20.15. Schematic drawing of Hertz apparatus for generating and receiving electromagnetic waves.

■ THE ELECTROMAGNETIC SPECTRUM

Electromagnetic radiation encompasses all the effects from radio to x rays. In one sense, the only difference between these very different phenomena is the frequency of the wave motion. That factor makes such a large difference, however, that we should divide the whole range of electromagnetic radiation into various realms. The organizing variable for this analysis could be either the frequency or the wavelength. These two variables are related by the basic equation for wave motion: $\lambda \nu = c$. [It is customary to use ν (nu) for the frequency of electromagnetic radiation.]

There are other factors that characterize regions of the electromagnetic spectrum. Visible light seems to interact with matter in a bunched or quantized way.

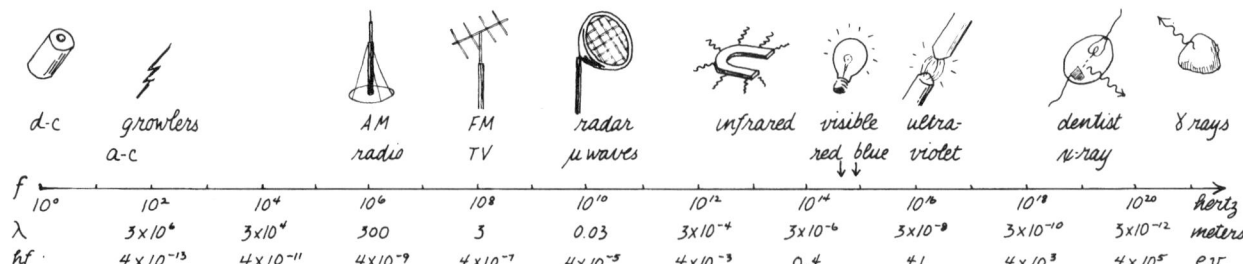

Figure 20.16. The electromagnetic spectrum.

Each photon of light delivers an amount of energy $h\nu$, where h is Planck's constant. For visible-light frequencies and higher, the photon energy is often the most important parameter.

While we are organizing the electromagnetic spectrum, we should also point out particular features of the transmitters or receivers in each region. The antenna size is linked to wavelength. Another allied factor is the absorption of a particular frequency. Some materials are transparent for certain wavelengths and opaque to others.

Fig. 20.16 presents a chart of the electromagnetic spectrum, indicating the ranges that have been given particular names. The scale is necessarily logarithmic since the frequencies and wavelengths extend over such a large range. Notice in particular the very small region for visible light.

■ GROWLERS AND WHISTLERS

From the earliest days of radio, certain types of receivers have been plagued by a type of noise called "growlers" or "whistlers." These are short-lived signals with frequencies in the audio range—a few tens to a few hundreds of cycles per second. The pitch varies during each brief signal. The resulting wavelengths are of the order of 1000 km, almost the width of continents. The waves cling to the surface of the earth, circling it many times before being dissipated. Since radiated power is proportional to the fourth power of frequency, it must take an enormous amount of current in the source to produce sizable signals at these low frequencies. Apparently these signals are caused by lightning, which provides sufficient power so that the signals can travel around the earth.

Human-made radio signals have been used with frequencies in the audio range. In the 1920s and 1930s a 25 000-Hz system was occasionally used for overseas transmission from the RCA center in Long Island, New York. The wavelength of this signal was 12 km. The antenna consisted of a horizontal wire stretching for 3 km on high towers. The system was designed before high-power radio tubes were manufactured. A motor generator set with many commutators was built in order to provide sufficient radiated power. The advantage of a transmitter using such long wavelengths was that the signals could creep around the surface of the circular earth rather than flying tangentially off into space. Ordinary radio waves for overseas transmission are directed up to the layer of ionization called the Kennelly–Heaviside layer, high above the atmosphere. The short-wavelength signals are then reflected back down to earth again and may make several reflections or skips before arriving at a place where they can be detected. Solar magnetic storms can destroy the Kennelly–Heaviside layer temporarily, but communications between continents could be maintained with the very-long-wavelength system.

Ultralong wavelengths of electromagnetic radiation can penetrate ocean depths or even the earth itself. For military purposes in contacting submarines, transmitters have been proposed for frequencies in the range of 100 Hz. The transmitting antennas for such systems would require formidable land space.

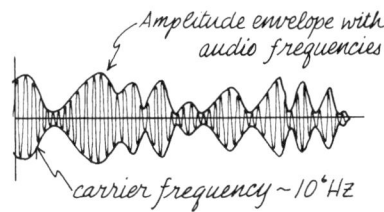

Figure 20.17. An amplitude modulated radio signal.

■ AM RADIO

AM stands for amplitude modulation. The transmitter uses one particular frequency for the carrier signal but changes the amplitude of the signal at the much lower frequencies of the audio range. A diagram of such amplitude modulation is shown in Fig. 20.17. The standard AM band in the United States extends from 535 to 1605 kHz. The center of this range is at 1 MHz. The wavelength corresponding to this frequency is 300 m. Radiation of this wavelength can penetrate ordinary nonmetallic buildings and can also creep around most human-made structures. (Diffraction reduces shadow effects if $\lambda/a > 1$.)

We seldom have occasion to consider the photon nature of radio waves but let us calculate the energy involved in such a photon:

$$h\nu = (6.6 \times 10^{-34} \text{ J s})(1 \times 10^6 \text{ Hz})$$

$$= 6.6 \times 10^{-28} \text{ J} = \frac{6.6 \times 10^{-28} \text{ J}}{1.6 \times 10^{-19} \text{ J/eV}} \approx 4 \times 10^{-9} \text{ eV}.$$

This energy is small even by atomic standards. It takes about 1 eV to produce a chemical interaction.

The transmitting antennas for AM radio are vertical towers. Sometimes these are placed on the top of a very high building in the center of a city, although sometimes an array of several towers built up from the ground will be placed on the outskirts of a populated region. By controlling the phase of the signals in the several towers, the direction of the radiation can be controlled so that most of the energy is delivered to the populated region. The towers are usually about 1/4 wavelength high with the signals being sent in at the ground level. Note that for a 1-MHz signal, the 1/4-wavelength antenna would have to be 75 m high. The complete *effective* antenna is actually half a wavelength long. The other 1/4-wavelength is provided by the ground reflection of the actual antenna. Sometimes the field around the antenna is covered with a lightly buried layer of wires to create the ground mirror. To provide a satisfactory signal, the electric field strength at a receiver must be anywhere from 100 μV in electrically quiet rural areas to 10 mV in the heart of an electrically noisy city. Most commercial radio stations have a range of from 25 to 100 miles. Reception within this region is due to the so-called ground wave that comes directly from the antenna and is refracted in such a way that it clings to the curved surface of the earth. Some of the radiated energy, however, shoots up to the Kennelly–Heaviside layers (the ionosphere) and then

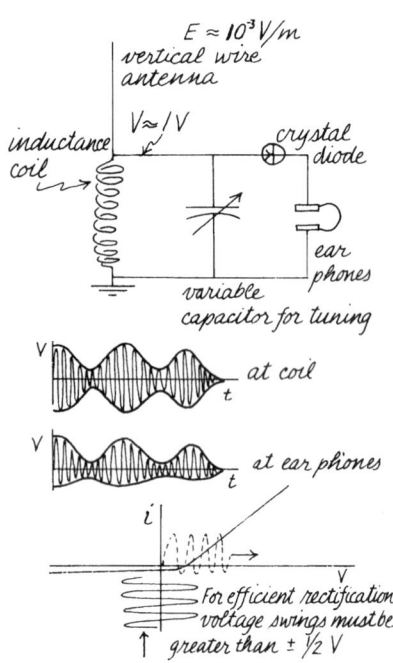

Figure 20.18. Circuit and operation of a crystal model radio.

bounces back down again many hundreds of miles away. The bounce effect is usually small for AM radio but becomes increasingly important for shorter wavelengths and at night. (The Kennelly–Heaviside layer is produced by ionizing radiation from the sun, and so descends and rises from day to night. During the day when the ionosphere is thickest, the lowest layer tends to absorb radio waves rather than refract them. At night the ions of this layer recombine, leaving only the higher layers where little absorption takes place.)

The generation and reception of radio waves are crucially dependent on the resonant characteristics of LRC circuits. The final stage of the transmitter circuit and antenna forms an LC circuit tuned to the particular frequency of the station. Because of the resonance, large currents and high voltages can be produced in the transmitting antenna.

The receiving antenna and the first stage of the receiving circuit must also be sharply tuned to the frequency of the desired signal. The main reason for this is obvious. The space around the antenna is filled with electromagnetic radiation of many frequencies coming from many sources. If you tune the antenna and receiver to one particular frequency, the signal at that frequency will set the circuit into resonant oscillation. Radiation of other frequencies will not be able to build up such an oscillation.

The magnitude of the voltage buildup in a receiver is surprising. If the electric-field strength of the signal is only about 1 mV/m, and the antenna is only a few meters long, then the input voltage to the receiver is only a few millivolts. Nevertheless, crystal radios, without any tubes or transistors, can produce audible signals even though they need inputs of the order of a volt.

Let us analyze a typical circuit of a simple crystal radio, as shown in Fig. 20.18. The purpose of the crystal (silicon) diode is to rectify the high-frequency radio signal; otherwise the loudspeaker or earphones could not respond to the high-frequency signal that exerts alternate positive and negative forces. By rectifying the signal, the circuit produces short pulses of varying magnitude and all in the same direction. The audio part of the circuit can then respond to the slowly varying unidirectional current.

The trouble with such a scheme is that a solid-state diode rectifier (or tube diode, for that matter) changes its forward to backward resistance only over a range of several tenths of a volt. If a signal of a few millivolts is impressed across such a rectifier, the resulting signal is scarcely rectified at all. The signal across the diode is not just a few millivolts, however. The resonant circuit of antenna and input coil is driven into oscillation by the weak radio signal. The resonant oscillation builds up a potential hundreds of times greater than the driving voltage. (The resonant Q of the circuit must have a value of at least 100.) Consequently, a high-frequency signal with an amplitude of at least a volt is imposed across the diode so that the resulting rectified signal looks like the one in the diagram. To get a high-Q circuit, the resistance must be small. Not only is the resonant voltage amplification equal to Q, but the discrimination against other frequencies depends on Q. The full width of the response curve (at half amplitude) divided by the resonant frequency is equal to Q. For a Q value of 100 at a resonant frequency of 1 MHz, the full width (at half amplitude) would be 10 kHz, which indeed is about the frequency band allocated to an AM station. (Note the further implication that the width of the audio band can therefore be only 5 kHz, which is not exactly high fidelity.) For a further description of Q, see 13.11 and 13.19.

■ TELEVISION AND FM RADIO

Each of the television channels is allocated a width of 6 MHz starting with channel 2, which goes from 54 to 60 MHz. Channel 6 extends from 82 to 88 MHz. The frequency range from 88 to 108 MHz is reserved for FM. Each FM station can use a channel that is 150 kHz wide. Television channels 7 through 13 occupy the range from 174 to 216 MHz, while the UHF (ultrahigh-frequency) channels from 14 through 83 use the range from 470 to 890 MHz.

There is a frequency change of about a factor of 15 between channel 2 and channel 83. Let us calculate the wavelengths corresponding to the extremes of this region. The wavelength for channel 2 is

$$\lambda = (3 \times 10^8 \text{ m/s})/(54 \times 10^6 \text{ Hz}) \approx 6 \text{ m}.$$

The middle of the FM band has a wavelength of about 3 m and channel 83 in the UHF has a wavelength of about 1/3 m.

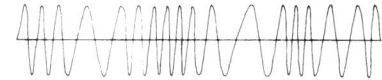

Figure 20.19. Frequency modulated (FM) radio signal.

Antennas for TV and FM look very different from the simple vertical wires for AM radio. If the transmitter is located at the center of the city, it must be designed to radiate equally well in all horizontal directions. The receivers, however, should be directional and aimed at the transmitting tower. Notice the TV antennas on the roofs of houses in a suburb. They are all pointed in the same general direction, usually toward the center of the metropolis. Furthermore, the receiving antennas are cut to match the approximate size of the wavelength. A "yagi" antenna has a row of horizontal cross bars, each with a length of $\frac{1}{2}\lambda$. For FM and TV the electric field is generated to be horizontal, or at 45° to the vertical. The active bar to which the antenna lead is fastened is called a "folded dipole." The other bars, which are not electrically connected, serve to reflect and reinforce the signal. A "yagi" antenna is very directional and has high gain for the particular frequency for which the length of its arms is one half the wavelength. Sensitivity to other wavelengths is low. The more general type of TV antenna will resonate to a broad range of frequencies but has lower gain for any one in particular. Many different configurations of wires and shapes can be made to resonate at a particular frequency. In all cases, however, the size of the antenna is approximately 1/2 a wavelength. Notice that the loop antenna used for UHF is less than 1/5 the size of the antennas designed for the lower channels.

Since the frequency of FM is about 100 times that of AM, the energy of an FM photon is about 100 times that an AM photon. That factor raises the energy of an FM photon to less than one millionth of an electron volt.

Although the video part of television is amplitude modulated (notice how a lightning stroke affects the picture), the audio part of the signal is frequency modulated (FM). Figure 20.19 shows schematically the appearance of an FM signal. The audio input controls the *frequency* of the carrier radio signal, which can vary by 75 kHz on either side of the central frequency. If there is no audio signal, the FM signal remains at its central frequency and the receiver remains quiet. The louder the original audio signal, the further the radio signal departs from its central frequency and the louder the resulting tone in the receiver. The frequency of the original audio signal determines the frequency with which the radio signal departs from its central frequency and therefore the frequency of the rectified signal from the receiver. Ordinary static, whether human-made or from lightning, is not detected as a signal by such a system.

The wavelength of FM and TV is smaller than the size of many human-made structures such as buildings and airplanes. Consequently, the TV or FM signals frequently reflect from such objects and arrive at your antenna by two different routes. The path difference between two such routes creates a phase difference, which appears on your TV set as "ghosts." Suppose there is a path difference of 1 km between the two signals. The travel time difference would be about 3 μs. On a TV screen the ghost would be separated from the main picture by several centimeters. Sometimes there is interference from signals coming from the same channel in a city several hundred miles distant. The high-frequency signal in such a case did not refract around the curve of the earth but instead shot up to the ionized layer and then bounced back down. As one program fades out and the other surges in you are watching the effect of the ion layers rising and falling.

■ RADAR AND MICROWAVES

In a radar system the transmitter is also the receiver. A short burst of electromagnetic radiation is sent out in the form of a beam from a reflecting-type transmitter. The same system then listens for any fraction of the pulse that may be reflected from some object. The time interval between sending the pulse and receiving the reflection is proportional to the distance between the radar set and the object. If the object is moving, the reflected pulse will be Doppler shifted—to a higher frequency if the object is approaching and to a lower frequency if the object is retreating, and in both cases the shift is roughly proportional to the speed.

In order for the transmitter to produce a beam, the wavelength of the radiation must be small compared with the size of the reflecting transmitter. The angular spread of the beam cannot be less than the diffraction pattern of the circular transmitter. The half-angular width, from the center of the diffraction pattern to the first minimum, is $\theta = 1.2\lambda/d$. The intensity is down by a factor of one-half at about half this angle. Consequently, we can take the *full* angular width of the diffraction pattern to be about $1.2\lambda/d$. If λ/d equals 1/10, then $\theta = 1.2/10 = 0.12$ rad $\approx 7°$. At a distance of 100 m, the width of the beam would be (100 m)(0.12 rad) = 12 m. Such a beam could detect objects at a distance of 100 m providing they were not closer together than 12 m.

Various frequencies are used for radar depending on the size of the object being tracked. Police radar uses the smallest wavelength. You can estimate what this wavelength must be by observing the small portable systems that police can aim from their car. The diameter of the radar gun may be as small as 10 cm. If the resulting beam is to be narrow enough to have any discrimination between cars, then the wavelength used must be only about 1 cm. Wavelengths much less than this are absorbed by water vapor in the air.

The frequency of 1-cm radiation is 3×10^{10} Hz or 3×10^4 MHz. The energy of a photon of this frequency is

$$E = h\nu = (6.6 \times 10^{-34} \text{ J s})(3 \times 10^{10} \text{ Hz}) \approx 20 \times 10^{-24} \text{ J} = 1.2 \times 10^{-4} \text{ eV}.$$

A photon energy of 1-cm radiation is still well below the energy of molecular binding or the average vibrational or translational kinetic energy of molecules at room temperature. However, microwave photons are in the energy range of molecular *rotations*. The atoms in a molecule cannot only vibrate but can also rotate around each other. A rotating system can change its angular momentum only in integral multiples of a basic unit; angular momentum is quantized. The basic unit is Planck's constant. When the angular frequency is given in radians per second, the magnitude of this unit of angular momentum change is

$$\hbar = h/2\pi = 1 \times 10^{-34} \text{ kg m}^2/\text{s}.$$

Let us calculate the energy involved with the simple hydrogen molecule when the two revolving protons are given one extra unit of angular momentum:

$$\Delta(\text{angular momentum}) = \Delta(I\omega) = \hbar,$$

$$(\text{moment of inertia of hydrogen molecule}) = I = 2m_p r^2,$$

$$(\text{rotational kinetic energy}) = E_r = \tfrac{1}{2} I \omega^2 = \tfrac{1}{2}(I\omega)^2/I = \tfrac{1}{2}\hbar^2/I$$

$$= \frac{1}{2} \frac{(1 \times 10^{-34})^2}{2(1.7 \times 10^{-27} \text{ kg})(0.37 \times 10^{-10} \text{ m})^2}$$

$$\approx 1 \times 10^{-21} \text{ J} = 1 \times 10^{-21} \text{ J} \left(\frac{1 \text{ eV}}{1.6 \times 10^{-19} \text{ J}}\right)$$

$$\approx 1 \times 10^{-2} \text{ eV}.$$

(The actual measured value is 1.5×10^{-2} eV.)

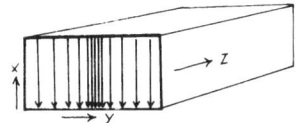
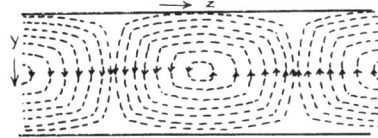
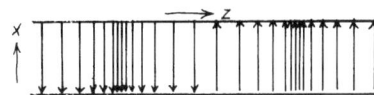

Figure 20.20. Electric and magnetic fields in a pipe carrying microwaves.

As you can see, when a hydrogen molecule changes its angular momentum by one unit, the change of angular momentum is very small, though still larger by a factor of about 100 than the energy in a photon with a wavelength of 1 cm. For larger molecules, the change in rotational energy can be much smaller than that for the hydrogen molecule. In the expression for the rotational energy in terms of angular momentum, notice that the moment of inertia term is in the denominator. For a molecule with larger mass and larger radius, the moment of inertia is larger and the rotational kinetic energy for a given angular momentum is less. Consequently, microwave radiation can be used to study molecular structure. Radiation with these wavelengths can be generated with very precisely known frequencies. If a beam of such radiation is absorbed by a material at a particular frequency, then the molecules in the material must be changing their motions using amounts of energy corresponding to the photon energy.

Microwaves can be generated by vacuum tubes, such as the "magnetron," that use bunches of electrons to excite standing waves of electromagnetic radiation in small cavities. Instead of sending this high-frequency radiation on wires, we can literally pipe it from one place to another with very little loss. The electric and magnetic fields inside a pipe carrying microwaves are shown in Fig. 20.20. For practical reasons this technique can be used only for radiation of wavelengths in the centimeter region since the dimensions of the pipe must be about the same as the wavelength of the radiation.

A radar beam raises the temperature of anything that absorbs it. As a matter of fact, many kitchens have devices that depend on this effect. The radiation in the standard microwave oven has a wavelength of about 10 cm. In popular folklore there is a legend that microwaves cook food from the inside out. This is not true, as you can readily test. The penetration depth into food is large, but the outer region absorbs more. It is also not true that the frequency is tuned to water absorption bands. The heating is simply Ohmic, depending on conductivity of the material. Tap water or moisture in food has good conductivity and so absorbs the most microwave energy.

A serious proposal for harnessing solar energy envisions an enormous field of silicon photocells in a geosynchronous orbit 22 000 miles above the surface of the earth. The photocells would absorb sunlight and provide dc voltage. This energy would be converted by magnetron tubes, like those in the microwave ovens, but without any need for glass envelopes, to 10-cm radiation. The microwaves would be beamed to a rectifying field on the earth close to the city that would use it. Would this radiation cook any birds or airplanes flying through it? Suppose that the beam supplying energy for a major city has a total power of 10 000 MW (10^{10} W). If the average power density in the beam is 100 W/m^2, how large a rectifier field would be required? (The current safety level for microwave oven radiation leakage in the United States is 10 mW/cm^2.) With a power density of 100 W/m^2 (which would meet the safety standards), the rectifier field would have to have an area of 10^8 m^2. The diameter of the field would therefore have to be a little larger than 10 km. The land underneath the rectifying arrays could be used for many purposes, including farming or grazing. Birds roosting on the antenna might get warm. The metal skin of airplanes would reflect the microwaves with little absorption.

■ THE INFRARED REGION

The low-frequency end of this region has no sharp boundary as it merges with the microwave region. Wavelength, rather than frequency, determines this boundary since for wavelengths less than 1 mm, tubes and hardware of ordinary electronics become too small to be fabricated or used. For a wavelength of 1 mm, the corresponding frequency is 3×10^{11} Hz. The upper frequency limit of the infrared is determined by the beginning of the visible range. That frequency is about 4×10^{14} Hz with a wavelength of 7×10^{-7} m. Notice that this wavelength is a little smaller than 1 μm (0.7×10^{-6} m = 0.7 μm).

The photon energy at the low frequency range is

$$h\nu = \frac{(3 \times 10^{11} \text{ Hz})(6.6 \times 10^{-34} \text{ J s})}{1.6 \times 10^{-19} \text{ J/eV}} \approx 1 \times 10^{-3} \text{ eV}.$$

That energy, as we have seen, corresponds to the changes of rotational energy of molecules. At the high-frequency end of the infrared, the photons have enough energy to break apart some types of molecules:

$$h\nu = \frac{hc}{\lambda} = \frac{(6.6 \times 10^{-34} \text{ J s})(3 \times 10^8 \text{ m/s})}{(7 \times 10^{-7} \text{ m})(1.6 \times 10^{-19} \text{ J/eV})} = 1.8 \text{ eV}.$$

Because the photons in the infrared have enough energy to disturb molecules, infrared radiation is a prime tool for chemists. Infrared beams are sent through materials and the absorption is measured as a function of frequency. Each frequency where there is absorption corresponds to a particular energy that changed the rotational or vibrational motion of the molecules. From such evidence the actual structure of the molecule can be figured out.

One way to produce infrared radiation is simply to raise the temperature of some solid material. As the temperature of the solid rises, two things happen. The total amount of radiation given off increases rapidly (proportional to T^4); also the radiation shifts to higher frequencies. This latter observation agrees with everyday experience. As the temperature of a coal or glowing wire increases, its color changes from dull red to bright white.

At any temperature above absolute zero, atomic and molecular motion gives off electromagnetic radiation. We notice the effect with objects hotter than our skin temperature. Glowing electric coils or fireplace coals can warm or even cook us. The effect does not depend on the visible radiation; you can easily feel the ratiation from a hot flat iron or a hot water radiator even though they are not hot enough to glow.

Many types of lasers can produce radiation in the infrared. Radiation from these sources can be generated in a narrow frequency range and also is coherent.

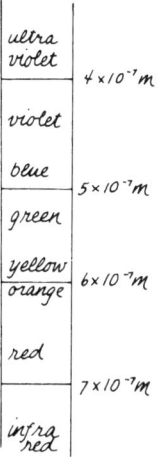

Figure 20.21. Wave length boundaries in the visible EM spectrum.

■ VISIBLE LIGHT

The visible region of the electromagnetic spectrum is very small and has fairly sharp boundaries determined by the characteristics of human eyes. The high frequency in the blue-violet has a wavelength of 4×10^{-7} m. The low-frequency limit in the red has a wavelength of 7×10^{-7} m. The wavelengths and frequencies of various colors are shown in Fig. 20.21.

We have already figured out the energy corresponding to the red–infrared border—1.8 eV. The energy of a blue-violet photon is 3 eV, almost a factor of 2 larger.

Notice how many different phenomena are necessarily linked together by the magnitude of the energy of visible-light photons. Since atoms are bound together with energies of about 1 eV, voltaic cells produce voltages of about 1 V (the standard flashlight cell is $1\frac{1}{2}$ V; the lead storage cell is 2 V). Each time that a molecular combination changes in such a cell, the energy involved is of the order of an electron volt. That much energy is sufficient to shove an electron up a potential hill of about 1 V. Since visible light can clearly produce chemical changes (photosynthesis, photography, bleaching, tanning), the energy of visible photons must be of the order of an electron volt.

Visible light can be generated by many different processes. Most of these involve bound atomic or molecular systems. An electron in such a system is pulled away from its ground state by thermal agitation or by electron bombardment. When the disturbed electron falls back to its original condition, the energy is released in the form of a photon. As we have seen, a hot solid emits a continuous spectrum of photons of different frequencies. If the atoms are isolated, as they are in a low-pressure gas, then the photons that are emitted have only certain energies and hence wavelengths that are characteristic of that particular type of atom.

■ ULTRAVIOLET AND X RAYS

Our main experience with ultraviolet rays comes in the summer when we spend too much time in the sun. Our hair is bleached and our skin is tanned or burned. (If you want a skin like leather, let the sun tan your hide!) The ultraviolet region starts at the blue-violet end of the visible spectrum and merges without distinction into the x-ray spectrum. For these photons, energy is a more important parameter than frequency. For a wavelength of 1×10^{-10} m, the atomic radius, the photon energy is about 1×10^4 eV. X rays with this energy and above "see" individual atoms since their wavelengths are smaller than the atom. X rays for medical diagnosis usually have energies between 20 000 and 100 000 eV. For radiological destruction of tissue, photons with energies of 1×10^6 eV (1 MeV) or more are used.

A dental x-ray machine is a classic example of the phenomenon of electromagnetic radiation from an accelerated charge. Electrons are accelerated across a voltage between the filament and a high-density target. The acceleration through the vacuum is relatively small and produces little radiation. However, when the electrons strike the high-density target (frequently tungsten), they are stopped abruptly, with very large acceleration. Radiated photons stream off the target with energies up to the maximum energy of the bombarding electrons.

Since x rays can be used to probe atomic structure, it is important to have high-intensity sources of them. During the 1980s a new source of high intensity, polarized x rays, was developed. This is the electron synchrotron. When high-energy electrons (with energies in the 10^9-eV range) are forced by a magnetic field to travel in an orbit of small radius, they must experience large centripetal acceleration. Consequently they radiate photons in a narrow beam mostly in the direction of their motion. This synchrotron radiation can be very intense and extends from the visible to x rays with wavelengths of atomic dimensions.

■ X RAYS AND γ RAYS

There is no upper limit to photon energies. Even as there is no real boundary between the ultraviolet and soft x rays, there is no physical difference between x rays and γ rays. Historically, photons coming from energy changes in *atoms* were called x rays, while those coming from disturbances in atomic *nuclei* were called γ rays. When the photon energies get above 1 MeV they can under certain circumstances produce a pair of electrons, one negative and one positive. In electron pair production, the photon energy provides the rest mass energies of the electron and positron as well as their kinetic energies. It is not a case, however, of "energy turning into matter." A high-energy photon is best considered as a subatomic particle. All such particles can transform into each other if certain conservation laws are satisfied.

1. Woodruff, Arthur E, "The Radiometer and how it does not work," Phys. Teach. **6**, 358 (1968).

21
Microstructure

To begin our description of the microworld, it might seem reasonable to list the various subatomic particles and then describe how they form nuclei, atoms, molecules, and us. It is not that simple. First, there are some new rules or laws that we do not ordinarily meet in introductory physics. But these rules dominate the microworld. Second, the particles cannot be described except in terms of their interactions, and the interactions must be explained in terms of the properties of the particles. Finally, the phenomena of the microworld require description in terms that challenge our everyday view of the universe and our interaction with it. To describe these phenomena, we must raise primitive questions about the nature of reality. We are in the position of trying to describe a play in which the roles of the actors cannot be understood without knowing the plot, which cannot be explained without describing the roles of the actors. Furthermore, some of the stage rules are different from those of the everyday world, which turns out to be an illusion. The play is the real world.

Let us proceed, warily, by taking a closer look at a familiar particle. In describing electricity, light, heat, and other phenomena, we have frequently referred to electrons. Now we describe the parameters, or attributes, of the electron. In the process, we will learn about many of the strange features of the microworld. With this background as a guide, we can then describe the other subatomic particles and how they combine.

■ THE ELECTRON

The first particle to be actually identified as such was the electron. In 1897, J. J. Thomson measured the charge-to-mass ratio (e/m) of the particles coming from a negative terminal inside a vacuum system. The existence of "cathode rays" had been known for some years and it had been shown that they carried negative charge. It was conceivable, however, that the rays were simply a stream of electric current. Thomson's experiment of deflecting the path of the rays with electric and magnetic fields was most easily explained in terms of the motion of individual particles.

Production

The electron is the easiest particle to isolate. Experimenters 100 years ago were intrigued with the way electrons seemed to leap out of any substance with only the mildest encouragement. Certain minimum conditions are necessary, of course. Electrical instruments are needed to detect most of the effects, and these effects

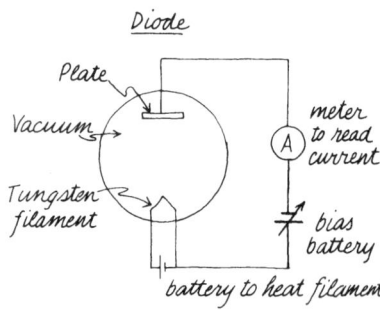

Figure 21.1. Current is produced only when filament is hot. Therefore, charges must be coming from filament. The current can be diminished or stopped by imposing a negative bucking field with the bias battery. Therefore, charges must be negative.

mostly take place in high vacuum. To drive an electron out of a material, it is necessary to give it sufficient energy to escape from its atomic or molecular bonds. With some metals this can be done by heating the whole system and boiling off the electrons. They can also be knocked out by light, as in the photoelectric effect. As we saw earlier, the amount of extra escape energy needed is a few electron volts. Ordinary blue light can provide about 3 eV per encounter, which is sufficient for electron emission from the alkali metals. Ultraviolet light, which can deliver energy of 6–10 eV to each electron, could knock electrons out of chicken fat, were that desirable (and assuming that the ultraviolet light could get through the air without being absorbed by knocking electrons out of the air molecules).

The average vibrational thermal energy of a bound atom at room temperature is about $\frac{1}{25}$ eV (0.04 eV). A tungsten filament in a vacuum tube may be heated to 3000 K, 10 times as hot as the 300 K room temperature. Even at a temperature of 3000 K, the average kinetic energy of a bound atom is only 0.4 eV. How can the tungsten emit electrons if their binding energy is several electron volts? An energy of only 0.4 eV would not be enough to cause an electron to be emitted from tungsten, but that is only the average energy. The high-energy end of the distribution can give sufficiently energetic kicks to electrons frequently enough to make the white-hot filament a very copious electron emitter.

Charge

The diagram of the diode in Fig. 21.1 shows a way to demonstrate that something negative is coming from the filament to the receiving plate. In spite of the success of Thomson's particle model, however, it is still conceivable that this negative charge is flowing as some kind of fluid instead of traveling in individual packages. A single electron would not make even the most sensitive electrical detector in this circuit respond with a signal greater than the natural fluctuations of the instrument. The individual charge was first measured accurately by Millikan in 1903. On p. 401 we described his method of measuring the electric charge on tiny oil droplets. The amount of electric charge that can cling to a droplet is quantized (exists in multiples of unit quantities). Millikan's experiment yielded a quantitative value for the magnitude of the electron charge. It is negative and equal to 1.6×10^{-19} C.

■ MASS

The most obvious attribute of the electron is its electric charge. The next question is, does it have any mass? To measure the mass of a baseball we have two methods. First, we can measure the weight or, in other words, the attractive force between baseball and earth. Because the weight is proportional to the mass, $F = G(mM/r^2)$, we can compare the weights (and so the masses) of the baseball and a standard mass on a balance. The second method depends on the inertial properties of mass, its reluctance to have its state of motion changed. When a force is applied to an object, the object accelerates: $F \propto a$. The constant of proportionality is called mass: $F = ma$. The mass of the baseball could be found by applying a known force to it and measuring the acceleration produced.

The first method is completely impractical for finding the mass of the electron. No experiment has been performed where the gravitational effect on the electron could be observed. The force—the weight—is just too small. But because the electron has charge, electromagnetic forces can be exerted on it with much larger effects than the gravitational. The bubble chamber tracks in Fig. 21.2 clearly show the result of electron acceleration. The electron velocity was continually being changed by a force perpendicular to the path. Such a perpendicular or centripetal force does not change the speed or energy of the electron but does make it move in a circular path. In the bubble-chamber picture, the tracks are spirals because the electrons were constantly losing energy in their passage through the gas. The perpendicular force caused by the movement of the electron in a magnetic field is equal to Bev, where B is the strength of the magnetic field, e is the electron charge, and v is its velocity. A centripetal acceleration must thus be produced with magnitude v^2/r, where r is the radius of the circular path. Since $F=ma$, it follows that

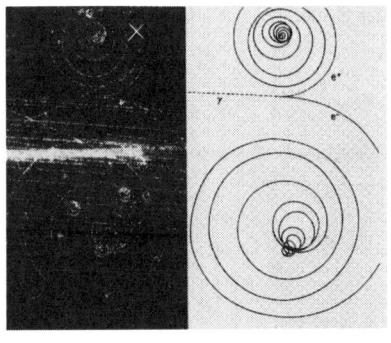

Figure 21.2. Photon production of an electron-positron pair.

$$eBv = m\frac{v^2}{r}.$$

Solving for m, we have

$$m_{\text{electron}} = \frac{Ber}{v}.$$

If the velocity of the electron is known, its mass can be determined. There are many ways to combine magnetic and electric forces acting on electrons or other particles so that the velocity can be measured and the mass determined. For the electron, the value turns out to be 9×10^{-31} kg. How could you find the velocity of an electron by measuring the electric and magnetic fields that influence it? Here is one way to produce a beam of electrons with known velocity. If the electron is accelerated through a known potential difference, its kinetic energy becomes $\frac{1}{2}mv^2 = eV$. Such a measurement would yield a value for v in terms of e/m: $v = \sqrt{2Ve/m}$. If the electron is then sent through a magnetic field, it travels in a circle with radius r: $m = Ber/v = Ber/\sqrt{2Ve/m}$. The magnitude of m can be found by substituting into this equation the measured values of B, r, V, and e.

There is yet another way of measuring the mass of the electron, depending on an effect that appears quite different from the inertial ones, although really not independent of inertia. Instead of forcing an existing electron to change its motion, create a new electron! This takes energy, of course, in an amount equal to mc^2, where c is the velocity of light and m is the mass of the created electron. The bubble-chamber picture in Fig. 21.2 shows such an event. A photon came from the left and in the vicinity of a particular atom, turned all its energy into the mass and kinetic energy of two electrons. In the magnetic field the two electrons circled in opposite senses; one was negative, our standard electron, and one was positive, the so-called positron. If only one electron were created, a unit of electric charge would have come from nowhere. This never happens; the total charge remains constant. There was none to begin with and the net amount ($+$plus$-$) afterward is also zero. It takes an x ray of at least 1 MeV to produce such a pair creation. Each electron uses half of this for its mass. If any is left over, the electrons can share it to provide forward motion. So in this other way of measuring the mass of an electron we find that the mass is 0.51 MeV. The arithmetic of course works out:

$$mc^2 = 9 \times 10^{-31} \text{ kg} \times (3 \times 10^8 \text{ m/s})^2 = 81 \times 10^{-15} \text{ J}.$$

Since 1.6×10^{-19} J = 1 eV, $mc^2 = 0.51 \times 10^6$ eV $\approx \frac{1}{2}$ MeV.

Spin

The electron has yet another attribute that is vital to its interactions. Besides having mass and electric charge, it acts as if it were spinning on its own axis. Such a spin can make itself known through three main effects.

1. *Magnetic moment.* (a) First, a spinning electric charge becomes a small magnet. In a uniform magnetic field, the electrons should align themselves like compass needles all pointing together. If for any reason some of them are forced out of alignment, they will be in a different energy state from the stable ones. When they swing back into line, their energy will be released in some form, since energy had to be provided to get them out of line in the first place. Furthermore, in a nonuniform magnetic field, the electron magnets should not only rotate but should move toward or away from the region of strong field. These effects can be detected with a bit of trickery.

    Figure 21.3 shows magnet pole pieces that provide a very nonuniform magnetic field. As you can see, a magnet with its north end up will move down in the diagram; the force down on the south end is greater than the force up on the north. If a beam of electrons is shot through such a magnetic field, we might expect the emerging beam to spread out in a vertical line. The deviation of each electron would depend on how its spin was oriented when it entered. We assume that the spin of each would not flip into line while going through the magnet, but instead would precess like a top, still maintaining its original angle to the vertical. But, unfortunately, the experiment cannot be done quite so simply. Why not? Aside from the vertical motion of the electrons due to the nonuniform magnetic field, what else would happen to electrons traveling in such a field? Not only would the electrons spread out slightly vertically but, because they are charged particles moving in a magnetic field, they would be forced sideways in a direction perpendicular to their velocity and to the magnetic field.

    To avoid this problem, the actual experiment is done by having the electrons ride along on an atom. Silver atoms were used by Otto Stern and Walther Gerlach in 1924. Silver has just one valence electron, existing in a state such that the only angular momentum or spin must be that associated with the electron itself. All the other electrons in the closed shells produce effects that cancel. The whole atom is electrically neutral so that it will not be influenced by a magnetic field unless the atom is also behaving like a small magnet, in which case it will be deviated up or down. Figure 21.4 shows the experimental method and results. Some atoms were indeed forced up and others down, but only into two groups! Apparently, the electron spins were either in line with the field or 180° out of line. The effect is *quantized.* If the whole system were rotated on its axis 90°, we would find that the deviations were then horizontal, but still divided into two definite groups with nothing in between. Thus the directional deviation is not a function of the way the atomic beam was produced. The silver atoms are boiled out of the oven with completely random orientation. We are forced by this phenomenon (and many others) to conclude that the electron spin can exist only parallel or antiparallel

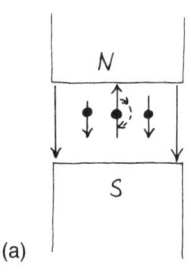

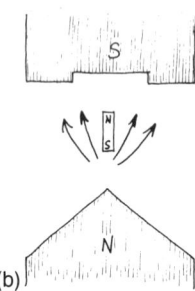

Figure 21.3. (a) Alignment of magnets in a magnetic field. (b) A magnetic dipole experiences a translational force in an inhomogeneous magnetic field. The force is stronger on the bottom than on the top of the dipole.

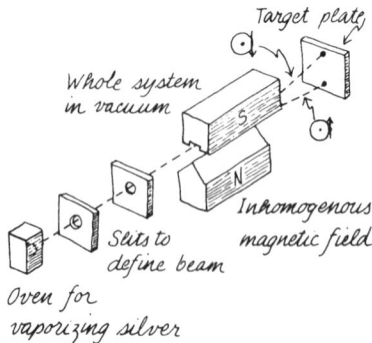

Figure 21.4. Stern–Gerlach experiment using neutral silver atoms to demonstrate discrete alignment of magnetic moments.

with the magnetic field with which we detect it. One is tempted to insist that in the original beam the spins must have had any orientation and somehow arranged themselves in this peculiar way just when they entered the magnetic field. Perhaps so, but of course that is metaphysics and meaningless. If an effect can never be detected, directly or indirectly, then it is rather useless to ponder it. In any experimental arrangement where electron spin is determined, some standard direction or axis is established. The electrons are always found to be lined up either in that particular direction or opposite to it, but never in between.

The convenient unit of angular momentum for particles is Planck's constant divided by 2π. In the literature of physics this is written as the symbol $\hbar = h/2\pi$. It has the value of 1×10^{-34} J s and the proper dimensions to be angular momentum [(mass) (length)^2]/(time).

Chapter 8 dealt with the subject of angular momentum. As we pointed out there, it is rather startling that there is a natural unit of angular momentum in this world. We know of very few other quantities that have basic units which exist naturally. There are none for mass, length, or time—as far as we know. Electrical charge has a natural unit, and now we find one for angular momentum. In terms of this unit, the electron has spin 1/2. As we will see, this is a very significant value. Particles cannot have just any value of spin, but only certain ones. Like electric charge, spin is quantized and can exist in multiples of $\frac{1}{2}\hbar$, but never with values in between, such as $0.65\hbar$. Furthermore, the measured values for a given system, corresponding to the different directions of the spin, must always be different from each other by one whole unit, $\hbar$. Therefore the valence electron of silver is found in only two orientations, with the direction-giving field $(+1/2)$ or against $(-1/2)$.

(b) The magnetic-moment effect of the electron spin can also be determined in *uniform* magnetic fields. For this it is necessary to use atoms or molecules with unpaired electrons; the goal is to deal with electrons that are relatively free and yet bound sufficiently so that a large number can be held in one place. The conduction electrons in a metal are plentiful enough, but they are by no means independent agents. A conduction electron could not orient itself in a magnetic field without interacting with all the others. Most of the electrons in material are paired off with electrons of opposite spin existing in the same energy level. In paramagnetic substances, however, there are unpaired electrons that do not have to interact with the neighboring atoms to form molecular bonds.

If one of these substances is placed in a magnetic field, the electron spins will point either with or against the field. It would seem at first that all of them would flip around so as to align themselves and thus be in the lowest possible energy state. But actually, energy is being fed into the electron-spin system by the random thermal motion of the molecules. It is as though a group of compass needles were trying to line up in a magnetic field but were constantly being jiggled into random directions.

The amount of energy that it takes to flip one electron in a particular magnetic field is a very definite amount:

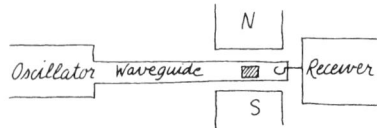

Figure 21.5. Schematic of microwave apparatus to detect flipping of electron magnetic moments.

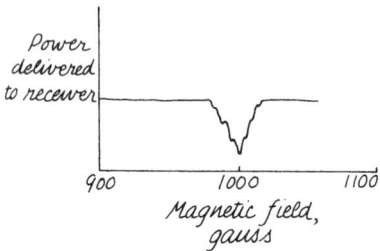

Figure 21.6. Absorption of radar-frequency energy due to electron magnetic resonance. The jagged subpeaks along the sides of the major dip correspond to absorption at various local magnetic fields in the structure.

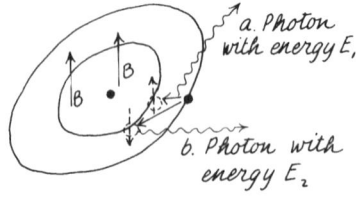

Figure 21.7. Electron spin splitting of spectral lines. $E_1 > E_2$ since in (b) part of the energy is still retained by the electron when it is oriented against the magnetic field.

$$2\mu_e B \approx (2 \times 10^{-23} B) \text{ J} \approx (10^{-4} B) \text{ eV} = 10^{-5} \text{ eV}$$

for $B = 0.1$ T or 1000 G,

where μ_e is the magnetic moment of the electron. (Magnetic moment is the measure of strength of a magnetic dipole, and is best defined and measured in terms of this equation. See p. 438.) It would take a chunk of electromagnetic energy of just this amount to flip the electron, and, of course, if the electron flips back, a photon of that energy is emitted. Photon energy is directly related to frequency. Thus, $E = h\nu$ where h is Planck's constant and ν is frequency in cycles per second. In this particular example, a magnetic field of 0.1 T would require photons or radio signals with a frequency of about 3000 MHz. What is the wavelength of a 3000-MHz signal?

$$\lambda = c/\nu = (3 \times 10^8 \text{ m/s})/(3 \times 10^9 \text{ Hz}) = 1 \times 10^{-1} \text{ m} = 10 \text{ cm}.$$

This is the wavelength used in microwave ovens.

Normally, such a signal is not heard from the sample. There are as many electrons flipping one way as the other to maintain the particular balance determined by the temperature. If, however, a radio coil driven by a transmitter is placed around the sample, as shown in Fig. 21.5, and the frequency is gradually raised, in one particular narrow frequency range the coil will appear to absorb extra energy from the transmitter. This energy has produced the flipping of many electrons out of the magnetic field direction. How long it takes for them to jiggle back depends on the temperature and how easily the substrate transmits its thermal agitation to the electrons.

Figure 21.6 shows a typical plot obtained in such an experiment. In this case, the frequency remained constant and the magnetic field was changed.

(c) The different energy states of the electron as it assumes the two orientations (either parallel or antiparallel to the magnetic field) give rise to spectroscopic effects that were observed over a hundred years ago. However, we had to wait until 1924 for the interpretation, which was made by Samuel Goudsmit and George Uhlenbeck. Strong magnetic fields of the order of 10 T can exist naturally inside atoms. The electronic circulation (in the older picture, the orbital motion of electrons) produces fields that interact with the magnetic effects of the intrinsic spin of each electron. Since light is produced by the jump of an outside electron into a vacant position that has been momentarily created in the orbits, the energy of the light pulse will be equal to the difference in energy levels of the jumping electron.

Each energy level really has two possible values, corresponding to the two orientations the electron can assume with respect to the local fields. As can be seen in Fig. 21.7, this will give fine structure to most transitions. Instead of producing a photon of light with one particular energy, and so a particular wavelength, each transition produces a double or triple wavelength in the spectrum.

2. *Angular momentum conservation—another role of electron spin.* Electron spin must also be taken into account in spectroscopy and in particle decays where electrons are emitted, simply because spin means angular momentum, and angular momentum is one of the few things in this world that must be

rigorously conserved. If an atomic system emits a photon of light, the photon carries off one whole unit of angular momentum. (The fact that photons possess angular momentum has been experimentally demonstrated). The remaining system must then be one unit different from what it was to begin with. If the photon carries off clockwise angular momentum, the remaining system must recoil counterclockwise. Within the system various parts are producing angular momentum—some associated with certain electron configurations (as though the electrons were actually orbiting around the nucleus). Also, of course, we must take into account the spin, or intrinsic angular momentum, of each of the electrons. The nucleus has a spin, too. All of these must be added together, but the addition is vectorial; that is, many cancel. The spectral lines can be understood only when the description of total angular momentum includes the spins. See Fig. 21.8.

In particle decays, total angular momentum must also be conserved. When a radioactive nucleus decays to emit an electron, a β ray, there is an apparent breakdown of this rule. The parent nucleus recoils, but it can do so with no change in its own spin. Yet the electron goes tearing off, spinning as it goes. It is as surprising as if a person on a friction-free table were able to get a bicycle wheel spinning without recoiling in the opposite sense himself. The very same problem arises when the muon (one of the main parts of cosmic rays at sea level) decays. A single electron comes spinning out.

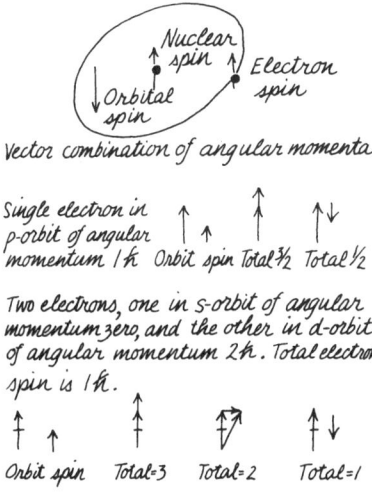

Figure 21.8. Vector combination of angular momenta.

There is a way out of this apparent breakdown of angular momentum conservation. In all such cases, at least one other particle, the neutrino, ν, is also emitted. The neutrino has no electric charge and so is not easily detected, but it does have spin. The person on the friction-free table would have no trouble getting two bicycle wheels spinning, as long as one went clockwise and the other counterclockwise. The neutrino and its problems will be dealt with later when we define all of the subatomic particles.

3. *The exclusion law—the third role of spin.* The electron has intrinsic angular momentum, and so it acts like a magnet, but its spin of $\frac{1}{2}\hbar$ has an even more profound effect. All particles with half odd-integral spin (1/2, 3/2, 5/2,...), and this includes protons and neutrons as well as electrons, must obey a special exclusion law. The great physicist Wolfgang Pauli first framed this principle in 1925. Two other great names in physics are associated with the particular rules resulting from Pauli's principle, Enrico Fermi and Paul Dirac. In scientific literature, particles with spin $\frac{1}{2}\hbar$, $\frac{3}{2}\hbar$, and so forth, are referred to as *fermions*. No two fermions can possess identical properties while inhabiting the same region. It is this saturation restriction that is responsible for the particular arrangement of electrons in the atoms and so determines the nature of the periodic table. The nucleons—protons and neutrons—are fermions, and so are the heavier *hyperons* that decay into the proton and mesons. The meson family has integral spin: 0 or 1. The agents of the interactions, such as the photon (electromagnetic), the graviton (gravity), the W (weak), and gluons (strong nuclear) also have integral spin. They follow Bose–Einstein statistics and are called *bosons*. In an ensemble of these, the statistical distribution of energy allows any number of particles to have the same energy. Therefore there is no exclusion principle, and also there is no conservation principle requiring that they be created or absorbed in pairs of particle and antiparticle.

Any number of photons can be churned out of a charged particle or system and similarly with the other bosons as long as the other conservation principles such as electric charge, energy, and angular momentum are satisfied.

In Fig. 21.9 we show electron-spin arrangements for the first few atoms in the periodic table. Hydrogen has one electron existing in a region close to the proton nucleus. Helium has two electrons in the same region with all the same properties except that the spins of the two electrons are in opposite directions. Lithium has three electrons, but only two can exist in the lowest allowed energy region. The third electron in this region could have its spin aligned in only one of two ways and, in either case, this would duplicate all of the

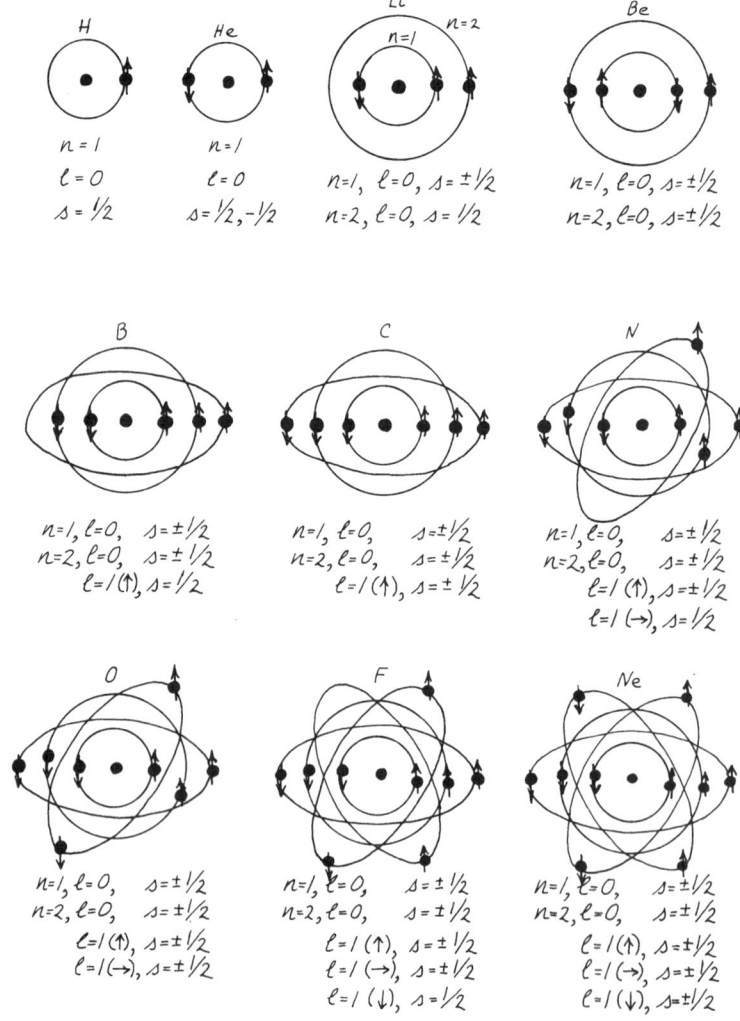

Figure 21.9. Electron spin arrangements for first ten elements in periodic table.

properties of an electron already there. If there were two kinds of electrons, red and blue, or if the electron had some other kind of variable property—for example, lopsidedness, then more than two could fit into the same energy region. But only two do so. Thus, it must be that the electron has a limited number of attributes, and with spin we have exhausted the list. The third electron of lithium must exist in a higher energy level, farther from the nucleus. The fourth electron of beryllium can, with opposing spin, associate with number three. The fifth electron of boron is almost in the same second energy level, but not quite because it can have one unit of angular momentum (besides its intrinsic spin). The existence of this extra unit of angular momentum slightly changes the energy of the fifth electron and thus makes it distinguishable from electron number four. There can be three directions of angular momentum of magnitude $1\hbar$. In any experimental situation that sets up a preferred direction (e.g., a magnetic field), the axis of a system with one unit of angular momentum can either point with the field, against it, or perpendicular to it.

Why can a system with an angular momentum of $1\hbar$ assume three different orientations with respect to magnetic field? For a spin of $\frac{1}{2}\hbar$, there are only two possibilities. The quantization law requires that the angular momentum of a system can change by only whole units of $\hbar$. If the intrinsic angular momentum is $1\hbar$, then the direction along an axis can change from in-line to perpendicular (a change of one unit from $+1\hbar$ to $0\hbar$ in the original direction), or from perpendicular to antiparallel with the original axis (a change of one unit from $0\hbar$ to $-1\hbar$).

If the intrinsic spin is $\frac{1}{2}\hbar$, then the angular momentum in a particular direction can change by $1\hbar$ only if the particle flips from parallel to antiparallel: $+\frac{1}{2}\hbar$ to $-\frac{1}{2}\hbar$.

The second main level of an atom's electron system can thus contain four possible configurations: orbital angular momentum 0 and three orientations of orbital angular momentum 1. Each of these can hold two electrons, with spin "up" and "down." Altogether, eight electrons can fit into the second level. Even without an external field, the internal magnetic fields in an atom cause each of the eight electrons to have slightly different energies. These differences are represented in spectra by photons of slightly different wavelengths, producing "fine structure" or "splitting" of the spectral lines.

Size of the Electron—The Wave–Particle Problem

So far we have assigned only three attributes to the familiar electron: mass, charge, and spin. If we were describing a marble, we would already have inquired about its "looks": color, size, and so on. (Actually, we have been describing the "looks" of the electron; i.e., we have described the experimental responses to probes that can detect electrons.) Of course, the electron has no color, simply because the shortest wavelength of visible light (4×10^{-7} m for blue) is certainly larger than the size of an electron. One cannot determine the structure of an object with a probe larger than the object. There is a special problem concerned with the definition of particle size. In some cases, the problem is as simple as the one we would face in measuring the "size" of a small bar magnet. With a meter stick, and using light and our eyes, we would get one result. Using a magnetic probe, we

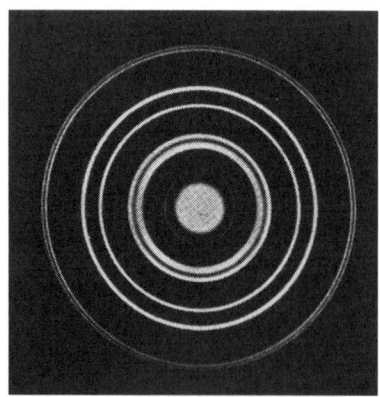

Figure 21.10. Interference pattern produced by an electron beam going through thin foil. *(Courtesy of Dr. Lester Germer.)*

would apparently have a larger object with rather fuzzy edges. If a particle is responsive to several forces, it is possible that its boundaries may be different for each one.

But there is another size problem more subtle than this. Under certain conditions, the behavior of electrons must be described in terms of the behavior of waves. The wavelength is related to the momentum by the following formula: $\lambda = h/p$. Here λ is the wavelength in meters, h is the ubiquitous Planck's constant, and p is the standard symbol used in the literature to represent momentum. At low velocities, $p = mv$ (mass times velocity), the usual definition. At velocities close to the speed of light, the momentum is a more complicated function, as we saw in Chap. 6:

$$p = \frac{mv}{\sqrt{1-(v^2/c^2)}}.$$

Thus experimentally momentum assumes a more important role than either the mass or velocity separately.

Observe now how the size of the electron changes. Surely it is as large as one wavelength and presumably its influence would extend over several wavelengths. For an electron accelerated through 10 V, perhaps in a radio tube, the velocity $\approx 1.6 \times 10^6$ m/s. Then

$$\lambda = \frac{6.6 \times 10^{-34} \text{ J s}}{(9 \times 10^{-31} \text{ kg})(1.6 \times 10^6 \text{ m/s})} \approx 5 \times 10^{-10} \text{ m}.$$

(The outer electrons in atoms have kinetic energy in the range from 1 to 10 eV. Their wavelengths, therefore, are about the size that is indicated by the equation.) Since this wavelength is larger than the size of an atom, the point is emphasized that an electron cannot really be considered to be orbiting about a nucleus. In this sense, the electron is as large as the atom, and indeed there is no other sense. By any experiment that can be performed, the atomic electrons are found to exist over regions comparable in size to the whole atom. It is meaningless to ask how large the electrons are or where they are going during the time when we do not measure them.

There is, of course, good experimental evidence for the wave *behavior* of electrons. For our present arguments, we deliberately avoid the use of the words "wave nature." The similar wave behavior of light is most clearly seen in interference or diffraction patterns. The photograph in Fig. 21.10 shows a diffraction pattern produced by a beam of electrons passing through holes produced by the spacing of atoms in a crystal. The pattern is very similar to that produced by light passing through a small hole. The spacing of the rings depends in both cases on the ratio of wavelengths to hole diameter. With red light, the rings are spaced farther apart than with blue light. With slow electrons, the spacing is larger than with fast electrons. The analysis of the optical phenomenon is in terms of waves originating in different regions of the hole, traveling different distances to the screen, and interfering with each other there, sometimes canceling and sometimes reinforcing. The ring geometry arises solely because with a round hole there must be circular symmetry. Since exactly the same phenomenon is observed with electrons, we use the same language and the mathematics of wave motion to describe the situation.

We need not concern ourselves as to whether an electron *is* a wave. It *is* an electron. In situations where the electron interacts with objects larger than the wavelength given by $\lambda = h/p$, it acts like a hard spinning sphere. When the interaction is with objects comparable in size to λ, or smaller, the effect has to be described in terms of the interactions of waves. There is a similar situation with sound. The behavior of high-frequency sound in a room is very much like that of "rays" bouncing about. The low-frequency sounds, however, curl around corners like waves.

In mathematical descriptions of processes, this dual state of affairs is very common. We may have a particular formula or equation that is complicated and difficult to compute or solve, but that is completely accurate at all times. This is exactly the case for our description of photons and any electromagnetic interactions. To obtain a solution for a particular case, the usual technique is to make certain approximations. An exact solution to most physical problems is never possible, and part of the skill of the mathematician or scientist lies in choosing the right approximation, and knowing to what accuracy the solution should be carried.

For example, consider the equation describing a marble falling in air:

$$ma = mg - Kv^2$$

The product of mass and acceleration of the marble is equal to its weight minus the air friction force. (We assume that the marble is large enough and falling fast enough so that the air resistance is proportional to the square of the velocity.) A graph of the solution to this equation was shown in Chap. 4. The exact solution to the equation is complicated and requires the use of calculus. For two extreme cases, however, approximations can be made that yield simple solutions. First, at the beginning of the fall of the marble, when the velocity is small, we get the simple equation of free-fall in a vacuum: $a = g$, and $x = \frac{1}{2}at^2$. In the second extreme case, the friction is very large: $mg = Kv^2$, and $\Delta x = v \Delta t$.

Needless to say, we have not produced a paradox about the nature of marbles just because the same equation yields two different solutions. For the same reason, there is no duality paradox about the nature of subatomic particles. Under certain experimental conditions, the best approximation to the equations describing particles makes the equations similar to wave equations. In other experimental situations, the proper approximation produces an equation similar to the kind describing the action of solid balls. We do not claim that the subatomic particle *is* a wave or *is* a hard ball; it merely behaves in this or that fashion under particular conditions. Nor should it be imagined that a subatomic particle (or a photon) moves with a wiggly, wavelike motion. *It is the solution to the equation that has the wave character. This solution is a probability function that gives the chances of finding particular values of position or velocity of a given particle at a particular time.*

The equation for the position of a falling marble gives the exact position of the marble as a function of time: $x(t)$. Doesn't the quantum theory yield the same information about subatomic particles? Not exactly. The quantum-mechanics equations depend on the geometry, the nature of the particle, and the location of other particles. The solutions to these equations (which mathematically can have wavelike characteristics in space and time) are then combined or operated on to yield the *probability* of finding the particle in a particular position, or with a particular momentum. The probabilistic nature of these predictions corresponds with the actual experimental situation. Repeated measurements of the same atomic system usually yields a distribution of values, centered around a most probable value.

■ HEISENBERG'S UNCERTAINTY PRINCIPLE

There is yet another important fact of life concerned with particle measurements. It is not always possible to measure—and so know—certain pairs of related properties of a particle, both with infinite precision. This is not just a matter of failing to have a sensitive enough meter of some kind. There is in principle, as well as in practice, a limit to the precision with which we can measure the position of a particle at the same time that we measure its momentum. Of course, every time any measurement is made of the position of some object a certain amount of error is involved. For example, if you judge by eye the position of a scribe mark on the floor of a room, you might guess the distance between it and a wall to within half a foot: $x=3$ ft, $\Delta x=\pm \frac{1}{2}$ ft, where Δx describes not any mistake made but simply your assigned limits of accuracy. With a meter stick, the precision could be much greater: $x=1$ m, $\Delta x=\pm 2$ mm. If there were some reason to attain greater precision it could be done. The distance between two scribe marks on the standard meter stick in Paris was measured by Michelson in terms of the number of wavelengths of a particular spectral line of visible light. His precision with this technique was better than one part in a million: $x=2\times 10^6\lambda$, $\Delta x=\pm 1\lambda$.

Even though practical considerations might rule out the extension of such efforts, it would seem that there should be no limit in principle to attaining greater precision. The measuring probe, of course, must be smaller than the Δx required. To do better than Michelson, it would be necessary to use light (or particles!) with a smaller wavelength than visible light. Blue light with $\lambda=4\times 10^{-7}$ m would be a little better than red with $\lambda=7\times 10^{-7}$ m. In some microscopes, this color effect is taken advantage of by using blue light to increase the resolution. Ultraviolet microscopes have been constructed, but the most widely used type is the electron microscope. The wavelength of the electrons in such a machine is about 2×10^{-9} m, 100 times better than can be obtained with visible light. Actually, resolution of even the best electron microscope is not so good as this, because of practical considerations of construction. To get even finer resolution, we need particles or light of yet smaller wavelength. But to get particles with smaller wavelength, we must increase their momentum: $\lambda=h/p$. The same problem faces us when we use light. The amount of energy in each photon of light is $E=h\nu=hc/\lambda$, where E is the energy in joules, h is Planck's constant, ν is the frequency of light in cycles per second, and λ is the wavelength in meters. As we go from radio to radar to infrared to visible to ultraviolet to x rays to γ rays, we deal with photons with greater and greater energy and with more of the behavior we normally associate with particles. Light does, of course, carry momentum. (Comet tails are blown away from the sun by photons and other particles emitted from the sun.) The amount of momentum is

$$p=\frac{E}{c}=\frac{h\nu}{c}=\frac{h}{\lambda}.$$

But this is just the expression we find for any other kind of particle!

If the measuring probe has higher momentum as we decrease the wavelength, we can still measure the position of a tiny object as accurately as we choose, but in doing so we knock it away with the probe, as shown in Fig. 21.11. In measuring

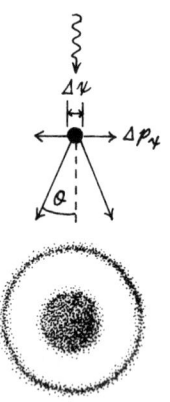

Figure 21.11. Heisenberg uncertainty principle: The diffraction pattern produced by a light beam with wavelength λ has an opening angle of $\theta \approx \lambda/\Delta x$. The indeterminate momentum in the x direction, delivered by a probe glancing off at an angle θ, is $\Delta p_x = p_{probe} \sin \theta \approx p\theta = (h/\lambda)(\lambda/\Delta x) = (h/\Delta x)$. Then, $\Delta p_x \Delta x \approx h$.

position, we add an unknown amount of momentum to the object. The *unknown* momentum is *perpendicular* to the original direction of the wave probe. Instead of casting a sharp shadow of the object, which would determine the position of the object exactly, the wave probe necessarily falls somewhere in a diffraction pattern. Any particular photon (or other probe particle) can fall anywhere within the diffraction pattern in a completely indeterminate way. The opening angle for the maximum of the diffraction pattern is $\theta \approx \lambda/\Delta x$. The wavelength of the probe is λ, and the indeterminate size of the object (either the vagueness of its edge or, more usually, the whole diameter) is Δx. Of course, the probe particle can also be deflected at an angle two or three times this large and land in one of the outer fringes of the diffraction pattern. Most, however, fall in the central maximum. A probe with small wavelength produces a small diffraction pattern. But, as a result of the small wavelength, the probe carries higher momentum. This compensates for the smaller angle of deflection, producing a sideways impulse independent of the wavelength.

The dilemma is summarized by Heisenberg's uncertainty relationship. A more careful analysis of the uncertainties leads to a slightly lower limit for the product than the one derived in the diagram.

$$\Delta x \Delta p_x \geq \frac{h}{2\pi} = \hbar \approx 1 \times 10^{-34} \text{ J s}.$$

In words, the product of the uncertainty in position and the uncertainty in momentum must always be equal to or greater than Planck's constant divided by 2π. This rule actually follows from our formula for the momentum of a particle, $p = h/\lambda$. In a glancing collision of a probe particle with an object, an appreciable fraction of its momentum p may be given to the object in a direction perpendicular to the original probe direction. Then, that fraction, $\Delta p \approx p$, is the uncertainty in momentum of the object after the measurement. The position uncertainty Δx cannot be much less than the measuring wavelength λ. Since for the particle $p = h/\lambda$, for the object after the collision, $\Delta p_x \approx h/\Delta x$. If angular momentum were not quantized, that is, if Planck's constant were 0, both position and momentum could be measured simultaneously with infinite precision. This world would not be this world at all in that case.

Heisenberg's principle also applies to other pairs of quantities that in classical science can, in principle, be known together with infinite precision. Instead of linear position x, choose angular position θ. Then $\Delta \theta \Delta L \geq \hbar$, where $\Delta \theta$ is the uncertainty in angle of a rotating system and ΔL is the uncertainty in *angular momentum*.

Another pair of variables is of importance in particle physics: $\Delta E \Delta t \geq \hbar$, where ΔE is the uncertainty in the energy of a system *at the time we measure it* and Δt is the uncertainty in the time duration of the measurement. One is tempted to say that ΔE is the uncertainty with which we *know* the energy E, implying that actually the system has a definite energy. This is a deluding thought. Once again, if in principle something cannot be measured, then the question as to whether it exists is meaningless.

How can we measure the energy of an excited state of a system? The most general method is to let the system return to normal and see how much energy comes out. A collision, perhaps, throws an atom into an excited state, with one of the electrons in a nonstable position. The uncertainty of the energy of the system presents no problem, however, until the time when we receive the information. As the atom returns to normal, as the electron hops down into place, a photon is emitted. It carries energy $E = h\nu$, and this implies that it has a definite wavelength, $\lambda = c/\nu$. However, the photon is emitted in a time of about 10^{-8} s for an ordinary atomic transition. Before then there is no photon, and after that the electromagnetic vibrations stop coming from the atom. Figure 21.12 shows how a meter might record this situation. But a train of sine waves has a definite frequency (or is a pure tone), only if it has been emitted forever and will continue indefinitely. A short burst of oscillations must have overtones associated with it. These are higher frequencies that, when mixed with the main note, produce a beginning and end to the pulse. The shorter the duration, the more important the overtones. Instead of a single sharp wavelength, a narrow band of wavelengths is produced. Even sharp spectral lines have measurable widths.

In the case of visible light where $\Delta t \approx 10^{-8}$ s, we have

$$\Delta E \Delta t \geq \hbar \approx 10^{-34} \text{ J s,}$$

$$\Delta E \approx 10^{-26} \text{J} \approx 10^{-7} \text{ eV.}$$

This seems like a small uncertainty, indeed. Since the energy of a visible photon is about 2 eV, it appears that the energy is definite to 1 part in 10 million, but spectroscopy can attain precision such that this is observed as the natural linewidth. The uncertainty in frequency is $\Delta \nu \approx 10^8$ Hz, and the irreducible width of the spectral line is $\Delta \lambda \approx 10^{-13}$ m, or about 0.3 millionth of the length of the whole spectrum from red to blue.

We have occasion to deal with particles or excited states with lifetimes as small as 10^{-22} s. In this case, $\Delta E \approx 10^{-12}$ J ≈ 10 MeV. Uncertainties of this size can play dominant roles in certain particle reactions.

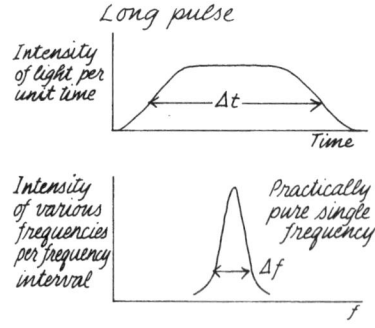

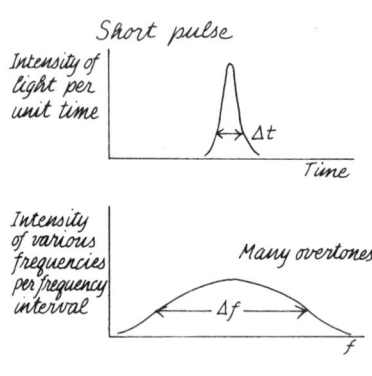

Figure 21.12. Relationship between duration of light pulse and spread of frequencies.

■ ATOMIC STRUCTURE

Our preliminary model for the atom is basically the one worked out by Niels Bohr (1885–1962) and Ernest Rutherford (1871–1937) during the second decade of this century. A central nucleus containing most of the mass of the atom has a positive electric charge. The amount of charge ranges from 1 unit for hydrogen to 92 for uranium. Each atom contains a corresponding number of negatively charged electrons, thus making the atom electrically neutral. Each chemical element is defined by the number of electrons in its atom, the *atomic number*, or Z. Note that for hydrogen the atomic number is 1, for carbon it is 6, for oxygen it is 8, for copper it is 29, for gold it is 79, and for lead it is 82. The numbers are integers, since electric charge is quantized.

Within the nucleus, there are neutrons and protons. Neutrons have about the same mass as protons, about 1800 times greater than the electron mass. Each proton has one unit of positive charge; the neutrons are neutral. Protons and neutrons are not point elementary particles. Among other internal features, they contain circulating electric currents, which produce magnetic moments. The atomic number of an element is also the number of protons in each nucleus. To a first approximation, each nucleus contains about as many neutrons as protons. In the next section, we will see how the actual ratio of neutrons to protons affects nuclear stability.

The radius of a nucleus is quite accurately given by the formula $r = r_0 A^{1/3}$, where $r_0 = 1.3 \times 10^{-15}$ m, and A is the total number of protons and neutrons in the nucleus. This is just the relationship expected if the nucleus consisted of tiny marbles packed together, each with radius, r_0.

Why should we expect this relationship if protons and neutrons pack like marbles? If a marble has a radius r_0, it has volume $V = \frac{4}{3}\pi r_0^3$. The volume of A marbles packed together would be $V_{tot} = KA\frac{4}{3}\pi r_0^3$, where K is the packing friction for fitted spheres, a fraction close to 1. The radius of the whole nucleus would therefore be $r = K^{1/3} r_0 A^{1/3}$.

For the uranium nucleus, this formula gives $r = (1.3 \times 10^{-15} \text{ m})(238)^{1/3} = 8 \times 10^{-15}$ m. The radius of the atom itself is about 1×10^{-10} m, about 10 000 times greater. Although the electrons in an atom contain only about 1/4000 the atomic mass, they evidently occupy most of the volume.

The Bohr–Rutherford model of the atom pictured the electrons circulating in planetary orbits about the nucleus. Electrostatic attraction provided the centripetal force. For the simplest of atoms, hydrogen, a single electron orbits a single proton.

$$\frac{mv^2}{r} = k\frac{q_1 q_2}{r^2}.$$

The undetermined variables are v and r. So far it would appear that an electron could have an orbit at any radius, and thus have the corresponding velocity. Bohr proposed, however, that the angular momentum of the electron must be quantized. (However, for the actual historical situation see the article by Hua-Xiang Liu.) For the electron in a circular orbit:

$$mvr = n\hbar, \quad \text{where } n \text{ is an integer}.$$

Let us combine the angular-momentum requirement with the centripetal-force formula.

Centripetal-force requirement:

$$\frac{mv^2}{r} = k\frac{q_1 q_2}{r^2} \rightarrow m^2 v^2 r^2 = mk q_1 q_2 r.$$

Angular-momentum quantization:

$$mvr \rightarrow m^2 v^2 r^2 = n^2 \hbar^2;$$

Therefore

$$r = \frac{n^2 \hbar^2}{mk q_1 q_2}.$$

Now we have a formula for only one variable, r. It appears that only certain radii are allowed, and that their size increases as squares of the integers: 1, 4, 9, 16, and so forth.

What is the size of the hydrogen atom in its unexcited state when $n = 1$? Since the model is crude, use only order-of-magnitude numbers:

$q_1 = q_2 = 10^{-19}$ C, $\hbar = h/2\pi = 10^{-34}$ J s, $k = 10^{10}$ N m^2/C^2; $m = 10^{-30}$ kg,

$$r = \frac{1 \times (10^{-34})^2}{(10^{-30})(10^{10})(10^{-19})^2} = 10^{-10} \text{ m}.$$

This is the order-of-magnitude size of any atom!

Why should atoms all be about the same size? For an element with a large atomic number, high up in the periodic table, there are more positive protons in the nucleus and more negative electrons in shells or orbits around the nucleus. The radius of the first electron shell, where $n=1$, is inversely proportional to the positive charge of the nucleus. As the atomic number Z increases, the first shell is drawn in toward the nucleus. There are other factors involved, but in general as the number of protons in the nucleus increases, the radius of each electron shell decreases. The electrons in outer shells are partially shielded from the central positive charge by the intervening negative charge on the inner shells. To first approximation, the single valence electron in the outer shell of an alkali-metal atom (lithium, sodium, potassium, rubidium, cesium, francium) sees only a single positive central charge, since the $Z-1$ electrons in the closed inner shells shield the Z protons in the nucleus. Therefore, to first approximation, the radius of the valence shell is about the same as that of the hydrogen atom.

You can see the validity of the approximation in the graph of atomic radius versus atomic number shown in Fig. 21.13. Note that the atoms with closed shells are smallest, and that the alkali-metal atoms are largest with their single valence electrons all by themselves in outer orbits. With the alkali-metal atoms, the larger the atomic number, the more complete is the shielding of the central charge and the less tightly bound is the single electron. Consequently the alkali-metal atom with high Z is more chemically active and is also larger. Even with these variations, it is still true that all atoms are the same size within a factor of 2, if r_{atom} is chosen to be 1.5×10^{-10} m.

Note, however, that Fig. 21.13 shows the radii of *free* atoms. In molecules, or in crystals, the outer electron shells are completely rearranged and most atoms are even closer to being the same size. In a metallic crystal, for instance, the valence

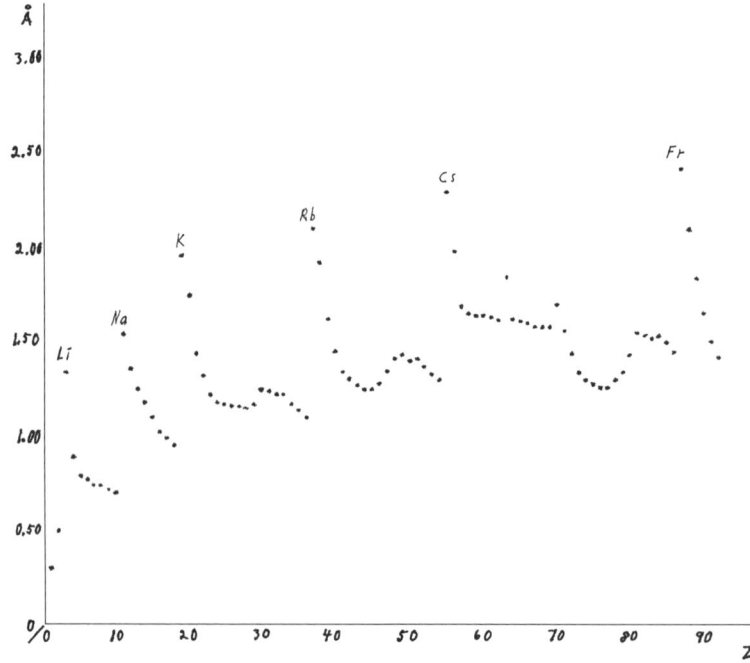

Figure 21.13. *(Single-bond covalent) Radii vs. Z.*

electron from each atom is separated from its original atom and becomes part of the electron sea of the crystal. Each remaining ion has a size almost independent of atomic number. You can test this for yourself (or it is a good exercise for students), by calculating the volume of a mole of any substance:

$$V_{mole} = \frac{(\text{mass of mole})}{(\text{density})} = \frac{A(\text{atomic "weight" in grams})}{\rho(\text{in g/cm}^3)}.$$

For carbon (graphite):

$$V_{mole} = \frac{12}{2.3} = 5.2 \text{ cm}^3, \quad V_{atom} = \frac{5.2 \text{ cm}^3/\text{mol}}{6.0 \times 10^{23} \text{ atoms/mol}} = 8.7 \times 10^{-24} \text{ cm}^3/\text{atom}.$$

Taking the cube root of this volume yields $d_{\text{carbon atom}} = 2.1 \times 10^{-8}$ cm. The same calculation for uranium, with $A = 238$ and $\rho = 19.1$, yields $d_{\text{uranium atom}} = 2.7 \times 10^{-8}$ cm.

The theory yields the same approximate size for the hydrogen atom that we have been using throughout the book. More important for the usefulness of the theory is its prediction concerning the energy of the system. Since the electron and proton are bound together, their electrostatic potential energy must be negative:

$$E_{pot} = -k\frac{q_1 q_2}{r}.$$

The electron has positive kinetic energy equal to $E_{kin} = \frac{1}{2}mv^2$. Since the centripetal-force requirement is

$$\frac{mv^2}{r} = \frac{kq_1 q_2}{r^2},$$

the kinetic energy must be

$$\tfrac{1}{2}mv^2 = \tfrac{1}{2}k\frac{q_1 q_2}{r}.$$

The total energy of the system is

$$E_{tot} = E_{kin} + E_{pot} = \tfrac{1}{2}k\frac{q_1 q_2}{r} - k\frac{q_1 q_2}{r} = -\tfrac{1}{2}k\frac{q_1 q_2}{r}.$$

Notice that the total energy is necessarily negative, because the electron is bound. Since only certain radii are allowed, only certain energies can exist:

$$E_{tot} = -\tfrac{1}{2}k\frac{q_1 q_2}{r} = -\tfrac{1}{2}k^2\frac{mq_1^2 q_2^2}{n^2 \hbar^2}.$$

Use order-of-magnitude numbers to calculate the binding energy of the lowest hydrogen orbit, where $n = 1$:

$$E_{tot} = -\tfrac{1}{2}(10^{10})^2 \frac{(10^{-30})(10^{-19})^4}{(1)(10^{-34})^2} \approx 10^{-18} \text{J} \approx -10 \text{ eV}.$$

The actual energies for the various radii are shown in the energy level diagram in Fig. 21.14.

Consider the experimental consequences of such a theory. It would take at least 13.6 eV to "ionize" a hydrogen atom; that is, to rip the electron away from the proton. Indeed, that is the measured ionization energy. If a hydrogen atom is bombarded by other atoms in a hot gas, or by photons, or by subatomic particles

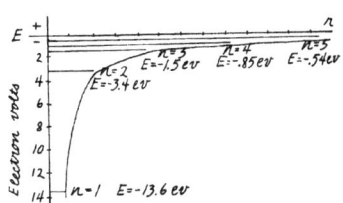

Quantized energy levels of Bohr orbits for hydrogen.

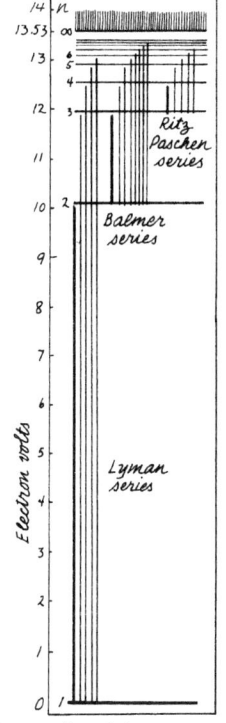

Spectra produced by the quantized energy levels of hydrogen.

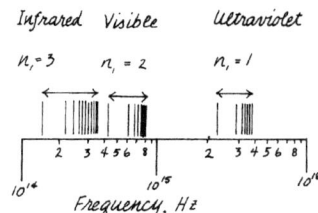

Hydrogen spectra in the infrared, visible, and ultraviolet.

Figure 21.14. *Quantized energy levels of Bohr orbits for hydrogen. Spectra produced by the quantized energy levels of hydrogen. Hydrogen spectra in the infrared, visible, and ultraviolet.*

passing by, it can accept amounts of energy smaller than 13.6 eV only if the energy is exactly the right amount to raise it to one of the quantized levels. The hydrogen atom cannot exist at energy levels in between. Furthermore, once the atom is excited to a higher energy level, it can lose only the exact energy required to change to one of the lower quantized levels.

In a very hot gas, such as in a fluorescent or spectrum tube, collisions provide the energy needed for an atom to go from its ground state to one of the excited states. Within 10^{-8} s, for most atoms and most transitions, the atom emits a photon and sinks into a lower energy level. Since the atomic energy changes can occur only between quantized levels, the photons emitted must have only certain, discrete energies. Each energy corresponds to a particular frequency and wavelength.

For hydrogen, the allowed energies of photons are shown in the Fig. 21.14. Note that there are families or series of frequencies, corresponding to the lower energy level of the transition. The series are known by the names of nineteenth-century physicists. In the diagram below, we see a frequency spectrum of three of these hydrogen series. Only the Balmer series is in the visible range. The experimental spectra agree very closely with the predictions of this crude model.

The series of frequencies produced by these energy transitions in the free atoms of gases, as opposed to bound atoms in molecules or solids, are called *line spectra*. When light from a hot gas is analyzed by a prism or a diffraction grating in a spectroscope, the light enters through a narrow slit. The image of that slit is a line. Its position depends on the frequency or wavelength of the light. Thus, the spectrum consists of a number of individual "lines."

The energy levels of atoms can also be observed in the *absorption* of light. A spectrum produced by sunlight, for example, appears at first to be a continuous spread of colors. On close examination, you can see the "Fraunhofer" or absorption lines. Light coming from the "surface" of the sun contains essentially all frequencies. As this radiation passes through the cooler gases in the solar atmosphere, particular frequencies are absorbed corresponding to the quantized energy levels of the cooler gases. The gases in the solar atmosphere have been identified by analysis of these spectra of missing frequencies. Helium was discovered this way in 1868. Its prominent absorption lines did not correspond to those of any known element and so it was given its name derived from *Helios*, the sun god.

So far we have derived quantitatively the energy levels only for hydrogen. What happens when there are many electrons in each atom? Even when there are two, as in helium, there are complications. To calculate the energy levels, we would have to take into account the repulsion between the two electrons as well as the attraction of both to the nucleus. Furthermore, Pauli's exclusion principle dominates the calculation. The two electrons cannot exist in the same level unless their spins are opposed. Also, transitions between energy levels can take place only if angular momentum is conserved. The emitted photon itself has one unit of angular momentum. The rearranged atomic system, including the electron spins, must account for the rest of the original angular momentum. Only certain transitions can accommodate these requirements. With helium, for instance, the complete spectrum consists of two separate arrays of lines, corresponding to two different sets of final levels. In the lower level, the two electrons are essentially in the same energy state, but have opposite spins. In the other series, one electron is in a higher state, but the electron spins are parallel. See Fig. 21.15.

Bohr's simple model agreed astonishingly well with the experimental data for

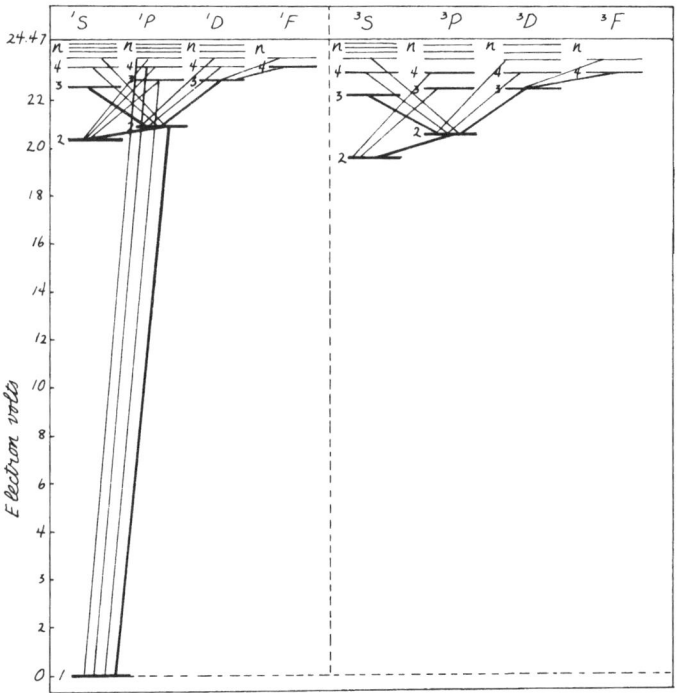

Figure 21.15. The two systems of spectra produced by the energy levels of helium and the prohibition of Pauli's exclusion principle.

hydrogen, gave qualitative agreement for certain hydrogenlike atoms, such as *ionized* helium or the alkali-metal atoms, but failed completely with more complicated systems, including helium. Consider all the facts about electrons that the simple model ignores. If electrons really traveled in tiny orbits, the calculated speeds would be large enough to require relativistic corrections. By 1920, a number of people had calculated the relativistic results, some of which agreed better with experiments. A more serious objection to the Bohr model was its assumption that electrons actually travel in orbits. As we have already seen, there is a relationship between the uncertainty of the momentum of a particle and the uncertainty of its position. To assume a certain orbit for a particle implies that both position and momentum can be measured precisely enough to determine the orbit.

One form of Heisenberg's uncertainty principle relates uncertainty of *angular* momentum and uncertainty of *angular* position: $\Delta L \Delta \theta \geq \hbar$. According to the Bohr model, the angular momentum of an electron in the first orbit ($n=1$) is *exactly* $\hbar$. (In the more complete quantum theory, the angular momentum in this first level is 0.) Since these orbital angular momenta are integral (including 0) multiples of the basic unit, the uncertainty in angular momentum, ΔL, is 0. Therefore, $\Delta \theta = \infty$. There is complete uncertainty about the angular position, θ. This situation is incompatible with the model of an electron actually traveling from position to position in an orbit.

As we saw in the last section, any particle has a wavelength associated with it, and wavelength is related to particle momentum:

$$\lambda = \frac{h}{p}.$$

Let us see how long this wavelength is for an electron in the ground state ($n=1$) of the Bohr model for hydrogen. Since

$$mvr = n\hbar, \quad mv = p = \frac{\hbar}{r}.$$

Therefore,

$$\lambda = \frac{h}{\hbar/r} = 2\pi r.$$

The wavelength is as large as the circumference of the orbit! Evidently the model has been carried too far. The electron cannot be considered to be a point particle orbiting the nucleus.

New laws for the mechanics of the microworld were discovered in the 1920s. The subject is called *quantum mechanics* or wave mechanics. Systems are described in terms of equations that have wavelike solutions. The solutions can be combined to give the probability of finding a particle in a particular location or the probability of a particle changing from one energy state to another. The quantum mechanics has been completely successful in predicting the results of experiments with atoms or subatomic particles—*if the calculations can be done*. Even the new quantum mechanics faces the old problems of complexity when dealing with more than two particles.

With atoms containing many electrons, approximations have to be made. For example, lithium has three electrons. Two of them can reside in the smallest region with $n=1$. The Pauli exclusion principle is satisfied because the electrons can have opposite spin. The third electron cannot be accommodated with $n=1$ or it would be identical with one of the electrons already there, which is forbidden. It has $n=2$ and acts almost as if the inner electrons formed a closed shell around the nucleus. Since this shell partially masks the positive nuclear charge of 3, the outer electron sees only an effective positive charge of 1. Since it is farther away from the center of that charge than is the electron in hydrogen, it is more loosely held. The ionization energy is only 5.4 eV. In its lowest energy state, the outer electron in lithium is not completely shielded from the nucleus by the two inner electrons. To a first approximation, however, the spectrum from lithium is similar to that from a hydrogen atom.

The analysis of more complicated atoms requires better approximations and lengthy computer calculations. The energy levels and the corresponding spectra of most atoms are very complex. For example, one of the best known spectral lines, and the easiest to produce without fancy apparatus, is the sodium D line. Sprinkle some salt in a gas flame and you can see the bright yellow color produced by the outer electron, jumping from an excited state to the ground state of $n=3$. When this light is analyzed in a spectrometer, it turns out that there are two lines close together, one with $\lambda = 5.896 \times 10^{-7}$ m and the other with $\lambda = 5.890 \times 10^{-7}$ m. Many of the spectral lines turn out to be doublets, or even higher multiplets.

Not only do the atomic energy levels determine the spectra, but they also control the chemistry of the elements. The periodic table was originally arranged by Dmitri Mendeleev (1834–1907) on the basis of chemical activity and similarities of the elements. The quantum theory now gives a complete explanation of chemistry (at least in principle—the detailed calculations are frequently too complicated). The alkali metals, for instance, such as lithium, sodium, and potassium, have an extra electron outside a filled shell of inner electrons. The spectra are hydrogenlike, and the chemical valence is +1. The halogens, such as fluorine, chlorine, bromine, and iodine, lack only one electron to fill a shell. They have a chemical valence of −1. The heavier the alkali-metal atom, the further its valence electron from the nucleus and, hence, the more loosely it is held. Consequently, the ionization energy of potassium is less than that of lithium, and potassium is more active chemically. With the halogens it is just the opposite. Fluorine, which is the lightest, lacks an electron in the second shell, with $n=2$. The attraction of the nucleus is strong, and fluorine is a strong oxidizing agent (it grabs electrons). With iodine, there is the same net positive charge of the central region, attracting an electron to fill a shell, but in this case $n=5$, and there is a greater distance to the nucleus. The attraction is weaker, and iodine is a weaker oxidizing agent.

■ MOLECULES

The simplest stable molecule is hydrogen, consisting of just two protons and two electrons. It is misleading to think of this molecule as a simple combination of two hydrogen atoms. The energy levels and regions of electron concentration are not related to those of the individual atoms. To calculate the structural details, we start out with twin positive centers and then ask how two electrons can establish themselves in the region. In the lowest energy level, there is a high probability of finding the two electrons (with spins opposed) between the two protons, thus producing binding. When electrons are interchanged like this and neither atom maintains its original outer structure, the binding is called *covalent*. At the other extreme, such as with sodium chloride, the valence electron of sodium transfers quite completely to the outer electron shell of chlorine. The resulting system consists of two spherical ions, one positive and one negative, attracting each other. This bonding is called *ionic*. Most molecules are held together by bonds in between pure covalent and pure ionic. In any case, those are merely names for first approximations of the actual energy states and electron configurations.

With a hydrogen molecule, there are four particles interacting, all influencing each other's motions. With sodium (atomic number 11) and chlorine (atomic number 17), there are 30 particles interacting. How is it possible to calculate the motions of 30 particles all at once? It is indeed difficult to calculate the interactions of more than two objects. However, complex groups can sometimes be considered, in first approximation, to consist of a main group that is motionless while a single other object moves. With the hydrogen molecule, for example, assume that the two protons are massive and stay fixed with respect to each other and the two electrons. The problem is thus reduced to calculating the behavior of two electrons moving around fixed centers. With sodium chloride, the inner electrons are largely shielded from what goes on at the atomic surface. In the first approximation, once the chlorine grabs the extra sodium electron, there is just a two-body interaction between ions.

Even when there can be no exchange or sharing of electrons between molecules, there may still be forces between them. The van der Waals's force between molecules, which requires a correction in the gas laws and explains physical adsorption, is caused by fluctuations in the electric charge centers of atoms or molecules. Although atoms and molecules are electrically neutral, they are composed of many moving electrically charged particles. There are momentary displacements of the symmetry of the components, producing fluctuating electric dipoles. These, in turn, induce similar displacements in adjacent atoms or molecules. The resulting forces are weak compared to chemical bonding and depend on the inverse sixth power of the distance between molecules.

The spectra of molecules are characterized by bands of colors that can be separated into individual lines only with high resolution. Besides having energy levels due to electron arrangements, molecules have energy levels due to vibration and rotation. These energies are also subject to the quantum conditions. Usually, however, the separation between vibration or rotation energy levels is small, and so there is only a small difference in the energy of photons emitted in transitions to neighboring levels. A typical molecular spectrum in the visible range consists not of lines, but of a multitude of lines, so close packed that they look like a continuous band of colors. Each transition between different electron energy levels has many slightly different energies possible, because of the various atomic vibration and rotation states associated with each electron state.

■ THE SOLID STATE

The complications that arise when a vast number of molecules are bound together are so great that much of the research in this field has been done only since World War II. Out of such studies has come a partial understanding of the art of metallurgy and such practical devices as transistors and lasers.

The most impressive feature of solids is the regularity of the atomic arrangements. Most solids are crystalline with the atoms bound in a lattice. In some cases, the same pattern repeats itself for only a few thousand atoms before running into a similar pattern that is oriented in a different direction, or perhaps the pattern is broken by an impurity atom. On the other hand, if the atoms are assembled very slowly, a single uniform crystal containing many moles of atoms may be formed. A few solid materials, such as glass, do not have crystalline structure and are most easily described as congealed liquids.

Imagine now what happens to the electron arrangements in a crystal in which the atomic centers are uniformly arrayed in all directions and are only about an atomic diameter apart. The inner electron shells will still remain attached to the parent atom and will be relatively undisturbed. The outer electrons, however, will find themselves in completely different energy levels. We saw that in the covalent bond, the electron was as likely to be with one atom as the other. When the bound atoms stretch out in all directions, some of the outer electrons may be shared with the whole crystal.

With an *isolated* atom, the energy levels and probability distributions of outer electrons are calculated in terms of a *central* attractive force, partially shielded by inner electron shells. For the outer electrons of each atom in a crystal, however, the problem is more like calculating the motions and positions of golf balls on a green containing numerous shallow holes. To make such a model a little more like the real situation, we would have to add a rim to the green so that the golf balls would be in a negative potential state compared to the outside and so would not escape. Furthermore, the entire green must be pictured as vibrating with thermal energy. This keeps the golf balls in constant motion and does not allow them to remain trapped for long in any one hole.

After what we have seen about the behavior of electrons, the golf green analogy can scarcely be satisfactory. The wavelength of the probability function describing an electron is larger than the distance between atoms. Two quantum effects dominate the real calculations. First, the Pauli exclusion principle demands that no more than two electrons within the same crystal have exactly the same energy. If we could take out of the crystal all those electrons not bound to their parent atoms and then start feeding them back in one at a time, the first one would drop to the lowest energy level possible. The next one could also exist at that lowest level, but with opposite spin. A third electron would have to go to the next energy level. Occasionally these electrons would be momentarily raised to higher energy levels by thermal collisions.

The second quantum effect results from the interactions of the atoms with each other. In an isolated atom, there are typically several electron volts separating energy levels. But if two atoms are joined their electrons cannot have the same energy, according to the exclusion principle. What happens is a splitting of energy levels. Each level becomes two, separated by a small amount of energy. Add a third atom, and each level becomes three. In any crystal, the millions of atoms are all joined, and the result is that there is an enormous number of energy levels, barely separated from each other; so the electron energies seem to form a continuum.

Note that in the case shown in Fig. 21.16, the energy levels do not form a completely continuous set. The levels are formed by the splitting of atomic levels that may be far apart. Each atomic level splits to form a more or less wide band of levels, but there are gaps between the bands. Within any band, an electron can easily gain or lose a small amount of energy (provided that it can find an unoccupied level to move to), but it must receive a substantial jolt to jump across the gap from a lower band to a higher one.

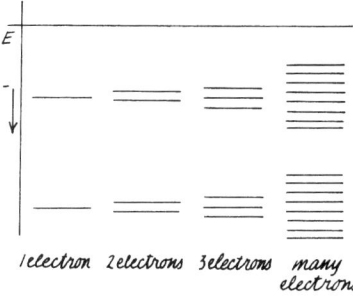

Figure 21.16. Energy of electrons trapped in a crystal. Each main energy level is subdivided into many others.

In some materials, all the energy levels in the lower band are filled right up to the top energy of several electron volts. At room temperature, the average kinetic energy of a free molecule, or an atom bound in a solid, is about $\frac{1}{25}$ eV. Most of the electrons in a solid cannot receive such small amounts of energy, because they would then be lifted to energy states that are already filled with two electrons. But this means that most of the electrons cannot give or receive any thermal energy!

This quantum model explains a number of experimental observations that could not be understood in terms of the classical theories. The older explanations had assumed that in a metal at least one electron per atom must be free to roam. That partially explained electrical conductivity. However, the free electrons should act like a gas enclosed in the crystal box. They would then have a major effect on the heat capacity. Here is another example of how gross measurements of something as old-fashioned as the thermal characteristics of a substance give us a clue about its microstructure.

The specific heat of a metal ought to be $3kT$, since atoms can vibrate in three dimensions, and the energy for each dimension is kT: $\frac{1}{2}kT$ potential and $\frac{1}{2}kT$ kinetic. This agrees very well with the observed value of 6 cal/mol °C. But what about all those electrons? Each of them ought to have $\frac{3}{2}kT$ of energy, just like the molecules of a gas, since they can bounce around freely in all directions. Every time heat is added to a metal with one conduction electron, the electron gas should absorb one-third of the energy.

The electron gas, then, ought to raise the specific heat of a metal to $\frac{9}{2}kT$, or 9 cal/mol °C. It does not. The quantum model explains why. The electrons cannot absorb energy because most of them are in the lower energy band, which is completely filled. The only way energy could be absorbed would be by giving a lot of energy to the electrons in the upper part of the band, so that they could jump into the higher energy band; but such large amounts of energy are not available. The electrons simply do not have the freedom of gas molecules, because they are bound in quantum states within the crystal.

■ ELECTRICAL CONDUCTIVITY IN SOLIDS

The quantum model also explains why some materials are electrical conductors and others are insulators. In Fig. 21.17 we show three different possibilities for the arrangement of energy-level bands. In part (a), there are just twice as many electrons as there are energy levels, and so every level is filled.

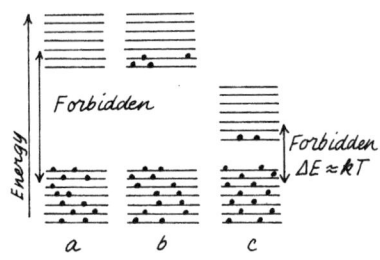

Figure 21.17. Electrons in the energy levels of: (a) insulator (b) conductor (c) semi-conductor.

How can there be twice as many electrons as energy levels? Each band is divided into as many closely spaced energy levels as there are atoms in the crystal. A particular type of atom may provide two electrons each to roam the crystal. Each separate level can hold two electrons with their spins opposed.

If an electric field is imposed on material of the type shown in part (a), an electron will feel a force, but cannot move and accept energy. To do so, an electron in a lower energy state would have to move into an energy level that is already filled. The electrons at the highest levels could accept energy only if they could be lifted abruptly to energy levels in the next higher band. In an ordinary electric field, this cannot happen because the electron as it is slowly accelerated slowly gains energy. It would have to pass through all the intermediate energy states, which in this type of material do not exist.

The energy-level diagram of an electrical conductor is shown in part (b) of the diagram. Some of the energy levels in one band are not occupied. An electron can accept energy from an imposed accelerating field. It might seem as if it would keep right on accelerating until it had reached the maximum energy available in that band. The electron, however, suffers frequent collisions that keep its speed low and, on the average, constant. Impurities in the crystal structure and the thermal vibrations of the atoms in the lattice are responsible for the collisions. They soak up some of the electron energy that is provided by the imposed electric field and turn it into thermal energy. In general, the hotter the lattice atoms, the more they absorb energy from the conduction electrons. The resistance of a conductor usually rises with increasing temperature.

The semiconductor situation is shown in part (c) of the diagram. All of the levels in one band are occupied, but only a small energy gap separates this band from a higher one with empty levels. Thermal energy or electromagnetic radiation can provide the small quantum of energy needed for a few electrons to make the jump into the open conduction band. The electrical resistance of such a material will depend not only on the terminal velocity of the charge carriers, which is determined by the nature of the collisions they suffer, but also on how many electrons are available to carry the current. In the case of carbon, in the form of graphite, the electrical resistance *decreases* with increasing temperature. The higher temperature sends more electrons into the upper conduction band.

If two solids are touching each other, the electrons at the junction may be able to move in ways not possible in either solid by itself. The energy levels and forbidden regions in the two materials may be different. An electron at the junction may be able to receive energy from an outside electric field if it moves into one region that contains open energy levels, but not the other region. This electron movement will then leave an open level in the region previously filled, so that another electron can move there. However, this current can only take place in the direction of the material with the open levels. Thus, we have a rectifier that can pass current in one direction, but not the other.

The theory of the solid state is now adding to our technology at an enormous rate. The junction diode, described above, was made according to specifications worked out first on paper, based on the quantum theory of crystals. From this has come the transistor, which works on a similar principle, and a great many other devices. The theory is being used to make alloys that are better conductors, that can become superconducting at higher temperatures, that have improved tensile strength, that respond to light and heat in various ways. In this field of the complex arrangement of simple particles, there are no boundary lines between theory and technology.

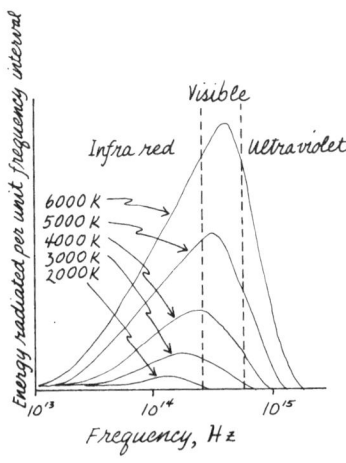

Figure 21.18. Rate of radiation of a blackbody as a function of frequency, for several temperatures.

■ SPECTRA PRODUCED BY HOT DENSE SOURCES

When isolated atoms are raised to high temperature in a gas, they yield line spectra. If atoms are combined into molecules, they produce band spectra, each band made up of many individual lines. If the atoms are closely packed into solids, or materials at high density, they give off light at essentially all wavelengths. A red-hot piece of iron produces such a continuous spectrum. So does the surface of the sun, which is not a solid but a plasma of ions and electrons with a density equal to that of a solid.

Since dense material is made up of atoms that have discrete energy levels, how can photons of all energies be produced? Isolated iron atoms yield a line spectrum. When the iron atoms are fastened together, the separate atomic energy levels must disappear. Indeed, in a solid metal the outer electrons are no longer part of the structure of individual atoms, but are part of a whole crystal. If any one atom is disturbed, the disturbance is shared with the whole crystal. Although the energy levels of the system are still quantized, there is now an enormous number of such levels, closely spaced.

The spectrum produced by a dense source is a function primarily of temperature and, to a lesser extent, of the surface color and conditions of the hot object. There is a way to eliminate completely the dependence on the surface characteristics. If a closed hollow box or tube has a small hole in its wall, the radiation coming from the hole will be independent of the material out of which the box is made. Furthermore, the radiation coming from the hole will be more intense than that coming from the wall. The hole is brighter! Because a hole in a cavity absorbs all of the light striking it, the arrangement is called a "blackbody." The radiation coming from a blackbody depends only on temperature. Explaining this dependence became a crucial stumbling block for classical physics and led to the beginnings of quantum mechanics.

Figure 21.18 shows the intensity as a function of frequency for blackbody radiation at several temperatures. There are two regularities in the spectra. First, there is a simple relationship between the temperature and ν_m, the frequency of maximum radiation; that is, $\nu_m = \text{const} \times T$. Second, the total energy radiated is proportional to the fourth power of the temperature. Both of these laws agree qualitatively with everyday observations. As the temperature of a light-bulb filament rises, its color changes from red to white hot. Furthermore, the amount of light produced by a hot filament increases dramatically as its temperature rises—recall the bright flash as an incandescent bulb burns out.

Before 1900, classical physics could explain many electromagnetic phenomena. Maxwell's equations successfully predicted radiation through space. Those equations and the basic principles of thermodynamics ought to have been able to explain a type of radiation whose characteristics depended only on the temperature of the source and not on the material from which it was made. The model for blackbody radiation pictured the solid walls of a hollow box as containing an enormous number of electron oscillators. These could oscillate at any frequency. Because of the equipartition of energy, each mode or frequency of the oscillators should share equally in the thermal energy available. Each mode should have energy equal to $3kT$, since each oscillator has $\frac{3}{2}kT$ for average potential energy and $\frac{3}{2}kT$ for average kinetic energy. However, not all modes or frequencies are possible. The electron oscillators radiate their energy into the hollow region of the box, but they also absorb that radiation. At any given temperature, there is an equilibrium between the electromagnetic radiation in the enclosed space and the oscillations of the electrons in the walls. However, the radiation can exist only at frequencies corresponding to standing waves in the cavity. The situation is analogous to that of the resonance conditions for sound waves in an organ pipe. The actual allowed frequencies depend on the dimensions of the cavity. At low frequencies (long wavelengths), there is the fundamental, then the next harmonic, and so on, each frequency separated by a considerable interval from the next. At higher frequencies, more and more waves are resonant. The higher the frequency, the more modes exist per frequency interval. If each mode has an energy $3kT$, we would expect blackbody radiation to be concentrated in the ultraviolet—or even x-ray—region! Indeed, since the number of modes (overtones) goes to infinity as the frequency goes to infinity, the energy required would be infinite if each mode had to be given $3kT$. Appropriately enough, this paradox of classical physics was called "the ultraviolet catastrophe."

In 1900, Max Planck (1858–1947) resolved the paradox by proposing that the energy exchange between oscillators and electromagnetic field was quantized. The oscillators could exchange energy only in amounts proportional to their frequency:

$$\Delta E = h\nu.$$

The proportionality constant is our old friend, Planck's constant, which has the very small value of 6.6×10^{-34} J s.

According to Planck's proposal, is radiant energy quantized like angular momentum or electric charge? Must radiant energy always exist in multiples of some basic unit of energy? No. Notice that the quantum of energy depends on the frequency. Infinitesimal amounts of radiation can be emitted or absorbed, but only at very low frequencies, corresponding to long wavelengths.

Planck's quantum requirement upsets the equipartition of energy in blackbody radiation. The equipartition law assumes that energy is transferred from one mode to another by processes that can exchange infinitesimal amounts of energy during each interaction. However, according to Planck's condition, a high-frequency oscillator can change energy only by a large amount, an unlikely event. Hence, the high-frequency modes are seldom excited. The detailed theory yielded a distribution of radiation energy versus frequency that matched exactly the curves shown in the graph. Note the way the distribution falls off at low frequencies—there are not many resonant modes. Note the way the distribution falls off at high frequencies—many modes are possible, but few are populated because of the large jump in energy needed to excite a high-frequency mode.

Planck was uneasy about the implications of his theory. For a long time he preferred to think that the quantization was an artificial peculiarity of the oscillators. Electromagnetic radiation must surely consist of waves, whose energy could assume any value. Albert Einstein (1879–1955), however, made a complete break with the classical theory of radiation. In one of his three great papers of 1905, he explained the photoelectric effect in terms of quanta of radiation. He assumed that electromagnetic radiation acts like chunks of energy when it interacts with matter. Each "photon" carries energy $h\nu$ and, when it interacts with matter, it does so at a point—not spread out in a wave front. Consequently, a photon's energy can be delivered to a single electron bound in an atom. If the energy is sufficient, the atom will be ionized and the electron freed.

For instance, blue light has a frequency of 7.5×10^{14} Hz [$\nu=c/\lambda=(3\times10^8$ m/s)/(4×10^{-7} m)]. How much energy can such a photon deliver to an atom? Is the energy sufficient to ionize an atom?

$$E = h\nu = (6.6\times10^{-34} \text{ J s})(7.5\times10^{14} \text{ Hz}) = 4.95\times10^{-19} \text{ J} = 3.1 \text{ eV}.$$

As we have seen, the valence electrons in the alkali metals are bound with energies in this region. Evidently, visible-light photons can free electrons and cause chemical changes in material—people can be sunburned, colors bleach, photosynthesis occurs, and photography works.

Einstein's equation for the photoelectric effect is

$$(\tfrac{1}{2}mv^2)_{\text{emitted electron}} = h\nu - E_{\text{escape}}.$$

In most photoelectric effects, the emitted electron does not escape from the material, and so its kinetic energy cannot be measured. After 1905, several physicists demonstrated the qualitative agreement between Einstein's simple law and experiments. However, it was not until 1916 that Robert Millikan (1868–1953) performed the delicate measurements that confirmed Einstein's photoelectric law and thus confirmed the reality of photons. The measurements had to be done in high vacuum, with emitting surfaces freshly exposed in the vacuum. The kinetic energy of the electrons was determined by measuring the retarding voltage that they could overcome. The binding energy of a metal corresponds to a threshold frequency of the light. As the frequency of the light increased above the threshold, the kinetic energy of the electrons increased.

■ THE DUALITY PROBLEM

Is light a wave or a particle? All of the evidence of diffraction, interference, and polarization calls for a wave model of light. Furthermore, we know that light is simply a short-wavelength version of electromagnetic radiation, such as radio waves, where the wave nature is observed both in production and detection. On the other hand, the *formation* of light involves features of quantized energy values. The spectra of isolated atoms contain only certain frequencies. The complicated radiation from hot solids is explained if, and only if, it is assumed that the micro-oscillators can exist only in quantized energy states and thus emit only quantized packages of radiant energy. When light interacts with matter, it acts as if it were a particle delivering packages of energy to specific points.

Notice, however, that the two models of light are not completely independent. The package of light, the photon, has an energy proportional to its *frequency*:

$$E_{\text{photon}} = h\nu.$$

But frequency is a wave concept. How can a particle have a frequency?

When the implications of this situation were realized in the early decades of this century, various philosophical escapes were sought. For instance, it was suggested that the advancing wave fronts might contain photons, much as a breaking water wave could carry a group of surfboards. No quantitative predictions of such a model were successful.

Experiments were performed in attempts to force light to reveal its true nature as either a particle or a wave. The most revealing type of experiment involves the creation of an interference pattern of light at intensities so low that only one photon at a time is likely to exist in the interference region. Such experiments were tried in the 1920s and again, with greatly improved accuracy, in the early 1950s.

The diagram in Fig. 21.19 illustrates one such experiment, which was first performed in Budapest in 1957. The apparatus is similar to the Michelson interferometer, which was used in the 1880s to calibrate wavelengths in terms of the standard meter and to demonstrate that the speed of light is independent of the motion of the source or receiver. Its operation for that purpose was described on p. 136, Chapt. 6. Light of a single wavelength comes from the left through filters that control the intensity. The half-silvered surface of mirror A transmits 50% of the light and reflects the other 50%. Each beam takes the path shown, and they recombine at the screen. The geometry of the interference pattern depends on the

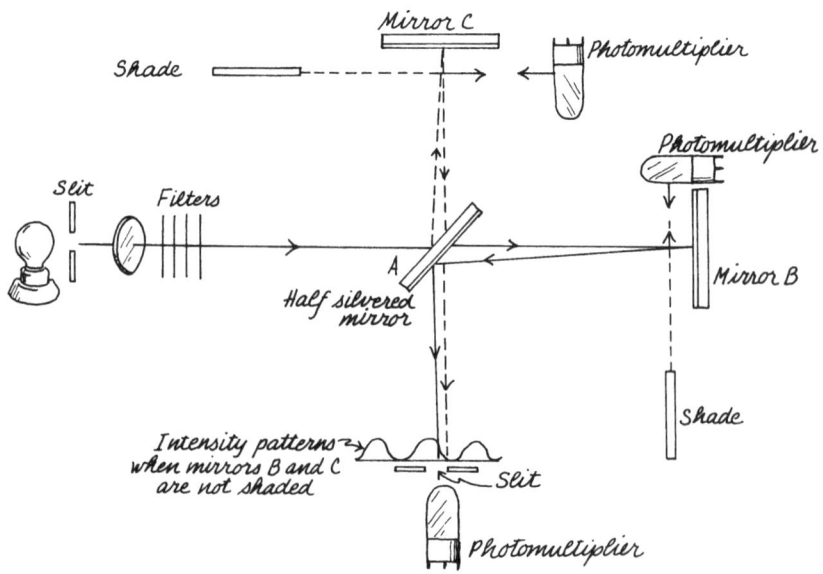

Figure 21.19. Interference pattern built up by one photon at a time in the Michelson interferometer.

original slit geometry. In this case, alternate bands of reinforcement and cancellation are formed on the screen. In effect, looking back from the screen, one sees the single source as two sources at slightly different distances, assuming that the two path lengths are almost but not quite the same.

If the slit opening of a sensitive photomultiplier tube is positioned at a dark band of the interference pattern, no signal will be produced except for the intrinsic electrical noise of the tube. (A photomultiplier tube contains a light-sensitive surface that emits electrons by the photoelectric effect. These electrons are then accelerated from electrode to electrode within the tube, generating more electrons at each stage. Millionfold amplification can be obtained.) The light intensity is now reduced by the filters to the level where only a million photons per second are being transmitted. [This number could be determined by measuring the original energy of the beam, dividing by the energy of a single photon ($h\nu$), and then multiplying by the filter reduction factor which can be separately calibrated.] To check the reduction factor, other photomultipliers can be used at positions B and C, and the number of photons actually counted. The geometry is such that the travel time for a photon through the whole system is about

$$t_{\text{transit}} = \frac{1 \text{ m}}{3 \times 10^8 \text{ m/s}} = 3 \times 10^{-9} \text{ s}.$$

Since about 10^6 photons per second are coming through at random, the total time *per second* that there will be a photon in the apparatus is

$$t_{\text{occupied}} = 10^6 \times 3 \times 10^{-9} \text{ s} = 3 \times 10^{-3} \text{ s}.$$

Since the apparatus is occupied by photons for only 3×10^{-3} s out of each second, there is only about 0.3% chance that two photons will be there during the same time. It would seem that the chance for interference between photons, constructive or destructive, would be practically nil.

Perhaps, though, a photon can divide at mirror A, half of it going one way and half the other. Eventually both halves might arrive at the photomultiplier and interfere with each other. This possibility can be checked experimentally. When the mirrors B and C are replaced by the photomultipliers, we would expect coincidence of pulses if the photons divide. Modern electronic circuits linked with photomultipliers can detect whether or not pulses are coincident to within 10^{-9} s. These pulses from B and C are usually not in coincidence. Each photomultiplier records half a million counts per second, but only a very few arrive in coincidence, and that number agrees with the probability prediction for pulses arriving at random.

The observation is that, although there is seldom more than one photon at a time in the apparatus, the photomultiplier scanning the destructive interference region at the screen still sees nothing. *However*, if mirror B or mirror C is blocked by a shade, the photomultiplier immediately starts registering counts. With one mirror covered, a photographic plate placed at the screen would not record a pattern, but only the continuous exposure expected from illumination by the central maximum of a single slit pattern. As soon as both mirrors are uncovered, the image of an interference pattern starts building up on the photographic plate, and the photomultiplier records no counts from a cancellation band, *in spite of the fact that only one photon at a time can be in the apparatus.*

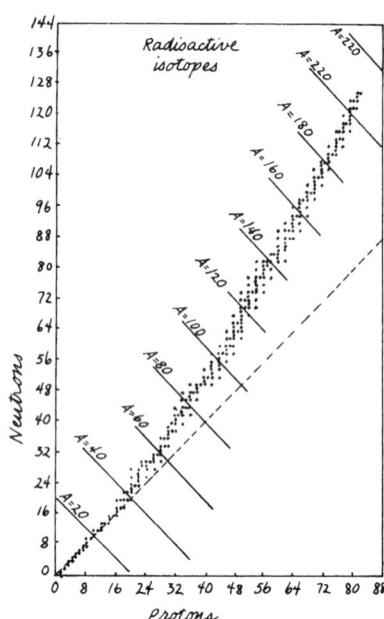

Figure 21.20. Number of neutrons and protons in stable nuclei.

Such a stark paradox probably means that we have not yet seen the light and must somehow change our whole viewpoint. Although we usually think that the solutions to Maxwell's equations describe traveling waves of electric and magnetic fields, we actually use the solutions to predict the probability of a meter response or photographic grain development at a particular location. Photons do not travel in or on the waves; *the mathematical functions (which are those of waves) yield the probability of a photon interaction.*

■ THE ATOMIC NUCLEUS

Each element is characterized by a certain number of electrons per atom. There is an equal number of protons in the nucleus, leaving the atom electrically neutral. There are also neutrons in the nucleus, usually about as many as there are protons. For a given element, however, the number of neutrons can vary, giving rise to isotopes. Isotopes of the same element have almost identical chemical characteristics, but slightly different masses.

Hydrogen has three isotopes, two of which have special names. *Deuterium*, sometimes called heavy hydrogen, has a nucleus consisting of one proton and one neutron. It is stable but exists naturally only to the extent of one part per 7000. The nucleus of *tritium* contains two neutrons and one proton. It is radioactive, with a half-life of 12.5 years, and exists naturally only with an abundance of 10^{-18} of that of ordinary hydrogen. (It is slowly but continuously being produced by cosmic radiation and has also been produced in sizable quantities in nuclear reactors.) Because of the large mass ratios of these hydrogen isotopes, their chemical behavior differs sufficiently so that they are easily separated from each other.

Probably the most famous isotope is uranium-235. Its nucleus contains 92 protons (the atomic number of uranium) and 143 neutrons. Its abundance is only 1/144 that of uranium-238. Elaborate production facilities have been created to separate ^{235}U because it is a fuel for atomic reactors.

Most elements exist naturally as a mixture of several isotopes, usually with one or two predominating. Figure 21.20 shows all the stable isotopes, and many of the unstable ones, up to $Z=82$ (lead). Note that for the light elements, nuclear stability is achieved when the number of neutrons is about the same as the number of protons (along the 45° line of the graph). For nuclei with larger atomic number, however, extra neutrons are needed for stability. This effect is caused by several factors. Evidently there must be a nuclear binding force between proton and proton, neutron and neutron, and proton and neutron. This force is a "contact" force, which goes essentially to 0 if the protons and neutrons are not touching each other. However, the Coulomb repulsion between protons is relatively long range. Although only adjacent nuclear particles attract each other, all of the protons repel each other. Nuclear stability is achieved by interlacing more and more neutrons. For nuclei above lead ($Z=82$), there are so many protons that complete stability is impossible.

The nuclear binding is tightest when proton–neutron pairs can be formed. If a proton and neutron combine to form a deuterium nucleus, their mass-energy is less than that of the original particles:

$m_{\text{proton}} = 1.6724 \times 10^{-27}$ kg,

$m_{\text{neutron}} = 1.6748 \times 10^{-27}$ kg,

(sum of separate masses) $= 3.3472 \times 10^{-27}$ kg,

$m_{\text{deuteron}} = 3.3431 \times 10^{-27}$ kg,

(mass difference) $= 0.0041 \times 10^{-27}$ kg.

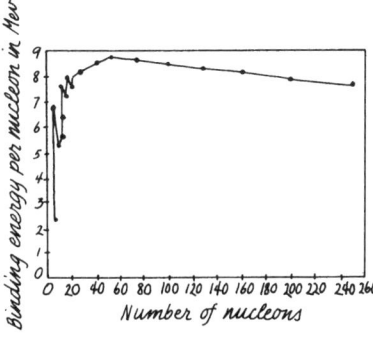

Figure 21.21. The binding energy per nucleon vs the number of nucleons.

This lost mass is given off in the form of a high-energy photon (γ ray) when the proton and neutron combine. The energy of the γ ray, which is the negative binding energy of the deuteron, is

$$E = mc^2 = (4.1 \times 10^{-30} \text{ kg})(3.0 \times 10^8 \text{ m/s})^2 = 3.69 \times 10^{-13} \text{ J} = 2.3 \times 10^6 \text{ eV} = 2.3 \text{ MeV}.$$

Can we find the binding energy of a water molecule this way? In theory, yes; in practice, no. The mass deficit is much too small to measure. Chemical binding energies are in the electron-volt range; nuclear binding energies are larger by a factor of a million.

In Fig. 21.21 we show the binding energy per nucleon (neutron or proton) as a function of the number of nucleons. The mass deficit for each element is calculated using the same method that we just used for deuterium. The deficit is then divided by the total number of nucleons in the nucleus of that element. Note that the binding energy per nucleon generally increases as the atomic number increases, although there are some special effects for certain combinations.

The general shape of the binding energy curve can be explained by assuming that the nuclear force is due to actual contact between nucleons. Greater binding can therefore be achieved by completely surrounding a nucleon with other nucleons. Surface nucleons are not so tightly bound as are interior ones. Increasing the total number of nucleons increases the volume-to-surface ratio. However, for the elements containing more than about 50 nucleons, the increasing Coulomb repulsion reduces the total binding energy.

■ FUSION

If two protons and two neutrons could be combined into a helium nucleus, the mass defect would yield 28.4 MeV. One way to accomplish this might be to shoot one deuteron at another. To be sure, the mass defect of each deuteron is 2.3 MeV, but we would still end up with 23.8 MeV. Unfortunately for our human energy needs, it is very hard to get the deuterons to combine. Both are positively charged and so would have to collide with sufficient energy to overcome the repulsive electrostatic barrier.

Why not just shoot a beam of deuterons into a block of deuterium ice? When a charged particle travels through matter, it loses energy to the atoms through which it passes. A direct hit on the nucleus of a target atom is rare, because the nucleus is so small. Most of the energy of the bombarding deuterons would be lost in ionizing the atoms in the deuterium ice.

Most of the attempts now being made to yield fusion energy depend on bringing deuterium or tritium together at high densities and at temperatures in the range of 100 million degrees. These are the conditions existing in the centers of stars where fusion of hydrogen into helium provides stellar energy. In the so-called hydrogen bomb, the fusing elements are compacted by the preliminary explosion of a uranium or plutonium fission bomb.

■ FISSION

At the high end of the atomic numbers, energy can be gained by splitting up large nuclei into smaller ones. Of course, it would still take energy to break uranium into 235 protons and neutrons, but energy is yielded when ^{235}U breaks down into two smaller nuclei, such as barium and krypton. The binding energy per nucleon of uranium is about 7.6 MeV. For medium-mass nuclei, the binding energy per nucleon is about 8.6 MeV. This difference of about 1 MeV per nucleon produces an average of 200 MeV for every uranium nucleus that is split into two smaller nuclei.

On a human scale, 200 MeV is not much energy. How much energy is released if a mole of uranium undergoes fission? A mole of ^{235}U has a mass of 235 g, weighing about half a pound:

$$(6 \times 10^{23} \text{ nuclei})(200 \times 10^6 \text{ eV/nucleus})(1.6 \times 10^{-19} \text{ J/eV}) = 2 \times 10^{13} \text{ J}.$$

This is about the energy produced in three hours by the power plants at Niagara Falls.

The heavy nuclei do not spontaneously split up, even though all that energy can be released by doing so. It takes a trigger to start the process. With ^{235}U, the capture of a low-energy neutron distorts the nucleus and throws it into oscillations that soon cause it to split in half (or sometimes into three pieces). In the process, two or three neutrons are emitted. If there is more ^{235}U surrounding the original fission site, the new neutrons can trigger more fission. The reaction is thus self-sustaining. With the right geometry of uranium and neutron absorbers, the process can be controlled to produce steady energy output in a reactor. With a different geometry, the process proceeds exponentially, producing a nuclear explosion.

The most common form of uranium, ^{238}U, can undergo fission, but only when bombarded by a high-energy nucleon. Consequently, it cannot produce a self-sustaining reaction.

■ RADIOACTIVITY

Nuclei with an imbalance of protons or neutrons can, under certain conditions, change into other nuclei. There are three main routes.

1. Some very heavy nuclei can emit a whole helium nucleus—a tightly bound combination of two protons and two neutrons. Historically this was called an α particle. Because of its double positive charge, an α particle ionizes heavily, expending its kinetic energy in a relatively short path. Most α particles are emitted with energies of several MeV and can travel only a few centimeters in air.

    When a nucleus emits an α particle, its atomic number drops by 2 and its nucleon number (or atomic mass) drops by 4. For instance, the decay of

uranium-238 to thorium-234 is written

$$^{238}_{92}U \rightarrow ^{234}_{90}Th + ^{4}_{2}\alpha.$$

The half-lives for most α-particle decays are in the range of billions of years. If the parent nucleus is unstable, why should it wait so long to decay? First, there is experimental evidence that large nuclei are to some extent composed of α-particle units of four nucleons. The α particle is bound, however, just as if it were behind a high wall (an electrostatic one in this case). We might wonder if occasionally it gets enough energy to sail over the wall, even as a gas atom occasionally gets enough energy to escape from earth. However, all of the emitted α particles from a particular element have exactly the same kinetic energy, *and* the energy is considerably less than the height of the Coulomb barrier required to get back into the nucleus. Apparently the α particles do not go over the wall. Instead, they go through it! According to the Heisenberg uncertainty principle, there is a small probability that the position of the α particle will at some time be outside the nucleus. Once outside, it keeps going. The detailed theory explains the experimental facts perfectly. See Fig. 21.22.

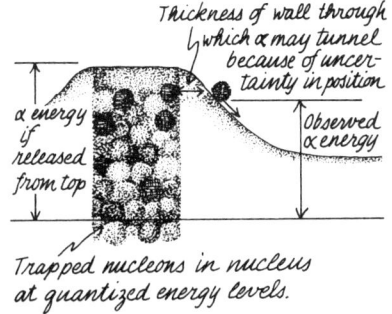

Figure 21.22. The quantum mechanical explanation of α-particle radioactivity.

2. Another way for a nucleus to change its ratio of neutrons to protons is to emit an electron, either negative or positive. The process is called β *decay*. The prototype for β decay is the decay of the neutron itself. The neutron, in free space, is unstable, decaying to a proton and electron with a half-life of about 11 minutes:

$$^{1}_{0}n \rightarrow ^{1}_{-1}p + ^{0}_{-1}e \quad (+ \text{ antineutrino}).$$

One of the best known β decays is from the naturally occurring isotope, carbon-14. The decay has a half-life of only 5580 years, so any of it formed during the origin of the universe must have long since disappeared. It is created continuously, however, by the bombardment of nitrogen with neutrons, which are generated by cosmic rays in the upper atmosphere:

$$^{14}_{7}N + ^{1}_{0}n \rightarrow ^{14}_{6}C + ^{1}_{1}H.$$

The carbon-14 eventually decays according to the following scheme:

$$^{14}_{6}C \rightarrow ^{14}_{7}N + ^{0}_{-1}e \quad (+ \text{ antineutrino}).$$

In some artificially radioactive elements, there are too many protons. In these cases, positive electrons (positrons) may be emitted. For instance,

$$^{13}_{7}N \rightarrow ^{13}_{6}C + ^{0}_{+1}e \quad (+ \text{ neutrino}).$$

Although a definite amount of mass-energy difference exists between the mother and daughter nuclei of a β decay, the electrons are emitted with varying energies up to the maximum. See Fig. 21.23. It turns out that in every β decay, two particles are emitted. Along with the electron there is a neutrino—a particle with 0 (or very small) mass, having no electric charge

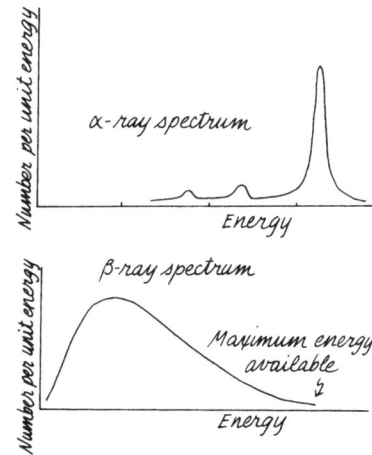

Figure 21.23. An α particle emitted from a particular nucleus must have one of several discrete energies. An electron emitted from a nucleus can have any energy up to a particular maximum.

but with a spin of $\frac{1}{2}\hbar$, like the electron. The electron (β particle) and the neutrino share the available energy.

When a nucleus undergoes a change, such as the emission of an α particle or a β ray, it is frequently left in an excited, energetic state. When atoms are similarly excited, they emit light as they settle into their ground state. Nuclei also emit photons, although the energies are apt to be in the MeV range. These are called γ rays.

■ RADIOACTIVE SEQUENCES

The radioactive heavy nuclei decay in series of reactions of α and β decays. Each α decay reduces the atomic number by 2, and each negative β decay increases it by 1. The series continues until a stable isotope of lead is reached. There are three main series. One starts with ^{238}U and ends at lead-206 (^{206}Pb). The sequence is shown in the Fig. 21.24. Another series starts with ^{235}U and ends with ^{207}Pb. The third starts with thorium-232 (^{232}Th) and ends at ^{208}Pb. There is a similar series starting with neptunium-237 and ending with bismuth-209, but the half-life for neptunium-237 is so short that these nuclei no longer exist in nature.

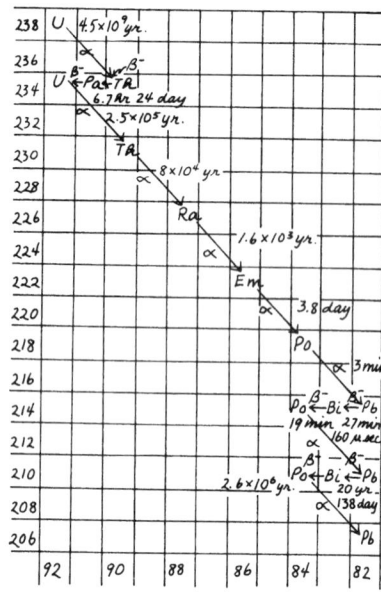

Figure 21.24. The radioactive decay sequence starting with ^{238}U.

■ THE FUNDAMENTAL INTERACTIONS

We know about any object only because of its interactions with other objects. It might seem that the world is as complicated as it is because there are so many different kinds of forces or interactions. There are forces exerted by springs, muscles, wind, expanding gas, gravity, physical contact, magnetism, electricity, and on and on. But actually we know of only four types of forces and all those others are merely manifestations of the four basic kinds. Even the four are not completely separate. Theory has now united two of them, although the union would be observed only at very high energy, and it is possible that there is also a unifying connection with the third.

Gravity is the first interaction that we experience. It pulls us down. Newton gave a successful description of gravity over 300 years ago. Although the interpretation of this force on a cosmological scale was altered by Albert Einstein's general theory of relativity, Newton's original formula is still satisfactory for most purposes here on earth. All particles are subject to gravitational attraction, and so far as we know the force between any two objects is always attractive—even antiparticles do not have negative mass. In the sense that mass warps space geometry, mass might be considered the "charge" or source of the gravitational field. Gravitational field radiation might be generated by rapidly rotating twin stars or by the collapse of a nova. Such gravitational waves have not yet been detected, though much sought after. According to theory, the "agent" of gravitational radiation would be a particle (called the graviton) with mass zero, and spin of $2\hbar$. Compared with the other interactions, gravity is extremely weak. The gravitational interaction between two protons that are touching is smaller than their nuclear interaction by a factor of 10^{39}. Why then do we feel so heavy at times? There is no cancellation of gravitational effects due to opposite charges. All the atoms in the earth pull down on all the atoms in our body. Also because there is no cancellation of charges, the velocity dependence of the gravitational force is completely masked by the primary effect of attraction. (See p. 447 for the relativistic explanation of the velocity dependence of electromagnetism, which yields the relationship between electricity and magnetism.)

Electricity and magnetism are aspects of the same interaction. These two were unified by Maxwell over 100 years ago. The "agent" of the interaction is the photon, a "particle" with 0 mass, 0 charge, and spin of $1\hbar$. The source of the interaction is the electric charge, which can be either positive or negative. As we have seen, the charge is quantized. The charge on particles that can be isolated is always an integral multiple of the basic unit, e. For two protons touching each other, the electrostatic repulsive force is smaller than the strong nuclear attractive force by a factor of about 100. Almost all the forces of everyday life are caused by the electromagnetic interaction. It binds electrons in atoms, and atoms within molecules, and thus is the interaction underlying all of chemistry and biology. The quantum view of an electromagnetic field is that a source charge emits clouds of "virtual" photons. The energy required to emit a photon can be supplied by a fluctuation in the energy of the source, as long as Heisenberg's uncertainty principle is satisfied: $\Delta E \Delta t \geq \hbar$. For a very short time, there can be a fluctuation ΔE in the source, which is the energy of the virtual photon that can be emitted and reabsorbed within that time. While emitted, the photon can interact with another charge, thus exerting a force. Note that electromagnetism does not interact with particles that have 0 electric charge *and* 0 magnetic moment. (The neutron has 0 charge, but is influenced by the electromagnetic interaction because it has a magnetic moment.)

The strong nuclear interaction is complicated because it manifests itself in different ways for different particles. First, it has no effect on electrons or the other members of the *lepton* family—electrons, heavy electrons, and their neutrinos. Second, the primary cause of the interaction is the behavior of *gluons* on *quarks*, which are the constituents of the heavy particles such as protons, neutrons, and mesons. Gluons have 0 electric charge and spin of 0. They are exchanged by the quarks to produce a binding force that *increases* with separation distance. The nucleons—protons and neutrons—are compounds of the quarks and can exchange *mesons* that are also quark compounds. This latter exchange is responsible for what is usually considered nuclear binding. The force between neutron and neutron, proton and proton, or neutron and proton is very short range, essentially a contact force. It is stronger than the electromagnetic force (for nucleons that are touching) by a factor of about 100. In terms of binding energy, the nucleon–nucleon interaction is about 7 MeV, whereas an electron is held in an atom by only a few electron volts. Although, as we shall see in succeeding sections, the quarks can be detected and measured, they cannot be isolated and knocked out of their compound status in nucleons or mesons. If a bombarding projectile hits a quark, the quark bond stretches but gets stronger with separation rather than weaker. Eventually, rather than knocking out the quark, the bombarding energy is turned into quark pairs, which in turn can be ejected as jets of heavy particles which are once again compounds of quarks. An isolated quark is never produced.

The weak interaction is responsible for β decay. It is weaker (at low energies) than the strong nuclear interaction (the one between nucleons) by a factor of 10^{13}. All particles are subject to the weak interaction. The agent is a set of massive particles: W^+, W^-, ($m=83$ GeV), and Z^0 ($m=93$ GeV). These have spin 1, and are extremely short-lived, which also signifies that the interaction range is very short, essentially the size of a nucleon. During the 1980s a successful theory has shown that at energies in the region about a tera-electron-volt (10^{12} eV), the weak interaction and electromagnetism merge into a common mechanism, called the electroweak.

■ OTHER PARTICLES

Let us take a census of the subatomic particles that we have met so far. They and others are organized in order of their masses in the chart in Fig. 21.25. Down

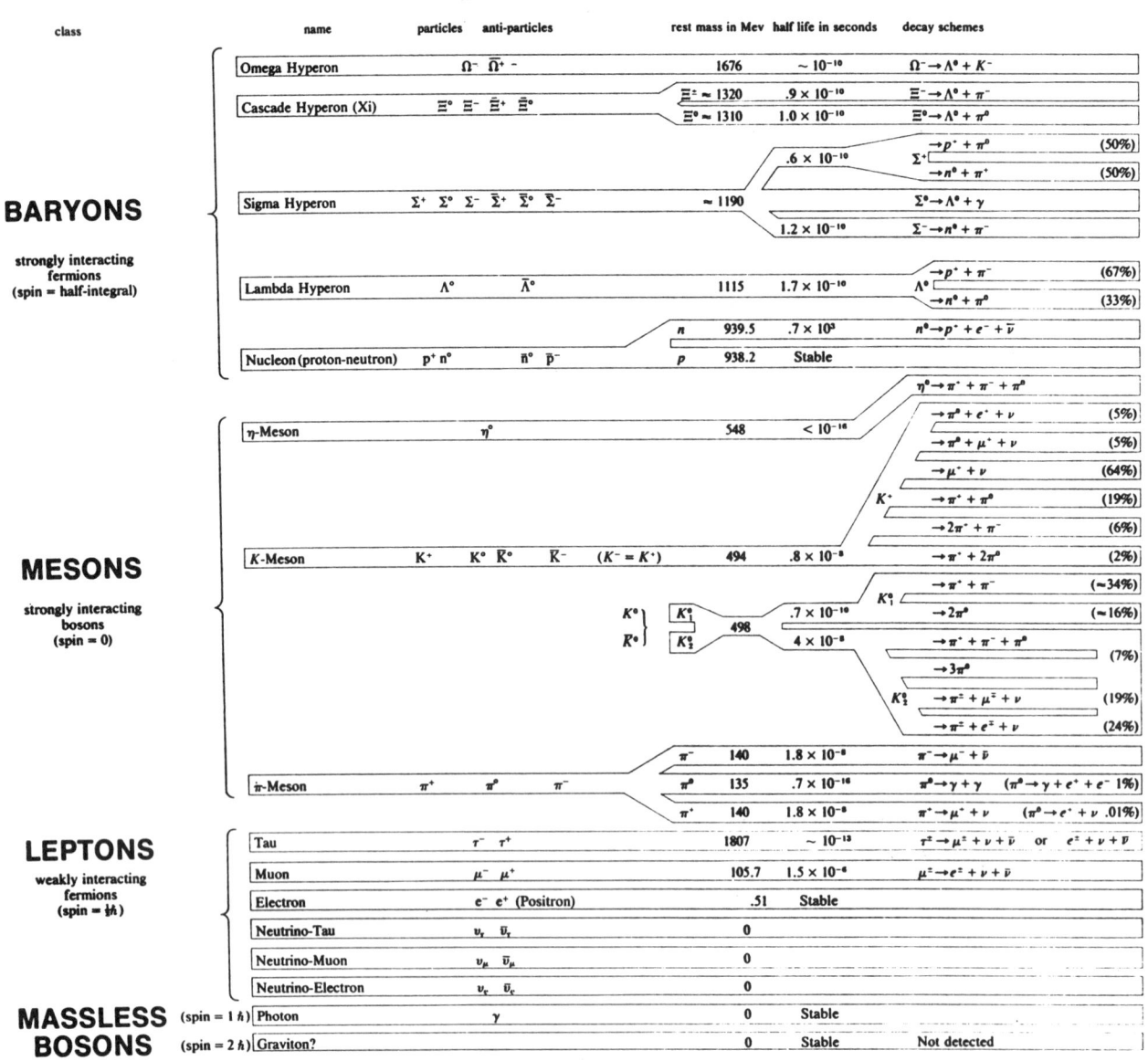

Figure 21.25. Particles stable against strong nuclear decay.

near the bottom of the chart is the familiar electron. One particle has no mass at all; photons travel at the speed of light and therefore cannot have mass. Neutrinos have mass less than 16 eV, and perhaps their mass is zero. Near the top of the chart are the nucleons—proton and neutron.

Note that even with these familiar particles there are novel complications. Each has an antiparticle, a particle with identical mass and spin but with opposite electric charge. Particles and antiparticles can be created in pairs, as long as energy and momentum are conserved. For instance, a γ ray with an energy of 1.02 MeV or greater can create an electron pair. The rest mass of electron and positron is 0.51 MeV each. It might seem that neutrinos have so few nonzero attributes that there is nothing to be "anti" about. Nevertheless, the neutrino accompanying negative β decay is experimentally different from its antiparticle that accompanies positive β decay.

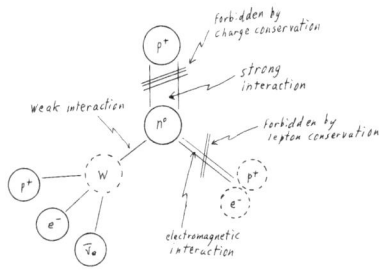

Figure 21.26. Decay of free neutron.

Actually, there are at least three electronlike particles with different masses. The muon was first detected in cosmic rays in the 1930s, although its nature was not understood until 10 years later. In every respect it is a heavy electron, having its own antimuon and matching neutrinos. In 1978, a yet-heavier electron was found, dubbed the tau particle (τ). The electrons and neutrinos seem to form a separate family of particles, called *leptons*, having no internal structure—or at least, none detected so far. Each member of the family has spin $\frac{1}{2}\hbar$. There is a conservation law within the family. No member can be created or destroyed without the cooperation of an appropriate antiparticle, such as one of the neutrinos.

The nucleons are the base of a large family of heavier particles called the *baryons*—Λ lambda, Σ sigma, Ξ xi, Ω omega, and so on. All are unstable, except the proton, and decay into the proton with half-lives in the 10^{-10} s range. (The neutron is an exception, which we will explain later. Its half-life is 11 min.) These particles have half-integral spins—$\frac{1}{2}\hbar$, or $\frac{3}{2}\hbar$, and so forth. They show internal structure and are currently pictured as composed of three members, each of a more fundamental group of particles called *quarks*, which we will describe in the next section. The theory based on quark arrangements has been highly successful in explaining the complicated spectrum of the many baryons—their mass, charge, and spin relationships.

There is another family of particles called *mesons*. These interact strongly with the hyperons and, to a first approximation, act as the nuclear glue holding nucleons together. They have integral spins of 0, 1, and so on (which makes them bosons), and are not subject to a conservation law of particles. As long as the appropriate energy and charge is available, mesons can be created or destroyed one at a time. In the quark theory, mesons are composed of a quark and an antiquark. All the particles subject to the strong nuclear interaction, the baryons and the mesons, are called *hadrons*.

Electrons and neutrinos take no part in the strong interactions, but are subject to the weak interaction as well as gravitation and electromagnetism. There are special conservation laws that must be obeyed by some of the four basic interactions, but not by others—although all must obey conservation of energy, momentum, and angular momentum. When a particle decays, the sequence must follow a path obeying all the conservation laws that apply to that interaction.

Consider, for example, the decay of a free neutron, as shown in Fig. 21.26. Each of the three main interactions may be considered to be a possible decay route, as shown in the diagram. The decay will occur through whichever route allows the greatest energy difference and the smallest number of participants. Consider first the strong nuclear interaction. Although the neutron is more massive than the proton, it cannot simply transform into a proton, because that would violate charge conservation. There is not enough mass-energy to produce a proton and a negative meson. There is enough energy to produce a proton and electron, but the process cannot take place through the strong nuclear interaction because the electron is not influenced by that interaction. *If* the strong interaction were possible, the decay would happen in times of the order of 10^{-23} s.

Since the strong interaction is forbidden, perhaps the neutron could decay through the electromagnetic interaction. It is weaker than the strong nuclear interaction by a factor of more than 100 and so the decay would take longer. Although the electron is influenced by electromagnetism, the neutrino is not. Because of the conservation of leptons, the neutron cannot decay simply to a proton and electron. Therefore, the electromagnetic route is blocked.

The remaining route is the weak interaction. (Gravitation is weaker than the strong nuclear by a factor of 10^{39} and can be ignored for individual particle interactions.) The neutron, proton, electron, and neutrino are all subject to the weak interaction, and there is sufficient energy to permit the decay. However, the energy difference between the neutron and the products is small, and the number of final particles is large (three). Furthermore, the weak interaction is smaller than the strong nuclear interaction by a factor of 10^{13}. We should expect that it would take a long time for the participants to be arranged with the necessary combination of momenta so that the decay can take place. Indeed, the half-life is 11 min. As we shall see in the next section, the agent of the weak interaction is the W boson, a particle with a mass of about 80 GeV. Through the uncertainty relationship, this massive particle can be created for a very brief time. The virtual W boson then decays to a proton, electron, and antineutrino.

Quarks and Gluons

During the 1950s and 1960s a large number of new "particles" were found. In these experiments high-energy protons and mesons from accelerators such as the Cosmotron and Alternating Gradient Synchrotron (AGS) were sent through bubble chambers. The collisions of the high-energy particles with the nuclei of the liquid in the chambers yielded V-shaped or forked tracks, such as those shown in Fig. 21.27. The new particles had lifetimes of the order of 10^{-10} s, allowing them to travel measureable distances before decaying. These particles evidently were stable against the strong nuclear and electromagnetic interactions but were decaying through the weak interaction.

More and more particles were also found with even shorter half-lives, whose existence could not be seen in measureable tracks but in the resonancelike appearance of production cross sections as a function of energy. What was observed was not the particle itself, but its decay products. As the bombarding energy was increased, there would be a sudden peak in the production of these specific decays. The narrower the width of the peak in the production curve, the more definite the mass of the new particle, and the longer its lifetime ($\Delta E \Delta t \geq \hbar$). In Fig. 21.28 we show such a production graph that accompanied the announcement of what is now known as the J/ψ particle. In this case, the J/ψ particle was produced in a proton–proton collision, and was detected by its decay into an electron pair. There was a peak in this production when the combined mass of the electron pair was equal to 3.1 GeV. It was soon discovered that there was a

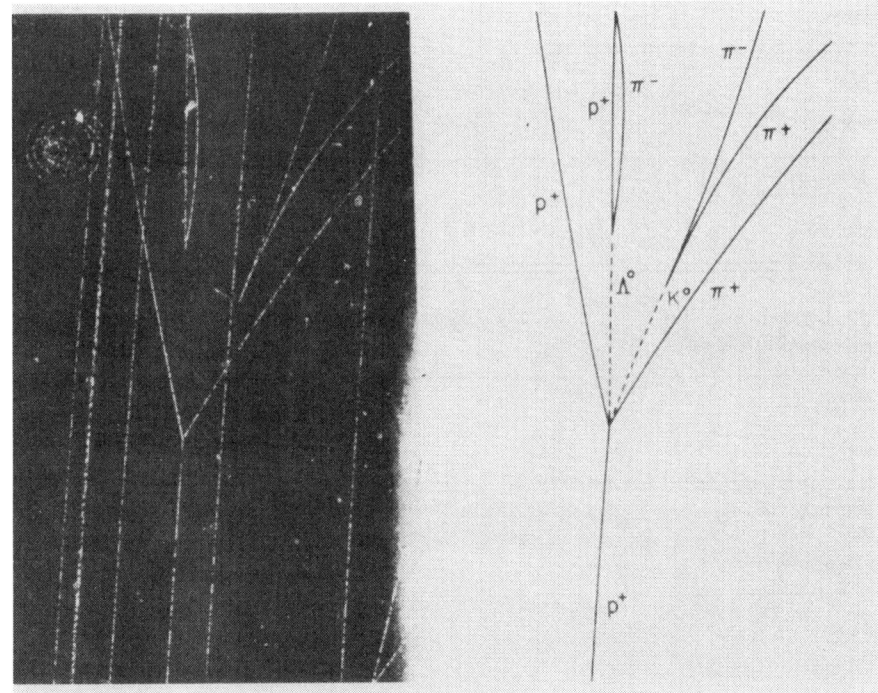

Figure 21.27. This is a bubble chamber picture of tracks of a proton-proton collision. Two neutral particles (Λ_0 and K_0) were produced, flew forward almost at the speed of light, and then decayed. Their decay times of about 10^{-10} s can be measured by measuring the decay distances: $t = x/v$.

resonance at higher energy corresponding to an excited state of the J/ψ particle. The J/ψ particle was also produced in the inverse reaction of electron–positron collisions (from counter-rotating beams) when the combined energy was 3.1 GeV.

It was not clear whether these events represented new particles or resonances of the older ones. The method of observing the enhanced production of an event at a particular energy is known as the "missing mass" or "invariant mass" technique. In the case of the J/ψ particle produced by proton–proton collisions, many particles are produced in the collision. In the $p-p$ experiments one baryon is colliding with another. Out of this must come two baryons, usually accompanied by other hadrons such as mesons. The appearance of two leptons, the electron pair, might have been explained in terms of π^0 decay into two photons and subsequent pair production. The peculiar feature, however, was that a graph of the sum of the energies of the positron and electron showed a sharp maximum at 3.1 GeV. This could happen only if the electron pair were the decay product of a particle with that mass. The lifetime of the particle is too short to be seen in a bubble-chamber picture, hence shorter than about 10^{-12} s, corresponding to a flight path of 0.3 mm. However, the lifetime can be deduced from the width of the resonance production curve. In the case of the J/ψ particle, the width is very narrow—about 5 keV for the reaction, leading to an electron pair. From the uncertainty principle: $\Delta t \approx \hbar/\Delta E = (6 \times 10^{-16} \text{ eV s})/(5 \times 10^3 \text{ eV}) \approx 1 \times 10^{-19}$ s. This is a time longer by a factor of 10^4 than the lifetime of a particle decaying through the strong nuclear reaction.

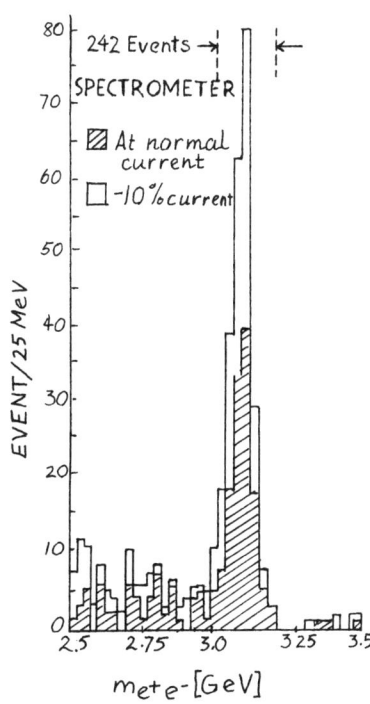

Figure 21.28. Production evidence for the J/ψ particle.

Name	Symbol	Mass	Charge	Spin
up	u	≈5 MeV	$+\frac{2}{3}$	$\frac{1}{2}$
anti-up	$\bar{u}$		$-\frac{2}{3}$	
down	d	≈10 MeV	$-\frac{1}{3}$	$\frac{1}{2}$
anti-down	$\bar{d}$		$+\frac{1}{3}$	
strange	s	≈100 MeV	$-\frac{1}{3}$	$\frac{1}{2}$
anti-strange	$\bar{s}$		$+\frac{1}{3}$	
charm	c	1.5 GeV	$+\frac{2}{3}$	$\frac{1}{2}$
anti-charm	$\bar{c}$		$-\frac{2}{3}$	
bottom (or beauty)	b	4.7 GeV	$-\frac{1}{3}$	$\frac{1}{2}$
anti-bottom	$\bar{b}$		$+\frac{1}{3}$	
top (or truth)	t	174 GeV	$+\frac{2}{3}$	$\frac{1}{2}$
ant-top	$\bar{t}$		$-\frac{2}{3}$	

Figure 21.29. *The quarks.*

The J/ψ particle caused great excitement not just because it appeared to be a new particle, but because it confirmed a prediction of the theory linking the weak and the electromagnetic interactions. The fantastic multiplicity of particles and/or resonances implied that the particles such as protons and neutrons must have structure. There must be more fundamental building blocks. Various schemes were tried, leading to what is now known as the standard model of quarks. In Fig. 21.29 we present an organizational scheme of the quarks, showing their properties and how they combine to form baryons (such as the nucleons) and mesons. In general, the baryons are compounds of three quarks (antibaryons are compounds of three antiquarks). Mesons are compounds (or mixtures of compounds) of a quark and an antiquark. That leaves out the leptons—the electron, muon, τ, and their neutrinos—which are shown in Fig. 21.30. The leptons are apparently without internal structure. There appears to be a parallelism between the organizations and properties of the leptons and the quarks.

In Fig. 21.31 we list the arrangements of quarks for some of the particles that we have already mentioned. The proton and neutron are compounds of the "up" and "down" quarks. In one sense they are merely different states of the same basic particle, called an "isospin (isotopic-spin) doublet." In the proton and neutron the quarks are in a ground state with no angular momentum. Since each quark has spin of 1/2, two of the quarks must be antiparallel, leaving the third to provide spin 1/2 to the nucleon. In such a compound of three particles, there are obviously many other ways to combine, either in the type of quark [up (u) or down (d)] or angular momentum. There are indeed many excited states of the nucleons, a whole spectrum of resonances at different energies. For instance, the first resonance was found by Fermi in an experiment scattering π mesons off

Name	Symbol	Mass	Charge	Spin	Lifetime
electron	e^-		-1		stable
		0.511 MeV		$\frac{1}{2}$	
positron	e^+		$+1$		stable
electron neutrino	ν_e		0		stable ?
		0 ? ($<$50 eV)		$\frac{1}{2}$	
electron anti-$\bar{\nu}$	$\bar{\nu}_e$		0		
muon	μ^-		-1		2×10^{-6} s
		105.6 MeV		$\frac{1}{2}$	
anti-muon	μ^+		$+1$		
muon neutrino	ν_μ		0		stable ?
		0 ? ($<$0.5 MeV)		$\frac{1}{2}$	
muon anti-ν	$\bar{\nu}_\mu$		0		
tau	τ^-		-1		3×10^{-13} s
		1.784 GeV		$\frac{1}{2}$	
anti-tau	τ^+		$+1$		
tau neutrino	ν_τ		0		stable ?
		0 ? ($<$70 MeV)		$\frac{1}{2}$	
tau anti-ν	$\bar{\nu}_\tau$		0		

Figure 21.30. Leptons.

protons. This is what is now called the $\Delta_{(1232)}$ baryon. It is composed of three up quarks with an electric charge of $2e$.

The baryons stable against strong nuclear decay, such as the Λ^0, Σ, and Ω were assigned "strangeness" numbers, S, to indicate another parameter that must be conserved in strong interactions, but which can be violated in the weak. We now attribute the strangeness to an s quark, so that, for instance, the Λ^0 is a compound consisting of uds. Yet another attribute associated with a conservation law was called "charm," with quark c. The famous J/ψ particle is a meson compound of $c\bar{c}$. As we have already seen, the J/ψ particle is stable against strong nuclear decay and so lives long enough to have a compound existence analogous to that of the positron–electron "atom." This latter is called positronium and so the $c\bar{c}$ is called charmonium. Each has a spectrum of excited states, paralleling the familiar hydrogen spectrum. Of course, with the hydrogen atom or with positronium the energy levels are separated by a few electron volts, while with charmonium the separations are in the 100-MeV range. There is also a difference between the two spectra because in positronium the electrons exist in a r^{-1} potential well. The relative spacing of levels in charmonium is different, corresponding to a linear potential and a constant force, independent of separation distance.

There is a similar meson, called upsilon (Y), requiring creation energy of over 10 GeV. It consists of two bottom (or beauty) quarks, $b\bar{b}$, which can also exist in excited states like charmonium.

Besides the quarks and leptons, there are particles that are agents of the interactions. The properties of these are listed in Fig. 21.32. The best known is the photon, γ, the agent of the electromagnetic force. The weak interaction is caused

Name	Symbol	Charge	Spin	Quark Arrangement
Pi-zero	π^0	0	0	$u\bar{u}$ or $d\bar{d}$
Pi-plus	π^+	+1	0	$u\bar{d}$
Pi-minus	π^-	−1	0	$d\bar{u}$
K-zero	K^0	0	0	$d\bar{s}$
K-plus	K^+	+1	0	$u\bar{s}$
K-minus	K^-	−1	0	$s\bar{u}$
J/Psi	J/ψ	0	1	$c\bar{c}$
D-zero	D^0	0	0	$c\bar{u}$
D-plus	D^+	+1	0	$c\bar{d}$
Upsilon	Y	0	1	$b\bar{b}$
Proton	p	+1	$\frac{1}{2}$	uud
Anti-proton	$\bar{p}$	−1	$\frac{1}{2}$	$\bar{u}\bar{u}\bar{d}$
Neutron	n	0	$\frac{1}{2}$	ddu
Anti-neutron	$\bar{n}$	0	$\frac{1}{2}$	$\bar{d}\bar{d}\bar{u}$
Lambda	Λ	0	$\frac{1}{2}$	uds
Anti-lambda	$\bar{\Lambda}$	0	$\frac{1}{2}$	$\bar{u}\bar{d}\bar{s}$
Sigma-plus	Σ^+	+1	$\frac{1}{2}$	uus
Sigma-minus	Σ^-	−1	$\frac{1}{2}$	dds
Sigma-zero	Σ^0	0	$\frac{1}{2}$	uds
Xi-minus	Ξ^-	−1	$\frac{1}{2}$	dss
Xi-zero	Ξ^0	0	$\frac{1}{2}$	uss
Omega-minus	Ω^-	−1	$\frac{3}{2}$	sss

Figure 21.31. Quark compounds.

by the exchange of one of three quanta, W^+, W^-, and Z^0. In the unification of electromagnetism and the weak interaction, all four of these quanta are considered related, even though the photon has zero mass and the W and Z bosons (sometimes called the intermediate bosons) have mass in the 80-GeV/c^2 range. The agents of the strong nuclear interaction are the gluons with 0 mass and spin of 1. Theory calls for the existence of quanta of the gravitational field, which should also have 0 mass but spin 2.

Name	Symbol	Mass	Charge	Spin	Lifetime	Interaction
Photon	γ	0	0	1	stable	electromagnetic
W-plus W W-minus	W^+ W^-	83 GeV	$+1$ -1	1	10^{-25} s	weak
Z	Z	93 GeV	0	1	10^{-25} s	weak
Gluon	g	0 ?	0	1	stable	strong
Gravitation (not yet observed)		0	0	2	stable	gravity

Figure 21.32. The "agents" of interactions—gauge bosons.

Questions Raised by the Quark Model

1. The model seems to be a complete and consistent organizational scheme to explain the multiplicity of observed particles and their excited states. Can quarks be detected?

   The atomic nucleus was detected by Rutherford in a scattering experiment. α particles passing through thin foils were occasionally scattered at large angles. This could happen only if the mass of the atom were concentrated in a small region rather than being spread over the whole volume of the atom. In similar experiments, when nucleons are bombarded with electrons and neutrinos in the many-GeV region, the scattering matches predictions based on the quark (sometimes called the parton) model. In particular, the electron or neutrino interacts with only one of the quarks in a nearly elastic collision. Because the quarks are not static in the nucleon, the deflected electron has a spread of energy at a given angle. This distribution yields information about the quark dynamics within the nucleon. The recoiling quark converts the energy it got from the collision into two oppositely directed jets of hadrons. The angular spread of these jets also provides information about the original quark arrangement.

2. When atomic nuclei are struck by high-energy particles, the constituent protons and neutrons can be knocked out and examined. Why aren't the quarks knocked out of the nucleons?

   It is indeed the case that isolated quarks have never been detected, but the model explains why. In the binding of nucleons to each other, the force is short range. The energy needed to release a proton or neutron from an average nucleus is of the order of 7 MeV. The electromagnetic binding of an electron in an atom is longer range, but the escape energy is still finite—only a few electron volts for the valence electrons and some keV for electrons in inner shells of large atoms. In contrast, the attractive force between quarks becomes constant as the separation distance increases. Consequently, complete separation of quarks would require infinite energy. Instead, when a quark receives a large amount of collision energy and starts to separate from its neighbor, like two bobs connected by a spring, the stretching energy can suddenly convert into another quark pair. With enough energy for multiple production of these extra pairs, new hadrons are created which are emitted in opposite directions of the stretched quark axis.

3. In the chart, gluons are listed as the agents of the force between quarks. Are gluons detectable?

The properties assigned to the gluons provide a consistent part of the standard model. They must be massless bosons with no electric charge and spin 1. They are not subject to the electromagnetic or weak interaction and interact only with quarks *or themselves*. (Photons do not interact with each other.) Because they are continually being interchanged between the quarks, they must contribute to the internal energy of baryons and mesons, and scattering experiments confirm this.

4. The quarks are characterized in terms of "flavors," and are assigned an extra attribute called "color." Why are these categories necessary?

The words themselves have no relationship to their everyday meaning. The term "flavors" refers to the six forms of quarks (up, down, strange, charmed, top, and bottom) and six forms of leptons (electron, muon, τ, and their associated neutrinos).

It was necessary to add another attribute to the quarks, besides electric charge (which is $+2/3$ or $-1/3$), mass (which we will discuss next), and spin. Because the spin is $1/2$ the quarks are fermions, subject to the Pauli exclusion principle. As we have seen in the case of the Δ however, the quarks can combine as uuu in the ground state. Only two identical fermions can occupy the same energy level, and then only if their spins are opposed. If three up quarks are bound together, they must have some additional attribute to make them nonidentical. The theory calls for such an attribute to exist in three forms and three antiparticle forms. To emphasize certain combinatory properties, the attribute was named color, with forms red, green, and blue. As for the combinatorial requirements, any free particles must be colorless. The proton, for instance, can contain a red and a green up quark and a blue down quark (thus becoming white), but not two red up quarks. A π meson can contain any color u as long as the $\bar{d}$ has the same anticolor.

The color is more than an organizational tag, however; it is the "charge" of the hadronic interaction, mediated by gluons. The theory of this strong interaction between quarks is analogous to quantum electrodynamics (QED) with its photon agent. Therefore the theory is called quantum chromodynamics (QCD). The gluons are the agents of this interaction, but both quarks and gluons possess "color charge." In fact, the gluons are bicolored, made up of a color and anticolor. In the π^+ meson, for example, a red up quark can emit a red–antiblue gluon, becoming blue up. (The red leaves; an antiblue leaving is the same as a blue arriving.) Its antired, antidown quark partner receives the gluon and becomes antiblue. (The red arrives, canceling the antired; the arriving antiblue remains.)

Notice how the organizational scheme is consistent in specifying that quarks and gluons cannot be isolated, by asserting that isolated particles must be color-free. Since each quark has color, an isolated quark would have color and so would a gluon. Color-free combinations of quarks must consist of a quark–antiquark pair (containing color–anticolor pair), or a trio of quarks or antiquarks, each containing one of three different colors.

5. How do you measure the mass of a quark?

   The masses of the quarks are not known precisely because they interact so strongly with the gluon and virtual quark pairs that surround them that independent mass has little meaning. In the case of a proton and neutron forming a deuteron, the product mass is less than the sum of the components because the binding energy has been emitted in the form of a γ ray. In the case of three quarks forming a nucleon, however, the binding energy is essentially infinite. The mass of the nucleon is the sum of the masses of the three quarks plus the mass-energy of the gluons and the bound kinetic energy of the whole assembly. The mass assignments in Fig. 21.29 are based on specific theories about what is going on inside the hadrons, and on experimental thresholds for the production of the higher mass quarks, such as c and $\bar{c}$ in the J/ψ particle.

6. Besides the success of the organizational scheme, the production of jets in high-energy collisions, and the "parton" type of e–nucleon and neutrino–nucleon scattering (paralleling Rutherford scattering), is there other experimental evidence of quarks and gluons?

   In very-high-energy e^+e^- collisions, the particles can turn into $\mu^+\mu^-$ pairs or can produce quark–antiquark pairs. The quark pairs become manifest in the form of hadrons. Well beyond the mass threshold for production of these, there should be equal probability of producing either muons or quarks modulated by the ratio of the square of electric charges of the products. (The interaction is electromagnetic.) Thus the ratio of hadrons to muon pairs should depend on (a) the number of flavors of quarks that can be produced at that energy (only 2 if only up and down can be produced, or 3 if there is also high enough energy for s, etc.), (b) the square of electric charge on each quark (1/3 or 2/3), and (c) the number of colors (additional quantum numbers) of each flavor of quark. The experimental results are consistent with the assignments of fractional electric charge and three color types for each flavor.

7. What happened to the idea that mesons were the nuclear glue?

   The nuclear strong force is provided by the exchange of gluons. In the continual emission and absorption of gluons and quarks, virtual quark pairs can be produced by nucleons and exchanged with other nucleons. A proton, for instance, can produce a virtual π^+ which will be absorbed by an adjacent neutron. Thus the proton and neutron exchange places. This produces the binding characteristic of the nucleus which is less than that binding the quarks in the nucleons. The situation is analogous to the van der Waals binding of molecules, which is electromagnetic but weaker than the direct Coulomb force binding electrons to the nucleus.

1. Liu, Hua-Xiang, "The Correspondence Principle and the Founding of the Atomic Quantum Theory," Phys. Teach. **33**, 396 (1993).

■ EPILOGUE

In Fig. 21.33 we present a logarithmic map of all the realms in the universe. One of the exciting features of our human condition is that we have learned how to explore all these realms using only a few principles and laws. There are four basic interactions; there are fewer than a dozen conservation laws. As we have seen, the motions of balls and pendulums turn out to be models for the motions of atoms and galaxies.

In this final chapter we have explored worlds within worlds. The chair you sit on, the water you drink, the air you breathe, and you yourself are all made of molecular units composed of groups of atoms. These in turn consist mostly of empty space populated by electrons and positive nuclei. The nucleons also have structure, acting like tiny molecules made up of combinations of quarks. However, we are obviously more than a simple collection of point objects. Organization is as real an entity as the particles. The ordered groupings are equivalent to information, and, in certain complex molecules, the organization has led to self-awareness and intelligence.

Although we have learned to probe the microworld and map the universe, we still do not know how or why the whole system began. We do not know why humans on this planet have developed the intelligence to learn these things. We do not know whether we are the only creatures in the universe who measure and try to comprehend it. Whether we are alone or not, and whether there is some great drama of creation, it seems peculiarly appropriate for humans to explore and understand their universe. Perhaps it is our proper role. But even if there is no drama, and even if we have no part, still the adventure is fun and noble in itself.

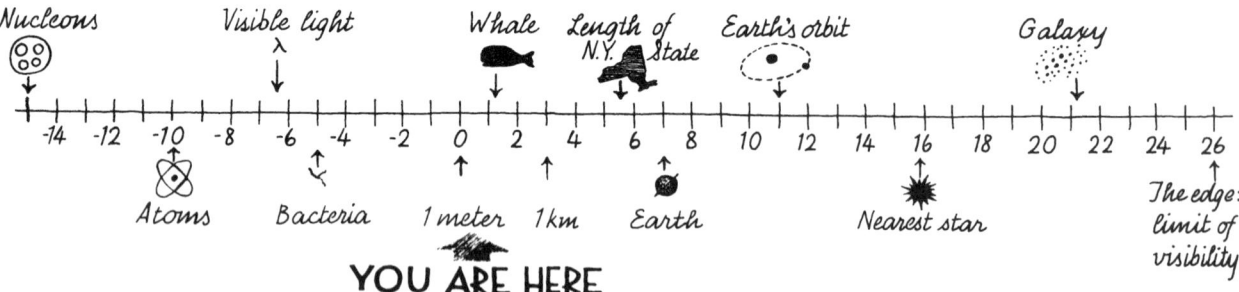

Figure 21.33. Map of our universe.

Index

A
2.7 K E-M radiation, 157
Absolute and percentage errors, 59
Absolute temperature, 226, 271
Action-reaction, 159
Adhesion, 289
Adiabatic process, 236
Acceleration due to gravity, 104
Accuracy and precision, 66
Alnico magnet, 446
Alpha decay, 535
Alternating current, 469
Ammeter, 420
Ampere, definition, 404, 428
Ampere's law, 430, 478
AM radio, 495
Angular momentum:
   bicycle wheel, 178, 179, 180, 186
   cat, falling, 182
   conservation, 181
   demonstrations, 178, 185, 186
   direction of ang. mom., 186
   earth, 180, 188
   electron in Bohr orbit, 180
   electron spin, 506
   monent of inertia, 174, 177
   moment of inertia tensor, 194
   photon spin, 189
   physical pendulum, 190
   planet in elliptical orbit, 181
   precession, 186, 188
   principal axes, 195
   quantization, 188
   radius of gyration, 190
   radius of percussion, 191
   solar system, 183
   spin and axis dependent, 179
   star formation, 183
   $\tau = dL/dt \approx d(I\omega)/dt$, 180
   twirling skater, 182
   wobbly rotations, 193
Antennas, 495–498
Anti-particles, 505, 539
Archimedes' principle, 283
Atmosphere:
   grav binding energy, 112
   pressure f(height), 243
Atomic model:
   gas, 237
   liquids and surface, 287, 289
   nucleus (see Microstructure)
   size, 237, 518
   spectra, 519
   structure, 516
Atwood's machine, 170
Audio sound, table, 368, 370
Aurora, 437
Avogadro's number, 237

B
Balmer series, 519, 520
Band spectra, 524
Barkhausen effect, 445
Barometer, 283
Baryons, 539
Baseball bat, 192
Batteries, 417
Beats, 351
Bernoulli's equation, 292
Beta decay, 535
Bicycle wheel:
   demonstration, 178
   direction of L, 186
   gears, 208
   magnitude of L, 180
   moment of inertia, 179
   precession, 187
Binding energy:
   atmosphere, 244
   covalent, 523
   escape from Earth, 211
   molecular, 212
   nuclear, 532
   photoelectric, 529
Biot–Savart law, 431
Black body radiation, 527
Blue sky, 492
Bohr angular momentum, 180
Bohr atom, 516
Boiling temperature table, 227
Boltzmann's constant, 235, 240, 274
Bose–Einstein statistics, 509
Brownian motion, 246
BTU, 230
Bubble chamber tracks, 164, 505, 541
Bulk modulus, 282, 326
Buoyancy, 279, 284

C
Calorie, 230
Capacitance, electrical, 386, 398
Capacitive reactance, 319, 471
Capacitors:
   charge decay, 452
   construction, 400
   energy storage, 400
Capacitors, practical, 400
Capillary, 289
Capstan, 208
Carnot cycle, (see Heat engines)
Cat, falling, 182
Cavendish, 102, 412
Celsius–Fahrenheit, 224
Center of mass:
   collision, 123
   definition, 168
Center of weight, 168
Centrifugal force, 131, 216
Centripetal acceleration, 109, 174, 306
Charge flow in wire, 405
Charge of electron, 401, 504
Charge surface density: $E = \sigma/\epsilon$, 396
   surface curvature, 401
Chords, 379
Circular dynamics, 173
Circular polarization, photons, 189
Class discussions:
   good topics, 6
   reasons, 5
   technique, 8
Coefficient of friction, 87, 88, 90, 91
Coefficient of thermal expansion, 250
Coherence, 362, 367
Cohesion, 286
Collisions:
   elastic, 197
   damage done, 203
   equal mass particles, 199
   space station, 122
Components, vectors, 79
Complex plane vectors, 317
Conduction, electrical, 401–416, 526

Conduction, thermal, 258
Conservation:
   angular momentum, 181
   energy in collisions, 197
   energy as heat, 230
   energy in dynamics problems, 217
   momentum in collisions, 162, 197
   particle decays, 539
Continuous spectra, 527
Convection, thermal, 257, 262
Coriolis force:
   space colony, 134
   washbowls, 133
   on Earth, 132
Coulomb's law, 385
Coulomb, unit, 385
Covalent binding, 523
Crooke's radiometer, 490
Crystal model radio, 496
Curie plot, magnetism, 443
Currents and fields that change:
   alternating current, 469
   capacitor charge decay, 452
   conductor motion in B, 454
   eddy currents, 461
   electric generators, 464
   electric motors, 462
   emf, 457, 460
   energy in capacitor, 400
   energy in inductor, 467
   impedance, 319, 472, 474
   inductance, 464
   induction motor, 463
   interdependence of E and B, 459
   Faraday, 451
   Faraday's induction law, 457
   force on moving charge, 454
   Lenz's law, induction, 455
   Lorentz equation, 460
   LRC circuits, 313, 468
   LR circuits, 466
   nonconservative potential, 460
   Oersted, 451
   peak voltage, 470
   phasor representation, ac, 319, 472
   RC time constant, 453
   reactance, 319, 471
   three phase, 470
   transformers, 473
Cyclotron, 435

D

Damped oscillations, 313
Degrees of freedom, 253

Demonstrations:
   3-D illustrations, 13
   coat hanger chime, 342
   dc and ac, 404
   electrostatics, 382
   electromagnetic radiation, 477
   failure, 9
   fluids, 278
   helium change of pitch, 374
   Poisson spot, 360
   reasons, 8
   sound diffraction, 360
   sources, 14
   technique, 10
Density:
   conduction electrons, 408
   gas, 238
   table for solids, 238
Deuterium, 532, 533
Diamagnetism, 442
Dielectric, 400
Diffraction:
   electron, 512
   pinhole, 360
   Poisson-Arago spot, 360
   Rayleigh's criterion, 36
   resolving power, 361
   single slit, 357
Dimensional analysis, 72, 323
Dimensional constraints of graphs, 78
Dimensionless numbers:
   fine structure constant, 77
   Reynolds number, 74
   "Specific" quantities, 74, 231
Diode, 496
Dipole:
   electron magnetic moment, 440, 506
   energy in B, 439
   field, 390, 398
   magnetic moment, 439
   strength, electric, 390
Dispersion:
   light, 364
   phase–group speed, 365
   rainbow, 338
   water ripples, 329
Distribution of molecular energy, 241
Doppler shift, 124
Dot product, 201
Double slit interference, 362
Drift speed, electrons, 408
Driven oscillations, 316
Drude, 405
Dry cell, 417
Duality, 512, 529

DuLong and Petit, 232

E

$e^{i\theta}$ representation, 307, 350
E and V, source geometry, 398
Ear, 369
Earth:
   angular momentum, 180
   magnetic field, 431
   potential well, 211, 244
   precession, 188
Eddy currents, 461
Effective voltage in ac, 470
Efficiency of heat engine, 267, 270, 272
Einstein:
   Brownian motion, 246
   gravitation and general theory, 153
   inertial–gravitational mass, 107
   photoelectric effect, 529
   special theory of relativity, 135
Elastic collisions, 197, 199
Electric currents:
   ammeters and voltmeters, 420
   ampere, definition, 404
   amps and watts in familiar circuits, 404
   batteries, car and flashlight, 417
   bridge circuits, 423
   conduction, 405–416, 526
   demonstrations, dc and ac, 404
   drift speed of charges, 408
   electric field in wire, 406
   galvanometer, 420
   ground, 423
   guide charges on wire, 407
   insulators, 526
   internal resistance, 419
   I(V) for circuit elements, 410
   Kirchhoff's rules, 424
   mobility of electrons, 410
   model of charge flow, 405
   null reading devices, 423
   Ohm's law, 412
   potentiometer, 423
   power, 405, 420
   problems with simple model, 409
   resistance mental models, 413
   resistivity table, 413
   resistivity as f(T), 415
   resistors in parallel, 418
   semiconductors, 526
   superconductivity, 415
   thermal speed of electrons, 408
   use of common terms, 405
   voltaic cells, 416

voltage divider, 422
watts, definition, 405
Wheatstone bridge, 423, 425
Electric batteries, 416
Electric generators, 464
Electric motors, 462
Electromagnetic radiation:
  accelerated charge radiation, 484
  AM radio, 495
  black body radiation, 527
  blue sky scattering, 492
  Crooke's radiometer, 490
  crystal model radio, 496
  demonstrations, 477
  derivation of radiation, 481
  ∇ operator derivation, 483
  E-M spectrum chart, 494
  energy density radiated, 486
  growlers and whistlers, 494
  Hertz generation of E-M, 493
  infrared, 500
  localization of E-M energy, 488
  Maxwell's equations, 478, 479
  momentum carried by waves, 489
  polarization of E-M waves, 491
  Poynting vector, 487
  pressure, 490
  propagation in media, 364
  radar and microwaves, 498
  stream from hose analogy, 485
  thermal, 259
  TV and FM, 497
  visible light region, 501
  ultraviolet and x rays, 501
  x rays and γ rays, 502
Electromotive force (emf), 457, 460
Electron:
  arrangements in low Z atoms, 510
  Bohr orbit angular momentum, 180
  charge, 401, 504
  magnetic moment, 440, 506
  mass, 504
  pair production, 505
  Pauli's exclusion principle, 440, 509
  production, 503
  size, 511
  spin, 506
  wave-particle duality, 512
Electron volt, eV, 212
Electrostatics:
  background, 381
  capacitance, 386, 398
  capacitor energy storage, 400
  capacitors, practical, 400
  charge density, curvature, 401

charge on balloon, 386
coulomb, size, 385
Coulomb's law, 385
demonstrations, 381
dielectric, microstructure, 400
dipole, 390
E and surface charge, 396, 407
E and V for charged cylinder, 395
E and V for parallel plates, 396
E and V for point charges, 389
E and V for two charges, 390
E and V summary geometry, 398
faraday, charge on mole, 418
Faraday cage, 394
Faraday's field line model, 391
field, 387, 388
Gauss's law, 393
industrial uses, 384
lightning, 383
Millikan oil drop, 401
permittivity, 385
potential, 387
relationship between E and V, 389
Van de Graaf generator, 384
Ellipsoidal mirror, 336
Elliptical orbits:
  actual planetary orbits, 115
  planetary orbits, 181
  precession of Mercury, 154
Energy:
  1771 Encyclopedia Brittanica, 196
  binding, 211, 212, 244, 532
  capacitor storage, 400
  density, magnetic, 467
  first law Thermo, 230
  flywheel storage, 205
  inductor storage, 467
  joule unit, 197
  kinetic energy, 197
  mass equivalence, 149, 151
  potential energy, 209 (see Pot. E.)
  problem of definition, 196, 201
  reference frame, 122
  rotational kinetic, 204
  surface free energy, 287
  vis viva, 197, 202
  wave transmission, 339
Energy units:
  British Thermal Unit, BTU, 230
  calorie, 230
  electron volt, eV, 212
  joule, 197
  kilowatt-hour (kWh), 217, 404, 417
Entropy:
  arrangement probability, 273

coin heads distribution, 274
equivalence of definitions, 276
$\Delta S = Q/T$ definition, 275
$S = k \ln W$ definition, 275
Equal mass collisions, 199
Equal-tempered scale, 380
Equilibrium, thermal, 229
Equipartition of energy, 253, 528
Error analysis:
  absolute and percentage, 59
  accuracy and precision, 66
  background, 53
  exponential notation, 54
  Fermi questions, 56
  graphical representation, 65
  propagation, 59, 63
  significant figures, 57
  standard deviation, 63, 68
  standard deviation of means, 68
  statistical, 62
Escape velocity, 111, 244
Euler's $e^{i\theta}$, 307, 350
Evaluation:
  background, 32
  cheating, 46
  difficulty, 46
  frequency, 43
  grading, 47
  reasons, 33, 50
  test construction, 36
  types, 34
Exponential notation, 54
Extremum of paths, 335

F

$F = dp/dt$, 159
Fahrenheit–Celsius, 224
farad, 386, 398
Faraday cage, 394
Faraday induction, 451, 457, 478
Faraday's field lines, 391
Faraday's homopolar generator, 464
Feedback, 321
Fermat's principle, 335
Fermi-Dirac statistics, 509
Fermi questions, 55
Fermions, 509
Ferromagnetism, 441, 443
Feynman's phase construction, 355
Field, electric:
  cylinder, 395, 398
  dipole, 390, 398
  Faraday's model, 397
  Gauss's law, 393

inside sphere, 398
inside conducting wire, 407
parallel plates, 396
point charge, 389, 398
sphere, 398
surface charges, 396, 407
Field, magnetic:
dipole, 427, 434
loop, 433
qualitative sketches, 429
range of strength, 431
straight wire, 430
Filament, I(V), 411
Fine structure constant, 77
First law of thermodynamics, 230
First law, Newton's dynamics, 156
Fission, 534
Fluids:
adhesion, 289
Archimedes' principle, 283
atomic model, surface, 287, 289
barometer, 283
Bernoulli's equation, 292
bulk modulus, 282
buoyancy, 279, 284
capillary, 289
cohesion, 286
curve ball, 296
demonstrations, 279
flow in pipe, 297, 300
Galileo and Archimedes, 285
hydrometer, 285
hydrostatic paradox, 282
lift due to flow, 296
meniscus, 289
negative pressure, 291
Pascal's vases, 282
Pitot tube, 293
Poise, 298
Poiseuilles's equation, 300, 303
pressure, 281, 297
Reynolds' number, 302
siphon, 280, 283
soap bubbles, 280, 288
solid fluids, 300
stream lines, 291
super fluid–helium, 227, 300
surface free energy, 287
surface tension, 279, 286
turbulence, 97, 296, 302
vena contracta, 294
Flux, electric, 393
Flux, magnetic, 455
FM, 497
Fundamental interactions, 536, 545

Focusing particle beams, 436
Force:
as interaction, 536
B on I and q(v) (Lorentz), 434, 435
buoyant, 284
centrifugal, 131, 216
centripetal, 174
Coriolis, 132
definition, 158
exerted by table, 159
gravitational, 102
Hooke, 158, 307
restoring in potential well, 213
inside homogeneous sphere, 108
molecular, 214
normal, 88
shear, 90, 298, 300
units, 161
van der Waals, 523
Fourier series, 353
Franklin, Benjamin, 382, 385
Fraunhofer absorption lines, 520
Fraunhofer-Fresnel diffraction, 357
Free energy, surface, 352
Frequencies:
E-M wave, 494
oscillation, chpt 13
sound, 369, 370
Friction:
coefficient, 87, 88, 90, 91
Coulomb's theory, 87
fluid, 96
inclined plane method, 88
Reynolds number, 97
rolling, 93
sliding dry, 88
sliding, ice, 93
static coefficient, 91
stick-slip, 91
Stoke's law, 98
surface physics, 89
Teflon, 93
terminal velocity, 98
velocity dependent, 98, 171
Fundamental frequency, 353
Fusion, heat of, 232, 233
Fusion, nuclear, 533

G

$\gamma = c_P/c_V$:
adiabatic expansion, 236
speed of sound in gas, 326
γ rays, 501
Galilean transformation, 118

Galvanometer, 420
Gas:
average molecular energy, 240
Brownian motion, 246
constant, 235
E and v distribution, 241
helium leakage, 245
kinetic model derivation, 239
law, ideal, 235
molecular model, 238
speed of sound, 245
thermometer, 223, 229
van der Waals's equation, 248
gauss, unit of B, 430
Gaussian pulse, 329
Gaussian surface, 393
Gauss's law, 108, 393, 478
Gears, 207
General theory of relativity, 153
Generators, electric, 464
Geometrical functional dependence, 86
Geosynchronous satellite, 110
Gluons, 541
Grades (see Evaluation)
Graphs:
dimensional constraints, 78
error bars, 65
Gravitation:
Cavendish balance, 102
elliptical orbits, 115, 181
escape velocity, 111
Gauss's law, 108
g inside sphere, 108
general theory of relativity, 153
gravitational field strength, 104
hole through Earth, 108
hovering satellite, 110
implications of inverse square, 103
loss of atmosphere, 112, 244
mass is energy is mass, 108, 148
measurements of G, 102
nature of mass, 106
Newton's gravitation law, 101
orbits, circular, 109
orbits, elliptical, 115, 181
potential energy, 110
precession, 114
problems of weight definition, 104
tides, 112, 185
unit of mass, 107
universal constant, 102
variation of g, 105
weightlessness, 105
Ground, electric, 423
Group and phase velocity, 365

Growlers and whistlers, 494

H
Hall effect, 437
Harmonics–overtones, 354, 377
Harmony, 377
Heat:
  atomic model for specific heat, 252
  capacities, 231
  c_p and c_V, 235
  DuLong and Petit, 232
  expansion of solids, 250
  first law, 230
  heat, average energy, T, 262
  molar specific heats, 234
  of fusion, 232, 233
  of vaporization, 232, 233
  regularities in specific heats, 232
  specific heat of hydrogen, 234
  specific heat of water, 231
  specific heats, table, 232
Heat engines:
  actual efficiencies, 272
  adiabatic expansion, 266
  background, 264
  Carnot cycle, 267, 268
  Carnot, maximum efficiency, 270
  efficiency of reversible cycle, 267, 268
  isothermal expansion, 265
  proof that $Q_1/Q_2 = T_1/T_2$, 269
  refrigerator, 267
  refrigerator quality factor, 272
  reversibility, 264
  temperature definition, 270
  thermal pollution, 271
Heat transmission:
  definition of R, 258
  electromagnetic radiation, 260
  Newton's law of cooling, 257
  R values, 260
  thermal conductivity, table, 258
Heisenberg uncertainty principle, 514
Helium:
  change of pitch, 374
  Helium–Neon laser, 362, 367
  spectra, 521
  superfluid, 300
henry, unit of inductance, 465
Henry, Joseph, 337, 465
Hero of Alexandria, 335
hertz, frequency unit, 173, 351, 369, 370, 494
Hertz, generation of E-M, 493
Hole through Earth, 108

Holograms, 367
Hooke's law:
  simple harmonic motion, 307
  in defining force, 158
Horns–bells, mouthpieces, 376
Horsepower, 217
Hull speed, 84
Huygens's construction, 354
Hydrogen:
  isotopes, 532
  specific heat, 234
  spectra, 519
  structure, 517
Hydrometer, 285
Hydraulic lever, 207
Hydrostatic paradox, 282
Hysteresis, 446

I
Ice skating friction, 93
Ideal gas law, 235
Impedance:
  ac circuits, 319, 345, 472, 474
  coat hanger chime, 342
  definition, 340, 344
  E-M waves, 343
  matching in ear, 343, 371
  musical instruments, 343
  reflection coefficients, 346
  sound in vacuum, 343
  transmission between media, 345
Impulse, 164
Inclined plane, 210
Inductance, electrical, 464, 467
Induction, electrical, 451, 457, 478
Induction motor, 463
Inductive reactance, 319, 471
Inductor, 465, 467
Inertial fields:
  accelerated reference frame, 128
  elevator, 130
Inertial–gravitational mass, 107
Inertial reference frame, 128, 157
Infrared, 500
Insulation, thermal, 257
Insulators, electrical, 526
Intensity:
  E-M radiation (irradiance), 487
  sound, 341
Interactions, fundamental, 536, 545
Interference of waves, 357, 361
Interferometer, Michelson's, 136, 362, 530

Internal energy:
  atomic model of gases, 253
  atomic model of solids, 249, 255
  degrees of freedom, 253
  equipartition of energy, 253, 528
  first law of thermo, 230
  Maxwell–Boltzmann distribution, 241
  rotation energy, 254
  van der Waals's equation, 248
  vibration energy, 255
  water, 256
Internal resistance, 419
Invariant interval, 143
Inverse square law, 103
Ionic binding, 523
Iron, electron arrangement, 443
Irradiance, 487
Isothermal process, 236
Isotopes, 532

J
Joule, James, 197, 221
joule, unit, 197

K
Kelvin temperatures, 226, 271
kelvin, unit, 226
Kepler, 182
Kilocalorie (diet calorie), 230
Kilogram, 107, 161
Kilowatt-hour, 217, 404, 417
Kinematic viscosity, 299
Kinetic energy:
  rotational, 204
  translational, 197
Kinetic theory of gases, 239
Kirchhoff's rules, 424

L
Laboratory:
  background, 17
  evaluation, 20
  homely, 24, 30
  open-end, 18
  projects, 27
  report, 23
  scheduling, 19
Laser, 362, 367
Lead storage battery, 418
Least action:
  Lagrangian, 218
  variational principles, 220
Lenz's law, induction, 455

Leptons, 539, 543
Levers, 206
Lift, fluid, 295
Light:
   diffraction, 357
   dispersion, 364
   E-M relativity → speed of light, 449
   extremum paths, 335
   holograms, 367
   interference, 361
   invariance of speed, 135
   Maxwell's derivation of c, 482
   polarization, 189, 491
   rainbow, 338
   Rayleigh's resolution criterion, 361
   refraction and reflection, 355
   speed of light, 135, 482
   visible range, 501
   wave speed relation, 327
Lightning, 383, 394
Liquids: (see Fluids)
   atomic model, 251
Local fields, 128
Longitudinal wave, 325
Lorentz force equation, 460, 478
Lorentz transformation:
   compared with Galilean, 135
   derivation, 141
   E-M fields, 153
   invariance of proper time, 143
   mass-energy, 152
   momentum and energy, 148
LR circuit, 466
LRC oscillations, 313, 468

M

Mach, Ernst, 129
Machines, simple, 205
Magnetic dipole, 434, 438
Magnetic moment electron, 440
Magnetism:
   alnico, 446
   Ampere's law, 430
   auroras, 437
   B at center of loop, 433
   B from a line current, 430
   B defined, $N/A \cdot m$, 428
   B field lines, sketches, 429
   Barkhausen effect, 445
   B in solenoid, 465
   Biot–Savart law, 431
   Curie plot, 443
   cyclotron, 435
   diamagnetism, 442
   dipole field, 426
   dipole, magnetic, 434, 438
   dipole or current sources, 426
   domains, 445
   Earth's field strength, 431
   electron magnetic moment, 440
   ferromagnetism, 441, 443
   flux, 455
   force between parallel currents, 427
   force on current in B, 434
   force on moving q in B, 435
   gauss, defined, 430
   Hall effect, 437
   hysteresis, 446
   in materials, 440
   magnetic dipole moment, 434, 440
   magnetic field strengths table, 431
   monopole, 426
   mu-metal, 446
   Oersted, B from I, 451
   paramagnetism, 441
   permeability, 327, 428
   potential energy of dipole, 439
   relativity and E and B, 447
   solenoid, 465, 467
   susceptibilities table, 442
   tesla, defined, 430
   torque on loop in B, 438
Mass:
   "charge" of gravitational field, 106
   electron, 504
   identity of gravitational and inertial, 107
   mass *is* energy *is* mass, 149, 151
   quantity of matter, 160, 533
   units, 107, 161
Maxwell-Boltzmann distribution, 241
Maxwell's equations, 478, 479
Mechanical advantage, 206, 208
Melting temperatures, table, 227
Meniscus, 289
Mercury thermometer, 222, 226
Mesons, 539
Michelson interferometer:
   independence of light speed, 136
   interference pattern, 362
   duality of light problem, 530
Microstructure:
   agents of interactions, 545
   alpha decay, 535
   anti-particles, 505, 539
   atomic radius, 517
   atomic structure, 516
   baryons, 539
   beta decay, 535
   Bohr, 516
   Bose-Einstein statistics, 509
   conductivity in solids, 526
   duality problem of light, 529
   electron (see Electron)
   electron atomic arrangements, 510
   equipartition of energy, 528
   Fermi-Dirac statistics, 509
   fission, 534
   fundamental interactions, 536, 545
   fusion, 533
   gluons, 541
   Heisenberg's uncertainty principle, 514
   helium spectra, 521
   hydrogen atom systematics, 519
   isotopes, 532
   leptons, 539, (chart, 543)
   mesons, 539
   Michelson interferometer, 530
   molecules, 523
   muon and time dilation, 140
   neutrino, 535, 539
   neutron, 532, 535
   neutron decay, 539
   nuclear binding energy, 533
   nuclear radius, 517
   nuclear structure, 532
   particle chart, 538
   periodic table of elements, 522
   Planck's quantum proposal, 528
   proton, 532, 535
   quark compounds chart, 544
   quarks, 541, (chart, 542)
   questions raised by quarks, 545
   radioactive sequences, 536
   radioactivity, 534
   Rutherford, 516
   solid state, 524
   spectra of dense sources, 527
   strong nuclear interaction, 537
   weak interaction, 537
Microwaves, 498
Middle A, 368
Millikan oil drop, 98, 401
Minimum deviation, 338
Mobility, electron, 410
Model, boats, planes, 83
Molar specific heat, 234
Molar volume, 238
Mole, 231, 234
Molecular:
   beams, 242
   energy distribution, 241
   forces, 214
   potential wells, 212

speeds in gas, 240, 242
structure, 523
Moments of inertia:
approximation calculation, 174
parallel axis theorem, 175
perpendicular axis theorem, 176
table of values, 177
Moment of inertia tensor, 194
Momentum:
angular, (see Angular Momentum)
carried by E-M waves, 489
conservation, 159, 162
linear, 159, 162
relativistic invariant, 148
Monopole, 426
Motors, electric, 462
Mu-metal, 446
Muon and time dilation, 140
Musical instruments, 371
Music chords and scales, 377

N
Neutrino, 535, 539
Neutron, 532, 535, 539
newton, force unit, 158, 161
Newton's laws of dynamics:
2.7 K E-M background, 157
action-reaction, 159
Atwood's machine, 170
center of mass, 123, 168
centripetal acceleration, 174
circular dynamics, 173
$F = dp/dt \approx d(mv)/dt$, 159
first law, 156
force definition, 158
force exerted by table, 169
friction as f(v), 98, 171
Hooke's law, 158
impulse, 164
inertial mass, 106, 160
moments of inertia, 174, 177
momentum conservation, 159, 162
parallel axis theorem, 175
perpendicular axis theorem, 176
rocket dynamics, 167
"quantity of matter", 160, 533
second law, 157
terminal velocity, 84, 98, 172
third law, 159
units of force and mass, 161
Newton's law of cooling, 257
Newton's law of gravitation, 101
Normal force, 88
Nuclear:
binding energy, 533
fission, 534
fusion, 533
radius, 517
structure, 532
Nucleons, 539
Null reading devices, 423

O
Oersted, Hans Christian, 451
Ohm's law, 412
Orbits, gravitational, 109, 115, 181
Oscillations (see Vibrations)
Oxidation and reduction, 416

P
Pair production, 505,
π, μ, e decay, 164
Parabolic mirror, 337
Parallel axis theorem, 175
Paramagnetism, 441
Particle chart, 538
Pascal's vases, 282
Pauli's exclusion principle, 440, 509
Peak voltage, 470
Pedal note, 373, 376
Pendulum:
physical, 190, 312
simple, 310
Period:
oscillation, (see Vibrations)
rotation, 174
wave, 330
Periodic table structure, 522
Permeability, 327, 428
Permittivity, 327, 385
Phase and group velocities, 365
Phase angle in ac, 472
Phase diagrams, CO_2 and H_2O, 228
Phase transitions,
atomic model, 252
compared with molecular bonds, 252
heats of fusion, 232, 233
heats of vaporization, 232, 233
Phasors, 319, 359, 472
Photoelectric effect, 529
Photon:
energy, 494, 528
mass–energy, 151
momentum, 490
spin, 189
Physical pendulum, 190, 312
Piano frequencies chart, 370

Pin hole diffraction, 360
Pitch, musical, 349, 373, 378
Pitot tube, 293
Planck's constant, 188, 490, 498, 507, 528
Plumb line:
direction of "down", 118
"down" in accelerated frame, 130
Poise, 298
Poiseuille's equation, 300, 303
Poisson's spot, 360
Polarization of E-M waves, 491
Polarization of photons, 189
Potential energy:
arbitrary zero point, 210
centrifugal force in rotating system, 216
dipole, 439
escape energy, Earth, 211
gravitation, 110, 210
gravitational well, 244
molecular forces, 214
molecular wells, 212
oscillation frequency in well, 215
potential wells, 211
restoring force, 213
spring, 202, 213
Potentials:
arbitrary zero point, 210
conservative, 209
electric, 387
Potentiometer, 423
Pound, force unit, 161
Power:
definition and units, 217
electrical, 404
electrical, ac, 470, 472
horsepower, 217
longitudinal waves, 340
magnitude, lightning and car, 218
Poynting vector, 487
Precession:
derivation of period, 186
locked phases, 114
orbit of Mercury, 154
spin of Earth, 188
Pressure:
atmospheric, 243
longitudinal waves, 341
negative, 291
P–V and P–T graphs, 228
in pipe with viscosity, 297
static liquid, 281
Principal axes, 195
Probability, molecular arrangement, 273
Proper time, 143

Proton, 532, 535
Pseudovectors, 8
Pulleys, 208

Q

Quality factor, Q (oscillations), 315, 321, 496
Quantity of matter:
  mass of deuteron, 533
  unsatisfactory mass definition, 160
Quantization:
  angular momentum, 188
  electric charge, 403
  Stern-Gerlach, 506
Quantum phenomena:
  angular momentum, 188
  atomic structure, 509
  black body radiation, 528
  electric current, 406
  electron spin, 506
  energy levels in solids, 525
  hydrogen specific heat, 234, 254
  magnetism, 440
  potential well levels, 249
  Schrödinger wave equation, 332
  uncertainty principle, 514
  wave–particle duality, 511
Quark compound chart, 544
Quarks, 541, 545

R

Radar, 498
Radiation (see Electromagnetic radiation)
Radioactive sequence, 536
Radioactivity, 534
Radius of gyration, 190
Radius of percussion, 191
Rainbow, 338
Reactance, 319, 471
RC time constant, 453
Recoil of projectile, 162, 164
Red shift, 128
Reference frames:
  accelerated frames, 128
  center of mass, 123
  centrifugal force, 131, 216
  collision and energy, 122
  Coriolis force, 132
  Doppler shift, 124
  elevator field, 130
  familiar examples, 117
  Galilean transformation, 118
  local inertial fields, 128, 130

Newton's first law, 156
path of falling stone, 119
plumb line, 118, 130
simple harmonic motion, 120
space colony fields, 134
velocity addition, 121
washbowl problem, 133
Reflection:
  between media, 346
  Huygens's construction, 355
  in one-D, 334
  in two-D, 334
Refraction:
  Huygens's construction, 355
  mirages, 337
  Snell's law, 335
  sound, 337
Refrigerators, 269, 272
Relativity:
  Einstein, 135
  experimental evidence, 139
  general theory, 153
  invariance of light speed, 135
  invariants, mass, 152
  invariants, E-M, 153
  length relationships, 140
  Lorentz transformation, 135, 141
  mass-energy relationships, 149
  Michelson-Morley experiment, 136, 530
  momentum, energy, mass, 148
  muon paradox, 140
  proper time, 143
  synchronization of clocks, 138
  time dilation, 137
  twin paradox, 145
  uniting E and B, 447
  velocity addition, 144
Resistivity, f(T), 415
Resistivity table, 413
Resolving power, 361
Resonance, 314
Restoring force and potential energy, 213
Reversible process, 264
Reynolds number:
  definition, 75
  differential equation, 171
  fluid friction, 96
  scaling regimes, 301
  model scaling, 83
Right hand rule:
  magnetic fields, 429
  vectors, 82
Ripple tank, 323, 328
Rocket dynamics, 167

Rotational kinetic energy, 204
Rotational molecular energy, 499
Rutherford, 516

S

Satellites:
  energy, 111
  orbits, 109, 115, 181
Scalar product, 201
Scalars, 80
Scales, music, 377
Scaling, 82
Schrödinger wave equation, 332
Second law of thermodynamics, 264, 270
Seat belts, 166
Semiconductor, 526
Shear stress, 90, 298, 300
Significant figures, 57
Simple harmonic motion:
  analyzed and graphed, 305
  reference frame transformation, 120
Simple machines:
  arm as lever, 207
  bicycle gears, 208
  capstan, 208
  impedance match, 206
  impulse machines, 209
  inclined plane, 210
  gears, 207
  lever, 206
  mechanical advantage, 206, 208
  pneumatic lever, 207
  pulley, 208
  three classes of lever, 207
  wheel and axle, 207
  work in and out, 206
Simple pendulum approximation, 310
Single slit diffraction, 358
Siphon, 280, 283
Skater:
  ice friction, 93
  twirling, 182
Slug, mass unit, 161
Snell's law, 335
Soap bubbles, 280, 288
Solar system angular momentum, 183
Solenoid, 465
Solids:
  atomic model, 249
  solid fluids, 300
  solid state, 524
Sound:
  coat hanger chime, 342
  diffraction, 360

impedance, 342, 347
in pipes, 371
power-intensity, 342
pressure, 342
refraction, 337
speed as f(gas), 245
speed, table, 327
standing waves, strings, 372
Space colonies, 134
Special theory of relativity, 135
"Specific" quantities:
  dimensionless number, 74
  heats, (see entries under Heat)
  usage problem, 231
Spectra:
  dense sources, 527
  helium, 521
  hydrogen, 519
Spectrum chart, E-M, 494
Speed of light, 135, 327, 482
Spider webs, 289
Spin, 179, 406, 506
Springs:
  Hooke's law, 158
  potential energy, 202, 213
  simple harmonic motion, 309
  spring constants and mass, 310
Stable isotopes, table, 532
Standard deviation:
  definition, 63
  of means, 68
Standing waves, 350
Star formation, 183
Stern, Gerlach, electron spin, 506
Stern, Otto, molecular speeds, 241
Stoke's law, 98
Storage battery, 418
Stream lines, 291
Strong interaction, 537
Subatomic particles, Chpt. 21
Superconductivity, 415
Superfluids, 227, 300
Surface tension, 279, 286, 328
Susceptibility table, 442
Synchronization of clocks, 138

T
Temperature:
  boiling and melting table, 227
  critical point, 226, 228
  dependence of resistivity, 415
  difficulty of definition, 228
  equilibrium requirement, 229
  heat engine definition, 270

Kelvin scale, 226, 271
  relationship to internal energy, 221, 262
  response for molecular activity, 247
  scales, 224
  triple point, 226
  zeroth law, 229
Tension, 171, 283, 291
Terminal velocity:
  calculation, 172
  dimensions derivation, 84
  friction as f(v)—values, 98
tesla, unit of B, 430
Testing (see Evaluation)
Textbooks:
  background, 1
  choosing, 2
  using, 2
Thermal:
  conductivity, 258
  equilibrium, 229
  expansion, 250
  pollution, 271
  radiation, 260
Thermistor, 223
Thermocouple, 222
Thermodynamics:
  entropy, 273
  first law, 230
  heat engines, 264
  reversible process, 264
  second law, 264, 270
  third law, 277
  zeroth law, 229
Thermometers:
  bimetallic, 223
  boiling and melting table, 227
  calibration, 224
  Fahrenheit and Celsius, 224
  gas, 223, 229
  human, 221
  Kelvin scale, 226, 270
  liquid, 222
  mercury, 222, 226
  nonlinear effects, 225
  phase changes as f(T), 227
  resistance, 223
  thermistor, 223
  thermocouple, 222
  triple point, 226
Thermos bottle, 262
Thin film interference, 363
Three phase electricity, 470
Thrust, of rocket, 167
Tides:

effect on planets, 185
  locked phases, 114
  on Earth by moon and Sun, 112
Time:
  constant, 453 (RC), 466(L/R)
  dilation, 137
  synchronization, 138
Torque, 174, 180, 181, 186, 438
Trajectories, 119
Transformers, 473
Transient response, 317
Transverse wave, 324
Triple point of water, 226
Tritium, 532
Tsunami, 323, 329
Turbulence, 97, 296, 302
TV, 497
Twin paradox, 145

U
Ultraviolet, 501
Ultraviolet catastrophe, 261, 528
Uncertainty principle, 514
Units (converting), 71
Unit vectors, 80
Universal gas constant, 235
Universal gravitational constant, 102
Universe, map, 548
Uranium-235, 534

V
Van de Graaff generator, 384
van der Waals's equation, 248
van der Waals's force, 523
Vaporization, heat of, 232, 233
Vectors:
  addition, 79
  complex plane, 317
  components, 79
  displacements, 79
  pseudovectors, 81
  quantities, 80
  right hand rule, 82
  scalar (dot) product, 82, 201, 439
  unit vectors, 80
  vector (cross) product, 81, 82, 174, 179, 181, 186, 432, 438, 487
Velocity:
  Addition, 121, (relativistic) 144
  charge flow, 408
  escape, 111, 244
  gas molecules, 241
  group and phase, 365

invariance of light speed, 135
light, 135, 449, 482
sound, 245
terminal, 84, 98, 172
water ripples, 328
Vena Contracta, 294
Venturi tube, 293
Vibrations:
   ac (electrical) phasor notation, 319
   approximation to potential well, 309
   damped oscillations, 313
   demonstration of driven, 317
   differential equation for SHM, 307
   driven oscillations, 316
   $e^{i\theta}$ representation, 307
   energy and SHM, 308
   feedback, 321
   Hooke's law and SHM, 307
   Huygens' analysis of SHM, 306
   LRC oscillations, 313
   oscillations without inductors, 321
   physical and other pendulums, 312
   plausibility solution of SHM, 305
   quality factor, Q, 315
   quantitative solutions, 317
   resonance, 316
   simple harmonic motion, 305
   simple pendulum approximation, 310
   spring constants and mass, 310
   springs and SHM, 309
   summary of definitions of Q and γ, 321
Violin string waves, 373, 375
VisViva, 197, 202
Viscosity, 75
Viscosity:
   centipoise, table, 298
   effect on flow, 301
   function of P and T, 299
   ketchup, water, air, 291
   kinematic viscosity, 299
   fluid friction, 96
   low Reynolds regime, 83
   Reynolds derivation, 75
   shear stress, 90, 298
Visible light range, 501
Volt, 388
Voltage divider, 422
Voltaic cells, 416
Voltmeter, 420
Volume expansion coefficient, 250

W

Water:
   boiling and freezing points, 225
   internal energy, 256
   phase diagram, 228
   ripples, 328
   specific heat, 231
   triple point, 226
   viscosity, 298
Watt, definition, 405
Wave interactions:
   addition of sine waves, 350
   audio sound, table, 368, 370
   beats, 351
   chords, 379
   coherence, 362
   diffraction, 357
   dispersion of light, 36
   dispersion, rainbow, 338
   dispersion water ripples, 329
   double slit, 362
   $e^{i\theta}$ notation, 350
   ear, 369
   equal-tempered scale, 380
   Feynman's phase construction, 355
   Fourier series, 353
   Fraunhofer–Fresnel, 357
   harmonics–overtones, 354, 377
   harmonics in music, 377
   harmony, 377
   helium change of pitch, 374
   holograms, 367
   horns–bells, mouthpieces, 376
   Huygens's principle, 354
   interference requirements, 361
   Michelson interferometer, 136, 362, 530
   musical instruments, 371
   phase and group velocity, 365
   phasor analysis, 359
   piano frequencies chart, 370
   Poisson spot, 360
   production of sounds, 373
   reflection–refraction, 355
   resolving power, 361
   response of instruments, 374
   single slit, 358
   sound diffraction, 360
   standing wave, 350
   standing waves in pipes, 371
   standing waves in strings, 372
   strings with pulses or bowed, 373, 375
   thin film by phasors, 363
Wave–particle duality, 512, 529
Wave transmission:
   coat hanger chime, 342
   dimensional analysis, 323
   energy transmission, 339
   equations, 330
   extremum of paths, 335
   Fermat's principle, 335
   functions, 329
   impedance, 340, 342, 344
   longitudinal–pressure, 341
   longitudinal–power, 341
   mirages, 337
   phase and focusing, 336
   power delivered, 340
   rainbow, 338
   reflection between media, 346
   reflection in one-D, 334
   reflection, refraction, 2-D, 334
   refraction of sound, 337
   ripple tanks, 328
   Schrödinger wave equation, 332
   Snell's law, 335
   sound sensitivity, 342
   speed, E-M waves, 327, 482, 449
   speed in rope, 324, 331, 345
   speed, longitudinal, 325
   speed, sound, table, 327
   speed, water waves, 328
   spread of wave energy, 347
   transmission of pulses, 332
   traveling sine wave, 330
   tsunami, 323, 329
Weak interaction, 537
Weight, 104
Weightlessness, 105
Wheatstone bridge, 423, 425
Wheel and axle, 207
Wobbly rotations, 193
Work:
   braking distances, 203
   $\int \mathbf{F} \cdot \mathbf{dx}$ definition, 197, 201
   damage done in collision, 203
   problem of definition, 196, 201
   producing kinetic energy, 200
   who does the work?, 203

X

X rays, 501

Y

Young's modulus, 90, 326

Z

Zeroth law of thermodynamics, 229